**W9-CSM-862**

## Chapter 8

## Chapter 9

## Chapter 10

## Chapter 11

## Chapter 12

DUXBURY

# Mind on Statistics

*Third Edition*

**Jessica M. Utts**
*University of California, Davis*

**Robert F. Heckard**
*Pennsylvania State University*

**THOMSON**
™
**BROOKS/COLE**

Australia • Brazil • Canada • Mexico • Singapore
Spain • United Kingdom • United States

**THOMSON**

**BROOKS/COLE**

*Mind on Statistics,* **Third Edition**
**Jessica M. Utts and Robert F. Heckard**

Senior Acquisitions Editor: Carolyn Crockett

Development Editor: Danielle Derbenti

Senior Assistant Editor: Ann Day

Technology Project Manager: Fiona Chong

Marketing Manager: Joseph Rogove

Marketing Assistant: Brian R. Smith

Marketing Communications Manager: Darlene Amidon-Brent

Project Manager, Editorial Production: Sandra Craig

Creative Director: Rob Hugel

Art Director: Lee Friedman

Print Buyer: Barbara Britton

Permissions Editor: Kiely Sisk

Production Service: Martha Emry

Text Designer: tani hasegawa

Photo Researcher: Stephen Forsling

Copy Editor: Barbara Willette

Illustrator: Lori Heckelman

Cover Designer: Lee Friedman

Cover Image: © Jack Hollingsworth/Corbis

Cover Printer: Phoenix Color Corp

Compositor: G&S Book Services

Printer: R.R. Donnelley/Willard

Library of Congress Control Number: 2005931910

ISBN 0-534-99864-X

**Thomson Higher Education**
**10 Davis Drive**
**Belmont, CA 94002-3098**
**USA**

For more information about our products, contact us at:
**Thomson Learning Academic Resource Center**
**1-800-423-0563**

For permission to use material from this text or product, submit a request online at **http://www.thomsonrights.com.**

Any additional questions about permissions can be submitted by e-mail to **thomsonrights@thomson.com.**

*To Bill Harkness—energetic, generous, and innovative educator, guide, and friend—who launched our careers in statistics and continues to share his vision.*

# Brief Contents

# Contents

# 8  Random Variables   278

# 9  Understanding Sampling Distributions: Statistics as Random Variables   330

## 17   Turning Information Into Wisdom   704

*The Supplemental Topics are available on the Student's Suite CD, or print copies may be custom published.*

## SUPPLEMENTAL TOPIC 1   Additional Discrete Random Variables   S1-1

## SUPPLEMENTAL TOPIC 2   Nonparametric Tests of Hypotheses   S2-1

SUPPLEMENTAL TOPIC **3** Multiple Regression     S3-1

SUPPLEMENTAL TOPIC **4** Two-Way Analysis of Variance     S4-1

SUPPLEMENTAL TOPIC **5** Ethics     S5-1

# Preface

## A Challenge

Before you continue, think about how you would answer the question in the first bullet, and read the statement in the second bullet. We will return to them a little later in this Preface.

- What do you *really know* is true, and how do you know it?
- The diameter of the moon is about 2160 miles.

## What Is Statistics and Who Should Care?

Because people are curious about many things, chances are that your interests include topics to which statistics has made a useful contribution. As written in Chapter 17, "information developed through the use of statistics has enhanced our understanding of how life works, helped us learn about each other, allowed control over some societal issues, and helped individuals make informed decisions. There is almost no area of knowledge that has not been advanced by statistical studies."

Statistical methods have contributed to our understanding of health, psychology, ecology, politics, music, lifestyle choices, business, commerce, and dozens of other topics. A quick look through this book, especially Chapters 1 and 17, should convince you of this. Watch for the influences of statistics in your daily life as you learn this material.

## How Is this Book Different?
## Two Basic Premises of Learning

We wrote this book because we were tired of being told that what statisticians do is boring and difficult. We think statistics is useful and not difficult to learn, and yet the majority of college graduates we've met seemed to have had a negative experience taking a statistics class in college. We hope this book will help to overcome these misguided stereotypes.

Let's return to the two bullets at the beginning of this Preface. Without looking, do you remember the diameter of the moon? Unless you already had a pretty good idea, or have an excellent memory for numbers, you probably don't remember. One premise of this book is that **new material is much easier to learn and remember if it is related to something interesting or previously known.** The diameter of the moon is about the same as the air distance between Atlanta and Los Angeles, San Francisco and Chicago, London and Cairo, or Moscow and Madrid. Picture the moon sitting between any of those pairs of cities, and you are not likely to forget the size of the moon again. Throughout this book, new material is presented in the context of interesting and useful examples. The first and last chapters (1 and 17) are exclusively devoted to examples and case studies, which illustrate the wisdom that can be generated through statistical studies.

Now answer the question asked in the first bullet: What do you really know is true and how do you know it? If you are like most people, you know because it's something you have experienced or verified for yourself. It is not likely to be something you were told or heard in a lecture. The second premise of this book is that **new material is easier to learn if you actively ask questions and answer them for yourself.** *Mind on Statistics* is designed to help you learn statistical ideas by actively thinking about them. Throughout most of the chapters there are boxes entitled *Thought Questions*. Thinking about the questions in those boxes will help you to discover and verify important ideas for yourself. We encourage you to think and question, rather than simply read and listen.

# New to this Edition

The biggest changes have been made in **Chapters 9 to 13,** containing the core material on sampling distributions and statistical inference. The new organization presents the material in a modular, more flexible format. There are six modules for each of the topics of sampling distributions, confidence intervals, and hypothesis testing. The first module presents an introduction and the remaining five modules each deal with a specific parameter, such as one mean, one proportion, or the difference in two means. Chapter 9 covers sampling distributions, Chapters 10 and 11 cover confidence intervals, and Chapters 12 and 13 cover hypothesis testing.

In response to reviewer feedback, we made these changes for two reasons: pedagogy and practicality. **The new structure emphasizes the similarity among the inference procedures for the five parameters discussed.** It allows instructors to illustrate that each procedure covered is a specific instance of the same process. We recognize that instructors have different preferences for the order in which to cover inference topics. For instance, some prefer to first cover all topics about proportions and then cover all topics about means. Others prefer to first cover everything about confidence intervals and then cover everything about hypothesis testing. **With the new modular format, instructors can cover these topics in the order they prefer.**

To aid in the navigation through these modular chapters, we have added **color-coded, labeled tabs that correspond to the introductory and parameter modules.** The table below, also found in Chapter 9, lays out the color-coding system as well as the flexibility of these new chapters. In addition, the table is a useful course planning tool.

## Organization of Chapters 9 to 13

| Parameter | Chapter 9: Sampling Distributions (SD) | Chapter 10: Confidence Intervals (CI) | Chapter 11: Confidence Intervals (CI) | Chapter 12: Hypothesis Tests (HT) | Chapter 13: Hypothesis Tests (HT) |
|---|---|---|---|---|---|
| 0. Introductory | SD Module 0 Overview of sampling distributions | CI Module 0 Overview of confidence intervals | | HT Module 0 Overview of hypothesis testing | |
| 1. Population Proportion ($p$) | SD Module 1 SD for one sample proportion | CI Module 1 CI for one population proportion | | HT Module 1 HT for one population proportion | |
| 2. Difference in two population proportions ($p_1 - p_2$) | SD Module 2 SD for difference in two sample proportions | CI Module 2 CI for difference in two population proportions | | HT Module 2 HT for difference in two population proportions | |
| 3. Population mean ($\mu$) | SD Module 3 SD for one sample mean | | CI Module 3 CI for one population mean | | HT Module 3 HT for one population mean |
| 4. Population mean of paired differences ($\mu_d$) | SD Module 4 SD for sample mean of paired differences | | CI Module 4 CI for population mean of paired differences | | HT Module 4 HT for population mean of paired differences |
| 5. Difference in two population means ($\mu_1 - \mu_2$) | SD Module 5 SD for difference in two sample means | | CI Module 5 CI for difference in two population means | | HT Module 5 HT for difference in two population means |

In response to feedback from users, some other chapters have been expanded and reorganized. Most notably, **Chapters 3 and 4 have been essentially reversed,** so that random samples and surveys are presented before the more complicated studies based on randomized experiments and observational studies.

Furthermore, to add to the flexibility of topic coverage, **Supplemental Topics 1 to 5** on discrete random variables, nonparametric tests, multiple regression, two-way ANOVA, and ethics **are now available for use in both print and electronic formats.** Instructors, please contact your sales representative to find out how these chapters can be custom published for your course.

# Student Resources: Tools for Expanded Learning

There are a number of tools provided in this book and beyond to enhance your learning of statistics.

## Tools for Conceptual Understanding

*Updated!* **Thought Question** boxes, previously called "Turn on Your Mind," appear throughout each chapter to encourage active thinking and questioning about statistical ideas. **Hints** are provided at the bottom of the page to help you develop this skill.

> ***thought question 3.5*** You have now learned that survey results have to be interpreted in the context of *who* responded and *to what questions* they responded. When you read the results of a survey, for which of these two areas do you think it would be easier for you to recognize and assess possible biases? Why?*
>
> ***HINT:** What information would you need in each case, and what information is more likely to be included in a description of the survey?

**Updated! Skillbuilder Applet** sections, previously called "Turn on Your Computer," provide opportunities for in-class or independent hands-on exploration of key statistical concepts. The applets that accompany this feature can be found on the Student's Suite CD or at http://1pass .thomson.com.

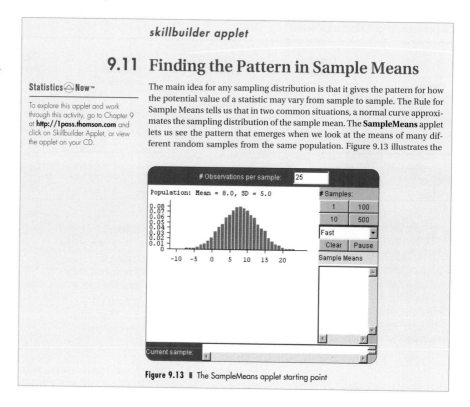

*skillbuilder applet*

## 9.11  Finding the Pattern in Sample Means

Statistics ⏏ Now™

To explore this applet and work through this activity, go to Chapter 9 at **http://1pass.thomson.com** and click on Skillbuilder Applet, or view the applet on your CD.

The main idea for any sampling distribution is that it gives the pattern for how the potential value of a statistic may vary from sample to sample. The Rule for Sample Means tells us that in two common situations, a normal curve approximates the sampling distribution of the sample mean. The **SampleMeans** applet lets us see the pattern that emerges when we look at the means of many different random samples from the same population. Figure 9.13 illustrates the

**Figure 9.13 ▪** The SampleMeans applet starting point

**Updated! Technical Notes** boxes, previously called "Tech Notes," provide additional technical discussion of key concepts.

*technical note* | **The Number of Units Per Block**

Some statisticians argue that the number of experimental units in a block should equal the number of treatments so that each treatment is assigned only once in each block. That allows as many sources of known variability as possible to be controlled. For the example of the effect of caffeine on swimming speed, there are two treatments, so blocks of size 2 (matched pairs) would be created. This could be accomplished by matching people on known variables such as sex, initial swim speed, usual caffeine consumption, age, and so on.

# Investigating Real-Life Questions

**Updated! Relevant Examples** form the basis for discussion in each chapter and walk you through real-life uses of statistical concepts.

Example 9.1 | **The "Freshman 15"**  Do college students really gain weight during their freshman year? The lore is that they do, and this phenomenon has been called "the freshman 15" because of speculation that students typically gain as much as 15 pounds during the first year of college. How can we turn our curiosity about the freshman 15 into a question about a parameter? There are several ways in which we could investigate whether or not students gain weight during the first year. We might want to know what proportion of freshmen gain weight. A related question would be whether most students gain weight. Or we might want to know what the average weight gain is across all first-year students. We might want to know whether women gain more weight than men or vice versa. Here are two ideas for satisfying our curiosity, along with the standard notation that we use for the relevant parameters:

- Parameter $= p =$ proportion of the population of first-year college students who weigh more at the end of the year than they did at the beginning of the year.

***Updated!* Case Studies** apply statistical ideas to intriguing news stories. As the Case Studies are developed, they model the statistical reasoning process.

---

### *case study 7.2*    Doin' the iPod Random Shuffle

The ability to play a collection of songs in a random order is a popular feature of portable digital music players. As an example, an Apple iPod Shuffle player with 512 megabytes of memory can store about 120 songs. Players with larger memory can store and randomly order thousands of songs. When the shuffle function is used, the stored songs are played in a random order.

We mention the iPod because there has been much grumbling, particularly on the Internet, that its shuffle might not be random. Some users complain that a song might be played within the first hour in two or three consecutive random shuffles. A similar complaint is that there are clusters of songs by the same group or musician within the first hour or so of play. *Newsweek* magazine's technology writer, Steven Levy (31 January 2005), wrote, "From the day I first loaded up my first Pod, it was as if the little devil liked to play favorites. It had a particular fondness for Steely Dan, whose songs always seemed to pop up two or three times in the first hour of play. Other songs seemed to be exiled to a forgotten corner" (p. 10). Conspiracy theorists even accuse Apple of playing favorites, giving certain musicians a better chance to have their songs played early in the shuffle.

more songs from the same album will be among the first four songs played in a random shuffle? We can find this by first determining the probability that all of the first four songs of the shuffle are from different albums and then subtracting that probability from 1. Define

Event A = first song is anything.
Event B = second song is from different album than first.
C = third song is from different album than first two picked.
D = fourth song is from different album than first three picked.

The probability that all four are from different albums is

$$P(A \text{ and } B \text{ and } C \text{ and } D) = \frac{120}{120} \times \frac{108}{119} \times \frac{96}{118} \times \frac{84}{117} = .53$$

Thus, the probability that all four songs are not from different albums, meaning that at least two songs are from the same album, is

$$P(\text{at least two of first four are from the same album}) = 1 - .53 \\ = .47$$

---

### *case study 10.3*    What a Great Personality

Students in a statistics class at Penn State were asked, "Would you date someone with a great personality even though you did not find them attractive?" By gender, the results were

- 61.1% of 131 women answered "yes."
- 42.6% of 61 men answered "yes."

There clearly is a difference between the men and the women in these samples. Can this difference be generalized to the populations represented by the samples? This question brings up another question: What are the populations that we are comparing? These men and women weren't randomly picked from any particular population, but for this example, we'll assume that they are like a random sample from the populations of all American college men and women.

A comparison of the 95% confidence intervals for the population percentage will help us to make a generalization:

- For men, the approximate 95% confidence interval is 30.2% to 55%.

make any conclusions about whether there is a difference in the population proportions. Instead, we can find a confidence interval for the *difference* in the proportions of men and women who would answer yes to the question.

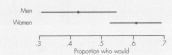

**Figure 10.5** ▍ 95% confidence intervals for proportions who would date someone with a great personality who wasn't attractive

A 95% confidence interval for the difference is .035 to .334. It is entirely above 0, so we can conclude that the proportion of women in the population who would answer "yes" is probably higher than the proportion of men who would do so.

---

***New!*** Original **Journal Articles** for 19 Examples and Case Studies can be found on the Student's Suite CD-ROM. By reading the original, you are given the opportunity to learn much more about how the research was conducted, what statistical methods were used, and what conclusions the original researchers drew.

---

**Example 5.3**

Statistics⬤Now™

Read the original source on your CD.

**The Development of Musical Preferences**  Will you always like the music that you like now? If you are about 20 years old, the likely answer is "yes," according to research reported in the *Journal of Consumer Research* (Holbrook and Schindler, 1989). The researchers concluded that we tend to acquire our popular music preferences during late adolescence and early adulthood.

In the study, 108 participants from 16 to 86 years old each listened to 28 hit songs that had been on *Billboard*'s Top 10 list for popular music some time between 1932 and 1986. Respondents rated the 28 songs on a 10-point scale, with 1 corresponding to "I dislike it a lot" and 10 corresponding to "I like it a lot." Each individual's ratings were then adjusted so that the mean rating for each participant was 0. On this adjusted rating scale, a positive score indicates a rating that was above average for a participant, whereas a negative score indicates a below-average rating.

# Getting Practice

**Updated! Basic Exercises,** comprising 25% of all exercises found in the text, focus on practice and review; these exercises, indicated by a green circle and appearing toward the beginning of each exercise section, complement the conceptual and data-analysis exercises. Basic exercises give you ample practice for these key concepts.

Relevant conceptual **Exercises** have been added and updated throughout the text. All exercises are found at the end of each chapter, with corresponding exercise sets written for each section and chapter. You will find **over 1500 exercises,** allowing for ample opportunity to practice key concepts.

## Exercises

● Denotes basic skills exercises

◆ Denotes dataset is available in StatisticsNow at **http:// 1pass.thomson.com** or on your CD but is **not required** to solve the exercise.

**Bold-numbered exercises** have answers in the back of the text and fully worked solutions in the Student Solutions Manual.

**Statistics ⟳ Now™**   Go to the StatisticsNow website at **http://1pass.thomson.com** to:
● Assess your understanding of this chapter
● Check your readiness for an exam by taking the Pre-Test quiz and exploring the resources in the Personalized Learning Plan

### Section 7.1

7.1  ● According to a U.S. Department of Transportation website (http://www.dot.gov/airconsumer), 76.1% of domestic flights flown by the top ten U.S. airlines from June 1998 to May 1999 arrived on time. Represent this in terms of a random circumstance and an associated probability.

7.2  ● Jan is a member of a class with 20 students. Each day for a week (Monday to Friday), a student in Jan's class is randomly selected to explain how to solve a homework problem. Once a student has been selected, he or she is not selected again that week. If Jan was not one of the four students selected earlier in the week, what is the probability that she will be picked on Friday? Explain how you found your answer.

7.3  Identify three random circumstances in the following story, and give the possible outcomes for each of them:

It was Robin's birthday and she knew she was going to have a good day. She was driving to work, and when she turned on the radio, her favorite song was playing. Besides, the traffic light at the main intersection she crossed to get to work was green when she arrived, something that seemed to happen less than once a week. When she arrived at work, rather than having to search as she usually did, she found an empty parking space right in front of the building.

7.4  Find information on a random circumstance in the news. Identify the circumstance and possible outcomes, and assign probabilities to the outcomes. Explain how you determined the probabilities.

### Section 7.2

7.5  ● Suppose you live in a city that has 125,000 households with telephones and a polling organization randomly selects 1000 of them to phone for a survey. What is the probability that your household will be selected?

7.6  ● Is each of the following values a legitimate probability value? Explain any "no" answers.
a. .50
b. .00
**c.** 1.00
**d.** 1.25
e. −.25

● Basic skills      ◆ Dataset available but not required      **Bold-numbered** exercises answered in the back

---

**Answers to Selected Exercises,** indicated by bold numbers in the Exercise sections, have complete or partial solutions found in the back of the text for checking your answers and guiding your thinking on similar exercises.

# Answers to Selected Exercises

*The following are partial or complete answers to the exercises numbered in **bold** in the text.*

**Chapter 1**
1.2   a. .00043
1.5   a. 400
1.7   c. Randomized experiment.   d. Observational study.
1.11  189/11,034, or about 17/1000, based on placebo group.
1.15  a. 150 mph.   b. 55 mph.   c. 95 mph.   d. 1/2   e. 51
1.19  No.
1.22  The base rate for that type of cancer.
1.26  a. 212/1525 = .139.   b. $1/\sqrt{1525}$ = .026.
      c. .113 to .165.
1.28  a. Self-selected or volunteer.   b. No. Readers with strong opinions will respond.

**Chapter 2**
2.1   a. 4   b. State in the United States.   c. $n = 50$.
2.3   a. Whole population.   b. Sample.
2.5   a. Population parameter.   b. Sample statistic.
      c. Sample statistic.
2.11  a. Categorical.   b. Quantitative.
      c. Quantitative.   d. Categorical.

2.34  a. Median is greater for males.   b. Spread between extremes is about the same; spread between quartiles is slightly greater for males.
2.37  a. Skewed to the right.   b. 13 ear pierces may be an outlier.   c. 2 ear pierces; about 45 women had this number.   d. About 32 or so.
2.39  a. Roughly symmetric.   b. Highest = 92.
2.44  Yes. Values inconsistent with the bulk of the data will be obvious.
2.46  a. Shape is better evaluated by using a histogram.
2.48  Skewed to the left.
2.50  a. Median = (72 + 76)/2 = 74; mean = 74.33.
      b. Median = 7; mean = 25.
2.52  a. Range = 225 − 123 = 102.   b. IQR = 35.   c. 50%
2.53  a. Median = 12.
2.54  d. There are no outliers.
2.59  The median is 16.72 inches. The data values vary from 6.14 to 37.42 inches. The middle 1/2 of the data is between 12.05 and 25.37 inches, so "typical" annual rainfall covers quite a wide range.

# Technology for Developing Concepts and Analyzing Data

**Technology manuals,** written specifically for *Mind on Statistics,* Third Edition, walk you through the statistical software and graphing calculator—step by step. You will find manuals for:

- **SPSS,** written by Brenda K. Gunderson and Kirsten T. Namesnik at the University of Michigan at Ann Arbor
- **MINITAB,** written by Edith Seier and Robert M. Price, East Tennessee State University
- **Excel,** written by Tom Mason, University of St. Thomas
- **TI-83/84,** written by Roger E. Davis, Pennsylvania College of Technology
- *New!* **JMP IN,** written by Jerry Reiter and Christine Kohnen, Duke University
- *New!* **R,** written by Mark A. Rizzardi, Humboldt State University

*Note:* These technology manuals are **available in both print and electronic formats.** Instructors, contact your sales representative to find out how these manuals can be custom published for your course.

**Datasets for examples and exercises** are formatted for MINITAB, Excel, SPSS, JMP, SAS, R, TI-83/84, and ASCII.

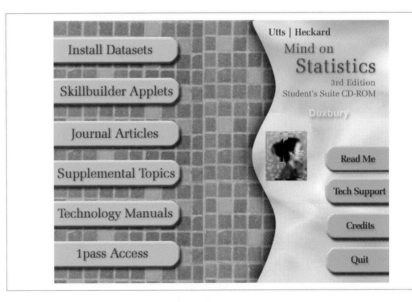

*New and Updated!* **Minitab, Excel, TI-84, and SPSS Tips** offer key details on the use of technology. TI-84 and SPSS Tips are new to this edition.

| MINITAB *tip* | **Calculating a Confidence Interval for a Proportion** |
|---|---|
| | • To compute a confidence interval for a proportion, use **Stat>Basic Statistics>1 Proportion.** (This procedure is not in versions earlier than Version 12.) |
| | • If the raw data are in a column of the worksheet, specify that column. If the data have already been summarized, click on "Summarized Data," and then specify the sample size and the count of how many observations have the characteristic of interest. |
| | *Note:* To calculate intervals in the manner described in this chapter, use the **Options** button, and click on "Use test and interval based on normal distribution." Note also that the confidence level can be changed by using the **Options** button. |

# Tools for Review

**Updated!** **In Summary** boxes serve as a useful study tool, appearing at appropriate points to enhance key concepts and calculations. A complete list of these can be found in StatisticsNow.

---

*in summary*    **Possible Reasons for Outliers and Reasonable Actions**

- *The outlier is a legitimate data value and represents natural variability for the group and variable(s) measured.* Values may not be discarded in this case—they provide important information about location and spread.
- *A mistake was made while taking a measurement or entering it into the computer.* If this can be verified, the values should be discarded or corrected.
- *The individual in question belongs to a different group than the bulk of individuals measured.* Values may be discarded if a summary is desired and reported for the majority group only.

---

**Key Terms** at the end of each chapter, organized by section, can be used as a "quick-finder" and as a review tool.

# Key Terms

**Section 10.1**
statistical inference, 401
confidence interval, 401, 403, 405, 408
sampling distribution, 403

**Section 10.2**
unit, 403
population, 403
universe, 403
sample, 404
sample size, 404
population parameter, 404
sample statistic, 404, 408
sample estimate, 404, 408
point estimate, 404

Fundamental Rule for Using Data for
  Inference, 404, 414–415
interval estimate, 405, 408
confidence level, 405, 406
multiplier, 408
standard error of the sample statistic,
  408

**Section 10.3**
confidence interval for a population
  proportion $p$, 412, 414
standard error of a sample proportion,
  412, 414, 417
margin of error, 420
conservative margin of error, 420

margin of error for a sample proportion,
  420
approximate 95% confidence interval
  for a proportion $p$, 420
conservative 95% confidence interval
  for $p$, 421
conservative estimate of the margin of
  error, 422

**Section 10.4**
confidence interval, difference between
  two population proportions, 423, 424
confidence interval for $p_1 - p_2$, 424

**Section 10.5**
confidence intervals and decisions, 428

---

*New!*

Statistics⬡Now™ is a personalized learning companion that helps you gauge your unique study needs and makes the most of your study time by building focused, chapter by chapter, Personalized Learning Plans that reinforce key concepts.

- **Pre-Tests,** developed by Deborah Rumsey of The Ohio State University, give you an initial assessment of your knowledge.
- **Personalized Learning Plans,** based on your answers to the pre-test questions, outline key elements for review.
- **Post-Tests,** also developed by Deborah Rumsey of The Ohio State University, assess your mastery of core chapter concepts; results can be emailed to your instructor.

*Note:* StatisticsNow also serves as a one-stop portal for many of your *Mind on Statistics* resources which are also found on the Student's Suite CD, as well as the Interactive Video Skillbuilder CD. **Throughout the text, StatisticsNow icons have been thoughtfully placed to direct you to the resources you need when you need them.**

*New!*   **Interactive Video Skill Builder CD-ROM** contains hours of helpful, interactive video instruction presented by Lisa Sullivan of Boston University. Watch as she walks you through key examples from the text, step by step—giving you a foundation in the skills you need to know. Each example found on the CD, as well as StatisticsNow, is identified by the StatisticsNow icon located in the margin. Think of it as portable office hours!

**vMentor™** allows you to talk (using your own computer microphone) to tutors who will skillfully guide you through a problem using an interactive whiteboard for illustration. Up to 40 hours of live tutoring a week is available with every new book and can be accessed through **http://1pass.thomson.com.**

The **Book Companion Website** offers book- and course-specific resources, such as tutorial quizzes for each chapter and datasets for exercises. You can access the website through the Student's Suite CD or through **http://1pass.thomson.com.**

The **Student Solutions Manual,** prepared by Jessica M. Utts and Robert F. Heckard, provides worked-out solutions to selected problems in the text. This is available for purchase through your local bookstore, as well as at **http://1pass .thomson.com.**

## Tools for Active Learning

*Updated!*   **Skillbuilder Applet** sections, previously called "Turn on Your Computer," provide opportunities for in-class or independent hands-on exploration of key statistical concepts. The applets that accompany this feature can be found on the Student's Suite CD or at **http://1pass.thomson.com.**

*New!*   The **Activities Manual,** written by Jessica M. Utts and Robert F. Heckard, includes a variety of activities for students to explore individually or in teams. These activities guide students through key features of the text, help them understand statistical concepts, provide hands-on data collection and interpretation team-work, include exercises with tips incorporated for solution strategies, and provide bonus dataset activities.

*New!*   **JoinIn™ on Turning Point®** offers instructors text-specific JoinIn content for electronic response systems, prepared by Brenda K. Gunderson and Kirsten T. Namesnik at the University of Michigan at Ann Arbor. You can transform your classroom and assess students' progress with instant in-class quizzes and polls. Turning Point software lets you pose book-specific questions and display students' answers seamlessly within Microsoft PowerPoint lecture slides, in conjunction with a choice of "clicker" hardware. Enhance how your students interact with you, your lecture, and each other.

**Internet Companion for Statistics,** written by Michael Larsen of Iowa State University, offers practical information on how to use the Internet to increase students' understanding of statistics. Organized by key topics covered in the introductory course, the text offers a brief review of a topic, listings of appropriate websites, and study questions designed to build students' analytical skills. This can be accessed through **http://1pass.thomson.com.**

## Tools for Online Learning

**CyberStats** contains complete online content for your introductory statistics course. It promotes learning through interaction on the Web and can be used as the sole text for a course or in conjunction with a traditional text. Students internalize the behavior of statistical concepts by interacting with hundreds of applets (simulations and calculations) and receiving immediate feedback on practice items. Effective for both distance and on campus courses, CyberStats provides a learning opportunity that cannot be delivered in print. Instructors interested in using this for their course should contact their local sales representative. CyberStats is available for purchase by students through **http://1pass .thomson.com.**

# Instructor Resources: Tools for Assessment

**iLrn Statistics** is your system for homework, integrated testing, and course management on the Web. Using iLrn, instructors can easily set up online courses, assign tests, quizzes, and homework, as well as monitor student progress, enabling them to mentor students on the right points at the right time. Student responses are automatically graded and entered into the iLrn grade book, making it easy for you to assign and collect homework or offer testing over the Web. Accessed through **http://1pass.thomson.com** and available with each new text, iLrn Statistics is comprised of two parts:

*Updated!*
- **iLrn Testing,** containing **algorithmically generated test items,** can be used for testing, homework, or quizzing. You choose! Instructors, contact your sales representative to find out how to get access to this valuable resource.

- **iLrn Homework,** which contains the exercises from the book, facilitates classroom management and assesses students through homework, on quizzes or on exams, in the process of doing real data analysis on the Web.

*Updated!*
The **Test Bank,** written by Brenda K. Gunderson and Kirsten T. Namesnik, of the University of Michigan at Ann Arbor, includes test questions by section and is made up of a combination of multiple-choice and free-response questions.

# Student Access Information

To summarize, many of the items mentioned above are available either on a CD or at **http://1pass.thomson.com** via your 1pass access code, which is bound in every new text. If you did not purchase this book new or you purchased a Basic Select text—containing no media resources—but would like to use any of the items described above, visit us online at **http://1pass.thomson .com** to purchase any of these valuable resources. The following chart summarizes the access information.

| Tools for Expanded Learning | http:1pass.thomson.com | Student's Suite CD | Interactive Video Skillbuilder CD |
|---|---|---|---|
| iLrn Statistics | ✔ | | |
| StatisticsNow | ✔ | | |
| InfoTrac® College Edition | ✔ | | |
| Michael Larsen's Internet Companion | ✔ | | |
| vMentor™ | ✔ | | |
| Link to Book Companion Website | ✔ | | ✔ |
| MINITAB Manual | ✔* | ✔ | |
| Excel Manual | ✔* | ✔ | |
| TI-83/84 Graphing Calculator Manual | ✔* | ✔ | |
| SPSS Manual | ✔* | ✔ | |
| JMP IN Manual | ✔* | ✔ | |
| R Manual | ✔* | ✔ | |
| Supplemental Topics 1 to 5 | | ✔ | |
| Datasets | ✔* | ✔ | |
| Skillbuilder Applets | ✔* | ✔ | |
| Link to 1pass | | ✔ | |
| Skillbuilder Videos | ✔ | | ✔ |
| Journal Articles | ✔ | | |

*Found in StatisticsNow.

# A Note to Instructors

The entire *Mind on Statistics* learning package has been informed by the recommendations put forth by the ASA/MAA Joint Curriculum Committee and GAISE (Guidelines for Assessment and Instruction in Statistics Education). Each of the pedagogical features and ancillaries listed in the section entitled "Student Resources: Tools for Expanded Learning" and "Instructor Resources: Tools for Assessment" have been categorized by suggested use to provide you with options for designing a course that best fits the needs of your students.

In addition to these tools you will also have access to the **Instructor's Suite CD-ROM,** which includes everything featured on the Student's Suite CD-ROM plus the following:

- Course Outlines and Syllabi
- Class Projects
- Instructor's Additional Resources
- Suggested discussions for the Thought Questions located throughout the text
- Multimedia manager containing all of the figures from the book in Power-Point and as .jpg files
- Complete Solutions Manual
- Test Bank in Microsoft Word format
- Supplemental Topic Solutions
- List of Applications and Methods

# Acknowledgments

We thank William Harkness, Professor of Statistics at Penn State University, for continued support and feedback throughout our careers and during the writing of this book, and for his remarkable dedication to undergraduate statistics education. Preliminary editions of *Mind on Statistics,* the basis for this text, were used at Penn State, the University of California at Davis, and Texas A & M University, and we thank the many students who provided comments and suggestions on those and on subsequent editions. Thanks to Dr. Melvin Morse (Valley Children's Clinic and University of Washington) for suggesting the title for Chapter 17 and to Deb Niemeier, University of California, Davis, for suggesting that we add a supplemental chapter on Ethics (on the CD). We are indebted to Neal Rogness, Grand Valley State University, for help with the SPSS Tips, and Larry Schroeder and Darrell Clevidence, Carl Sandburg College, for help with the TI-84 Tips; these tips are new to this edition. At Penn State, Dave Hunter, Steve Arnold, and Tom Hettmansperger have provided many helpful insights. At the University of California at Davis, Rodney Wong has provided insights as well as material for some exercises and the Test Bank. We extend special thanks to Phyllis Curtiss, Grand Valley State University, who provided hundreds of valuable suggestions for improving this edition of the book.

The following reviewers offered valuable suggestions for this and previous editions:

Patti B. Collings, Brigham Young University

Elizabeth Clarkson, Wichita State University

James Curl, Modesto Junior College

Wade Ellis, West Valley College

Linda Ernst, Mt. Hood Community College

Joan Garfield, University of Minnesota

Jonathan Graham, University of Montana

Jay Gregg, Colorado State University

Donnie Hallstone, Green River Community College

Donald Harden, Georgia State University

Rosemary Hirschfelder, University Sound

Sue Holt, Cabrillo Community College

Mark Johnson, University of Central Florida

Tom Johnson, North Carolina University

Andre Mack, Austin Community College

D'Arcy Mays, Virginia Commonwealth

Megan Meece, University of Florida

Mary Murphy, Texas A & M University

Emily Murphree, Miami University

Helen Noble, San Diego State University

Thomas Nygren, Ohio State University

Nancy Pfenning, University of Pittsburgh

David Robinson, St. Cloud State University

Neal Rogness, Grand Valley State University

Kelly Sakkinen, Lansing Community College

Heather Sasinouska, Clemson University

Kirk Steinhorst, University of Idaho

Gwen Terwilliger, University of Toledo

Robert Alan Wolf, University of San Francisco

Finally, our sincere appreciation and gratitude goes to Carolyn Crockett, Danielle Derbenti, and the staff of Duxbury, without whom this book could not have been written, and to Martha Emry who kept us on track throughout the editing and production of all three editions of the book. Finally, for their support, patience, and numerous prepared dinners, we thank our families and friends, especially Candace Heckard, Molly Heckard, Wes Johnson, Claudia Utts-Smith, and Dennis Smith.

*Jessica Utts*
*Robert Heckard*

# 1

DiMaggio/Kalish/Corbis

***Is a male or a female more likely to be behind
the wheel of this speeding car?***

See Case Study 1.1 *(p. 2)*

# Statistics Success Stories and Cautionary Tales

The seven stories in this chapter are meant to bring life to the term *statistics*. When you are finished reading these stories, if you still think the subject of statistics is lifeless or gruesome, check your pulse!

Let's face it. You're a busy person. Why should you spend your time learning about a subject that sounds as dull as statistics? In this chapter, we give seven examples of situations in which statistics either provided enlightenment or misinformation. With these examples, we hope to convince you that learning about this subject will be interesting and useful.

Each of the stories in this chapter illustrates one or more concepts that will be developed throughout the book. These concepts are given as "the moral of the story" after a case is presented. Definitions of some terms used in the story also are provided following each case. By the time you read all of these stories, you already will have an overview of what statistics is all about. ▮

## 1.1 What Is Statistics?

When you hear the word *statistics* you probably think of lifeless or gruesome numbers, such as the population of your state or the number of violent crimes committed in your city last year. The word *statistics*, however, actually is used to mean two different things. The better-known definition is that statistics are numbers measured for some purpose. A more complete definition, and the one that forms the substance of this book, is the following:

> *definition*  **Statistics** is a collection of procedures and principles for gathering data and analyzing information to help people make decisions when faced with uncertainty.

The stories in this chapter are meant to bring life to this definition. When you are finished reading them, if you still think the subject of statistics is lifeless or gruesome, check your pulse!

# 1.2  Seven Statistical Stories with Morals

The best way to gain an understanding of some of the ideas and methods used in statistical studies is to see them in action. Each of the seven stories presented in this section includes interesting lessons about how to gain information from data. The methods and ideas will be expanded throughout the book, but these seven stories will give you an excellent overview of why it is useful to study statistics. To help you understand some basic statistical principles, each case study is accompanied by a "moral of the story" and by some definitions. All of the ideas and definitions will be discussed in greater detail in subsequent chapters.

## *case study* 1.1    Who Are Those Speedy Drivers?

A survey taken in a large statistics class at Penn State University contained the question "What's the fastest you have ever driven a car? _____ mph." The *data* provided by the 87 males and 102 females who responded are listed here.

*Males:* 110 109 90 140 105 150 120 110 110 90 115 95 145 140 110 105 85 95 100 115 124 95 100 125 140 85 120 115 105 125 102 85 120 110 120 115 94 125 80 85 140 120 92 130 125 110 90 110 110 95 95 110 105 80 100 110 130 105 105 120 90 100 105 100 120 100 100 80 100 120 105 60 125 120 100 115 95 110 101 80 112 120 110 115 125 55 90

*Females:* 80 75 83 80 100 100 90 75 95 85 90 85 90 90 120 85 100 120 75 85 80 70 85 110 85 75 105 95 75 70 90 70 82 85 100 90 75 90 110 80 80 110 110 95 75 130 95 110 110 80 90 105 90 110 75 100 90 110 85 90 80 80 85 50 80 100 80 80 80 95 100 90 100 95 80 80 50 88 90 90 85 70 90 30 85 85 87 85 90 85 75 90 102 80 100 95 110 80 95 90 80 90

From these numbers, can you tell which sex tends to have driven faster and by how much? Notice how difficult it is to make sense of the *data* when you are simply presented with a list. Even if the numbers had been presented in numerical order, it would be difficult to compare the two sexes.

Your first lesson in statistics is how to formulate a simple summary of a long list of numbers. The **dotplot** shown in Figure 1.1 helps us see the pattern in the data. In the plot, each dot represents the response of an individual student. We can see that the men tend to claim a higher "fastest ever driven" speed than do the women.

The graph shows us a lot, and calculating some statistics that summarize the data will provide additional insight. There are a variety of ways to do so, but for this example, we examine a **five-number summary** of the data for males and females. The five numbers are the lowest value; the cut-

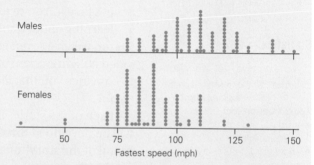

**Figure 1.1** ▮ Responses to "What's the fastest you've ever driven?"

off points for 1/4, 1/2, and 3/4 of the data; and the highest value. The three middle values of the summary (the cutoff points for 1/4, 1/2, and 3/4 of the data) are called the *lower quartile, median,* and *upper quartile,* respectively. Five-number summaries can be represented like this:

|  | **Males** (87 Students) | | **Females** (102 Students) | |
|---|---|---|---|---|
| **Median** | | 110 | | 89 |
| **Quartiles** | 95 | 120 | 80 | 95 |
| **Extremes** | 55 | 150 | 30 | 130 |

Some interesting facts become immediately obvious from these summaries. By looking at the medians, you see that half of the men have driven 110 miles per hour or more, whereas the halfway point for the women is only 89 miles per hour. In fact, 3/4 of the men have driven 95 miles per hour or more, but only 1/4 of the women have done so. These facts were not at all obvious from the original lists of numbers.

**Moral of the Story:** *Simple summaries of data can tell an interesting story and are easier to digest than long lists.*

**Definitions: Data** is a plural word referring to numbers or nonnumerical labels (such as male/female) collected from a set of entities (people, cities, and so on). The **median** of a numerical list of data is the value in the middle when the numbers are put in order. For an even number of entities, the median is the average of the middle two values. The **lower quartile** and **upper quartile** are (roughly) the medians of the lower and upper halves of the data.

---

## *case study 1.2*   Safety in the Skies?

Air travelers already have enough concerns, with the potential for delays, lost baggage, and so on. So if you do a lot of air travel or know anyone who does, you may have been disturbed by the headline in *USA Today* that read, "Planes get closer in midair as traffic control errors rise" (Levin, 1999). You may have been even more disturbed by the details: "Errors by air traffic controllers climbed from 746 in fiscal 1997 to 878 in fiscal 1998, an 18% increase." Don't cancel your next vacation yet. A look at the statistics indicates that the news is actually pretty good! And there is some reassurance when we are told that "most [errors] involve planes passing so far apart that there is no danger of collision."

The headline and details do sound ominous — all those errors and an increase of almost 20%! If things continue at that rate, won't your next flight be quite likely to suffer from air traffic controller error? The answer is a resounding "no," which becomes obvious when we are told the *base rate* or *baseline risk* for errors. "The errors per million flights handled by controllers climbed from 4.8 to 5.5." So the original *rate* of errors in 1997, from which the 18% increase was calculated, was fewer than 5 errors per million flights.

Fortunately, the rates for the two years were provided in the story. This is not always the case in news reports of increases in rates. For instance, an article may say that the rate of a certain type of cancer is doubled if you eat a certain unhealthful food. But what good is that information unless you know the actual risk? Doubling your chance of getting cancer from 1 in a million to 2 in a million is trivial, but doubling your chance from 1 in 50 to 2 in 50 is not.

**Moral of the Story:** *When discussing the change in the rate or risk of occurrence of something, make sure you also include the base rate or baseline risk.*

**Definitions:** The **rate** at which something occurs is simply the number of times it occurs per number of opportunities for it to occur. In fiscal year 1998, the rate of air traffic controller errors was 5.5 per million flights. The **risk** of a bad outcome in the future can be estimated by using the past rate for that outcome, if it is assumed the future is like the past. Based on 1998 data, the estimated risk of an error for any given flight in 1999 would be 5.5/1,000,000, or .0000055. The **base rate** or **baseline risk** is the rate or risk at a beginning time period or under specific conditions. For instance, the base rate of air traffic controller errors was 4.8 per million flights in fiscal year 1997.

---

## *case study 1.3*   Did Anyone Ask Whom You've Been Dating?

"According to a new *USA Today*/Gallup Poll of teenagers across the country, 57 percent of teens who go out on dates say they've been out with someone of another race or ethnic group" (Peterson, 1997). That's over half of the dating teenagers, so of course it was natural for the headline in the *Sacramento Bee* to read, "Interracial dates common among today's teenagers." The article contained other information as well, such as "In most cases, parents aren't a major obstacle. Sixty-four percent of teens say their parents don't mind that they date interracially, or wouldn't mind if they did."

There are millions of teenagers in the United States whose experiences are being reflected in this story. How could the polltakers manage to ask so many teenagers these questions? The answer is that they didn't. The article states that "the results of the new poll of 602 teens, conducted Oct. 13–20, reflect the ubiquity of interracial dating today." They asked only 602 teens? Could such a small sample possibly tell us anything about the millions of teenagers in the United States? The answer is "yes" if those teens constituted a *random sample* from the *population* of interest.

*(continued)*

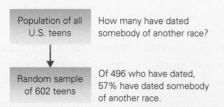

**Figure 1.2** ■ Population and sample for the survey

The featured statistic of the article is that "57 percent of teens who go out on dates say they've been out with someone of another race or ethnic group." Only 496 of the 602 teens in the poll said that they date, so the 57% value is actually a percentage based on 496 responses. In other words, the pollsters were using information from only 496 teenagers to estimate something about all teenagers. Figure 1.2 illustrates this situation.

How accurate could this sample possibly be? The answer may surprise you. The results of this poll are accurate to within a *margin of error* of about 4.5%. As surprising as it may seem, the true percentage of all teens in the United States who date interracially is reasonably likely to be within 4.5% of the reported percentage that's based only on the 496 teens asked! We'll be conservative and round the 4.5% margin of error up to 5%. The percent of all teenagers in the United States who date that would say they have dated someone of another race or ethnic group is likely to be in the range 57% ± 5%, or between 52% and 62%. (The symbol ± is read "plus and minus" and means that the value on the right should be added to and subtracted from the value on the left to create an interval.)

Polls and *sample surveys* are frequently used to assess public opinion and to estimate population characteristics such as the percent of teens who have dated interracially. Many sophisticated methods have been developed that allow pollsters to gain the information they need from a very small number of individuals. The trick is to know how to select those individuals. In Chapter 3, we examine a number of other strategies that are used to ensure that sample surveys provide reliable information about populations.

**Moral of the Story:** *A representative sample of only a few thousand, or perhaps even a few hundred, can give reasonably accurate information about a population of many millions.*

**Definitions:** A **population** is a collection of all individuals about which information is desired. The "individuals" are usually people but could also be schools, cities, pet dogs, agricultural fields, and so on. A **random sample** is a subset of the population selected so that every individual has a specified probability of being part of the sample. In a **sample survey,** the investigators gather opinions or other information from each individual included in the sample. The **margin of error** for a properly conducted survey is a number that is added to and subtracted from the sample information to produce an interval that is 95% certain to contain the truth about the population. In the most common types of sample surveys, the margin of error is approximately equal to 1 divided by the square root of the number of individuals in the sample. Hence, a sample of 496 teenagers who have dated produces a margin of error of about $1/\sqrt{496} = .045$, or about 4.5%.

## *case study* 1.4    Who Are Those Angry Women?

A well-conducted survey can be very informative, but a poorly conducted one can be a complete disaster. As an extreme example, Moore (1997, p. 11) reports that for her highly publicized book *Women and Love,* Shere Hite (1987) sent questionnaires to 100,000 women asking about love, sex, and relationships. Only 4.5% of the women responded, and Hite used those responses to write her book. As Moore notes, "The women who responded were fed up with men and eager to fight them. For example, 91% of those who were divorced said that they had initiated the divorce. The anger of women toward men became the theme of the book." Do you think that women who were angry with men would be likely to answer questions about love relationships in the same way as the general population of women?

The Hite sample exemplifies one of the most common problems with surveys: The sample data may not represent the population. Extensive *nonresponse* from a random sample, or the use of a *self-selected* (i.e., *all-volunteer) sample,* will probably produce biased results. Those who voluntarily respond to surveys tend to care about the issue and therefore have stronger and different opinions than those who do not respond.

**Moral of the Story:** *An unrepresentative sample, even a large one, tells you almost nothing about the population.*

**Definitions: Nonresponse bias** can occur when many people who are selected for the sample either do not respond at all or do not respond to some of the key survey questions. This may occur even when an appropriate random sample is selected and contacted. The survey is then based on a nonrepresentative sample, usually those who feel strongly about the issues. Some surveys don't even attempt to contact a random sample but instead ask anyone who wishes to respond to do so. Magazines, television stations, and Internet websites routinely conduct this kind of poll, and those who respond are called a **self-selected sample** or a **volunteer sample.** In most cases, this kind of sample tells you nothing about the larger population at all; it tells you only about those who responded.

## case study 1.5    Does Prayer Lower Blood Pressure?

**Statistics⊘Now™** Read the original source on your CD.

Newspaper headlines are notorious for making one of the most common mistakes in the interpretation of statistical studies: jumping to unwarranted conclusions. A headline in *USA Today* read, "Prayer can lower blood pressure" (Davis, 1998). The story that followed continued the possible fallacy it began by stating, "Attending religious services lowers blood pressure more than tuning into religious TV or radio, a new study says." The words "attending religious services lowers blood pressure" imply a direct cause-and-effect relationship. This is a strong statement, but it is not a statement that is justified by the research project described in the article.

The article was based on an *observational study* conducted by the U.S. National Institutes of Health, which followed 2391 people aged 65 or older for six years. The article described one of the study's principal findings: "People who attended a religious service once a week and prayed or studied the Bible once a day were 40% less likely to have high blood pressure than those who don't go to church every week and prayed and studied the Bible less" (Davis, 1998). So the researchers did observe a relationship, but it's a mistake to think that this justifies the conclusion that prayer actually *causes* lower blood pressure.

When groups are compared in an observational study, the groups usually differ in many important ways that may contribute to the observed rela-tionship. In this example, people who attended church and prayed regularly may have been less likely than the others to smoke or to drink alcohol. These could affect the results because smoking and alcohol use are both believed to affect blood pressure. The regular church attendees may have had a better social network, a factor that could lead to reduced stress, which in turn could reduce blood pressure. People who were generally somewhat ill may not have been as willing or able to go out to church. We're sure you can think of other possibilities for *confounding variables* that may have contributed to the observed relationship between prayer and lower blood pressure.

**Moral of the Story:** *Cause-and-effect conclusions cannot generally be made on the basis of an observational study.*

**Definitions:** An **observational study** is one in which participants are merely observed and measured. Comparisons based on observational studies are comparisons of naturally occurring groups. A **variable** is a characteristic that differs from one individual to the next. It may be numerical, such as blood pressure, or it may be categorical, such as whether or not someone attends church regularly. A **confounding variable** is a variable that is not the main concern of the study but may be partially responsible for the observed results.

## case study 1.6    Does Aspirin Reduce Heart Attack Rates?

**Statistics⊘Now™** Read the original source on your CD.

In 1988, the Steering Committee of the Physicians' Health Study Research Group released the results of a five-year *randomized experiment* conducted using 22,071 male physicians between the ages of 40 and 84. The purpose of the experiment was to determine whether taking aspirin reduces the risk of a heart attack. The physicians had been *randomly assigned* to one of the two *treatment* groups. One group took an ordinary aspirin tablet every other day, while the other group took a *placebo*. None of the physicians knew whether he was taking the actual aspirin or the placebo.

The results, shown in Table 1.1, support the conclusion that taking aspirin does indeed help to reduce the risk of having a heart attack. The rate of heart attacks in the group taking aspirin was only about half the rate of heart attacks in the placebo group. In the aspirin group, there were 9.42 heart attacks per 1000 participating doctors, while in the placebo group, there were 17.13 heart attacks per 1000 participants.

**Table 1.1** The Effect of Aspirin on Heart Attacks

| Treatment | Heart Attacks | Doctors in Group | Attacks per 1000 Doctors |
|-----------|---------------|------------------|--------------------------|
| **Aspirin** | 104 | 11,037 | 9.42 |
| **Placebo** | 189 | 11,034 | 17.13 |

Because the men in this experiment were randomly assigned to the two conditions, other important risk factors such as age, amount of exercise, and dietary habits should have been similar for the two groups. The only important difference between the two groups should have been whether they took aspirin or a placebo. This makes it possible to conclude that taking aspirin actually *caused* the lower rate of heart attacks for that group. In a later chapter, you will learn how to determine that the difference seen in this sample is *statistically significant*. In other words, the observed sample difference probably reflects a true difference within the population.

*(continued)*

To what population does the conclusion of this study apply? The participants were all male physicians, so the conclusion that aspirin reduces the risk of a heart attack may not hold for the general population of men. No women were included, so the conclusion may not apply to women at all. More recent evidence, however, has provided additional support for the benefit of aspirin in broader populations.

**Moral of the Story:** *Unlike with observational studies, cause-and-effect conclusions can generally be made on the basis of randomized experiments.*

**Definitions:** A **randomized experiment** is a study in which treatments are randomly assigned to participants. A **treatment** is a specific regimen or procedure assigned to participants by the experimenter. A **random assignment** is one in which each participant has a specified probability of being assigned to each treatment. A **placebo** is a pill or treatment designed to look just like the active treatment but with no active ingredients. A **statistically significant** relationship or difference is one that is large enough to be unlikely to have occurred in the sample if there was no relationship or difference in the population.

## case study 1.7   Does the Internet Increase Loneliness and Depression?

It was big news. Researchers at Carnegie Mellon University had found that "greater use of the Internet was associated with declines in participants' communication with family members in the household, declines in size of their social circle, and increases in their depression and loneliness" (Kraut et al., 1998, p. 1017). An article in the *New York Times* reporting on this study was entitled "Sad, lonely world discovered in cyberspace" (Harmon, 1998). The study included 169 individuals in 73 households in Pittsburgh, Pennsylvania, who were given free computers and Internet service in 1995. The participants answered a series of questions at the beginning of the study and either one or two years later, measuring social contacts, stress, loneliness, and depression. The *New York Times* reported:

> In the first concentrated study of the social and psychological effects of Internet use at home, researchers at Carnegie Mellon University have found that people who spend even a few hours a week online have higher levels of depression and loneliness than they would if they used the computer network less frequently. . . . it raises troubling questions about the nature of "virtual" communication and the disembodied relationships that are often formed in cyberspace.
>
> Source: "Sad, Lonely World Discovered in Cyberspace," by A. Harmon, New York Times, August 30, 1998, p. A3. Reprinted with permission of the New York Times Company.

Given these dire reports, one would think that using the Internet for a few hours a week is devastating to one's mental health. But a closer look at the findings reveals that the changes were actually quite small, though statistically significant. Internet use averaged 2.43 hours per week for participants. The number of people in the participants' "local social network" decreased from an average of 23.94 people to an average of 22.90 people, hardly a noticeable loss. On a scale from 1 to 5, self-reported loneliness decreased from an average of 1.99 to 1.89 (lower scores indicate greater loneliness). And on a scale from 0 to 3, self-reported depression dropped from an average of 0.73 to an average of 0.62 (lower scores indicate higher depression).

The *New York Times* did report the magnitude of some of the changes, noting for instance that "one hour a week on the Internet was associated, on average, with an increase of 0.03, or 1 percent on the depression scale." But the attention the research received masked the fact that the impact of Internet use on depression, loneliness, and social contact was actually quite small.

As a follow-up to this study, in July 2001, *USA Today* (Elias, 2001) reported that in continued research, the bad effects had mostly disappeared. The article, titled "Web use not always a downer: Study disputes link to depression," began with the statement "Using the Internet at home doesn't make people more depressed and lonely after all." However, the article noted that the lead researcher, Robert Kraut of Carnegie Mellon University, believes that the earlier findings were correct but that "the Net has become a more social place since the study began in 1995." His explanation for the change in findings is that "either the Internet has changed, or people have learned to use it more constructively, or both." Whether the link ever existed will never be known, but it is not surprising, given the small magnitude of the original finding, that it subsequently disappeared.

**Moral of the Story:** *A "statistically significant" finding does not necessarily have practical importance. When a study reports a statistically significant finding, find out the magnitude of the relationship or difference. A secondary moral to this story is that the implied direction of cause and effect may be wrong. In this case, it could be that people who were more lonely and depressed were more prone to using the Internet. And as the follow-up research makes clear, remember that "truth" doesn't necessarily remain fixed across time. Any study should be viewed in the context of society at the time it was done.*

# Save time.
# Improve understanding.
# Get a better grade.

One password is all you need to access the study tools designed to help you get a better grade.

## 1pass™ gives you access to:

 • **StatisticsNow**™—the robust, personalized online learning companion, complete with videos and a host of learning tools

 • **iLrn**™ **Statistics Homework**—a powerful online homework system

 • **InfoTrac® College Edition**—the online university library

• Michael Larsen's **Internet Companion for Statistics**— a guide to using the Internet more efficiently to learn statistics

 • **vMentor**™—live, online tutoring with a statistical expert

 • The **Companion Website**—with tutorial quizzes and datasets

 • Datasets, applets, and technology manuals found on the **Student's Suite CD**

The **1pass** code on the card at left is all you need—simply register for **1pass** by following the directions on the card.

# Statistics ⬤ Now™

## You determine what you need to learn... NOW! Log on today using 1pass access

This student website greatly improves your chances of success. Dynamic, interactive, and instructive, **StatisticsNow**™ allows you to create a personalized study plan for each chapter of **Mind on Statistics**—and gain a better understanding of statistical concepts.

## What does **StatisticsNow's** integrated learning system do?

It allows you to build a learning plan for yourself…based on what you know and what you need to know to get a better grade in your course. Icons within the text suggest opportune times to access **StatisticsNow**.

Just log on to **StatisticsNow** by using the 1pass access code packaged in front of this card. Once you log on, you will immediately notice the system's simple, browser-based format—as easy to use as surfing the web. Just a click of the mouse gives you the freedom to enter and explore the system at any point.

You can build a complete, personalized learning plan for yourself by taking advantage of all three of **StatisticsNow's** powerful components:

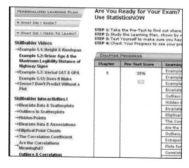

• The **Pre-Test**, authored by Deborah Rumsey of The Ohio State University, helps determine how well you've learned the chapter material

• **The Personalized Learning Plan**, based on your answers to the Pre-Test, lets you review chapter content according to your specific needs, use applets to interact with statistics, view video examples, and think critically about statistics

• The **Post-Test**, also authored by Deborah Rumsey, lets you assess your mastery of core chapter concepts

# 1.3 The Common Elements in the Seven Stories

The seven stories were meant to bring life to our definition of statistics. Let's consider that definition again:

> **Statistics** is a collection of procedures and principles for gathering data and analyzing information to help people make decisions when faced with uncertainty.

Think back over the seven stories. In every story, *data are used to make a judgment about a situation.* This common theme is what statistics is all about.

## The Discovery of Knowledge

Each story illustrates part of the process of discovery of new knowledge, for which statistical methods can be very useful. The basic steps in this process are as follows:

1. *Asking the right question(s).*
2. *Collecting useful data,* which includes deciding how much is needed.
3. *Summarizing and analyzing data,* with the goal of answering the questions.
4. *Making decisions and generalizations* based on the observed data.
5. *Turning the data and subsequent decisions into new knowledge.*

We'll explore these five steps throughout the book, concluding with a chapter on "Turning Information into Wisdom." We're confident that your active participation in this exploration will benefit you in your everyday life and in your future professional career.

In a practical sense, almost all decisions in life are based on knowledge obtained by gathering and assimilating data. Sometimes the data are quantitative, as when an instructor must decide what grades to give on the basis of a collection of homework and exam scores. Sometimes the information is more qualitative and the process of assimilating it is informal, such as when you decide what you are going to wear to a party. In either case, the principles in this book will help you to understand how to be a better decision maker.

---

*thought question* **1.1**   Think about a decision that you recently had to make. What "data" did you use to help you make the decision? Did you have as much information as you would have liked? If you could freely use them, how would you use the principles in this chapter to help you gain more useful information?[*]

---

**\*HINT:** As an example, how did you decide to live where you are living? What additional data, if any, would have been helpful?

*in summary*     ## Some Important Statistical Principles

The "moral of the story" items for the case studies presented in this chapter give a good overview of many of the important ideas covered in this book. Here is a summary:

- Simple summaries of data can tell an interesting story and are easier to digest than long lists.
- When discussing the change in the rate or risk of occurrence of something, make sure you also include the base rate or baseline risk.
- A representative sample of only a few thousand, or perhaps even a few hundred, can give reasonably accurate information about a population of many millions.
- An unrepresentative sample, even a large one, tells you almost nothing about the population.
- Cause-and-effect conclusions cannot generally be made on the basis of an observational study.
- Cause-and-effect conclusions can generally be made on the basis of randomized experiments.
- A statistically significant finding does not necessarily have practical significance or importance. When a study reports a statistically significant finding, find out the magnitude of the relationship or difference.

# Key Terms

Every term in this chapter is discussed more extensively in later chapters, so don't worry if you don't understand all of the terminology that has been introduced here. The following list indicates the page number(s) where the important terms in this chapter are introduced and defined.

# Exercises

●   Denotes basic skills exercises

◆   Denotes dataset is available in StatisticsNow at **http://1pass.thomson.com** or on your CD but is **not required** to solve the exercise.

**Bold-numbered exercises** have answers in the back of the text and fully worked solutions in the Student Solutions Manual.

**Statistics ◯ Now**™  Go to the StatisticsNow website at **http://1pass.thomson.com** to:
- Assess your understanding of this chapter
- Check your readiness for an exam by taking the Pre-Test quiz and exploring the resources in the Personalized Learning Plan

Note: *Many of these exercises will be repeated in later chapters in which the relevant material is covered in more detail.*

**1.1** ● A five-number summary for the heights in inches of the women who participated in the survey in Case Study 1.1 is as shown.

| | Female Heights (Inches) | |
|---|---|---|
| Median | 65 | |
| Quartiles | 63.5 | 67.5 |
| Extremes | 59 | 71 |

  a. What is the median height for these women?
  b. What is the range of heights, that is, the difference in heights between the shortest and the tallest women?
  c. What is the interval of heights containing the shortest 1/4 of the women?
  d. What is the interval of heights containing the middle 1/2 of the women?

**1.2** ● In the year 2000, Vietnamese American women had the highest rate of cervical cancer in the country. Suppose that among 200,000 Vietnamese American women, 86 developed cervical cancer.
  **a.** Calculate the rate of cervical cancer for these women.
  b. What is the estimated risk of developing cervical cancer for Vietnamese American women in the next year?
  c. Explain the conceptual difference between the rate and the risk, in the context of this example.

**1.3** ● Using Case Study 1.6 as an example, explain the difference between a population and a sample.

**1.4** ● A telephone survey of 2000 Canadians conducted March 20–30, 2001, found that "Overall, about half of Canadians in the poll say the right number of immigrants are coming into the country and that immigration has a positive effect on Canadian communities. Only 16 percent view it as a negative impact while one-third said it had no impact at all" (*The Ottawa Citizen*, August 17, 2001, p. A6.).
  a. What is the population for this survey?

  b. How many people were in the sample used for this survey?
  c. What is the approximate margin of error for this survey?
  d. Provide an interval of numbers that is 95% certain to cover the true percentage of Canadians who view immigration as having a negative impact.

**1.5** ● In Case Study 1.3, the margin of error for the sample of 496 teenagers was about 4.5%. How many teenagers should be in the sample to produce each of the following as the approximate margin of error?
  **a.** Margin of error = .05 or 5%.
  b. Margin of error = .30 or 30%.

**1.6** ● A proposed study design is to leave 100 questionnaires by the checkout line in a student cafeteria. The questionnaire can be picked up by any student and returned to the cashier. Explain why this volunteer sample is a poor study design.

**1.7** ● For each of the examples given here, decide whether the study was an observational study or a randomized experiment.
  a. A group of 100 students was randomly divided, with 50 assigned to receive vitamin C and the remaining 50 to receive a placebo, to determine whether vitamin C helps to prevent colds.
  b. All patients who received a hip transplant operation at Stanford University Hospital during 1995 to 2005 will be followed for 10 years after their operation to determine the success (or failure) of the transplant.
  **c.** A group of students who were enrolled in an introductory statistics course were randomly assigned to take a Web-based course or to take a traditional lecture course. The two methods were compared by giving the same final examination in both courses.
  **d.** A group of smokers and a group of nonsmokers who visited a particular clinic were asked to come in for a physical exam every 5 years for the rest of their lives to monitor and compare their health status.

**1.8** ● Suppose that an observational study showed that students who get at least 7 hours of sleep performed better on exams than students who didn't. Which of the following are possible confounding variables, and which are not? Explain why in each case.
  a. Number of courses the student took that term.
  b. Weight of the student.
  c. Number of hours the student spent partying in a typical week.

**1.9** ● Explain the distinction between statistical significance and practical significance. Can the result of a study be statistically significant but not practically significant?

**1.10** A headline in a major newspaper read, "Breast-Fed Youth Found to Do Better in School."

a. Do you think this statement was based on an observational study or a randomized experiment? Explain.

b. Given your answer in part (a), which of these two alternative headlines do you think would be preferable: "Breast-Feeding Leads to Better School Performance" or "Link Found Between Breast-Feeding and School Performance"? Explain.

**1.11** Refer to Case Study 1.6, in which the relationship between aspirin and heart attack rates was examined. Using the results of this experiment, what do you think is the base rate of heart attacks for men like the ones in this study? Explain.

**1.12** A random sample of 1001 University of California faculty members taken in December 1995 was asked, "Do you favor or oppose using race, religion, sex, color, ethnicity, or national origin as a criterion for admission to the University of California?" (Roper Center, 1996). Fifty-two percent responded "favor."

a. What is the population for this survey?

b. What is the approximate margin of error for the survey?

c. Based on the results of the survey, could it be concluded that a majority (over 50%) of *all* University of California faculty members favor using these criteria? Explain.

**1.13** In this chapter, you learned that cause and effect can be concluded from randomized experiments but generally not from observational studies. Why don't researchers simply conduct all studies as randomized experiments rather than observational studies?

**1.14** Give an example of a question you would like to have answered, such as whether eating chocolate helps to prevent depression. Then explain how a randomized experiment or an observational study could be done to study this question.

**1.15** Refer to the data and five-number summaries given in Case Study 1.1. Give a numerical value for each of the following.

**a.** The fastest speed driven by anyone in the class.

**b.** The slowest speed driven by a male.

**c.** The speed for which 1/4 of the women had driven at that speed or faster.

**d.** The proportion of females who had driven 89 mph or faster.

**e.** The number of females who had driven 89 mph or faster.

**1.16** Why was the study described in Case Study 1.5 conducted as an observational study instead of an experiment?

**1.17** Students in a statistics class at Penn State were asked, "About how many minutes do you typically exercise in a week?" Responses from the *women* in the class were

60, 240, 0, 360, 450, 200, 100, 70, 240, 0, 60, 360, 180, 300, 0, 270

Responses from the *men* in the class were

180, 300, 60, 480, 0, 90, 300, 14, 600, 360, 120, 0, 240

a. Compare the women to the men using a dotplot. What does your plot show you about the difference between the men and the women?

b. For each gender, determine the median response.

c. Do you think there's a "significant" difference between the weekly amount that men and women exercise? Explain.

**1.18** Refer to Exercise 1.17.

a. Create a five-number summary for the men's responses. Show how you found your answer.

b. Use your five-number summary to describe in words the exercise behavior of this group of students.

**1.19** Case Study 1.6 reported that the use of aspirin reduces the risk of heart attack and that the relationship was found to be "statistically significant." Does either of the cautions in the "moral of the story" for Case Study 1.7 apply to this result? Explain.

**1.20** An article in the magazine *Science* (Service, 1994) discussed a study comparing the health of 6000 vegetarians and a similar number of their friends and relatives who were not vegetarians. The vegetarians had a 28% lower death rate from heart attacks and a 39% lower death rate from cancer, even after the researchers accounted for differences in smoking, weight, and social class. In other words, the reported percentages were the remaining differences after adjusting for differences in death rates due to those factors.

a. Is this an observational study or a randomized experiment? Explain.

b. On the basis of this information, can we conclude that a vegetarian diet causes lower death rates from heart attacks and cancer? Explain.

c. Give an example of a potential confounding variable and explain what it means to say that this is a confounding variable.

**1.21** Refer to Exercise 1.20, comparing vegetarians and non-vegetarians for two causes of death. Were base rates given for the two causes of death? If so, what were they? If not, explain what a base rate would be for this study.

**1.22** An article in the *Sacramento Bee* (March 8, 1984, p. A1) reported on a study finding that "men who drank 500 ounces or more of beer a month (about 16 ounces a day) were three times more likely to develop cancer of the rectum than nondrinkers." In other words, the rate of cancer was triple in the beer-drinking group when compared with the nonbeer drinkers in this study. What important numerical information is missing from this report?

**1.23** Dr. Richard Hurt and his colleagues (Hurt et al., 1994) randomly assigned volunteers wanting to quit smoking to wear either a nicotine patch or a placebo patch to determine whether wearing a nicotine patch improves the chance of quitting. After 8 weeks of use, 46% of those wearing the nicotine patch but only 20% of those wearing the placebo patch had quit smoking.

a. Was this a randomized experiment or an observational study?

---

● Basic skills     ◆ Dataset available but not required     **Bold-numbered** exercises answered in the back

b. The difference in the percentage of participants who quit (20% versus 46%) was statistically significant. What conclusion can be made on the basis of this study?

c. Why was it advisable to assign some of the participants to wear a placebo patch?

1.24 Refer to the study in Exercise 1.23, in which there was a statistically significant difference in the percentage of smokers who quit using a nicotine patch and a placebo patch. Now read the two cautions in the "moral of the story" for Case Study 1.7. Discuss each of them in the context of this study.

1.25 Refer to the study in Exercises 1.23 and 1.24, comparing the percentage of smokers who quit using a nicotine patch and a placebo patch. Refer to the definition of statistics given on page 1, and explain how it applies to this study.

**1.26** The Roper Organization conducted a poll in 1992 (Roper, 1992) in which one of the questions asked was whether or not the respondent had ever seen a ghost. Of the 1525 people in the 18- to 29-year-old age group, 212 said "yes."

 **a.** What is the risk of someone in this age group having seen a ghost?

 **b.** What is the approximate margin of error that accompanies the proportion in part (a)?

 **c.** What is the interval that is 95% certain to contain the actual proportion of people in this age group who have seen a ghost?

1.27 Refer to Exercise 1.26. The Roper Organization selected a random sample of adults in the United States for this poll. Suppose listeners to a late-night radio talk show were asked to call and report whether or not they had ever seen a ghost.

 a. What is this type of sample called?

 b. Do you think the proportion reporting that they had seen a ghost for the radio poll would be higher or lower than the proportion for the Roper poll? Explain.

**1.28** A popular Sunday newspaper magazine often includes a yes-or-no survey question such as "Do you think there is too much violence on television?" or "Do you think parents should use physical discipline?" Readers are asked to mail their answers to the magazine, and the results are reported in a subsequent issue.

 **a.** What is this type of sample called?

 **b.** Do you think the results of these polls represent the opinions of all readers of the magazine? Explain.

1.29 Refer to Case Study 1.6. Go through the five steps listed under "The Discovery of Knowledge" in Section 1.3, and show how each step was addressed in this study.

1.30 Refer to Case Study 1.5. Explain what mistakes were made in the implementation of steps 4 and 5 of "The Discovery of Knowledge" when *USA Today* reported the results of this study.

**Statistics⏿Now™** Preparing for an exam? Assess your progress by taking the post-test at **http://1pass.thomson.com.**

**ⓋMentor**

Do you need a live tutor for homework problems? Access vMentor at **http://1pass.thomson.com** for one-on-one tutoring from a statistics expert.

# 2

*Why might the average weight of the men in this boat be lower than you think?*

See Example 2.16 (p. 48)

# Turning Data Into Information

> Looking at a long list of numbers is about the same as looking at a scrambled set of letters. To get information from numerical data, you have to organize it in ways that allow you to answer questions of concern to you.

In Case Study 1.1, we analyzed the responses that 189 college students gave to the question "What's the fastest you've ever driven a car?" The "moral of the story" for that case study was that *simple summaries of data can tell an interesting story and are easier to digest than long lists.* In this chapter, you will learn how to create simple summaries and pictures from various kinds of raw data. ∎

## 2.1  Raw Data

**Raw data** is a term used for numbers and category labels that have been collected but have not yet been processed in any way. For example, here is a list of questions asked in a large statistics class and the "raw data" given by one of the students:

| | |
|---|---|
| 1. What is your sex (m = male, f = female)? | Raw data: m |
| 2. How many hours did you sleep last night? | Raw data: 5 hours |
| 3. Randomly pick a letter—*S* or *Q*. | Raw data: *S* |
| 4. What is your height in inches? | Raw data: 67 inches |
| 5. Randomly pick a number between 1 and 10. | Raw data: 3 |
| 6. What's the fastest you've ever driven a car (mph)? | Raw data: 110 mph |
| 7. What is your right handspan in centimeters? | Raw data: 21.5 cm |
| 8. What is your left handspan in centimeters? | Raw data: 21.5 cm |

**Statistics △ Now™**

Throughout the chapter, this icon introduces a list of resources on the StatisticsNow website at **http://1pass.thomson.com** that will:
- Help you evaluate your knowledge of the material
- Allow you to take an exam-prep quiz
- Provide a Personalized Learning Plan targeting resources that address areas you should study

For questions 7 and 8, a centimeter ruler was provided on the survey form, and handspan was defined as the distance covered on the ruler by a stretched hand from the tip of the thumb to the tip of the small finger. For question 3, about one half of the students saw the choice of letters in the reverse order, so their question was "Randomly pick a letter—*Q* or *S*." This was done to learn whether students might be more likely to pick the first choice offered, regardless of whether it was the *S* or the *Q*. If you do Exercise 2.27 at the end of this chapter, you will learn the result. You may be wondering why question 5 was asked. Your curiosity will be satisfied as you keep reading this chapter.

## Datasets, Observations, and Variables

A **variable** is a characteristic that can differ from one individual to the next. Students in the statistics class provided raw data for eight variables: sex, hours of sleep, choice of a letter, height, choice of a number, fastest speed ever driven, right handspan, and left handspan. The instructor imposed a ninth variable: the order of listing *S* and *Q* in question 3.

An **observational unit** is a single individual who participates in a study. More simply, an individual may be called an **observation.** The word *observation* might also be used to describe the value of a single measurement, such as height = 67 inches. The **sample size** for a study is the total number of observational units. The letter $n$ is used to represent the sample size; one hundred and ninety students participated in the class survey, so the sample size is $n = 190$.

A **dataset** is the complete set of raw data, for all observational units and variables, in a survey or experiment. When statistical software or a spreadsheet program will be used to summarize the raw data, the dataset typically is organized so that each row of the dataset gives the data for one observational unit and each column gives the raw data for a particular variable. For the statistics class survey, the following display shows the first five rows of a dataset created for the Minitab statistical software program. These five rows give the raw data for five of the 190 students in the total dataset. Notice that each column label gives a clue to which variable is in the column. (T after column number indicates text.) The final column indicates the order of presenting letters in question 3.

**pennstate1.mtw ***

| → | C1-T | C2 | C3-T | C4 | C5 | C6 | C7 | C8 | C9-T |
|---|------|----|------|----|----|----|----|----|------|
|   | Sex | HrsSleep | SQpick | Height | RandNumb | Fastest | RtSpan | LftSpan | Form |
| 1 | M |  5.00 | S | 67.000 | 3 | 110 | 21.50 | 21.50 | SorQ |
| 2 | M |  7.00 | S | 75.000 | 9 | 109 | 22.50 | 22.50 | SorQ |
| 3 | M |  6.00 | S | 73.000 | 7 |  90 | 23.50 | 24.00 | SorQ |
| 4 | F |  7.50 | S | 64.000 | 8 |  80 | 20.00 | 21.00 | SorQ |
| 5 | F |  7.00 | S | 63.000 | 7 |  75 | 19.00 | 19.00 | SorQ |

## Data from Populations and Samples

In Case Study 1.3 in Chapter 1, we learned that data from a representative sample can provide information about the larger population represented by the data. When measurements are taken from a subset of a population, they represent **sample data.** When all individuals in a population are measured, the measurements represent **population data.** Sometimes the reason for collecting the data is important in making this distinction. For instance, data measured in a statistics class are population data if we use them to describe only that class but are sample data if we use them to represent a larger collection of students.

A major theme of this book is that sample data can be used to make inferences about a larger population, so it is important to determine whether raw data were obtained from a sample or from the complete population. However, the material in this chapter explains how to summarize data, and these descriptive methods are the same for both sample and population data. Therefore, in this chapter we will distinguish between sample and population data only when the notation differs for the two situations.

## Parameters and Statistics

The generic names used for summary measures from sample and population data also differ. A summary measure computed from sample data is called a **statistic,** while a summary measure for an entire population is called a **parameter.** However, this distinction is often overlooked when numerical summaries are the only information of interest from either a population or a sample. In that case, the summary numbers for either a population or a sample are simply called **descriptive statistics.** We will begin to emphasize the distinction between populations and samples in Chapter 9.

2.1 Exercises are on page 58.

*thought question 2.1*   There were almost 200 students who answered the survey questions. Formulate four interesting questions that you would like to answer using the data from these students. What kind of summary information would help you answer your questions?*

## ✳ 2.2  Types of Variables ✳

Different types of summaries are appropriate for different types of data. It makes sense, for example, to calculate the average number of hours of sleep last night for the members of a group, but it doesn't make sense to calculate the average sex (male, female) for the group. For gender data, it makes more sense to determine the proportion of the group that is male and the proportion that is female.

We learned in the previous section that a **variable** is a characteristic that differs from one individual to the next. A variable may be a *categorical* characteristic, such as a person's sex, or a *numerical* characteristic, such as hours of sleep last night. To determine what type of summary might provide meaningful information, you first have to recognize which type of variable you want to summarize.          *w/ examples*

The raw data from a **categorical variable** consists of group or category names that don't necessarily have any logical ordering. Each individual falls into one and only one category. For a categorical variable, the most fundamental summaries are how many individuals and what percent of the group fall into each category. Here are a few examples of categorical variables and their possible categories:

| Variable | Possible Categories |
|---|---|
| Dominant hand | Left-handed, Right-handed |
| Regular church attendance | Yes, No |
| Opinion about marijuana legalization | Yes, No, Not sure |
| Eye color | Brown, Blue, Green, Hazel |

**\*HINT:** An example is "What was the average amount of sleep for these students?" Case Study 1.1 could be utilized to generate another example.

The term **ordinal variable** may be used to describe the data when a categorical variable has ordered categories. For example, suppose that you are asked to rate your driving skills compared to the skills of other drivers, using the codes 1 = better than average, 2 = average, and 3 = worse than average. The response is an ordinal variable because the response categories are ordered perceptions of driving skills. Another example of an ordinal variable is the response to the instruction "On a scale of 1 to 7, where 1 = poor and 7 = excellent, rate your teacher."

Raw data from a **quantitative variable** are recorded as numerical values, and the data are either measurements or counts taken on each individual. All individuals can be meaningfully ordered according to these values, and averaging and other arithmetic operations make sense for these data. Some examples of quantitative variables are height, weight, hours of sleep last night, years of education, and number of classes missed this week. Not all numbers fit the definition of a quantitative variable. For instance, Social Security numbers or student identification numbers may carry some information (such as region of the country where the Social Security number was obtained), but it is not generally meaningful to put them into numerical order or to determine the average Social Security number.

Quantitative variable, **measurement variable,** and **numerical variable** are synonyms for quantitative data. The term **continuous variable** can also be used for quantitative data when every value within some interval is a possible result. For example, height is a continuous quantitative variable because any height within some range is possible. The limitations of measuring tapes, however, don't allow us to measure heights accurately enough to find that a person's actual height is 66.5382617 inches. Even if we could measure that accurately, we would usually prefer to round such a height to 66.5 inches.

A data type can be a function of how something is measured. For instance, household income is a numerical value with two digits after the decimal place, and if it is recorded this way, it is a quantitative variable. Researchers, however, often collect household income data using ordered categories such as 1 = less than $25,000, 2 = $25,000 to $50,000, 3 = $50,000 to $75,000, and so on. When household income is recorded by using categories like these, it becomes a categorical variable or, more specifically, an ordinal variable. In some situations, household income could be categorized very broadly, as when it is used to determine whether or not someone qualifies for a loan. In that case, income may be either "high enough" to qualify for the loan or "not high enough."

The way we summarize an ordinal variable can depend on the purpose. As an example, consider the self-rating of driver skill with 1 = better than average, 2 = average, and 3 = worse than average. We can summarize the responses in the way we usually do for a categorical variable, which is to find the number and percentage of the sample who responded in each category. It could also be informative to treat the variable as quantitative data and find the average of the responses to see whether it is close to 2, which it should be if all respondents give an honest appraisal of their abilities. On the other hand, it would not make sense to talk about "average household income" using the numerical codes attached to the income categories described in the previous paragraph.

*in summary*    Types of Variables

- Raw data from **categorical variables** consist of group or category names that don't necessarily have a logical ordering. Examples: eye color, country of residence.
- Categorical variables for which the categories have a logical ordering are called **ordinal variables.** Examples: highest educational degree earned, T-shirt size (S, M, L, XL).
- Raw data from **quantitative variables** consist of numerical values taken on each individual. Examples: height, number of siblings.

*thought question 2.2*    Review the data collected in the statistics class, listed in Section 2.1, and identify a type for each variable. The only one that is ambiguous is question 5. That question asks for a numerical response, but as we will see later in this chapter, it is more interesting to summarize the responses as if they are categorical.*

## Asking the Right Questions

As with most situations in life, the information you get when you summarize a dataset depends on how careful you are about asking for what you want. Here are some examples of the types of questions that are most commonly of interest for different kinds of variables and combinations of variables.

### One Categorical Variable

*Example:* What percentage of college students favors the legalization of marijuana, and what percentage of college students opposes legalization of marijuana?

Opinion about the legalization of marijuana is a categorical variable with two possible response categories (favor or oppose). In general, for one categorical variable, it is useful to ask what percentage of individuals fall into each category.

### Two Categorical Variables

*Example:* In Case Study 1.6, we asked if the risk of having a heart attack was different for the physicians who took aspirin than for those who took a placebo. Does the likelihood of a male physician having a heart attack depend on whether he has been taking aspirin or a placebo?

This example involves two categorical variables: whether a physician had a heart attack or not and whether a physician took aspirin or a placebo. For a combination of two categorical variables, we may ask if there is a relationship between the two variables. Specifically, does the category into which an indi-

*****HINT:** For each variable, consider whether the raw data are meaningful quantities or category names.

vidual falls for one variable seem to depend on which category they are in for the other variable?

### One Quantitative Variable

*Example:* What is the average handspan measurement for college women, and how much variability is there in their handspan measurements?

Handspan is a quantitative variable. To summarize one quantitative variable, we typically ask about the interesting summary measures, such as the average or the range of values, that help us understand the collection of individuals who were measured.

### One Categorical and One Quantitative Variable

*Example:* Do men and women drive at the same "fastest speeds" on average?

In this example, we are considering how a quantitative variable (fastest ever driven) is related to a categorical variable (sex). A general question about this type of situation is whether the quantitative measurements are similar across the categories, or do they differ? This question might be approached by examining whether the average measurement (such as fastest ever driven) is the same or different for the categories (men and women). We might also ask whether the range of measurements is the same or different across the categories.

### Two Quantitative Variables

*Example:* Do taller people also tend to have larger handspans?

Height and handspan, the two variables in this example, are both quantitative variables. A question we can ask about two quantitative variables is whether they are related so that when measurements are high (low) on one variable the measurements for the other variable also tend to be high (low).

## Explanatory and Response Variables

Three of the questions just listed were questions about the relationship between two variables. In these instances, it is useful to identify one variable as the **explanatory variable** and the other variable as the **response variable.** In general, the value of the *explanatory variable* for an individual is thought to partially explain the value of the *response variable* for that individual. For example, in examining the relationship between smoking and lung cancer, whether or not an individual smokes is the explanatory variable, and whether or not he or she develops lung cancer is the response variable. If we note that people with higher education levels generally have higher incomes, education level is the explanatory variable and income is the response variable. The identification of one variable as "explanatory" and the other as "response" does not imply that there is a *causal* relationship. It simply implies that knowledge of the value of the explanatory variable may help to provide knowledge about the value of the response variable for an individual.

2.2 Exercises are on page 59.

## 2.3 Summarizing One or Two Categorical Variables

### Numerical Summaries

To summarize a categorical variable, the first step is to count how many individuals fall into each possible category. Percentages usually are more informative than counts, so the second step is to calculate the percentage in each category. These two easy steps can also be used to summarize a combination of two categorical variables.

**Example 2.1**

Statistics◯Now™

For software help, download your Minitab, Excel, TI-83, SPSS, R, and JMP manuals from **ttp://1pass .thomson.com**, or find them on your CD.

**Table 2.1** Seatbelt Use by Twelfth-Graders When Driving

| Response | Count | Percent |
|----------|-------|---------|
| Always | 1686 | 55.4% |
| Most times | 578 | 19.0% |
| Sometimes | 414 | 13.6% |
| Rarely | 249 | 8.2% |
| Never | 115 | 3.8% |
| **Total** | **3042** | **100%** |

*Source:* http://www.cdc.gov/HealthyYouth/ yrbs/index.htm.

**Seatbelt Use by Twelfth-Graders** How often do you wear a seatbelt when driving a car? This is one of many questions asked in a biennial nationwide survey of American high school students. The survey, conducted as part of a federal program called the Youth Risk Behavior Surveillance System, is sponsored and organized by the U.S. Centers for Disease Control. Survey questions concern potentially risky behaviors such as cigarette smoking, alcohol use, and so on. For the question about seatbelt use when driving, possible answers were Always, Most times, Sometimes, Rarely, and Never. An additional choice allowed respondents to say that they don't drive, which often was the case because many survey participants were under the minimum legal driving age.

Table 2.1 summarizes responses in the 2003 survey given by twelfth-grade students who said that they drive. The total sample size for the table is $n = 3042$ students. Notice that a majority, $1686/3042 = .554$, or 55.4%, said that they always wear a seatbelt when driving, while just $115/3042 = .038$, or 3.8%, said that they never wear a seatbelt. Because 55.4% said that they always wear a seatbelt, we can calculate the percentage who don't always wear a seatbelt as $100\% - 55.4\% = 44.6\%$. Alternatively, the percentage saying that they don't always wear a seatbelt could be determined as 19.0% + 13.6% + 8.2% + 3.8%, the sum of the percentages for all categories other than Always.

One stereotype about males and females is that males are more likely to engage in risky behaviors than females are. Do you think the percentage who never wear a seatbelt is greater for males than it is for females? Are females more likely to say that they always wear a seatbelt? Table 2.2 summarizes seatbelt use for twelfth-grade males and females in the sample. To facilitate comparison, percentages are given within each sex. Using these percentages, we see that females were more likely than males to respond Always. Among females, 62.4% said that they always wear a seatbelt ($915/1467 = .624$) compared to 49.0% of the males ($771/1575 = .490$). Males were more likely than females to

**Table 2.2** Gender and Seatbelt Use by Twelfth-Graders When Driving

| | Always | Most Times | Sometimes | Rarely | Never | Total |
|--------|--------|-----------|-----------|--------|-------|-------|
| **Female** | 915 (62.4%) | 276 (18.8%) | 167 (11.4%) | 84 (5.7%) | 25 (1.7%) | 1467 (100%) |
| **Male** | 771 (49.0%) | 302 (19.2%) | 247 (15.7%) | 165 (10.5%) | 90 (5.7%) | 1575 (100%) |

*Source:* http://www.cdc.gov/HealthyYouth/yrbs/index.htm.

rarely or never use seatbelts. Adding the percentages for Rarely and Never gives 10.5% + 5.7% = 16.2% for the males and 5.7% + 1.7% = 7.4% for the females.

The results in Table 2.2 describe how gender and seatbelt use are related for this particular sample of twelfth-grade drivers. Do these sample data provide enough information for us to infer that gender and seatbelt use are related variables in the larger population of all U.S. twelfth-grade drivers? We will learn how to answer this type of question in Chapters 6 and 15. More generally, using sample data to answer population questions is what statistics is all about. ■

## Frequency and Relative Frequency

**Frequency** is a synonym for the count of how many observations fall into a category. The proportion or percentage in a category is a type of **relative frequency,** the count in a category relative to the total count over all categories. A **frequency distribution** for a categorical variable is a listing of all categories along with their frequencies (counts). A **relative frequency distribution** is a listing of all categories along with their relative frequencies (given as proportions or percentages, for example). It is commonplace to give the frequency and relative frequency distributions together, as was done in Table 2.1 on page 19.

Example 2.2

Statistics⌂Now™

Read the original source on your CD.

**Lighting the Way to Nearsightedness** A survey of 479 children found that those who had slept with a nightlight or in a fully lit room before the age of 2 had a higher incidence of nearsightedness (myopia) later in childhood (*Sacramento Bee,* May 13, 1999, pp. A1, A18). The raw data for each child consisted of two categorical variables, each with three categories. Table 2.3 gives the categories and the number of children falling into each combination of them. The table also gives percentages (relative frequencies) falling into each eyesight category, where percentages are computed within each nighttime lighting category. For example, among the 172 children who slept in darkness, about 90% (155/172 = .90) had no myopia.

**Table 2.3** Nighttime Lighting in Infancy and Eyesight

| Slept with: | No Myopia | Myopia | High Myopia | Total |
|---|---|---|---|---|
| **Darkness** | 155 (90%) | 15 (9%) | 2 (1%) | 172 (100%) |
| **Nightlight** | 153 (66%) | 72 (31%) | 7 (3%) | 232 (100%) |
| **Full Light** | 34 (45%) | 36 (48%) | 5 (7%) | 75 (100%) |
| **Total** | 342 (71%) | 123 (26%) | 14 (3%) | 479 (100%) |

The pattern in Table 2.3 is striking. As the amount of sleeptime light increases, the incidence of myopia also increases. However, this study does not prove that sleeping with light actually *caused* myopia in more children. There are other possible explanations. For example, myopia has a genetic component, so those children whose parents have myopia are more likely to suffer from it themselves. Maybe nearsighted parents are more likely to provide light while their children are sleeping. ■

*thought question* **2.3**   Can you think of possible explanations for the observed relationship between use of nightlights and myopia, other than direct cause and effect? What additional information might help to provide an explanation?*

## Explanatory and Response Variables for Categorical Variables

In both Example 2.1 and Example 2.2, percentages were given across rows. For instance, Table 2.3 in Example 2.2 shows that 90% (100% × 155/172) of children who had slept in darkness were free of myopia but only 45% of those who had slept in full light were free of myopia. The table does not provide "column percentages" such as the fact that 34/342 or about 10% of those who were free of myopia had slept in full light.

In summarizing two categorical variables, it may be possible to identify one variable as a response variable (or **outcome variable**), and the other as an explanatory variable. In Example 2.1, the outcome variable is how often a student wears a seatbelt when driving, and the explanatory variable is gender (male or female). In Example 2.2, the outcome variable is degree of myopia, and the explanatory variable is the amount of sleeptime lighting.

Notice that in both examples, the categories of the explanatory variable defined the rows of the table and the categories of the response variable defined the columns. Tables often are formed this way, and when they are, the row percentages are of more interest than the column percentages. No matter how the table is constructed, you should determine whether one variable is a response variable and the other is an explanatory variable. Within each explanatory group, we are interested in knowing what percentage fell into each response or outcome category.

---

| MINITAB *tip* | **Numerically Describing One or Two Categorical Variables** |
|---|---|

- To determine how many and what percentage fall into the categories of a single categorical variable, use **Stat>Tables>Tally Individual Variables.** In the dialog box, specify a column containing the raw data for a categorical variable. Click on any desired options for counts and percentages under "Display."

- To create a two-way table for two categorical variables, use **Stat>Tables> Crosstabulation and Chi-Square.** Specify a categorical variable in the "For rows" box and another categorical variable in the "For columns" box. Select any desired percentages (row, column, and/or total) under "Display."

---

***HINT:** Reread Example 2.2, in which one possible explanation is mentioned. What data would we need to investigate the possible explanation mentioned there?

> **SPSS *tip***  **Numerically Describing One or Two Categorical Variables**
>
> - To create a frequency table for one categorical variable, use **Analyze> Descriptive Statistics>Frequencies.**
> - To create a two-way table for two categorical variables, use **Analyze> Descriptive Statistics>Crosstabs.** Use the *Cells* button to request row and/or column percentages.

## Visual Summaries for Categorical Variables

There are two simple visual summaries that are used for categorical data:

- **Pie charts** are useful for summarizing a single categorical variable if there are not too many categories.
- **Bar graphs** are useful for summarizing one or two categorical variables and are particularly useful for making comparisons when there are two categorical variables.

Both of these simple graphical displays are easy to construct and interpret, as the following examples demonstrate.

**Example 2.3**

**Statistics⬡Now™**

For software help, download your Minitab, Excel, TI-83, SPSS, R, and JMP manuals from **ttp://1pass .thomson.com**, or find them on your CD.

**Humans Are Not Good Randomizers** Question 5 in the class survey described in Section 2.1 asked students to "Randomly pick a number between 1 and 10." The pie chart shown in Figure 2.1 illustrates that the results are not even close to being evenly distributed across the numbers. Notice that almost 30% of the students chose 7, while only just over 1% chose the number 1.

Figure 2.2 illustrates the same results with a bar graph. This bar graph shows the actual frequencies of responses on the vertical axis. The display makes it even more obvious that the number of students who chose 7 was more than double that of the next most popular choice. We also see that very few students chose either 1 or 10. ■

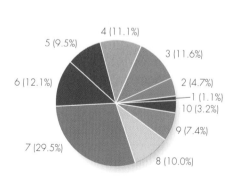

**Figure 2.1** ▮ Pie chart of numbers picked

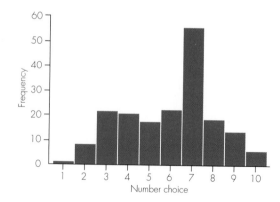

**Figure 2.2** ▮ Bar graph of numbers picked

**Example 2.4**

**Revisiting Nightlights and Nearsightedness**    Figure 2.3 illustrates the data presented in Example 2.2 with a bar chart showing, for each lighting group, the percentage that ultimately had each level of myopia. This bar chart differs from the one in Figure 2.2 in two respects. First, it is used to present data for two categorical variables instead of just one. Second, the vertical axis represents percentages instead of counts, with the percentages for myopia status computed separately within each lighting category.

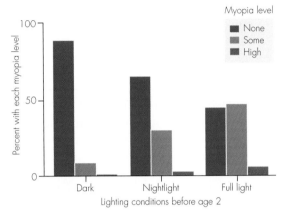

2.3 Exercises are on page 59.    **Figure 2.3** ▋ Bar chart for myopia and nighttime lighting in infancy    ■

---

***thought question 2.4***    Redo the bar graph in Figure 2.3 using counts instead of percentages. The necessary data are given in Table 2.3. Would the comparison of frequency of myopia across the categories of lighting be as easy to make using the bar graph with counts? Generalize your conclusion to provide guidance about what should be done in similar situations.*

---

*in summary*    ## Bar Graphs for Categorical Variables

In a bar graph for *one categorical variable,* you can choose one of the following to display as the height of a bar for each category, indicated by labeling the vertical axis:

- Frequency or count
- Relative frequency = number in category/overall number
- Percentage = relative frequency × 100%

*(continued)*

---

**\*HINT:** Which graph makes it easier to compare the percentage with myopia for the three groups? What could be learned from the graph of counts that isn't apparent from the graph of percentages?

In a bar graph for *two categorical variables,* if an explanatory and response variable can be identified, it is most common to:

- Draw a separate group of bars for each category of the explanatory variable.
- Within each group of bars, draw one bar for each category of the response variable.
- Label the vertical axis with percentages and make the heights of the bars for the response categories sum to 100% within each explanatory category group. It can sometimes be useful to make the heights of the bars equal the counts in the category groups instead of percentages.

---

**MINITAB *tip***    **Graphically Describing One or Two Categorical Variables**

- To draw a *bar graph,* use **Graph>Bar Chart.** In the resulting display, select **Simple** to graph one variable or select **Cluster** to graph the relationship between two variables. Then, in the "Categorical variables" box, specify the column(s) containing the raw data for the variable(s). To graph percentages rather than counts, use the **Bar Chart Options** button.
- To draw a *pie chart,* use **Graph>Pie Chart.** Use the **Multiple Graphs** button to create separate pie charts for subgroups within the dataset.

---

## 2.4  Exploring Features of Quantitative Data with Pictures

Looking at a long, disorganized list of data values is about the same as looking at a scrambled set of letters. To begin finding the information in quantitative data, we have to organize it using visual displays and numerical summaries. In this section, we focus on interpreting the main features of quantitative variables. More specific details will be given in the following sections.

Table 2.4 displays the raw data for the right handspan measurements (in centimeters) made in the student survey described in Section 2.1. The measurements are listed separately for males and females but are not organized in any other way. Imagine that you know a female whose stretched right handspan

**Table 2.4** Stretched Right Handspans (cm) of 190 College Students

---

**Males (87 students):**
21.5, 22.5, 23.5, 23, 24.5, 23, 26, 23, 21.5, 21.5, 24.5, 23.5, 22, 23.5, 22, 22, 24.5, 23, 22.5, 19.5, 22.5, 22, 23, 22.5, 20.5, 21.5, 23, 22.5, 21.5, 25, 24, 21.5, 21.5, 18, 20, 22, 24, 22, 23, 22, 22, 23, 22.5, 25.5, 24, 23.5, 21, 25.5, 23, 22.5, 24, 21.5, 22, 22.5, 23, 18.5, 21, 24, 23.5, 24.5, 23, 22, 23, 23, 24, 24.5, 20.5, 24, 22, 23, 21, 22.5, 21.5, 24.5, 22, 22, 21, 23, 22.5, 24, 22.5, 23, 23, 23, 21.5, 19, 21.5

**Females (103 students):**
20, 19, 20.5, 20.5, 20.25, 20, 18, 20.5, 22, 20, 21.5, 17, 16, 22, 22, 20, 20, 20, 20, 21.7, 22, 20, 21, 21, 19, 21, 20.25, 21, 22, 18, 20, 21, 19, 22.5, 21, 20, 19, 21, 20.5, 21, 22, 20, 20, 18, 21, 22.5, 22.5, 19, 19, 19, 22.5, 20, 13, 20, 22.5, 19.5, 18.5, 19, 17.5, 18, 21, 19.5, 20, 19, 21.5, 18, 19, 19.5, 20, 22.5, 21, 18, 22, 18.5, 19, 22, 17, 12.5, 18, 20.5, 19, 20, 21, 19, 19, 21, 18.5, 19, 21.5, 21.5, 23, 23.25, 20, 18.8, 21, 21, 20, 20.5, 20, 19.5, 21, 21, 20

is 20.5 cm. Can you see how she compares to the other females in Table 2.4? That will probably be hard because the list of data values is disorganized.

In Case Study 1.1, we graphed the "fastest ever driven" responses with a simple **dotplot.** We also summarized the data using a **five-number summary,** which consists of the median, the quartiles (roughly, the medians of the lower and upper halves of the data), and the extremes (low, high). Let's use those methods to organize the handspan data in Table 2.4.

**Right Handspans**  In Figure 2.4, each dot represents the handspan of an individual student, with the value of the measurement shown along the horizontal axis. From this dotplot, we learn that a majority of the females had handspans between 19 and 21 centimeters and a good number of the males had handspans between 21.5 and 23 cm. We also see that there were two females with unusually small handspans compared to those of the other females.

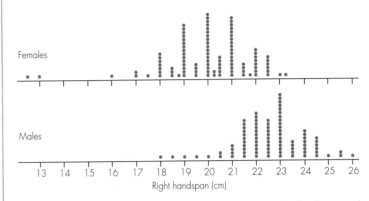

**Figure 2.4** ▌ Stretched right handspans (in centimeters) of college students

Here are five-number summaries for the male and females handspan measurements given in Table 2.4 and graphed in Figure 2.4:

|  | Males (87 Students) | | Females (103 Students) | |
|---|---|---|---|---|
| Median |  | 22.5 |  | 20.0 |
| Quartiles | 21.5 | 23.5 | 19.0 | 21.0 |
| Extremes | 18.0 | 26.0 | 12.5 | 23.25 |

Remember that the five-number summary approximately divides the dataset into quarters. For example, about 25% of the female handspan measurements are between 12.5 and 19.0 centimeters, about 25% are between 19 and 20 cm, about 25% are between 20 and 21 cm, and about 25% are between 21 and 23.25 cm. The five-number summary, along with the dotplot, gives us a good idea of where our imagined female with the 20.5-centimeter handspan fits into the distribution of handspans for females. She is in the third quarter of the data, slightly above the median (the middle value). ▌

## Summary Features of Quantitative Variables

The **distribution** of a quantitative variable is the overall pattern of how often the possible values occur. For most quantitative variables, three summary characteristics of the overall distribution of the data tend to be of the most interest. These are the **location** (center, average), the **spread** (variability), and the **shape** of the data. We also will be interested in whether there are any **outliers** in the data, values that are unusual compared to the bulk of the other values.

## Location (Center, Average)

The first concept for summarizing a quantitative variable is the idea of the "center" of the distribution of values, also called the *location* of the data. What is a typical or average value? The **median,** approximately the middle value in the data, is one estimate of location. The **mean,** which is the usual arithmetic average, is another. Details about how to compute these are given in Section 2.5.

## Spread (Variability)

Essentially, statistics is the study of **variability**—the variability (spread) among individual measurements and, as you will learn later in the text, the variability among different samples taken from the same population. The amount of variability among the individual measurements is an important feature of any dataset. How spread out are the values? Are all values about the same? Are most of them together but with a few that are unusually high or low?

In a five-number summary, we can assess the amount of spread (variability) in the data by looking at the difference between the two extremes (called the *range*) and the difference between the two quartiles (called the *interquartile range*). Later in this chapter, you will learn about the *standard deviation,* another important measure of variability.

The idea of variability is particularly important in interpreting data. Much of our work in statistics involves comparing an observed difference to what we expect if the difference is due solely to natural variability. For instance, to know whether the amount of rainfall during a year at a location is unusual, we have to know how much rainfall amounts naturally vary from year to year. To determine whether a one-year-old child might be growing abnormally, we need to know how much heights of one-year-old children naturally vary.

## Shape

The third interesting feature of a quantitative variable to consider is the **shape** of how the values are distributed. Using the appropriate visual displays, we can address questions about *shape* such as the following: Are most of the values clumped in the middle, with values tailing off at each end (like the handspan measurements shown in Figure 2.4)? Are most of the values clumped together on one end (either high or low), with the remaining few values stretching relatively far toward the other end? Are there two distinct groupings of values? We will discuss *shape* more completely later in this section, on page 33.

# Outliers

Another interesting feature to consider is whether any individual values are **outliers.** There is no precise definition for an outlier, but in general, an outlier is a data point that is not consistent with the bulk of the data. For a single variable, an outlier is a value that is unusually high or low. When two variables are considered, an outlier is an unusual combination of values. For instance, if you look back at Example 2.5, you will see that a female with a handspan of 24.5 cm would be an outlier because this handspan is well past the largest of the measurements made by the 103 females. On the other hand, a male with a handspan of 24.5 cm is not an outlier because this measurement is clearly within the normal range for males.

It is important to note that in every dataset, there will be extreme values in each direction (low and high) unless all values are exactly the same. The extremes do not necessarily qualify as outliers. The next example illustrates that sometimes the extreme points are the most interesting features of a dataset, even if they might not be outliers.

**Example 2.6** | **Ages of Death of U.S. First Ladies** Much has been written about ages of U.S. Presidents when elected and at death, but what about their wives? Do these women tend to live short lives or long lives? Table 2.5 lists the First Ladies of the United States and their approximate ages at death (to within one year) as listed in late 2002 at the White House website. It is not completely accurate to label all of these women "First Ladies" if the strict definition is "the wife of a President while in office." For example, Harriet Lane served socially as "First Lady" to President James Buchanan, but he was unmarried and she was his niece. A few of the women listed died before their husband's term in office. Nonetheless, we will use the data as provided by the White House and summarize the ages at death for these women. Here is a five-number summary for these ages:

|  | First Ladies' Ages at Death | |
| --- | --- | --- |
| Median | | 70 | |
| Quartiles | 60 | | 81.5 |
| Extremes | 34 | | 97 |

If you are at all interested in history, this summary will make you curious about the extreme points. Who died at 34? Who lived to be 97? The extremes are more interesting features of this dataset than is the summary of ages in the middle, which tend to match what we would expect for ages at death. From Table 2.5, you can see that Thomas Jefferson's wife Martha died in 1782 at age 34, almost 20 years before he entered office. He reportedly was devastated, and he never remarried, although historians believe that he may have had other children in his relationship with Sally Hemings. At the other extreme, Bess Truman died in 1982 at age 97; her husband Harry preceded her in death by ten years, but he too lived a long life—he died at age 88.

**Table 2.5** The First Ladies of the United States of America

| Name | Born–Died | Age at Death |
|------|-----------|--------------|
| Martha Dandridge Custis Washington | 1731–1802 | 71 |
| Abigail Smith Adams | 1744–1818 | 74 |
| Martha Wayles Skelton Jefferson | 1748–1782 | 34 |
| Dolley Payne Todd Madison | 1768–1849 | 81 |
| Elizabeth Kortright Monroe | 1768–1830 | 62 |
| Louisa Catherine Johnson Adams | 1775–1852 | 77 |
| Rachel Donelson Jackson | 1767–1828 | 61 |
| Hannah Hoes Van Buren | 1783–1819 | 36 |
| Anna Tuthill Symmes Harrison | 1775–1864 | 89 |
| Letitia Christian Tyler | 1790–1842 | 52 |
| Julia Gardiner Tyler | 1820–1889 | 69 |
| Sarah Childress Polk | 1803–1891 | 88 |
| Margaret Mackall Smith Taylor | 1788–1852 | 64 |
| Abigail Powers Fillmore | 1798–1853 | 55 |
| Jane Means Appleton Pierce | 1806–1863 | 57 |
| Harriet Lane | 1830–1903 | 73 |
| Mary Todd Lincoln | 1818–1882 | 64 |
| Eliza McCardle Johnson | 1810–1876 | 66 |
| Julia Dent Grant | 1826–1902 | 76 |
| Lucy Ware Webb Hayes | 1831–1889 | 58 |
| Lucretia Rudolph Garfield | 1832–1918 | 86 |
| Ellen Lewis Herndon Arthur | 1837–1880 | 43 |
| Frances Folsom Cleveland | 1864–1947 | 83 |
| Caroline Lavinia Scott Harrison | 1832–1892 | 60 |
| Ida Saxton McKinley | 1847–1907 | 60 |
| Edith Kermit Carow Roosevelt | 1861–1948 | 87 |
| Helen Herron Taft | 1861–1943 | 82 |
| Ellen Louise Axson Wilson | 1860–1914 | 54 |
| Edith Bolling Galt Wilson | 1872–1961 | 89 |
| Florence Kling Harding | 1860–1924 | 64 |
| Grace Anna Goodhue Coolidge | 1879–1957 | 78 |
| Lou Henry Hoover | 1874–1944 | 70 |
| Anna Eleanor Roosevelt Roosevelt | 1884–1962 | 78 |
| Elizabeth Virginia Wallace Truman | 1885–1982 | 97 |
| Mamie Geneva Doud Eisenhower | 1896–1979 | 83 |
| Jacqueline Lee Bouvier Kennedy Onassis | 1929–1994 | 65 |
| Claudia Taylor Johnson | 1912– | |
| Patricia Ryan Nixon | 1912–1993 | 81 |
| Elizabeth Bloomer Ford | 1918– | |
| Rosalynn Smith Carter | 1927– | |
| Nancy Davis Reagan | 1923– | |
| Barbara Pierce Bush | 1925– | |
| Hillary Rodham Clinton | 1947– | |
| Laura Welch Bush | 1946– | |

*Source:* http://www.whitehouse.gov/WH/glimpse/firstladies/html/firstladies.html.

Should we attach the label "outlier" to either of the most extreme points in the list of ages at death for the First Ladies? To study this issue, we have to examine all of the data to see whether the two extremes clearly stand apart from the other values. If you look over Table 2.5, you may be able to form an opinion about whether Martha Jefferson and Bess Truman should be called outliers by comparing them to the other First Ladies. Making sense of a list of numbers, however, is difficult. The most effective way to look for outliers is to graph the data, which we will learn more about in the remainder of this section.

## Pictures of Quantitative Data

There are three similar types of pictures that are used to represent quantitative variables, all of which are valuable for assessing center, spread, shape, and outliers. **Histograms** are similar to bar graphs and can be used for any number of data values, although they are not particularly informative when the sample size is small. **Stem-and-leaf plots** and **dotplots** present all individual values, so for very large datasets, they are more cumbersome than histograms. A fourth kind of picture, called a **boxplot** or **box-and-whisker plot,** displays the information given in a five-number summary. It is especially useful for comparing two or more groups and for identifying outliers. We will examine boxplots at the end of this section, after covering the first three types.

Figures 2.5 to 2.7 illustrate a histogram, a stem-and-leaf plot, and a dotplot, respectively, for the females' right handspans displayed in Table 2.4. Figure 2.7 is merely a portion of the dotplot shown previously in Figure 2.4 on page 25. Examine the three figures. Notice that if the stem-and-leaf plot were turned on its side, all three pictures would look similar. Each picture shows the *distribution* of the data—the pattern of how often the various measurements occurred.

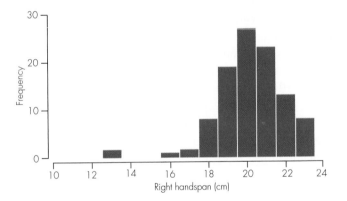

**Figure 2.5** ▮ Histogram of females' right handspans

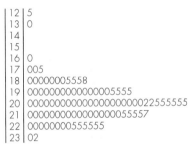

```
12 | 5
13 | 0
14 |
15 |
16 | 0
17 | 005
18 | 0000000555 8
19 | 00000000000000005555
20 | 000000000000000000000022555555
21 | 00000000000000000055557
22 | 00000000555555
23 | 02
```

Example:   | 12 | 5 = 12.5

**Figure 2.6** ▮ Stem-and-leaf plot of females' right handspans

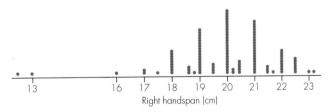

**Figure 2.7** ▮ Dotplot of females' right handspans

## Interpreting Histograms, Stem-and-Leaf Plots, and Dotplots

Each of these pictures is useful for assessing the location, spread, and shape of a distribution, and each is also useful for detecting outliers. For the data presented in Figures 2.5 to 2.7, notice that the values are *centered* at about 20 cm, which we learned in Example 2.5 is indeed the median value. There are two possible *outlier* values that are low in comparison to the bulk of the data. These are identifiable in the stem-and-leaf plot as 12.5 and 13.0 cm but are evident in the other two pictures as well. Except for those values, the handspans have a *range* of about 7 cm, extending from about 16 to 23 cm. They tend to be clumped around 20 and taper off toward 16 and 23.

There are many computer programs that can be used to create these pictures. Figures 2.5, 2.6, and 2.7, for instance, are slight modifications of pictures created using Minitab. We will go through the steps for creating each type of picture by hand, but keep in mind that statistical software such as Minitab automates most of the process.

## Creating a Histogram

A histogram is a bar chart of a quantitative variable that shows how many values are in various intervals of the data. The steps in creating a histogram are as follows:

**Step 1:** Decide how many *equally spaced* intervals to use for the horizontal axis. The experience of many researchers is that somewhere between 6 and 15 intervals is a good number. Consider using intervals that make the range of each interval convenient.

**Step 2:** Decide whether to use *frequencies* or *relative frequencies* on the vertical axis. Frequencies simply represent the actual number of individuals who fall into each interval. Relative frequencies represent either the proportion or the percent that are in an interval.

**Step 3:** Draw the appropriate number of equally spaced intervals on the horizontal axis, making sure you cover the entire range of the data. Determine the frequency or relative frequency of data values in each interval and draw a bar with corresponding height.

**Example 2.7** | **Histograms for Ages of Death of U.S. First Ladies** Figures 2.8 and 2.9 show two different histograms for the ages of death for the First Ladies of the United States. The raw data were given in Table 2.5 on page 28. In each histogram, the

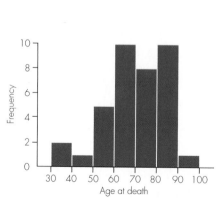

**Figure 2.8** ▌ Histogram of ages of death of U.S. First Ladies using seven ten-year intervals

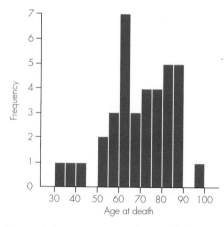

**Figure 2.9** ▌ Histogram of ages of death of U.S. First Ladies using fourteen five-year intervals

horizontal axis gives age at death and the vertical axis gives the frequency of how many First Ladies died within the age interval represented by any particular bar.

Figure 2.8 is drawn using seven ten-year age intervals, beginning at 30 and ending at 100. Thus the heights of the bars show how many First Ladies died in their 30s, in their 40s, and so on, up to 90s. Figure 2.9 gives more detail by using fourteen five-year age intervals, beginning with the interval 30 to 35 and ending with the interval 95 to 100.

A First Lady with an age of death falling on the boundary between two age intervals is counted in the interval that begins with her age. For instance, there were two First Ladies who died at 60 years old. In Figure 2.8, they are counted among the ten First Ladies in the interval 60 to 70, and in Figure 2.9 they are counted into the interval 60 to 65. ■

## Creating a Dotplot

The dotplot in Figure 2.7 was created using the Minitab Statistical Package, and the online help feature in Minitab gives a good description of how it's done: "A dotplot displays a dot for each observation along a number line. If there are multiple occurrences of an observation, or if observations are too close together, then dots will be stacked vertically. If there are too many points to fit vertically in the graph, then each dot may represent more than one point" (Minitab, Release 12.1, 1998).

## Creating a Stem-and-Leaf Plot

A stem-and-leaf plot is created much like a histogram, except every individual data value is shown. This type of plot is a quick way to summarize small datasets and is also a useful tool for ordering the data from lowest to highest. The basic structure of the plot is that the row "stem" contains all but the last digit of a number and the "leaf" within the row stem is the last digit of the number, regardless of whether it falls before or after a decimal point.

To simplify the work, data values sometimes are truncated or rounded off. To truncate a value, simply drop digits. The number 23.58 is truncated to 23.5, but it is rounded off to 23.6. In Figure 2.6, the handspan values are truncated to one digit after the decimal point. The largest handspan is 23.25 cm, which is truncated to 23.2. This number is shown on the "23" stem along with the value of 23.0 cm. These two handspans are displayed as |23| 02. In other words, this stem has two leaves, each representing a different individual.

Here are the steps required to create a stem-and-leaf plot:

**Step 1:** Determine the stem values. Remember that the "stem" contains all but the last of the displayed digits of a number. As with histograms, it is reasonable to have between 6 and 15 stems (where each stem defines an interval of values). The stems should define equally spaced intervals.

**Step 2:** For each individual, attach a "leaf" to the appropriate stem. A "leaf" is the last of the displayed digits of a number. It is standard, but not mandatory, to put the leaves in increasing order at each stem value.

*Note:* There will be more than one way to define equally spaced stems. For example, the ages at death of First Ladies range from 34 to 97. We could have stem values representing the decades (3, 4, . . . , 9) for a total of seven stems, *or* we could allow two stem values for each decade, for a total of 14 stem values. With two stems for each decade, the first instance of each stem value would receive leaves of 0 to 4, and the second would receive leaves of 5 to 9. So the two deaths in the thirties, at ages 34 and 36, could be represented in two different ways:

$$|3|46 \quad \text{or} \quad \begin{array}{c|c} 3 & 4 \\ 3 & 6 \end{array}$$

The first method, with fewer stem values, is generally preferable for small datasets, while the second, with more stem values, is preferable for larger datasets.

---

***thought question 2.5*** For the First Ladies data, could you use three stems for each decade? Why or why not? Could you use five stems for each decade? Why or why not?*

---

**Example 2.8**

**Big Music Collections** About how many music CDs do you own? Responses to this question for 24 students in a senior-level statistics course in 1999 at Penn State University were

> 220, 20, 50, 450, 300, 30, 20, 50, 200, 35, 25, 50,
> 250, 100, 0, 100, 20, 13, 200, 2, 125, 150, 90, 60

Here is a stem-and-leaf plot of these data:

Reported music CDs owned for *n* = 24 Penn State students (stem unit = 100s, leaf unit = 10s)

```
0 | 001222233
0 | 55569
1 | 002
1 | 5
2 | 002
2 | 5
3 | 0
3 |
4 |
4 | 5
```

*Source: Class data collected by Robert Heckard.*

Notice that the numbers range from 0 to about 450. The stem unit is "100s," and the leaf unit within a row is "10s." The final digit of the response is truncated. For instance, the row $|1|002$ represents the data values 100, 100, and 125 (but would be read from the stemplot as 100, 100, 120). There is a clear outlier, with someone reporting that he or she owns 450 CDs. Even without the outlier, the data have a shape called "skewed to the right," which we will define following this example. The median number of CDs owned is 55. The arithmetic average or mean is about 107. The outlier of 450 is a legitimate part of the data but (along with the skewness) causes a big difference between the mean and the median. ■

---

**\*HINT:** Is the number of possibilities for the second digit of age evenly divisible by 3?

| MINITAB *tip* | **Drawing a Histogram, Dotplot, or Stem-and-Leaf Plot** |

- To draw a *histogram* of a quantitative variable, use **Graph>Histogram.** In the resulting display, select *Simpie.* In the "Variables" box, specify the column containing the raw data for the variable. To graph percents rather than counts, use the *Scale* button and then select the "Y-Scale Type" tab.

- To draw a *dotplot* of a quantitative variable, use **Graph>Dotplot.** Select *Simple* under "One Y" to create a dotplot showing the entire sample, as in Figure 2.6. Select *With Groups* under "One Y" to compare subgroups within the sample, as in Figure 1.1 in Chapter 1.

- To draw a *stem-and-leaf plot* of a quantitative variable, use **Graph>Stem-and-Leaf.**

## Describing Shape

The *shape* of a dataset is usually described as either **symmetric,** meaning that it is similar on both sides of the center, or **skewed,** meaning that the values are more spread out on one side of the center than on the other. If it is **skewed to the right,** the higher values (toward the right on a number line) are more spread out than the lower values. If it is **skewed to the left,** the lower values (toward the left on a number line) are more spread out than the higher values. The shape in Figures 2.5 to 2.7 is called "skewed to the left" because the values trail off to the left. Were it not for the two low values, the shape would be relatively *symmetric.*

A symmetric dataset may be **bell-shaped** or another symmetric shape. Without the two low values, the data set in Figures 2.5 to 2.7 would be considered bell-shaped because the pictures would look somewhat like the shape of a bell. We will learn much more about bell-shaped curves in Section 2.7.

The **mode** of a dataset is the most frequent value. The shape is called **unimodal** if there is a single prominent peak in a histogram, stemplot, or dotplot, as in Figures 2.5 to 2.7. The shape is called **bimodal** if there are two prominent peaks in the distribution. Figure 2.10 (on the next page) shows a histogram with a bimodal shape. The data are eruption durations (in minutes) for $n = 230$ eruptions of the Old Faithful geyser in Yellowstone Park. There is a prominent peak around 2.0 minutes and another prominent peak around 4.5 minutes.

Some datasets will be described best with a combination of these terms, while others will require only one of them. The description "bell-shaped" already conveys the information that the shape is symmetric and that it is unimodal. On the other hand, knowing that a shape is symmetric or skewed does not tell us about the number of modes. The dataset could be unimodal, bimodal, or neither. For instance, Figure 2.10 might be described as bimodal and slightly skewed to the left, since the larger bulk of the data is on the right.

## Boxplots: Picturing Location and Spread for Group Comparisons

We have already seen how to create a simple and informative five-number summary using the extremes, the quartiles, and the median. A **boxplot,** also called a box-and-whisker plot, is a simple way to picture the information in one or more

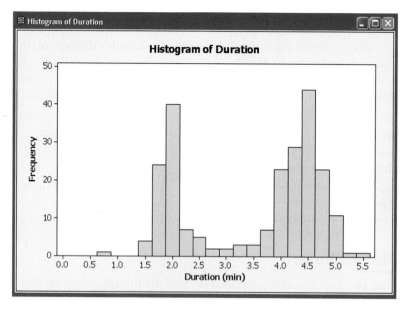

**Figure 2.10** ▌ Histogram of eruption durations (minutes) for Old Faithful geyser
*Source: Hand et al., 1994.*

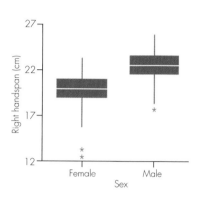

**Figure 2.11** ▌ Boxplots for right handspans of men and women

five-number summaries. This type of graph is particularly useful for comparing two or more groups and is also an effective tool for identifying outliers.

Figure 2.11 uses boxplots to compare the spans of the right hands of males and females. For each group, the box covers the middle 50% of the data, and the line within a box marks the median value. With the exception of possible outliers, the lines extending from a box reach to the minimum and maximum data values. Possible outliers are marked with an asterisk. In Figure 2.11, the vertical axis is used for the quantitative variable (handspan), but the graph could also be drawn so that the horizontal axis is used for the quantitative variable.

In Figure 2.11, we see that several features of each group are immediately obvious. The comparison between the two groups is simplified as well. The only feature of a dataset that is not obvious from a boxplot is the shape, although there is information about whether the values tend to be clumped in the middle or tend to stretch more toward one extreme or the other.

Statistical software can be used to draw boxplots. Figure 2.11, for instance, was drawn using Minitab. We will learn how to draw a boxplot by hand in Section 2.5, in which we cover the details of determining numerical summaries such as the median and quartiles.

> ***thought question 2.6*** Using the boxplots in Figure 2.11, what can you say about the respective handspans for males and females? Are there any surprises, or do you see what you would expect?*

**\*HINT:** Is there a difference in the location of the data for the two graphs? Is either group more spread out than the other?

| MINITAB *tip* | **Drawing a Boxplot** |
|---|---|

- To draw a *boxplot* of a quantitative variable, use **Graph>Boxplot.** In the resulting display, select **Simple** under "One Y." In the "Graph Variables" box, specify the column containing the raw data for the variable.

- To create a comparative boxplot as in Figure 2.11, use **Graph>Boxplot** and select **With Groups** under "One Y." Specify the quantitative response variable in the "Graph Variables" box; specify the categorical explanatory variable that creates the groups in the "Categorical variables for grouping" box.

## Strengths and Weaknesses of the Four Visual Displays

Histograms, stem-and-leaf plots, dotplots, and boxplots organize quantitative data in ways that let us begin finding the information in a dataset. As to the question of which type of display is the best, there is not a unique answer. The answer to that question depends on what feature of the data may be of interest and, to a certain degree, on the sample size. Let's consider some strengths and weaknesses of each type of plot.

### Histograms

Histograms are an excellent tool for judging the shape of a dataset, although with a small sample size, we won't get a very good picture of shape. In this case, a histogram may not "fill in" sufficiently well to show the shape of the data. The number of bars (intervals) used in this display also affects our ability to judge shape. With too few bars or too many, we may not see the true shape of the data. Somewhere between 6 and 15 intervals is usually a good number to use. We can examine location (center), spread (variability), and outliers but we will not be able to judge these features exactly using a histogram.

### Stem-and-Leaf Plots

A stem-and-leaf plot is an excellent tool for sorting data, and with a sufficient sample size, it can be used to judge shape. However, with a large sample size, the plot may be too cluttered because the display shows all individual data points. As with a histogram, we can use this type of display to make a somewhat vague judgment about location, spread, and outliers. One disadvantage of this plot, compared to a histogram, is that the intervals for the stem are restricted to be of length 1, 2, 5, or 10 and to start at multiples of those numbers. This may cause a problem for certain datasets.

### Dotplots

A dotplot shows us individual data values and can be used to help us form rather vague descriptions of location (center), spread (variability), and outliers. Dotplots tend not to be as useful for judging shape as histograms and stem-and-leaf plots. For large datasets, especially, we can judge shape better by summarizing the data into intervals as we do when creating histograms and stem-

and-leaf plots. With moderate sample sizes, comparative dotplots can be useful for comparing two or more groups, as we did in Figure 2.4 on page 25.

### Boxplots

Generally, boxplots give a direct look at location (center), spread (variability), and outliers; and they are also an excellent tool for comparing two or more groups. Unfortunately, boxplots are not entirely useful for judging shape. For instance, it is not possible to judge whether a dataset is bell-shaped using a boxplot, nor is it possible to judge whether a dataset may be bimodal.

2.4 Exercises are on page 61.

---

*in summary*   **Using Visual Displays to Identify Interesting Features of Quantitative Data**

Here is a summary of which pictures are useful for various purposes:

- To illustrate **location** and **spread,** any of the pictures work well. A boxplot marks the median, so it may be best for identifying location.
- To illustrate **shape,** histograms and stem-and-leaf plots are best.
- To see **individual values,** use stem-and-leaf plots and dotplots.
- To **sort the values,** use stem-and-leaf plots. This can be a useful first step for creating a five-number summary.
- To **compare groups,** use side-by-side boxplots, or create versions of one of the other pictures using the same scale for each group.
- To **identify outliers** using the standard definition, use a boxplot (see p. 43). Any of the other pictures will enable you to identify possible outliers.

---

## 2.5 Numerical Summaries of Quantitative Variables

In this section, we learn how to compute numerical summaries of these features for quantitative data. Recall from Section 2.1 that the numbers in a data set are called *raw data*. To write formulas for some of the summaries in this section, we need notation for the raw data.

---

*formula*   **Notation for Raw Data**

$n$ = the number of individuals in a dataset

$x_1, x_2, x_3, \ldots, x_n$ represent the individual raw data values

*Example:* A dataset consists of handspan values in centimeters for six females; the values are 21, 19, 20, 20, 22, and 19. Then,

$n = 6$

$x_1 = 21, x_2 = 19, x_3 = 20, x_4 = 20, x_5 = 22,$ and $x_6 = 19$

# Describing the Location of a Dataset

The word *location* is used as a synonym for the "middle" or "center" of a dataset. There are two common ways to describe this feature.

- The **mean** is the usual numerical average, calculated as the sum of the data values divided by the number of values. It is nearly universal to represent the mean of a sample with the symbol $\bar{x}$, read as "*x*-bar."

- The **median** of a sample is the middle data value for an odd number of observations, after the sample has been ordered from smallest to largest. It is the average of the middle two values, in an ordered sample, for an even number of observations. We will use the letter *M* to represent the median of a sample.

---

*formula* | **Determining the Mean and Median**

**The Mean**

The symbol $\bar{x}$ is nearly always used to represent the mean of a sample. Notation for calculating the sample mean of a list of values is

$$\bar{x} = \frac{\sum x_i}{n}$$

The capital Greek letter sigma, written as $\sum$, is the universal symbol meaning "add up whatever follows." Therefore, for a dataset with individual values $x_1, x_2, x_3, \ldots, x_n$, the notation $\sum x_i$ is the same as saying "add together all the values."

**The Median**

It would require more notation than is convenient to write a formula for the median, so we simply write the rule:

- If *n* is odd, the median *M* is the middle of the ordered values. Find *M* by counting $(n + 1)/2$ up from the bottom or down from the top of the ordered list.

- If *n* is even, the median *M* is the average of the middle two of the ordered values. Find *M* by averaging the values that are $(n/2)$ and $(n/2) + 1$ from the top or the bottom of the ordered list.

*Note:* If you are determining a median "by hand," your first step should be to put the data in order from lowest to highest.

---

The first of the next two examples illustrates how to find the median when the sample size is odd; the second shows how to find the median when the sample size is even. The second example also demonstrates how an outlier can cause the values of the mean and median to differ.

**Example 2.9** | **Median and Mean Quiz Scores** Suppose that scores on a quiz for $n = 7$ students in a class are

91   79   60   94   89   93   86

To find the median score, first write the data in order from smallest to largest. The ordered scores are

60   79   86   **89**   91   93   94

Because the number of data values is odd ($n = 7$), the median is the middle value of the ordered scores. The median, $M = 89$, is underlined and in bold typeface in the ordered list. To find the mean, add the seven data values, and then divide that sum by $n = 7$. The sum of the seven scores is 592, so the mean is $\bar{x} = 592/7 = 84.57$. The low score of 60 pulls the mean down slightly in comparison to the median. ∎

**Example 2.10**

**Median and Mean Number of CDs Owned**  In Example 2.8 on page 32, we gave data for the number of music CDs owned by $n = 24$ college students in a statistics class. The ordered list of data values is

0    2    13    20    20    20    25    30    35    50    50    50
60    90    100    100    125    150    200    200    220    250    300    450

The sample size is even ($n = 24$), so the median $M$ is the average of the middle two values in the ordered data. Using the list of ordered data, we see that the median is located between 50 and 60, the 12th- and 13th-highest values in the dataset, the two middle values among 24 observations. The median is the average of these two middle values, so $M = (50 + 60)/2 = 55$. Notice that one half of the students owned fewer than 55 CDs and the other one half of the students owned more than 55 CDs. The mean, the sum of the values divided by the sample size, is $\bar{x} = 2560/24 = 106.7$. The large outlier, 450, along with the overall skew to the right (toward big values) causes the mean to be decidedly greater than the median. ∎

It is not necessarily correct to say that the mean and median measure the "typical" value in a dataset. This is true only when data values are clumped within a central region. In some situations, values near the mean and median may be rare. For instance, suppose that one half of a sample has the value 10 and the other one half has the value 50. The mean is 30, but no values in the sample are close to 30. Also, statements such as "the value is way above normal" do not have a clear meaning unless accompanied by a description of the overall variability in a dataset. This point is made in the next example.

**Example 2.11**

**Will "Normal" Rainfall Get Rid of Those Odors?**  A company (that will remain unnamed) located near Davis, California, was having an odor problem in its wastewater facility. Blame it on the weather:

> [According to a company official] "Last year's severe odor problems were due in part to the 'extreme weather conditions' created in the Woodland area by El Niño." She said Woodland saw 170 to 180 percent of its normal rainfall. "Excessive rain means the water in the holding ponds takes longer to exit for irrigation, giving it more time to develop an odor." (Amy Goldwitz, *The Davis Enterprise,* March 4, 1998, p. A1)

This wording is typical of weather-related stories in which it is often remarked that rainfall is vastly "above normal" or "below normal." In fact, these stories occur so frequently that one wonders if there is *ever* a normal year. The rainfall (in inches) for Davis, California, for 47 years is shown in Table 2.6. A histogram is shown in Figure 2.12.

**Table 2.6**  Annual Rainfall for Davis, California
(July 1–June 30), in Inches

| Year | Rainfall | Year | Rainfall | Year | Rainfall | Year | Rainfall |
|------|----------|------|----------|------|----------|------|----------|
| 1951 | 20.66 | 1963 | 11.2  | 1975 | 6.14  | 1987 | 16.3  |
| 1952 | 16.72 | 1964 | 18.56 | 1976 | 7.69  | 1988 | 11.38 |
| 1953 | 13.51 | 1965 | 11.41 | 1977 | 27.69 | 1989 | 15.79 |
| 1954 | 14.1  | 1966 | 27.64 | 1978 | 17.25 | 1990 | 13.84 |
| 1955 | 25.37 | 1967 | 11.49 | 1979 | 25.06 | 1991 | 17.46 |
| 1956 | 12.05 | 1968 | 24.67 | 1980 | 12.03 | 1992 | 29.84 |
| 1957 | 28.74 | 1969 | 17.04 | 1981 | 31.29 | 1993 | 11.86 |
| 1958 | 10.98 | 1970 | 16.34 | 1982 | 37.42 | 1994 | 31.22 |
| 1959 | 12.55 | 1971 | 8.6   | 1983 | 16.67 | 1995 | 24.5  |
| 1960 | 12.75 | 1972 | 27.69 | 1984 | 15.74 | 1996 | 19.52 |
| 1961 | 14.99 | 1973 | 20.87 | 1985 | 27.47 | 1997 | 29.69 |
| 1962 | 27.1  | 1974 | 16.88 | 1986 | 10.81 |      |       |

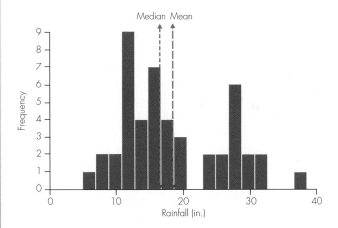

**Figure 2.12** ▌ Annual rainfall in Davis, California

What is the "normal" annual rainfall in Davis? The mean in this case is 18.69 inches, and the median is 16.72 inches. For the year under discussion in the article (1997–98), rainfall was 29.69 inches, hence the comment that it was "170 to 180 percent of normal." But 29.69 inches of rain is within the range of rainfall values over this 47-year period, and more rain occurred in 4 of the other 46 years. The next time you hear about weather conditions that are not "normal," pay attention to whether they are truly outliers or just different from the *average* value. In this case, the company's excuse for the problem just doesn't hold water. ▪

## The Influence of Outliers on the Mean and Median

Outliers have a larger influence on the mean than on the median. Outliers at the high end will increase the mean, while outliers at the low end will decrease it. For instance, suppose there had been only five First Ladies, with ages at death of 70, 72, 74, 76, and 78. Then the median and the mean age of death would both be 74. But now suppose that the youngest age was 35 instead of 70. The median would *still* be 74 years old. But the mean would be

$$\bar{x} = \frac{35 + 72 + 74 + 76 + 78}{5} = 67 \text{ years}$$

You should remember this when you hear statistics about "average life expectancy." Those values are calculated on the basis of averaging the anticipated age at death for all babies born in a given time period. The majority of individuals will live to be older than this "average," but those who die in infancy are the outliers that pull down the average. Datasets that have large outliers at the high end generally have higher means than medians. Examples of data that might have large outliers include annual incomes of executives and sales prices of homes for a large area.

## The Influence of Shape on the Mean and Median

In a perfectly symmetric dataset, the mean and the median are equal; but in a skewed dataset, they differ. When the data are skewed to the left, the mean will tend to be smaller than the median. When the data are skewed to the right, with extreme high values, the mean will tend to be larger than the median. Example 2.8, about the number of music CDs owned, illustrates an extreme case with data strongly skewed to the right, and the mean of about 107 CDs is much larger than the median of 55.

Figures 2.13 and 2.14 show two more examples of how shape affects the relative sizes of the mean and median. Figure 2.13, a histogram of hours of sleep the previous night for $n = 173$ college students, is more or less symmetric. The mean of the sample data is $\bar{x} = 6.94$ hours, and the median is nearly the same at $M = 7$ hours. Having similar values for the mean and median is characteristic of a symmetric dataset. Figure 2.14 is a histogram of self-reported hours spent per

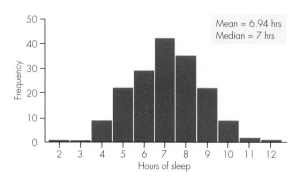

**Figure 2.13** ▮ A symmetric shape—hours of sleep the previous night for $n = 173$ college students

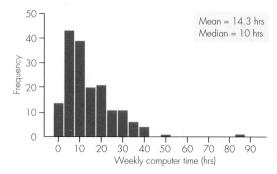

**Figure 2.14** ▮ Data skewed to the right—weekly hours using the computer for $n = 173$ college students

week using a computer for the same 173 students who were used for Figure 2.13. The histogram in Figure 2.14 is skewed to the right. The mean weekly time of 14.3 hours is decidedly greater than the median time of 10 hours. There is an outlier at 84 hours, but deleting it decreases the mean only to 13.9 hours which is still clearly greater than the median.

## Describing Spread: Range and Interquartile Range

Three summary measures that describe the *spread* or *variability* of a dataset are

- **Range** = high value − low value
- **Interquartile range** = upper quartile − lower quartile. The notation **IQR** is often used to represent the interquartile range.
- **Standard deviation**

The standard deviation is easiest to interpret in the context of bell-shaped data, so we postpone a description of it until Section 2.7, in which bell-shaped distributions are discussed. In this section, we describe how to calculate the range and the interquartile range.

Before going into detail about how to compute these measures of spread, let's revisit part of Case Study 1.1. This example shows how informative it is to reduce a dataset to a few simple summary values, providing information about location, spread, and outliers.

**Example 2.12**

**Statistics⬠Now™**

For software help, download your Minitab, Excel, TI-83, SPSS, R, and JMP manuals from **ttp://1pass .thomson.com**, or find them on your CD.

**Range and Interquartile Range for Fastest Speeds Ever Driven** In Case Study 1.1, we summarized responses to the question "What's the fastest you've ever driven a car?" The five-number summary for the 87 males surveyed is as follows:

|  |  | Males (87 Students) |  |
|---|---|---|---|
| Median |  | 110 |  |
| Quartiles | 95 |  | 120 |
| Extremes | 55 |  | 150 |

This summary provides substantial information about the location, spread, and possible outliers. Remember that the median, 110 mph in this case, measures the *center* or *location* of the data. The other four numbers in the five-number summary can be used to describe how *spread out* or *variable* the responses are.

- The two *extremes* describe the spread over 100% of the data. Here, the responses are spread from 55 mph to 150 mph.
- The two *quartiles* describe the spread over approximately the middle 50% of the data. About 50% of the men gave responses between 95 and 120 mph.

Given the values of the extremes and the quartiles, it's simple to calculate the range and the interquartile range (IQR). For the fastest speed reportedly driven by males, values for these two measures of variability are

- Range = high − low = 150 − 55 = 95 mph
- IQR = upper quartile − lower quartile = 120 − 95 = 25 mph

While the range from the smallest to the largest data point is 95 mph, the middle 50% of the data fall in a relatively narrow range of only 25 mph. In other words, the responses are more densely clumped near the center of the data and are more spread out toward the extremes. ■

## Finding Quartiles and Five-Number Summaries

To find the quartiles, first put the data values from lowest to highest and then determine the median. The **lower quartile** ($Q_1$) is the median of the data values that are located below the median. The **upper quartile** ($Q_3$) is the median of the data values that are located above the median. These values are called **quartiles** because, along with the median and the extremes, they divide the ordered data approximately into quarters.

| *formula* | **Finding Quartiles** |
|---|---|

$Q_1 =$ **lower quartile** = median of lower half of the ordered data values

$Q_3 =$ **upper quartile** = median of upper half of the ordered data values

***Note:*** The Minitab program uses a different procedure. When the sample size is even, its quartile estimates will differ slightly from those determined by the procedures described here. Other software packages may do this as well.

**Example 2.13**

**Fastest Driving Speeds for Men**   Here are the 87 males' responses to the question about how fast they have driven a car, as given in Case Study 1.1, except now the data are in numerical order. To make them easier to count, the data are arranged in rows of ten numbers:

```
 55   60   80   80   80   80   85   85   85   85
 90   90   90   90   90   92   94   95   95   95
 95   95   95  100  100  100  100  100  100  100
100  100  101  102  105  105  105  105  105  105
105  105  109  110  110  110  110  110  110  110
110  110  110  110  110  112  115  115  115  115
115  115  120  120  120  120  120  120  120  120
120  120  124  125  125  125  125  125  125  130
130  140  140  140  140  145  150
```

The median is the middle value in an ordered list, so for 87 values, the median is the $(87 + 1)/2 = 88/2 = 44$th value in the list. The 44th value is 110, and this value is shown in bold in the data list. Aside from the middle value of 110, there were 43 values at or below 110 and another 43 values at or above 110. Notice that there are many responses of 110, which is why we are careful to say that 43 of the values are *at or above* the median.

There are 43 values on either side of the median. To find the *quartiles,* simply find the median of each set of 43 values. The lower quartile is the $(43 + 1)/2 = 22$nd value from the bottom of the data, and the upper quartile is the 22nd value from the top. These values are in bold and italics in the data list; $Q_1 = 95$ and $Q_3 = 120$. Notice that the 87 values have been partitioned as follows:

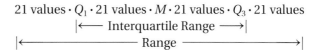

The median and quartiles divide the data into equal numbers of values but do not necessarily divide the data into equally wide intervals. For example, the lowest 1/4 of the males had responses ranging over the 40-mph interval from 55 mph to 95 mph, while the next 1/4 had responses ranging over only a 15-mph interval, from 95 to 110. Similarly, the third 1/4 had responses in only a 10-mph interval (110 to 120), while the top 1/4 had responses in a 30-mph interval (120 to 150). It is common to see the majority of values clumped in the middle and the remainder tapering off into a wider range. ∎

---

***thought question 2.7*** A **resistant statistic** is a numerical summary of the data that is "resistant" to the influence of outliers. In other words, an outlier is not likely to have a major influence on its numerical value. Two of the summary measures from the list *mean, median, range,* and *interquartile range* are *resistant,* while the other two are not. Explain which two are resistant and which two are not.*

---

### How to Draw a Boxplot and Identify Outliers

Now that you know how to find a five-number summary you can draw a boxplot.

**Step 1:** Label either a vertical axis or a horizontal axis with numbers from the minimum to the maximum of the data.

**Step 2:** Draw a box with the lower end of the box at the lower quartile (denoted as $Q_1$) and the upper end at the upper quartile ($Q_3$).

**Step 3:** Draw a line through the box at the median.

**Step 4:** Calculate IQR = $Q_3 - Q_1$.

**Step 5:** Draw a line that extends from the lower quartile end of the box to the smallest data value not smaller than the value of ($Q_1 - 1.5 \times$ IQR). Also draw a line that extends from the upper quartile end of the box to the largest data value that is not greater than the value of ($Q_3 + 1.5 \times$ IQR).

**Step 6:** Mark the location of any data points smaller than ($Q_1 - 1.5 \times$ IQR) or larger than ($Q_3 + 1.5 \times$ IQR) with an asterisk. In other words, values more than one and a half IQRs beyond the quartiles are considered to be *outliers.*

The following two examples illustrate the process of finding five-number summaries, identifying outliers, and presenting the information in a boxplot. Notice that boxplots can be displayed vertically, as in Figure 2.15, or horizontally, as in Figure 2.16.

**Example 2.14** | **Five-Number Summary and Outlier Detection for the Cambridge University Crew Team** The weights (in pounds) for nine men on the Cambridge crew

---

***HINT:** Which of the statistics incorporate all data values at the extremes in the calculations?*

team were as follows (*The Independent,* March 31, 1992; also Hand, D. J. et al., 1994, p. 337):

188.5, 183.0, 194.5, 185.0, 214.0, 203.5, 186.0, 178.5, 109.0

The first step in finding a five-number summary is to write the data in order from smallest to largest. Here, the ordered values are

109.0　178.5　183.0　185.0　**186.0**　188.5　194.5　203.5　214.0

The median, 186.0, is underlined and in bold typeface in the ordered list. Because the number of data values is odd ($n = 9$), the median value is the middle of the ordered values. The lower quartile is the median of the four values below the median (186.0), which are 109.0, 178.5, 183.0, and 185.0. Thus, the lower quartile is $Q_1 = (178.5 + 183)/2 = 180.75$. The upper quartile is the median of the four values above the median. These are 188.5, 194.5, 203.5, and 214.0, so the value of the upper quartile is $Q_3 = (194.5 + 203.5)/2 = 199.0$.

The resulting five-number summary and a boxplot for these data are as follows:

| | Weights of Cambridge Crew Members | | |
|---|---|---|---|
| Median | | 186 | |
| Quartiles | 180.75 | | 199 |
| Extremes | 109 | | 214 |

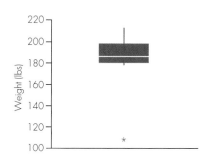

**Figure 2.15** ▌ Boxplot of Cambridge crew members

In Figure 2.15, we see that the value 109 is marked as an outlier. A value below the lower quartile is marked as an outlier if it is more than $1.5 \times \text{IQR}$ below $Q_1$. The interquartile range for these data is $\text{IQR} = Q_3 - Q_1 = 199.0 - 180.75 = 18.25$, and $1.5 \times \text{IQR} = 1.5 \times 18.25 = 27.375$. Any value smaller than $Q_1 - 1.5 \times \text{IQR} = 180.75 - 27.375 = 153.375$ would be marked as an outlier on a boxplot, the case for the weight 109.0. No other values would be identified as outliers in this example. For instance, a value above the upper quartile would have to be larger than $Q_3 + 1.5 \times \text{IQR} = 199.0 + 27.375 = 226.375$ to be identified as an outlier. None of the crew team members was that heavy. ▄

**Example 2.15**　**Five-Number Summary and Outlier Detection for Music CDs** For the example about the number of music CDs owned, started on page 32 and continued earlier in this section, the ordered list of $n = 24$ data values is

0　2　13　20　20　20　25　30　35　50　50　50
60　90　100　100　125　150　200　200　220　250　300　450

Previously we determined that the median of this sample is $M = 55$, the average of 50 and 60, the middle two values in the ordered list. The lower quartile $Q_1$ is the median of the lower half of the data, the 12 values that are smaller than

$M = 55$. These are the values in the first row of the list of ordered values above, ranging from 0 to 50. Thus, the lower quartile is $Q_1 = (20 + 25)/2 = 22.5$, the average of the middle two values in the lower half of the data. The upper quartile is $Q_3 = (150 + 200)/2 = 175$, the average of the middle two values of the 12 observations that are greater than the median. The resulting five-number summary for the number of music CDs owned is as follows:

|  | Number of Music CDs Owned | |
|---|---|---|
| Median |  | 55 |  |
| Quartiles | 22.5 | 175 |
| Extremes | 0 | 450 |

Notice by inspecting the list of ordered data that the five-number summary divides the number of data values into quarters. There are six data values between the low extreme (0) and $Q_1$ (22.5), six values between $Q_1$ (22.5) and $M$ (55), six values between $M$ and $Q_3$ (175), and six values between $Q_3$ and the high extreme (450). Notice, though, that the five-number summary does not necessarily divide the data into equally wide intervals. For example, the lowest 1/4 of the students had responses ranging from 0 to 22.5, while the highest 1/4 ranged over a much wider interval, from 175 to 450.

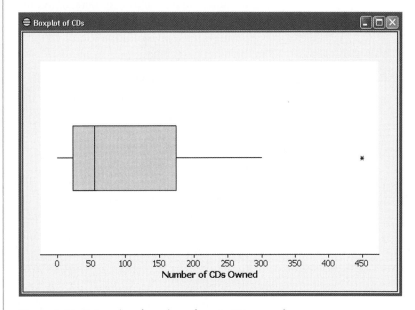

**Figure 2.16** ▌ Boxplot of number of music CDs owned

Figure 2.16 shows a boxplot of the data, drawn with a horizontal axis for the number of CDs owned. The plot shows that the data are skewed to the right, as there is a far greater stretch to the right of the median than to the left. The highest value, 450, is marked as an outlier. Remember that outliers are values that are more than $(1.5 \times \text{IQR})$ beyond the quartiles. In this example, IQR $=$ $Q_3 - Q_1 = 175 - 22.5 = 152.5$, so $(1.5 \times \text{IQR}) = (1.5 \times 152.5) = 228.75$. On the high side of the data, the boundary for marking outliers is $Q_3 + 228.75 = 175 + 228.75 = 403.75$. The value 450 is greater than 403.75, so it is marked with an asterisk. Outliers on the low side would be below $Q_1 - 228.75 = 22.5 - 228.75 = -206.25$, a value substantially below 0. So obviously, there could not be any outliers on the low side. ■

## Percentiles

The quartiles and the median are special cases of **percentiles** for a dataset. In general, the $k$th *percentile* is a number that has $k$% of the data values at or below it and $(100 - k)$% of the data values at or above it. The lower quartile, median, and upper quartile are also the 25th percentile, 50th percentile, and 75th percentile, respectively. If you are told that you scored at the 90th percentile on a standardized test (such as the SAT), it indicates that 90% of the scores were at or below your score, while 10% were at or above your score.

2.5 Exercises are on page 63.

---

**EXCEL *tip***

Suppose the dataset has been stored in a range of cells, which we represent by the word *list* in what follows. For instance, if the dataset is in column A, rows 1 to 30, then "list" is A1:A30. You can also "list" the actual numerical values themselves, rather than the range of cells containing them. All of these commands are part of the "statistical functions" provided by Excel. You can insert them directly into a cell by preceding the command with the symbol @. Some values have multiple options. For instance, there are many commands that give the minimum value.

> *Average(list)* = mean
>
> *Quartile(list, 0)* = *Min(list)* = minimum value
>
> *Quartile(list, 1)* = lower quartile
>
> *Quartile(list, 2)* = *Median(list)* = median
>
> *Quartile(list, 3)* = upper quartile
>
> *Quartile(list, 4)* = *Max(list)* = maximum value
>
> *Small(list,k)* gives the $k$th-smallest value, for example, *small(list,1)* = minimum
>
> *Large(list,k)* gives the $k$th-largest value, for example, *large(list,n)* = minimum
>
> *Percentile(list,p)* gives the $k$th percentile, where $p = k/100$. In other words, you must express the desired percentile as a proportion rather than a percent. For instance, to find the 90th percentile, use *percentile(list,.9)*.
>
> *Count(list)* = $n$, the number of values in the dataset.

***Note:*** As with most computer programs, Excel uses a more precise algorithm to find the upper and lower quartiles than the one we recommend using if you are finding them "by hand," so your values may differ slightly.

| MINITAB *tip* | **Numerical Summaries of a Quantitative Variable** |

- To determine summary statistics for a quantitative variable, use **Stat> Basic Statistics>Display Descriptive Statistics.** In the dialog box, specify one or more columns containing the raw data for a quantitative variable(s). Use the ***Statistics*** button to select or deselect summary statistics that will be displayed. The ***Graphs*** button provides options for several different graphs.

- If you wish to compare numerical summaries of a quantitative variable across categories (for example, to compare the handspans of men and women), specify the categorical variable that defines the groups in the "By Variables" box.

## 2.6  How to Handle Outliers

Outliers need special attention because they can have a big influence on conclusions drawn from a dataset and because they can lead to erroneous conclusions if they are not treated appropriately. Outliers can also cause complications in some statistical analysis procedures, as you will learn throughout this book. As a result, some researchers wrongly discard them rather than treating them as legitimate data. Outliers should never be discarded without justification. The first step in deciding what to do with outliers is determining why they exist so that appropriate action can be taken.

*in summary*       Possible Reasons for Outliers
and Reasonable Actions

- *The outlier is a legitimate data value and represents natural variability for the group and variable(s) measured.* Values may not be discarded in this case—they provide important information about location and spread.

- *A mistake was made while taking a measurement or entering it into the computer.* If this can be verified, the values should be discarded or corrected.

- *The individual in question belongs to a different group than the bulk of individuals measured.* Values may be discarded if a summary is desired and reported for the majority group only.

Let's consider each of these possible reasons for outliers and what to do about them.

### The Outlier Is a Legitimate Data Value and Represents Natural Variability for the Group and Variable(s) Measured

The characterization of natural variability is one of the most important themes in statistics and data analysis. We should not discard legitimate values that inherently occur, unless the goal is to study only a partial range of the possible values. In the handspan data given in Table 2.4 and Figure 2.4, the two smallest fe-male handspan measurements, 12.5 cm and 13.0 cm, were well below the other female measurements. If they are legitimate measurements, they provide im-portant information about the spectrum of possibilities for female handspans, and they should be retained in the dataset. Discarding them would result in an erroneous depiction of female handspans, with a measure of variability that is too low and a mean that is too high.

### A Mistake Was Made While Taking a Measurement or Entering It into the Computer

Faulty measuring equipment, unclear instructions on a survey, or typing errors when data are entered into a computer cause many of the outliers that occur in datasets. For example, a stopwatch with a dying battery might give the wrong value for the time needed to do a task. Or the handspan measurement for a woman whose span is 20 cm would be recorded as 8 if she misunderstood the instructions and reported the measurement in inches instead of centimeters. A typing mistake could cause a height of 68 inches to be recorded as 86 inches. Fortunately, outliers caused by these kinds of problems are often easy to iden-tify. If possible, the values for outliers caused by mistakes should be corrected and retained. If it is not possible to correct them, they should be discarded.

### The Individual in Question Belongs to a Different Group Than the Bulk of Individuals Measured

A group sometimes includes a few individuals that are different from the others in an important way. For instance, a college class might include mostly tradi-tional-age college students (perhaps aged 18 to 22), and a few returning stu-dents who are much older. Measurements for the older students are likely to be outliers for any variables related to age, such as dollar value of assets owned or length of longest romantic relationship.

If we know that outliers are individuals that are different from the others for a specific reason, our reason for studying the data should be considered in de-ciding whether to discard them or not. For example, in measuring assets owned by college students, we should not include the older, returning students if we want to study the assets owned by traditional-age students. If we are interested in the value of assets owned by all college students, we should retain all mea-surements. In that case, the relationship between age and assets may also be of interest.

**Example 2.16**

Statistics◬Now™

Watch a video example at **http://1pass.thomson.com** or on your CD.

**Tiny Boatmen**  Here are the weights (in pounds) of 18 men who were on the crew teams at Oxford and Cambridge universities (*The Independent,* March 31, 1992; also Hand, D. J. et al., 1994, p. 337):

*Cambridge:*    188.5, 183.0, 194.5, 185.0, 214.0, 203.5, 186.0, 178.5, 109.0
*Oxford:*    186.0, 184.5, 204.0, 184.5, 195.5, 202.5, 174.0, 183.0, 109.5

Read over the list. Do you notice anything unusual? The last weight given in each list is very different from the others. In fact, those two men were the coxswains for their teams, while the other men were the rowers. What is the mean weight

2.6 Exercises are on page 64.

for the crew team members? If all members are included, it is 181 lb. If only the rowers are included, it is 190 lb. Different questions, different answers. ■

# 2.7 Features of Bell-Shaped Distributions

Nature seems to follow a predictable pattern for many kinds of measurements. Most individuals are clumped around the center, and the greater the distance a value is from the center, the fewer individuals have that value. Except for the two outliers at the lower end, that pattern is evident in the females' right hand-span measurements in Figures 2.5 to 2.7. If we were to draw a smooth curve connecting the tops of the bars on a histogram with this shape, the smooth curve would resemble the shape of a symmetric bell.

Numerical variables that follow this pattern are said to follow a **bell-shaped curve,** or to be "bell-shaped." A special case of this distribution of measurements is so common it is also called a **normal distribution** or **normal curve.** There is a precise mathematical formula for this smooth curve, which we will study in more depth in Chapter 8, but in this chapter we will limit ourselves to a few convenient descriptive features for data of this type. Most variables with a bell shape do not fit the mathematical formula for a normal distribution exactly, but they come close enough that the results in this section can be applied to them to provide useful information.

**Example 2.17**    **The Shape of British Women's Heights**  A representative sample of 199 married British couples, taken in 1980, provided information on five variables: height of each spouse (in millimeters), age of each spouse, and husband's age at the time they were married (Hand et al., 1994). Figure 2.17 displays a histogram of the wives' heights, with a normal curve superimposed. The particular normal curve shown in Figure 2.17 was generated using Minitab statistical software, and of all the possible curves of this type, it was chosen because it is the best match for the histogram. The mean height for these women is 1602 millimeters, and the median at 1600 millimeters is very close. Although it is difficult to tell precisely in Figure 2.17, the normal curve is centered at the mean of 1602. For bell-shaped curves, there is a useful measure of spread called the *standard deviation,* which we describe next.

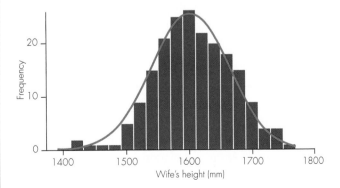

**Figure 2.17** ▪ Histogram of wife's height and normal curve

| MINITAB *tip* | **Superimposing a Normal Curve onto a Histogram** |
|---|---|
| | To draw a normal curve onto the histogram of a quantitative variable, use **Graph>Histogram.** In the resulting display, select ***With Fit.*** In the "Graph Variables" box, specify the column containing the raw data for the variable. |

## Describing Spread with Standard Deviation

Because normal curves are so common in nature, a whole set of descriptive features has been developed that apply mostly to variables with that shape. In fact, two summary features uniquely determine a normal curve, so if you know those two summary numbers, you can draw the curve precisely. The first summary number is the mean, and the bell shape is centered on that number. The second summary number is called the **standard deviation,** and it is a measure of the spread of the values. The symbol $s$ is used to represent the standard deviation of a sample. The squared value of the standard deviation is called the **variance,** and it is represented by $s^2$.

## The Concept of Standard Deviation

You can think of the standard deviation as *roughly the average distance values fall from the mean.* Put another way, it measures variability by summarizing how far individual data values are from the mean. Consider, for instance, the standard deviations for the following two sets of numbers, both with a mean of 100:

| Set | Numbers | Mean | Standard Deviation |
|---|---|---|---|
| 1 | 100, 100, 100, 100, 100 | 100 | 0 |
| 2 | 90, 90, 100, 110, 110 | 100 | 10 |

In the first set of numbers, all values equal the mean value, so there is no variability or spread at all. For this set, the standard deviation is 0, as will *always* be the case for such a set of numbers. In the second set of numbers, one number equals the mean, while the other four numbers are each 10 points away from the mean, so the average distance away from the mean is about 10, the standard deviation for this set of data.

## Calculating the Standard Deviation

The formula for calculating the standard deviation is a bit more involved than the conceptual interpretation that we just discussed. This is the first instance of a summary measure that differs based on whether the data represent a sample or an entire population. The version given here is appropriate when the dataset is considered to represent a sample from a larger population. This distinction will become clear later in the book.

| | |
|---|---|
| *formula* | **Formulas for Standard Deviation and Variance** |

The formula for the (sample) standard deviation is

$$s = \sqrt{\frac{\sum(x_i - \bar{x})^2}{n - 1}}$$

The value of $s^2$, the squared standard deviation, is called the (sample) **variance.** The formula for the (sample) variance is

$$s^2 = \frac{\sum(x_i - \bar{x})^2}{n - 1}$$

In practice, statistical software such as Minitab or a spreadsheet program such as Excel typically is used to find the standard deviation for a dataset. For situations in which you have to calculate the standard deviation by hand, here is a step-by-step guide to the steps involved:

**Step 1:** Calculate $\bar{x}$, the sample mean.

**Step 2:** For each observation, calculate the difference between the data value and the mean.

**Step 3:** Square each difference calculated in step 2.

**Step 4:** Sum the squared differences calculated in step 3, and then divide this sum by $n - 1$. The answer for this step is called the variance.

**Step 5:** Take the square root of the variance calculated in step 4.

**Example 2.18**

Statistics ⟁ Now™

Watch a video example at **http:// 1pass.thomson.com** or on your CD.

**Calculating a Standard Deviation** Calculate the standard deviation of the four pulse rates 62, 68, 74, 76.

**Step 1:** The sample mean is

$$\bar{x} = \frac{62 + 68 + 74 + 76}{4} = \frac{280}{4} = 70$$

**Steps 2 and 3:** For each observation, calculate the difference between the data value and the mean. Then square this difference. The results of these two steps are shown here:

| Data Value | Step 2 Value − Mean | Step 3 (Value − Mean)² |
|---|---|---|
| 62 | $62 - 70 = -8$ | $(-8)^2 = 64$ |
| 68 | $68 - 70 = -2$ | $(-2)^2 = 4$ |
| 74 | $74 - 70 = 4$ | $4^2 = 16$ |
| 76 | $76 - 70 = 6$ | $6^2 = 36$ |

**Step 4:** The sum of step 3 quantities is $64 + 4 + 16 + 36 = 120$. Divide this sum by $n - 1 = 4 - 1$ to get the variance:

$$s^2 = \frac{120}{4 - 1} = \frac{120}{3} = 40$$

**Step 5:** Take the square root of the variance computed in step 4:

$$s = \sqrt{40} = 6.3 \ \blacksquare$$

---

**technical note**    **Population Mean and Standard Deviation**

For reasons that will become clear later in this book, datasets are commonly treated as if they represent a sample from a larger population. However, in situations in which the dataset includes measurements for an entire population, the notations for the mean and standard deviation are different, and the formula for the standard deviation is also slightly different. A **population mean** is represented by the Greek letter $\mu$ ("mu"), and a **population standard deviation** is represented by the Greek letter $\sigma$ ("sigma"). The formula for the population standard deviation is

$$\sigma = \sqrt{\frac{\Sigma(x_i - \mu)^2}{n}}$$

Notice that the difference between this formula and the sample version is that the denominator is now $n$ instead of $n - 1$. Also, the appropriate notation for the mean (population) is used.

---

**EXCEL *tip***    The **Excel** commands for the standard deviation and variance are

*Stdev(list)* = sample standard deviation

*Stdevp(list)* = population standard deviation

*Var(list)* = sample variance

*Varp(list)* = population variance

## Interpreting the Standard Deviation for Bell-Shaped Curves: The Empirical Rule

Once you know the mean and standard deviation for a bell-shaped curve, you can also determine the approximate proportion of the data that will fall into any specified interval. We will learn much more about how to do this in Chapter 8, but for now, here are some useful benchmarks.

---

*definition*    The **Empirical Rule** states that for any bell-shaped curve, approximately

- 68% of the values fall within 1 standard deviation of the mean in either direction
- 95% of the values fall within 2 standard deviations of the mean in either direction
- 99.7% of the values fall within 3 standard deviations of the mean in either direction

Combining the Empirical Rule with knowledge that bell-shaped variables are symmetric allows the "tail" ranges to be specified as well. The first statement of the Empirical Rule implies that about 16% of the values fall more than 1 standard deviation *below* the mean and 16% fall more than 1 standard deviation *above* the mean. Similarly, about 2.5% fall more than 2 standard deviations below the mean, and so on.

**Example 2.19**

**Statistics⬡Now™**

Watch a video example at **http://1pass.thomson.com** or on your CD.

**Women's Heights and the Empirical Rule** The mean for the 199 British women's heights is 1602 millimeters, and the standard deviation is 62.4 millimeters. Figure 2.18 illustrates how the Empirical Rule would apply if the distribution exactly followed a normal curve.

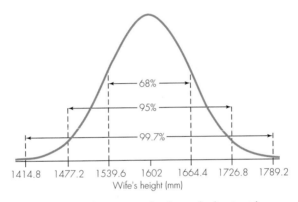

**Figure 2.18** ▌ The Empirical Rule applied to British women's heights

For instance, about 68% of the 199 heights would fall into the range 1602 ± 62.4, or 1539.6 to 1664.4 mm. (The symbol "±" is read "plus or minus" and indicates that you form an interval by first subtracting and then adding the value that follows the symbol from the value that precedes it.) About 95% of the heights would fall into the interval 1602 ± (2 × 62.4), or 1477.2 to 1726.8 mm. And about 99.7% of the heights would be in the interval 1602 ± (3 × 62.4), or 1414.8 to 1789.2 mm. In fact, these intervals work well for the actual data. Here is a summary of how well the Empirical Rule compares with the actual numbers and percents of heights falling within 1, 2, and 3 standard deviations (s.d.) of the mean:

| Interval | Numerical Interval | Empirical Rule % and Number | Actual Number | Actual Percent |
|---|---|---|---|---|
| Mean ± 1 s.d. | 1539.6 to 1664.4 | 68% of 199 = 135 | 140 | 140/199 or 70% |
| Mean ± 2 s.d. | 1477.2 to 1726.8 | 95% of 199 = 189 | 189 | 189/199 or 95% |
| Mean ± 3 s.d. | 1414.8 to 1789.2 | 99.7% of 199 = 198 | 198 | 198/199 or 99.5% |

Notice that the women's heights, although not perfectly bell-shaped, follow the Empirical Rule quite well. ▌

## The Empirical Rule, the Standard Deviation, and the Range

The Empirical Rule implies that the range from the minimum to the maximum data values equals about 4 to 6 standard deviations. For relatively large samples, you can get a rough idea of the value of the standard deviation by dividing the range of the data values by 6. In other words, the standard deviation can be approximated as

$$s \approx \frac{\text{Range}}{6}$$

This approximation works reliably only for bell-shaped data with a sample size of about 200 or more observations. It does not work reliably for skewed data or for smaller sample sizes.

In Example 2.17, about British women's heights, the data are approximately bell-shaped. The sample size of $n = 199$ is large enough to use the formula to estimate the standard deviation. The minimum height was 1410 mm, while the maximum was 1760, for a range of $1760 - 1410 = 350$ mm. Therefore, a reasonable guess for the standard deviation is Range$/6 = 350/6 = 58.3$ mm. This is indeed close to the actual standard deviation of 62.4 mm.

## Standardized z-Scores

The standard deviation is also useful as a "yardstick" for measuring how far an individual value falls from the mean. Suppose you were told that scores on your last statistics exam were bell-shaped (they often are) and that your test score was 2 standard deviations above the mean for your class. Without even knowing your score or the class mean score, you would know that only about 2.5% of the students had scores exceeding yours. From the Empirical Rule, we know that the scores for about 95% of the class are within 2 standard deviations of the mean. Of the remaining 5% of the scores, about half, or 2.5%, will be more than 2 standard deviations above the mean.

The **standardized score** or **z-score** is a useful measure of the relative value of any observation in a dataset. The formula for this score is simple:

$$z = \frac{\text{Observed value} - \text{Mean}}{\text{Standard deviation}}$$

Notice that a $z$-score is simply the distance between the observed value and the mean, measured in terms of number of standard deviations. Data values below the mean have negative $z$-scores, and values above the mean have positive $z$-scores.

As an example, suppose that the mean resting pulse rate for adult men is 70 beats per minute, the standard deviation is 8 beats per minute, and we calculate the standardized score for a resting pulse rate of 80 beats per minute. The calculation is

$$z = \frac{80 - 70}{8} = 1.25$$

The value $z = 1.25$ indicates that a pulse rate of 80 is 1.25 standard deviations above the mean pulse rate for adult men.

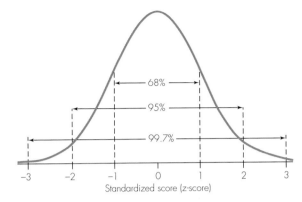

**Figure 2.19** ▌ The Empirical Rule applied to standardized scores (*z*-scores)

| definition | The **Empirical Rule** for bell-shaped data can be restated as follows: |

- About 68% of values have *z*-scores between −1 and +1
- About 95% of values have *z*-scores between −2 and +2
- About 99.7% of values have *z*-scores between −3 and +3

Figure 2.19 illustrates this version of the Empirical Rule.

Many computer programs and calculators will find the approximate proportion of a bell-shaped variable falling below any *z*-score you specify. For instance, the function NORMSDIST(z) in Excel does this. Remember that a common special case of a bell-shaped distribution is called the "normal distribution," and it is this special case that is used by Excel. As an example, NORMSDIST(−1) = 0.158655, or 15.8655%, corresponding to the information from the Empirical Rule that about 16% of values fall more than 1 standard deviation below the mean. In Chapter 8, you will learn a more precise interpretation for *z*-scores.

2.7 Exercises are on page 65.

*thought question* **2.8**   Why do you think measurements with a bell-shaped distribution are so common in nature? For example, why do you think women's heights are distributed in this way rather than, for instance, being equally spread out from about 5 feet tall to 6 feet tall?*

| TI-84 *tip* | **Numerical Summaries of a Quantitative Variable** |

- First store the data values into a list, say L1.
- Press STAT. Scroll horizontally to **CALC,** then scroll vertically to **1:1-Var Stats** and press ENTER. Assuming the data are in list L1, complete the expression as **1-Var Stats L1** followed by ENTER. The display will show the mean, standard deviation (both sample and population), the five-number summary, the sample size, the sum of *x*-values, and the sum of $x^2$ values.

***HINT:** What factors contribute to a person's adult height? Considering these factors, why would it be more likely that heights are close to the mean than far from the mean?

*skillbuilder applet*

# 2.8  The Empirical Rule in Action

Statistics⊗Now™

To explore this applet and work through this activity, go to Chapter 2 at **http://1pass.thomson.com** and click on Skillbuilder Applet, or view the applet on your CD.

The **Empirical** applet on the CD accompanying this book can be used to explore how well the Empirical Rule works for each of eight variables, some with bell-shaped distributions and some with skewed distributions. The data are from the **UCDavis1** and **pennstate1** datasets on the CD for this book. A general description of the eight variables is given at the top of the Web page that includes the applet. For each variable, the applet will display a histogram along with information about the intervals *mean±s* and *mean±2s.* The percent of the sample contained in each interval is reported. When the Empirical Rule applies, these two percents should be about 68% and 95%, respectively.

## What to Do

Open the **Empirical** applet. Figure 2.20 shows the initial applet display, a summary of the hours of sleep the previous night for *n* = 173 students in a UC Davis statistics class. A histogram of the hours of sleep data is displayed with superimposed vertical lines indicating the intervals *mean±s* and *mean±2s.* Notice that the histogram has approximately a bell shape. Below the histogram, we see that the sample mean = 6.935 hours and the standard deviation is *s* = 1.705. Notice also that the interval *mean±s* = (5.23, 8.64) contains 114 of the 173 data values, which is 65.9%. The interval *mean±2s* = (3.525, 10.35) contains 166/173 = 95.95% of the data values. These percents are consistent with the Empirical Rule —not surprising since the distribution is approximately bell-shaped.

After examining the results for the hours of sleep variable, click on *TV Hours,* the second variable in the menu at the left of the applet display. Figure 2.21 displays the result, a summary of self-reported weekly hours of watching television for the same 173 students in the hours of sleep example. Notice that the distribution is skewed to the right, and there is an extreme outlier at 100 hours, so the Empirical Rule won't work well. Here, the interval *mean±s,* given as −1.484 to 19.26 hours, contains about 89% of the dataset, much more than the (approximate) 68% that would be in this interval if the Empirical Rule applied. Another

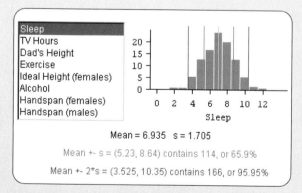

Mean = 6.935  s = 1.705

Mean +- s = (5.23, 8.64) contains 114, or 65.9%

Mean +- 2*s = (3.525, 10.35) contains 166, or 95.95%

**Figure 2.20** ■ The Empirical applet display of hours of sleep reported by *n* = 173 students

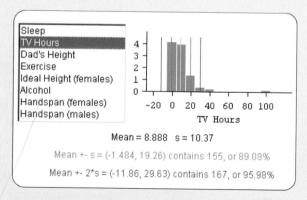

Mean = 8.888  s = 10.37

Mean +- s = (-1.484, 19.26) contains 155, or 89.08%

Mean +- 2*s = (-11.86, 29.63) contains 167, or 95.98%

**Figure 2.21** ■ The Empirical applet display of weekly hours watching television reported by *n* = 173 students

difficulty is that the lower value of the interval is negative, an impossible value for weekly hours of watching television. This also is the case for the interval $mean \pm 2s = (-11.86, 29.63)$.

Click on each of the other six variables to further explore the connection between the shape of the histogram and the applicability of the Empirical Rule. For each variable, judge the shape of the distribution and take note of the percents in the two intervals given. Which of the variables are well described by the Empirical Rule? Typically, what is the approximate shape of the distributions of these variables? Which variables are not well described by the Empirical Rule? What shape do the distributions of these variables typically have?

### Lessons Learned

You'll see that the Empirical Rule works well when the distribution is more or less bell-shaped. But when the distribution is skewed or an extreme outlier is present, you will see that the interval $mean \pm s$ tends to include noticeably more than 68% of the dataset, and the values in the interval $mean \pm 2s$ generally don't match the characteristics of the actual data.

2.8 Exercises are on page 66.

# Key Terms

**Section 2.1**
raw data, 13
variable, 14, 15
observational unit, 14
observation, 14
sample size, 14
dataset, 14
sample data, 14
population data, 14
statistic, 15
parameter, 15
descriptive statistics, 15

**Section 2.2**
categorical variable, 15, 17
ordinal variable, 16, 17
quantitative variable, 16, 17
measurement variable, 16
numerical variable, 16
continuous variable, 16
questions for variable types, 17–18
explanatory variable, 18–21
response variable, 18, 21

**Section 2.3**
frequency, 20
relative frequency, 20

frequency distribution, 20
relative frequency distribution, 20
outcome variable, 21
pie chart, 22
bar graph, 22–24

**Section 2.4**
dotplot, 25, 29, 31
five-number summary, 25, 42
distribution, 26
location, 26
spread, 26
shape, 26, 33–34, 40
outlier, 26, 27–28, 40, 43, 47
median, 26, 37–38, 40
mean, 26, 37–38, 40
variability, 26
histogram, 29
stem-and-leaf plot, 29, 31–32
boxplot, 33–34, 43–45
symmetric (shape), 33
skewed (shape), 33
skewed to the right, 33
skewed to the left, 33
bell-shaped dataset, 33
mode, 33

unimodal, 33
bimodal, 29

**Section 2.5**
range, 41, 54
interquartile range (IQR), 41
standard deviation (sample), 41, 50–52, 54
lower quartile, 42
upper quartile, 42
quartiles, 42
resistant statistic, 43
percentile, 46

**Section 2.6**
reasons for outliers, 47

**Section 2.7**
bell-shaped curve, 49
normal distribution (or curve), 49
variance (sample), 51
population mean, 52
population standard deviation, 52
Empirical Rule, 52, 54, 55
standardized score, 54
z-score, 54

# Exercises

● Denotes basic skills exercises

◆ Denotes dataset is available in StatisticsNow at **http://1pass.thomson.com** or on your CD but is **not required** to solve the exercise.

**Bold-numbered exercises** have answers in the back of the text and fully worked solutions in the Student Solutions Manual.

**Statistics⚠Now**™ Go to the StatisticsNow website at **http://1pass.thomson.com** to:

• Assess your understanding of this chapter
• Check your readiness for an exam by taking the Pre-Test quiz and exploring the resources in the Personalized Learning Plan

## Section 2.1

**2.1** ● A sociologist assembles a dataset consisting of the poverty rate, per capita income, serious crime rate, and teen birth rate for the 50 states of the United States.
   **a.** How many variables are in this dataset?
   **b.** What is an observational unit in this dataset?
   **c.** What is the sample size for the dataset?

**2.2** ● Suppose that in a national survey of 620 randomly selected adults, each person is asked how important religion is to him or her (very, fairly, not very) and whether the person favors or opposes stricter regulation of what can be broadcast on network television.
   a. How many variables are measured in this survey?
   b. What is an observational unit in this study?
   c. What is the sample size for this survey?

**2.3** ● In each situation, explain whether it would be more appropriate to treat the observed data as a sample from a larger population or as data from the whole population.
   **a.** An instructor surveys all the students in her class to determine whether students would prefer a take-home exam or an in-class exam.
   **b.** The Gallup Organization polls 1000 individuals to estimate the percent of American adults who approve of the president's job performance.

**2.4** ● In each situation, explain whether it would be more appropriate to treat the observed dataset as sample data or as population data.
   a. A historian summarizes the ages at death for all past presidents of the United States.
   b. A nutritionist wants to determine which of two weight-loss programs is more effective. He assigns 25 volunteers to each program and records each participant's weight loss after two months.

**2.5** ● For each of the following statistical summaries, explain whether it is a population parameter or a sample statistic.
   **a.** In the 2000 census of the United States, it was determined that the average household size was 2.59 persons per household (www.census.gov).
   **b.** In an ABC News poll done June 9–13, 2004, 36% of $n = 500$ persons surveyed said that they supported replacing the portrait of Alexander Hamilton on the U.S. $10 bill with a portrait of Ronald Reagan (www.pollingreport.com/news.htm).
   **c.** To estimate average normal body temperature of all adults, a doctor measures the temperatures of 100 healthy adults. The average temperature for that group is 98.2 degrees Fahrenheit.

**2.6** ● For each of the following statistical summaries, explain whether it is a population parameter or a sample statistic.
   a. A highway safety researcher wants to estimate the average distance at which all drivers can read a highway sign at night. She measures the distance for a sample of 50 drivers; the average distance for these drivers is 495 feet.
   b. The average score on the final exam is 76.8 for $n = 83$ students in a statistics class. The instructor is only interested in describing the performance of this particular class.
   c. Case Study 1.3 (p. 3) reported that in a Gallup poll, 57% of $n = 496$ teens who date said they have been out with someone of another race or ethnic group.

**2.7** Case Study 1.1 (p. 2) was about the fastest speeds that students in a statistics class claimed they have ever driven. *Variables*
   a. What variables are described in Case Study 1.1?
   b. What are the observational units in the study?
   c. Explain whether you think it would be more appropriate to treat the data as sample data or as population data.

**2.8** Read Case Study 1.5 (p. 5) about prayer and blood pressure.
   a. What was the sample size for the observational study conducted by the National Institutes of Health?
   b. Describe the observational units in this study.
   c. What variables are mentioned in Case Study 1.5?
   d. Explain whether you think the researchers treated the observed data as sample data or as population data.

**2.9** Read Case Study 1.6 (p. 5) about aspirin and heart attack rates.
   a. What two variables are measured on each individual in Case Study 1.6?
   b. Describe the observational units in this study.
   c. What was the sample size for the study?
   d. Explain whether you think the researchers treated the observed data as sample data or as population data.

**2.10** Case Study 1.2 (p. 3) gave the information that the rate of errors made by air traffic controllers in the United States during fiscal year 1998 was 5.5 errors per million flights. Discuss whether this summary value is a population summary (a parameter) or a sample summary (a statistic).

● Basic skills    ◆ Dataset available but not required    **Bold-numbered** exercises answered in the back

explantory var. is a var. we believe
that it can explain why the response
var. change.

Turning Data Into Information   59

## Section 2.2

**2.11** ● For each of the following variables, indicate whether the variable is categorical or quantitative.
  **a.** Importance of religion to respondent (very, somewhat, or not very important). *ordinal*
  **b.** Hours of sleep last night. *discrete*
  **c.** Weights of adult women, measured in pounds. *discrete*
  **d.** Favorite color for an automobile. *Nominal*

**2.12** ● For each of the following, indicate whether the variable is ordinal or not. If the variable is not ordinal, indicate its variable type.
  **a.** Opinion about a new tax law (favor or oppose).
  **b.** Letter grade in a statistics course (A, B, and so on).
  c. Whether the person believes in love at first sight.
  d. Student rating of teacher effectiveness on a 7-point scale where 1 = not at all effective and 7 = extremely effective.

2.13 ● For each of the following characteristics of an individual, indicate whether the variable is categorical or quantitative.
  a. Length of forearm from elbow to wrist (in centimeters).
  b. Whether the person has ever been the victim of a crime.
  c. Number of music CDs owned.
  d. Feeling about own weight (overweight, about right, underweight).

**2.14** ● For each of the following quantitative variables, explain whether the variable is continuous or not.
  **a.** Number of classes a student misses in a week.
  b. Head circumference (in centimeters).
  c. Time it takes students to walk from their dorm to a classroom.
  d. Number of coins presently in someone's pockets and/or purse.

**2.15** ● For each pair of variables, specify which variable is the explanatory variable and which is the response variable in the relationship between them.
  **a.** Amount person walks or runs per day and performance on a test of lung function.
  b. Feeling about importance of religion and age of respondent.
  c. Score on the final exam and the final course grade in a psychology course.
  d. Opinion about the death penalty (favor or oppose), and gender (male or female).

2.16 For each of the following situations reported in the news, specify what variable(s) were measured on each individual and whether they are best described as categorical, ordinal, or quantitative.
  **a.** A *Los Angeles Times* survey found that 60% of the 1515 adult Californians polled supported a recent state law banning smoking in bars (*Sacramento Bee*, May 28, 1998, p. A3).
  **b.** According to the *College Board News* (Dec. 1998, p. 1), "Students using either one of two major coaching programs [for the SAT] were likely to experience an average gain of 5 to 19 points on verbal and 5 to 38 points on math."
  c. According to the *Associated Press* (June 19, 1998),

"Smokers are twice as likely as lifetime nonsmokers to develop Alzheimer's disease and other forms of dementia . . . [according to a study that] followed 6,870 men and women ages 55 and older."

2.17 A physiologist records the pulse rates of 30 men and 30 women.
  a. Specify the two variables measured in this situation.
  b. For each variable, explain whether it is categorical or quantitative.
  c. Using the examples under the "Asking the Right Questions" heading in Section 2.2 (p. 17) as a guide, write a question that would be helpful for comparing the pulse rates of men and women. What summary information would be useful for making this comparison?

2.18 Give an example of an ordinal variable that is likely to be treated as a categorical variable because numerical summaries like the average would not make much sense.

**2.19** Give an example of an ordinal variable for which a numerical summary like the average would make sense.

2.20 To answer the following questions, researchers would need to measure two variables for each individual unit measured or observed in the study. In each case, specify the two variables, and whether each one would most likely be categorical, ordinal, or quantitative. Then, specify which of the two variables you would call the explanatory variable and which you would designate as the response variable.
  a. Is the average IQ of left-handed people higher than the average IQ of right-handed people?
  b. For married couples, is there a relationship between owning a pet and whether or not they get divorced?
  c. For college students, is there a relationship between grade point average (GPA) and number of hours spent studying each week?
  d. Individuals in the United States fall into one of a small number of tax brackets based on level of income. Is there a relationship between a person's tax bracket and the percent of income donated to charities?

2.21 Find an example of a study in a magazine, newspaper, or website. Determine what variables were measured, and, for each variable, determine its type. Which of the questions listed under "Asking the Right Questions" were addressed in this study? Describe the question(s) in the context of the study, then explain what answer was found.

## Section 2.3

**2.22** ● Table 2.1 (p. 19) summarized frequency of seatbelt use while driving for twelfth-grade participants in the 2003 Youth Risk Behavior Surveillance System (YRBSS) survey. In 2001, YRBSS survey students were asked the same question. For the 2001 survey, a summary of responses given by 2530 students in the twelfth grade grade who said that they drive is as follows.

| Wears Seatbelt | Frequency |
|---|---|
| Never | 105 |
| Rarely | 248 |
| Sometimes | 286 |
| Most times | 464 |
| Always | 1427 |

*Source:* http://www.cdc.gov/nccdphp/dash/yrbs.

**a.** What percent of the twelfth-grade students who drive said that they always wear a seatbelt when driving?

**b.** What percent of the twelfth-grade students who drive said that they do not always wear a seatbelt when they drive?

c. Find the percentage in each of the five response categories.

d. Draw a bar graph of the percentages found in part (c).

**2.23** ● Refer to Exercise 2.22. Students also were asked what grades they usually get in school. For twelfth-grade students who responded to this question and the question about how often they wear seatbelts when driving, a summary of frequency counts for combinations of responses to the two questions is as follows:

| Wears Seatbelt | Usual School Grades | | | |
|---|---|---|---|---|
| | A and B | C | D and F | Total |
| Never | 52 | 32 | 18 | 102 |
| Rarely | 128 | 93 | 22 | 243 |
| Sometimes | 166 | 104 | 8 | 278 |
| Most times | 298 | 128 | 24 | 450 |
| Always | 1056 | 300 | 41 | 1397 |
| Total | 1700 | 657 | 113 | 2470 |

**a.** The total number of students in the table is 2470. What percentage of these 2470 students said that they usually get A's and B's in school?

**b.** What percentage of the 1700 students who said that they usually get A's and B's said that they always wear a seatbelt when driving?

c. What percentage of the 657 students who said that they usually get C's said that they always wear a seatbelt when driving?

d. What percentage of the 113 students who said that they usually get D's and F's said that they always wear a seatbelt when driving?

**2.24** ● ◆ A sample of college students was asked how they felt about their weight. Of the 143 women in the sample who responded, 38 women said that they felt overweight, 99 felt that their weight was about right, and 6 felt that they were underweight. Of the 78 men in the sample, 18 men felt that they were overweight, 35 felt that their weight was about right, and 25 felt that they were underweight. (*Data source:* **pennstate3** dataset on the CD for this book.)

a. In the relationship between feelings about weight and gender, which variable is the explanatory variable and which is the response variable?

b. Summarize the observed counts by creating a table similar to Table 2.3.

c. For the 143 women, find the percentage responding in each category for how they felt about their weight.

d. For the 78 men, find the percentage responding in each category for how they felt about their weight.

e. Using the percentages found in parts (c) and (d), summarize how the women and men differed in how they felt about their weight.

**2.25** ● Refer to Exercise 2.24 about gender and feelings about weight. To compare the men and women, draw a bar graph of the percents found in parts (c) and (d). Use Figure 2.3 for guidance.

**2.26** ● For each of the following situations, which is the explanatory variable and which is the response variable?

**a.** The two variables are whether or not someone smoked and whether or not the person developed Alzheimer's disease.

b. The two variables are whether or not somebody voted in the last election and the person's political party (Democrat, Republican, Independent, or Other).

c. The two variables are income level and whether or not the person has ever been subjected to a tax audit.

**2.27** In the sample survey described in Section 2.1, there were 92 students who responded to "Randomly pick a letter—*S* or *Q*." Of these 92 students, 61 picked *S* and 31 picked *Q*. The order of the letter choices was reversed for another 98 students who responded to "Randomly pick a letter—*Q* or *S*." Of these 98 students, 45 picked *S* and 53 picked *Q*.

a. Construct a two-way table of counts summarizing the relationship between the letter listed first in the survey question and the letter picked by the student.

b. For the 92 students who saw *S* listed first in the question, determine the percents who picked *S* and *Q*.

c. For the 98 students, who had *Q* listed first, determine the percents who picked *S* and *Q*.

d. Draw a bar chart of the percents found in parts (c) and (d) to show the relationship between the letter listed first and the letter picked.

e. Explain whether you think the letter listed first in the question affected the choice of letter.

**2.28** Refer to Exercise 2.27.

a. Reconstruct the table using the two categorical variables "letter listed first (*S* or *Q*)" and "ordering of letter chosen (listed first or second)."

b. Draw an appropriate picture to accompany your numerical summary.

c. Explain whether you think the variables used in Exercise 2.27 or the variables used in this exercise were more appropriate for illustrating the point of this dataset.

**2.29** Each of the following quotes is taken from an article entitled "Education seems to help in selecting husbands" (*Sacramento Bee*, Dec. 4, 1998, p. A21), which reported on new data in the *Statistical Abstract of the*

*United States.* Draw an appropriate graph to represent each situation.

a. "The data show that 3.8 percent of women who didn't complete high school had four or more husbands. For high school graduates, the share with four or more partners drops to 3 percent. Among those who attended college 2 percent had four or more husbands, and that fell to 1 percent for those with college degrees."

b. "From 1997 on, 5.5 percent of children lived with their grandparents, a share that has been rising steadily. It was only 3.6 percent in 1980, and by 1990 it was 4.9 percent."

c. "The center said 20.1 percent of Americans took part in some regular activity—21.5 percent of men and 18.9 percent of women."

**2.30** According to Krantz (1992, p. 190) the percentage of women having their first child at various ages is as follows (for all women who have children):

| Under 20 | 20–24 | 25–29 | 30–34 | 35 and Over |
|---|---|---|---|---|
| 25% | 33% | 25% | 12.5% | 4% |

(The numbers do not sum to 100% owing to being rounded off.) Draw a pie chart and a bar graph to represent the data. Explain which picture you think is more informative.

## Section 2.4

**2.31** ● *This is the same as Exercise 1.1.* A five-number summary for the heights in inches of the women who participated in the survey described in Section 2.1 is as follows:

| | Female Heights (Inches) | |
|---|---|---|
| Median | 65 | |
| Quartiles | 63.5 | 67.5 |
| Extremes | 59 | 71 |

a. What is the median height for these women?

b. What is the range of heights, that is, the difference in heights between the shortest and the tallest women?

c. What is the interval of heights containing the shortest 1/4 of the women?

d. What is the interval of heights containing the middle 1/2 of the women?

**2.32** ● Refer to Exercise 2.31.

a. Give a value from the five-number summary that characterizes the *location* of the data.

b. Describe the spread of the data using values from the five-number summary.

**2.33** ● *This is the same as Exercise 1.15.* The five-number summaries of the fastest ever driven data given in Case Study 1.1 (p. 2) were as follows:

| | Males (87 students) | | Female (102 Students) | |
|---|---|---|---|---|
| Median | 110 | | 89 | |
| Quartiles | 95 | 120 | 80 | 95 |
| Extremes | 55 | 150 | 30 | 130 |

Give a numerical value for each of the following.

a. The fastest speed driven by anyone in the class.

b. The slowest speed driven by a male.

c. The speed for which 1/4 of the women had driven at that speed or faster.

d. The proportion of females who had driven 89 mph or faster.

e. The number of females who had driven 89 mph or faster.

**2.34** ● Refer to the five-number summaries given in Exercise 2.33.

a. ● Using the appropriate summary value, compare the *location* of the fastest ever driven response for males to the location for females.

b. Explain whether the *spread* is greater for one sex than the other or whether it is about the same.

**2.35** In an experiment, one female and one male restaurant server drew happy faces on the checks of randomly chosen dining parties. The figure for this exercise is a dotplot comparing tip percentages for the female ($n = 22$ checks) to the tip percentages for the male ($n = 23$ checks).

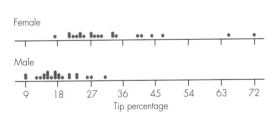

a. Compare the two servers with respect to the approximate centers (locations) of their tip percentages.

b. Compare the two servers with respect to the variation (spread) among tip percentages.

c. Explain whether you think there are any outliers in the dataset or not. If you think there are outliers, give their approximate values.

**2.36** ● Hand et al. (1994, p. 148) provide data on the number of words in each of 600 randomly selected sentences from the book *Shorter History of England* by G. K. Chesterton. They summarized the data as follows:

| Number of Words | Frequency | Number of Words | Frequency |
|---|---|---|---|
| 1–5 | 3 | 31–35 | 68 |
| 6–10 | 27 | 36–40 | 41 |
| 11–15 | 71 | 41–45 | 28 |
| 16–20 | 113 | 46–50 | 18 |
| 21–25 | 107 | 51–55 | 12 |
| 26–30 | 109 | 56–60 | 3 |

a. Create a histogram for the number of words in sentences in the *Shorter History of England*.

b. Provide a summary of the dataset based on your histogram.

c. Explain why you could not create a stem-and-leaf plot for this dataset.

d. Count the number of words in the first 20 sentences in Chapter 1 of this book (not including headings), and create a histogram of sentence lengths. Compare the sentence lengths to those in the *Shorter History of England*.

**2.37** ● ◆ The figure for this exercise is a histogram summarizing the responses given by 137 college women to a question asking how many ear pierces they have. (*Data source:* **pennstate2** dataset on the CD for this book.)

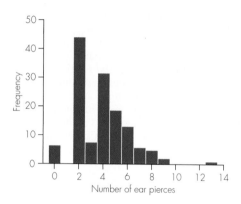

Number of ear pierces

**a.** Describe the shape of the dataset. Explain whether it is symmetric or skewed.

**b.** Are there any outliers? For any outlier, give a value for the number of ear pierces, and explain why you think the value is an outlier.

**c.** What number of ear pierces was the most frequently reported value? Roughly, how many women said they have this number of ear pierces?

**d.** Roughly, how many women said they have four ear pierces?

**2.38** ● ◆ The figure for this exercise is a histogram summarizing the responses given by 116 college students to a question asking how much they had slept the previous night. (*Data source:* **sleepstudy** dataset on the CD for this book.)

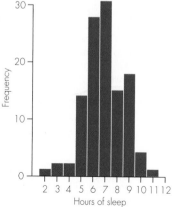

Hours of sleep

a. Describe the shape of the dataset. Explain whether it is symmetric or skewed.

b. Are there any outliers? For any outlier, give an approximate value for the amount of sleep, and explain why you think the value is an outlier.

c. What was the most frequently reported value (approximately) for the amount of sleep the previous night?

d. Roughly, how many students said that they slept 8 hours the previous night?

**2.39** ● ◆ The following stem-and-leaf plot is for the mean August temperatures (Fahrenheit) in 20 U.S. cities. The "stem" (row label) gives the first digit of a temperature, while the "leaf" gives the second digit. (*Data source:* **temperature** dataset on the CD for this book.)

```
6 | 44
6 | 89
7 | 01124
7 | 56667
8 | 1223
8 | 5
9 | 2
```

**a.** Describe the shape of the dataset. Is it skewed or is it symmetric?

**b.** What is the highest temperature in the dataset?

**c.** What is the lowest temperature in the dataset?

**d.** What percent of the 20 cities have a mean August temperature in the 80s?

**2.40** A set of exam scores is as follows:

75, 84, 68, 95, 87, 93, 56, 87, 83, 82, 80, 62, 91, 84, 75

a. Draw a stem-and-leaf plot of the scores.

b. Draw a dotplot of the scores.

**2.41** ◆ Case Study 1.1 (p. 2) presented data on the fastest speed men and women had driven a car, and dotplots were shown for each sex. Data for the men are also in the **pennstate1M** dataset on the CD for this book.

a. Create a stem-and-leaf plot for the male speeds.

b. Create a histogram for the male speeds.

c. Compare the pictures created in (a) and (b) and the dotplot in Case Study 1.1. Comment on which is more informative, if any of them are, and comment on any other differences you think are important.

d. How would you describe the shape of this dataset?

**2.42** ◆ Annual rainfall for Davis, California, for 1951 to 1997 is given in Table 2.6 in Section 2.5 and in the **rainfall** dataset on the CD for this book. A histogram is shown in Figure 2.12 (p. 39).

a. Create a stem-and-leaf plot for the rainfall data, rounded (not truncated) to the nearest inch.

b. Create a dotplot for the rainfall data, rounded to the nearest inch.

c. Describe the shape of the rainfall data.

**2.43** ◆ Here are the ages, arranged in order, for the CEOs of the 60 top-ranked small companies in America in 1993. (*Data source: Forbes,* "America's Best Small Companies," Nov. 8, 1993; also at http://lib.stat.cmu.edu/DASL/Datafiles/ceodat.html.) These data are part of the **ceodata** dataset on the CD for this book.

32, 33, 36, 37, 38, 40, 41, 43, 43, 44, 44, 45, 45, 45, 45,
46, 46, 47, 47, 47, 48, 48, 48, 48, 49, 50, 50, 50, 50, 50,
50, 51, 51, 52, 53, 53, 53, 55, 55, 55, 56, 56, 56, 56, 57,
57, 58, 58, 59, 60, 61, 61, 61, 62, 62, 63, 69, 69, 70, 74

a. Create a histogram for these ages.
b. Create a stem-and-leaf plot for these ages.
c. Create a dotplot for these ages.
d. Describe the shape of this dataset.
e. Are there any outliers in this dataset?
f. In general, would outliers be more likely to occur in the salaries of heads of companies or in the ages of heads of companies? Explain.

**2.44** Does a stem-and-leaf plot provide sufficient information to determine whether a dataset contains an outlier? Explain.

2.45 For the following situations, would you be most interested in knowing the average value, the spread, or the maximum value for each dataset? Explain. If you think it would be equally useful to know more than one of these summaries, explain that as well. (Answers may differ for different individuals. It is your reasoning that is important.)
a. A dataset with the annual salaries for all employees in a large company that has offered you a job.
b. You need to decide from which of two statistics instructors you will take a class. You have two datasets, with previous final exam scores given by each of the two instructors.
c. A dataset with ages at death for 20 of your relatives who died of natural causes.

**2.46** Histograms and boxplots are two types of graphs that were discussed in Section 2.4.
**a.** Explain what features of a dataset are best identified using a histogram.
**b.** Explain what features of a dataset are best identified using a boxplot.

2.47 Construct an example and sketch a histogram for a measurement that you think would be bimodal.

**2.48** About 75% of the students in a class score between 80 and 100 on a quiz. The other 25% of the students have scores spread out between 35 and 79. Characterize the shape of the distribution of quiz scores. Explain.

2.49 The figure for this exercise is a boxplot comparing tip percentages for a male and a female restaurant server, each of whom drew happy faces on the checks of randomly selected dining parties. A dotplot of the data was given as the figure for Exercise 2.35. Discuss the ways in which the tip percentages for the two servers differed.

*this boxplot clearly states that female restaurant server drew more happy faces on random checks.*

## Section 2.5

**2.50** Find the mean and the median for each list of values:
   **a.** 64, 68, 72, 76, 80, 86
   **b.** 10, 6, 2, 7, 100
   **c.** 30, 10, 40, 30

2.51 Refer to part (b) of Exercise 2.50. Explain why there is such a large difference between the mean and median values.

**2.52** ● ◆ Sixty-three college men were asked what they thought was their ideal weight. A five-number summary of the responses (in pounds) is as shown.

| Median | | 175 | |
|---|---|---|---|
| Quartiles | 155 | | 190 |
| Extremes | 123 | | 225 |

(*Data source:* **idealwtmen** dataset on the CD for this book)
**a.** Find the value of the range for these data.
**b.** Find the value of the interquartile range (IQR).
**c.** About what percent of the men gave a response that falls in the interval 155 to 190 pounds?

**2.53** ● ◆ Students in a statistics class wrote as many letters of the alphabet as they could in 15 seconds using their nondominant hand. The figure for this exercise is a boxplot that compares the number of letters written by males and females in the sample. (*Data source:* **letters** dataset on the CD for this book.)
**a.** What is the median number of letters written by females?
b. What is the median for males?
c. Explain whether the interquartile range is larger for males or for females.
d. Find the value of the range for males.
e. Find the value of the range for females.

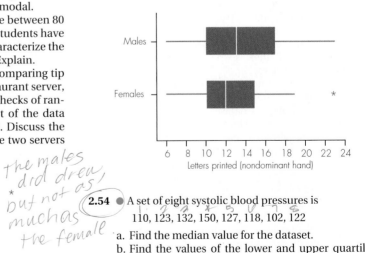

*the males did drew, but not as much as the female*

**2.54** ● A set of eight systolic blood pressures is
110, 123, 132, 150, 127, 118, 102, 122
a. Find the median value for the dataset.
b. Find the values of the lower and upper quartiles.
c. Find the value of the interquartile range (IQR).
**d.** Identify any outliers in the dataset. Use the criterion that a value is an outlier if it is either more than $1.5 \times$ IQR above $Q_3$ or more than $1.5 \times$ IQR below $Q_1$.
e. Draw a boxplot of the dataset.

● Basic skills     ◆ Dataset available but not required     **Bold-numbered** exercises answered in the back

**2.55** ◆ Create side-by-side boxplots for the "fastest ever driven a car" described in Case Study 1.1 (p. 2): one for males and one for females. Compare the two sexes based on the boxplots. (Five-number summaries are given in Exercise 2.33 and in Case Study 1.1. The raw data are on page 2 and in the **pennstate1** dataset on the CD for this book.)

**2.56** *This is the same as parts (a) and (b) of Exercise 1.17.* Students in a statistics class were asked, "About how many minutes do you typically exercise in a week?" Responses from the *women* in the class were

60, 240, 0, 360, 450, 200, 100, 70, 240, 0, 60, 360, 180, 300, 0, 270

Responses from the *men* in the class were

180, 300, 60, 480, 0, 90, 300, 14, 600, 360, 120, 0, 240

a. Compare the women to the men using a dotplot. What does your plot show you about the difference between the men and the women?
b. For each gender, determine the median response.

**2.57** *Parts (a) and (b) are the same as Exercise 1.18.* Refer to Exercise 2.56.
a. Create a five-number summary for the men's responses. Show how you found your answer.
b. Use your five-number summary to describe in words the exercise behavior of this group of students.
c. Draw a boxplot of the men's responses.

**2.58** ◆ Describe the data on First Ladies' ages at death given in Table 2.5 (p. 28) and also in the dataset **first-ladies** on the CD for this book. Compute whatever numerical summaries you think are appropriate, then write a narrative summary based on the computed information. Include pictures if appropriate.

**2.59** ◆ Refer to Exercise 2.58. Repeat that exercise to describe the rainfall data given in Table 2.6 (p. 39) and also in the dataset **rainfall** on the CD for this book.

**2.60** Refer to Example 2.11, Table 2.6, and Figure 2.12 (p. 39) for the rainfall data. Specify whether the shape is skewed to the left or to the right, and explain whether the relationship between the mean and the median (which one is higher) is what you typically expect for data with that shape.

**2.61** ◆ Create a five-number summary for the rainfall data in Example 2.11, Table 2.6 (p. 39). Write a few sentences describing the dataset. The data are in the **rainfall** dataset on the CD for this book.

**2.62** Refer to the sentence-length dataset in Exercise 2.36. Notice that you cannot compute exact summary values. Provide as much information as you can about the median, interquartile range, and range for sentence lengths in the *Shorter History of England*.

**2.63** The football team at the school of one of the authors won four of eleven games it played during the 2004 college football season. Point differences between teams in the eleven games were

+38, −14, +24, −13, −9, −7, −2, −11, −7, +4, +24

A positive difference indicates that the author's school won the game, and a negative difference indicates that the author's school lost.

a. Find the value of the mean point difference and the value of the median point difference for the eleven games.
b. Explain which of the two summary values found in part (a) is a better summary of the team's season.

**2.64** In an experiment conducted by one of this book's authors, 26 students were asked to estimate (in millions) the population of Canada, which was about 30 million at that time. Before they made their estimates, ten of the students (group 1) were told that the population of the United States was about 290 million at that time. Nine of the students (group 2) were told that the population of Australia was roughly 20 million at that time. The other seven students (group 3) were not given any information. The estimates (millions) made by the students in these three groups were

Group 1: 2, 30, 35, 70, 100, 120, 135, 150, 190, 200
Group 2: 8, 12, 16, 29, 35, 40, 45, 46, 95
Group 3: 5, 23, 50, 55, 150, 150, 300

**a.** Find the median estimate in group 1.
b. Find the median estimate in group 2.
c. Find the median estimate in group 3.
d. Compare the values of the range for the three groups. Which group had the largest range? Which group had the smallest range?

**2.65** Refer to Exercise 2.64.
a. Find values of the lower and upper quartiles for the data in group 1.
b. Find values of the lower and upper quartiles for the data in group 2.
c. Find values of the lower and upper quartiles for the data in group 3.
d. Draw a boxplot that compares the three groups. Refer to Figure 2.11 for guidance.

**2.66** Refer to Exercise 2.31, which gives a five-number summary of heights for college women. Draw a boxplot displaying the information in this five-number summary.

**2.67** ◆ Create a five-number summary for the ages of CEOs of small companies listed in Exercise 2.43 and on the **ceodata** dataset on the CD for this book. Write a few sentences describing the dataset.

**2.68** ◆ Create a boxplot for the ages of CEOs of small companies listed in Exercise 2.43 and on the **ceodata** dataset on the CD for this book.

**2.69** ◆ Find the mean and median for the ages of CEOs of small companies listed in Exercise 2.43 and on the **ceodata** dataset on the CD for this book. Is the relationship between them what you would typically expect for data with the shape of this dataset? Explain.

## Section 2.6

**2.70** ● In the data discussed in Section 2.1, one student reported having slept 16 hours the previous night.
a. What additional information do you need to determine whether or not this value is an outlier?
b. If you do determine that the data value of 16 hours of sleep is an outlier, what additional information do you need to decide whether or not to discard it before summarizing the data?

● Basic skills ◆ Dataset available but not required **Bold-numbered** exercises answered in the back

**2.71** ● A male whose height is 78 inches might be considered to be an outlier among males in a statistics class but not among males who are professional basketball players. Give another example in which the same measurement taken on the same individual would be considered to be an outlier in one dataset but not in another dataset.

**2.72** ● One of the authors of this book (one male, one female) has a right handspan measurement of 23.5 cm. Would you consider this value to be an outlier? What additional information do you need to make your decision?

**2.73** Refer to the rainfall data given in Table 2.6 (p. 39). Discuss whether there are any outliers, and, if so, whether to discard them or not.

**2.74** Give an example, not given in Section 2.6, of a situation in which a measurement is an outlier because the individual belongs to a different group than the bulk of individuals measured. Be specific about the variable measured and the way in which the individual may differ from the others who were measured.

**2.75** In a statistics class survey, students reported their heights in inches. The instructor entered the data into a computer file and used statistical software to find separate five-number summaries for men and women in the class. In the five-number summary for men, the minimum height was 17 inches, an obvious outlier. What is a possible reason for this outlier, and what should the instructor do about it?

## Section 2.7

**2.76** ● The typical amount of sleep per night for college students has a bell-shaped distribution with a mean of 7 hours and a standard deviation equal to 1.7 hours. Use the Empirical Rule to complete each sentence:

**a.** About 68% of college students typically sleep between ____ and ____ hours per night.

**b.** About 95% of college students typically sleep between ____ and ____ hours per night.

**c.** About 99.7% of college students typically sleep between ____ and ____ hours per night.

**2.77** ● Refer to Exercise 2.76 about hours of sleep per night for college students. Draw a picture of the distribution. Indicate the locations of the three intervals found in the previous exercise. Use Figure 2.18 for guidance.

**2.78** ● Suppose that the mean weight for men 18 to 24 years old is 170 pounds, and the standard deviation is 20 pounds. In each part, find the value of the standardized score ($z$-score) for the given weight:

**a.** 200 pounds.        **c.** 170 pounds.
**b.** 140 pounds.        **d.** 230 pounds.

**2.79** ● Find the mean and standard deviation for each sample of values:

**a.** 18, 19, 20, 21, 22        **c.** 1, 5, 7, 8, 79
**b.** 20, 20, 20, 20, 20

**2.80** ● The scores on the final exam in a course have approximately a bell-shaped distribution. The mean score was 70, the highest score was 98, and the lowest score was 41.

**a.** Find the value of the range for the exam scores.

**b.** Refer to part (a). Use the value of the range to estimate the value of the standard deviation.

**2.81** ● A sample of $n = 500$ individuals is asked how many hours they typically spend using a computer in a week. The mean response is $\bar{x} = 8.3$ hours, and the standard deviation is $s = 7.2$ hours. Find values for the interval $\bar{x} \pm 2s$, and explain why the result is evidence that the distribution of weekly hours spent using the computer is not bell-shaped.

**2.82** The mean for the women's right handspans is about 20 cm, with a standard deviation of about 1.8 cm. Using the stem-and-leaf plot in Figure 2.6, determine how well this set of measurements fits with the Empirical Rule.

**2.83** The data for Exercise 2.54 was this set of systolic blood pressures:

110, 123, 132, 150, 127, 118, 102, 122

**a.** Find values of the mean and standard deviation for these data.

**b.** What is the value of the variance for these data?

**2.84** Scores on the Stanford-Binet IQ test have a bell-shaped distribution with mean = 100 and standard deviation = 16.

**a.** Use the Empirical Rule to specify the intervals into which 68%, 95%, and 99.7% of Stanford-Binet IQ scores fall.

**b.** Draw a picture similar to Figure 2.18 illustrating the intervals found in part (a).

**c.** What is the value of the variance of Stanford-Binet IQ scores?

**2.85** Suppose verbal SAT scores for students admitted to a university are bell-shaped with a mean of 540 and a standard deviation of 50.

**a.** Draw a picture of the distribution of these verbal SAT scores, indicating the cutoff points for the middle 68%, 95%, and 99.7% of the scores.

**b.** What is the variance of verbal scores for students admitted to the university?

**2.86** Head circumferences of adult males have a bell-shaped distribution with a mean of 56 centimeters and a standard deviation of 2 centimeters.

**a.** Explain whether it would be unusual for an adult male to have a 52-centimeter head circumference.

**b.** Explain whether it would be unusual for an adult male to have a 62-centimeter head circumference.

**2.87** Refer to Exercise 2.86. What is the variance of head circumferences of adult males?

**2.88** Both of the following lists of $n = 8$ data values have a mean of 20:

*List 1:*    10, 10, 10, 10, 30, 30, 30, 30
*List 2:*    10, 15, 19, 20, 20, 21, 25, 30

Draw a dotplot comparing the two data lists and explain how the plot shows that the standard deviation for list 1 is greater than the standard deviation for list 2.

**2.89** Do you think the "ages at death" in Table 2.5 are likely to fit a bell-shaped curve? Explain why or why not.

**2.90 a.** Would the "ages at death" in Table 2.5 be considered a population of measurements or a sample from some larger population? Explain.

**b.** Find the appropriate standard deviation (sample or population) for the "ages at death" data in Table 2.5.

2.91 The interquartile range and the standard deviation are two different measures of spread. Which measure do you think is more affected by outliers? Explain.

2.92 Explain why women's heights are likely to have a bell shape but their ages at marriage do not.

2.93 Can a categorical variable have a bell-shaped distribution? Explain.

2.94 The data for 103 women's right handspans are shown in Figures 2.5 to 2.7, and a five-number summary is given in Example 2.5.

   **a.** Examine Figures 2.5 to 2.7 and comment on whether or not the Empirical Rule should hold.

   **b.** The mean and standard deviation for these measurements are 20.0 cm and 1.8 cm, respectively. Determine whether the range of the data (found from the five-number summary) is about what would be expected using the Empirical Rule.

2.95 Refer to the women's right handspan data. As you can see in Figures 2.5 to 2.7, there are two apparent outliers: at 12.5 cm and 13.0 cm.

   **a.** If these values are removed, do you think the mean will increase, decrease, or remain the same? What about the standard deviation? Explain.

   **b.** With the outliers removed, the mean and standard deviation for the remaining 101 values are 20.2 cm and 1.45 cm, respectively. The range is 7.25, from 16.0 to 23.25. Determine whether the Empirical Rule for mean ± 3 standard deviations appears to hold for these values.

   **c.** Refer to Figures 2.5 to 2.7. Based on those figures, do you think the Empirical Rule should hold when the outliers are removed? How about when the outliers have not been removed? Explain.

   **d.** Is there any justification for removing the two outliers? Explain.

2.96 Heights for traditional college-age students in the United States have means and standard deviations of approximately 70 inches and 3 inches for males and 65 inches and 2.5 inches for females. Specify your own height and find the z-score for a college-age student of your sex and height.

2.97 **a.** If a data value has a z-score of 0, the value equals one of the summary measures discussed in this chapter. Which summary is that?

   **b.** Verify that a data value having a z-score of 1.0 is equal to the mean plus 1 standard deviation.

2.98 Using a computer or calculator that provides proportions falling below a specified z-score, determine the approximate proportion for each of the following situations. In each case, assume the values are approximately bell-shaped.

   **a.** The proportion of SAT scores falling below 450 for a group with a mean of 500 and a standard deviation of 100.

   **b.** The proportion of boys with heights below 36.5 inches for a group with mean height of 34 inches and standard deviation of 1 inch.

   **c.** The proportion of a large class that scored below you on a test for which the mean was 75, the standard deviation was 8, and your score was 79.

   **d.** The proportion of a large class that scored below you on a test for which the mean was 75, the standard deviation was 4, and your score was 79.

2.99 If you learn that your score on an exam was 80 and the mean was 70, would you be more satisfied if the standard deviation was 5 or if it was 15? Explain.

**2.100** Write a set of seven numbers with a mean of 50 and a standard deviation of 0. Is there more than one possible set of numbers? Explain.

2.101 Remember that a resistant statistic is a numerical summary whose value is not unduly influenced by an outlier of any magnitude. Is the standard deviation a resistant statistic? Justify your answer by giving an example of a small dataset, then adding a very large outlier and noting how the standard deviation is affected.

2.102 For a bell-shaped dataset with a large number of values, approximately what z-score would correspond to a data value equaling each of the following?

   a. The median.          c. The highest value.
   b. The lowest value.    d. The mean.

**2.103** ◆ Refer to Exercise 2.43, in which the ages for the 60 CEOs of America's best small companies of 1993 were given. These data are in the **ceodata** dataset on the CD for this book.

   **a.** Find the mean and standard deviation for these ages.

   **b.** Recall that the range should be equivalent to 4 to 6 standard deviations for bell-shaped data. Determine whether that relationship holds for these ages.

   **c.** Find the z-scores for the youngest and oldest CEO. Are they about what you would expect? Explain.

## Section 2.8: Skillbuilder Applet Exercises

**Statistics ⚘ Now™** Test yourself on these questions and explore the applet at **http://1pass.thomson.com** or on your CD.

*For Exercises 2.104 to 2.108, use the **Empirical** applet described in Section 2.8. It is on the CD for this book.*

2.104 Examine the results given by the applet for each of the eight variables.

   **a.** Among the eight variables, which variables are best described by the Empirical Rule, and which are not well described by the Empirical Rule?

   **b.** Generally, what is the shape of the histogram for the variables that are well described by the Empirical Rule? What is the shape of the histogram for the variables that are not well described by the Empirical Rule?

2.105 The parts of this exercise concern the variable **Dad's Height,** which is data for father's height as reported by n = 167 college students.

   **a.** What is the shape of the histogram given for the variable **Dad's Height**? Are there any outliers?

   **b.** Refer to part (a). Based on the shape of the histogram, explain whether the Empirical Rule will apply or not.

● Basic skills     ◆ Dataset available but not required     **Bold-numbered** exercises answered in the back

c. What numerical values are given by the applet for the interval **mean±s**? What percentage of the data values are in this interval? Compare this percentage to the percentage that would be expected if the Empirical Rule applies.

d. What numerical values are given by the applet for the interval **mean±2s**? What percentage of the data values are in this interval? Compare this percentage to the percentage that would be expected if the Empirical Rule applies.

2.106 The parts of this exercise concern the variable **Alcohol,** which is data for number of alcoholic beverages consumed in a typical week as reported by $n = 167$ college students.

a. What is the shape of the histogram given for **Alcohol**? Are there any outliers?

b. Refer to part (a). Based on the shape of the histogram, explain whether the Empirical Rule will apply or not.

c. What numerical values are given by the applet for the interval **mean±s**? What percentage of the data values are in this interval? Compare this percentage to the percentage that would be expected if the Empirical Rule applies.

d. What numerical values are given by the applet for the interval **mean±2s**? Explain why this interval is not a good description of possible values for the variable **Alcohol.**

2.107 The parts of this exercise concern the variable **Ideal Height,** which is the respondent's desired height as reported by $n = 149$ college women.

a. Use information given by the applet to explain whether the Empirical Rule applies for **Ideal Height.**

b. Assuming the Empirical Rule applies, give numerical values for the interval that will contain about 99.7% of the data values.

c. Compare the interval found in part (b) to the histogram of the data. Explain whether or not the interval is a reasonable description of the data.

2.108 The data for each of the following variables includes one or more outliers. In each case, identify the outlier(s). Then explain whether the Empirical Rule would apply to the remaining data if the outlier(s) were removed from the dataset.

a. **TV Hours.**

b. **Handspan (females).**

c. **Dad's Height.**

## Chapter Exercises

2.109 ◆ A question in the 2002 General Social Survey (GSS) conducted by the National Opinion Research Center asked participants how long they spend on e-mail each week. A summary of responses (hours) for $n = 1881$ respondents follows. (The raw data are in the file **GSS-02** on the CD for the book.)

| Mean | StDev | Minimum | Q1 | Median | Q3 | Maximum |
|------|-------|---------|----|--------|----|---------|
| 4.14 | 7.235 | 0 | 0 | 2 | 5 | 70 |

a. Explain how the summary statistics shows us that at least 25% of the respondents said that they do not use e-mail.

b. What is the interval that contains the lower 50% of the responses?

c. What is the interval that contains the upper 50% of the responses?

d. Explain whether the maximum value, 70 hours, would be marked as an outlier on a boxplot.

e. Calculate Range/6 and compare the answer to the value of the standard deviation. What feature(s) of the data do you think causes the values to differ?

f. Compare the mean to the median. What feature(s) of the data do you think causes the values to differ?

2.110 For each of the following situations, would you prefer your value to be average, a low outlier, or a high outlier? Explain.

**a.** Number of children you have.

**b.** Your annual salary.

c. Gas mileage for your car.

d. Crime rate in the city or town where you live.

2.111 Specify the type (categorical, ordinal, quantitative) for each of the following variables recorded in a survey of telephone usage in student households:

**a.** Telephone exchange (first three numbers after area code).

**b.** Number of telephones in the household.

c. Dollar amount of last month's phone bill.

d. Long-distance phone company used.

2.112 In the same survey for which wives' heights are given in Example 2.19, husbands' heights were also recorded. A five-number summary of husbands' heights (mm) is as follows:

| | Husbands' Heights ($n = 199$) | |
|------|:---:|------|
| Median | | 1725 |
| Quartiles | 1691 | 1774 |
| Extremes | 1559 | 1949 |

a. Construct a boxplot for the husbands' heights.

b. Use the range to approximate the standard deviation for these heights.

c. What assumption did you need to make in part (b) to make the approximation appropriate?

d. The mean and standard deviation for these heights are 1732.5 mm and 68.8 mm. Use the Empirical Rule to construct an interval that should cover 99.7% of the data, and compare your interval to the extremes. Does the interval cover both extremes?

2.113 Can a variable be *both* of the following types? If so, give an example.

**a.** An explanatory variable and a categorical variable.

**b.** A continuous variable and an ordinal variable.

c. A quantitative variable and a response variable.

d. A bell-shaped variable and a categorical variable.

e. A bell-shaped variable and a response variable.

2.114 Reach into your wallet, pocket, or wherever you can find at least ten coins, and sort all of the coins you have by type.

a. Count how many of each kind of coin you have (pennies, nickels, and so on, or the equivalent for your country). Draw a pie chart illustrating the distribution of your coins.

b. In part (a), "kind of coin" was the variable of interest for each coin. Is that a categorical, ordinal, or quantitative variable?

c. Now consider the total monetary value of all of your coins as a single data value. What type of variable is it?

d. Suppose you had similar data for all of the students in your statistics class. Write a question for which the variable "kind of coin" is the data of interest for answering the question. Then write a question for which the variable "total monetary value of the coins" is the data of interest for answering the question.

2.115 Look around your living space or current surroundings, and find a categorical variable for which there are at least three categories and for which you can collect at least 20 observations (example: color of the shirts in your closet). Collect the data.
   a. Draw a pie chart for your data.
   b. Draw a bar graph for your data.
   c. Is one of the pictures more informative than the other? Explain. (Your answer may depend on the variable you chose.)

2.116 Look around your living space or current surroundings and find a quantitative variable for which you can collect at least 20 observations (example: monetary amounts of the last 25 checks you wrote). List the data with your response.
   a. Create a five-number summary.
   b. Draw a boxplot.
   c. Draw your choice of a histogram, stem-and-leaf plot, or dotplot.
   d. Refer to your picture in part (c) and comment on the shape and presence or absence of outliers.
   e. Compute the mean and compare it to the median. Explain whether the relationship between them is what you would expect based on the information you discussed in part (d).

**2.117** For each of the following two sets of data, explain which one is likely to have a larger standard deviation:
   **a.** Set 1: Heights of the children in a kindergarten class.
   Set 2: Heights of all of the children in an elementary school.
   b. Set 1: Systolic blood pressure for a single individual taken daily for 30 days.
   Set 2: Systolic blood pressure for 30 people who visit a health clinic in 1 day.
   c. Set 1: SAT scores (which range from 200 to 800) for the students in an honors class.
   Set 2: Final examination scores (which range from 0 to 100) for all of the students in the English classes at a high school.

2.118 For each of the following datasets, explain whether you would expect the mean or the median of the observations to be higher:
   a. In a rural farming community, for each household the number of children is measured.

b. For all households in a large city, yearly household income is measured.
   c. For all students in a high school (not just those who were employed), income earned in a job outside the home in the past month is measured.
   d. For the coins in someone's pocket that has 1/3 pennies, 1/3 nickels, and 1/3 quarters, the monetary value of each coin is recorded.

**2.119** For each of the following research questions, explain what two variables would have to be measured for each individual in the study.
   **a.** Is there a relationship between the amount of beer people drink and their systolic blood pressure?
   **b.** Is there a relationship between calories of protein consumed per day and incidence of colon cancer?
   c. Is there a relationship between eye color and whether or not corrective lenses are needed by age 18?
   d. For women who are HIV-positive when they get pregnant, is there a relationship between whether or not the HIV is transmitted to the infant and the length of time the woman had been infected before getting pregnant?

2.120 Refer to Exercise 2.119. In each case, specify which of the two variables is the explanatory variable and which is the response variable. If it is ambiguous, explain why.

*Exercises 2.121 to 2.124 each describe one or two variables and the individuals for whom they were measured. For each exercise, state an interesting research question about the situation. Use the examples under the "Asking the Right Questions" heading (p. 17) in Section 2.2 as a guide.*

2.121 Individuals are all of the kindergarten children in a school district.
   *One variable:* Adult(s) with whom the child lives (both parents, mother only, father only, one or both grandparents, other).

**2.122** Individuals are all mathematics majors at a college.
   *Two variables:* Grade point average and hours spent studying last week.

2.123 Individuals are a representative sample of adults in a large city.
   *Two variables:* Ounces of coffee consumed per day and marital status (currently married or not).

2.124 Individuals are a representative sample of college students.
   *Two variables:* Male or female and whether the person dreams in color (yes or no).

## Dataset Exercises

Statistics⏾Now™ Datasets **are required** to solve these exercises and can be found at **http://1pass.thomson.com** or on your CD.

2.125 Use the **oldfaithful** dataset on the CD for this book; it gives data for $n = 299$ eruptions of the Old Faithful geyser.
   a. The variable **Duration** is the duration (in minutes) of an eruption. Draw a boxplot of this variable. Figure 2.10 (p. 34) is a histogram of the same data.

What important feature of the data can be seen in histogram, but not in the boxplot?

b. The variable *TimeToNext* is the time until the next eruption after the present eruption. Draw a histogram of this variable. Describe the shape of the histogram.

2.126 The data for this exercise are in the **GSS-02** dataset on the CD for this book. The variable *gunlaw* is whether a respondent favors or opposes stronger gun control laws.

a. Determine the percentage of respondents who favor stronger gun control laws and the percentage of respondents who oppose stronger gun control laws. (*Note:* Not all survey participants were asked the question about gun laws, so the sample size for *gunlaw* is smaller than the overall sample size.)

b. Draw a graphical summary of the *gunlaw* variable.

c. Create a two-way table of counts that shows the relationship between gender (variable name *sex*) and opinion about stronger gun control laws. From looking at this table of counts, are you able to judge whether the two variables are related? Briefly explain.

d. What percentage of females favors stronger gun control laws? What percentage of males favors stronger gun control laws?

e. Based on the percentages found in part (d), do you think that gender and opinion about gun control are related? Briefly explain.

2.127 Use the **pennstate1** dataset on the CD for this book for this exercise. The data for the variable *HrsSleep* are responses by $n = 190$ students to the question "How many hours did you sleep last night?"

a. Draw a histogram of the data for the *HrsSleep* variable. Describe the shape of this histogram, and comment on any other interesting features of the data.

b. Determine the five-number summary for these data.

c. What is the range of the data? What is the interquartile range?

2.128 Use the **pennstate2** dataset on the CD for this book for this exercise. The variable *CDs* is the approximate number of music CDs owned by a student.

a. Draw a stem-and-leaf plot for the *CDs* variable.

b. Draw a histogram for the *CDs* variable.

c. Draw a dotplot for the *CDs* variable.

d. Describe the shape of the data for the *CDs* variable, and comment on any other interesting features of the data.

e. Calculate the mean number and the median number of CDs. Compare these two values.

f. For these data, do you think the mean or the median is a better description of the location of the data? Briefly explain.

2.129 For this exercise, use the **GSS-02** dataset on the CD for this book. The variable *cappun* is the respondent's opinion about the death penalty for persons convicted of murder, and the variable *polparty* is the respondent's political party preference (Democrat, Republican, Independent, Other).

a. In this dataset, what percentage favors the death penalty? What percentage opposes it?

b. Create a table that displays the relationship between political party and opinion about the death penalty. Calculate an appropriate set of conditional percentages for describing the relationship.

c. Are the variables *polparty* and *cappun* related? Explain.

2.130 Use the **cholest** dataset on the CD for this book for this exercise. The dataset contains cholesterol levels for 30 "control" patients and 28 heart attack patients at a medical facility. For the heart attack patients, cholesterol levels were measured 2 days, 4 days, and 14 days after the heart attack.

a. Calculate the mean, the standard deviation, and the five-number summary for the control patients.

b. Calculate the mean, the standard deviation, and the five-number summary for the heart attack patients' cholesterol levels 2 days after their attacks.

c. Generally, which group has the higher cholesterol levels? How much difference is there in the *location* of the cholesterol levels of the two groups?

d. Which group of measurements has a larger *spread*? Compare the groups with regard to all three measures of spread introduced in Sections 2.6 and 2.7.

e. Compare the controls and the heart attack patients using a comparative dotplot (as in Case Study 1.1 on p. 2). Briefly explain what this plot indicates about the difference between the two groups.

2.131 Use the **pennstate1** dataset on the CD for this book.

a. Draw a histogram of the *height* variable.

b. What is the shape of this histogram? Why do you think it is not a bell shape?

c. Draw a boxplot of the height variable.

d. Which graph, the histogram or the boxplot, is more informative about this dataset? Briefly explain.

2.132 Use the **GSS-02** dataset on the CD for this book. The variable *degree* indicates the highest educational degree achieved by a respondent.

a. Is the *degree* variable quantitative, categorical, or ordinal? Explain.

b. Determine the number and percentage falling into each degree category.

c. What percentage of the sample has a degree that is beyond a high school degree?

d. The variable *tvhours* is the self-reported number of hours of watching television in a typical day. Find the mean number of television-watching hours for each of the five degree groups.

e. Is there a relationship between self-reported hours of watching television and educational degree? Explain.

f. Draw any visual summary of the variable *tvhours* (for the whole sample). What are the interesting features of your graph?

**Statistics⌂Now™** Preparing for an exam? Assess your progress by taking the post-test at **http://1pass.thomson.com.**

**⌄Mentor™**
Do you need a live tutor for homework problems? Access vMentor at **http://1pass.thomson.com** for one-on-one tutoring from a statistics expert.

● Basic skills    ◆ Dataset available but not required    **Bold-numbered** exercises answered in the back

# 3

Roy Morsch/zefa/Corbis

*How many people would eat a genetically altered tomato?*

See Example 3.4 *(p. 77)*

# Sampling: Surveys and How to Ask Questions

Here is a fact that may astound you. If you use proper methods to sample 1500 people from a population of millions and millions, you can almost certainly gauge the percentage of the entire population who have a certain trait or opinion to within 3%. The tricky part is that you have to use a proper sampling method.

People are curious about each other. When a newsworthy event occurs, people want to know what others think. In the midst of election campaigning, people want to know which candidates are in the lead. We devour news stories about the behavior and opinions of others. For this reason, news organizations such as *USA Today* and CNN often conduct sample surveys. Many recent surveys can be found on the Web. In this chapter, we learn how to conduct surveys, how to make sure they are representative, and what can go wrong. ∎

## 3.1 Collecting and Using Sample Data Wisely

There are two major categories of statistical techniques that can be applied to data. The first is **descriptive statistics,** in which we use numerical and graphical summaries to characterize a dataset. We partially covered descriptive statistics in Chapter 2, and we will introduce additional descriptive techniques in Chapters 5 and 6. The second important category of statistical techniques is **inferential statistics,** in which we use sample data to make conclusions about a broader range of individuals than just those who are observed. For example, in Case Study 1.6 about aspirin use and the risk of heart disease, the data from a sample of 22,071 physicians was used to *infer* that taking aspirin helps to prevent heart attacks for all men who are similar to the participants.

In Chapters 3 and 4, you will learn how to collect representative data. In these chapters, you will learn that the data collection method used affects the extent to which sample data can be used to make inferences about a larger population. Descriptive summaries such as the mean and the standard deviation, as well as graphical techniques, can be used whether the data are from a sample or from an entire population, but inferential methods can be used only when the data in hand are from a *representative* sample for the question being asked about a larger population. When you use inferential methods, a key concept is that you have to think about *both* the source of the data *and* the question(s) of interest. A dataset may contain representative information for some questions but not for others.

| | |
|---|---|
| *definition* | The **Fundamental Rule for Using Data for Inference** is that available data can be used to make inferences about a much larger group *if the data can be considered to be representative with regard to the question(s) of interest.* |

**Example 3.1**

**Do First Ladies Represent Other Women?**    If we want to know how old American women live to be, it would not make sense to use the ages of death given in Chapter 2 for the First Ladies. Even if you ignore the fact that most of the First Ladies did not live in modern times with modern health care, you can probably think of many reasons why those women do not represent the full population of American women. In fact, there is probably no larger group that is represented by these women. Past First Ladies are not even likely to represent future First Ladies on the question of age at death because medical, social, and political conditions keep changing in ways that may affect their health.  ■

**Example 3.2**

Statistics△Now™
_____
Watch a video example at **http://1pass.thomson.com** or on your CD.

**Do Penn State Students Represent Other College Students?**    One dataset in Chapter 2 was collected at Penn State University. Do Penn State students represent all college students? This probably depends on the question. If we want to know the average handspan of female college students in the United States, we could convincingly argue that the data collected in statistics classes at Penn State University are indeed representative. We could even argue that the Penn State women represent all females in the same age group, not just college students. On the other hand, Penn State students may not represent all college students when questioned about how fast they have ever driven a car. Penn State is in rural Pennsylvania, where there are many country roads and little traffic. Students in larger cities, even nearby ones such as New York or Philadelphia, may not have access to the same open spaces and may not have had the opportunity to drive a car as fast as students in rural areas. Many students in larger cities may not even drive cars at all.  ■

These examples show that it may require knowledge of the subject matter to determine whether a sample represents the larger group for the question of interest. There are, however, a small number of common strategies that researchers use to improve the likelihood that the data they collect can be used to make inferences about a larger population. We will examine some of those strategies in the remainder of this chapter and the next. Also, we will discuss the strengths of those strategies and some problems that may arise when they are used.

## Populations, Samples, and Simple Random Samples

In most statistical studies, the objective is to use a small group of units to make an inference about a larger group. The larger group of units about which inferences are to be made is called the **population.** The smaller group of units actually measured is called the **sample.** Sometimes measurements *are* taken on the whole group of interest, in which case *these measurements comprise a **census** of the whole population.* Occasionally, you will see someone make the mistake of trying to use census data to make inferences to some hypothetical "larger group" when there isn't one.

| | |
|---|---|
| *definition* | • A **population** is the entire group of units about which inferences will be made.<br>• A **sample** is the group of units that are actually measured or surveyed.<br>• A **census** is taken when every unit in the population is measured or surveyed. |

Remember the fundamental rule for making valid inferences about the group represented by the sample for which the data were measured: *The data must be representative of the larger group with respect to the question of interest.* The principal way to guarantee that sample data represents a larger population is to use a **simple random sample** from the population.

| | |
|---|---|
| *definition* | With a **simple random sample,** every conceivable group of units of the required size from the population has the same chance to be the selected sample. |

An ideal data collection method is to obtain a simple random sample of the population of interest or to collect sample data by using one of the more complex random sampling methods described later in this chapter. In some research studies, however, random sampling is not possible for practical or ethical reasons (or both). For instance, suppose researchers want to study the effect of using marijuana to reduce pain in cancer patients. It would be neither practical nor ethical to select a random sample of all cancer patients to participate. Instead, the researchers would use volunteers who want to take part and hope that those volunteers represent the larger population of all cancer patients. The use of volunteers will be discussed more fully in Chapter 4, when we cover randomized experiments.

Simple random samples and related sampling methods *are* typically used for one type of statistical study: sample surveys or polls. Remember from Chapter 1 that in a *sample survey,* the investigators gather opinions or other information from each individual included in the sample. Because this gathering of information is usually not time-consuming or invasive, it is often both practical and ethical to contact a large random sample from the population of interest. Throughout this chapter, we will learn more about how to select simple random samples and how to conduct sample surveys.

In a **sample survey,** a subgroup of a large population is questioned on a set of topics. The investigator simply asks the participants to answer some questions. There is no manipulation of a respondent's behavior in this type of research. The results from a sample survey are used as if they represent the larger population, which they will if the sample is chosen correctly and if those who are selected to participate cooperate in responding.

## Advantages of a Sample Survey over a Census

### When a Census Isn't Possible

Suppose you need a laboratory test to find out whether or not your blood has too high a concentration of a certain substance. Would you prefer that the lab measure the entire population of your blood, or would you prefer to give the lab a sample? Similarly, suppose that a manufacturer of firecrackers wants to know what percentage of its products are duds. The company would not make much of a profit if it tested them all, but it could get a reasonable estimate of the desired percentage by testing a properly selected sample. As these examples illustrate, there are situations in which measurements destroy the units being tested and therefore a census is not feasible.

### Speed

Another advantage of a sample survey over a census is speed. For instance, it takes several years to successfully plan and execute a census of the entire population of the United States, something that is done every ten years. Getting figures such as monthly unemployment rates would be impossible with a census; they would be quite out of date by the time they were released. It is much faster to collect a sample than a census if the population is large.

### Accuracy

A final advantage to a sample survey is that you can devote your resources to getting the most accurate information possible from the sample you have selected. It is easier to train a small group of interviewers than a large one, and it is easier to track down a small group of nonrespondents than the larger one that would inevitably result from trying to conduct a census.

## Bias: How Surveys Can Go Wrong

The purpose of most surveys is to use a characteristic of the sample (such as the proportion with a certain opinion) as an estimate of the corresponding characteristic of the population from which the sample was selected. While it is unlikely that the sample value will equal the population value precisely, the goal of a good survey is to get an **unbiased** sample value. Survey results are **biased** if the method used to obtain those results would produce values that are either consistently too high or consistently too low. When the fundamental rule for inference does not apply, it's likely that the results will be biased.

There are three common types of bias that might occur in surveys:

- **Selection bias** occurs if the method for selecting the participants produces a sample that does not represent the population of interest. For instance, if shoppers at a mall are surveyed to determine attitudes about raising the sales tax, the results are not likely to represent all area residents.
- **Nonresponse bias** (nonparticipation bias) occurs when a representative sample is chosen for a survey but a subset cannot be contacted or does not respond. For instance, a survey conducted by phoning people at home in the evening would omit people who work during that time. Their opinions would not be represented, so if their opinions were much different, the sample results would be biased.

- **Response bias** occurs when participants provide incorrect information. The way questions are worded, the way the interviewer behaves, as well as many other factors might lead an individual to answer untruthfully. For instance, surveys about socially unacceptable behavior such as heavy smoking or drinking, abusive behavior, and so on must be worded and conducted carefully to minimize the possibility of response bias.

3.1 Exercises are on page 106.

---

**thought question 3.1** The terms *nonresponse bias* and *response bias* sound very similar. Do they refer to two aspects of the same kind of problem? Explain.*

---

## 3.2 Margin of Error, Confidence Intervals, and Sample Size

Surprisingly, a sample does not have to include a large portion of the population to be reasonably accurate. National polling organizations such as the Gallup Organization often question only slightly over 1000 individuals. How can a survey of such a small number of individuals possibly obtain accurate information about the opinions of millions?

Here is some information that may astound you. If you use commonly accepted methods to sample 1500 people from an entire population of millions and millions, you can almost certainly gauge the percentage of the entire population who have a certain trait or opinion to within 3%. Even more amazing is the fact that this result does not depend on how big the population is. It depends only on how many are in the sample! A sample of 1500 would do about equally well for a population of 4 billion as it would for a population of 10 million. Of course, the tricky part is that you have to use a proper sampling method.

You can see why researchers are content to rely on public opinion polls rather than trying to ask everyone for their opinions. It is much cheaper to ask 1500 people than several million, especially when you can get an answer that is almost as accurate. It also takes much less time to conduct a sample survey than it does to conduct a census, in which everyone in the population is measured. And because fewer interviewers are needed for a survey than for a census, there may be better quality control of interviewing procedures when only a small sample of individuals is measured.

### The Margin of Error: The Accuracy of Sample Surveys

Sample surveys are often used to estimate the *proportion* or *percentage* of people who have a certain trait or opinion. For example, the Nielsen ratings, used to determine the percentage of American television sets tuned to a particular show, are based on a sample of a few thousand households. Newspapers and magazines routinely survey only one or two thousand people to determine public

---

**\*HINT:** In each case, think about whether the bias has to do with *who* responds or with *how* they respond.

opinion on current topics of interest. As we have said, if properly conducted, these surveys are amazingly accurate.

When a survey is used to find a proportion based on a sample of only a few thousand individuals, the obvious question is how close that proportion comes to the truth for the entire population. The **margin of error** is a measure of the accuracy of a sample proportion. It provides an upper limit on the difference between the sample proportion and the population proportion that holds for at least 95% of simple random samples of a specific size. In other words, the difference between the sample proportion and the unknown value of the population proportion is less than the margin of error at least 95% of the time, or in at least 19 of every 20 situations. A conservative estimate of the margin of error for a sample proportion is calculated as $1/\sqrt{n}$, where $n$ is the sample size. To express results in terms of percentages instead of proportions, simply multiply everything by 100.

---

*definition*   The **conservative margin of error** for a sample proportion is calculated by using the formula $1/\sqrt{n}$, where $n$ represents the **sample size,** the number of people in the sample. The amount by which the sample proportion differs from the true population proportion is less than this quantity in at least 95% of all random samples. Survey results reported in the media are usually expressed as percentages. The *conservative margin of error for a sample percentage* is

$$\frac{1}{\sqrt{n}} \times 100\%$$

---

In Chapter 10, we will learn a more complicated and precise formula for the margin of error that depends on the actual sample proportion. It will never give an answer larger than the formula given here, so we call this formula a *conservative* margin of error. For simplicity, many media reports use the conservative formula. For example, with a sample of 1600 people, we will usually get an estimate that is accurate to within $1/\sqrt{1600} = 1/40 = 0.025 = 2.5\%$ of the truth.

## Confidence Intervals

You might see results such as "Fifty-five percent of respondents support the President's economic plan. The margin of error for this survey is plus or minus 2.5 percentage points." This means that it is almost certain that between 52.5% and 57.5% of the entire population supports the plan. In other words, add and subtract the margin of error to the sample value (55% in this example) to create an interval. If you were to follow this method every time you read the results of a properly conducted survey, the interval would miss covering the truth only about 1 in 20 times (5%) and would cover the truth the remaining 95% of the time.

A **confidence interval** is an interval of values that *estimates* an unknown population value. In the preceding paragraph, the percentage of *all* U.S. adults favoring the President's economic policy is the unknown value we wish to estimate. The confidence interval estimate of this value is 52.5% to 57.5%, calculated as *sample percentage ± margin of error.*

The word *confidence* refers to the fraction or percentage of random samples for which a confidence interval procedure gives an interval that includes the unknown value of a population parameter. For the procedure described here, the confidence is at least 95%, meaning that intervals calculated in this manner will capture the population proportion for at least 95% of all properly selected random samples.

---

*definition*    **95% Confidence Interval for Population Proportion**

For about 95% of properly conducted sample surveys, the interval

$$sample\ proportion - \frac{1}{\sqrt{n}} \text{ to } sample\ proportion + \frac{1}{\sqrt{n}}$$

will contain the actual population proportion. This interval is called an **approximate "95% confidence interval"** for the population proportion. In Chapter 10, a more precise formula will be given.

---

Another way of writing the approximate 95% confidence interval is

$$sample\ proportion \pm \frac{1}{\sqrt{n}}$$

To report the results in percentages instead of proportions, multiply everything by 100%.

Here are three examples of polls, with the margin of error and an approximate 95% confidence interval for each one. In each case, the poll is based on a nationwide random sample of Americans aged 18 and older.

**Example 3.3** | **The Importance of Religion for Adult Americans** For a CNN/*USA Today*/ Gallup Poll conducted on September 2 to 4, 2002, a random sample of $n = 1003$ adult Americans was asked, "How important would you say religion is in your own life: very important, fairly important, or not very important?" (http://www.pollingreport.com/religion.htm, September 17, 2002). The percentages that selected each response were as follows:

| | |
|---|---|
| Very important | 65% |
| Fairly important | 23% |
| Not very important | 12% |
| No opinion | 0% |

Conservative margin of error:  $\frac{1}{\sqrt{1003}} = .03$ or $.03 \times 100\% = 3\%$

Approximate 95% confidence intervals for the proportion and percentage of *all* adult Americans who would say religion is very important are

Proportion:   $.65 \pm .03$ or [$.65 - .03$ to $.65 + .03$] or .62 to .68
Percentage:   $65\% \pm 3\%$ or [$65\% - 3\%$ to $65\% + 3\%$] or 62% to 68% ■

**Example 3.4** | **Would You Eat Those Modified Tomatoes?** An ABCNews.com poll conducted by TNS Intersearch on June 13 to 17, 2001, posed the following question

Statistics⬡Now™

Watch a video example at **http://1pass.thomson.com** or on your CD.

to a random sample of $n = 1024$ adult Americans: "Scientists can change the genes in some food crops and farm animals to make them grow faster or bigger and be more resistant to bugs, weeds and disease. Do you think this genetically modified food, also known as bio-engineered food, is or is not safe to eat?" (http://www.pollingreport.com/science.htm, September 17, 2002). The percentages that selected each response were as follows:

Is safe          35%
Is not safe      52%
No opinion       13%

Conservative margin of error:  $\dfrac{1}{\sqrt{1024}} = .03$ or $.03 \times 100\% = 3\%$

Approximate 95% confidence intervals for the proportion and percentage of *all* adult Americans who would say genetically modified food is *not* safe to eat are

Proportion:   $.52 \pm .03$ or $[.52 - .03$ to $.52 + .03]$ or .49 to .55
Percentage:   $52\% \pm 3\%$ or $[52\% - 3\%$ to $52\% + 3\%]$ or 49% to 55% ∎

**Example 3.5**

**Cloning Human Beings** A Pew Research Center for the People & the Press and Pew Forum on Religion & Public Life survey conducted by Princeton Survey Research Associates between February 25 and March 10, 2002, asked a random sample of $n = 2002$ adult Americans, "Do you favor or oppose scientific experimentation on the cloning of human beings?" (http://www.pollingreport.com/science.htm, September 17, 2002). The percentages that selected each response were as follows:

Favor                 17%
Oppose                77%
Don't know/Refused     6%

Conservative margin of error:  $\dfrac{1}{\sqrt{2002}} = .022$ or $.022 \times 100\% = 2.2\%$

Approximate 95% confidence intervals for the proportion and percentage of *all* adult Americans who would say they *favor* this experimentation are

Proportion:   $.17 \pm .022$ or $[.17 - .022$ to $.17 + .022]$ or .148 to .192
Percentage:   $17\% \pm 2.2\%$ or $[17\% - 2.2\%$ to $17\% + 2.2\%]$ or
              14.8% to 19.2% ∎

### Interpreting the Confidence Intervals in Examples 3.3 to 3.5

In each of the three examples of surveys, an approximate 95% confidence interval was found. There is no way to know whether all, some, or none of these intervals actually covers the population value of interest. For instance, the interval from 62% to 68% may or may not capture the percentage of adult Americans who considered religion to be very important in their lives in September 2002. But in the long run, this procedure will produce intervals that capture the unknown population values about 95% of the time, as long as it is used with properly conducted surveys. This long-run performance is usually expressed after an interval is computed by saying that we are 95% confident that the population value is covered by the interval. We will learn more about how to interpret a confidence interval and the accompanying confidence level (such as 95%) in Chapter 10.

## Choosing a Sample Size for a Survey

**Table 3.1** Relationship Between Sample Size and Margin of Error for 95% Confidence

| Sample Size $n$ | Margin of Error $= 1/\sqrt{n}$ |
|---|---|
| 100 | .10 (10%) |
| 400 | .05 (5%) |
| 625 | .04 (4%) |
| 1000 | .032 (3.2%) |
| 1600 | .025 (2.5%) |
| 2500 | .02 (2%) |
| 10,000 | .01 (1%) |

When surveys are planned, the **choice of a sample size** is an important issue. One commonly used strategy is to use a sample size that provides a **desired margin of error** for a 95% confidence interval. Table 3.1 displays the margin of error for several different sample sizes. The margin of error calculations were done by using the "conservative" formula $1/\sqrt{n}$. This is commonly done by polling organizations because, before the sample is observed, there is no way to know what sample proportion to use in the more exact margin of error formula. With a table like Table 3.1, researchers can pick a sample size that provides suitable accuracy for any sample proportion, within the constraints of the time and money available for the survey.

Two important features of Table 3.1 are

1. When the sample size is *increased*, the margin of error *decreases*.
2. When a large sample size is made even larger, the improvement in accuracy is relatively small. For example, when the sample size is increased from 2500 to 10,000, the margin of error decreases only from 2% to 1%. In general, cutting the margin of error in half requires a fourfold increase in sample size.

Polling organizations determine a sample size that is accurate enough for their purposes and is also economical. Many national surveys use a sample size of about 1000, which, as you can see from Table 3.1, makes the margin of error roughly 3%. This is a reasonable degree of accuracy for most questions asked in these surveys.

Some federal government surveys utilize much larger sample sizes, sometimes as large as $n = 120,000$, to make accurate estimates of quantities such as the unemployment rate. Also, when researchers want to make accurate estimates for subgroups within the population, they have to use a very large overall sample size. For instance, to get information from approximately 1000 African-American women in the 18- to 29-year-old age group, a random sample of 120,000 Americans might be necessary.

## The Effect of Population Size

You might wonder how the number of people in the population affects the accuracy of a survey. The surprising answer is that for most sample surveys, the number of people in the population has almost no influence on the accuracy of sample estimates. The margin of error for a sample size of 1000 is about 3% whether the number of people in the population is 30,000 or 200 million.

The formulas for margin of error in this chapter were derived by assuming that the number of units in the population is essentially infinite. In practice, as long as the population is at least ten times as large as the sample, we consider only how sample size affects accuracy and we ignore the specific size of the population. For small populations, a "finite population correction" is used. We will not discuss it in this book, but you can consult any book on survey sampling for details.

3.2 Exercises are on page 108.

## 3.3  Choosing a Simple Random Sample

The ability of a relatively small sample to accurately reflect a huge population does not happen haphazardly. It happens only if proper sampling methods are used. A **probability sampling plan** is one in which everyone in the population has a specified probability to be selected for the sample. The basic idea is that everyone in the population must have a specified chance of making it into the sample.

The most basic probability sampling plan is to use a **simple random sample.** Remember that with a simple random sample, every conceivable group of units of the required size has the same chance of being the selected sample. In this section, we discuss how to choose a simple random sample. A variety of other probability sampling methods will be discussed in Section 3.4.

Choosing a simple random sample is somewhat like choosing the winning numbers in many state lotteries. For instance, in the Pennsylvania Match 6 lottery game, six numbers are randomly selected from the choices 1, 2, . . . , 49. Every possible set of six numbers is equally likely to be the winning set. There are actually 13,983,816 different possible sets, which is why the odds of any specific individual guessing the winning set are so small!

Similarly, the chances of any *particular* group of units getting selected to be the random sample from a large population is quite small, but whatever group is selected is likely to be *representative*. For instance, in a simple random sample of 1000 people in your state or country, it is extremely unlikely that you and your next-door neighbor would both be selected. But it is extremely likely that someone in the sample will be representative of each of you, having similar opinions to yours.

By the way, if there were 100 million people in a population, the number of different possible samples of 1000 individuals would be incomprehensively large. It would take about one and a half pages of this book to show the value. Approximately, the number of possible different samples of 1000 individuals selected from a population of 100 million is 247 followed by 5430 zeros.

To actually produce a simple random sample, you need only two things. First, you need a list of the units in the population. For instance, in drawing the winning lottery numbers, the list of units is the numbers 1, 2, . . . , 49. In selecting a simple random sample of students from your school, the list of units is all students in the school (which the registrar can usually produce).

Second, you need a source of **random numbers.** Computer programs such as Minitab and Excel or the right calculator can be used to generate random numbers. Random numbers can also be found in tables called **tables of ran-**

*HINT: The sample size is $n = 400$. Is the number of students at your school more than ten times the sample size?

**dom digits.** If the population isn't very large, physical methods can be used, such as in lotteries in which the numbers are written on small, hollow plastic balls and six of them are physically selected.

Table 3.2 illustrates a portion of a table of random digits. It is organized into numbered rows to make it easier to find specific sections of the table. There are only ten rows in Table 3.2, so they are labeled by using the single digits 0 to 9. A larger table would have to use longer identifying numbers for the rows. The digits are grouped into columns of five for easier reading. These tables are generated by the equivalent of writing the digits from 0 to 9 on slips of paper, mixing them well, choosing one, and then repeating this process over and over again with replacement. No single digit, pair of consecutive digits, triplet of digits, and so on is any more or less likely to occur than any others.

**Table 3.2**  A Table of Random Digits

| Row | | | | | | | | |
|---|---|---|---|---|---|---|---|---|
| 0 | 00157 | 37071 | 79553 | 31062 | 42411 | 79371 | 25506 | 69135 |
| 1 | 38354 | 03533 | 95514 | 03091 | 75324 | 40182 | 17302 | 64224 |
| 2 | 59785 | 46030 | 63753 | 53067 | 79710 | 52555 | 72307 | 10223 |
| 3 | 27475 | 10484 | 24616 | 13466 | 41618 | 08551 | 18314 | 57700 |
| 4 | 28966 | 35427 | 09495 | 11567 | 56534 | 60365 | 02736 | 32700 |
| 5 | 98879 | 34072 | 04189 | 31672 | 33357 | 53191 | 09807 | 85796 |
| 6 | 50735 | 87442 | 16057 | 02883 | 22656 | 44133 | 90599 | 91793 |
| 7 | 16332 | 40139 | 64701 | 46355 | 62340 | 22011 | 47257 | 74877 |
| 8 | 83845 | 41159 | 67120 | 56273 | 67519 | 93389 | 83590 | 12944 |
| 9 | 12522 | 20743 | 28607 | 63013 | 60346 | 71005 | 90348 | 86615 |

## Using Table 3.2 to Choose a Simple Random Sample

Here are the steps for selecting a simple random sample using Table 3.2.

1. Number the units in the population, using the same number of digits for each one.
   *Example:* Suppose there are 270 students in a class and the teacher wants to choose a simple random sample of 10 of them to call on in class. Number the students from 001 to 270.
2. Choose a starting point in Table 3.2. You can close your eyes and point or use any other method as long as you have not studied the table and then chosen numbers that give a favorable sample for your purpose.
   *Example:* Start in row 3, column 2 (10484 . . .).
3. From the starting point, read across the row to get numbers with the correct number of digits to identify a unit. Continue with consecutive rows. For instance, if the units are numbered 001 to 270, read three-digit numbers.
   *Example:* Reading three-digit numbers starting with row 3, column 2 results in 104, 842, 461, 613, 466, 416, 180, 855, 118, 314, 577, 002, 896, and so on.
4. If a number is in the range of the unit numbers, select that unit number. Otherwise, continue along the row, choosing more potential unit numbers until you have a sample of the desired size. (But see step 5 for a more efficient method.) Because each unit can be used only once, if a unit number occurs that has already been selected, simply ignore that number and continue.

*Example:* Unit numbers can only be 001 to 270, so most of the numbers that are chosen are simply ignored. Select units 104, 180, 118, 002, then keep going until the required 10 students have been selected.

5. Step 4 is very inefficient. To make it more efficient, reassign some of the higher numbers onto the range of unit numbers in a way that still ensures that each unit number has an equal chance of selection. For instance, suppose the units are numbered 001 to 270. If a number between 301 and 570 is chosen, subtract 300 and use it. If a number between 601 and 870 is chosen, subtract 600 and use it. As in step 4, if a unit number occurs more than once, simply ignore subsequent occurrences after the first one. For instance, in the scenario just given, if unit 017 has already been selected, then any subsequent occurrence of 017, 317, or 617 would be discarded.

*Example:* Now the string generated earlier is more useful. Let's see what decision follows each three-digit entry:

| Three-Digit Number | Using Step 4, Choose Unit | Using Step 5, Choose Unit |
|---|---|---|
| 104 | 104 | 104 |
| 842 | Discard | 242 (subtracted 600) |
| 461 | Discard | 161 (subtracted 300) |
| 613 | Discard | 013 (subtracted 600) |
| 466 | Discard | 166 (subtracted 300) |
| 416 | Discard | 116 (subtracted 300) |
| 180 | 180 | 180 |
| 855 | Discard | 255 (subtracted 600) |
| 118 | 118 | 118 |
| 314 | Discard | 014 (subtracted 300) |
| 577 | Discard | Not needed, 10 already selected |
| 002 | 002 | Not needed |

Using the method in step 4, the first four students who are selected would be those numbered 104, 180, 118, and 002, and the process would continue until six more were selected. Using the method in step 5, the sample of ten units would include 104, 242, 161, 013, 166, 116, 180, 255, 118, and 014, and unit 002 would not be needed.

---

**MINITAB *tip*** | **Picking a Random Sample**

- To create a column of ID numbers, use **Calc>Make Patterned Data> Simple Set of Numbers.** In the dialog box, specify a column for storing the ID numbers, and specify the first and last possible ID number for the population.

- To sample values from a column, use **Calc>Random Data>Sample from Columns.** In the dialog box, specify how many items (rows) will be selected from a particular column, and specify a column where the sample will be stored.

*Note:* Items can be randomly selected from a column of names or data values, so it may not be necessary to assign ID numbers to the units in the population in order to select a sample.

# Not Reaching the Individuals Selected

Even if a sample of units is selected properly, researchers might not reach the desired units. For example, *Consumer Reports* magazine mails a lengthy survey to its subscribers to obtain information on the reliability of various products. If you received such a survey and had a close friend who had been having trouble with a highly rated automobile, you may very well have decided to pass the questionnaire on to your friend. That way, he would get to register his complaints about the car, but *Consumer Reports* would not have reached the intended recipient.

Telephone surveys tend to reach a disproportionate number of women because they are more likely to answer the phone. Researchers sometimes ask to speak to the oldest adult male at home to try to counter that problem. Surveys are also likely to have trouble contacting people who work long hours and are rarely home and those who tend to travel extensively. Some people screen their calls or simply refuse to cooperate even when they do answer the phone.

In recent years, there has been pressure on news organizations to produce surveys of public opinions quickly. When a controversial story breaks, people want to know how others feel about it. This pressure results in what *Wall Street Journal* reporter Cynthia Crossen calls "**quickie polls.**" As she notes, these are "most likely to be wrong because questions are hastily drawn and poorly pretested, and it is almost impossible to get a random sample in one night" (Crossen, 1994, p. 102). Even if the computer can randomly generate phone numbers for the sample, there will be groups of people who are not likely to be home that night, and they may have different opinions from those who are likely to be home. Most responsible reports about polls include information about the dates during which the poll was conducted. For instance, the millennium poll in Example 3.7 was conducted on a Saturday and Sunday, so people who frequently travel on weekends may be underrepresented.

Failing to contact or measure the individuals who were selected in the sampling plan leads to *nonresponse bias.* It is important, once a sample has been selected, that those individuals are the ones who are actually measured. It is better to put resources into getting a smaller sample than to get one that has been biased by moving on to the next person on the list when someone is initially unavailable.

# Nonresponse or Nonparticipation

Even the best surveys are not able to contact everyone on the list of selected units, and not everyone who is contacted will respond. The General Social Survey (GSS), run by the prestigious National Opinion Research Center (NORC) at the University of Chicago, noted in its September 1993 *GSS News:*

> In 1993 the GSS achieved its highest response rate ever, 82.4%. This is five percentage points higher than our average over the last four years. Given the long length of the GSS (90 minutes), the high response rates of the GSS are testimony to the extraordinary skill and dedication of the NORC field staff.

The GSS is now conducted biennially, and the response rate was 70% in each of the years 2000, 2002, and 2004, according to the NORC website (http://www

.norc.uchicago.edu/projects/gensoc.asp). Beyond having a dedicated staff, not much can be done about getting everyone in the sample to respond.

Response rates should be reported in research summaries. As a reader, remember that the lower the response rate, the less the results can be generalized to the population as a whole. Responding to a survey (or not) is voluntary, and those who respond are likely to have stronger opinions than those who do not respond. As we mentioned earlier, this type of **nonresponse bias** can lead to systematically overestimating or underestimating the truth about a population.

With mail surveys, it may be possible to compare those who respond immediately with those who need a second prodding; and in telephone surveys, you could compare those who are home on the first try with those who require numerous call-backs. If those groups differ on the measurement of interest, then those who were never reached are probably different as well.

In a mail survey, researchers should not just accept that those who did not respond the first time can't be cajoled into responding. Often, sending a reminder with a brightly colored stamp or following up with a personal phone call will produce the desired effect. Surveys that simply use those who respond voluntarily are sure to be biased in favor of those with strong opinions or with time on their hands.

---

***thought question* 3.4**   Suppose you want to know how students at your school feel about the computer services that are offered. You are able to obtain the list of e-mail addresses for all students who are taking statistics classes, so you send a survey to a simple random sample of 100 of those students and 65 respond. Using the difficulties discussed so far in this section, explain to whom you could extend the results of your survey and why. *

---

**Example 3.9**

**Statistics⟁Now™**

Watch a video example at **http://1pass.thomson.com** or on your CD.

**Which Scientists Trashed the Public?**  According to a poll taken among scientists and reported in the prestigious journal *Science* (Mervis, 1998), scientists don't have much faith in either the public or the media. The article reported that based on the results of a "recent survey of 1400 professionals" in science and in journalism, 82% of scientists "strongly or somewhat agree" with the statement, "The U.S. public is gullible and believes in miracle cures or easy solutions." Eighty percent agreed that "the public doesn't understand the importance of federal funding for research." Eighty-two percent also trashed the media, agreeing with the statement, "The media do not understand statistics well enough to explain new findings." It isn't until the end of the article that we learn who responded: "The study reported a 34% response rate among scientists, and the typical respondent was a white, male physical scientist over the age of 50 doing basic research." Remember that those who feel strongly about the issues in a survey are the most likely to respond. With only about a third of those contacted responding, it is inappropriate to generalize these findings and conclude that most scientists have so little faith in the public and the media. This is especially true since we were told that the respondents represented only a narrow subset of scientists. ■

---

**\*HINT:** What type of bias is represented and how is it likely to affect the results?

# Not Using a Probability Sample

Basing a sample survey on a **self-selected sample** (also called a **volunteer sample**) or a **convenience sample** is usually so problematic that the results cannot be extended to anyone beyond the sample.

## Self-Selected Sample

While relying on volunteer responses from a properly selected random sample presents somewhat of a difficulty, relying on a self-selected sample consisting of only those who pick themselves to participate is usually a complete waste of time. When a television or radio station, magazine, or Internet site presents a survey and asks any readers or viewers who are interested to respond, the results reflect the opinions of only those listeners or readers who decide to take the trouble to respond. And people who weren't watching or listening, or reading the magazine or Internet site, would not even know the poll existed.

As we noted earlier, those who have a strong opinion about the question are more likely to respond than are those who do not have a strong opinion. Thus, the self-selected responding group is simply not representative of any larger group. Most media outlets now acknowledge that such polls are "unscientific" when they report the results, but most readers are not likely to understand how misleading the results can be. The next example illustrates the contradiction that can result between a scientific poll and one relying solely on a self-selected sample.

**Example 3.10** | **A Meaningless Poll** On February 18, 1993, shortly after Bill Clinton became president of the United States, a television station in Sacramento, California, asked viewers to respond to the question "Do you support the President's economic plan?" The next day the results of a properly conducted study that asked the same question were published in the newspaper. Here are the results from these two different surveys:

|  | Television Poll | Survey |
| --- | --- | --- |
| Yes (support plan) | 42% | 75% |
| No (don't support plan) | 58% | 18% |
| Not sure | 0% | 7% |

As you can see, those who were dissatisfied with the president's plan were much more likely to respond to the television poll than those who supported it, and no one who was "Not sure" called the television station, since they were not invited to do so. Trying to extend those results to the general population is misleading. It is irresponsible to publicize such studies, especially without a warning that they result from an unscientific survey and are not representative of general public opinion. You should never interpret such polls as anything other than a count of who bothered to respond. ■

## Convenience or Haphazard Sample

Another problematic sampling technique for surveys is either to use the most convenient group available or to decide haphazardly on the spot whom to sample. It is almost always the case that these types of samples break the Fundamental Rule of Using Data for Inference given in Section 3.1. The responses from a **convenience sample** or **haphazard sample** rarely represent any larger population for the question of interest.

**Example 3.11**

Statistics⬡Now™

Watch a video example at **http://1pass.thomson.com** or on your CD.

**Haphazard Sampling**  A few years ago, the student newspaper at a California university announced as a front-page headline "Students ignorant, survey says." The article explained that a "random survey" indicated that American students were less aware of current events than international students were.

The article quoted the undergraduate researchers, who were international students themselves, as saying "the students were randomly sampled on the quad." The quad is a wide expanse of lawn where students relax, eat lunch, and so on. There is simply no proper way to collect a random sample of students by selecting them in an area like that. In such situations, the researchers are likely to approach people who they think will support the results they intended for their survey. Or they are likely to approach friendly-looking people who look as though they will easily cooperate. This is called a haphazard sample, and it cannot be expected to be representative at all.  ■

You have seen the proper way to collect a sample and have been warned about the many difficulties and dangers inherent in the process. Here is a famous example that helped researchers learn some of these pitfalls.

## *case study* 3.1    The Infamous *Literary Digest* Poll of 1936

Before the presidential election of 1936, a contest between Democratic incumbent Franklin Delano Roosevelt and Republican Alf Landon, the magazine *Literary Digest* had been extremely successful in predicting the results in U.S. presidential elections. But 1936 turned out to be the year of its downfall, when it predicted a 3-to-2 victory for Landon. To add insult to injury, young pollster George Gallup, who had just founded the American Institute of Public Opinion in 1935, not only correctly predicted Roosevelt as the winner of the election, but also predicted that the *Literary Digest* would get it wrong. He did this before they had even conducted their poll! And Gallup surveyed only 50,000 people, while the *Literary Digest* sent questionnaires to 10 million people (Freedman, Pisani, Purves, and Adhikari, 1991, p. 307).

The *Literary Digest* made two classic mistakes. First, the lists of people to whom they mailed the 10 million questionnaires were taken from magazine subscribers, car owners, telephone directories, and, in a few instances, lists of registered voters. In 1936, those who owned telephones or cars or subscribed to magazines were more likely to be wealthy individuals who were not happy with the Democratic incumbent.

Despite what many accounts of this famous story conclude, the bias produced by the more affluent list was not likely to have been as severe as the second problem (Bryson, 1976). *The main problem was volunteer response.* The *Literary Digest* received 2.3 million responses, a response rate of only 23%. Those who felt strongly about the outcome of the election were most likely to respond. That included a majority of those who wanted a change: the Landon supporters. Those who were happy with the incumbent were less likely to bother to respond.

Gallup, on the other hand, knew the value of random sampling. He not only was able to predict the election, but also predicted what the results of the *Literary Digest* poll would be to within 1%. How did he do this? According to Freedman et al. (1991, p. 308), "he just chose 3,000 people at random from the same lists the *Digest* was going to use, and mailed them all a postcard asking them how they planned to vote.

Case Study 3.1 illustrates the beauty of random sampling and the idiocy of trying to base conclusions on nonrandom and biased samples. The *Literary Digest* went bankrupt the following year, so it never had a chance to revise its

3.5 Exercises are on page 110.

methods. The organization founded by George Gallup has flourished, although it has made a few sampling blunders as well. (See Exercise 3.50.)

# 3.6 How to Ask Survey Questions

You may be surprised at how much the answers to questions can change on the basis of simple changes in wording. Here is one example. Loftus and Palmer (1974, quoted in Plous, 1993, p. 32) showed college students films of an automobile accident, after which they asked the students a series of questions. One group was asked the question, "About how fast were the cars going when they contacted each other?" The average response was 31.8 miles per hour. Another group was asked, "About how fast were the cars going when they collided with each other?" In that group, the average response was 40.8 miles per hour. Simply changing from the word *contacted* to the word *collided* increased the estimates of speed by 9 miles per hour, or 28%, even though the respondents had witnessed the same film.

The wording and presentation of questions can significantly influence the results of a survey. The seven pitfalls discussed in this section are all possible sources of response bias in surveys.

## Possible Sources of Response Bias in Surveys

Many pitfalls leading to response bias can be encountered when asking questions in a survey or experiment. Here are seven of them, each of which will be discussed in turn:

- Deliberate bias in questions
- Unintentional bias in questions
- Desire of respondents to please
- Asking the uninformed
- Unnecessary complexity
- Ordering of questions
- Confidentiality and anonymity concerns

### Deliberate Bias in Questions

Sometimes, if a survey is being conducted to support a certain cause, questions are deliberately worded in a biased manner (**deliberate bias**). Be careful about survey questions that begin with phrases such as "Do you agree that . . . ?" Most people want to be agreeable and will be inclined to answer "yes" unless they have strong feelings the other way.

For example, suppose that an anti-abortion group and a pro-choice group each wanted to conduct a survey in which it would find the best possible agreement with its position. Here are two questions that would each produce an estimate of the proportion of people who think abortion should be completely illegal. The questions are almost certain to produce different estimates:

1. Do you agree that abortion, the murder of innocent beings, should be outlawed?

2. Do you agree that there are circumstances under which abortion should be legal, to protect the rights of the mother?

The wording of a question should not indicate a desired answer. For instance, a Gallup Poll conducted in June 1998 contained the question, "Do you think it was a good thing or a bad thing that the atomic bomb was developed?" Notice that the question does not indicate which answer is preferable (61% of the respondents said "bad," while 36% said "good," and 3% were undecided).

### Unintentional Bias in Questions

Sometimes questions are worded in such a way that the meaning is misinterpreted by a large percentage of the respondents (**unintentional bias**). For example, if you were to ask people whether or not they use drugs, you would need to specify whether you mean prescription drugs, illegal drugs, over-the-counter drugs, or common substances such as caffeine. If you were to ask people to recall the most important date in their life, you would need to clarify whether you meant the most important calendar date or the most important social engagement with a potential partner. (It is unlikely that anyone would mistake the question as being about the shriveled fruit, but you can see that the same word can have multiple meanings.)

**Example 3.12**

**Laid Off or Fired?**  Plewes (1994) described a number of changes implemented at the beginning of 1994 to the Current Population Survey (CPS), which is a monthly survey used to determine government unemployment figures and other government statistics in the United States. One change involved clarification and expansion of questions about being laid off. Plewes explained that people were answering that they had been laid off when in reality they had been fired: "The CPS concept of 'on layoff' differs from the everyday usage by respondents. CPS defines 'on layoff' as embodying an expectation of recall to the job; common usage considers it a euphemism for 'fired'" (p. 38). ∎

### Desire of Respondents to Please

Most survey respondents have a desire to please the person who is asking the question. They tend to understate their responses about undesirable social habits and opinions and vice versa. For example, in recent years, estimates of the prevalence of cigarette smoking based on surveys do not match those based on cigarette sales. Either people are not being completely truthful or lots of cigarettes are ending up in the garbage.

**Example 3.13**

Statistics⬨Now™

Watch a video example at **http:// 1pass.thomson.com** or on your CD.

**Most Voters Don't Lie but Some Liars Don't Vote**  Abelson, Loftus, and Greenwald (1992) discuss how the National Election Surveys always find that more people claim to vote than actually did. They explain that researchers have tried to make it more socially acceptable to admit to not voting by adding a preamble to questions about voting. The preamble reads, "In talking to people about elections, we often find that a lot of people were not able to vote because they weren't registered, they were sick, or they just didn't have time. How about you—did you vote in the elections this November?" (p. 139). In a survey six to seven months after the November 1986 election, the authors of the report surveyed people for whom they knew whether or not they had actually voted. The question started with the preamble and finished with "did you vote in the 1986 elections for United States Congress?" (p. 141). They found that of the 211 respondents who had actually voted, 203 (96%) accurately reported that they did. But of the 98 respondents who had *not* voted, 39 (40%) claimed that they *did* vote. ■

### Asking the Uninformed

People do not like to admit that they don't know what you are talking about when you ask them a question. Crossen (1994, p. 24) gives an example: "When the American Jewish Committee studied Americans' attitudes toward various ethnic groups, almost 30% of the respondents had an opinion about the fictional Wisians, rating them in social standing above a half-dozen other real groups, including Mexicans, Vietnamese and African blacks."

Political pollsters, who are interested in surveying only those who will actually vote, learned long ago that it is useless to simply ask people whether they plan to vote. Most of them will say "yes." Instead, they ask questions to establish a history of voting, such as "Where did you go to vote in the last election?"

### Unnecessary Complexity

If questions are to be understood, they must be kept simple. A question such as "Shouldn't former drug dealers not be allowed to work in hospitals after they are released from prison?" is sure to lead to confusion. Does a "yes" answer mean they should or should not be allowed to work in hospitals? It would take a few readings to figure that out.

Another way in which a question can be unnecessarily complex is by actually asking more than one question at once. An example would be a question such as "Do you support the president's health care plan, since it would ensure that all Americans receive health coverage?" If you agree with the idea that all Americans should receive health coverage but disagree with the remainder of the plan, do you answer "yes" or "no"? Or what if you support the president's plan, but not for that reason?

**Example 3.14**   **Why Weren't You at Work Last Week?**   In the work leading to the revision of the Current Population Survey discussed in Example 3.12, researchers discovered that some questions were too complex to be understood by many respondents. For instance, as Plewes (1994, p. 38) described, one such question on the old version was "Did you have a job from which you were temporarily absent or on layoff LAST WEEK?" In the revised version of the questionnaire, people are first asked whether they were absent in the last week, and if they were, further questions about temporary absence are asked. A new and separate series of questions asks about layoffs. ■

### Ordering of Questions

If one question requires respondents to think about something that they may not have otherwise considered, then the order in which questions are presented can change the results. For example, suppose a survey were to ask, "To what extent do you think teenagers today worry about peer pressure related to drinking alcohol?" and then ask, "Name the top five pressures you think face teenagers today." It is quite likely that respondents would use the idea they had just been given and name peer pressure related to drinking alcohol as one of the five choices.

Example 3.15

Statistics⊘Now™

Watch a video example at **http://1pass.thomson.com** or on your CD.

**Is Happiness Related to Dating?**   Clark and Schober (1992, p. 41) report on a survey that asked the following two questions:

1.  How happy are you with life in general?
2.  How often do you normally go out on a date? about _____ times a month

When the questions were asked in this order, there was almost no relationship between the two answers. But when question 2 was asked first, the answers were highly related. Clark and Schober speculate that in that case, respondents interpreted question 1 to mean "Now, considering what you just told me about dating, how happy are you with life in general?"   ■

## Confidentiality and Anonymity Concerns

People will often answer questions differently based on the degree to which they believe they are anonymous. Because researchers often need to perform follow-up surveys, it is easier to try to ensure confidentiality than anonymity. In ensuring **confidentiality,** the researcher promises not to release identifying information about respondents. In an **anonymous** survey, the researcher does not know the identity of the respondents.

Questions on issues such as sexual behavior and income are particularly difficult because people consider those to be private matters. A variety of techniques have been developed to help ensure confidentiality, but surveys on such issues are hard to conduct accurately.

The problems discussed in this section are well known to professional pollsters, and most reputable polling agencies go to great lengths to overcome them to the extent possible. The survey questions in Examples 3.3 to 3.5 in Section 3.2 are excellent illustrations of how questions should be worded. For instance, the question in Example 3.5, "Do you favor or oppose scientific experimentation on the cloning of human beings?" is worded so as not to suggest which answer the interviewer would prefer. Here are two examples that illustrate attempts to overcome bias but also illustrate how difficult that can be.

Example 3.16

Statistics⊘Now™

Watch a video example at **http://1pass.thomson.com** or on your CD.

Read the original source on your CD.

**When Will Adolescent Males Report Risky Behavior?**   It is particularly difficult to measure information on behavior that is sensitive, risky, or illegal. In an article in *Science*, Turner et al. (1998) describe the use of audio, computer-assisted self-interviews as part of the 1995 National Survey of Adolescent Males (aged 15 to 19). The respondents were randomly assigned to answer the survey using the traditional paper form ($n = 368$) or using a laptop computer ($n = 1361$). The paper questionnaire was filled out and sealed in an envelope to return to the surveyor but was accompanied by identifying code numbers. The computer method included listening to questions through headphones, then recording the answers on a laptop computer. The authors believed that this method would allow respondents to feel that their responses were more private. In particular, respondents who could not read would need to have the paper version read to them, whereas the audio-computer method allowed complete self-administration of the survey.

Results indicated that the additional perception of privacy increased the reported incidence of certain behaviors while decreasing the reporting of others. The behaviors with increased reported incidence for the audio-computer version tended to be less socially acceptable, while those with higher incidence for the paper version tended to be more socially acceptable (for the adolescent males in this survey). Here are some examples reported in the article:

| Question: | Estimated Prevalence per 100 Males in Population | |
|---|---|---|
| *Have you ever* . . . | *Paper Version* | *Audio-Computer Version* |
| Had sex with a prostitute | 0.7 | 2.5 |
| Had intercourse with a female | 68.1 | 63.9 |
| Made a girl pregnant | 7.9 | 6.5 |
| Had sex with another male | 1.5 | 5.5 |
| Taken street drugs using a needle | 1.4 | 5.2 |

**Example 3.17**

**Politics Is All in the Wording**   Crossen (1994, p. 112) described three very different sets of responses to a question about whether the president of the United States should be able to veto specific items in the federal budget. The first set came from a questionnaire published in *TV Guide* in which founder of the Reform Party and presidential candidate Ross Perot asked, "Should the president have the line item veto to eliminate waste?" Respondents were volunteers who mailed in the survey. Of those who responded, 97% said "yes." The second set of responses came from asking a properly selected random sample the same question, and 71% said "yes." But clearly, the question is written to elicit a "yes" answer. When the question was reworded to read, "Should the president have the line item veto, or not?" and a proper random sample was queried, only 57% answered yes.

Cohn and Cope (2001, pp. 145–146) report that Perot also asked the question "Should laws be passed to eliminate all possibilities of special interests giving huge sums of money to candidates?" Again the "yes" responses were an overwhelming 99%. When a properly selected random sample was asked the same, biased version of the question, 80% said "yes." But when a properly selected random sample was asked, "Should laws be passed to prohibit interested groups from contributing to campaigns, or do groups have a right to contribute to the candidate they support?" only 40% said that such laws should be passed. ∎

*thought question* **3.5**   You have now learned that survey results have to be interpreted in the context of *who* responded and *to what questions* they responded. When you read the results of a survey, for which of these two areas do you think it would be easier for you to recognize and assess possible biases? Why?*

*****HINT:** What information would you need in each case, and what information is more likeˡ be included in a description of the survey?

# Be Sure You Understand What Was Measured

Sometimes words mean different things to different people. When you read about survey results, you should get a precise definition of what was actually asked or measured. Here are two examples that illustrate that even common terminology may mean different things to different people.

**Example 3.18**

**Teenage Sex** A letter to advice columnist Ann Landers stated, "According to a report from the University of California at San Francisco ... sexual activity among adolescents is on the rise. There is no indication that this trend is slowing down or reversing itself." The letter went on to explain that these results were based on a national survey (*Davis Enterprise*, Feb. 19, 1990, p. B-4). On the same day, in the same newspaper, an article entitled "Survey: Americans Conservative with Sex" reported that "teenage boys are not living up to their reputations. [A study by the Urban Institute in Washington] found that adolescents seem to be having sex less often, with fewer girls and at a later age than teenagers did a decade ago" (*Ibid.*, p. A-9).

Here we have two apparently conflicting reports on adolescent sexuality, both reported on the same day in the same newspaper. One indicated that teenage sex was on the rise, while the other indicated that it was on the decline. Although neither report specified exactly what was measured, the letter to Ann Landers proceeded to note that "national statistics show the average age of first intercourse is 17.2 for females and 16.5 for males." The article stating that adolescent sex was on the decline measured it in terms of *frequency*. It was based on interviews with 1880 boys between the ages of 15 and 19, in which "the boys said they had had six sex partners, compared with seven a decade earlier. They reported having had sex an average of three times during the previous month, compared with almost five times in the earlier survey."

Thus, it is not enough to know that both surveys were measuring adolescent or teenage sexual behavior. In one case, the author was, at least partially, discussing the *age* of first intercourse, whereas in the other case, the author was discussing the *frequency* of intercourse. ∎

**Example 3.19**

**Statistics⬡Now**™

Watch a video example at **http:// 1pass.thomson.com** or on your CD.

**The Unemployed** Ask people whether or not they know anyone who is unemployed and, invariably, they will say that they do. Most people, however, don't realize that to be officially unemployed and included in the unemployment statistics given by the U.S. government, you must meet very stringent criteria. The government excludes "discouraged workers" who are "identified as persons who, though not working or seeking work, have a current desire for a job, have looked for one within the past year [but not within the past four weeks], and are currently available for work" (U.S. Dept. of Labor, 1992, p. 10). If you knew someone who fit that definition, you would undoubtedly think of him or her as unemployed. But they are not included in the official statistics. You can see that the true number of people who are not working is higher than the government statistics would lead you to believe. ∎

# Some Concepts Are Hard to Precisely Define

Sometimes it is not the language but the concept itself that is ill-defined. For example, there is still not universal agreement on what should be measured with intelligence tests. IQ tests were developed at the beginning of the 20th century to determine the mental level of schoolchildren. The intelligence quotient (IQ) of a child was found by dividing the child's "mental level" by his or her chronological age. The "mental level" was determined by comparing the child's performance on the test with that of a large group of "normal" children to find the age group the individual's performance matched. Thus, if an 8-year-old child performed as well on the test as a "normal" group of 10-year-old children, he or she would have an IQ of $100 \times (10/8) = 125$.

IQ tests have been expanded and refined since the early days, but they continue to be surrounded by controversy. One reason is that it is very difficult to define what is meant by *intelligence*. It is hard to measure something if you can't even agree on what it is you are trying to measure. If you are interested in knowing more about these tests and the surrounding controversies, you can find numerous books on the subject. Anastasi (1988) provides a detailed discussion of a large variety of psychological tests, including IQ tests.

# Measuring Attitudes and Emotions

Similar problems exist with trying to measure attitudes and emotions such as self-esteem and happiness. The most common method for trying to measure such things is to have respondents read statements and determine the extent to which they agree with the statement. For example, a test for measuring happiness might ask respondents to indicate their level of agreement, from "strongly disagree" to "strongly agree," with statements such as "I generally feel optimistic when I get up in the morning." To produce agreement on what is meant by characteristics such as "introversion," psychologists have developed standardized tests that claim to measure those attributes.

# Open or Closed Questions: Should Choices Be Given?

An **open question** is one in which respondents are allowed to answer in their own words; a **closed question** is one in which respondents are given a list of alternatives from which to choose their answer. Usually, the latter form offers a choice of "other" in which the respondent is allowed to fill in the blank.

### Problems with Closed Questions

To show the limitation of closed questions, Schuman and Scott (1987) asked about "the most important problem facing this country today." Half of the sample, 171 people, were given this as an open question. The most common responses were

Unemployment (17%)
General economic problems (17%)
Threat of nuclear war (12%)
Foreign affairs (10%)

In other words, one of these four choices was volunteered by over half of the respondents.

The other half of the sample was given the question as a closed question, with the list of choices and the percentages that chose them being

The energy shortage (5.6%)
The quality of public schools (32.0%)
Legalized abortion (8.4%)
Pollution (14.0%)

These choices combined were mentioned by only 2.4% of respondents in the open question survey, yet they were selected by 60% when they were the only specific choices given. Don't think that respondents had no choice; in addition to the list of four, they were told, "If you prefer, you may name a different problem as most important." On the basis of the closed-form questionnaire, policymakers would have been seriously misled about what is important to the public. It is possible to avoid this kind of astounding discrepancy. If closed questions are preferred, they first should be presented as open questions to a test sample before the real survey, and the most common responses should then be included in the list of choices for the closed question. This kind of exercise is usually done as part of what's called a "pilot survey," in which various aspects of a study design can be tried before it's too late to change them.

### Problems with Open Questions

The biggest problem with open questions is that the results can be difficult to summarize. If a survey includes thousands of respondents, it can be a major chore to categorize their responses. Another problem, found by Schuman and Scott (1987), is that the wording of the question might unintentionally exclude answers that would have been appealing had they been included in a list of choices (such as in a closed question). To test this, they asked 347 people to "name one or two of the most important national or world event(s) or change(s) during the past 50 years." The most common choices and the percentages that mentioned them were

World War II (14.1%)
Exploration of space (6.9%)
Assassination of John F. Kennedy (4.6%)
The Vietnam war (10.1%)
Don't know (10.6%)
All other responses (53.7%)

The same question was then repeated in closed form to a new group of 354 people. Five choices were given: the preceding ones and "invention of the computer." Of the 354 respondents, the percentages that selected each choice were

World War II (22.9%)
Exploration of space (15.8%)
Assassination of John F. Kennedy (11.6%)
The Vietnam war (14.1%)
Invention of the computer (29.9%)

**a.** The heights of women in a psychology class will be used to estimate the average height of all women at a college.

**b.** Parents of children attending a daycare facility are surveyed about whether the state government should increase funding to daycare providers to provide lower rates for low-income families. The data are used to estimate statewide support for increased state funding of daycare facilities.

**3.4** ● A political scientist surveys 400 voters randomly selected from the list of all registered voters in a community. The purpose is to estimate the proportion of registered voters who will vote in an upcoming election.
a. What is the population of interest in this survey?
b. What is the sample in this survey?

**3.5** ● In a class of 20 students (John, Maria, Inez, Bill, etc.), a simple random sample of two people will be selected from the class. Use the definition of a simple random sample to compare the chance that John and Maria will be the group selected with the chance that Maria and Bill will be the group selected.

**3.6** ● To estimate the percentage of households in the United States that use a DVD player, a researcher surveys a randomly selected sample of 500 households and asks about DVD player usage.
a. What is the population of interest in this survey?
b. What is the sample in this survey?

**3.7** ● Briefly explain the difference between a sample survey and a census.

**3.8** ● Briefly explain what it means to say that a survey method is biased.

**3.9** For each definition, identify the correct term for the type of bias being defined. Possible answers are *selection bias, nonresponse bias,* and *response bias.*
a. Participants respond differently from how they truly feel.
b. The method for selecting the participants produces a sample that does not represent the population of interest.
**c.** A representative sample is chosen, but a subset of the sample cannot be contacted or does not respond.

3.10 In each of the following situations, indicate whether the potential bias is a selection bias, a nonresponse bias, or a response bias.
a. A survey question asked of unmarried men was "What is the most important feature you consider when deciding whether to date somebody?" The results were found to depend on whether the interviewer was male or female.
b. In a study of women's opinions about community issues, investigators randomly selected a sample of households and interviewed a woman from each selected household. When no woman was present in a selected household, a next-door neighbor was interviewed instead. The survey was done during daytime hours, so working women might have been disproportionately missed.
c. A telephone survey of 500 residences is conducted. People refused to talk to the interviewer in 200 of the residences.

3.11 Give an example of a situation in which a sample must be used because a census is not possible.

3.12 The U.S. government gathers numerous statistics based on random samples, but every ten years, it conducts a census of the U.S. population. What can it learn from a census that cannot be learned from a sample?

**3.13** Refer to the three types of bias given in Section 3.1. Which type of bias do you think would be introduced in each of the following situations?
**a.** A list of registered automobile owners is used to select a random sample for a survey about whether people think homeowners should pay a surtax to support public parks.
**b.** A survey is mailed to a random sample of residents in a city asking whether they think the current mayor is doing an acceptable job.
c. In a college town, college students are hired to conduct door-to-door interviews, based on a multistage cluster sample, to determine whether city residents think there should be a law forbidding loud music at parties.
d. A magazine sends a survey to a random sample of its subscribers asking them whether they would like the frequency of publication reduced from biweekly to monthly or would prefer that it remain the same.
e. A random sample of registered voters is contacted by phone and asked whether or not they are going to vote in the upcoming presidential election.

3.14 For each of the following examples from the text, explain whether there would be more interest in descriptive statistics (for the acquired data set only) or in inferential statistics (extending results to a larger population).
a. Ages at death of First Ladies.
b. The relationship between nicotine patch use and cessation of smoking.
c. Annual rainfall data for Davis, California.

**3.15** For each of the following situations, explain whether or not the Fundamental Rule for Using Data for Inference holds.
**a.** *Available Data:* Salaries for a random sample of male and female professional basketball players. *Research Question:* Are women paid less than men who are in equivalent jobs?
**b.** *Available Data:* Pulse rates for smokers and nonsmokers in a large statistics class at a major university. *Research Question:* Do college-age smokers have higher pulse rates than college-age nonsmokers?
c. *Available Data:* Opinions on whether or not the legal drinking age should be lowered to 19 years old, collected from a random sample of 1000 adults in the state. *Research Question:* Does a majority of adults in the state support lowering the drinking age to 19?
d. *Available Data:* Opinions on whether or not the legal drinking age should be lowered to 19 years old collected from a random sample of parents of school students in the state.

*Research Question:* Does a majority of adults in the state support lowering the drinking age to 19?

3.16 A friend has recommended the work of a musician who has recorded five CDs, each containing ten selections. You decide to visit a music store and listen to a simple random sample of four songs. Explain how you could select the four songs.

**3.17** Refer to Exercise 3.16. Suppose you randomly chose four of the CDs, then randomly chose one song from each one. Would this be considered a simple random sample of the musician's songs? Explain.

## Section 3.2

3.18 ● A survey is planned to estimate the proportion of voters in a community who plan to vote for candidate Y. Calculate the conservative margin of error, as a proportion and as a percentage, for each of the following possible sample sizes:
   a. $n = 100$.
   b. $n = 400$.
   c. $n = 900$.

**3.19** ● The Nielsen survey produces the rankings of television shows. The survey is based on a random sample of about 5000 households with TV sets. Calculate a conservative margin of error for this survey as a proportion and as a percentage.

3.20 ● In a random sample of 90 students at a university, 72 students (80% or .80 of the sample) say that they use a laptop computer.
   a. Calculate the conservative margin of error for the survey.
   b. Compute an approximate 95% confidence interval for the population *proportion* that uses a laptop computer.

3.21 ● *This is also Exercise 1.4.* For a survey based on 2000 adults, what is the approximate margin of error?

3.22 ● *This is also Exercise 1.5.* What sample size produces each of the following as the approximate margin of error?
   a. Margin of error = .05 or 5%.
   b. Margin of error = .30 or 30%.

3.23 ● In a CNN/*Time* poll conducted December 17–18, 1998, a sample of $n = 1031$ adults in the United States was asked, "Do you think the police should or should not be allowed to collect DNA information from suspected criminals, similar to how they take fingerprints?" Of those sampled, 66% answered "should" (http://www.pollingreport.com/crime.htm).
   a. Calculate the conservative margin of error for the survey, as a percentage.
   b. Compute an approximate 95% confidence interval for the percentage of all American adults who think police should be allowed to collect DNA information on suspected criminals.

News Poll conducted between January 21 ... 0, a random sample of $n = 1006$ adult vas asked, "Compared to buying things or in a store, do you think that buying e Internet poses more of a threat to your cy, less of a threat, or about the same?"

(http://www.pollingreport.com/computer.htm, September 17, 2002). The percentages that selected each response were

| More of a threat | 40% |
|---|---|
| Less of a threat | 7% |
| About the same | 47% |
| No opinion | 6% |

   a. What is the margin of error for this poll? Give your answer as both a proportion and a percentage.
   b. What is a conservative 95% confidence interval for the population proportion that would have responded "More of a threat?" Give your answer for the population percentage as well.
   c. Write a sentence or two interpreting the interval that you found in part (b).
   d. Based on the interval you found in part (b), could you conclude that in January 2000 fewer than half of all adult Americans perceived Internet shopping as posing more of a threat than buying things by mail order or in a store? Explain.

3.25 Suppose a national polling agency conducted 100 polls in a year, using proper random sampling, and reported a 95% confidence interval for each poll. About how many of those confidence intervals would be wrong, that is, would not cover the true population value?

3.26 The question about the importance of religion in Example 3.3 was asked in a survey of elementary statistics students at Penn State University in the fall of 2001 and spring of 2002. The results were as follows:

| Very important | 460 = 27% |
|---|---|
| Fairly important | 703 = 41% |
| Not very important | 543 = 32% |
| Total | 1706 = 100% |

Because these students were not a random sample from a larger population, and because statistics students may not represent other students on questions of religion, treat these 1706 students as a population. Therefore, we know that for the population of elementary statistics students at Penn State in the 2001–2002 academic year, 27% felt that religion was very important in their lives.
   a. Based on the results in Example 3.3, can we conclude that this percentage differs from the percentage of the population of all adult Americans who felt that religion was very important in their lives in September 2002? Explain.
   b. Based on the preceding results and those in Example 3.3, answer the question in part (a) for the percentage who felt religion was fairly important and for the percentage who felt it was not very important.
   c. Write a short news article comparing the two surveys as if you were writing it for the student newspaper.
   d. Suppose we believe that the Penn State sample is representative of all students in the academic programs that require the elementary statistics course. Calculate an approximate 95% confidence interval

for the population proportion that would say that religion is very important in their lives.

3.27 An Internet report on a 1999 Gallup Poll (August 2, 1999, http://www.gallup.com/poll/index.asp) included the following statement: "The results below are based on telephone interviews with a randomly selected national sample of 1,021 adults, 18 years and older, conducted July 22–25, 1999. For results based on this sample, one can say with 95 percent confidence that the maximum error attributable to sampling and other random effects is plus or minus 3 percentage points." What is the margin of error for this survey? Give two separate ways in which you know this is the answer.

**3.28** Refer to Exercise 3.27. One of the questions asked was "Do you think there will or will not come a time when Israel and the Arab nations will be able to settle their differences and live in peace?" The choices and percentage choosing them were "Yes, will be a time" (49%), "No, will not" (47%), "No opinion" (4%).

**a.** Give the interval of values that is likely to cover the true proportion of the population who would answer, "Yes, will be a time."

**b.** Give the interval of values that is likely to cover the true proportion of the population who would answer, "No, will not."

c. Compare your answers in parts (a) and (b). Is there a clear majority of the population for either opinion? Explain.

3.29 A poll commissioned by the Student Aid Alliance sampled 1022 American adults in May 1999 and was reported to have a margin of error of 3.1%. Respondents were asked, "How important do you think it is to increase the current level of federal funding for each of the following things?" The programs and the percentages of the sample who answered "Very important" or "Somewhat important" were "Health care" (90%), "Social Security" (89%), "Financial aid for college students" (87%), "Defense" (77%), "Welfare" (56%), and "Foreign aid" (47%). (*Source:* NASULGC[1] *Newsline,* July–August 1999, p. 7.) Write a short news article explaining these results. Address the question of which program(s) the highest percentage of *all* adult Americans think it is at least somewhat important to fund at increased levels.

3.30 Suppose that a researcher is designing a survey to estimate the proportion of adults in your state who oppose a proposed law that requires all automobile passengers to wear a seat belt.

**a.** What would be the approximate margin of error if the researcher randomly sampled 400 adults?

**b.** What sample size would be needed to provide a margin of error of about 2%?

3.31 Explain whether the width of a confidence interval would increase, decrease, or remain the same as a result of each of the following changes.

a. Increase the sample size from 1000 to 2000.

b. Decrease the sample size from 1000 to 500.

c. Increase the population size from 10 million to 20 million.

3.32 In which of the following three samples will the margin of error be the smallest? Explain. Assume that each sample is a random sample.

*Sample A:* sample of $n = 1000$ from a population of 10 million.

*Sample B:* sample of $n = 2500$ from a population of 200 million.

*Sample C:* sample of $n = 400$ from a population of 50,000.

3.33 A college dean plans a student survey to estimate the percentage of currently enrolled students who plan to take classes during the next summer session. She wants the 95% margin of error for the estimate to be at most 5%. What is the necessary sample size?

3.34 An epidemiologist plans a survey to estimate the prevalence of Alzheimer's disease in the population of adults aged 65 years or older. It is desired that the 95% margin of error for the sample estimate be no more than 3%. What sample size is needed?

## Section 3.3

3.35 ● Define each of the following terms:
a. Probability sampling plan.
b. Simple random sample.

3.36 ● There are 8000 items in a population, and these items are labeled by using four-digit numbers ranging from 0000 to 7999. Use the following stream of random digits to select four items from the population. Explain how you determined your answer.

76429   69730   23395   12694   43387

**3.37** A lottery game is played by choosing six whole numbers between 1 and 49. The grand prize is won if all six numbers chosen match the winning numbers drawn.

**a.** Use Table 3.2 to draw six numbers. Start at the beginning of the table, explain your process, and write down the six numbers chosen.

**b.** George plays this game weekly. Assume that the winning numbers are drawn fairly, so that any number has the same chance of occurrence. Would George have a better chance of winning if he chose the same numbers every week or chose different numbers every week, or doesn't it matter?

3.38 A radio station has a daily contest in which the DJ randomly selects one birthday (month and day, not year) and announces it on the air. The first person with that birthday who calls the station wins a prize. Each day, the station selects a new birthday, and it doesn't mind if the same birthday is selected multiple times over the course of the contest. Explain how it could use Table 3.2 to select the birthdays. Start at the beginning of the row labeled 9 and write down the first five birthdays the station would use.

3.39 ◆ The right handspan measurements for 103 female college students in Table 2.4 in Chapter 2 are given again here and in the **pennstate1F** dataset on the CD for this book.

---

[1] NASULGC is the National Association of State Universities and Land Grant Colleges.

*Females (103 Students):* 20, 19, 20.5, 20.5, 20.25, 20, 18, 20.5, 22, 20, 21.5, 17, 16, 22, 22, 20, 20, 20, 20, 21.7, 22, 20, 21, 21, 19, 21, 20.25, 21, 22, 18, 20, 21, 19, 22.5, 21, 20, 19, 21, 20.5, 21, 22, 20, 20, 18, 21, 22.5, 22.5, 19, 19, 19, 22.5, 20, 13, 20, 22.5, 19.5, 18.5, 19, 17.5, 18, 21, 19.5, 20, 19, 21.5, 18, 19, 19.5, 20, 22.5, 21, 18, 22, 18.5, 19, 22, 17, 12.5, 18, 20.5, 19, 20, 21, 19, 19, 21, 18.5, 19, 21.5, 21.5, 23, 23.25, 20, 18.8, 21, 21, 20, 20.5, 20, 19.5, 21, 21, 20

a. Draw three simple random samples of ten measurements each from this data set (an individual can be in more than one of your samples). Explain how you chose the samples and list the ten handspan measurements in each sample.
b. Find the median for each of your three samples. Compare them to the median for the full data set (20.0 cm).
c. Find the mean for each of your three samples. Compare them to the mean for the full data set (also 20.0 cm).
d. There were 21 women whose recorded handspan measurements were 20.0 cm. Would it be possible to have a random sample of ten measurements with a mean of 20.0 and a standard deviation of 0? Explain.

## Section 3.4

**3.40** ● A class of 200 students is numbered from 1 to 200, and a table of random digits is used to choose 60 students from the class. Is the group of students selected a simple random sample, a stratified random sample, a cluster sample, or a systematic sample?

3.41 ● In each part, identify whether the sample is a stratified random sample or a cluster sample.
a. A class of 200 students is seated in 20 rows of 10 students per row. Three students are randomly selected from every row.
b. An airline company randomly chooses one flight from a list of all flights taking place that day. All passengers on that selected flight are asked to fill out a survey on meal satisfaction.

3.42 ● In a factory producing television sets, every 100th set produced is inspected. Is the collection of sets inspected a simple random sample, a stratified random sample, a cluster sample, or a systematic sample?

**3.43** Suppose a state has 10 universities, 25 four-year colleges, and 50 community colleges, each of which offer multiple sections of an introductory statistics class each year. Researchers want to conduct a survey of students taking introductory statistics in the state. Explain a method for collecting each of the following types of samples:
**a.** A stratified sample.
b. ...er sample.
c. ...e random sample.
...xercise 3.43. Give one advantage of each
...ethod in the context of the problem.
...nple of a survey routinely conducted by the

U.S. government. (The Internet is a good source; for instance, try http://www.fedstats.gov.) Explain how the survey is conducted, and identify the type(s) of sampling used.

**3.46** Is a sample that is found by using random-digit dialing more like a stratified sample or a cluster sample? Explain.

## Section 3.5

3.47 ● A group of biologists wants to estimate the abundance of barrel cactus in a desert. They divide the desert into a grid of 100 rectangular areas but exclude 10 of those areas because they are difficult to access. The biologists then measure the density of cactus in a randomly selected sample of 40 of the 90 accessible areas.
a. What was the sampling frame in this study, and how did it differ from the population of interest?
b. Explain why "using the wrong sampling frame" might lead to a biased estimate of the abundance of cactus in the desert.

**3.48** ● A local government wants to determine whether taxpayers support increasing local taxes to provide more public funding to schools. They randomly select 500 schoolchildren from a list of all children enrolled in local schools and then survey the parents of these children about possible tax increases.
**a.** What is the population of interest for the local government?
**b.** What sampling frame did the government use for the survey?
c. Explain why "using the wrong sampling frame" might lead to a biased estimate of taxpayer support for increasing taxes.

3.49 ● In each part, indicate whether the sample should be called a self-selected sample or a convenience sample.
a. To assess passenger satisfaction, an airline distributed questionnaires to 100 passengers in the airline's frequent flyer lounge. All 100 individuals responded, and 95 respondents said that they had a high degree of satisfaction with the airline.
b. A magazine contains a survey about sexual behavior. Readers are asked to mail in their answers, and $n = 2000$ readers do so.
c. A political scientist surveys the 80 people in a class he teaches to evaluate student political views.

3.50 Despite his success in 1936, George Gallup failed miserably in trying to predict the winner of the 1948 U.S. presidential election. His organization, as well as two others, predicted that Thomas Dewey would beat incumbent Harry Truman. All three used what is called **quota sampling.** The interviewers were told to find a certain number, or quota, of each of several types of people. For example, they might have been told to interview six women under age 40, one of whom was black and the other five of whom were white. Imagine that you are one of their interviewers, trying to follow these instructions. Whom would you ask? Now explain why you think these polls failed to predict the true winner and why quota sampling is not a good method.

3.51 Explain why the main problem with the *Literary Digest* poll is described as "volunteer response" and not "volunteer sample."

**3.52** Gastwirth (1988, p. 507) describes a court case in which Bristol Myers was ordered by the Federal Trade Commission to stop advertising that "twice as many dentists use Ipana as any other dentifrice" and that more dentists recommended it than any other dentifrice. Bristol Myers had based its claim on a survey of 10,000 randomly selected dentists from a list of 66,000 subscribers to two dental magazines. They received 1983 responses, with 621 saying they used Ipana and only 258 reporting that they used the second most popular brand. As for the recommendations, 461 respondents recommended Ipana, compared with 195 for the second most popular choice.

   **a.** Specify the sampling frame for this survey, and explain whether or not you think "using the wrong sampling frame" was a difficulty here, based on what Bristol Myers was trying to conclude.

   **b.** Of the remaining four "Difficulties and Disasters in Sampling" listed in Section 3.5 (other than "using the wrong sampling frame"), which do you think was the most serious in this case? Explain.

   **c.** What could Bristol Myers have done to improve the validity of the results after it had mailed the 10,000 surveys and received 1983 of them back? Assume it kept track of who had responded and who had not.

3.53 Find an example of a poll based on a self-selected sample. Report the wording of the questions and the number who chose each response. Comment on whether or not you think the results can be extended to any population.

3.54 *This is also Exercise 1.28.* A popular Sunday newspaper magazine often includes a yes-or-no survey question such as "Do you think there is too much violence on television?" or "Do you think parents should use physical discipline?" Readers are asked to phone their answers to the magazine, and the results are reported in a subsequent issue.

   **a.** What is this type of sample called?

   **b.** Do you think the results of these polls represent the opinions of all readers of the magazine? Explain.

3.55 *This is also Exercise 1.6.* A proposed study design is to leave 100 questionnaires by the checkout line in a student cafeteria. The questionnaire can be picked up by any student and returned to the cashier. Explain why this volunteer sample is a poor study design.

3.56 *This is a modification of Exercise 1.27.* Suppose listeners to a late-night radio talk show were asked to call and report whether or not they had ever seen a ghost.

   a. What is this type of sample called?

   b. Do you think the proportion reporting that they had seen a ghost for the radio poll would be higher or lower than the proportion for a poll done using a random sample of adults in the United States? Explain.

## Section 3.6

3.57 ● A survey question will be asked to determine whether people think smoking should be banned on all airline flights.

   a. Write a version of the question that is as neutral and unbiased as possible.

   b. Write a version of the question that is likely to get people to respond that smoking should be forbidden on all flights.

   c. Write a version of the question that is likely to get people to respond that smoking should be allowed on airline flights.

3.58 ● An example of an unnecessarily complex survey question is "Shouldn't former drug dealers not be allowed to work in hospitals after they are released from prison?" Restate this question so that it is clearer.

3.59 ● In a planned survey about moviegoing, two questions that will be asked are

   • How many times per month do you go to the movies?
   • Do you consider yourself to be well-informed about recent movies or not?

The questions could be asked in either order. Briefly explain how you think the question order for asking them might affect how people answer them.

3.60 ● In presidential election years in the United States, a Gallup poll is conducted in which the first survey question asks which presidential candidate the voter prefers. Subsequent questions concern other political, social, and election issues. Explain why this question order might be the best order for estimating how people may vote in the presidential election.

**3.61** ● Medical tests, such as those for detecting HIV, sometimes concern such sensitive information that people do not want to give their names when they take the test. In some instances, a person taking such a medical test is given a number or code that he or she can use later to learn the test result by phone. Explain whether this procedure is an example of anonymous testing or confidential testing.

3.62 An Internet poll sponsored by a site called About.com asked Internet users to pick one of two choices in response to the question, "Should jurors opposed to gun control laws refuse to convict defendants even if they have clearly broken gun laws?" The two choices and the number and percentage choosing them were

   • Yes, that's an effective way to defeat unjust laws (16,864, 23%).
   • No, that undermines the legal system (55,519, 77%).

Discuss this poll, including whether or not you think the results are representative of all adults and whether you think the wording is appropriate.

3.63 A Gallup poll that was released on July 9, 1999, included a series of questions about possible religious activities in public schools. The poll was based on telephone interviews with a randomly selected sample of 1016 U.S. adults conducted June 25–27, 199

The questions, asked in random order, included the following:

- Do you favor or oppose teaching creationism ALONG WITH evolution in public schools? Favor—68%, Oppose—29%, No opinion—3%.
- Do you favor or oppose teaching creationism IN-STEAD OF evolution in public schools? Favor—40%, Oppose—55%, No opinion—5%.

a. What is the margin of error for this poll?
b. Write a few sentences describing the results of the survey, being truthful but biasing the story in favor of teaching creationism.
c. Repeat part (b), but bias the story in favor of not teaching creationism.
d. Repeat part (b), but write an unbiased story.
e. Explain which of the pitfalls in Section 3.6 is illustrated by this example.

**3.64** Give an example of two survey questions for which you think the results would be substantially different depending on which order they were asked.

3.65 Explain which of three methods—a door-to-door interview, a telephone interview, or a mail survey—would be *most* likely and which would be *least* likely to suffer from each of the following problems:
a. Bias due to desire to please the interviewer.
b. Volunteer response.
c. Bias due to perceived lack of confidentiality.

**3.66** Refer to Example 3.16, "When Will Adolescent Males Report Risky Behavior?" Explain which two of the seven "Possible Sources of Response Bias in Surveys" are illustrated by this example.

3.67 An advertiser of a certain brand of aspirin (let's call it Brand B) claims that it is the preferred painkiller for headaches, on the basis of the results of a survey of headache sufferers. But further investigation reveals that the choices given to respondents were Tylenol, Extra-Strength Tylenol, Brand B aspirin, and Advil.
a. Is this an open- or closed-form question? Explain.
b. Comment on the choices given to respondents.
c. Comment on the advertiser's claim.
d. What choices do you think should have been given? Explain.

3.68 An NBC News/*Wall Street Journal* poll conducted at the end of the 20th century (September 9–12, 1999) asked a random sample of $n = 1010$ adult Americans, "Which one of the following do you consider to be the best American movie of the 20th century?" The choices given and the percentages who chose them were: *Gone With the Wind* 28%, *Schindler's List* 18%, *Titanic* 11%, *Star Wars* 11%, *Casablanca* 8%, *The Godfather* 7%, *Citizen Kane* 6%, *The Graduate* 1%, Not sure 4%. In addition, 2% volunteered the answer "all of them," and 4% volunteered an answer not listed ̶̶̶̶w.pollingreport.com/hollywoo.htm, Sep-̶̶̶̶2002).
̶̶̶̶asked as an open-form or closed-form

̶̶̶̶ink the form that was used was appropri-

ate for trying to ascertain what movies adult Americans thought were the best of the 20th century? Explain.
c. Give one advantage and one disadvantage each for open-form and closed-form questions in this situation.

**3.69** Rock singer Elvis Presley died on August 16, 1977, and his life received substantial publicity as the 25th anniversary of his death approached on August 16, 2002. On August 7–11, 2002, an ABC News poll asked a random sample of $n = 1023$ adult Americans, "Who do you think is the greatest rock 'n' roll star of all time?" There were 128 different stars volunteered by respondents. The top responses and the percentages they received were Elvis Presley 38%, Jimi Hendrix 4%, John Lennon 2%, Mick Jagger 2%, Bruce Springsteen 2%, Paul McCartney 2%, Eric Clapton 2%, Michael Jackson 2%, Other 25%, None 7%, No opinion 15% (http://abc news.go.com/sections/us/DailyNews/elvis_poll.html, Sept 29, 2002).
**a.** Was this an open- or closed-form question?
**b.** Do you think the percentage who responded "Elvis Presley" was influenced by the timing of the poll? Explain.
**c.** Do you think the percentage who responded "Elvis Presley" would have been higher, lower, or about the same if the question had given specific choices?

3.70 Refer to Exercise 3.69. Another question that was asked in the poll was "Do you consider yourself a fan of Elvis Presley, or not?" The responses were almost equally split, with 49% saying "yes" and 51% saying "no" (http://www.pollingreport.com/music.htm, September 29, 2002). Do you think the order in which this question and the one in Exercise 3.69 were asked would influence the results to either question? Explain.

## Section 3.7: Skillbuilder Applet Exercises

**Statistics⟨△⟩Now**™  Test yourself on these questions and explore the applet at **http://1pass.thomson.com** or on your CD.

*For Exercises 3.71 to 3.75, use the* **Sampling** *applet described in Section 3.7. The applet is on the CD for this book.*

3.71 Choose one stick figure to represent yourself, and identify which one it was in your answers to this exercise by giving the row number (1 to 10) and column number (1 to 10) where you reside.
a. Suppose you are in a class of 100 individuals, represented by these stick figures. Your teacher randomly selects 10 people each class period and asks them to answer a question about the reading assignment. What are your chances of being in the sample each day?
b. Repeat the sample selection process 20 times, each time noting whether you made it into the sample. How many times did you make it into the sample? Does that surprise you, or is it about what you expected?

3.72 Repeat the sampling process 50 times without pressing the **Start Over** button. Use **Show Results** to display the results of the 50 samples. Cut and paste the results into Minitab or another program.
  a. Display a histogram of the 50 means of the chosen samples.
  b. Describe the histogram in part (a), including shape, center, and spread.
  c. Display a histogram of the 50 values of "percent female."
  d. Describe the histogram in part (c).

3.73 Repeat the sampling process 20 times without pressing the **Start Over** button. Use **Show Results** to display the results of the 20 samples. Write down the results, or cut and paste them into a computer program.
  a. What is the median of the 20 sample means?
  b. What is the range of the 20 sample means?
  c. Display a boxplot of the 20 sample means.
  d. We know that the population has a mean height of 68 inches. Based on your results in parts (a), (b), and (c), discuss how useful the mean of a sample of size 10 is for estimating the population mean in this situation.

3.74 Repeat the sampling process 20 times without pressing the **Start Over** button. Use **Show Results** to display the results of the 20 samples. Write down the results, or cut and paste the results into a computer program. Find the mean of the 20 sample means. This value represents the mean of all 200 sample values. Is this combined mean a better estimate of the population mean of 68 than most of the individual sample means? Is that what you would expect, or not? Explain.

3.75 Generate one random sample, and print the display showing the applet and your result. Show what stream of numbers from a table of random digits would have produced your sample, numbering the stick figures from 00 to 99, starting at the top and going across rows.

**3.76** Refer to Figure 3.7, showing the results of taking a simple random sample of ten stick figures from Figure 3.6. Label the stick figures in Figure 3.6 from 00 to 99, going across rows. Suppose a table of random digits had been used to select the sample in Figure 3.7. Write down a stream of random digits that would have led to the selection of the sample shown.

3.77 Refer to Exercise 3.76 and write down a stream of random digits that would have led to the selection of the sample shown in Figure 3.8.

3.78 The **Sampling** applet described in Section 3.7 drew a simple random sample of 10 individuals from a population of 45 men and 55 women and measured their heights. The average height for the sample could be used as an estimate of the average height in the population. Another possibility for estimating the average height would be to use stratified sampling, in which men and women would be the two strata, then estimate the heights of men and women separately and use the knowledge that 45% of the population is male to combine the two estimates. Do you think stratified sampling would be a more accurate or less accurate method than simple random sampling for estimating the mean height in the population? Explain.

## Chapter Exercises

3.79 Suppose a community group wants to convince city officials to put more trash containers on the city streets. They decide to conduct a survey, but rather than a scientifically valid result, they want their results to show that as many citizens as possible want additional trash containers. Give an example of how they could use each of the following to their advantage:
  **a.** Self-selected sample.
  **b.** Deliberate bias.
  c. Ordering of questions.
  d. Desire to please.

3.80 Find an example of a poll conducted by a reputable scientific polling organization. Write a few paragraphs describing how the poll was conducted, what was asked, and the results. Explain whether or not any of the problems discussed in Sections 3.5 and 3.6 were likely to have biased the results of the poll.

3.81 Refer to Example 3.17, "Politics Is All in the Wording."
  a. Explain which one of the "Difficulties and Disasters" listed in Section 3.5 is illustrated by this example.
  b. Explain which one of the seven "Possible Sources of Response Bias in Surveys" is illustrated by this example.

3.82 Refer to Example 3.5, in which a survey question was "Do you favor or oppose scientific experimentation on the cloning of human beings?" Suppose an agency that advocated cloning of human beings wanted to conduct a survey that would show support for their cause. Explain how it might use each of the following to do so:
  a. Selection bias.
  b. Deliberate bias.
  c. Desire to please.
  d. Ordering of questions.

3.83 Suppose you wanted to estimate the proportion of adults who write with their left hands and decide to watch a sample of $n$ people signing credit card receipts at a mall.
  a. Do you think this sample would be representative for the question of interest? Explain.
  b. Assuming that the sample was representative, how many people would you have to include to estimate the proportion with a margin of error of 5%?
  c. Suppose you observed 500 people and 60 of them signed with their left hand. What interval of values would you give as being likely to contain the true proportion of people in the larger population who would sign with their left hand?
  d. Is the sampling plan that was used a probability sampling plan? Explain.

3.84 The faculty senate at a large university wanted to know what proportion of the students thought that a foreign language should be required for everyone. The statistics department offered to cooperate in conducting a survey, and a simple random sample of 500 students was selected from all students enrolled in statistics classes. A survey form was sent by e-mail to the 500 students. Discuss the extent to which each three types of bias would be likely to occur in this survey:

a. Selection bias.

b. Nonresponse bias.

c. Response bias.

d. Which of these three types of bias do you think would be the most serious? Explain.

**3.85** Refer to Exercise 3.84.

**a.** What is the population of interest to the faculty senate?

**b.** What is the sampling frame?

**c.** What is the sample?

d. Is the sample representative of the population of interest? Explain.

3.86 A large medical professional organization with membership consisting of doctors, nurses, and other medical employees wanted to know how its members felt about HMOs (health maintenance organizations). Name the type of sampling plan they used in each of the following scenarios:

a. They randomly selected 500 members from each of the lists of all doctors, all nurses, and all other employees and surveyed those 1500 members.

b. They randomly selected ten cities from all cities in which its members lived, then surveyed all members in those cities.

c. They randomly choose a starting point from the first 50 names in an alphabetical list of members, then chose every 50th member in the list starting at that point.

**3.87** An article in the *Sacramento Bee* was headlined "Drop found in risky behavior among teens over last decade" (June 7, 2000, p. A6; reprinted from the *Los Angeles Times*). The article reported that "the findings show that for the general population [not a specific ethnic group], the share of teens abstaining from all 10 risky behaviors jumped from 20 percent to 25 percent [from the beginning to the end of the 1990s]." The article did not say who was measured but was obviously based on *samples* taken at each end of the decade. What additional information would you need to know to determine whether or not this drop represents a real drop in the percentage of the *population* of teens abstaining? Explain.

3.88 One evening around 6 P.M., authorities in a major metropolitan area received hundreds of phone calls from people reporting that they had just seen an unidentified flying object (UFO). A local television station would like to report the story on their 10 P.M. news and wants to include an estimate of the proportion of area residents who witnessed the UFO. Explain how each of the following methods of obtaining this information would be biased, if at all:

a. A "quickie poll" taken between 7 P.M. and 9 P.M. that eveni using random-digit dialing for residents of

c. A door-to-door survey taken between 7 P.M. and 9 P.M., based on a multistage sampling plan, in which ten city blocks (or equivalent area) were randomly selected and interviewers knocked on the doors of all homes in those areas and surveyed those who answered the door.

d. An e-mail survey sent at 7 P.M. to all e-mail addresses for local residents served by a local Internet service provider in the area, using responses received by 9 P.M.

3.89 Refer to Exercise 3.88. Acknowledging that some bias is inevitable when information is required on such short notice, which of the four methods would you recommend that the television station use to get the information it desires? Explain.

3.90 Refer to Exercises 3.88 and 3.89. Suppose that a local newspaper conducted a survey during the next week, based on random-digit dialing of 1400 residents of the area, and found that 20% of them reported having seen a UFO on the evening in question.

a. What is the margin of error for the survey?

b. Give an interval of values that probably covers the true percentage of the population that saw the UFO.

**3.91** Refer to the survey described in Example 3.9, "Which Scientists Trashed the Public?" Explain which type of bias (selection, nonresponse, or response) was the most problematic in that survey.

3.92 Suppose an (unscrupulous!) organization wanted to conduct a survey in which the results supported its position that drinking coffee in public should be illegal except in designated coffee bars. Explain how the organization could use each of the following sources of bias to help produce the results it wants:

a. Selection bias.

**b.** Nonresponse bias.

c. Response bias.

3.93 Suppose that a survey reported that 55% of respondents favored gun control, with a margin of error of ±3 percentage points.

a. What was the approximate size of the sample?

b. What is an approximate 95% confidence interval for the percentage of the corresponding population who favor gun control?

c. Based on the material in this chapter, write a statement interpreting the interval you found in part (b) that could be understood by someone with no training in statistics. (In Chapter 10, you will learn a more precise way to interpret this type of interval.)

3.94 Refer to the *Literary Digest* poll in Case Study 3.1. Discuss the extent to which each type of bias (selection, nonresponse, and response) played a role in producing the disastrous results, if at all.

3.95 ◆ For this exercise, use the data in Case Study 1.1 (p. 2) or in the file **pennstate1** on the CD for this book.

a. Select a random sample of ten males and a random sample of ten females, and write down their answers to the question "What's the fastest you've ever driven a car?"

b. Compare the males' and females' responses, using appropriate numerical and/or graphical summary information.

c. Recall from Chapters 1 and 2 that there was an obvious difference between male and female responses for the set of all students. Is this difference obvious in your samples? Explain.

d. It is true in statistics that the larger the actual difference between two groups is in the population, the smaller the sample sizes needed to detect the difference. Discuss your answer to part (c) in the context of this statement.

*Exercises 3.96 to 3.99 are based on a report of a Field Poll printed in the* Sacramento Bee *on July 7, 2000 (Dan Smith, p. A4), with the headline* "Ratings up for Boxer, Feinstein in new poll." *The results were reported with a graphical display containing the information that* "results are based on a state-wide survey of 750 adults conducted June 9–18. The sample includes 642 voters considered likely to vote in the Nov. 7 general election. The margin of error is 3.8 percentage points."

**3.96**  The text of the article said that the 750 adults surveyed were actually 750 *registered voters.* Based on the information about who was surveyed, if the poll results are to be used to predict the proportion of voters who were likely to vote for Senator Feinstein in November, should the results of all 750 voters be used? Explain, using the Fundamental Rule for Using Data for Inference in your explanation.

3.97  The report stated that the margin of error was 3.8 percentage points. This value is based on the 642 voters who were considered likely to vote and uses the precise formula you will learn in Chapter 10.

a. Compute the conservative margin of error for this poll based on those likely to vote, and compare it to the precise margin of error presented.

b. For this sample, 57% reported that they approve of Dianne Feinstein's job performance. What is an approximate 95% confidence interval for the percent of all likely voters who approve of her performance?

3.98  Read the headline that accompanied the article, which stated that ratings were "up" for both senators. For the June 2000 poll, 57% approved of Feinstein's performance, compared with 56% and 53%, respectively, for similar polls taken in February 2000 and October 1999. Presumably, the headline refers to all registered voters in the state, not just those in the samples. Considering the margin of error accompanying each poll, do you agree with the headline? Explain.

**3.99**  The article also reported that "among GOP voters, 39 percent give Feinstein high marks." Would the margin of error accompanying this result be larger, smaller, or the same as the margin of error for the entire sample? Explain.

*Exercises 3.100 to 3.102 refer to a survey of University of California faculty members on affirmative action policies for the university (Roper Center, 1996). The survey was based on telephone interviews with a random sample of 1001 faculty members conducted in December 1995.*

3.100  One of the questions asked was "Do you favor or oppose using race, religion, sex, color, ethnicity, or national origin as a criterion for admission to the University of California?" Of the 804 male respondents, 47% said "Favor," and of the 197 female respondents, 73% said "Favor."

a. Find the conservative margin of error for the males, and use it to compute an approximate 95% confidence interval for the percent of males in the population who favor this policy.

b. Repeat the analysis in part (a) for females.

c. Compare the approximate 95% confidence intervals you found in parts (a) and (b), and comment on whether you think male and female faculty members in the population differed on this question.

**3.101**  Refer to Exercise 3.100. Of the 166 arts and humanities faculty members, 66% favored the policy, while of the 229 engineering, mathematical, and physical sciences faculty, 38% favored it.

**a.** Find the conservative margin of error for the arts and humanities faculty, and use it to compute an approximate 95% confidence interval for the percent of them in the population who favor this policy.

**b.** Repeat the analysis in part (a) for the engineering and sciences faculty.

**c.** Compare the approximate 95% confidence intervals you found in parts (a) and (b), and comment on whether you think faculty members in these disciplines in the population differed on this question.

3.102  One question asked was as follows: "The term affirmative action has different meanings to different people. Please tell me which statement best describes what you mean by the term: First, affirmative action means granting preferences to women and certain racial and ethnic groups. Second, affirmative action means promoting equal opportunities for all individuals without regard to their race, sex, or ethnicity." The options "Neither" or "Both" were not given and were recorded only if the respondent offered them. The first statement was chosen by 37%, the second one by 43%, "Neither" by 13%, and "Both" by 2%. The remainder said, "Don't know." The organization that commissioned the poll, the California Association of Scholars, issued a press release that said in reference to this question, "These findings demolish the claim that the faculty at UC wants preferential policies." Refer to the "Possible Sources of Response Bias" in Section 3.6, and discuss them in the context of this question and resulting press release.

**Statistics ⏚ Now™**  Preparing for an exam? Assess your progress by taking the post-test at **http://1pass.thomson.com.**

**⏚Mentor™**
Do you need a live tutor for homework problems? Access vMentor at **http://1pass.thomson.com** for one-on-one tutoring from a statistics expert.

● Basic skills   ◆ Dataset available but not required   **Bold-numbered** exercises answered in the back

# 4

*Royalty-free/Corbis*

*Does prayer lower blood pressure?*

See Example 4.1 *(p. 120)*

# Gathering Useful Data for Examining Relationships

> Probably the biggest misinterpretation made when reporting studies is to imply that a *cause-and-effect* relationship can be concluded on the basis of an observational study. Groups that naturally differ for the explanatory variable of interest are almost certain to differ in other ways, any of which may contribute to differences in the responses.

In this chapter, we learn about ways to collect data in order to examine relationships between variables. We have already seen several examples that involved possible links between variables. In Chapter 2, Example 2.1 was about the connection between gender and seat belt use for twelfth-grade students. Example 2.2 was about a possible connection between the use of nightlights in infancy and nearsightedness. Case Study 1.6 described a study that demonstrated a link between taking an aspirin a day and a decreased risk of heart attacks for men.

In studies like these, we want to know whether a **cause-and-effect relationship** exists. That is, we want to know whether changing the value of one variable *causes* changes in another variable. We will learn in this chapter that the way in which a study is conducted affects our ability to infer that a cause-and-effect relationship exists. ▮

## 4.1 Speaking the Language of Research Studies

Although there are a number of different strategies for collecting meaningful data, there is common terminology used in most of them. Statisticians tend to borrow words from common usage and apply a slightly different meaning, so be sure that you are familiar with the special usage of a word in a statistical context.

### Types of Research Studies

Two basic types of statistical research studies are conducted to detect relationships between variables:

- *Observational studies*
- *Experiments*

In an **observational study,** the researchers simply observe or question the participants about opinions, behaviors, or outcomes. Participants are not asked to do anything differently. For example, Case Study 1.5 described an observational study in which blood pressure and frequency of certain types of religious activity (such as prayer and church attendance) were measured. The goal was to determine whether people with higher frequency of religious activity had lower blood pressure. Researchers simply measured blood pressure and frequency of religious activity. They did not ask participants to change how often they prayed or went to religious services or to change any other aspect of their lives.

In an **experiment,** researchers manipulate something and measure the effect of the manipulation on some outcome of interest. **Randomized experiments** are experiments in which the participants are *randomly assigned* to participate in one condition or another. The different "conditions" are called **treatments.**

As an example, Case Study 1.6 described a randomized experiment in which physicians were randomly assigned to take an aspirin or a placebo every other day. The goal was to determine whether regular intake of aspirin resulted in lower risk of heart attacks than that experienced by the group taking only a placebo. The two possible treatments were taking aspirin or taking a placebo. The random assignment should have ensured that the groups were similar in all other respects, so any difference in heart attack rates could actually be attributed to the difference in the pill they took (aspirin or placebo).

## Observational Study or Randomized Experiment?

A major theme of this chapter will be that a randomized experiment provides stronger evidence of a cause-and-effect relationship than an observational study does. In many situations, however, there are practical and ethical issues that force researchers to conduct an observational study rather than an experiment. For instance, in an investigation of the effect of oral contraceptive usage on blood pressure, it would be unreasonable to randomly assign some women to use oral contraceptives and others to use a placebo. Also, some variables, such as handedness, are inherent traits and cannot be randomly assigned. A study done to find out whether handedness affects the risk of a workplace accident, for example, would have to be done as an observational study.

In some cases, researchers can choose between the strategies for collecting data and must weigh the advantages and disadvantages of each strategy. For instance, suppose researchers want to study whether people lose more weight by exercising or by limiting their fat intake. They could conduct an *observational study,* in which they question people about their exercise and eating habits, as well as their history of weight loss, and compare people who exercise with those who eat limited fat. Or they could conduct a *randomized experiment* in which they randomly assign people to either exercise or eat a specific diet and then measure and compare weight loss. As we will see later in this chapter, each type of study has advantages and disadvantages.

This method of giving each participant two "treatments" to ensure that the experiment is blind is called **double dummy.** Sometimes a third group is used and given placebos of both methods to compare the results from each treatment method to what would happen with no active treatment at all. In our example, a third group would receive a placebo patch and placebo gum.

Case Study 4.3 describes a well-designed experiment that was done to assess the effectiveness of a nicotine patch as an aid to quitting smoking. The researchers used randomization, a placebo group, and a double-blind method.

## case study 4.3    Quitting Smoking with Nicotine Patches

**Statistics△Now**™  Read the original source on your CD.

There is no longer any doubt that smoking cigarettes is hazardous to your health and the health of those around you. Yet for someone who is addicted to smoking, quitting is no simple matter. One promising technique is to apply a patch to the skin that dispenses nicotine into the blood. In fact, these nicotine patches have become one of the most frequently prescribed medications in the United States.

To test the effectiveness of these patches on the cessation of smoking, Dr. Richard Hurt and his colleagues (1994) recruited 240 smokers at Mayo Clinics in Rochester, Minnesota; Jacksonville, Florida; and Scottsdale, Arizona. Volunteers were required to be between the ages of 20 and 65, have an expired carbon monoxide level of 10 ppm or greater (showing that they were indeed smokers), be in good health, have a history of smoking at least 20 cigarettes per day for the past year, and be motivated to quit.

Volunteers were randomly assigned to receive either 22-mg nicotine patches or placebo patches for eight weeks. They were also provided with an intervention program recommended by the National Cancer Institute, in which they received counseling before, during, and for many months after the eight-week period of wearing the patches.

After the eight-week period of patch use, almost half (46%) of the nicotine group had quit smoking, while only one fifth (20%) of the placebo group had. Having quit was defined as "self-reported abstinence (not even a puff) since the last visit and an expired air carbon monoxide level of 8 ppm or less" (p. 596). After a year, rates in both groups had declined, but the group that had received the nicotine patch still had a higher percentage who had successfully quit than did the placebo group, 27.5% versus 14.2%.

The study was double-blind, so neither the participants nor the nurses taking the measurements knew who had received the active nicotine patches. The study was funded by a grant from Lederle Laboratories and was published in the *Journal of the American Medical Association.* Because this was a well-designed randomized experiment, it is likely that the nicotine patches actually *caused* the additional percentage who had quit in the nicotine patch group compared with the control group.

## Pairing and Blocking

It is sometimes easier and more efficient to have each person in a study serve as his or her own control. That way, differences that inherently exist among individuals don't obscure the treatment effects. We encountered this idea when we discussed how to compare driving ability under the influence of alcohol and marijuana and when sober.

Sometimes, instead of using the same individual for the treatments, researchers will match people on traits that are likely to be related to the outcome, such as age, IQ, or weight. They will then randomly assign each of the treatments to one member of each matched pair or grouping. For example, in a study comparing chemotherapy to surgery to treat cancer, patients might be matched by sex, age, and level of severity of the illness. One from each pair would then be randomly chosen to receive the chemotherapy, and the other would receive surgery. (Of course, such a study would be ethically feasible only if there were no prior knowledge that one treatment was superior to the other. Patients in such cases are always required to sign an informed consent.)

### Matched-Pair Designs

Experimental designs that use either two matched individuals or the same individual to receive each of two treatments are called **matched-pair designs.** The important feature of these designs is that randomization is used to assign the order of the two treatments. Of course, it is still important to try to conduct the experiment in a double-blind fashion so that neither the participant nor the researcher knows which *order* was used.

### Block Designs

The matched-pair design is a special case of a **block design.** Experimental units are divided into homogeneous groups called **blocks,** and each treatment is randomly assigned to one or more units in each block. Sometimes the "blocks" are individuals, and the "experimental units" are the repeated time periods in which they receive the varying treatments. The method that we discussed for comparing drivers under three conditions was a block design. Each driver is a block, and the experimental units are the three time periods during which they drove sober, with alcohol, and with marijuana. This somewhat peculiar terminology results from the fact that these ideas were first used in agricultural experiments, in which the experimental units were plots of land that had been subdivided into homogeneous blocks. In the social sciences, designs such as these, in which the same participants are measured repeatedly under differing conditions, are referred to as **repeated-measures designs.**

Block designs are sometimes used when the experimental units are different enough on an important variable that experimenters group them on that variable, then assign all treatments within each group. For instance, an experiment to compare two methods for increasing memorization skills might first block participants into age groups (young, middle-aged, elderly) and then randomly assign the two methods to half of the individuals in each age group. Comparison of the two treatments would be done separately in each age group so that differences in memorization capacity with age would not mask the differences in the two methods. Figure 4.4 illustrates the steps that would be taken in such an experiment.

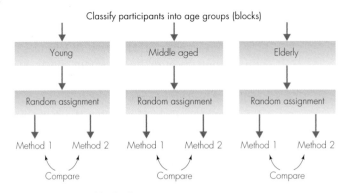

**Figure 4.4** ▮ A blocked experiment to compare two memorization methods. Within each of three age groups (blocks), individuals are randomly allocated to the two methods.

Blocks or matched pairs are used to reduce known sources of variability in the response variable across experimental units and should be used when such sources are known and can be determined for each experimental unit. For example, if you wanted to study the relationship between exercise and weight loss, it would make sense to separate people based on whether they were initially underweight, overweight, or about right. Initial weight would be used to create matched pairs or blocks.

---

***thought question 4.4*** Students are sometimes confused by the reasons for blocking and for randomization. One method is used to control *known* sources of variability among the experimental units, and the other is used to control *unknown* sources of variability. Explain which is which, and provide examples illustrating these ideas.*

---

## Design Terminology and Examples

If the treatments in an experiment are randomly assigned to experimental units without using matched pairs or blocks, the experiment has what is called a **completely randomized design.** When matched pairs are used, the experiment has a **matched-pair design;** and when blocks are used, the experiment has a **randomized block design.**

Notice that there is some overlap in the terminology that is used for these designs because a matched-pair design could also be considered to be a randomized block design, with each pair representing one block. Therefore, matched-pair designs are a subset of randomized block designs, in which there are only two treatments, each assigned only once within each block.

To help you distinguish these three types of experimental design, here are two examples of experiments that might be done by using them and how each design would be accomplished.

### Experiment 1: The Effect of Caffeine on Swimming Speed

Suppose a researcher wants to know whether taking caffeine an hour before swimming affects the time it takes swimmers to complete a one-mile swim and that 50 volunteers are available for the study.

- *Completely randomized design.* Randomly assign 25 volunteers to take a caffeine pill and the remaining 25 to take a placebo, have them all swim a mile, and compare the times for the two groups as the analysis.
  - *Matched-pair design.* There are two ways in which this could be done.
    - *Method 1:* Have each of the 50 swimmers take both treatments, with adequate time, perhaps a week, between them. In week 1, randomly assign swimmers to take either a caffeine pill or a placebo. Have them swim a mile and record the time. The next week, give each swimmer the treatment he or she did not have the previous week,

---

*****HINT:** Consider the example in Figure 4.4. Two reasons for differing ability to memorize might be age and individual brain structure. Which one can be measured and used to assign people to blocks?

have them swim a mile, and record the time. For each swimmer, take the difference in the two times, and use those measurements for the analysis.

*Method 2:* Create 25 pairs of swimmers based on speed. Before assigning treatments, measure how long it takes each one to swim a mile. Match the two fastest swimmers as one pair, the next two as the second pair, and so on. A week later, randomly assign one swimmer in each pair to take a caffeine pill and the other to take a placebo. Measure their times to swim a mile. For each pair, take the difference in the two times, and use those measurements for the analysis.

• *Randomized block design.* This would be most appropriate if there were different types of swimmers. For instance, suppose that 10 of the volunteers were males who swim competitively, 20 were males who don't swim competitively, 8 were females who swim competitively, and the remaining 12 were females who don't swim competitively. Then each of these four groups would constitute a block, and half of each block would be randomly assigned to take a caffeine pill while the other half would take a placebo. Each person would swim a mile. The times for caffeine versus placebo would be compared within each block as well as across the four blocks combined.

## Experiment 2: Advertising Strategies for Selling Coffee Near Colleges

Suppose that a national chain of cafes wants to compare three methods for increasing sales at their cafes near college campuses. Each method involves placing a weekly advertisement in the campus newspaper. With method 1, a coupon for a free cup of coffee is included in the ad. With method 2, a coupon for a free cookie with any coffee drink is included in the ad. The third method is to place the ad without any coupons. They place the ads for one full semester and measure how much sales change from the same semester the previous year.

• *Completely randomized design.* For this design, they would randomly choose one third of the campuses in the study to receive each of the treatments, and the analysis would compare the change in sales for the three treatment groups.

• *Matched-pair design.* This design would not be possible if they want to compare all three treatments.

• *Randomized block design.* Decide what known factors contribute to differing sales across campuses. Group or "block" the campuses on the basis of those factors. For instance, they might categorize the campuses on the basis of size of the student body (small or large) and population of the town or city (small, medium, large) for a total of six blocks. Once the campuses have been blocked, randomly assign one third of the campuses within each block to receive each treatment. Compare change in sales within each block, then across blocks. Other blocking variables might include initial sales, distance from the campus, and whether or not there is a competing cafe in the neighborhood.

4.2 Exercises are on page 143.

---

*technical note*  **The Number of Units per Block**

Some statisticians argue that the number of experimental units in a block should equal the number of treatments so that each treatment is assigned only once in each block. That allows as many sources of known variability as possible to be controlled. For the example of the effect of caffeine on swimming speed, there are two treatments, so blocks of size 2 (matched pairs) would be created. This could be accomplished by matching people on known variables such as sex, initial swim speed, usual caffeine consumption, age, and so on.

---

# 4.3 Designing a Good Observational Study

For the purpose of establishing cause-and-effect links, observational studies start with a distinct *disadvantage* in comparison to randomized experiments. In observational studies, the researchers observe, but do not control, the explanatory variables. However, these studies have the possible *advantage* that participants are measured in their natural setting. The following case study, about a possible association between baldness and heart disease, is an example of a well-designed observational study.

## *case study 4.4*  Baldness and Heart Attacks

**Statistics⊖Now**™  Read the original source on your CD.

On March 8, 1993, *Newsweek* announced, "A really bad hair day: Researchers link baldness and heart attacks" (p. 62). The article reported that "men with typical male pattern baldness . . . are anywhere from 30 to 300 percent more likely to suffer a heart attack than men with little or no hair loss at all." (Male pattern baldness is the type affecting the crown or vertex and not the same thing as a receding hairline; it affects approximately one third of middle-aged men.)

The report was based on an observational study conducted by researchers at Boston University School of Medicine. They compared 665 men who had been admitted to the hospital with their first heart attack to 772 men in the same age group (21 to 54 years old) who had been admitted to the same hospitals for other reasons. There were 35 hospitals involved, all in eastern Massachusetts and Rhode Island. Lesko, Rosenberg, and Shapiro reported the full results in the *Journal of the American Medical Association* (1993).

The study found that the percentage of men who showed some degree of male pattern baldness was substantially higher for those who had had a heart attack (42%) than for those who had not (34%). Furthermore, when they used sophisticated statistical tests to ask the question in the reverse direction, they found that there was an increased risk of heart attack for men with any degree of male pattern baldness. The analysis methods included adjustments for age and other heart attack risk factors. The increase in risk was more severe with increasing severity of baldness, after adjusting for age and other risk factors.

The authors of the study speculated that a third variable, perhaps a male hormone, might simultaneously increase the risk of heart attacks and the propensity for baldness. With an observational study such as this one, scientists can establish a connection, and they can then look for causal mechanisms in future investigations.

## Types of Observational Studies

### Retrospective and Prospective Studies

Observational studies can be classified according to the relevant time frame of the data. In a **retrospective study,** the data are from the past. Participants may be asked to recall past events, or the researchers may use information that already has been recorded about the participants (medical records, for instance). In a **prospective study,** researchers follow participants into the future and record relevant events and variables. The prospective approach generally is a better procedure because people often do not remember past events accurately. A possible difficulty with a prospective study, however, is that participants may change their behavior because they know that it is being recorded.

### Case-Control Studies

In a **case-control study,** "cases" who have a particular attribute or condition are compared to "controls" who do not. The idea is to compare the cases and controls to see how they differ on explanatory variables of interest. In medical settings, the cases usually are individuals who have been diagnosed with a particular disease. Researchers then identify a group of controls who are as similar as possible to the cases except that they don't have the disease. To achieve this similarity, researchers often use patients who have been hospitalized for other causes as the controls.

Case Study 4.4 about baldness and heart disease is an example of a case-control study. In this example, men who had been admitted to the hospital with a heart attack were the cases, and men who had been admitted for other reasons were the controls. The cases and controls were compared on a variable of interest, which in Case Study 4.4 was the degree of male pattern baldness. In this example, the response variable is whether a participant is a case or a control (heart attack or not), and the explanatory variable is the measurement of interest (degree of baldness). Figure 4.5 illustrates the steps that were taken in this study.

Sometimes cases are matched with controls on an individual basis. That type of design is similar to a matched-pair experimental design. The analysis proceeds by first comparing the pair, then summarizing over all pairs. Unlike a matched-pair experiment, the researcher does not randomly assign treatments within pairs but is restricted to how they occur naturally. For example, to identify whether or not left-handed people die at a younger age, researchers might match each left-handed case with a right-handed sibling as a control and compare their ages at death. Handedness could obviously not be randomly as-

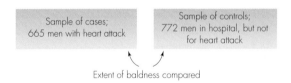

**Figure 4.5** ▮ The case-control design in Case Study 4.4: Samples of male heart attack patients (cases) and other male hospital patients (controls) were compared for the extent of baldness

signed to the two individuals, so confounding factors (such as the fact that most tools are made for right-handed people) might be responsible for any observed differences.

## Advantages of Case-Control Studies

Case-control studies have become increasingly popular in medical research, with good reason. They do not suffer from the ethical considerations that are inherent in the random assignment of potentially harmful or beneficial treatments. In addition, case-control studies can be efficient in terms of time, money, and the inclusion of enough people with the disease.

### Efficiency

In an exercise given later in this book, researchers were interested in whether or not owning a pet bird is related to the incidence of lung cancer. Imagine trying to design an experiment to study this problem. You would randomly assign people to either own a bird or not and then wait to see how many in each group contracted lung cancer. The problem is that you would have to wait a long time, and even then, you would observe very few cases of lung cancer in either group. In the end, you may not have enough cases for a valid comparison.

A case-control study, in contrast, would identify a large group of people who had just been diagnosed with lung cancer (response variable), then ask them whether or not they had owned a pet bird (explanatory variable). A similar control group would be identified and asked the same question. A comparison would then be made between the proportion of cases (lung cancer patients) who had birds and the proportion of controls who had birds.

### Reducing Potential Confounding Variables

Another advantage of case-control studies over other observational studies is that the controls are chosen to try to reduce potential confounding variables. For example, in Case Study 4.4, suppose that it was simply the case that balding men were less healthy than other men and were therefore more likely to get sick in some way. An observational study that recorded only whether or not someone had baldness and whether or not he had experienced a heart attack would not be able to control for that fact. By using other hospitalized patients as the controls, the researchers were able to partially account for general health as a potential confounding factor.

You can see that careful thought is needed to choose controls that reduce potential confounding factors and do not introduce new ones. For example, suppose we want to know whether heavy exercise induces heart attacks, and as cases, we use people who were recently admitted to the hospital for a heart attack. We would certainly not want to use other newly admitted patients as controls. People who are sick enough to enter the hospital (for anything other than sudden emergencies) would probably not have been recently engaging in heavy exercise. When you read the results of a case-control study, you should pay attention to how the controls were selected.

4.3 Exercises are on page 145.

*in summary*    Types of Observational Studies

- A **retrospective study** uses data from the past.
- A **prospective study** follows participants into the future and uses data from that future time.
- A **case-control study** compares individuals with a condition (cases) and individuals without the condition (controls) to determine whether cases and controls differ with respect to explanatory variables of interest.

# 4.4 Difficulties and Disasters in Experiments and Observational Studies

A number of complications can arise if researchers are not careful when they conduct and report experiments and observational studies. Some difficulties described here apply only to observational studies, some apply only to experiments, and some apply to both. We first list the possible problems and then describe each one in detail.

### Difficulties and Disasters in Experiments and Observational Studies

- Confounding variables and the implication of causation in observational studies
- Extending results inappropriately
- Interacting variables
- Hawthorne and experimenter effects
- Ecological validity and generalizability
- Using the past as a source of data

## Confounding Variables and the Implication of Causation in Observational Studies

Probably the most common mistake made in media reports of research studies is to imply that a *cause-and-effect* relationship can be concluded from an observational study. When a link between the explanatory and response variables is demonstrated, it is tempting to assume that the differences in the explanatory variable *caused* the differences in the response variable. A conclusion of cause and effect may be justified when a randomized experiment is done but not when the data are from an observational study. With an observational study, it is difficult, perhaps impossible, to separate the effects of confounding variables from the effects of the main explanatory variable(s) of interest. Consequently, we are not able to know the extent to which a specific explanatory variable causes changes in the response variable.

<table>
<tr><td>definition</td><td>The **Rule for Concluding Cause and Effect** is that cause-and-effect relationships can be inferred from randomized experiments but not from observational studies.</td></tr>
</table>

A partial solution for researchers who are carrying out an observational study is to measure possible confounding variables and take them into account in the analysis. For instance, if researchers are studying the connection between oral contraceptive use and blood pressure in women, they should measure and take into account other factors that are known to affect blood pressure, such as age, diet, and smoking habits. Another partial solution can be achieved when a case-control study is conducted by choosing the controls to be as similar as possible to the cases. There also may be other considerations that can help you to determine whether a causal link is plausible, such as a reasonable explanation of why cause and effect could occur in the particular situation. One part of the solution is up to you as a consumer of information: Don't be misled into thinking that a causal relationship exists when it may not.

**Example 4.4**    **Will Preventing Artery Clog Prevent Memory Loss?**    An observational study based on a random sample of over 6000 older people (average age 70 when the study began) found that 70% of the participants did not lose cognitive functioning over time. One important finding, however, was that "those who have diabetes or high levels of atherosclerosis in combination with a gene for Alzheimer's disease are eight times more likely to show a decline in cognitive function." This finding led one of the researchers to comment that "that has implications for prevention, which is good news. If we can prevent atherosclerosis, we can prevent memory loss over time, and we know how to do that with behavior changes—low-fat diets, weight control, exercise, not smoking, and drug treatments." (*Source:* "Study: Age doesn't sap memory," *Sacramento Bee,* Kathryn Doré Perkins, July 7, 1999, pp. A1 and A10.)

But notice that this conclusion assumes that the memory loss is *caused* by atherosclerosis. This conclusion is not justified on the basis of an observational study. Perhaps certain people have a genetic predisposition to artherosclerosis and to memory loss, in which case preventing the former condition with lifestyle changes will not reduce the memory loss. Or perhaps people who suffer from depression tend to eat poorly, not exercise well, and also suffer from memory loss. There are simply too many possible confounding factors that could be related to both artherosclerosis and memory loss to conclude that preventing artherosclerosis would prevent memory loss.  ■

## Extending Results Inappropriately

In interpreting both randomized experiments and observational studies, keep in mind the Fundamental Rule for Using Data for Inference: Available data can be used to make inferences about a much larger group *if the data can be considered to be representative with regard to the question(s) of interest.* Some observational studies are based on random samples, in which case the results can readily be extended to the population from which the sample was drawn. Example 4.4 is one such example; the study on aging and memory loss was part of the large Cardiovascular Health Study, sponsored by the National Heart, Lung and Blood Institute, part of the National Institutes of Health.

Many observational studies and almost all randomized experiments use convenience samples or volunteers. It is up to you to determine whether or not the results can be extended to any larger group for the question of interest. For example, the weight-lifting experiment in Case Study 4.2 used 11 girls and 32 boys who volunteered to participate in the study. There is no reason to suspect that these volunteers would differ from other children in terms of the differential effects of no training, moderate lifting, and heavy lifting measured in the study. Thus, although the sample consisted solely of volunteers, the results can probably be extended to other children in this age group.

## Interacting Variables

Sometimes a second explanatory variable *interacts* with the principal explanatory variable in its relationship with the response variable. This can be a problem if the results of a study are reported without taking the **interacting variable** into account. Although the reported results may be true on average, they may not hold for specific subgroups.

For example, in Case Study 4.3, there was an interaction between the treatment (nicotine or placebo patch) and whether there were other smokers at home. The researchers measured and reported this interaction, which is illustrated in Figure 4.6. When there were no other smokers in the participant's home, the eight-week success rate was actually 58% for the nicotine patch users compared to 20% for the placebo patch group. When there were other smokers in the home, the eight-week success rate for the nicotine patch was only 31%, but the success rate for the placebo users remained at about 20%. In other words, the amount by which the nicotine and placebo groups differed was affected by whether other smokers were present in the home. Therefore, it would be mis-

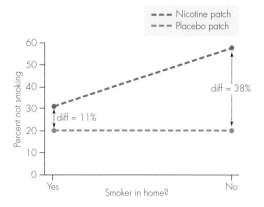

**Figure 4.6** ▌ Interaction in Case Study 4.3: The difference between the nicotine and placebo patches is greater when there are no smokers in the home than when there are smokers in the home.

leading to merely report that 46% of the nicotine recipients had quit without also providing the information about the interaction.

## Hawthorne and Experimenter Effects

We have already discussed the fact that a placebo can have a very strong effect on experimental outcomes because the power of suggestion is somehow able to affect the result. A similar difficulty is that participants in an experiment respond differently than they otherwise would, just because they are in the experiment. This is called the **Hawthorne effect** because it was first detected in the late 1920s in studying factory workers at the Western Electric Company's Hawthorne plant in Cicero, Illinois. (The phrase was not actually coined until much later; see French, 1953.)

The Hawthorne effect is a common problem in medical research. Many treatments have been observed to have a higher success rate in clinical trials than they do in actual practice. This may occur because patients and researchers are highly motivated to correctly carry out a treatment protocol in a clinical trial.

Related to these effects are numerous ways in which the experimenter can bias the results. These **experimenter effects** include recording the data erroneously to match the desired outcome, treating subjects differently on the basis of which condition they are receiving, and subtly making the subjects aware of the desired outcome. Most of these problems can be overcome by using double-blind designs and by including a placebo group or a control group that gets identical handling except for the active part of the treatment. Other problems, such as incorrect data recording, should be addressed by allowing data to be automatically entered into a computer when they are collected, if possible. Depending on the experiment, there may still be subtle ways in which experimenter effects can sneak into the results. You should be aware of these possibilities when you read the results of a study.

**Example 4.5** | **Dull Rats** In a classic experiment that was designed to test whether or not the experimenter's expectations could really influence the results, Rosenthal and Fode (1963) deliberately conned 12 experimenters. They gave each one five rats that had been taught to run a maze. They told six of the experimenters that the rats had been bred to do well (i.e., they were "maze bright") and told the other six that their rats were "maze dull" and should not be expected to do well. Sure enough, the experimenters who had been told that they had bright rats found learning rates far superior to those found by the experimenters who had been told they had dull rats. Hundreds of other studies have since confirmed the experimenter effect. ■

## Ecological Validity and Generalizability

Suppose you want to compare three assertiveness-training methods to find out which is most effective in teaching people how to say "no" to unwanted requests for their time. Would it be realistic to give participants the training, then measure the results by asking them to role-play situations in which they should say "no"? Probably not, since everyone involved would know that they were only role-playing. The usual social pressures to say "yes" would not be as present.

This is an example of an experiment with little **ecological validity.** The variables have been removed from their natural setting and are measured in the laboratory or in some other artificial setting. Thus, the results do not accurately reflect the impact of the variables in the real world or in everyday life. Whenever you read the results of a study, you should question its ecological validity and its generalizability.

**Example 4.6**

**Real Smokers with a Desire to Quit**  The researchers in Case Study 4.3, the nicotine patch study, did many things to help ensure ecological validity and generalizability. First, they used a standard intervention program available from and recommended by the National Cancer Institute, instead of inventing their own, so that other physicians could follow the same program. Next, they used participants at three different locations around the country rather than in one community only, and they involved a wide range of ages (20 to 65). They included individuals who lived in households with other smokers as well as those who did not. Finally, they recorded numerous other variables (sex, race, education, marital status, psychological health, etc.) and checked to make sure these were not related to the response variable or the patch assignment.  ■

## Using the Past as a Source of Data

Retrospective observational studies can be particularly unreliable because they ask people to recall past behavior. Some studies, in which the response variable is whether or not someone has died, can be even worse because they rely on the memories of relatives and friends rather than the actual participants to measure explanatory and other variables. Retrospective studies also suffer from the fact that variables that confounded things in the past may no longer be similar to those that would currently be confounding variables, and researchers may not think to measure them.

If at all possible, prospective studies should be used. That is not always possible. For example, researchers who first considered the potential causes of AIDS or toxic-shock syndrome had to start with those who were ill and try to find common factors from their pasts. If possible, retrospective studies should use authoritative sources such as medical records rather than relying on memory.

**Example 4.7**

**Statistics ◯ Now™**

Watch a video example at **http://1pass.thomson.com** or on your CD.

**Do Left-Handers Die Young?**  A few years ago, a highly publicized study pronounced that left-handed people did not live as long as right-handed people (Coren and Halpern, 1991). In one part of the study, the researchers had sent letters to next of kin for a random sample of recently deceased individuals, asking which hand the deceased had used for writing, drawing, and throwing a ball. They found that the average age of death for those who had been left-handed was 66, while for those who had been right-handed, it was 75.

What the researchers failed to take into account was that in the early part of the 20th century, many children were forced to write with their right hands,

4.4 Exercises are on page 145.

even if their natural inclination was to be left-handed. Therefore, people who died in their 70s and 80s during the time of this study were more likely to be right-handed than were those who died in their 50s and 60s. The confounding factor of how long ago one learned to write was not taken into account. A better study would be a prospective one, following current left- and right-handers to see which group survived longer, but you can imagine the practical difficulties of conducting such a study. The participants could well outlive the researchers. ■

# Key Terms

# Exercises

● Denotes basic skills exercises

◆ Denotes dataset is available in StatisticsNow at **http:// 1pass.thomson.com** or on your CD but is **not required** to solve the exercise.

**Bold-numbered exercises** have answers in the back of the text and fully worked solutions in the Student Solutions Manual.

**Statistics⬡Now™** Go to the StatisticsNow website at **http://1pass.thomson.com** to:
• Assess your understanding of this chapter
• Check your readiness for an exam by taking the Pre-Test quiz and exploring the resources in the Personalized Learning Plan

## Section 4.1

4.1 ● In each situation, indicate whether you think an observational study or a randomized experiment would be used. Explain why in each case.
   a. A teacher wants to compare the grade point averages of female students who are in sororities to the grade point averages of female students who are not in sororities.
   b. A doctor wants to compare the effectiveness of two allergy medications.
   c. A social psychologist wants to determine whether restaurant servers will get better tips if they introduce themselves by name to the people they serve.
   d. A child psychologist wants to find out whether there is an association between the amount of television 6-year-old children watch and how often they bully other children in school.

4.2 ● Refer to Exercise 4.1. In each part, specify the explanatory variable and the response variable.

**4.3** ● Suppose a statistics teacher wants to know whether the number of hours students spend studying in a group affects the final course grade. In each part, explain whether the research method described is a randomized experiment or an observational study.
   a. Each student keeps a log of the hours he or she spends studying in a group and reports the total after the course is completed.
   b. Students are randomly assigned to study groups. The teacher tells each group how often to meet. This varies from one hour the day before each exam to two hours per week.
   **c.** Students voluntarily join groups based on how often the groups will meet. The groups are designated as meeting weekly, meeting only before exams, or meeting whenever enough members feel that it is necessary.

**4.4** ● In each part, identify the response variable and the explanatory variable in the relationship between the two given variables.
   **a.** Daughter's height and mother's height.
   b. Age and weight of children between the ages of 3 and 10 years old.

   c. Opinion about the death penalty for persons convicted of murder and political party membership (Democrat, Republican, etc.)
   d. Smoking habits and performance on a test of lung function.

4.5 ● Remember that a confounding variable affects the response variable and is related to the explanatory variable. For each of the following situations, explain how the given confounding variable meets these criteria.
   a. *Response variable* = Math skills of children aged 6 to 12 years old; *explanatory variable* = shoe size; *confounding variable* = age.
   b. *Response variable* = Whether or not a student gets a cold during a school term; *explanatory variable* = whether or not the student procrastinates in doing assignments; *confounding variable* = number of hours per week the student spends socializing.
   c. *Response variable* = Child's IQ at age 10; *explanatory variable* = whether mother smokes or not; *confounding variable* = mother's educational level.
   d. *Response variable* = Number of missing or decayed teeth a child has; *explanatory variable* = amount of past exposure to lead; *confounding variable* = number of visits to a dentist the child has made in the past three years.

4.6 ● *This is also Exercise 1.8.* Suppose that an observational study showed that students who got at least seven hours of sleep performed better on exams than students who didn't. Which of the following are possible confounding variables, and which are not? Explain why in each case.
   a. Number of courses the student took that term.
   b. Weight of the student.
   c. Number of hours the student spent partying in a typical week.

**4.7** ● *This is also Exercise 1.7.* For each of the examples given here, decide whether the study was an observational study or a randomized experiment.
   a. A group of 100 students was randomly divided, with 50 assigned to receive vitamin C and the remaining 50 to receive a placebo, to determine whether vitamin C helps to prevent colds.
   b. All patients who received a hip transplant operation at Stanford University Hospital during 1995–2005 will be followed for ten years after their operation to determine the success (or failure) of the transplant.
   **c.** A group of students who were enrolled in an introductory statistics course were randomly assigned to take a Web-based course or a traditional lecture course. The two methods were compared by giving the same final examination in both courses.
   **d.** A group of smokers and a group of nonsmokers who visited a particular clinic were asked to come in for a physical exam every five years for the rest of their lives so that the researchers could monitor and compare their health status.

4.8    Specify what an individual unit is in each of the following studies. Then specify what two variables were measured on each unit.

    a. A study finds that college students who often procrastinate tend to be sick more often than students who do not procrastinate.

    b. A study finds that sport utility vehicles (SUVs) made by one car manufacturer tend to be more heavily damaged in a crash test than SUVs made by a second car manufacturer.

**4.9** For each of the following research scenarios, explain whether a randomized experiment could be used.

    **a.** To study the relationship between long-term practice of meditation and blood pressure.

    **b.** To determine whether a special training program improves scores on a standard college admissions test.

    c. To compare two programs to reduce the number of commuters who drive to work: providing discount coupons for the bus, or providing shuttle service for people who need to run errands during the workday.

    d. To study the relationship between age and opinion about whether marijuana should be legal.

**4.10** Refer to Exercise 4.9. In each case, state the explanatory variable and the response variable.

**4.11** Refer to Exercise 4.9. In each case, give an explanation of a possible confounding variable that could be present if an observational study were done.

**4.12** Refer to the first observational study described in Case Study 4.1 (p. 122). In the study, a link was found between tooth decay and exposure to lead for 24,901 children. Give an example of a possible confounding variable in addition to those described in the text. Explain why it could be a confounding variable.

**4.13** A news article on the *Reuters Health* website, Dec. 18, 1998 reported that

> A study of over 1200 people over age 65 showed that the "owls"—those who go to sleep after 11 p.m. and rise after 8 a.m.—tend to be as healthy and intelligent as "larks"—those who go to bed before 11 p.m. and rise before 8 a.m.—according to a report in the December 19/26 issue of the *British Medical Journal* [Vol. 317, pp. 1675–1677]. The researchers also found that owls tend to have higher average income than the early birds.

    a. Do you think this is based on an observational study or a randomized experiment? Explain.

    b. It is mentioned that the "owls" have higher average income than the "larks." Explain whether or not income is a possible confounding variable in this study.

    c. Give an example of a possible confounding variable that is not mentioned.

    d. Draw a figure similar to Figure 4.2 (p. 123), illustrating the steps in this study.

**4.14** A news article by *Reuters* on October 6, 1998 reported:

> In a new study, Dr. Matti Uhari and colleagues at the University of Oulu in Finland, randomly gave 857 healthy children in daycare centers xylitol in syrup, gum or a lozenge form, or a placebo gum or syrup in five doses per day for 3 months. According to a report in the October issue of the journal *Pediatrics* [Vol. 102, pp. 879–884, 971–972, 974–975], the incidence of ear infections was reduced by 40% in children given xylitol chewing gum, 30% in those given syrup and 20% in those given lozenges when compared to children given a placebo.

    a. Is this an observational study or a randomized experiment? Explain.

    b. What are the explanatory and response variables?

    c. Are confounding variables likely to be a problem in this study? Explain.

**4.15** *This is a modified version of Exercise 1.16.* Case Study 1.5 was about the association between blood pressure and frequency of participating in religious activities. Why was this study conducted as an observational study instead of an experiment?

## Section 4.2

**4.16** ● Suppose that we want to determine whether people who want to lose weight would lose more in three months by dieting or by exercising. Sixty people have volunteered to participate. Briefly describe how a randomized experiment would be done in this situation.

**4.17** ● An experiment will be done to determine whether college students learn information better when listening to rap music, when listening to classical music, or when not listening to music. In the study, each participant will read a chapter from a history textbook and then take a test about the material. Ninety students will participate.

    a. Briefly describe how a randomized experiment would be performed in this situation.

    b. For the experiment you described in part (a), draw a figure similar to Figure 4.3 (p. 125) illustrating the steps of the study.

**4.18** ● Two weight-training regimens designed to improve arm strength will be compared. The response variable will be the improvement in arm strength at the end of six weeks of training. Forty individuals, 20 males and 20 females, will participate in the study.

    a. Describe how a completely randomized experiment would be done in this situation.

    b. Describe how a matched-pair experiment could be done in this situation. Be specific about what constitutes a matched pair and how treatments would be assigned.

    c. Describe how a block experiment could be done in this situation. Be specific about how blocks would be formed and how weight-training methods would be assigned.

**4.19** ● Case Study 4.3 (p. 129) was about an experiment done to determine whether nicotine patches helped smokers to quit smoking. For that experiment, draw a figure similar to Figure 4.3 (p. 125) illustrating the steps of the study.

**4.20** ● Twenty students agreed to participate in a study on colds. Ten were randomly assigned to receive vitamin C, and the remaining ten received a tablet that looked and tasted like vitamin C but in fact contained only sugar and flavoring. The students did not know whether they were taking vitamin C or not, but the investigators did. The students were followed for two months to see who came down with a cold and who didn't. Explain whether each of the following terms applies to this randomized experiment.

a. Placebo.
b. Single-blind.
c. Double-blind.
d Matched pairs.
e. Repeated measures.

4.21 ● Researchers want to design a study to compare topical cream (applied to the skin) and a drug taken in capsule form for the treatment of a certain type of skin rash. Sixty volunteers who have the rash agree to participate in the study.
a. If each volunteer is given only one treatment, can the study be double-blind? Explain.
b. If each volunteer is given only one treatment, can the study be single-blind? Explain.
c. Could a double-blind, double-dummy study be done? If not, explain why not. If so, explain how it would be done.

4.22 ● Ten grocery stores, numbered from 0 to 9, will participate in an experiment to compare the effectiveness of two methods for displaying a product. Five of these stores will be randomly selected to use the first method; the other five stores will use the second method. Use the following row of random digits to select the five stores that will use the first display method.

68538   70621   32205   23593   61396

4.23 A school district has 20 elementary schools, and each school has 12 classes that can be used for a study.
a. Describe a randomized block design to compare three teaching methods, using schools as blocks.
b. Explain why schools should be used as blocks.

4.24 Twenty volunteers aged 40 to 65 years old will participate in an experiment to compare two methods of memorizing information. The age variation in the participants concerns the researchers because memorization skills decrease with age.
a. Describe how the study could be done using a matched-pair design.
b. Explain why a matched-pair design is a good idea in this situation.

4.25 Refer to Case Study 4.2 (p. 124), "Kids and Weight Lifting." Did the experiment described there use a completely randomized design, a matched-pair design, or a randomized block design? Explain.

4.26 One hundred volunteers agree to participate in an experiment. There are four treatments, each of which will be assigned to 25 participants. Explain how you could use a table of random digits to assign the treatments.

4.27 Refer to Case Study 4.3 (p. 129), "Quitting Smoking with Nicotine Patches," which used a completely randomized design. Explain how the experiment could have been done by using a randomized block design instead.

4.28 A fast-food chain sells its burgers alone or as part of a "value meal," which includes fries and a drink. They

know that some customers are health-conscious. They want to do an experiment to determine whether the proportion of customers choosing the meal would increase if they offered baby carrots with the meal as an alternative to fries. They will compare three treatments: (1) status quo, (2) offering carrots as an alternative but with no advertising, and (3) offering carrots as an alternative and advertising this option in the local area. They have restaurants in three types of areas: cities, suburban areas, and along major highways.
a. Explain how they would do this experiment using a completely randomized design.
b. Explain how they would do this experiment using a randomized block design.
c. Explain why it makes sense to use the type of area in which a restaurant is located as a block determinant in this experiment.
d. Explain why a matched-pair design could not be used for this experiment.

4.29 Explain whether each of the following experiments was single-blind, double-blind, or neither.
a. An electric company wanted to know whether residential customers would use less electricity during peak hours if they were charged more during those hours. One hundred customers were randomly selected to receive the special time-of-day rates, and they were matched with 100 customers who were similar in terms of past electrical use and household size. Special meters were installed for all 200 customers, and at the end of three months, the usage during peak hours was compared for each pair. The technician who read the meters did not know who had which rate plan.
b. To test an herbal treatment for depression, 100 volunteers who suffered from mild depression were randomly divided into two groups. Each person was given a month's supply of tea bags. For one group, the tea contained the herb mixed with a spice tea; for the other group, the bags contained only the spice tea. Participants were not told which type of tea they had and were asked to drink one cup of the tea per day for a month. At the end of the month, a psychologist evaluated them to determine whether their mood had improved. The psychologist did not know who had the tea with the herbal ingredient added.
c. To compare three packaging designs for a new food, a manufacturer randomly selected three grocery stores in each of 50 cities. The three designs were randomly assigned to the three stores in each city, and sales were compared for a two-month period.

4.30 Refer to Exercise 4.29. In each case, explain whether the experiment was a matched-pair design, a block design, or neither.

4.31 Refer to Exercise 4.29. In each case, designate the explanatory variable and the response variable.

4.32 Refer to Exercise 4.29. In each case, explain whether a control group was used and whether a placebo treatment was used.

4.33 Echinacea is an herb that may help to prevent colds and flu. Explain how researchers could conduct a double-blind, double-dummy experiment to compare echinacea tea with chewable vitamin C pills for effectiveness in preventing colds.

## Section 4.3

4.34 ● A teacher wants to determine how class attendance during the term affects grades on the final exam.
   a. Describe how the data would be collected in a prospective observational study.
   b. Describe how the data would be collected in a retrospective observational study.
   c. Refer to parts (a) and (b). Give one disadvantage for each method of collecting the data in this situation.

4.35 ● Suppose you want to know whether men or women spend more time talking on a cell phone during a week's time.
   a. Describe how you would collect the data in a retrospective observational study.
   b. Describe how you would collect the data in a prospective observational study.
   c. Refer to parts (a) and (b). Give one disadvantage for each method of collecting the data in this situation.

4.36 ● A medical researcher wants to know how lifetime sun exposure affects the risk of skin cancer for persons over 50 years old. Describe how data might be collected for a retrospective case-control study.

4.37 ● Students who had meningitis were matched with students without meningitis on gender, undergraduate (or graduate) status, and college. The students' recent activities were examined to discover risk factors for meningitis. Explain whether each of the following terms applies to this observational study.
   a. Prospective study.
   b. Retrospective study.
   c. Case-control study.

**4.38** ● Students whose undergraduate major was economics were followed for 10 years after graduation to study the number of different jobs they took. Explain whether each of the following terms applies to this observational study.
   **a.** Prospective study.
   **b.** Retrospective study.
   **c.** Case-control study.

4.39 Suppose researchers were interested in determining the relationship, if any, between the use of cellular telephones and the incidence of brain cancer. Would it be better to use a randomized experiment, a case-control study, or an observational study that did not use cases and controls? Explain.

4.40 Find an example of an observational study in the news. Determine whether or not it was a case-control study and whether it was prospective or retrospective. Specify the explanatory and response variables, and explain whether confounding variables were likely to be a major problem in interpreting the results. Be sure to include the news article with your response.

4.41 Reuters (June 24, 1997) reported on a study published in the *Journal of the American Medical Association* (1997, Vol. 277, pp. 1940–1944) in which researchers recruited 276 volunteers (aged 18 to 55) and used nose

drops to infect them with a cold virus. The volunteers were then quarantined and observed to see whether they came down with a cold or infection. They also answered questions about their social lives, including the number of different types of contacts they had (family, work, community, religious groups, and so on). *Reuters* reported that "those with one to three types of social contacts were 4.2 times as likely to come down with cold symptoms and signs of infection compared to those with six or more contacts."
   a. Explain why this is an observational study, not a randomized experiment.
   b. What were the explanatory and response variables of interest to the researchers?
   c. The researchers in this study took an unusual step to help reduce the impact of confounding variables. Explain what they did and why it might have helped to reduce the impact of confounding variables.

**4.42** A story at ABCNEWS.com ("Pet Contact" by Rita Rubin, March 17, 1998) reported that Karen Allen, a researcher at the University of Buffalo, found that couples who own cats or dogs have more satisfying marriages and are less stressed-out than those who don't own pets. Pet owners also have more contact with each other and with other people. Allen compared 50 couples who owned either cats or dogs with 50 pet-free couples. The volunteers completed a standard questionnaire assessing their relationship and how attached they were to their pets. They also kept track of their social contacts over a two-week period. To see how the couples responded to stress, Allen monitored their heart rates and blood pressure while they discussed sore subjects. Pet-owning couples started out with lower blood pressure readings than the others, and their numbers didn't rise as much when they argued.
   **a.** Was this a case-control study, a randomized experiment, or an observational study that was not case-controlled?
   **b.** What are the explanatory and response variables in this study?
   **c.** Give an example of a possible confounding variable.
   d. Draw a figure illustrating the steps in this study, similar to Figure 4.2 (p. 123).

## Section 4.4

4.43 ● *This is also Exercise 1.10.* A headline in a major newspaper read, "Breast-Fed Youth Found to Do Better in School."
   a. Do you think this statement was based on an observational study or a randomized experiment? Explain.
   b. Given your answer in part (a), which of these two alternative headlines do you think would be preferable: "Breast-Feeding Leads to Better School Performance" or "Link Found Between Breast-Feeding and School Performance"? Explain.

4.44 ● Explain why it is preferable, whenever possible, to conduct a randomized experiment rather than an observational study.

● Basic skills     ◆ Dataset available but not required     **Bold-numbered** exercises answered in the back

4.45 ● *This is also Exercise 1.20.* An article in the magazine *Science* (Service, 1994) discussed a study comparing the health of 6000 vegetarians and a similar number of their friends and relatives who were not vegetarians. The vegetarians had a 28% lower death rate from heart attacks and a 39% lower death rate from cancer, even after the researchers accounted for differences in smoking, weight, and social class. In other words, the reported percentages were the differences remaining after adjusting for differences in death rates due to those factors.

   a. Is this an observational study or a randomized experiment? Explain.

   b. On the basis of this information, can we conclude that a vegetarian diet causes lower death rates from heart attacks and cancer? Explain.

   c. Give an example of a potential confounding variable in this situation, and explain what it means to say that this is a confounding variable.

4.46 ● For each of the following, explain whether the study summary describes an interacting variable or a confounding variable.

   **a.** A comparison of the mean grade point averages of male and female students finds more difference for fraternity and sorority members than for students who are not fraternity and sorority members.

   b. A study finds that wine drinkers tend to have better health than beer drinkers but also finds that wine drinkers tend to smoke less than beer drinkers.

   c. It is found that drawing a happy face on a restaurant bill increases the mean tip percentage for female servers but does not increase the mean tip percentage for male servers.

4.47 ● Researchers have found that women who take oral contraceptives (birth control pills) are at higher risk of having a heart attack or stroke and that the risk is substantially higher if a woman smokes. In investigating the relationship between taking oral contraceptives (the explanatory variable) and having a heart attack or stroke (the response variable), would smoking be called a confounding variable or an interacting variable? Explain.

4.48 ● In a study on worker productivity done in 1927 in Cicero, Illinois, it was observed that productivity improved when the lighting was increased. However, it was observed that productivity also improved when the lighting was decreased. What term is used to describe the fact that productivity increased simply because the subjects knew they were in an experiment?

4.49 ● A new medication is observed to cause a weight loss in women but a weight gain in men. Identify what the response variable, explanatory variable, and interacting variable are in this situation.

4.50 An investigator believes that a new medication will help to reduce anxiety and that a placebo will not. In a single-blind study (patients are blind), she finds that the group taking the new medication had reduced anxiety while the placebo group did not. Explain how the experimenter effect may have played a role in the results.

4.51 In a sample of 500 college students aged 18 to 21, it was found that offering discount coupons for restaurant meals substantially increased the likelihood that the students would eat at those restaurants. Explain how "extending results inappropriately" may be a problem in the interpretation of this study.

4.52 An experiment on elephants in captivity was done to find out whether breeding would occur more often when the elephants were kept together in a herd or when a male was isolated with a female. Explain how ecological validity affects the ability to generalize these results to wild elephants.

4.53 A retrospective study of 517 veterans who never smoked was done to determine whether there was an association between lung disease and exposure to workplace gases, dust, and fumes (Clawson, *Stanford Report*, November 1, 2000). In this study, subjects were asked whether or not they remembered being exposed to gases, dust, or fumes in the workplace. A concern expressed by the investigator was that "patients who have histories of lung diseases may be more likely to recall a history of exposure—whether accurate or not—than subjects reporting good health." Discuss whether each of the following is likely to be a problem in this study.

   **a.** Relying on memory.

   **b.** Confounding variable.

   **c.** Hawthorne effect.

4.54 Refer to Exercise 4.42, describing a study about pet ownership and marriage. Explain whether each of the following is likely to be a problem for that study:

   a. Confounding variables and the implication of causation in observational studies.

   b. Hawthorne and experimenter effects.

   c. Ecological validity and generalizability.

4.55 Refer to Exercise 4.29. For each of the experiments described, pick one of the "Difficulties and Disasters" described in Section 4.4, and explain how it might be a problem in that experiment.

4.56 Is the experimenter effect more likely to be a problem in a study that is double-blind, single-blind, or not blind at all? Explain.

4.57 Pick the two "Difficulties and Disasters" that are most likely to be a problem in each of the following studies, and explain why they would be a problem:

   a. To compare marketing methods, a marketing professor randomly divides a large class into three groups and randomly assigns each group to watch one of three television commercials. The students are then asked to rate how likely they would be to buy the advertised product.

   b. Volunteers are recruited through a newspaper article to participate in a study of a new vaccine for hepatitis C, which is transmitted by contact with infected blood and is a particular problem for intravenous drug users. The article specifically requests both intravenous drug users and medical workers to volunteer, assuring them confidentiality and offering them free medical care for the life of the study. Volunteers are then randomly assigned to receive either a placebo or the vaccine, and a year later they are tested to see whether they have the disease.

c. A researcher at a large medical clinic wants to determine whether a high-fat diet in childhood is more likely to result in heart disease in later life. As a case-control study, she gives a questionnaire to 500 patients with heart disease and 500 other clinic patients matched by age and sex. The questionnaire asks about childhood diet.

4.58 Refer to part (c) of Exercise 4.57, describing a case-control study relating childhood diet and heart disease. Comment on the extent to which each of the following is likely to be a problem.
  a. Confounding variables and the implication of causation in observational studies.
  b. Extending results inappropriately.
  c. Interacting variables.
  d. Hawthorne and experimenter effects.
  e. Ecological validity and generalizability.
  f. Relying on memory or secondhand sources.

4.59 A categorical interacting variable defines subgroups for which the effect of the explanatory variable on the outcome variable differs. For instance, as explained in the text, for the nicotine patch experiment described in Case Study 4.3, one interacting variable was whether or not there were other smokers at home. When the difference in subgroups is thought to hold for the population, the subgroup variable is sometimes called an **effect modifier.** Explain why this term makes sense, using the context of the smokers at home in the nicotine experiment. For instance, what "effect" is being modified?

## Chapter Exercises

*Exercises 4.60 to 4.67 refer to a study published in the* Journal of the American Medical Association *and reported in the* Sacramento Bee *on Jan. 26, 2000 ("Study ties hormone replacement, cancer" by Shari Roan, p. A1). A sample of 46,355 postmenopausal women were studied for 15 years and asked about their use of hormone therapy and whether or not they had breast cancer; 2082 developed the disease. The article reported that "for each year of combined [estrogen and progestin] therapy, a woman's risk of breast cancer was found to increase by 8 percent compared to a 1 percent increase in women taking estrogen alone.*

**4.60** Do you think this research was an observational study or a randomized experiment? Explain.

4.61 Discuss whether this research is the following:
  a. A case-control study.
  b. A retrospective study or a prospective study.

4.62 What are the explanatory and response variables for this study? Are they quantitative or categorical? If they are categorical, what are the categories?

**4.63** To whom are the women taking combined therapy and estrogen alone being compared when the increased risks of 8% and 1% are computed?

4.64 Assuming that the women and their physicians made the decision about which treatment to pursue (combined hormones, estrogen alone, or no hormones), discuss the possibility of confounding variables in this study.

4.65 Near the end of the *Sacramento Bee* article, it was noted that "the NCI researchers analyzed their results by the woman's weight. . . . There was no increased breast cancer risk in heavier women." Which of the "Difficulties and Disasters" in Section 4.4 is represented by this statement? Explain.

4.66 One drawback of the study quoted in the *Sacramento Bee* was that "people who develop breast cancer are much more likely to remember whether they took hormones compared to women who don't develop the disease." Which of the "Difficulties and Disasters" in Section 4.4 is represented by this statement? Explain.

**4.67** Another drawback of the study quoted in the *Sacramento Bee* was that "the women in the study may not resemble women today, because many women now take lower doses of progestin than was used during the follow-up years of the study (1979–95)." Which of the "Difficulties and Disasters" in Section 4.4 is represented by this statement? Explain.

*Exercises 4.68 to 4.72 refer to a study described in a* Sacramento Bee *article titled "Much ado over those bad hairdos" (Jan. 26, 2000, p. A21). Here is part of the article:*

> Researchers surveyed 60 men and 60 women from 17 to 30, most of them Yale students. They were separated into three groups. One group was questioned about times in their lives when they had bad hair. The second group was told to think about bad product packaging, like leaky containers, to get them in a negative mind-set. The third group was not asked to think about anything negative. The groups were tested for self-esteem and self-judgment, and the bad hair group scored lower than the others.

4.68 Was this an observational study or a randomized experiment? Explain.

4.69 What were the explanatory and response variables for this study? Were they categorical variables or quantitative variables?

4.70 Discuss whether this study could have been blind or double-blind, and whether you think it was blind or double-blind.

4.71 Discuss each of the following "Difficulties and Disasters" in the context of this research.
  a. Ecological validity.
  b. Extending results inappropriately.
  c. Experimenter effect.

**4.72** The article also reported that "contrary to popular belief men's self-esteem may take a greater licking than women's when their hair just won't behave. Men were likely to feel less smart and less capable when their hair stuck out, was badly cut or otherwise mussed." In the context of this quote, is male/female a confounding variable, an interacting variable, or neither?

4.73 Refer to Example 4.3, in which 43 children were randomly assigned to three treatment groups. The example demonstrated how the children for group 1 were selected. Use the row of random digits below to select the 12 children (from the remaining 28) to go into the control group. Number the children from 01 to 28. Start at the beginning of the row. Explain your process, and list the unit numbers of the children selected.

27475 10484 24616 13466
41618 08551 18314 57700

4.74 In this chapter, you learned that cause and effect can be concluded from randomized experiments but generally not from observational studies. Why don't researchers simply conduct all studies as randomized experiments rather than observational studies?

4.75 Refer to the study reported in Example 2.2, relating the use of nightlights in childhood and the incidence of subsequent myopia.
  a. Was the research based on an observational study or a randomized experiment? Explain.
  b. Which one of the "Difficulties and Disasters" in Section 4.4 do you think is most likely to be a problem in this research? Explain.

4.76 Find an example of an observational study in the news. Answer the following questions about it. Be sure to include the news article with your response.
  a. What are the explanatory and response variables, or is this distinction not possible?
  b. Briefly describe how the study was done. For instance, was it a case-control study, a prospective or retrospective study?
  c. Discuss possible confounding variables. For instance, were any possible confounding variables included in the study and news report? Can you think of possible confounding variables that were not mentioned?
  d. Pick one of the "Difficulties and Disasters" in Section 4.4 and discuss how it applies to the study.

4.77 Find an example of a randomized experiment in the news. Answer the following questions about it. Be sure to include the news article with your response.
  a. What are the explanatory and response variables? What relationship was found, if any?
  b. What treatments were assigned? Was a control group or placebo used?
  c. Was the study a matched-pair design, a block design, or neither? Explain.
  d. Was the study single-blind, double-blind, or neither? Explain.

**4.78** Specify what an individual "unit" is in each of the following studies. Then specify what two variables were measured on each unit.
  **a.** A study found that tomato plants raised in full sunlight produced more tomatoes than did tomato plants raised in partial shade.
  b. A study found that gas mileage was higher for automobiles when the tires were inflated to their maximum possible pressure than when the tires were underinflated.
  c. Ten randomly selected classrooms in a school district were assigned to have a morning fruit snack break, and another ten classrooms of the same grade level in similar schools were measured as a

control group. The number of children who performed better than average on standardized tests was measured for each classroom.

4.79 Refer to Exercise 4.78 and answer these questions.
  a. For each study, on the basis of the information given, is it clearly a randomized experiment? If not, explain what additional information would make it clear that the study was a randomized experiment rather than an observational study.
  b. For each study, discuss whether or not matched pairs were used. If it is not clear, explain what additional information would make it clear.
  c. Discuss whether any of the studies would suffer from each of the following: the Hawthorne effect, interacting variables, ecological validity.

**4.80** Is it possible for each of the following to be used in the same study (on the same units)? Explain or give an example of such a study.
  **a.** A placebo and a double-blind procedure.
  **b.** A matched-pair design and a retrospective study.
  **c.** A case-control study and random assignment of treatments.

4.81 Refer to the study in Example 4.7, comparing ages at death for left-handed and right-handed people. Draw a figure similar to those in Figures 4.2 through 4.5, illustrating the steps for that study.

4.82 Refer to Case Study 1.6, in which physicians were randomly assigned to take aspirin or a placebo, and heart attack rates were compared. Draw a figure similar to Figures 4.2 through 4.5, illustrating the steps for that study.

4.83 A study was done (fictional) to compare the proportion of children who developed myopia after sleeping with and without a nightlight. The study found that the results differed based on whether at least one parent suffered from myopia by age 20. The percents of children suffering from myopia were as follows:

| Slept with: | Parent(s) Had Myopia | Parent(s) Did Not Have Myopia |
|---|---|---|
| | % with Myopia | % with Myopia |
| **No Light** | 10% | 10% |
| **Some Light** | 50% | 30% |

  a. Identify the explanatory variable, the response variable, and an interacting variable.
  b. Draw a figure similar to Figure 4.6 (p. 138), illustrating the interaction in this study.
  c. Write a few sentences that would be understood by someone with no training in statistics explaining the concept of interaction in this study.

4.84 Refer to Example 4.5 (p. 139), "Dull Rats," which was done by using a completely randomized design.
  a. What were the treatments in this experiment?
  b. Were the experimental units the 60 individual rats or the 12 individual experimenters? Explain.
  c. Explain how the experiment could have been done using a matched-pair design instead.

4.85 Give an example of an observational study, and explain the difference between a confounding variable and a lurking variable in the context of your example.

4.86 Explain whether a variable can be both
   a. A confounding variable and a lurking variable.
   b. A response variable and a confounding variable.
   c. An explanatory variable and a dependent variable.

**4.87** Refer to Exercise 4.29. In each case, explain the extent to which you think the results from the sample in the experiment could be extended to a larger population.

4.88 Explain why confounding variables are more of a problem in observational studies than in randomized experiments. Give an example.

**Statistics△Now™** Preparing for an exam? Assess your progress by taking the post-test at **http://1pass.thomson.com.**

**vMentor™**
   Do you need a live tutor for homework problems? Access vMentor at **http://1pass.thomson.com** for one-on-one tutoring from a statistics expert.

# 5

*Royalty-free/Corbis*

## *Does the driver's age affect the view?*

See Example 5.2 *(p. 153)*

> ***thought question 5.2***   Suppose you were to make a scatterplot of adult daughters' heights versus mothers' heights by collecting data on both variables from several of your female friends. You would now like to predict how tall your infant niece will be when she grows up. How would you use your scatterplot to help you make this prediction? What other variables, aside from her mother's height, might be useful for improving your prediction? How could you use these variables in conjunction with the mother's height?*

## 5.2  Describing Linear Patterns with a Regression Line

Scatterplots show us a lot about a relationship, but we often want more specific numerical descriptions of how the response and explanatory variables are related. Imagine, for example, that we are examining the weights and heights of a sample of college women. We might want to know what the increase in average weight is for each 1-inch increase in height. Or we might want to estimate the average weight for women with a specific height, such as 5′10″.

**Regression analysis** is the area of statistics that is used to examine the relationship between a quantitative response variable and one or more explanatory variables. A key element of regression analysis is the estimation of a **regression equation** that describes how, on average, the response variable is related to the explanatory variables. This regression equation can be used to answer the types of questions that we just asked about the weights and heights of college women.

A regression equation can be used to **predict** values of a response variable using known values of an explanatory variable. For instance, it might be useful for colleges to have an equation for the connection between verbal SAT score and college grade point average (GPA). They could use that equation to predict the potential GPAs of future students, based on their verbal SAT scores. Some colleges actually do this kind of prediction to decide whom to admit, but they use a collection of variables to predict GPA. The prediction equation for GPA usually includes high school GPA, high school rank, verbal and math SAT scores, and possibly other factors such as a rating of the student's high school or the quality of an application essay.

There are many types of relationships and many types of regression equations. *The simplest kind of relationship between two variables is a straight line,* and that is the only type we will discuss here. Straight-line relationships occur frequently in practice, so this is a useful and important type of regression equation. Before we use a straight-line regression model, however, we should always examine a scatterplot to verify that the pattern actually is linear. We remind you of the music preference and age example, in which a straight line definitely does not describe the pattern of the data.

*\*HINT:* Perhaps use the approximate average of daughters' heights for mothers who are about the same height as your niece's mother.

*definition* | A **regression line** is a straight line that describes how values of a quantitative response variable (*y*) are related, on average, to values of a quantitative explanatory variable (*x*). This line is used for two purposes:

- To *estimate the average value* of *y* at any specified value of *x*
- To *predict the value* of *y* for an individual, given that individual's *x* value

The term **simple linear regression** refers to methods used to analyze straight-line relationships.

**Example 5.5**

**Describing Height and Handspan with a Regression Line** In Figure 5.1 (p. 153), we saw that the relationship between handspan and height has a straight-line pattern. Figure 5.6 displays the same scatterplot as Figure 5.1, but now a regression line is shown that describes the average relationship between the two variables. We used statistical software (Minitab) to find the "best" line for this set of measurements. We will discuss the criterion for "best" later. For now, let's focus on what the line tells us about the data.

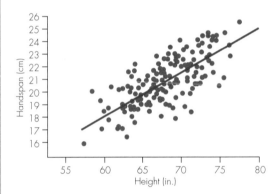

**Figure 5.6** ▌ Regression line describing height and handspan

The regression line drawn through the scatterplot describes how average handspan is linked to height. For example, when the height is 60 inches, the vertical position of the line is at about 18 centimeters. To see this, locate 60 inches along the horizontal axis (*x* axis), look up to the line, and then read the vertical axis to determine the handspan value. The result is that we can estimate that people who are 60 inches tall have an *average* handspan of about 18 centimeters (roughly 7 inches; 1 inch is 2.5 cm). We can also use the line to *predict* the handspan for an individual whose height is known. For instance, someone who is 60 inches tall is *predicted* to have a handspan of about 18 centimeters.

Let's use the line to estimate the average handspan for people who are 70 inches tall. We see that the vertical location of the regression line is somewhere between 21 and 22 centimeters, perhaps about 21.5 centimeters (roughly 8.5 inches). So when height is increased from 60 inches to 70 inches, average handspan increases from about 18 centimeters to about 21.5 centimeters.

The average handspan increased by 3.5 centimeters (about 1.5 inches) when the height was increased by 10 inches. This is a rate of 3.5/10 = 0.35 centimeter per 1-inch increase in height, which is the **slope** of the line. For each 1-inch difference in height, there is about a 0.35-centimeter average difference in handspan. ■

---

*formula*    **Equation of a Straight Line**

The equation for a straight line relating $y$ and $x$ is

$$y = b_0 + b_1 x$$

where $b_0$ is the "$y$ intercept" and $b_1$ is the slope. When $x = 0$, $y$ is equal to $y$ intercept. The letter $y$ represents the vertical direction, and $x$ represents the horizontal direction. The **slope** tells us how much the $y$ variable changes for each increase of one unit in the $x$ variable.

---

## Writing the Regression Equation

Statistical software will tell us the *regression equation,* the specific equation that is used to draw the line. For the handspan and height relationship, the regression equation determined by statistical software is

Handspan $= -3 + 0.35$ (Height)

When emphasis is on using the equation to estimate the average handspans for specific heights, we may write

Average handspan $= -3 + 0.35$ (Height)

When emphasis is on using the equation to predict an individual handspan, we might instead write

Predicted handspan $= -3 + 0.35$ (Height)

As examples, let's use the regression equation to estimate the average handspans for some specific heights.

For height $= 60$, average handspan $= -3 + 0.35(60) = -3 + 21 = 18$ cm
For height $= 67$, average handspan $= -3 + 0.35(67) = -3 + 23.45 =$
    20.45 cm
For height $= 70$, average handspan $= -3 + 0.35(70) = -3 + 24.5 = 21.5$ cm

The handspan values just calculated for heights of 60, 65, and 70 inches can also be used to predict the handspans of any individuals with those specific heights.

## Interpreting the Slope

In the handspan and height equation, the value 0.35 multiplies the height. This value is the *slope* of the straight line that links handspan and height. In general, the slope of a line measure how much the $y$ variable changes per each one-unit increase in the value of the $x$ variable. Consistent with our previous estimates, the slope in this example tells us that handspan increases by 0.35 centimeter, on average, for each increase of 1 inch in height. We can use the slope to estimate the average difference in handspan for any difference in height. If we consider two heights that differ by 7 inches, our estimate of the difference in handspans would be $7 \times 0.35 = 2.45$ centimeters, or approximately 1 inch.

## Statistical Relationships Versus Deterministic Relationships

In a **deterministic relationship,** if we know the value of one variable, we can exactly determine the value of the other variable. For example, the relationship between the volume and weight of water is deterministic. Every pint of water weighs 1.04 pounds, so we can determine exactly the weight of any number of pints of water.

In a **statistical relationship,** there is variation from the average pattern. You can see from Figure 5.6 that the regression line does not predict exactly what will happen for each individual. Most individuals do not have a handspan exactly equal to $-3 + 0.35$ (Height), the handspan that would be predicted from the regression equation.

Our ability to predict what happens for an individual depends on the amount of natural variability from the overall pattern. If most measurements are close to the regression line, we may be able to accurately predict what will happen for an individual. When there is substantial variation from the line, we will not be able to accurately predict what will happen for an individual.

## The Equation for the Regression Line

All straight lines can be expressed by the same formula in which $y$ is the variable on the vertical axis and $x$ is the variable on the horizontal axis. The equation for a regression line is

$$\hat{y} = b_0 + b_1 x$$

In any given situation, the sample is used to determine numbers that replace $b_0$ and $b_1$.

- $\hat{y}$ is spoken as "y-hat," and it is also referred to either as **predicted $y$** or **estimated $y$.**
- $b_0$ is the *intercept* of the straight line. The intercept is the value of $y$ when $x = 0$.
- $b_1$ is the *slope* of the straight line. The slope tells us how much of an increase (or decrease) there is for the $y$ variable when the $x$ variable increases by one unit. The sign of the slope tells us whether $y$ increases or decreases when $x$ increases.

*in summary*    Interpreting a Regression Line

- $\hat{y}$ estimates the average $y$ for a specific value of $x$. It also can be used as a prediction of the value of $y$ for an individual with a specific value of $x$.
- The **slope** of the line estimates the average increase in $y$ for each one-unit increase in $x$.

> • The **intercept** of the line is the value of $y$ when $x = 0$. Note that interpreting the intercept in the context of statistical data makes sense only if $x = 0$ is included in the range of observed $x$ values.

**Example 5.6**

Statistics⬦Now™

Watch a video example at **http:// 1pass.thomson.com** or on your CD.

For software help, download your Minitab, Excel, TI-83, SPSS, R, and JMP manuals from **ttp://1pass .thomson.com**, or find them on your CD.

**Regression for Driver Age and the Maximum Legibility Distance of Highway Signs**  Example 5.2 (p. 153) described a study in which researchers measured the maximum distance at which an automobile driver could read a highway sign. Thirty drivers participated. The regression line $\hat{y} = 577 - 3x$ describes how the maximum sign legibility distance (the $y$ variable) is related to driver age (the $x$ variable). Statistical software was used to calculate this equation and to create the graph shown in Figure 5.7. Earlier, we asked these two questions about distance and age:

- • How much does the distance decrease when age is increased?
- • For drivers of any specific age, what is the average distance at which the sign can be read?

The slope of the equation can be used to answer the first question. Remember that the slope is the number that multiplies the $x$ variable and the sign of the slope indicates the direction of the association. Here, the slope tells us that, on average, the legibility distance decreases 3 feet when age increases by one year. This information can be used to estimate the average change in distance for any difference in ages. For an age *increase* of 30 years, the estimated *decrease* in legibility distance is 90 feet because the slope is $-3$ feet per year.

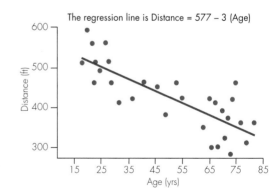

**Figure 5.7** ▌ Regression line for driver age and sign legibility distance

The question about estimating the average legibility distances for a specific age is answered by using the specific age as the $x$ value in the regression equation. To emphasize this use of the regression line, we write it as

$$\text{Average distance} = 577 - 3 \,(\text{Age})$$

Here are the results for three different ages:

| Age | Average Distance |
|-----|------------------|
| 20  | $577 - 3(20) = 517$ feet |
| 50  | $577 - 3(50) = 427$ feet |
| 80  | $577 - 3(80) = 337$ feet |

The equation can also be used to predict the distance measurement for an *individual* driver with a specific age. To emphasize this use of the regression line, we write the equation as

Predicted distance = 577 − 3 (Age)

For example, we can predict that the legibility distance for a 20-year-old will be 517 feet. For a 50-year-old the predicted legibility distance is 427 feet and for an 80-year-old it is 337 feet. ■

---

**MINITAB *tip*** | **Finding the Regression Line**

- To find a simple regression equation, use **Stat>Regression>Regression.** In the dialog box, specify the column containing the raw data for the response variable (Y) as the "Response," and specify the column containing the data for the explanatory variable (X) as a "Predictor."

- To find a regression line and also have Minitab draw this line onto a scatterplot of the data, use **Stat>Regression>Fitted Line Plot.** Specify the response variable (Y) and the predictor (X) in the dialog box.

---

**EXCEL *tip*** | **Finding the Regression Equation**

First, enter the *y* values and *x* values into separate columns. We refer to the range of cells as *y-range* and *x-range* in what follows:

*INTERCEPT(y-range,x-range)* = intercept of the regression equation.
*SLOPE(y-range,x-range)* = slope of the regression equation.

Notice that for the intercept and slope, the *y* range must be listed first.

---

## Prediction Errors and Residuals

For any given line, we can calculate the **predicted value** $\hat{y}$ for each point in the observed data. To do this for any particular point, we use the observed $x$ value in the regression equation. The **prediction error** for an observation is the difference between the observed $y$ value and the predicted value $\hat{y}$; the formula is $error = (y - \hat{y})$. The terminology "error" is somewhat misleading, since the amount by which an individual differs from the line is usually due to natural variation rather than "errors" in the measurements. A more neutral term for the difference $(y - \hat{y})$ is that it is the **residual** for that individual. In Chapter 14, we will learn that the standard deviation of the residuals for a dataset is a useful measure of the "typical" difference between actual and predicted values of the $y$ variable.

**Example 5.7** | **Prediction Errors for the Highway Sign Data** Examples 5.2 and 5.6 described a study in which $y$ = maximum distance at which a person can read a highway sign was related to $x$ = age. The regression equation for these data is $\hat{y} = 577 - 3x$. To calculate $\hat{y}$ for an individual, substitute his or her age for $x$ in the equation. For individuals in the sample, an observed value of $y$ is available, and the residual $(y - \hat{y})$ can then be found. For the first three individuals shown in Table 5.2, the *residuals,* or *prediction errors,* are calculated as follows:

| x = Age | y = Distance | ŷ = 577 − 3x | Residual = y − ŷ |
|---------|--------------|--------------|------------------|
| 18 | 510 | 577 − 3(18) = 523 | 510 − 523 = −13 |
| 20 | 590 | 577 − 3(20) = 517 | 590 − 517 =   73 |
| 22 | 560 | 577 − 3(22) = 511 | 560 − 511 =   49 |

This process could be carried out for any of the 30 observations in the dataset. The seventh individual in Table 5.2, for instance, has age = 27 years and distance = 560 feet. The predicted distance for this person is $\hat{y} = 577 - 3(27) = 496$ feet, so the residual is $(y - \hat{y}) = 560 - 496 = 64$ feet. A positive residual indicates that the individual had an observed value that was higher than what would be predicted for someone of that age. In this case, the 27-year-old in the study could see the sign at a distance 64 feet farther away than would be predicted for someone of that age. Figure 5.8 illustrates this residual by showing that the residual is the vertical distance from a data point to the regression line.

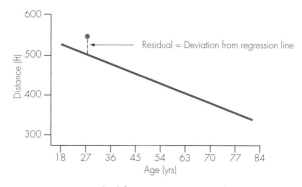

**Figure 5.8** ▌ Residual from regression line for 27-year-old who saw sign at a distance of 560 feet. The residual, also called the prediction error, is the difference between observed $y = 560$ feet and $\hat{y} = 496$ feet.

## The Least Squares Criterion

A mathematical criterion called **least squares** is nearly always the basis for estimating the equation of a regression line. The term *least squares* is a shortened version of "least sum of squared errors." A **least squares line** has the property that the sum of squared differences between the observed values of $y$ and the predicted values is smaller for that line than it is for any other line. Put more simply, the least squares line minimizes the sum of squared prediction errors for the observed data set. The notation **SSE,** which stands for **sum of squared errors,** is used to represent the sum of squared prediction errors. The least squares line (the regression line) has a smaller SSE than any other regression line that might be used to predict the response variable.

There is a mathematical solution that produces general formulas for computing the slope and intercept of the least squares line. These formulas are used by all statistical software, spreadsheet programs, and statistical calculators. To be complete, we include the formulas. In practice, however, regression analysis is done using a computer, so we don't include an example showing how to calculate the slope and intercept for the least squares line "by hand."

---

*formula*

**Formulas for the Slope and Intercept of the Least Squares Line**

$b_1$ is the slope and $b_0$ is the $y$ intercept.

$$b_1 = \frac{\sum_i (x_i - \bar{x})(y_i - \bar{y})}{\sum_i (x_i - \bar{x})^2}$$

$$b_0 = \bar{y} - b_1\bar{x}$$

$x_i$ represents the $x$ measurement for the $i$th observation.

$y_i$ represents the $y$ measurement for the $i$th observation.

$\bar{x}$ represents the mean of the $x$ measurements.

$\bar{y}$ represents the mean of the $y$ measurements.

---

**Example 5.8**

**Calculating the Sum of Squared Errors** Suppose that $x =$ score on exam 1 in a course and $y =$ score on exam 2 and that the first two rows in Table 5.3 (shown below) give $x$ values and $y$ values for $n = 6$ students. For these data, the least squares regression line is $\hat{y} = 20 + 0.8x$ (found using Minitab). Values of $\hat{y}$ for all observations are given in the third row of Table 5.3, and the fourth row gives the corresponding values of the prediction errors $(y - \hat{y})$. For instance, $x = 70$ and $y = 75$ for the first observation shown in Table 5.3, so $\hat{y} = 20 + 0.8(70) = 76$ and $(y - \hat{y}) = 75 - 76 = -1$. The sum of the squared prediction errors for the regression line is

$$\text{SSE} = (-1)^2 + (2)^2 + (-4)^2 + (2)^2 + (2)^2 + (-1)^2$$

$$= 1 + 4 + 16 + 4 + 4 + 1 = 30$$

**Table 5.3** Values of $x$, $y$, $\hat{y}$, and $(y - \hat{y})$ for Example 5.8

| $x =$ Exam 1 score | 70 | 75 | 80 | 80 | 85 | 90 |
|---|---|---|---|---|---|---|
| $y =$ Exam 2 score | 75 | 82 | 80 | 86 | 90 | 91 |
| $\hat{y} = 20 + 0.8x$ | 76 | 80 | 84 | 84 | 88 | 92 |
| $(y - \hat{y})$ | $-1$ | 2 | $-4$ | 2 | 2 | $-1$ |

The line $\hat{y} = 20 + 0.8x$ is the least squares line, so any other line will have a sum of squared errors greater than 30. As an example, if the line $\hat{y} = 4 + x$ were used to predict the values of $y$, the sum of squared values of $(y - \hat{y})$ would be

$$(75 - 74)^2 + (82 - 79)^2 + (80 - 84)^2 + (86 - 84)^2 + (90 - 89)^2$$

$$+ (91 - 94)^2 = 40 \quad \blacksquare$$

## Why Regression Is Called Regression

You may wonder why the word *regression* is used to describe the study of statistical relationships. Most of the vocabulary used by statisticians has at least some connection to the common usage of the words, but this doesn't seem to be true for *regression*. The statistical use of the word *regression* dates back to Francis Galton, who studied heredity in the late 1800s. (See Stigler, 1986 or 1989, for a detailed historical account.) One of Galton's interests was whether or not a man's height as an adult could be predicted by his parents' heights. He discovered that it could, but the relationship was such that very tall parents tended to have children who were shorter than they were and very short parents tended to have children who were taller than themselves. He initially described this phenomenon by saying that there was "reversion to mediocrity" but later changed the terminology to "regression to mediocrity." Thereafter, the technique of determining such relationships was called *regression*.

5.2 Exercises are on page 183.

---

***thought question 5.3***   Suppose the statistics community is having a contest to rename *regression* to something more descriptive of what it actually does. What would you suggest as a name for the whole procedure? As a name for the regression line?*

---

# 5.3 Measuring Strength and Direction with Correlation

The linear pattern is so common that a statistic was created to characterize this type of relationship. The statistical **correlation** between two quantitative variables is a number that *indicates the strength and the direction of a straight-line relationship.*

- The *strength* of the relationship is determined by the *closeness of the points to a straight line.*

- The *direction* is determined by whether one variable generally increases or generally decreases when the other variable increases.

As used in statistics, the meaning of the word *correlation* is much more specific than it is in everyday life. A statistical correlation describes only *linear relationships.* Whenever a correlation is calculated, a straight line (the regression line) is used as the frame of reference. When the pattern is nonlinear, as it was for the music preference data shown in Figure 5.3, a correlation is not an appropriate way to measure the strength of the relationship.

*Correlation* is represented by the letter *r*. Sometimes this measure is called the **Pearson product moment correlation** or the **correlation coefficient.** It doesn't matter which of the two variables is called the *x* variable and which is called the *y* variable. The value of the correlation is the same either way. For instance, the correlation between height and foot length is the same regardless of whether you use height as the *y* variable or use foot length as the *y* variable. Another useful feature of the correlation coefficient is that its value doesn't change

**\*HINT:** Two purposes for using regression are given in the definition box on page 158.

when the measurement units are changed for either or both of the variables. For instance, the correlation between weight and height is the same whether the measurements are in pounds and inches or in kilograms and centimeters.

The formula for calculating the correlation coefficient looks complicated, although it can be described rather simply in terms of standardized scores (introduced in Section 2.7). Approximately, the correlation value is the average product of standardized scores for variables $x$ and $y$. Calculating a correlation value by hand, however, generally involves much labor, so all statistical software programs and many calculators provide a way to easily calculate this statistic. In this section, we focus on how to interpret the correlation coefficient rather than how to calculate it.

---

*formula*    **A Formula for Correlation**

$$r = \frac{1}{n-1} \sum_i \left( \frac{x_i - \bar{x}}{s_x} \right) \left( \frac{y_i - \bar{y}}{s_y} \right)$$

$n$ is the sample size.

$x_i$ is the $x$ measurement for the $i$th observation.

$\bar{x}$ is the mean of the $x$ measurements.

$s_x$ is the standard deviation of the $x$ measurements.

$y_i$ is the $y$ measurement for the $i$th observation.

$\bar{y}$ is the mean of the $y$ measurements.

$s_y$ is the standard deviation of the $y$ measurements.

---

## Interpreting the Correlation Coefficient

Some specific features of the correlation coefficient are as follows:

- Correlation coefficients are always between $-1$ and $+1$.
- The magnitude of the correlation indicates the strength of the relationship, which is the overall closeness of the points to a straight line. The sign of the correlation does not tell us about the strength of the linear relationship.
- A correlation of either $+1$ or $-1$ indicates that there is a perfect linear relationship and all data points fall on the same straight line.
- The sign of the correlation indicates the direction of the relationship. A *positive* correlation indicates that the two variables tend to increase together (a positive association). A *negative* correlation indicates that when one variable increases, the other is likely to decrease (a negative association).
- A correlation of 0 indicates that the best straight line through the data is exactly horizontal, so knowing the value of $x$ does not change the predicted value of $y$.

The following examples illustrate these features.

**Example 5.9** | **The Correlation Between Handspan and Height** In Example 5.1 we saw that the relationship between handspan and height appears to be linear, so a

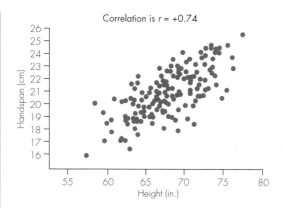

**Figure 5.9** ▌ Height and handspan

correlation is useful for characterizing the strength of the relationship. For these data, the correlation is $r = +0.74$, a value that indicates a somewhat strong positive relationship. Figure 5.9 (which is the same as Figure 5.1) shows us that average handspan definitely increases when height increases, but within any specific height there is some natural variation among individual handspans. ▪

**Example 5.10**

**The Correlation Between Age and Sign Legibility Distance**  For the data shown in Figure 5.10 (which is the same as Figure 5.2), relating driver age and sign legibility distance, the correlation is $r = -0.8$. This value indicates a somewhat strong negative association between the variables.

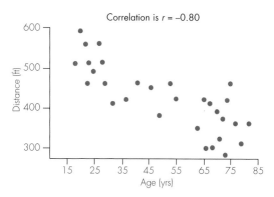

**Figure 5.10** ▌ Driver age and the maximum distance at which a highway sign was read

(Source: Adapted from data collected by Last Resource, Inc., Bellefonte, PA.)

**Example 5.11**

**Left and Right Handspans**  If you know the span of a person's right hand, do you think you could accurately estimate his or her left handspan? Figure 5.11 displays the relationship between the right and left handspans (in centimeters) of the 190 college students in the dataset of Chapter 2. In the plot, the points nearly fall into a straight line. The correlation coefficient for this strong positive association is +0.95. ▪

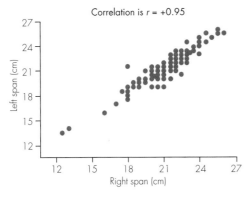

**Figure 5.11** ▮ Right handspan and left handspan

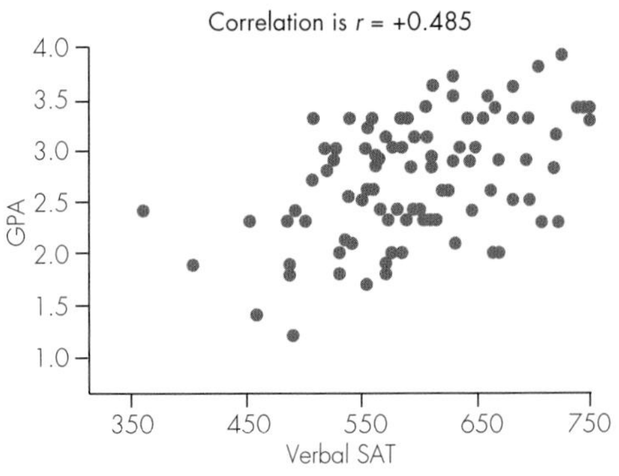

**Figure 5.12** ▮ Verbal SAT and grade point average

**Example 5.12**

**Statistics Now™**

Watch a video example at **http://1pass.thomson.com** or on your CD.

**Verbal SAT and GPA** The scatterplot in Figure 5.12 shows the grade point averages (GPAs) and verbal SAT scores for a sample of 100 students at a university in the northeastern United States. The correlation for the data in the scatterplot is $r = 0.485$, a value that indicates only a moderately strong relationship. ■

**Example 5.13**

**Age and Hours of Television Watching per Day** On a typical day, how many hours do you spend watching television? The National Opinion Research Center asks this question in its General Social Survey. In Figure 5.13, we see the relationship between respondent age and hours of daily television viewing for 898 respondents in the 2002 survey. There does not seem to be much of a relationship between age and television hours, and the correlation of only $+0.15$ confirms this weak connection between the variables. We also see some odd responses. A few respondents claim to watch television more than 20 hours per day! ■

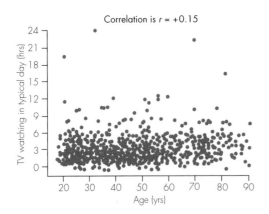

**Figure 5.13** ▮ Age and hours of watching television per day

*(Source: http://sda.berkeley.edu:7502/archive.htm and GSS-02 dataset on the CD for this book.)*

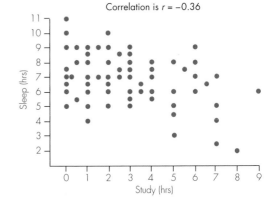

**Figure 5.14** ▮ Hours of study and hours of sleep

*(Source: Class data collected by one of the authors.)*

**Example 5.14**

**Hours of Sleep and Hours of Study** Figure 5.14 displays, for a sample of 116 college students, the relationship between the reported hours of sleep during the previous 24 hours and the reported hours of study during the same period. The correlation value for this scatterplot is $r = -0.36$, indicating a negative association that is not particularly strong. On average, the hours of sleep decrease as hours of study increase, but there is substantial variation in the hours of sleep for any specific hours of study. ■

### Interpreting the Squared Correlation, $r^2$

The squared value of the correlation is commonly used to describe the strength of a linear relationship. A **squared correlation, $r^2$,** always has a value between 0 and 1, although some computer programs will express its value as a percent between 0 and 100%. By squaring the correlation, we retain information about the strength of the relationship, but we lose information about the direction.

The phrase "**proportion of variation explained by $x$**" is sometimes used in conjunction with the squared correlation, $r^2$. For example, if a correlation has the value $r = 0.5$, the squared correlation is $r^2 = (0.5)^2 = 0.25$, or 25%, and a researcher may write that the explanatory variable explains 25% of the variation among observed values of the response variable. This interpretation stems from the use of the least squares line as a prediction tool.

Let's calculate and interpret $r^2$ for three of the examples given previously in this section.

- The correlation between height and stretched right handspan is $r = 0.74$ (Example 5.9, p. 166). The squared correlation is $r^2 = (0.74)^2 = 0.55$, or 55%. Height explains 55% of the variation among observed stretched right handspan values.

- For verbal SAT and college GPA (Example 5.11, p. 167), the correlation between the two variables is $r = 0.485$, so $r^2 = (0.485)^2 = 0.235$, or 23.5%. Verbal SAT scores explain 23.5% of the variation among observed GPAs.

- In Example 5.13 (p. 168), the correlation between television watching hours and age is only $r = 0.15$. The squared correlation is $r^2 = (0.15)^2 = 0.0225$, or 2.25%. Age explains just 2.25% of the variation among observed amounts of television watching. As we can see from Figure 5.13, knowing a person's age does not help us much in predicting how much television the person watches per day.

### Formula for $r^2$

The regression equation for the height and handspan example is Handspan = $-3 + 0.35$ (Height). We could use this equation to predict the handspan of any individual in the dataset. For instance, the predicted handspan of a student who is 70 inches tall is $-3 + 0.35(70) = 21.5$ cm.

For the 167 students in the dataset, $\bar{y}$, the average handspan, is about 20.8 centimeters. If we ignore the information given by the regression line, we could use this overall average to predict the handspan for any individual, regardless of his or her height. Our prediction "equation" is simply Handspan = 20.8.

For both this equation and the regression equation involving height, we can compute the sum of squared differences between the actual handspan values and the predicted values.

- The sum of squared differences between observed $y$ values and the sample mean $\bar{y}$ is called the *total variation* in $y$ or **sum of squares total** and is denoted by **SSTO.**
- The sum of squared differences between observed $y$ values and the predicted values based on the regression line is called the sum of squared errors and is denoted by *SSE.*

The squared correlation, $r^2$, can be calculated using SSTO and SSE as

$$r^2 = \frac{\text{SSTO} - \text{SSE}}{\text{SSTO}}$$

It can be shown (using algebra) that this quantity is exactly equal to the squared value of the correlation coefficient.

---

**MINITAB *tip***    **Finding the Correlation**

To calculate a correlation coefficient, use **Stat>BasicStatistics>Correlation.** Specify two or more columns as Variables.

---

## Reading Computer Results for Regression

Many statistical computer packages are available that will do all of the regression calculations for you. The box below illustrates the basic results of using the statistical package Minitab for the data in Figure 5.11 (p. 168). The explanatory variable is right handspan, and the response variable is left handspan.

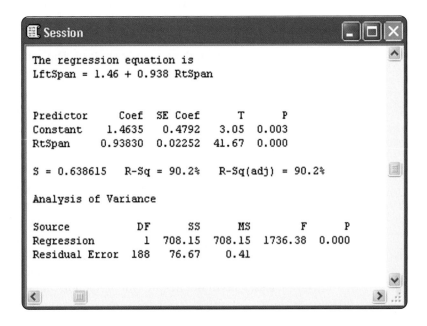

```
Session                                              _ □ X

The regression equation is
LftSpan = 1.46 + 0.938 RtSpan


Predictor      Coef    SE Coef        T        P
Constant     1.4635     0.4792     3.05    0.003
RtSpan      0.93830    0.02252    41.67    0.000

S = 0.638615   R-Sq = 90.2%   R-Sq(adj) = 90.2%

Analysis of Variance

Source           DF      SS       MS        F        P
Regression        1   708.15   708.15  1736.38    0.000
Residual Error  188    76.67     0.41
```

# Key Terms

**Section 5.1**

scatterplot, 152
explanatory variable, 152
response variable, 152
dependent variable, 152
y variable, 152
x variable, 152
positive association, 152, 153
linear relationship, 153
negative association, 153
nonlinear relationship, 154
curvilinear relationship, 154
outliers in regression, 156

**Section 5.2**

regression analysis, 157
regression equation, 157

prediction, 157
regression line, 158
simple linear regression, 158
slope of a straight line, 159, 160
intercept of a straight line, 159, 161
deterministic relationship, 160
statistical relationship, 160
predicted y, 160
estimated y, 160
predicted value, 162
prediction error, 162
residual, 162
least squares, 163
least squares line, 163
sum of squared errors (SSE), 163, 170

**Section 5.3**

correlation, 165–166
Pearson product moment correlation, 165
correlation coefficient, 165
squared correlation ($r^2$), 169
proportion of variation explained by x, 169
sum of squares total (SSTO), 170

**Section 5.4**

extrapolation, 171
influential observations, 172

**Section 5.5**

causation versus correlation, 176–177

# Exercises

●    Denotes basic skills exercises

◆    Denotes dataset is available in StatisticsNow at **http://1pass.thomson.com** or on your CD but is **not required** to solve the exercise.

**Bold-numbered exercises** have answers in the back of the text and fully worked solutions in the Student Solutions Manual.

**Statistics⊖Now™**   Go to the StatisticsNow website at **http://1pass.thomson.com** to:
• Assess your understanding of this chapter
• Check your readiness for an exam by taking the Pre-Test quiz and exploring the resources in the Personalized Learning Plan

## Section 5.1

**5.1** ● For each of the following pairs of variables, is there likely to be a positive association, a negative association, or no association? Briefly explain your reasoning.
  **a.** Amount of alcohol consumed and performance on a test of coordination.
  **b.** Height and grade point average for college students.
  **c.** Miles of running per week and time for a 5-kilometer run.
  **d.** Forearm length and foot length.

**5.2** ● ◆ The figure for this exercise is a scatterplot of $y =$ average math SAT score in 1998 versus $x =$ percent of graduating seniors who took the test that year for the 50 states and the District of Columbia. The data are from the **sats98** dataset on the CD for this book.

**Figure for Exercise 5.2**

  **a.** Does the plot show a positive association, a negative association, or no association between the two variables? Explain.
  **b.** Explain whether you think the pattern of the plot is linear or curvilinear.
  **c.** About what was the highest average math SAT for the 50 states and District of Columbia? Approximately, what percent of graduates took the test in that state?
  **d.** About what was the lowest average math SAT for the 50 states and District of Columbia? Approximately what percent of graduates took the test in that state?

●   Basic skills     ◆   Dataset available but not required     **Bold-numbered** exercises answered in the back

**5.3** ● Identify whether a scatterplot would or would not be an appropriate visual summary of the relationship between the following variables. In each case, explain your reasoning.

**a.** Blood pressure and age.

**b.** Region of country and opinion about stronger gun control laws.

**c.** Verbal SAT score and math SAT score.

**d.** Handspan and gender (male or female).

**5.4** ● ◆ The figure for this exercise is a scatterplot of $y$ = head circumference (centimeters) versus $x$ = height (inches) for the 30 females in the **physical** dataset on the CD for this book.

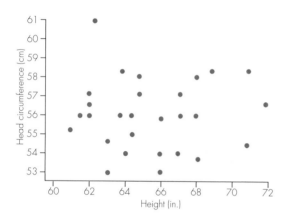

**a.** Does the plot show a positive association, a negative association, or no association between the two variables? Explain.

**b.** One data point appears to be an outlier. What are the approximate values of height and head circumference for that point?

**5.5** ● ◆ The figure for this exercise is a scatterplot of $y$ = pulse rate after marching in place for 1 minute versus $x$ = resting pulse rate measure before marching in place. (The data are in the **pulsemarch** dataset on the CD for this book.)

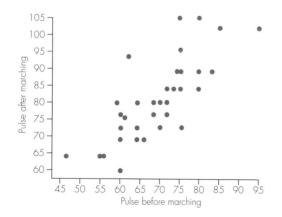

**a.** Does the plot show a positive association, a negative association, or no association between the two variables? Explain.

**b.** Explain whether you think the pattern of the plot is linear or curvilinear.

**c.** Explain whether there are any obvious outliers. If there are outliers, describe where they are located on the plot.

**5.6** ◆ The data in the following table are the geographic latitudes and the average August temperatures (Fahrenheit) for 20 cities in the United States. The cities are listed in geographic order from south to north. (These data are part of the **temperature** dataset on the CD for this book.)

**Geographic Latitude and Mean August Temperature**

| City | Latitude | Aug Temp |
|---|---|---|
| Miami FL | 26 | 83 |
| Houston TX | 30 | 82 |
| Mobile AL | 31 | 82 |
| Phoenix AZ | 33 | 92 |
| Dallas TX | 33 | 85 |
| Los Angeles CA | 34 | 75 |
| Memphis TN | 35 | 81 |
| Norfolk VA | 37 | 77 |
| San Francisco CA | 38 | 64 |
| Baltimore MD | 39 | 76 |
| Kansas City MO | 39 | 76 |
| Washington DC | 39 | 74 |
| Pittsburgh PA | 40 | 71 |
| Cleveland OH | 41 | 70 |
| New York NY | 41 | 76 |
| Boston MA | 42 | 72 |
| Syracuse NY | 43 | 68 |
| Minneapolis MN | 45 | 71 |
| Portland OR | 46 | 69 |
| Duluth MN | 47 | 64 |

*Data source: The World Almanac and Book of Facts, 1999, p. 220 and p. 456. Reprinted by permission.*

**a.** Draw a scatterplot of $y$ = August temperature versus $x$ = latitude.

**b.** Is the pattern linear or curvilinear? What is the direction of the association?

**c.** Are there any outliers? If so, which cities are outliers?

**5.7** The following table shows the relationship between the speed of a car (mph) and the average stopping distance (feet) after the brakes are applied:

| Speed (mph) | 0 | 10 | 20 | 30 | 40 | 50 | 60 | 70 |
|---|---|---|---|---|---|---|---|---|
| Distance (ft) | 0 | 20 | 50 | 95 | 150 | 220 | 300 | 400 |

*Source: Defensive Driving: Managing Time and Space, American Automobile Association, Pamphlet #3389, 1991.*

**a.** In the relationship between these two variables, which is the response variable ($y$) and which is the explanatory variable ($x$)?

**b.** Draw a scatterplot of the data. Characterize the relationship between stopping distance and speed.

5.8 ◆ The following table shows sex, height (inches), and mid-parent height (inches) for a sample of 18 college students. The variable mid-parent height is the average of mother's height and father's height. (These data are in the dataset **UCDchap5** on the CD for this book; they are sampled from the larger dataset **UCDavis2**.)

**Sex, Height, and Mid-Parent Height
for 18 College Students**

| Sex | Height | Mid-Parent Height |
| --- | --- | --- |
| M | 71 | 64.0 |
| F | 60 | 63.5 |
| F | 66 | 67.0 |
| M | 70 | 64.5 |
| F | 65 | 65.5 |
| F | 66 | 69.5 |
| M | 74 | 72.5 |
| F | 67 | 67.5 |
| F | 63 | 65.5 |
| M | 67 | 64.0 |
| F | 69 | 70.0 |
| M | 65 | 63.0 |
| M | 72 | 69.0 |
| M | 68 | 67.0 |
| F | 63 | 63.0 |
| F | 61 | 63.0 |
| M | 74 | 69.5 |
| F | 65 | 67.5 |

**a.** In the relationship between height and mid-parent height, which variable is the response variable ($y$) and which is the explanatory variable ($x$)?

**b.** Draw a scatterplot of the data for the $y$ and $x$ variables defined in part (a). Use different symbols for males and females.

**c.** Briefly interpret the scatterplot. Does the association appear to be linear? What are the differences between the males and females? Which points, if any, are outliers?

**d.** Calculate the difference between height and mid-parent height for each student, and draw a scatterplot of $y$ = difference versus $x$ = mid-parent height. Use different symbols for males and females. What does this graph reveal about the connection between height and mid-parent height?

## Section 5.2

5.9 ● Suppose that a regression equation for the relationship between $y$ = weight (pounds) and $x$ = height (inches) for men aged 18 to 29 years old is

$$\text{Average weight} = -250 + 6 \text{ (Height)}$$

**a.** Estimate the average weight for men in this age group who are 70 inches tall.

**b.** What is the slope of the regression line for average weight and height? Write a sentence that interprets this slope in terms of how much average weight changes when height is increased by one inch.

5.10 ● Refer to Exercise 5.9, in which a regression equation is given that relates average weight and height for men in the 18- to 29-year-old age group.

**a.** Suppose a man in this age group is 72 inches tall. Use the regression equation given in the previous exercise to predict the weight of this man.

**b.** Suppose this man, who is 72 inches tall, weighs 190 pounds. Calculate the residual (prediction error) for this individual.

5.11 ● Refer to the scatterplot for Exercise 5.2 showing the relationship between the average math SAT score and the percentage of high school graduates taking the test for the 50 states and District of Columbia. The regression line for these data is

$$\text{Average math} = 575 - 1.11 \text{ (Percent took)}$$

**a.** The slope of the equation is $-1.11$. Interpret this value in the context of how average math SAT changes when the percent of graduates taking the test changes.

**b.** In Missouri, only 8% of graduates take the SAT test. What is the predicted average math SAT score for Missouri?

**c.** In 1998, the average math SAT score for Missouri was 573. What is the residual (prediction error) for Missouri?

5.12 ● The equation for converting a temperature from $x$ = degrees Celsius to $y$ = degrees Fahrenheit is $y = 32 + 1.8x$. Does this equation describe a statistical relationship or a deterministic relationship? Briefly explain your answer.

5.13 ◆ The average August temperatures ($y$) and geographic latitudes ($x$) of 20 cities in the United States were given in the table for Exercise 5.6. The regression equation for these data is

$$\hat{y} = 113.6 - 1.01x$$

**a.** What is the slope of the line? Interpret the slope in terms of how mean August temperature is affected by a change in latitude.

**b.** Estimate the mean August temperature for a city with latitude of 32.

**c.** San Francisco has a mean August temperature of 64, and its latitude is 38. Use the regression equation to estimate the mean August temperature in San Francisco, and then calculate the prediction error (residual) for San Francisco.

5.14 A regression equation for $y$ = handspan (cm) and $x$ = height (in.) was discussed in Section 5.2. If the roles of the variables are reversed and only women are considered, the regression equation is Average height = 51.1 + 0.7 (Handspan).

**a.** Interpret the slope of 0.7 in terms of how height changes as handspan increases.

**b.** What is the estimated average height of women with a handspan of 20 centimeters?

**c.** Molly has a handspan of 20 centimeters and is

66.5 inches tall. What is the prediction error (residual) for Molly?

**5.15** Imagine a regression line that relates $y$ = average systolic blood pressure to $x$ = age. The average blood pressure for people 30 years old is 120, while for those 50 years old the average is 130.

a. What is the slope of the regression line?

b. What is the estimated average systolic blood pressure for people who are 34 years old?

**5.16** Iman (1994) reports that for professional golfers, a regression equation relating $x$ = putting distance (in feet) and $y$ = success rate (in percent) based on observations of distances ranging from 5 feet to 15 feet is

$$\text{Success rate} = 76.5 - 3.95 \,(\text{Distance})$$

a. What percentage of success would you expect for these professional golfers if the putting distance was 10 feet?

b. Explain what the slope of $-3.95$ means in terms of how success changes with distance.

**5.17** The figure for Exercise 5.5 is a scatterplot of pulse rate after marching in place for 1 minute ($y$) versus resting pulse rate measured before marching ($x$) for $n = 63$ individuals. The regression equation for these data is

$$\text{Pulse after marching} = 17.8 + 0.894 \,(\text{Resting pulse})$$

a. What is the slope of this equation? Write a sentence that interprets this slope in the context of this situation.

b. Predict the pulse rate after marching for somebody with a resting pulse rate of 50 beats per minute.

c. Predict the pulse rate after marching for somebody with a resting pulse rate of 90 beats per minute.

d. Use the results of parts (b) and (c) to draw the regression line. Clearly label the axes of your graph.

**5.18** Refer to Exercise 5.17.

a. Predict the pulse rate after marching for somebody with a resting pulse rate of 70.

b. Suppose the pulse rate after marching is 76 for somebody whose resting pulse rate is 70. What is the residual (prediction error) for this individual?

**5.19** The data for this exercise are as follows:

| $x$ | 1 | 2 | 3 | 4 |
|---|---|---|---|---|
| $y$ | 4 | 10 | 14 | 16 |

a. Determine the sum of squared errors (SSE) for each of the following two lines:

$$\text{Line 1:} \quad \hat{y} = 3 + 3x$$
$$\text{Line 2:} \quad \hat{y} = 1 + 4x$$

b. By the least squares criterion, which of the two lines is better for these data? Why is it better?

**5.20** The *least squares* regression equation for the data in the following table is $\hat{y} = 5 + 2x$.

| $x$ | 4 | 4 | 7 | 10 | 10 |
|---|---|---|---|---|---|
| $y$ | 15 | 11 | 19 | 21 | 29 |

a. Calculate the value of $\hat{y}$ for each data point.

b. Calculate the sum of squared errors for this equation.

## Section 5.3

**5.21** ● Which of the numbers 0, 0.25, $-1.7$, $-0.5$, 2.5 could not be values of a correlation coefficient? In each case, explain why.

**5.22** ● For $n = 188$ students, the correlation between $y$ = fastest speed ever driven and $x$ = number randomly picked between 1 and 10 is about $r = 0$. Describe what this correlation indicates about the association between the maximum speed driven and picking a number between 1 and 10.

**5.23** ● ◆ For 19 female bears, the correlation between $x$ = length of the bear (inches) and $y$ = chest girth (inches) is $r = +0.82$. (*Data source:* **bears-female** dataset on the CD for this book).

a. Describe how chest girth will change when length is increased.

b. Assuming that there are no outliers and the relationship is linear, explain what the correlation indicates about the strength of the relationship.

c. If the measurements were made in centimeters rather than inches, what would be the value of the correlation coefficient?

**5.24** ● Suppose the value of $r^2$ is 100% for the relationship between two variables.

a. What is indicated about the strength of the relationship?

b. What are the two possible values for the correlation coefficient for the two variables?

**5.25** ● Which implies a stronger linear relationship: a correlation of $+0.4$ or a correlation of $-0.6$? Briefly explain.

**5.26** ● In Figure 5.11 (p. 168), we observed that the correlation between the left and right handspans of college students was 0.95. The handspans were measured in centimeters. What would be the correlation if the handspans were converted to inches? Explain.

**5.27** Explain how two variables can have a perfect curved relationship yet have zero correlation. Draw a picture of a set of data meeting those criteria.

**5.28** Sketch a scatterplot showing data for which the correlation is $r = -1$.

**5.29** The figure for this exercise (see the next page) shows four graphs. Assume that all four graphs have the same numerical scales for the two axes.

a. Which graph shows the strongest relationship between the two variables? Which graph shows the weakest?

b. In scrambled order, correlation values for these four graphs are $-0.9$, 0, $+0.3$, $+0.6$. Match these correlation values to the graphs.

**5.30** In the 1996 General Social Survey, the correlation between respondent age and hours of daily television viewing for $n = 1913$ respondents was $r = +0.12$. Using this value, characterize the nature of the relationship between age and hours of television watching.

---

● Basic skills     ◆ Dataset available but not required     **Bold-numbered** exercises answered in the back

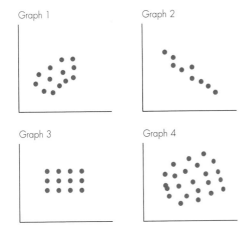

**Figure for Exercise 5.29**

**5.31** For each pair of variables, identify whether the pair is likely to have a positive correlation, a negative correlation, or no correlation. Briefly indicate your reasoning.
   a. Hours of television watched per day and grade point average for college students.
   b. Number of liquor stores and number of ministers in Pennsylvania cities.
   c. Performance on a strength test and age for people between 40 and 80 years old.
   d. Verbal skills and age for children under 12 years old.
   e. Height of husband and height of wife.

**5.32** ◆ Refer to Exercise 5.8 and the table for Exercise 5.8 in which heights and mid-parent heights are given for 18 college students (*Data source:* **UCDchap5** dataset on the CD for this book). Draw a scatterplot for the data, using different symbols for males and females as instructed in part (b) of Exercise 5.8. Based on the scatterplot, would you say that the correlation between height and mid-parent is higher for the females in the sample or for the males? Or are the correlation values about the same for males and females? Explain your reasoning.

**5.33** For Example 5.2, about highway sign reading distance and driver age, the correlation is $r = -0.8$. Calculate $r^2$, and write a sentence that interprets this value.

**5.34** ◆ The correlation between height and weight is $r = 0.40$ for twelfth-grade male respondents ($n = 1501$) in a survey done in 2003 by the U.S. Centers for Disease Control as part of the Youth Risk Behavior Surveillance System. (The raw data are in the dataset **YouthRisk03** on the CD for this book.)
   a. Calculate $r^2$, and write a sentence that interprets this value in the context of this situation.
   b. Heights and weights were recorded in meters and kilograms, respectively. What would be the value of the correlation if the measurements had instead been made in inches and pounds?

**5.35** Calculate $r^2$ for Example 5.9 in this chapter (about hours of sleep and hours of study) in which the correlation is $-0.36$. Write a sentence that interprets this value.

**5.36** In a regression analysis, the total sum of squares (SSTO) is 800, and the error sum of squares (SSE) is 200. What is the value for $r^2$?

**5.37** Suppose you know that the slope of a regression line is $b_1 = +3.5$. Based on this value, explain what you know and do not know about the strength and direction of the relationship between the two variables.

## Section 5.4

**5.38** ● An article in the *Sacramento Bee* (29 May 1998, p. A17) noted "Americans are just too fat, researchers say, with 54 percent of all adults heavier than is healthy. If the trend continues, experts say that within a few generations virtually every U.S. adult will be overweight." This prediction is based on extrapolation, which assumes that the current rate of increase will continue indefinitely. Is that a reasonable assumption? Do you agree with the prediction? Explain.

**5.39** ● ◆ The **physical** dataset on the CD for this book gives heights (inches) and head circumferences (cm) for a sample of college students. For females only, the correlation between the two variables is 0.05, while for males only, the correlation is 0.19. For the combined sample of males and females, however, the correlation is 0.42. Explain why the correlation in the combined sample is higher than the correlations in the separate samples of males and females. Refer to Example 5.11 for guidance.

**5.40** Refer back to Exercise 5.7 about stopping distance and vehicle speed. The least squares line for these data is

$$\text{Average distance} = -44.2 + 5.7 \text{ (Speed)}$$

   a. Use this equation to estimate the average stopping distance when the speed is 80 miles per hour. Do you think this is an accurate estimate? Explain.
   b. Draw a scatterplot of the data, as instructed in Exercise 5.7(b). Use the scatterplot to estimate the average stopping distance for a speed of 80 mph.
   c. Do you think the data on stopping distance and vehicle speed shown in Exercise 5.7 describe the relationship between these two variables for all situations? What are some other variables that should be considered when the relationship between stopping distance and vehicle speed is analyzed?

**5.41** When a correlation value is reported in research journals, there often is not an accompanying scatterplot. Explain why reported correlation values should be supported with either a scatterplot or a description of the scatterplot.

**5.42** A memorization test is given to ten women and ten men. The researchers find a negative correlation between scores on the test and height. Explain which of the reasons listed at the beginning of Section 5.4 for misleading correlations might explain this finding. Sketch a scatterplot for the relationship between the variables that is consistent with your explanation.

**5.43** Sketch a scatterplot in which the presence of an outlier decreases the observed correlation between the response and explanatory variables.

● Basic skills    ◆ Dataset available but not required    **Bold-numbered** exercises answered in the back

**5.44** Give an example of a prediction that is an extrapolation. Do not give an example that is already in this chapter.

**5.45** ◆ The data in the following table come from the time when the United States still had a maximum speed limit of 55 miles per hour. An issue of some concern at that time was whether lower speed limits reduce the highway death rate. (These data are called **speedlimit** on the CD for this book.)

**Highway Death Rates and Speed Limits**

| Country | Death Rate (per 100 million veh. miles) | Speed Limit (in miles per hour) |
|---------|------------------------------------------|----------------------------------|
| Norway | 3.0 | 55 |
| United States | 3.3 | 55 |
| Finland | 3.4 | 55 |
| Britain | 3.5 | 70 |
| Denmark | 4.1 | 55 |
| Canada | 4.3 | 60 |
| Japan | 4.7 | 55 |
| Australia | 4.9 | 60 |
| Netherlands | 5.1 | 60 |
| Italy | 6.1 | 75 |

Source: D. J. Rivkin, "Fifty-five mph speed limit is no safety guarantee," *New York Times* (letters to the editor), Nov. 25, 1986, p. 26.

a. In the relationship between death rate and speed limit, which variable is the response variable and which is the explanatory variable?
b. Plot the data in the table, and discuss the result. Does there appear to be an association? Are there any outliers? If so, what is their influence on the correlation?

**5.46** The table for Exercise 5.6 gave the average August temperature ($y$) and geographic latitude ($x$) for 20 cities in the United States. Exercise 5.13 gave the information that the regression equation relating these two variables is

$$\hat{y} = 113.6 - 1.01x$$

a. The latitude at the equator is 0. Using the regression equation, estimate the average August temperature at the equator.
b. Explain why we should not use this equation to estimate August temperature at the equator.

**5.47** In Exercise 5.16, a regression equation relating $x =$ putting distance (feet) to $y =$ success rate (in percent) for professional golfers was given as

$$\text{Success rate} = 76.5 - 3.95 \text{ (Distance)}$$

The equation was based on observations of distances ranging from 5 feet to 15 feet.
a. Use the equation to predict success rate for a distance of 2 feet and for a distance of 20 feet.
b. The original data included values beyond those used to determine the regression equation (5 feet to 15 feet). At a distance of 2 feet, the observed success rate was 93.3%, and at a distance of 20 feet, 15.8% of observed putts were successful. Compare your results in part (a) to the observed success rates for distances of 2 feet and 20 feet. Utilize your results from part (a) to explain why it is not a good idea to use a regression equation to predict information beyond the range of values used to determine the equation.
c. Draw a graph of what you think the relationship between putting distance and success rate would look like for the entire range from 2 feet to 20 feet.

## Section 5.5

**5.48** ● Explain why a strong correlation would be found between weekly sales of firewood and weekly sales of cough drops over a 1-year period.

**5.49** ● ◆ The **pennstate2** dataset on the CD for this book includes heights and the total number of ear pierces for each person in a sample of college students. The correlation between the two variables is $-0.495$. What third variable may explain this observed correlation? Explain how that third variable could create the negative correlation.

**5.50** ● Based on the data for the past 50 years in the United States, there is a strong correlation between yearly beer sales and yearly per capita income. Would you interpret this to mean that increasing a person's income will cause him or her to drink more beer? Explain.

**5.51** Suppose a positive relationship had been found between each of the following sets of variables. For each set, discuss possible reasons why the connection may not be causal. Refer to the list of possible reasons for an observed association in Section 5.5.
a. Number of deaths from automobiles and soft drink sales for each year from 1950 to 2000.
b. Amount of daily walking and quality of health for men over 65 years old.
c. Number of ski accidents and average wait time for the ski lift for each day during one winter at a ski resort.

**5.52** Suppose that in an observational study, it is observed that the risk of heart disease increases as the amount of dietary fat consumed increases. Write a paragraph discussing why this result does not imply that diets high in fat cause heart disease.

**5.53** Give an example of a situation in which it would be reasonable to conclude that an explanatory variable causes changes in a response variable.

**5.54** Suppose a medical researcher finds a negative correlation between amount of weekly walking and the incidence of heart disease for people over 50 years old; in other words, people who walked more had a lower incidence of heart disease. One possible explanation for this observed association is that increased walking reduces the risk of heart disease. What are some other possible explanations?

**5.55** It is said that a higher proportion of drivers of red cars are given tickets for traffic violations than drivers of any other car color. Does this mean that if you drive a red car rather than a car of some other color, it will cause you to get more tickets for traffic violations? Explain.

**5.56** Give an example not given elsewhere in this chapter of two variables that are likely to be correlated because they are both changing over time.

**5.57** Example 2.2 (p. 20) described an observational study in which it was found that children who slept with a nightlight or in a fully lit room before the age of 2 were more likely to be nearsighted than children who slept in darkness. Does this mean that sleeping with a light on as an infant causes nearsightedness? What are some other possible explanations?

**5.58** Researchers have shown that there is a positive correlation between average fat intake and the breast cancer rate across countries. In other words, countries with higher fat intake tend to have higher breast cancer rates. Does this correlation prove that dietary fat is a contributing cause of breast cancer? Explain.

## Section 5.6: Skillbuilder Applet Exercises

**Statistics⊘Now™** Test yourself on these questions and explore the applet at **http://1pass.thomson.com** or on your CD.

*For these exercises, use the **Correlation** applet described in Section 5.6. It is on the CD for this book. In each exercise, you are asked to sketch a facsimile of a graph you create with the applet. Alternatively, you might use "Print Screen" on your keyboard to copy the screen image; then paste it to a word-processing document.*

**5.59** Using the applet, create a plot for the target correlation $r = +0.5$. Don't include any outliers. Sketch an approximate facsimile of your resulting graph.

**5.60** Using the applet, create a plot for the target correlation $r = -0.8$. Don't include any outliers. Sketch an approximate facsimile of your resulting graph.

**5.61** Using the applet, create a plot for the target correlation $r = 0$. Don't include any outliers. Sketch an approximate facsimile of your resulting graph.

**5.62** Using the applet, create a plot for the target correlation $r = +0.5$ in which one point is an outlier that decreases the correlation. Make the plot such that if the outlier were removed, the correlation for the remaining points would be greater than $r = 0.7$. Sketch an approximate facsimile of your resulting graph.

**5.63** Using the applet, create a plot for the target correlation $r = -0.8$ in which one point is an outlier that inflates the correlation. Make the plot such that if the outlier were removed, the correlation for the remaining points would be between $-0.2$ and $+0.2$. Sketch an approximate facsimile of your resulting graph. *Hint:* Start by putting points in the upper left corner of the plot.

**5.64** Using the applet, with the target correlation $r = 0$, make a plot that has a curvilinear pattern for which the correlation is 0. Sketch an approximate facsimile of your resulting graph.

## Chapter Exercises

**5.65** The regression line relating verbal SAT scores and GPA for the data exhibited in Figure 5.12 is

$$\text{Average GPA} = 0.539 + 0.00362 \,(\text{Verbal SAT})$$

a. Estimate the average GPA for those with verbal SAT scores of 600.

b. Explain what the slope of 0.00362 represents in terms of the relationship between GPA and SAT.

c. For two students whose verbal SAT scores differ by 100 points, what is the estimated difference in college GPAs?

d. Explain whether the intercept has any useful interpretation in the relationship between GPA and verbal SAT score. Keep in mind that the lowest possible verbal SAT score is 200.

**5.66** ◆ The heights (inches) and foot lengths (centimeters) of 33 college men are shown in the following table. (These data are in the dataset **heightfoot** on the CD for this book.)

**Height (inches) and Foot Length (cm) for 33 College Students**

| Student | Height | Foot Length |
|---|---|---|
| 1 | 66.5 | 27.0 |
| 2 | 73.5 | 29.0 |
| 3 | 70.0 | 25.5 |
| 4 | 71.0 | 27.9 |
| 5 | 73.0 | 27.0 |
| 6 | 71.0 | 26.0 |
| 7 | 71.0 | 29.0 |
| 8 | 69.5 | 27.0 |
| 9 | 73.0 | 29.0 |
| 10 | 71.0 | 27.0 |
| 11 | 69.0 | 29.0 |
| 12 | 69.0 | 27.2 |
| 13 | 73.0 | 29.0 |
| 14 | 75.0 | 29.0 |
| 15 | 73.0 | 27.2 |
| 16 | 72.0 | 27.5 |
| 17 | 69.0 | 25.0 |
| 18 | 68.0 | 25.0 |
| 19 | 72.5 | 28.0 |
| 20 | 78.0 | 31.5 |
| 21 | 79.0 | 30.0 |
| 22 | 71.0 | 28.0 |
| 23 | 74.0 | 29.0 |
| 24 | 66.0 | 25.5 |
| 25 | 71.0 | 26.7 |
| 26 | 71.0 | 29.0 |
| 27 | 71.0 | 28.0 |
| 28 | 84.0 | 27.0 |
| 29 | 77.0 | 29.0 |
| 30 | 72.0 | 28.0 |
| 31 | 70.0 | 26.0 |
| 32 | 76.0 | 30.0 |
| 33 | 68.0 | 27.0 |

*Data source: William Harkness.*

a. Draw a scatterplot with $y =$ foot length (cm) and $x =$ height (inches). Does the relationship appear to be linear? Are there any outliers? If so, do you think the outliers are legitimate data values?

b. Use statistical software or a calculator to calculate the correlation between height and foot length.

If heights were converted to centimeters, what would be the correlation between height and foot length?

c. If there are any outliers, remove them and recalculate the correlation. Describe how the correlation changed from part (b).

5.67 ◆ Refer to Exercise 5.66 about $y$ = foot length and $x$ = height. (*Data source:* the **heightfoot** dataset on the CD for this book.) If the person who reportedly is 84 inches tall is excluded, the regression equation for the remaining 32 men is $\hat{y} = 0.25 + 0.384x$.

a. How much does average foot length increase for each 1-inch increase in height?

b. Predict the difference in the foot lengths of men whose heights differ by 10 inches.

c. Suppose Max is 70 inches tall and has a foot length of 28.5 centimeters. On the basis of the regression equation, what is the predicted foot length for Max? What is the value of the prediction error (residual) for Max?

**5.68** Refer to Case Study 5.1, in which regression equations are given for males and females relating ideal weight to actual weight. The equations are

Women:    Ideal = 44 + 0.6 (Actual)

Men:    Ideal = 53 + 0.7 (Actual)

**a.** Predict the ideal weight for a man who weighs 140 pounds and for a woman who weighs 140 pounds. Compare the results.

**b.** Do the intercepts have logical physical interpretations in the context of this example? Explain.

**c.** Do the slopes have logical interpretations in the context of this example? Explain.

5.69 The winning time in the Olympic men's 500-meter speed skating race over the years 1924 to 1992 can be described by the regression equation

Winning time = 255 − 0.1094 (Year)

a. Is the correlation between winning time and year positive or negative? Explain.

b. In 1994, the actual winning time for the gold medal was 36.33 seconds. Use the regression equation to predict the winning time for 1994, and compare the prediction to what actually happened.

c. Explain what the slope of −0.1094 indicates in terms of how winning times change from year to year.

d. Why should we not use this regression equation to predict the winning time in the 2050 Winter Olympics?

**5.70** ◆ The following table lists the number of pages and the price for 15 books. Eight of the books are hardcover and 7 are softcover. (These data are in the dataset **ProfBooks** on the CD for this book.)

a. Draw a scatterplot of $y$ = price versus $x$ = pages. Use different symbols for hardcover and softcover books.

b. For all 15 books, determine the correlation between price and pages.

**Pages versus Price for Books,**
**Type H = hardcover,**
**S = softcover**

| Pages | Price | Type |
|-------|-------|------|
| 104 | 32.95 | H |
| 188 | 24.95 | H |
| 220 | 49.95 | H |
| 264 | 79.95 | H |
| 336 | 4.50 | S |
| 342 | 49.95 | H |
| 378 | 4.95 | S |
| 385 | 5.99 | S |
| 417 | 4.95 | S |
| 417 | 39.75 | H |
| 436 | 5.95 | S |
| 458 | 60.00 | H |
| 466 | 49.95 | H |
| 469 | 5.99 | S |
| 585 | 5.95 | S |

c. Separate the books by type. Determine the correlation between price and pages for hardcover books only. Determine the correlation between price and pages for softcover books only.

d. Which of the reasons listed in Section 5.4 for misleading correlations is illustrated in this exercise?

5.71 ◆ U.S. Census Bureau estimates of the average number of persons per household in the United States for census years between 1850 and 2000 are shown in the following table. (These data are in the file **perhouse** on the CD for this book.)

**Persons per Household**
**in the United States**

| Year | Per House |
|------|-----------|
| 1850 | 5.55 |
| 1860 | 5.28 |
| 1870 | 5.09 |
| 1880 | 5.04 |
| 1890 | 4.93 |
| 1900 | 4.76 |
| 1910 | 4.54 |
| 1920 | 4.34 |
| 1930 | 4.11 |
| 1940 | 3.67 |
| 1950 | 3.37 |
| 1960 | 3.35 |
| 1970 | 3.14 |
| 1980 | 2.76 |
| 1990 | 2.63 |
| 2000 | 2.59 |

*Data source:* The World Almanac and Book of Facts, 1999, p. 383, and U.S. Bureau of the Census.

● Basic skills    ◆ Dataset available but not required    **Bold-numbered** exercises answered in the back

a. Draw a scatterplot for the relationship between persons per household and year. Is the relationship linear or curvilinear? Is the association between persons per household and year positive or negative?

b. On your scatterplot, add a line that you believe fits the data pattern. Extend this line to the year 2010. On the basis of this line, estimate the number of persons per household in the United States in the year 2010.

**5.72** ◆ Refer to Exercise 5.71 about the trend in number of persons per household.

   **a.** Using statistical software, determine the least squares line for these data. Use the equation of this line to estimate the number of persons per household in the year 2010 (*Data source:* **perhouse** dataset on the CD for this book).

   **b.** What is the slope of the line? Interpret the slope in the context of these variables.

   **c.** Based on the regression line, what would be the predicted persons per household in the year 2200? What is the lowest possible value of the persons per household number? How does the estimate for 2200 compare to this value?

   **d.** Part (c) illustrates that the observed pattern can't possibly continue in the same manner forever. Sketch the pattern for the trend in persons per household that you think might occur between now and the year 2200.

**5.73** ◆ For a statistics class project at a large northeastern university, a student examined the relationship between

$x$ = body weight (in pounds)

$y$ = time to chug a 12-ounce beverage (in seconds)

We'll leave it to you to imagine the beverage. The student collected data from 13 individuals, and those data are in the following table. (This dataset is named **chugtime** on the CD for this book.)

**Body Weight (pounds) and Chug Time (seconds) for 13 College Students**

| Person | Weight | Chug Time |
|--------|--------|-----------|
| 1 | 153 | 5.6 |
| 2 | 169 | 6.1 |
| 3 | 178 | 3.3 |
| 4 | 198 | 3.4 |
| 5 | 128 | 8.2 |
| 6 | 183 | 3.5 |
| 7 | 177 | 6.1 |
| 8 | 210 | 3.1 |
| 9 | 243 | 4.0 |
| 10 | 208 | 3.2 |
| 11 | 157 | 6.3 |
| 12 | 163 | 6.9 |
| 13 | 158 | 6.7 |

*Data source: William Harkness.*

a. Draw a scatterplot of the measurements. Characterize the relationship between chug time and body weight.

b. The heaviest person appears to be an outlier. Do you think that observation is a legitimate observation, or do you think an error was made in recording or entering the data?

c. Outliers should not be thrown out unless there's a good reason, but there are several reasons why it may be legitimate to conduct an analysis without them (for instance, see part (e)). Delete the data point for the heaviest person, and determine a regression line for the remainder of the data.

d. Use the regression line from part (c) to estimate the chug time for an individual who weighs 250 pounds. Do you think this time could be achieved by anybody?

e. Sometimes the relationship between two variables is linear for a limited range of $x$ values and then changes to a different line or curve. Using this idea, draw a sketch that illustrates what you think the actual relationship between weight and chug time might be for the range of weights from 100 to 300 pounds.

f. Discuss plausible reasons why the heaviest person appears to be an outlier with regard to his combination of weight and chug-time measurements.

**5.74** Measure the heights and weights of ten friends of the same sex.

a. Draw a scatterplot of the data, with weight on the vertical axis and height on the horizontal axis. Draw a line onto the scatterplot that you believe describes the average pattern. On the basis of two points on this line, estimate the slope of the relationship between weight and height.

b. Using statistical software, compute the least squares line, and compare the slope to your estimated slope from part (a).

## Dataset Exercises

**Statistics⌂Now™** Datasets **are required** to solve these exercises and can be found at **http://1pass.thomson.com** or on your CD.

**5.75** Use the dataset **poverty** on the CD for this book; it includes teenage mother birth rates and poverty rates for the 50 states and the District of Columbia. The variable **PovPct** is the percent of a state's population in 2000 living in households with incomes below the federally defined poverty level. The variable **Brth15to17** is the birth rate for females 15 to 17 years old in 2002, calculated as births per 1000 persons in this age group.

a. Plot **Brth15to17** ($y$) versus **PovPct** ($x$). Describe the direction and strength of the relationship, and comment on whether there are any outliers.

b. Determine the equation of the regression line relating $y$ = **Brth15to17** to $x$ = **PovPct**. Write the equation.

c. What is the value of the slope of the equation? Write a sentence that interprets the slope in the context of these variables.

d. Based on the equation, what is the estimated birth rate for females 15 to 17 years old in a state with a poverty rate of 15%?

5.76 Use the dataset **oldfaithful** on the CD for this book; it gives data for $n = 299$ eruptions of the Old Faithful geyser. The variable **Duration** is the duration (minutes) of an eruption, and the variable **TimeNext** is the time interval (minutes) until the next eruption.

a. Plot **TimeNext** ($y$) versus **Duration** ($x$). Describe the direction and strength of the relationship, and comment on whether there are any outliers.

b. Determine the equation of the regression line relating $y$ = **TimeNext** to $x$ = **Duration.** Write the equation.

c. What is the value of the slope of the equation? Write a sentence that interprets the slope in the context of these variables.

d. Estimate the interval of time until the next eruption following one that lasts 4 minutes.

5.77 Use the dataset **cholesterol** on the CD for this book. For $n = 28$ heart attack patients, the variables **2-Day** and **4-Day** are cholesterol levels measured two days and four days, respectively, after the attacks.

a. Plot **4-Day** ($y$) versus **2-Day** ($x$). Describe the direction and strength of the relationship, and comment on whether there are any outliers.

b. Determine the equation of the regression line relating $y$ = **4-Day** to $x$ = **2-Day.** Write the equation.

c. What is the value of the slope of the equation? Write a sentence that interprets the slope in the context of these variables.

d. Separately estimate cholesterol levels four days after the attack for patients with **2-Day** values of 200, 250, and 300.

e. Utilizing the result of part (d), describe how cholesterol levels for these patients generally changed in the time from two days to four days after their heart attacks.

5.78 Use the dataset **sats98** from the CD for this book for this exercise. The variable **Verbal** contains the average scores on the verbal SAT in 1998 for the 50 states and the District of Columbia. **PctTook** is the percent of high school graduates, in each state, who took the SAT that year.

a. Make a scatterplot showing the connection between average verbal SAT and the percent of graduates who took the SAT in a state. Describe the relationship between these two variables.

b. Compute the least squares regression line for the relationship between these two variables. Write a sentence that interprets the slope of this equation in a way that could be understood by people who don't know very much about statistics.

c. Based on the appearance of the scatterplot, do you think that a straight line is an appropriate mathematical model for the connection between **Verbal** and **PctTook**? Why or why not?

d. Explain why the intercept of the equation computed in part (b) would not have a sensible interpretation for these two variables.

5.79 Use the **sats98** dataset from the CD for this book.

a. Plot the relationship between average verbal (**Verbal**) and average math (**Math**) SAT scores in the 50 states. Describe the characteristics of the relationship.

b. What states are outliers? In what specific way are they outliers?

**5.80** Use the dataset **idealwtmen** from the CD for this book. It contains data for the men used for Case Study 5.1. The variable **diff** is the difference between actual and ideal weights and was computed as **diff = actual − ideal.**

a. Plot **diff** versus **actual** (actual weight). Does the relationship appear to be linear, or is it curvilinear?

**b.** Compute the equation of the regression line for the relationship between **diff** and **actual.** Estimate the average difference for men who weigh 150 pounds. On average, do 150-pound men want to weigh more or less than they actually do?

c. Repeat part (b) for men who weigh 200 pounds.

**d.** What is the value of $r^2$ for the relationship between **diff** and **actual**?

5.81 Use the dataset **ceodata** from the CD for this book. The ages of 60 CEOs of America's best small companies were given in Exercise 2.48 of Chapter 2. The annual salaries (in thousands of dollars) for 59 of these CEOs are in the dataset along with the ages.

a. Plot **Salary** versus **Age.**

b. Compute the correlation coefficient and $r^2$.

c. Characterize the relationship between annual salary and age. What is the pattern of the relationship? How strong is the association?

5.82 Use the dataset **UCDwomht** from the CD for this book. For a sample of college women, the variable **height** is student's height (in inches), and the variable **midparent** is the average height of the student's parents (in inches) as reported by the student.

a. Compute the regression equation for predicting a student's height from the average of her parents' heights.

b. Use the regression equation to predict the height for a college woman with parents who have an average height of 68 inches.

c. Use the regression equation to predict the height of a college woman whose mother is 62 inches tall and whose father is 70 inches tall.

d. What other summaries of the data should be done to determine the strength of the relationship between **height** and **midparent** height?

5.83 Use the dataset **temperature** from the CD for this book. A portion of this dataset was presented in Exercise 5.6, in which the relationship between mean August temperature and geographic latitude was analyzed. For predicting mean April temperature (*AprTemp*), which of these two variables in the dataset is a stronger predictor: geographic latitude (*latitude*) or mean January temperature (*JanTemp*)? Support your answer with relevant statistics and plots.

# 6

Simon Wilkinson / The Image Bank / Getty Images

*Do men and women have the same motivations for exercising?*

See Example 6.3 *(p. 197)*

# Relationships Between Categorical Variables

Whenever risk statistics are reported, there is a risk that they are misreported. Journalists often present risk data in a way that produces the best story rather than in a way that provides the best information. Very commonly, news reports either don't contain or don't emphasize the information you need to understand risk.

**Statistics △ Now™**

Throughout the chapter, this icon introduces a list of resources on the StatisticsNow website at **http:// 1pass.thomson.com** that will:
- Help you evaluate your knowledge of the material
- Allow you to take an exam-prep quiz
- Provide a Personalized Learning Plan targeting resources that address areas you should study

This chapter is about the analysis of the relationship between two categorical variables, so let's begin by recalling the meaning of the term *categorical variable*. The raw data from categorical variables consist of group or category names that don't necessarily have any ordering. Eye color and hair color, for instance, are categorical variables.

We can also use the methods of this chapter to examine *ordinal* variables. Ordinal variables can be thought of as categorical variables for which the categories have a natural ordering. For example, a researcher might define categories for quantitative variables, such as age, income, or years of education.

Although there are many questions that we can and will ask about two categorical variables, the principal question that we ask in most cases is "Is there a relationship between the two variables such that the category into which individuals fall for one variable seems to depend on the category they are in for the other variable?" ▮

> **thought question 6.1** Hair color and eye color are related characteristics. What exactly does it mean to say that these two variables are related? Suppose that you know the hair colors and the eye colors of 200 individuals. How would you assess whether the two variables are related for those individuals?*

## 6.1 Displaying Relationships Between Categorical Variables

We have already encountered several examples of the type of problem we will study in this chapter. In Chapter 2, for instance, we described a study of 479 children that found that children who slept either with a nightlight or in a

*HINT: Read the sentence just before Thought Question 6.1.

fully lit room before the age of 2 had a higher incidence of myopia (near-sightedness) later in childhood. The following table was used to examine this relationship:

| Slept with: | No Myopia | Myopia | High Myopia | Total |
|---|---|---|---|---|
| **Darkness** | 155 (90%) | 15 (9%) | 2 (1%) | 172 |
| **Nightlight** | 153 (66%) | 72 (31%) | 7 (3%) | 232 |
| **Full Light** | 34 (45%) | 36 (48%) | 5 (7%) | 75 |
| **Total** | 342 (71%) | 123 (26%) | 14 (3%) | 479 |

The table displays the number of children in each combination of the categories of two categorical variables: the sleep-time lighting condition and the child's eyesight classification at the time of the study. The table also gives row percentages—the percentages of children within each row who fall into the different eyesight categories. For example, 90% (155/172) of those who slept in darkness had no myopia, but only 45% (34/75) of those who slept in full light had no myopia.

Notice that the row percentages show us that the two variables are related. From these percentages, we learn that the incidence of myopia increases when the amount of sleep-time light increases. This result does not prove that sleeping with light *causes* myopia, but we can say that for some reason, the characteristics of eyesight and sleep-time lighting are *associated* characteristics.

Displays like the table above are often called **contingency tables** because they cover all contingencies for the combinations of the two variables. Because the categories of two variables are used to create the table, a contingency table is also referred to as a **two-way table.** Each row and column combination of the table is called a **cell.**

The first step in analyzing the relationship between two categorical variables is to count how many observations fall into each cell of the contingency table. It's difficult, however, to look at a table of counts and make useful judgments about a relationship. Usually, we need to consider the **conditional percentages** within either the rows or the columns of the table.

There are two types of conditional percentages that we can compute for a contingency table. **Row percentages,** like the percentages in the myopia table, are the percentages across a row of a contingency table. These percentages are based on the total number of observations in the row. **Column percentages** are the percentages down a column of a contingency table. These percentages are based on the total number of observations in the column.

In some cases, one variable can be designated as the *explanatory variable* and the other variable as the *response variable.* As we learned in Chapter 2, in these situations it is customary to define the rows using the categories of the explanatory variable and the columns using the categories of the response variable. When this is done, the row percentages can be used to examine the relationship because they tell us what percentage *responded* in each possible way. We can then see whether individuals responded equivalently for each category of the explanatory variable. For instance, we can see whether the myopia response was the same for different lighting conditions.

**Example 6.1**

Statistics⟨△⟩Now™

Watch a video example at **http://
1pass.thomson.com** or on your CD.

**Smoking and the Risk of Divorce**    Table 6.1 displays data on smoking habits and divorce history for the 1669 respondents who had ever been married in the 1991 and 1993 General Social Surveys done by the National Opinion Research Center at the University of Chicago.

**Table 6.1**  Smoking and Divorce,
GSS Surveys 1991–1993

|  | **Ever Divorced?** | | |
|---|---|---|---|
| **Smoke?** | *Yes* | *No* | *Total* |
| **Yes** | 238 | 247 | 485 |
| **No** | 374 | 810 | 1184 |
| **Total** | 612 | 1057 | 1669 |

*Data source: SDA archive at UC Berkeley website
(www.csa.berkeley.edu:7502/).*

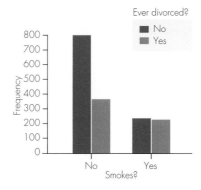

**Figure 6.1**  ❚ Frequency counts for smoking habits and divorce

   The frequency counts in the cells of this table are visually summarized in Figure 6.1, using a bar graph as described in Chapter 2. In the bar graph, we see evidence of an association between smoking habits and the likelihood of divorce because the set of bars for smokers has a different pattern from the set of bars for nonsmokers. Among smokers, the number who have ever been divorced is about equal to the number who have not been divorced. Among nonsmokers, the proportion who have ever been divorced is a minority.

   We can use row percentages to compare the divorce rates of smokers and nonsmokers. The first row of the table gives data for smokers. Among the 485 respondents who smoke, 49% (238/485) have been divorced, and 51% (247/485) have not been divorced. The second row of the table gives data for nonsmokers. Among the 1184 nonsmokers, only 32% (374/1184) have been divorced, and 68% (810/1184) have not. The difference between the two sets of row percentages indicates a relationship. Smokers are more likely to have been divorced.  ■

   In the next example, there is not an obvious way to designate one variable as an explanatory variable and the other as a response variable. In such situations, it may be interesting to describe the data using both row percentages and column percentages.

**Example 6.2**

**Table 6.2** Ear Pierces and Tattoos for Men (*n* = 1014)

| Ear Pierces | No Tattoo | Tattoo | Total |
|---|---|---|---|
| 0 | 699 | 68 | 767 |
| 1 | 87 | 27 | 114 |
| 2 or more | 87 | 46 | 133 |
| Total | 873 | 141 | 1014 |

**Tattoos and Ear Pierces**  From 1996 through 2002, students in several statistics classes at a large university in the Northeast provided responses to these two questions:

1. Do you have a tattoo?
2. How many total ear pierces do you have?

The responses from 1014 men are displayed in Table 6.2. In the table, the very few men who have more than two ear pierces are lumped with the men who have two ear pierces.

For this table, we could ask questions about the conditional percentages in either direction because there is no clear explanatory and response variable distinction. For instance, does the likelihood that a man has a tattoo differ on the basis of whether or not he has an ear pierce? Or we could ask whether men with tattoos are more likely to have an ear pierce. Let's consider the row percentages:

- Among men with no ear pierces, 68/767 = 8.9% have a tattoo.
- Among men with one ear pierce, 27/114 = 23.7% have a tattoo.
- Among men with two or more ear pierces, 46/133 = 34.6% have a tattoo.

Clearly, the percentage with a tattoo increases as the number of ear pierces increases, so the two characteristics are related. We also see the relationship when we consider the column percentages. Those percentages are shown in the bar chart in Figure 6.2. Men with tattoos are more likely to have an ear pierce.

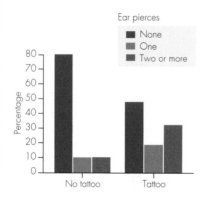

**Figure 6.2**  ▌ Column percentages for the ear-pierce and tattoo data

Percentages based on the total sample can also be interesting because they provide useful descriptions of the overall sample. For instance, 699/1014 = 68.9% of the men in the sample have neither a tattoo nor an ear pierce, while 46/1014 = 4.5% have a tattoo and also have at least two ear pierces. The percentage of the sample with a tattoo is 141/1014 = 13.9%, and the percentage with at least one ear pierce is (114 + 133)/1014 = 24.4%.  ▪

The next example describes a situation in which there might not be a relationship between the two categorical variables that form the two-way table. The distributions of row percentages are nearly the same for the two rows of the two-way table.

**Example 6.3**

**Gender and Reasons for Taking Care of Your Body** In a 1997 poll conducted by the *Los Angeles Times*, 1218 southern California residents were surveyed about their health and fitness habits. The respondents were selected using random-digit dialing methods. One of the questions was "What is the most important reason why you try to take care of your body: Is it mostly because you want to be attractive to others, or mostly because you want to keep healthy, or mostly because it helps your self-confidence, or what?"

The percent distribution of the responses is shown for men and for women in Table 6.3. Notice that the pattern of responses is more or less the same for men and women. It seems reasonable to conclude that the response to the question was not related to the gender of the respondent.

**Table 6.3** Gender and Reasons for Taking Care of Body

|  | Healthy | Self-Confidence | Attractive | Don't Know | Total |
|---|---|---|---|---|---|
| **Men** | 76% | 16% | 7% | 1% | 100% |
| **Women** | 74% | 20% | 4% | 2% | 100% |

6.1 Exercises are on page 217.    *Source:* www.latimes.com, *poll archives, study #401.*

---

***thought question 6.2***  In Example 6.1, the risk of divorce was associated with smoking habits. Were the data collected in an observational study or in an experiment? Do you think cigarette smoking may cause an increase in the divorce rate? Do you think getting divorced may cause a person to start smoking? How would you explain the association between smoking habits and the likelihood of divorce?*

---

*in summary*    ## Row Percentages, Column Percentages, and Relationships in Two-Way Tables

- A **row percentage** uses a total row count as the basis for computing a percentage. It is the percentage of the observations within a particular row category that are in a specified category of the column variable.

- A **column percentage** uses a total column count as the basis for computing a percentage. It is the percentage of observations within a particular column category that are in a specified category of the row variable.

- There is a **relationship** between the categorical variables that form a two-way table if two or more rows have different distributions of row percentages. Equivalently, there is a relationship if two or more columns have different distributions of column percentages.

---

***HINT:*** Do you think participants were randomly assigned to smoke or not? Think about possible confounding variables.

# 6.2 Risk, Relative Risk, and Misleading Statistics About Risk

When a particular outcome is undesirable, researchers and journalists may describe the *risk* of that outcome. The **risk** that a randomly selected individual within a group falls into the undesirable category is simply the proportion in that category.

$$\text{Risk} = \frac{\text{Number in category}}{\text{Total number in group}}$$

It is commonplace to express risk as a percentage rather than as a proportion. Suppose, for instance, that within a group of 200 individuals, asthma affects 24 people. In this group, the *risk* of asthma is $24/200 = 0.12$, or 12%.

## Relative Risk

We often want to know how the risk of an outcome relates to an explanatory variable. One statistic that is used for this purpose is **relative risk,** which is the ratio of the risks in two different categories of an explanatory variable.

$$\text{Relative risk} = \frac{\text{Risk in category 1}}{\text{Risk in category 2}}$$

Relative risk describes the risk in one group as a multiple of the risk in another group. For example, suppose that a researcher states that for those who drive while under the influence of alcohol, the relative risk of an automobile accident is 15. This means that the risk of an accident for those who drive under the influence is 15 times the risk for those who don't drive under the influence.

Some features of relative risk are as follows:

- When two risks are the same, the relative risk is 1.
- When two risks are different, the relative risk is different from 1; and when the category in the numerator has higher risk, the relative risk is greater than 1.
- The risk in the denominator (the bottom) of the ratio is often the **baseline risk,** which is the risk for the category in which no additional treatment or behavior is present.

---

***thought question* 6.3** Based on the study in Section 6.1, the relative risk of developing any myopia later in childhood is 5.5 for babies sleeping in full light compared with babies sleeping in darkness. Restate this information in a sentence that the public would understand.[*]

---

**Example 6.4**

Statistics⟁Now™

Watch a video example at **http://1pass.thomson.com** or on your CD.

**Smoking and the Relative Risk of Divorce**  Table 6.1 on page 195 gives data on the smoking habits (yes or no) and divorce history (yes or no) of 1669 people who had ever been married. In the dataset, there are 485 smokers, of whom 238 have been divorced. There are 1184 nonsmokers, and 374 of these people have

[*]**HINT:** See the interpretation of relative risk in Example 6.4.

been divorced. To compute the relative risk of divorce for smokers, we first find the risk of divorce in each smoking category:

- For those who smoke, the risk of divorce is 238/485 = 0.491, or about 49%.
- For nonsmokers, the risk of divorce is 374/1184 = 0.316, or about 32%. This can be considered to be the *baseline risk* of divorce.

The relative risk of divorce is the ratio of these two risks.

$$\text{Relative risk} = \frac{49\%}{32\%} = 1.53$$

In this sample, the risk of divorce for smokers is 1.53 times the risk of divorce for nonsmokers. ▪

## Percent Increase or Decrease in Risk

Sometimes an increase or decrease in risk is presented as a percent change instead of a multiple. The **percent increase** (or decrease) **in risk** can be calculated as follows:

$$\text{Percent increase in risk} = \frac{\text{Difference in risks}}{\text{Baseline risk}} \times 100\%$$

Equivalently, when the relative risk has already been determined, the percent increase in risk can be calculated by using the following relationship:

$$\text{Percent increase in risk} = (\text{relative risk} - 1) \times 100\%$$

When a risk is smaller than the baseline risk, the relative risk is less than 1, and the percent "increase" will actually be negative. In this situation, the term *percent decrease* should be used to describe the percent change in the risks.

**Example 6.5** | **Percent Increase in the Risk of Divorce for Smokers**  In Example 6.4, we found that the relative risk of divorce for smokers is 1.53. Given this information, the fastest way to determine the percent increase in the risk of divorce for smokers is

$$\text{Percent increase in risk} = (1.53 - 1) \times 100\% = 53\%$$

We can get the same answer for the percent increase in risk by calculating

$$\text{Percent increase in risk} = \frac{\text{Difference in risks}}{\text{Baseline risk}} \times 100\%$$

$$= \frac{49 - 32}{32} \times 100\% = 53\%$$

So the risk of divorce is 53% higher for smokers than it is for nonsmokers. ▪

## Odds Ratio

The **odds** of an event compare the chance that the event happens to the chance that it does not. For instance, suppose there is a 60% chance that it will rain tomorrow, so there is a 40% chance that it will not rain. The odds that it will rain tomorrow can be expressed as 60 to 40, a comparison of the chance that it rains to the chance that it does not. Dividing both sides of the odds of rain by 20 lets us express these odds as 3 to 2. Notice that odds are expressed using a phrase with the structure "*a* to *b*," so a ratio is implied but not actually computed.

In Example 6.1 about smoking and divorce history, among 1184 nonsmokers, there were 374 people who had divorced and 810 who had not. Thus, the odds of divorce for nonsmokers are 374 divorced to 810 not divorced or, equivalently, about 0.46 divorced to 1 not divorced (divide both counts by 810 to get this). For smokers, there were 238 divorced and 247 not divorced, so the odds of divorce are 238 to 247, or 0.96 to 1.

The **odds ratio** compares two different groups with regard to the odds of a certain behavior or event. For example, we may wish to compare the odds of a successful outcome for two different treatments of clinical depression. The formula for the odds ratio will be given in a moment, but a numerical example should make it easy for you to see how it is computed. For the smoking habits and divorce history data, the odds ratio comparing the odds of divorce for smokers and nonsmokers is

$$\frac{\text{Odds of divorce for smokers}}{\text{Odds of divorce for nonsmokers}} = \frac{238/247}{374/810} = \frac{0.96}{0.46} = 2.1$$

This odds ratio tells us that the odds of ending up divorced instead of still married for smokers are about double the odds of ending up divorced for nonsmokers.

A useful characteristic of the odds ratio is that its value stays the same if the roles of the response and explanatory variables are reversed. In the previous paragraph, we compared the odds of divorce for smokers and nonsmokers. If we had, instead, compared the odds of being a smoker for those who have been divorced and those who have never divorced, the answer would still be 2.1. The calculation is

$$\frac{\text{Odds of smoking for divorced}}{\text{Odds of smoking for never divorced}} = \frac{238/374}{247/810} = \frac{0.636}{0.305} = 2.1$$

The interpretation here is that the odds of being a smoker for those who have been divorced are about double the odds of being a smoker for those who have never been divorced.

*in summary*   Statistics on Risk, Relative Risk, Odds, and Odds Ratios

Let's summarize the various ways in which risk, relative risk, odds, and odds ratios are constructed from a two-way contingency table. These measures are usually employed when there is a definable explanatory and response variable and when there is a baseline condition, so we present them using those distinctions. In situations in which those distinctions cannot be made, simply be clear about which condition is in the numerator and which is in the denominator.

|  | **Response Variable** | | |
|---|---|---|---|
| **Explanatory Variable** | *Category 1* | *Category 2* | *Total* |
| **Category of Interest** | $A_1$ | $A_2$ | $T_A$ |
| **Baseline Category** | $B_1$ | $B_2$ | $T_B$ |

- Risk (of response 1) for category of interest $= A_1/T_A$
- Odds (of response 1 to response 2) for category of interest $= A_1$ to $A_2$
- Relative risk $= \dfrac{A_1/T_A}{B_1/T_B}$
- Odds ratio $= \dfrac{A_1/A_2}{B_1/B_2}$

# Misleading Statistics About Risk

> ***thought question 6.4***  Suppose a newspaper article claims that drinking coffee doubles your risk of developing a certain disease. Assume that the statistic was based on legitimate, well-conducted research. What additional information would you want about the risk before deciding whether or not to quit drinking coffee?*

Whenever risk statistics are reported, there is a risk that they are misreported. Unfortunately, journalists often present risk data in a way that produces the best story rather than in a way that provides the best information. Very commonly, news reports either don't contain or don't emphasize the information that you need to understand risk. You should always ask the following questions when you encounter statistics about risk:

1. What are the actual risks? What is the baseline risk?
2. What is the population for which the reported risk or relative risk applies?
3. What is the time period for this risk?

News reports sometimes give counts of how often a "bad" event has happened without giving information about how many people have been exposed to the risk. Without information on the amount of exposure, we don't know the actual risk and can't properly interpret the news report. The next two examples show that consideration of the amount of exposure to a risk is important to understanding the risk.

**Example 6.6**  **The Risk of a Shark Attack**  A terrifying risk of going to the beach, especially where the water is warm, is the possibility of being attacked by a shark. Media stories about shark attacks on humans often include counts of how many attacks have occurred within the past year or so, and those numbers can sound scary. For instance, the website of the Florida Museum of Natural History Ich-

**\*HINT:** Would your decision be the same if you knew that the disease was rare in noncoffee drinkers of your age as it would be if you knew that it was common?

thyology Department reports that in 2000, there were 54 unprovoked shark bites of humans in U.S. coastal waters, with one fatality. Fifty-four attacks sounds like a lot, but when estimated beach attendance is taken into account, we see that the risk of being attacked by a shark is extremely low. At the Florida Museum website, this risk is estimated to be only 1 in 11.5 million beach visits (http://www.flmnh.ufl.edu/fish/sharks/ISAF/ISAF.htm). ■

**Example 6.7**

Statistics⊘Now™

Watch a video example at **http:// 1pass.thomson.com** or on your CD.

**Disaster in the Skies? Case Study 1.2 Revisited**   Case Study 1.2 described a *USA Today* article that told us, "Errors by air traffic controllers climbed from 746 in fiscal 1997 to 878 in fiscal 1998, an 18% increase." To airplane travelers, this may sound frightening, but a look at the risk of controller error per flight should ease their fear. In 1998, there were only 5.5 errors per million flights compared to 4.8 errors per million flights in 1997. The risk of controller error did increase, but the actual risk is extremely small. ■

In the next example, we see that a reported risk value may be meaningless if we don't know what specific population is being described or the time period of exposure to the risk.

**Example 6.8**

**Dietary Fat and Breast Cancer**   "Italian scientists report that a diet rich in animal protein and fat—cheeseburgers, french fries, and ice cream, for example—increases a woman's risk of breast cancer threefold," according to *Prevention Magazine's Giant Book of Health Facts* (1991, p. 122). The statement attributed to the Italian scientists is nearly useless information for at least two reasons:

1. We don't how the data were collected, so we don't know what population these women represent.
2. We don't know the ages of the women studied, so we don't know the baseline rate of breast cancer for these women.

Age is a critical factor. A frequently stated statistic about breast cancer is that one in nine women will develop breast cancer, but this is actually an accumulated lifetime risk to the age of 85, so a woman who dies before she reaches age 85 does not have a one in nine risk. According to the *University of California at Berkeley Wellness Letter* (July 1992, p. 1), the accumulated lifetime risk of a woman developing breast cancer by certain ages is as follows:

by age 50: 1 in 50
by age 60: 1 in 23
by age 85: 1 in 9

Also, the annual risk of developing breast cancer is only about 1 in 3700 for women in their early 30s (Fletcher, Black, Harris, Rimer, and Shapiro, 1993, p. 1644). If the Italian study was done on very young women, the threefold increase in risk represents a small increase. Unfortunately, *Prevention Magazine's Giant Book of Health Facts* did not even give enough information to lead us to the original research report, so it is impossible to intelligently evaluate the claim. ■

6.2 Exercises are on page 219.

## *case study 6.1*    Is Smoking More Dangerous for Women?

"Higher heart risk in women smokers" was the headline of an April 3, 1998, article at the Yahoo!® Health news website. The article, from the Reuters news agency, described Danish research that was interpreted as evidence that smoking affects the risk of a heart attack more for women than for men. Here is part of the article:

> Women who smoke have a greater than 50% higher risk of a heart attack than male smokers according to a study from Denmark. The researchers suggest that this difference may be related to the interaction of tobacco smoke and the female hormone, estrogen.
>
> "Women may be more sensitive than men to some of the harmful effects of smoking," the team writes in the April 4th edition of the *British Medical Journal*.
>
> Analyzing data from nearly 25,000 Danish men and women, Dr. Eva Prescott of the University of Copenhagen and colleagues report that women who smoke have a 2.24 relative risk of myocardial infarction, or heart attack, compared with nonsmokers. This is significantly higher than the relative risk of male smokers compared with nonsmokers.

Unfortunately, the only information that we are given, for each sex, is the *relative risk* of a heart attack for smokers. What's missing from the article is any mention of the estimated risk of a heart attack for any group of interest. A look at those risks makes the research interpretation debatable.

The *British Medical Journal* article (Prescott, Hippe, Schnor, and Vestbo, 1998) provided more complete information about the risks. That information is displayed in Table 6.4. Based on the data in the table, do you agree with the conclusion stated in the Web article?

We see from the table that the risk of a heart attack is generally greater for men than it is for women. In particular, for men who smoke, the

risk of a heart attack is 10.62%, which is clearly higher than the risk of 5.88% for women who smoke. These numbers contradict the Web article's headline and first sentence, both of which imply that the risk of a heart attack is greater for women smokers than for men smokers.

The researchers focused on relative risk, which is the ratio of two risks. We can easily verify that the relative risks provided in the Web article are correct.

- For women who smoke, the relative risk of a heart attack is 5.88%/2.63% = 2.24.

- For men who smoke, the relative risk of a heart attack is 10.62%/7.42% = 1.43.

The first sentence of the article actually refers to a comparison of these two *relative* risks, which are not the *actual* risks for women and men smokers. In other words, the *relative risk* value for women, 2.24, is 50% higher than the *relative risk* for men, 1.43.

In a subsequent issue of the *British Medical Journal*, several letter writers argued that the correct interpretation of these data is as follows:

- For both smokers and nonsmokers, men have a higher risk of heart attack than women do.

- For both men and women, smoking *adds* about 3.2% to the risk experienced by nonsmokers. In other words, the effect of smoking is the same for the two sexes.

The researchers' response to these letters was that they believed that it was valid to consider the multiplicative effect of smoking on risk. Dr. Prescott and her colleagues believe it is important that smoking multiplies the risk of a heart attack by 2.24 for women but by only 1.43 for men.

The moral of the story here is that a *relative risk* is affected by the *baseline risk*, so it is important to know the baseline risk. If a risk increases to 3% from a baseline risk of 1%, the relative risk is 3. If a risk increases to 22% from a baseline risk of 20%, the relative risk is only 22/20 = 1.1, although the difference in risks is still 2%. Is an increase in risk from 1% to 3% more serious than an increase from 20% to 22%? That may depend on the situation.

**Table 6.4** Smoking, Gender, and the Risk of a Heart Attack

|  | Sample Size | Heart Attacks | Risk of Heart Attack |
|---|---|---|---|
| **Men** | | | |
| Smokers | 8490 | 902 | 10.62% (902/8490) |
| Nonsmokers | 4701 | 349 | 7.42% (349/4701) |
| **Women** | | | |
| Smokers | 6461 | 380 | 5.88% (380/6461) |
| Nonsmokers | 5011 | 132 | 2.63% (132/5011) |

*thought question* **6.5**  If you were a frequent beer drinker and were worried about getting colon cancer, would it be more informative to you to know the *risk* of colon cancer for frequent beer drinkers or the *relative risk* of colon cancer for frequent beer drinkers compared to nondrinkers? Which of those statistics would likely be of more interest to the media? Explain your responses.*

# 6.3  The Effect of a Third Variable and Simpson's Paradox

In previous chapters, you saw several examples in which a confounding or lurking variable may have affected the relationship between an explanatory variable and a response variable. In observational studies, a confounding variable might explain an apparent relationship between two variables or, in some instances, it can mask a relationship. Whenever the data are the product of an observational study, you should carefully consider the possibility that a third variable may affect the observed relationship.

**Example 6.9**

**Educational Status and Driving After Substance Use**  In December 1998, the U.S. government's Substance Abuse and Mental Health Services Administration released a report with information about drug use, alcohol use, and driving that was gathered in a 1996 nationwide survey of 11,847 individuals aged 16 and over (SAMSHA, 1998). A principal response variable throughout the report was a categorical variable that described whether a respondent had on any occasion in the previous year driven within two hours of using either alcohol or drugs. This response variable had three categories:

- Never drove while impaired
- Drove within two hours of alcohol use, but never after drug use
- Drove within two hours of drug use and possibly after alcohol use

Figure 6.3 displays the association between this variable and educational status. The vertical axis of the bar chart is the proportion within an educational category that falls into a particular category of the response variable. The bar chart clearly shows that as the amount of education increases, the proportion who drove within two hours of alcohol use also increases. The data suggest a possible way to reduce the problem of drinking and driving: Let's not give out any more college degrees. We hope, for many reasons, that you disagree with this solution.

What's going on here? This is an observational study, so we should consider the possibility of confounding. We should ask whether the various educational groups differ in ways that may affect the response variable. One likely difference between the educational groups has to do with age. In the survey, about 13% of the respondents were under 21 years old. It is probable that almost none of

*HINT:  Suppose the relative risk is 3.0. Does that mean the same to you as an individual if the disease is rare as it does if it is common?

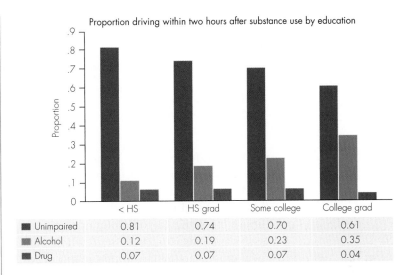

Figure 6.3 ▌ Education and incidence of driving after substance abuse

these younger respondents were in the college degree group. It is also possible that many members of the "less than high school group" were under 21.

Age differences could at least partially explain the results. Another table in the report showed that about 80% of those under 21 fell into the "never while impaired category," but only 70% of the 21 or over age group fell into this category. Unfortunately, the report didn't contain the information necessary to sort out the relative effects of age and educational status, so we'll have to classify age as a lurking variable.

The amount of alcohol consumed within two hours of driving should also be considered. The report indicates that compared to younger drivers, older drivers were more likely to say that they had never had more than one drink before driving. ▓

## Simpson's Paradox

Occasionally, the effect of a confounding factor is strong enough to produce a paradox known as **Simpson's Paradox.** The paradox is that the relationship appears to be in a different direction when the confounding variable is not considered than when the data are separated into the categories of the confounding variable.

**Example 6.10**

**Blood Pressure and Oral Contraceptive Use** We will illustrate Simpson's Paradox with a hypothetical example of an observational study that was done to examine the association between oral contraceptive use and blood pressure. Although the data are hypothetical, they are similar to data from several actual studies of the same problem. Suppose that 2400 women are categorized according to whether or not they use oral contraceptives and whether or not they have high blood pressure. The results shown in Table 6.5 indicate that the percentage with high blood pressure is about the same among oral contraceptive

**Table 6.5** Percentage with High Blood Pressure for Users and Nonusers of Oral Contraceptives

|  | Sample Size | Number with High B.P. | % with High B.P. |
|---|---|---|---|
| **Use Oral Contraceptives** | 800 | 64 | 64 of 800 = 8.0% |
| **Don't Use Oral Contraceptives** | 1600 | 136 | 136 of 1600 = 8.5% |

users as it is for nonusers. In fact, a slightly higher percentage of the nonusers have high blood pressure.

We're certain that you can think of many factors that affect blood pressure. If the users and nonusers of oral contraceptives differ with respect to one of these factors, that factor confounds the results in Table 6.5. As in Example 6.9, age is a critical factor. The users of oral contraceptives tend to be younger than the nonusers are. This is important because blood pressure increases with age.

**Table 6.6** Controlling for the Effect of Age

|  | Age 18–34 | | Age 35–49 | |
|---|---|---|---|---|
|  | Sample Size | n and % with High B.P. | Sample Size | n and % with High B.P. |
| **Use Oral Contraceptives** | 600 | 36 (6%) | 200 | 28 (14%) |
| **Don't Use Oral Contraceptives** | 400 | 16 (4%) | 1200 | 120 (10%) |

One way to control for the effect of age is to create separate contingency tables for women in different age groups. Table 6.6 divides the data from our hypothetical study into two age groups. In each age group, the percentage with high blood pressure is higher for the users than for the nonusers. This contradicts the direction of the relationship that we see when the age factor is not considered. In Table 6.5, not controlling for age differences masked the true nature of the relationship. Notice that the older women were less likely to be oral contraceptive users but more likely to have high blood pressure. ■

6.3 Exercises are on page 220.

# 6.4 Assessing the Statistical Significance of a 2 × 2 Table

Some instructors may wish to defer coverage of this section until later in the course, after hypothesis testing has been more formally introduced. That can be done with no loss of continuity in the material covered in the following chapters.

Most of the material that you have seen thus far involves *descriptive statistics*, which are methods for describing the data in hand. This section introduces a special case of *hypothesis testing*, one of the two most common statistical procedures making up *inferential statistics*. The other common inference procedure is the use of a *confidence interval* to estimate a population value.

Confidence intervals were introduced in Chapter 3. Hypothesis testing and confidence intervals will be covered in detail in Chapters 10 to 16, but this brief introduction will help you to become familiar with some basic concepts and will help to prepare you for those chapters.

Remember that observed data often represent a sample from a larger population. When this is the case, the purpose of collecting the data is usually to use the sample information to make generalizations about the population. Because this involves *inferring* something rather than being sure, the statistical methods that are used are called *inferential statistics*. When the data represent two categorical variables, the question of interest is whether a relationship that is observed in the sample data can also be inferred to hold in the population represented by the data.

When we see differences in the conditional distributions within the rows (or columns) of the contingency table, we say that the variables are related in the sample. A large difference between conditional distributions may convince us that the relationship is real. Small differences in conditional distributions, however, might not convince us. It is unlikely that any set of observed conditional distributions would ever be exactly the same for all categories of an explanatory variable. Small differences in observed conditional distributions could simply be the result of chance and may not represent a real relationship in the population.

*thought question 6.6* A random sample includes 110 women and 90 men. Of the women, approximately 9% are left-handed, while approximately 11% of the men are left-handed. Based on this observed data, do you think there is a relationship between gender and handedness in the population represented by this sample? Why or why not?*

Case Study 1.6 described the Physician's Health Study, in which about 22,000 physicians were randomized to take either aspirin or a placebo daily. Over a five-year period, the percentage of physicians experiencing heart attacks was lower in the aspirin group than in the placebo group. The difference was found to be **statistically significant,** and it was inferred that the observed difference reflected an actual difference in the population.

In the definitions following Case Study 1.6, the term *statistically significant* was defined as follows:

> A **statistically significant relationship** or difference is one that is large enough to be unlikely to have occurred in the observed sample if there is no relationship or difference in the population.

The practical upshot of calling an observed relationship statistically significant is that we are inferring that the relationship exists in the population. We make this inference by establishing that if there were really no relationship, it would be unlikely that we would have observed such a strong relationship in the sample.

**\*HINT:** Consider *how many* men and women in the samples are left-handed. Now consider how the percentages would change if one more or one fewer had been left-handed.

# The Five Steps to Determining Statistical Significance

Five steps are required in any hypothesis-testing situation. These steps lead to a decision about whether a statistically significant effect can be inferred about the population. The steps are as follows:

**Step 1:** Determine the null and alternative hypotheses.

**Step 2:** Verify necessary data conditions, and if they are met, summarize the data into an appropriate test statistic.

**Step 3:** Assuming that the null hypothesis is true, find the $p$-value.

**Step 4:** Decide whether or not the result is statistically significant based on the $p$-value.

**Step 5:** Report the conclusion in the context of the situation.

In this section, we describe how these steps are implemented for a $2 \times 2$ contingency table. In Chapter 15, you will learn how to implement them for any contingency table.

## Step 1: Null and Alternative Hypotheses

The question we are considering is whether or not the sample data permit us to conclude that two variables are related. Another way to describe our objective is that we are deciding between two possible hypotheses about the population. In statistical language, the possibilities are called the **null hypothesis** and the **alternative hypothesis,** and they can be stated as follows:

**Null hypothesis:** The two variables are not related.
**Alternative hypothesis:** The two variables are related.

The two hypotheses just given are specific to the context of two-way tables. In Chapters 12 and 13, you will learn how to write null and alternative hypotheses for a broad range of research questions.

## Step 2: The Chi-Square Statistic

The statistical significance of the association between two categorical variables can be examined using a value known as the **chi-square statistic.** This statistic measures the difference between the **observed counts** in the contingency table and the **expected counts,** which are the counts that would be expected if there were no relationship between the variables (the null hypothesis). A large difference between the observed and expected counts occurs when a relationship is present in the observed table. To simplify our discussion of the chi-square statistic, we will consider only $2 \times 2$ contingency tables so that each variable will have just two categories. The same principles apply to larger tables, and we will learn the details for larger tables in Chapter 15.

What follows is a brief discussion of the computation of the chi-square value. Don't worry too much about technical detail. Statistical software can be used to compute the chi-square statistic. You should, at this point, principally focus on learning about the general ideas and how to use the computer output to decide whether you can declare statistical significance.

The first step in the calculation of a chi-square statistic is to determine a table of expected counts. These are the counts that would be expected to fall into the cells of the contingency table if there were no relationship between the two variables. They are found by making the row percentages across columns equal for each row. To accomplish this, simply determine what percentage of the total sample falls into each column, then split the total for each row into columns according to those same percentages.

A table of expected counts has these two properties:

1. There is no association between the two variables. For instance, there are no differences between the conditional distributions in the rows (or columns) of a table of expected counts.
2. The expected counts in each row and column sum to the same totals as the observed numbers.

**Example 6.11**

**A Table of Expected Counts**  Suppose that 90 men and 110 women are asked about their handedness. Of the 200 total individuals, 20 are left-handed and 180 are right-handed. If gender and handedness are not related, what counts would we see in the cells of a contingency table? A strategy for determining these "expected counts" is first to calculate that $20/200 = 10\%$ of the overall sample is left-handed (falls into the first column) and then to make 10% of each gender be left-handed. The result is as follows:

|  | Left-Handed | Right-Handed | Total |
|---|---|---|---|
| **Men** | 9 (10%) | 81 (90%) | 90 |
| **Women** | 11 (10%) | 99 (90%) | 110 |
| **Total** | 20 (10%) | 180 (90%) | 200 |

Keep in mind that these counts are what we would see if there were no relationship between gender and handedness. The row percentages are the same for the two genders, and the total counts for each row and each column match the characteristics of the observed data. ■

*formula*  |  **Calculating Expected Counts**

The expected count for each cell can be calculated as

$$\frac{\text{Row total} \times \text{Column total}}{\text{Total } n \text{ for table}}$$

For instance, in Example 6.11, the expected count for the number of women who are left-handed could be calculated as

$$\frac{110 \times 20}{200} = 11$$

*technical note*  |  **Verifying Conditions for the Test**

The method described in this section won't work if the sample is too small. Proceed as long as at least three of the four expected counts are 5 or more and all of them are 1 or more.

### Calculating the Chi-Square Statistic

The specific way in which the chi-square statistic ($\chi^2$) measures the difference between the observed and expected counts is relatively simple:

- First, for each cell in the table, we compute

$$\frac{(\text{Observed count} - \text{Expected count})^2}{\text{Expected count}}$$

- Then we total these quantities over all cells of the table:

$$\chi^2 = \text{Sum of } \frac{(\text{Observed count} - \text{Expected count})^2}{\text{Expected count}}$$

### Step 3: The *p*-Value of the Chi-Square Test

Now we're faced with a bit of a mystery. A large chi-square value indicates a relationship, but how large should the value be for us to declare statistical significance? In practice, this question is transformed into a different but equivalent question: If there is actually no relationship in the population, what is the likelihood that the chi-square statistic could be as large as it is or larger?

*definition* | The ***p*-value** for a chi-square test is computed by assuming that the null hypothesis is true and then determining the likelihood of observing data that would produce a chi-square statistic as large as the one observed, or larger. It answers the question "How likely is it that a relationship of the magnitude observed, or one even stronger, would occur in the sample if there is no relationship in the population?"

The *p*-value is used to decide whether the relationship observed in the sample is statistically significant and can be inferred to hold in the population. The *p*-value will be part of the computer output provided by almost any statistical software. If you are computing the chi-square statistic "by hand" instead of using statistical software, you can use Excel to find the *p*-value; see the Excel Tip on page 212. If that isn't possible, you can still determine whether the result is statistically significant using a standard rule, but the reason it works will remain a mystery until you encounter it again in Chapter 15. The decision process is as follows.

### Steps 4 and 5: Making and Reporting a Decision

Generally, a large chi-square statistic and subsequently small *p*-value provide evidence that a real relationship exists in the population. To determine whether to decide in favor of the alternative hypothesis, that a real relationship exists, we will use the rule that most researchers commonly use:

- When the *p*-value is less than or equal to 0.05 (5%), we will assume that the observed relationship did not occur by chance. In that case, we *can* say that the relationship is statistically significant, and we reject the null hypothesis in favor of the alternative hypothesis. For a chi-square statistic based on a 2 × 2 contingency table, this is equivalent to finding a chi-square statistic of at least 3.84.

- When the *p*-value is greater than 0.05 (5%), we will say that the observed relationship could have occurred just by chance. In that case, we *cannot* say that the relationship is statistically significant, and we cannot reject the null hypothesis. For a chi-square statistic based on a 2 × 2 contingency table, this is equivalent to finding a chi-square statistic of less than 3.84.

**Example 6.12**

**Statistics⌂Now™**

For software help, download your Minitab, Excel, TI-83, SPSS, R, and JMP manuals from **http://1pass .thomson.com**, or find them on your CD.

**Does Order Influence Who Wins an Election?**  In an experiment done in a statistics class, 92 college students were given a form that read, "Randomly choose one of the letters *S* or *Q*." Another 98 students were given a form with the order of the letters reversed, which read, "Randomly choose one of the letters *Q* or *S*." The purpose was to determine whether the order of listing the letters might influence the choice of letters. The possible influence of the order of listing items is a concern in elections. Many election analysts believe that a candidate gains an advantage if he or she is the first candidate listed on the ballot.

In the observed data, the likelihood of choosing *S* (or *Q*) was related to the order of listing the letters. Of the students who were given the order "*S* or *Q*," 66% (61/92) picked *S*, the letter that was listed first. Only 46% (45/98) of the students who were given the order "*Q* or *S*" picked *S*. On the basis of the relationship observed in the sample, can we generalize that choice of letter and order of the letters on the form are related variables in the larger population represented by this sample of students?

Figure 6.4 displays computer output for testing the significance of the relationship for this experiment. The output was produced using the Minitab computer program. From the output, we learn that the *p*-value is 0.005. Because the *p*-value is less than 0.05, we can say that the relationship is statistically significant, and we reject the null hypothesis that there is no relationship. In other words, we infer that the observed result also holds in the population represented by these students.

Notice that the value of the chi-square statistic is 7.995. The *p*-value tells us that the chance is only 0.005 (which is 5 in 1000) that we would get a chi-square as large as 7.995 (or larger) if there really is no relationship between the order of the letters on the form and the letter that would be picked by people in this population.

```
Expected counts are printed below observed counts

                      Letter Picked
Form           S            Q          Total
S first       61           31           92
            51.33        40.67

Q first       45           53           98
            54.67        43.33

Total        106           84          190

Chi-Sq = 1.823 + 2.301 +
         1.712 + 2.160 = 7.995
DF = 1, P-Value = 0.005
```

**Figure 6.4** ▌ Minitab output for the *S* or *Q* experiment (Example 6.12)

> **MINITAB tip** | **Computing a Chi-Square Test for a Two-Way Table**
>
> - If the raw data are stored into columns of the worksheet, use **Stat> Tables>Cross Tabulation and Chi-Square.** Specify a categorical variable in the "For rows" box and a second categorical variable in the "For columns" box. Then click the **Chi-Square** button and select "Chi-Square analysis."
> - If the data are already summarized into counts, enter the table of counts into columns of the worksheet, and then use **Stat>Tables>Chi-Square Test (Table in Worksheet).** In the dialog box, specify the columns that contain the counts.

> **EXCEL tip** | The *p*-value can also be computed by using Microsoft Excel. The function CHIDIST(x,df) provides the *p*-value, where x is the value of the chi-square statistic and df is a number called "degrees of freedom," which will be explained later in this book. The formula for df is (#rows − 1)(#columns − 1). For instance, corresponding to the information in Figure 6.4, $x = 7.995$, df $= (2 − 1)(2 − 1) = 1$, and the *p*-value is CHIDIST(7.995,1) = .00469, or about .005 as given by Minitab. Excel will also calculate the *p*-value directly from the data, using the function CHITEST, but unlike the case with Minitab, the user must provide the expected counts as well as the observed counts. Use the command *CHITEST(array1,array2)*, where the observed counts are in table format covered by the cells in *array1* and the expected counts are in *array2*. For example, if the observed counts are in a $2 \times 2$ table in the first four cells of a worksheet, *array1* would be given as A1:B2. Unlike Minitab and other statistical software, Excel gives only the *p*-value and not the value of the chi-square statistic.

**Example 6.13**

Statistics⬡Now™

Read the original source on your CD.

**Breast Cancer Risk Stops Hormone Replacement Therapy Study** On July 17, 2002, the *Journal of the American Medical Association* published the results of a study that affected the lives of millions of women. The study was the first large randomized experiment to test the effects of combined estrogen and progestin, the most commonly prescribed hormone replacement therapy for postmenopausal women. The big news was that "On May 31, 2002, after a mean of 5.2 years of follow-up, the data and safety monitoring board recommended stopping the trial of estrogen plus progestin vs placebo because the test statistic for invasive breast cancer exceeded the stopping boundary for this adverse effect and the global index statistic supported risks exceeding benefits" (Writing Group for the Women's Health Initiative Investigators, 2002, p. 321).

The study measured several medical outcomes, but the two that received the most attention were breast cancer and coronary heart disease. Table 6.7 shows the results for breast cancer; results for coronary heart disease are given in Exercise 6.52. Is there a statistically significant effect of the hormone treatment on breast cancer?

**Table 6.7** Results of a Randomized Experiment Comparing Hormone Therapy and Placebo

|  | Invasive Breast Cancer? | | | Risk of Breast Cancer |
|---|---|---|---|---|
|  | Yes | No | Total |  |
| Hormones | 166 | 8,340 | 8,506 | .0195 |
| Placebo | 124 | 7,978 | 8,102 | .0153 |
| Total | 290 | 16,318 | 16,608 | .0175 |

### Step 1: The Hypotheses

*Null hypothesis:* Occurrence of breast cancer is unrelated to hormone therapy.

*Alternative hypothesis:* Occurrence of breast cancer is related to hormone therapy.

### Step 2: The Conditions and Test Statistic

The expected count for the "Hormones, Yes" cell is $[(8,506)(290)]/16,608 = 148.53$. The other expected counts can be found similarly, or they can be found by subtraction because the row and column totals are the same as they are for the observed counts. For instance, the expected count for the "Hormones, No" cell is $290 - 148.53 = 141.47$. The sample size condition is met, since all expected counts exceed 5. The chi-square statistic is

$$\frac{(166 - 148.53)^2}{148.53} + \frac{(8340 - 8357.47)^2}{8357.47} + \frac{(124 - 141.47)^2}{141.47} + \frac{(7978 - 7960.53)^2}{7960.53} = 4.288$$

### Steps 3, 4, and 5: The *p*-Value and Decision in Context

Using Excel, the *p*-value is found as CHIDIST(4.288,1) = .038382, or about .04. Using the standard convention, the result is statistically significant because .04 < .05. In the context of the problem, this means that there is a statistically significant relationship between hormone therapy and occurrence of breast cancer *in the population similar to the women in this experiment.* Because the data are from a randomized experiment, it can be concluded that hormone therapy *causes* a change in the incidence of breast cancer. It was for this reason that the study was terminated early.

Notice that the relative risk of breast cancer for the women in the study is .0195/.0153 = 1.27, so that women who took hormones had an increased risk of about 27%. The baseline risk was .0195, and the difference in risk for the two groups was .0042. This represents about four additional cases of breast cancer per 1000 women taking hormones. ■

## Factors That Affect Statistical Significance

In general, whether or not we can infer that an observed relationship represents a "real" relationship depends on two factors:

- The strength of the observed relationship
- How many people were studied

For $2 \times 2$ tables, the strength of the relationship can be measured by the difference between the two categories of the explanatory variable with respect to the percent falling into a particular category of the response variable. For instance, in Example 6.12, we saw that 66% of those with the "*S* or *Q*" order picked *S*, while only 46% of those with the "*Q* or *S*" form picked *S*. The difference between these two percentages reflects the strength of the observed relationship between the order of the letters on the form and letter choice.

The sample size of a study also affects the significance of the results. Imagine, for example, that a psychiatric researcher compares the effectiveness of two treatments for depression by conducting a randomized experiment in which only ten patients receive each treatment. The results are that after three months, eight of the ten patients using treatment A show improvement, but among those who used treatment B, only five of ten patients show improvement. Although the difference between the percentages showing improvement is large (80% − 50%), the study is too small to safely infer that treatment A is the better treatment for the larger population of patients. The chi-square statistic and *p*-value would be 1.978 and .16, respectively. On the other hand, if 100 patients were in each treatment group and the same percentages were observed (80% and 50%), the chi-square statistic would be 19.78, and the *p*-value would be $8.7 \times 10^{-8}$. We would almost certainly believe that treatment A was really better than treatment B in that case.

## Practical Versus Statistical Significance

Statistical significance does not mean that the two variables have a relationship that you would necessarily consider to have **practical significance.** A table based on a very large number of observations will have little trouble achieving the status of statistical significance even if the relationship between the two variables in the population is only minor.

On the other hand, an interesting relationship in a population may fail to achieve statistical significance if there are only a few observations in the sample. Whenever a study "fails to find a relationship" between two variables, this does not necessarily mean that a relationship was not observed or does not exist in the population. It means that whatever relationship was observed in the sample did not achieve statistical significance. It is hard to rule out chance unless you have either a very strong relationship or a sufficiently large sample.

In the following example, you will see that with a large sample size, a relatively small arithmetic difference can be statistically significant. Remember that "statistical significance" means that we are convinced that a real relationship exists in the larger population represented by the sample. It makes sense that if we have a very large sample and it is exhibiting a relationship, we would be more convinced that it is real than we would be if a very small sample exhibited that same relationship.

**Example 6.14** | **Aspirin and Heart Attacks** In Case Study 1.6, the possible categories for the explanatory and response variables were as follows:

Explanatory variable = took placebo or took aspirin
Response variable = heart attack or no heart attack

Statistics Now™

Watch a video example at **http://1pass.thomson.com** or on your CD.

Read the original source on your CD.

Expected counts are printed below observed counts

| | Heart Attack? Yes | No | Total |
|---|---|---|---|
| Aspirin | 104 | 10933 | 11037 |
| | 146.52 | 10890.48 | |
| Placebo | 189 | 10845 | 11034 |
| | 146.48 | 10887.52 | |
| Total | 293 | 21778 | 22071 |

Chi-Sq = 12.339 + 0.166 +
12.343 + 0.166 = 25.014
DF = 1, P-Value = 0.000

**Figure 6.5** ▮ Minitab output for aspirin study (Example 6.14)

Figure 6.5 shows the results of asking the Minitab program to compute the chi-square statistic. We see that the *p*-value is .000, so we can declare statistical significance. This *p*-value tells us that there is almost no chance that we could observe such a strong relationship if there is really no difference between the effects of aspirin and placebo. Because the *p*-value is less than .05, we can reject the null hypothesis of no relationship.

In the placebo group, 189/11,034 = 1.71% had a heart attack during the study period, while in the aspirin group, 104/11,037 = 0.94% had a heart attack. The difference has practical importance because the percent decrease in risk for those taking aspirin is (1.71 − 0.94)/1.71 = 45%. However, the arithmetic difference in the percentages who experienced a heart attack is 1.71% − 0.94% = 0.77%, or less than 1%. If a smaller study had been done, the researchers may not have been able to declare statistical significance, but since they used a sample with over 22,000 participants, they were able to detect this important relationship. ■

## Interpreting a Nonsignificant Result

When we cannot claim statistical significance, we have to be careful about how we state the conclusion. The correct interpretation of a **nonsignificant result** is that the sample results are not strong enough to safely conclude that there is a relationship in the population. A *p*-value that is too large for us to declare significance simply means that the observed relationship could have resulted by chance, even if there is no relationship in the population. This is not the same as saying that we believe that there is no relationship.

In the following case study, the difference between the relevant conditional percentages is larger than the difference we saw in the aspirin and heart attacks example (Example 6.14). The result of the chi-square test, however, will be that we cannot declare statistical significance. You will see that the sample size for the case study is much smaller than the sample size for the aspirin study. Remember that the sample size affects our ability to declare significance.

6.4 Exercises are on page 221.

**SPSS *tip***

**Computing a Chi-Square Test for a Two-Way Table**

- Use **Analyze>Descriptive Statistics>Crosstabs.** Enter variable names for "Row(s)" and "Column(s)" and then click ***Statistics*** to select "Chi-Square."
- Use the ***Cells*** button to request expected counts and/or conditional percentages of interest.

## case study 6.2   Drinking, Driving, and the Supreme Court

In the early 1970s, a young man challenged an Oklahoma state law that prohibited the sale of 3.2% beer to males under 21 but allowed its sale to females in the same age group. The case (*Craig v. Boren,* 429 U.S. 190, 1976) was ultimately heard by the U.S. Supreme Court.

Laws are allowed to use gender-based differences as long as they "serve important governmental objectives" and "are substantially related to the achievement of these objectives" (Gastwirth, 1988, p. 524). The defense argued that traffic safety was an important governmental objective and that data clearly show that young males are more likely to have alcohol-related accidents than young females.

The Supreme Court examined evidence from a "random roadside survey" that measured information on age, gender, and whether or not the driver had been drinking alcohol in the previous two hours. Although the survey was called a "random" survey of drivers, it probably was not. In roadside surveys, police tend to stop all drivers at certain locations at the time of the survey. This procedure does not really provide a random sampling of drivers in an area, but we'll treat it as though it does. Table 6.8 gives the results of the roadside survey for the drivers under 20 years of age.

Notice that the percentage of young men who had been drinking alcohol is slightly higher than the percentage of young women. The difference is 16% − 11.6% = 4.4%. However, we cannot rule out chance as a reasonable explanation for this difference. In other words, if there really is no difference between the percentages of young male and female drivers in the population who drink and drive, we could possibly see a difference as large as the one observed in a sample of this size.

In Figure 6.6, we present the results of asking the Minitab program to compute the chi-square statistic for this example. The chi-square summary statistic is 1.637, and the *p*-value for this statistic is .201. This *p*-value tells us that if there is really no association in the population (the null hypothesis), there is about a 20% chance that the sample would have a chi-square statistic as large as 1.637 or larger. In other words, the observed relationship could easily have occurred even if there is no relationship in the population represented by the sample.

The Supreme Court overturned the law, concluding that "the showing offered by the appellees does not satisfy us that sex represents a legitimate, accurate proxy for the regulation of drinking and driving" (Gastwirth, 1988, p. 527). Based on the chi-square analysis, you can see why the Supreme Court was reluctant to conclude that the difference in the sample represented sufficient evidence for a real difference in the population.

**Table 6.8**  Results of Roadside Survey for Young Drivers

| | **Drank Alcohol in Last Two Hours?** | | | **Percent Who Drank** |
|---|---|---|---|---|
| | *Yes* | *No* | *Total* | |
| **Males** | 77 | 404 | 481 | 16.0% |
| **Females** | 16 | 122 | 138 | 11.6% |
| **Total** | 93 | 526 | 619 | 15.0% |

*Source: Gastwirth, 1988, p. 526.*

Expected counts are printed below observed counts

| | Drank in Last 2 Hours? | | |
|---|---|---|---|
| | Yes | No | Total |
| Males | 77 | 404 | 481 |
| | 72.27 | 408.73 | |
| Females | 16 | 122 | 138 |
| | 20.73 | 117.27 | |
| Total | 93 | 526 | 619 |

Chi-Sq = 0.310 + 0.055 +
$\qquad$ 1.081 + 0.191 = 1.637
DF = 1, P-Value = 0.201

**Figure 6.6**  ▮  Minitab output for Case Study 6.2

# Key Terms

**Section 6.1**

contingency table, 194
two-way table, 194
cell, 194
conditional percentages, 194
row percentages, 194, 197
column percentages, 194, 197
relationship, 197

**Section 6.2**

risk, 198
relative risk, 198
baseline risk, 198
percent increase in risk, 199
odds, 199
odds ratio, 200

# Exercises

●   Denotes basic skills exercises

◆   Denotes dataset is available in StatisticsNow at **http://**
**1pass.thomson.com** or on your CD but is **not required** to
solve the exercise.

**Bold-numbered exercises** have answers in the back of the text and
fully worked solutions in the Student Solutions Manual.

**Statistics⬯Now™** Go to the StatisticsNow website at
**http://1pass.thomson.com** to:
● Assess your understanding of this chapter
● Check your readiness for an exam by taking the Pre-Test quiz and
exploring the resources in the Personalized Learning Plan

## Section 6.1

**6.1** ● ◆ The table below shows data for grades usually
achieved in school and how often the respondent puts
on sunscreen when going out in the sun for more than
one hour. Respondents are twelfth-grade participants
in the 2003 Youth Risk Behavior Surveillance System
survey. The survey, sponsored by the U.S. Centers for
Disease Control, is a national survey of high school students. (Raw data are in the **YouthRisk03** dataset on the
CD for this book.)

**Grades and Sunscreen Use for Twelfth-Graders**

| | Sunscreen Use | | | |
|---|---|---|---|---|
| Grade | Never or Rarely | Sometimes | Always or Most Times | All |
| As and Bs | 1322 | 450 | 285 | 2057 |
| Cs | 568 | 83 | 47 | 698 |
| Ds and Fs | 85 | 15 | 3 | 103 |
| All | 1975 | 548 | 335 | 2858 |

a. Among students who usually get As and Bs in school,
what percentage never or rarely uses sunscreen when
going out in the sun for more than one hour? Explain
whether this value is a row percentage or a column
percentage.
b. Among students who sometimes wear sunscreen,
what percentage usually gets Cs in school? Explain

whether this value is a row percentage or a column
percentage.
c. What percentage of the overall sample usually gets As
and Bs in school and also uses sunscreen always or
most times when going out in the sun for more than
one hour?

**6.2** ● ◆ Refer to the table in Exercise 6.1.
a. Determine a complete table of row percentages.
b. Briefly explain whether the percentages found in
part (a) indicate that there is a relationship between grades usually achieved and frequency of sunscreen use.

**6.3** ● Each fall, auditions for the band and orchestra are
held at a large university. Last fall, the numbers of males
and females in each class who auditioned were

| Class | Female | Male | Total |
|---|---|---|---|
| Freshman | 170 | 100 | 270 |
| Sophomore | 50 | 50 | 100 |
| Junior | 60 | 20 | 80 |
| Senior | 20 | 30 | 50 |
| Total | 300 | 200 | 500 |

a. Calculate the row percentage for freshman females
and explain what it means.
b. Calculate the column percentage for freshman females and explain what it means.
c. Which class had the highest percentage of female applicants? Support your answer with numbers.
d. Which gender had a higher percentage of sophomore
applicants? Support your answer with numbers.

**6.4** ● Anton and Edward often play a game together, so
they decide to see whether who goes first affects who
wins. They keep track of 50 games, with each going
first 25 times. Of the 25 times Anton went first, he won
15 times. Of the 25 times Edward went first, he won 12
times. In constructing a contingency table for these
results, the two variables are "Who went first?" and "Did
the person who went first win?"
a. What are the categories for each of the two variables?
b. If the explanatory variable is used to define the rows
of the contingency table, which variable would be the
row variable?

c. Construct a contingency table for the results.

d. Overall, what percent of the games were won by the person who went first?

e. Is there an advantage to going first? Explain.

6.5 ● For each pair of variables, indicate whether or not a two-way table would be appropriate for summarizing the relationship. In each case, briefly explain why or why not.

   **a.** Political party (Republican, Democrat, etc.) and opinion about new gun control law.

   b. Age group (under 20, 21–29, etc.) and rating of a song on 1 to 5 scale (1 = hate it, 5 = love it).

   **c.** Weight (pounds) and height (inches).

   d. Gender and opinion about capital punishment.

   e. Head circumference (centimeters) and gender.

6.6 ● Suppose a study on the relationship between gender and political party included 200 men and 200 women and found 180 Democrats and 220 Republicans. Is that information sufficient for you to construct a contingency table for the study? If so, construct the table. If not, explain why not.

6.7 In the 1996 General Social Survey, religious preference and opinion about when premarital sex might be wrong were among the measured variables. The contingency table of counts for these variables is as follows:

**Religious Preference and Opinion about Premarital Sex**

| | When Is Premarital Sex Wrong? | | | | |
|---|---|---|---|---|---|
| Religion | Always | Almost Always | Sometimes | Never | Total |
| Protestant | 355 | 117 | 227 | 384 | 1083 |
| Catholic | 62 | 37 | 120 | 226 | 445 |
| Jewish | 0 | 3 | 14 | 34 | 51 |
| None | 20 | 13 | 45 | 147 | 225 |
| Other | 15 | 13 | 23 | 40 | 91 |
| Total | 452 | 183 | 429 | 831 | 1895 |

Source: SDA archive at csa.berkely.edu:7502.

a. For each religious preference category, determine the percentage of respondents who think premarital sex is always wrong.

b. Do the percentages computed in part (a) indicate that there is a relationship between the two variables? Briefly explain why or why not.

6.8 Students in a class were asked whether they preferred an in-class or a take-home final exam and were then categorized as to whether or not they had received an A on the in-class midterm. Of the 25 A students, 10 preferred a take-home exam, while of the 50 non-A students, 30 preferred a take-home exam.

a. Display the data in a contingency table.

b. In the relationship between grade on the midterm and opinion about type of final, which variable is the response variable and which is the explanatory variable?

c. Determine an appropriate set of conditional percentages for determining whether there is a relationship between grade on the midterm and opinion about the type of final. Based on these percentages, does it appear that there is a relationship? Why or why not?

6.9 Do grumpy old men have a greater risk of having coronary heart disease than men who aren't so grumpy? Harvard Medical School researchers examined this question in a prospective observational study reported in the November 1994 issue of *Circulation* (Kawachi et al., 1994). For seven years, the researchers studied men between the ages of 46 and 90 years old. All study participants completed a survey of anger symptoms at the beginning of the study period. Among 199 men who had no anger symptoms, there were 8 cases of coronary heart disease. Among 559 men who had the most anger symptoms, there were 59 cases of coronary heart disease.

   **a.** Construct a contingency table for the relationship between degree of anger and the incidence of heart disease.

   **b.** Among those with no anger symptoms, what percentage had coronary heart disease?

   **c.** Among those with the most anger symptoms, what percentage had coronary heart disease?

   d. Draw a bar graph of these data. Based on this graph, does there appear to be an association between anger and the risk of coronary heart disease? Explain.

6.10 In a class survey, statistics students at a university were asked, "Regarding your weight, do you think you are: About right? Overweight? Underweight?" The following table displays the results by sex: .

**Gender and Perception of Weight**

| | Perception of Weight | | | |
|---|---|---|---|---|
| Sex | About Right | Overweight | Underweight | Total |
| Female | 87 | 39 | 3 | 129 |
| Male | 64 | 3 | 16 | 83 |
| Total | 151 | 42 | 19 | 212 |

Source: The authors.

a. Write a sentence that explains what would be measured by the row percentages for this table. Make your answer specific to this situation.

b. Determine the row percentages.

c. Draw a bar graph of the row percentages.

d. Briefly describe how males and females differ in their perceptions of weight.

e. An important objective of statistics is the use of sample information to make generalizations about a larger population. What population do you think is represented by this sample?

6.11 In a case-control study done in England, Voss and Mulligan (2000) collected data on height (short or not) and whether or not the student had ever been bullied in school for 209 secondary school students. The following table displays a contingency table of the data.

A student was categorized as short if he or she was below the third percentile for height on school entry, but because this was a case-control study, short people (cases) constitute almost half of the sample.

**Height and Bullying in School**

| Height | Ever Bullied | | |
|---|---|---|---|
| | Yes | No | Total |
| Short | 42 | 50 | 92 |
| Not Short | 30 | 87 | 117 |
| Total | 72 | 137 | 209 |

a. Among students in the "short" category, what percentage have ever been bullied?
b. Among students in the "not short" category, what percentage have ever been bullied?
c. Is there a relationship between height and the likelihood of having been bullied? Briefly justify your answer.

## Section 6.2

**6.12** ● For each of the following measures, give the value that implies no difference between the two groups being compared.
   **a.** Relative risk.
   b. Odds ratio.
   c. Percent increase in risk.

*For Exercises 6.13 to 6.16: A study is done to compare side effects for those taking a drug versus those taking a placebo. One hundred people are given the drug, and 100 are given the placebo. Results are as shown in the following table:*

| | Headache? | | Nausea? | | Insomnia? | |
|---|---|---|---|---|---|---|
| | Yes | No | Yes | No | Yes | No |
| Drug | 10 | 90 | 15 | 85 | 6 | 94 |
| Placebo | 5 | 95 | 5 | 95 | 4 | 96 |

**6.13** ● For each of the following side effects, compute the risk of the side effect when taking the drug.
   **a.** Headache.
   b. Nausea.
   c. Insomnia.
**6.14** ● For each of the following side effects, compute the relative risk of the side effect when taking the drug compared to when taking the placebo.
   **a.** Headache.
   b. Nausea.
   c. Insomnia.
**6.15** ● For each of the following side effects, compute the odds ratio for the side effect for those taking the drug compared to those taking the placebo.
   **a.** Headache.
   b. Nausea.
   c. Insomnia.

**6.16** ● For each of the following side effects, compute the percent increase (or decrease) in risk of side effect for those taking the drug compared to those taking the placebo.
   a. Headache.
   b. Nausea.
   c. Insomnia.
**6.17** ● If the baseline risk of a certain disease for nonsmokers is 1% and the relative risk of the disease is 5 for smokers compared to nonsmokers, what is the risk of the disease for smokers?
**6.18** ● For a relative risk of 2.1, what is the percent increase in risk?
**6.19** ● For an increase in risk of 40%, what is the relative risk?
**6.20** ● The relative risk of contracting a certain coronary disease is 2.0 for male smokers compared to male nonsmokers and 3.0 for female smokers compared to female nonsmokers. Is this enough information to determine whether male smokers or female smokers are more likely to contract the disease? If so, make that determination. If not, explain what additional information would be needed to make that determination.
**6.21** *Science News* (Feb. 25, 1995, p. 124) reported a study of 232 people, aged 55 or over, who had had heart surgery. The patients were asked whether their religious beliefs give them feelings of strength and comfort and whether they regularly participate in social activities. Of those who said yes to both, about 1 in 50 died within six months after their operation. Of those who said no to both, about 1 in 5 died within six months after their operation. What is the relative risk of death (within six months) for the two groups? Write your answer in a sentence or two that would be understood by someone with no training in statistics.
**6.22** Exercise 6.9 described a study in which men were classified according to how many anger symptoms they exhibit and whether they have coronary heart disease or not. Among 559 men with the most anger symptoms, 59 had coronary heart disease. Among 199 men with no anger symptoms, 8 had coronary heart disease.
   **a.** For those with the most anger symptoms, what is the relative risk of heart disease (compared to those with no anger symptoms)?
   **b.** For those with the most anger symptoms, what is the percent increase in the risk of heart disease?
**6.23** ◆ Exercise 6.11 gave data on height and whether or not a student had ever been bullied for 209 secondary school students in England. Among 92 short students, 42 had been bullied. Among 117 students who were not short, 30 had been bullied.
   a. For each height category, calculate the risk of having been bullied.
   b. What is the relative risk for short students of having been bullied? Write a sentence that interprets this relative risk.
   c. What is the increased risk of having been bullied for short students? Write a sentence that interprets this increased risk.

d. Calculate the odds ratio that compares the odds of having been bullied for the short students to the odds for students who are not short. Write a sentence that interprets this ratio.

6.24 Using the terminology of this chapter, what name applies to each of the boldface numbers in the following quotes (e.g., odds, risk, relative risk)?

a. "Fontham found increased risks of lung cancer with increasing exposure to secondhand smoke, whether it took place at home, at work, or in a social setting. A spouse's smoking alone produced an overall **30** percent increase in lung-cancer risk" (*Consumer Reports*, Jan. 1995, p. 28).

b. "What they found was that women who smoked had a risk [of getting lung cancer] **27.9** times as great as nonsmoking women; in contrast, the risk for men who smoked regularly was only **9.6** times greater than that for male nonsmokers" (Taubes, 1993, p. 1375).

c. "**One student in five** reports abandoning safe-sex practices when drunk" (*Newsweek*, Dec. 19, 1994, p. 73).

6.25 The Roper Organization (1992) conducted a study as part of a larger survey to ascertain the number of U.S. adults who had experienced phenomena such as seeing a ghost, "feeling as if you left your body," and seeing a UFO. A representative sample of adults (18 and over) in the continental United States were interviewed in their homes during July, August, and September 1991. The results when respondents were asked about seeing a ghost are shown in the following table:

|  | **Reportedly Has Seen a Ghost** | | |
|---|---|---|---|
|  | *Yes* | *No* | *Total* |
| **Aged 18 to 29** | 212 | 1313 | 1525 |
| **Aged 30 or over** | 465 | 3912 | 4377 |
| **Total** | 677 | 5225 | 5902 |

*Data source: The Roper Organization (1992), Unusual Personal Experiences, Las Vegas: Bigelow Holding Corp., p. 35.*

a. In each age group, find the percentage who reported seeing a ghost.

b. What is the *relative risk* of reportedly seeing a ghost for one group compared to the other? Write your answer in the form of a sentence that could be understood by someone who knows nothing about statistics.

c. Repeat part (b) using *increased risk* instead of relative risk.

d. What are the odds of reportedly seeing a ghost to not seeing one in the older group?

6.26 Suppose a newspaper article states that drinking three or more cups of coffee per day doubles the risk of gall bladder cancer. Before giving up coffee, what questions should be asked by a person who drinks this much coffee?

6.27 Suppose that you read in your hometown newspaper that there were 80 home burglaries in your town in 2005 compared to only 65 in 1990. Explain why this might not mean that your home is more at risk for a burglary now than it was in 1990.

6.28 Discuss how you might estimate your risk of being injured in the next year by falling down stairs.

6.29 The teacher of a class you're taking tells the class that he thinks students who often skip classes have twice the risk of failing the class compared to students who regularly attend class. To better understand his claim, what question(s) would be reasonable to ask the professor?

## Section 6.3

6.30 ● Case Study 1.5 (p. 5) was called "Does Prayer Lower Blood Pressure?" One of the results quoted in that study was "People who attended a religious service once a week and prayed or studied the Bible once a day were 40% less likely to have high blood pressure than those who don't go to church every week and prayed and studied the Bible less."

a. What is the explanatory variable in this study?

b. What is the response variable in this study?

c. Give an example of a third variable that might at least partially account for the observed relationship.

6.31 ● Refer to Exercise 6.30, in which one or more third variables are at least partially likely to account for the observed relationship between religious activities and reduced incidence of high blood pressure. Is this an example of Simpson's Paradox? Explain.

6.32 ● This exercise presents a real example of Simpson's paradox (Wagner, 1982). The total income and total taxes paid in each of five income categories are given for two years, 1974 and 1978, in the following table.

| Adjusted Gross Income | 1974 | | 1978 | |
|---|---|---|---|---|
|  | *Income* | *Tax* | *Income* | *Tax* |
| **Under $5,000** | 41,651,643 | 2,244,467 | 19,879,622 | 689,318 |
| **$5,000 to $9,999** | 146,400,740 | 13,646,348 | 122,853,315 | 8,819,461 |
| **$10,000 to $14,999** | 192,688,922 | 21,449,597 | 171,858,024 | 17,155,758 |
| **$15,000 to $99,999** | 470,010,790 | 75,038,230 | 865,037,814 | 137,860,951 |
| **$100,000 or more** | 29,427,152 | 11,311,672 | 62,806,159 | 24,051,698 |
| **Total** | 880,179,247 | 123,690,314 | 1,242,434,934 | 188,577,186 |

a. The "tax rate" for any income category is calculated as Tax divided by Income. Calculate the tax rates for each income bracket in each year.

b. Calculate the overall tax rate for each of the two years.

c. Did the tax rates within each income bracket increase or decrease from 1974 to 1978?

d. Did the overall tax rate increase or decrease from 1974 to 1978?

e. Compare your results in parts (c) and (d). Explain them in terms of Simpson's Paradox.

**6.33** In a 1997 Marist College Institute for Public Opinion survey of 995 randomly selected Americans, 31% of the men and 12% of the women surveyed said they have dozed off while driving (*Source:* www.mipo.marist.edu). Think of a third variable that might at least partially explain the observed relationship between gender and dozing when driving. Briefly explain.

6.34 Suppose two hospitals are willing to participate in an experiment to test a new treatment, and both hospitals agree to include 1100 patients in the study. Because the researchers who are conducting the experiment are on the staff of hospital A, they decide to perform the majority of cases with the new procedure. They randomly assign 1000 patients to the new treatment, with the remaining 100 receiving the standard treatment. Hospital B, which is a bit reluctant to try something new on too many patients, agrees to randomly assign 100 patients to the new treatment, leaving 1000 to receive the standard treatment. The following table displays the results:

**Survival Rates for Standard
and New Treatments at Two Hospitals**

| | Hospital A | | | Hospital B | | |
|---|---|---|---|---|---|---|
| **Treatment** | Survive | Die | Total | Survive | Die | Total |
| **Standard** | 5 | 95 | 100 | 500 | 500 | 1000 |
| **New** | 100 | 900 | 1000 | 95 | 5 | 100 |
| **Total** | 105 | 995 | 1100 | 595 | 505 | 1100 |

a. Which treatment was more successful in hospital A? Justify your answer with relevant percents.
b. Which treatment was more successful in hospital B? Justify your answer with relevant percents.
c. Combine the data from the two hospitals into a single contingency table that shows the relationship between treatment and outcome. Which treatment has the higher survival rate in this combined table? Justify your answer.
d. Explain how this exercise is an example of Simpson's paradox.

6.35 A researcher observes that, compared to students who do not procrastinate, students who admit to frequent procrastination are more likely to miss class due to illness. Does this mean that procrastinating increases illness? What is another explanation?

6.36 A well-known example of Simpson's Paradox, published by Bickel, Hammel, and O'Connell (1975), examined admission rates for men and women who had applied to graduate programs at the University of California at Berkeley. The actual data for specific programs is confidential, so we are using similar, hypothetical numbers. For simplicity, we assume there are only two graduate programs. The figures for acceptance to each program are shown in the following table:

| | Program A | | Program B | |
|---|---|---|---|---|
| | *Admit* | *Deny* | *Admit* | *Deny* |
| **Men** | 400 | 250 | 50 | 300 |
| **Women** | 50 | 25 | 125 | 300 |

a. Combine the data for the two programs into one table. What percentage of all men who applied was admitted? What percentage of all women who applied was admitted? Which sex was more successful in the admissions process?
b. What percentage of the men who applied did program A admit? What percentage of the women who applied did program A admit? Repeat the question for program B. Which sex was more successful in getting admitted to program A? Into program B?
c. Explain how this problem is an example of Simpson's Paradox. Provide a potential explanation for the observed figures by guessing what type of programs A and B might have been.

**6.37** The success rates of two treatments (A and B) for clinical depression are being compared. The research team included five doctors, and the participants were 200 patients with depression. The doctors were supposed to randomly assign treatments to patients, but two doctors didn't do this. Instead, they often assigned more severely depressed patients to use treatment A and less severely depressed patients to use treatment B.

**a.** Three variables measured by the investigators were the outcome of treatment (successful or not), method of treatment (A or B), and a rating of the initial severity of the depression (mild or severe). Which of these variables is the response variable in this investigation?
b. The doctors were surprised by the final result, which was that treatment B had a higher success rate than treatment A. Based on results attained by other investigators, they had expected the opposite to occur. Explain how the actions of the two doctors who did not always randomly assign treatments may have caused an unfair comparison of the two treatments.
c. Taking Example 6.10 (p. 205) into consideration, describe how the researchers might control for the effect of the initial severity of patient depression when they analyze the relationship between treatment method and outcome.

## Section 6.4

**6.38** ● ◆ The 2 × 2 contingency table on the next page shows data for gender and opinion on the death penalty for respondents in the 2002 General Social Survey in which a random sample of adults in the United States was surveyed. (Raw data are in the **GSS-02** dataset on the CD for this book.)

**Opinion on Death Penalty**

|        | Oppose | Favor | All  |
|--------|--------|-------|------|
| Male   | 156    | 475   | 631  |
| Female | 253    | 424   | 677  |
| All    | 409    | 899   | 1308 |

a. Write null and alternative hypotheses about the relationship between the two variables in this problem.

b. For this table, the value of the chi-square statistic is 24.31, and to three decimal places, the corresponding $p$-value is .000. Based on these results, state a conclusion about the two variables, and explain why you came to this conclusion.

c. For each sex separately, calculate the percentage who were opposed to the death penalty. Explain whether these percentages are evidence that opinion about the death penalty is related to gender.

**6.39** ● ◆ Calculate a table of expected counts for the data in Exercise 6.38.

**6.40** ● In a national survey, $n = 1000$ randomly selected adults are asked how important religion is in their own lives (very, fairly, or not very) and whether they approve or disapprove of same-sex marriage. Write null and alternative hypotheses about the two variables in this situation. Make your hypothesis statements specific to this situation.

**6.41** ● Researchers studied a random sample of North Carolina high school students who participated in interscholastic athletics to learn about the risk of lower-extremity injuries (anywhere between hip and toe) for interscholastic athletes (Yang et al., 2005). Of 999 participants in girls' soccer, 74 experienced lower-extremity injuries. Of 1667 participants in boys' soccer, 153 experienced lower-extremity injuries.

a. Write null and alternative hypotheses about gender and the risk of a lower-extremity injury while playing interscholastic soccer.

b. For these data, the value of the chi-square statistic is 2.51, and the $p$-value for the chi-square test is .113. Based on these results, state a conclusion about the two variables in this situation and explain how you came to this conclusion.

c. For each sex separately, calculate the percent of participants who had a lower-extremity injury. Explain how the difference between these percentages is consistent with the conclusion you stated in part (b).

**6.42** ● Refer to Exercise 6.41.

a. Write a two-way table of observed counts for gender and whether a participant had a lower-extremity injury or not.

b. Determine a two-way table of expected counts for these data.

c. Show calculations verifying that the value of the chi-square statistic is 2.51.

**6.43** ● For each of the following results, explain what conclusion can be made about the null hypothesis that

there is no relationship between two variables that form a 2 × 2 contingency table.

a. $p$-value = .001.
b. $p$-value = .101.
c. $p$-value = .900.
d. $p$-value = .049.
e. $p$-value = .755.
f. Chi-square statistic = 4.01.
g. Chi-square statistic = 2.98.

**6.44** ● If a relationship is statistically significant, does that guarantee that it also has practical significance? Explain.

**6.45** If a relationship has practical significance, does it guarantee that statistical significance will be achieved in every study that examines it? Explain.

**6.46** Explain whether each of the following is possible.

a. A relationship exists in the observed sample but not in the population from which the sample was drawn.

b. A relationship does not exist in the observed sample but does exist in the population from which the sample was drawn.

c. A relationship does not exist in the observed sample, but an analysis of the sample shows that there is a statistically significant relationship, so it is inferred that there is a relationship in the population.

**6.47** The table for Exercise 6.25 gave data for the relationship between age group and whether a person reports ever having seen a ghost.

a. Write null and alternative hypotheses about the possible relationship between the two variables.

b. The Minitab output for a chi-square test of the relationship follows. Based on this output, what can be concluded about the relationship? What is the basis for this conclusion?

Expected counts are printed below observed counts

|        | Yes    | No       | Total |
|--------|--------|----------|-------|
| 18–29  | 212    | 1313     | 1525  |
|        | 174.93 | 1350.07  |       |
| 30+    | 465    | 3912     | 4377  |
|        | 502.07 | 3874.93  |       |
| Total  | 677    | 5225     | 5902  |

Chi-Sq = 7.857 + 1.018 +
         2.737 + 0.355 = 11.967
DF = 1, P-Value = 0.001

c. Notice in the Minitab output that the $p$-value is .001. Write a sentence that interprets this number.

**6.48** Refer to the Minitab output for Exercise 6.47.

a. Demonstrate how the *expected count* was computed for the "age 18–29, yes" cell of the table.

b. Verify that the expected count for the "age 30+, yes" cell can be determined by subtracting the answer for part (a) from the total count for the "yes" column of the table.

c. In each row of the table, express the *expected count* for yes as a percent of the total count for that row. How do these two percentages compare to each other?

**6.49** Imagine that 50 men and 50 women are asked, "Do you favor or oppose capital punishment for those convicted of murder?" In the observed data, 38/50 = 76% of the men favor capital punishment compared to 32/50 = 64% of the women.

**a.** Write null and alternative hypotheses about the possible relationship between gender and opinion about capital punishment.

**b.** A chi-square test of the statistical significance of the relationship has a *p*-value of .19. Is this evidence that there is or is not a relationship? What is the justification for this conclusion?

**6.50** Refer to Exercise 6.49 about the relationship between gender and opinion about capital punishment.

**a.** Calculate the value of the chi-square statistic.

**b.** In Exercise 6.49, the *p*-value was given as .19. Write a sentence that interprets this number.

**c.** Suppose that 500 men and 500 women had been surveyed, rather than 50 of each sex. Further suppose that the proportions in favor of capital punishment remained the same, at 380/500 or 76% of the men and 320/500 or 64% of the women. Calculate the chi-square statistic in this case.

**d.** The *p*-value for the chi-square statistic in part (c) is .000035. Is this evidence that there is a relationship between gender and opinion about capital punishment?

**e.** Explain the reason for the discrepancy between the result in Exercise 6.49b and the result in part (d) of this exercise.

**6.51** Considering the effect of sample size on the chi-square test, explain why a finding that a relationship is "not statistically significant" should not be interpreted as absolute proof that there is no relationship in the population.

**6.52** In Example 6.13 (p. 212), a statistically significant relationship was found between hormone therapy and invasive breast cancer. In the same study, the following observed counts were found for death from coronary heart disease (CHD). Carry out the five steps leading to a conclusion about whether there is a statistically significant relationship between hormone therapy and death from CHD.

| | **Death from CHD?** | | |
|---|---|---|---|
| | Yes | No | Total |
| **Hormones** | 33 | 8473 | 8506 |
| **Placebo** | 26 | 8076 | 8102 |
| **Total** | 59 | 16,549 | 16,608 |

**6.53** ◆ The following 2 × 2 contingency table shows data for gender and opinion about the legalization of marijuana for respondents in the 2002 General Social Survey in which a random sample of U.S. adults was surveyed. (Raw data are in the **GSS-02** dataset on the CD for this book.)

| | **Opinion on Marijuana** | | |
|---|---|---|---|
| | Legalize | Don't Legalize | All |
| **Male** | 153 | 224 | 377 |
| **Female** | 153 | 321 | 474 |
| **All** | 306 | 545 | 851 |

a. Write null and alternative hypotheses about the relationship between the two variables in this problem.

b. For this table, the value of the chi-square statistic is 6.29, and the corresponding *p*-value is .01. Based on these results, state a conclusion about the two variables and explain how you came to this conclusion.

c. For each sex separately, calculate the percentage who were opposed to the legalization of marijuana. Explain why these percentages are evidence that the two variables in this exercise are related.

d. Verify that the value of the chi-square statistic is 6.29.

## Chapter Exercises

**6.54** Wechsler and Kuo (2000) used data from the 1999 College Alcohol Study to examine the relationship between student alcohol use and student definitions of binge drinking. In the study, approximately 14,000 college students from 119 schools answered questions about drinking habits. Using the definition that binge drinking is five consecutive drinks for men and four consecutive drinks for women, the researchers categorized students into four "type of drinker" categories on the basis of their answers to questions about personal alcohol use. Students in the survey were also asked how many drinks in a row they thought constituted binge drinking for men and for women. The following is a contingency table showing the relationship between the student definitions of binge drinking for a man and student alcohol use:

**Type of Drinker and Personal Definition of Binge Drinking**

| | **Definition of Binge Drinking for Men (Drinks in a Row)** | | | | |
|---|---|---|---|---|---|
| **Type of Drinker** | ≤4 | 5 | 6 | ≥7 | Total |
| **Abstainer** | 814 | 659 | 484 | 702 | 2659 |
| **Nonbinge** | 714 | 1129 | 997 | 2223 | 5063 |
| **Occasional Binge** | 231 | 471 | 480 | 1780 | 2962 |
| **Frequent Binge** | 179 | 442 | 367 | 2147 | 3135 |
| **Total** | 1938 | 2701 | 2328 | 6852 | 13,819 |

a. Among all students, what percentage defined binge drinking for men as being seven or more drinks in a row?

b. Write a sentence describing what would be measured by row percentages in the table. Make your answer specific to these variables.

c. Write a sentence describing what would be measured by column percentages in this table. Make your answer specific to these variables.

d. Calculate the row percentages in the table. In a few sentences, describe the relationship between type of drinker and the personal definition of binge drinking.

6.55 Example 6.2 presented data on how many college men have ear pierces and tattoos. The following table contains data on ear pierces and tattoos for a sample of 1375 college women. The ear-pierce response is the total number of ear pierces for a woman, and this has been categorized in the rows of the table.

**Ear Pierces and Tattoos,
1375 College Women**

| Pierces | No Tattoo | Have Tattoo |
|---------|-----------|-------------|
| 2 or less | 498 | 40 |
| 3 or 4 | 374 | 58 |
| 5 or 6 | 202 | 77 |
| 7 or more | 73 | 53 |

*Data source: One of the authors.*

a. For each ear-pierce category, determine the percentage with a tattoo. On the basis of these percentages, would you say there is a relationship between these two variables? Explain.

b. Draw a bar graph of the column percentages. What does the graph show us about the relationship between having a tattoo and the number of ear pierces?

c. Calculate the percentage with at least five ear pierces for women who have a tattoo and separately for women who don't have a tattoo.

d. What percentage of the total sample has a tattoo?

e. What percentage of the total has two or fewer ear pierces and also does not have a tattoo?

6.56 In a retrospective observational study, researchers asked women who were pregnant with planned pregnancies how long it took them to get pregnant (Baird and Wilcox, 1985; see also Weiden and Gladen, 1986). Length of time to pregnancy was measured according to the number of cycles between stopping birth control and getting pregnant. Women were also categorized on whether or not they smoked, with smoking defined as having at least one cigarette per day for at least the first cycle during which they were trying to get pregnant. The following table summarizes the observed counts:

**Time to Pregnancy for Smokers
and Nonsmokers**

| | Pregnancy Occurred After | |
|---|---|---|
| | First Cycle | Two or More Cycles |
| Smoker | 29 | 71 |
| Nonsmoker | 198 | 288 |

a. Among smokers, what percentage was pregnant after the first cycle?

b. Among nonsmokers, what percentage was pregnant after the first cycle?

c. Draw a bar graph that can be used to examine the relationship between smoking and how long it took to get pregnant.

d. Among those who were pregnant after the first cycle, what percentage smoked?

e. Among those who took two or more cycles to become pregnant, what percentage smoked?

f. Do you think there is a relationship between smoking and how long it takes to get pregnant? Explain.

g. Can these data be used as evidence of a causal connection between smoking and how long it takes to get pregnant? Explain.

6.57 A case-control study in Berlin, reported by Kohlmeier, Arminger, Bartolomeycik, Bellach, Rehm, and Thamm (1992) and by Hand et al. (1994) asked 239 lung-cancer patients and 429 controls (matched to the cases by age and sex) whether or not they had kept a pet bird during adulthood. Of the 239 lung-cancer cases, 98 said yes. Of the 429 controls, 101 said yes.

a. Construct a contingency table for the data.

b. Compute the risk of lung cancer for bird owners and for those who had never kept a bird.

c. Can the risks computed in part (b) be used as baseline risks for the populations of those who have and have not owned birds? Explain.

d. What is the relative risk of lung cancer for bird owners?

e. What additional information about the risk of lung cancer would you want before you made a decision about whether or not to own a pet bird?

f. Can these data be used to establish a causal connection between owning a bird and the risk of lung cancer? Explain.

g. For any study of lung cancer, the effect of smoking should be considered. How would you determine whether smoking might be a confounding variable in this study?

6.58 Compute the chi-square statistic, and assess the statistical significance for the relationship between bird ownership and lung cancer, based on the data in Exercise 6.57. State a conclusion about the relationship.

6.59 Pagano and Gauvreau (1993, p. 133) reported data for women participating in the first National Health and Nutrition Examination Survey (Carter, Jones, Schatzkin, and Brinton, 1989). The explanatory variable was whether or not the woman gave birth to her first child at the age of 25 or older, and the outcome variable was whether or not she developed breast cancer. Observed counts are in the following table.

a. Calculate the risk of breast cancer for women who were under 25 years old when they had their first child.
b. Calculate the risk of breast cancer for women who were 25 or older when they had their first child.

### Age at Birth of First Child and Breast Cancer

| Age at First Child | Breast Cancer | No Breast Cancer | Total |
|---|---|---|---|
| 25 or Older | 31 | 1597 | 1628 |
| Under 25 | 65 | 4475 | 4540 |
| Total | 96 | 6072 | 6168 |

Source: Pagano and Gauvreau (1993).

c. Calculate the relative risk of breast cancer for women who were 25 or older when they had their first child compared with women who were under 25. Write a sentence that interprets this relative risk in the context of this problem.
d. What is the percent increase in the risk of breast cancer for women who were 25 or older when they had their first child?
e. Can these data be used as evidence of a causal connection between age at first child and the risk of breast cancer? Explain.

6.60 In Example 6.13 (p. 212), a statistically significant relationship was found between hormone therapy and invasive breast cancer. There were 166 cases of invasive breast cancer out of 8506 women taking hormones and 124 cases of invasive breast cancer out of 8102 women taking placebo.

a. What is the relative risk of invasive breast cancer when taking hormones compared with taking placebo?
b. What is the baseline risk of invasive breast cancer, assuming that the placebo group is representative of the relevant population of women?
c. Explain why it is important to report the baseline risk found in part (b) when reporting the relative risk found in part (a).

**6.61** Compute the chi-square statistic, and assess the statistical significance for the relationship between smoking and time to pregnancy in Exercise 6.56. State a conclusion about the relationship.

6.62 Example 6.13 (p. 212), was about the relationship between hormone therapy and invasive breast cancer. One of the results reported in the paper was "Absolute excess risks per 10,000 person-years attributable to estrogen plus progestin were . . . 8 more invasive breast cancers" (Writing Group for the Women's Health Initiative Investigators, 2002, p. 321). In other words, for every 10,000 "person-years" of women taking the hor-

mones instead of a placebo, there would be 8 additional cases of invasive breast cancer. Discuss practical versus statistical significance in this situation.

6.63 Exercise 6.9 concerned the relationship between anger and the risk of coronary heart disease. Some computer output for a chi-square test is shown below. What do the results indicate about the relationship? Explain.

Expected counts are printed below observed counts

| | Heart Disease | | |
|---|---|---|---|
| | Yes | No | |
| No Anger | 8 | 191 | 199 |
| | 17.59 | 181.41 | |
| Most Anger | 59 | 500 | 559 |
| | 49.41 | 509.57 | |
| Total | 67 | 691 | 758 |

Chi-Sq = 5.228 + 0.507 +
$\qquad$ 1.861 + 0.180 = 7.777
DF = 1, P-Value = 0.0005

6.64 Refer to the data in Exercise 6.63, showing the frequency of coronary heart disease for men with no anger and men with the most anger. Show how the numbers in the following statement were calculated: The odds of remaining free of heart disease versus getting heart disease are about 24 to 1 for men with no anger, whereas those odds are only about 8.5 to 1 for men with the most anger.

**6.65** Refer to Exercises 6.63 and 6.64. Find the odds ratio for remaining free of heart disease for men with no anger compared with men with the most anger. Give the result in a sentence that someone with no training in statistics would understand.

6.66 "Saliva test predicts labor onset" was the headline of a Reuters Health story that appeared May 23, 2000, at the Yahoo! Health News website. The story described a medical test called SalEst. A positive SalEst test indicates an elevated estrogen level, and this knowledge may help to predict how soon a pregnant woman will give birth because estrogen level increases in the 2 to 3 weeks before delivery. Researchers employed by the manufacturers of SalEst tested 642 women who had been pregnant for 39 weeks. Among the 615 women who delivered before the 42nd week of the pregnancy, about 59% had a positive SalEst test. Among the 27 women who delivered in the 42nd week or later, 33% had a positive SalEst test.

a. Construct a contingency table of counts for the relationship between the SalEst test result and the time of delivery.
b. Which of the two variables that were used to make the contingency table is the explanatory variable, and which is the response variable?
c. Among all 642 women, what percentage delivered before 42 weeks?

d. The article states that if a test result is positive, there is a 98% chance that the woman will deliver before 42 weeks. Based on the contingency table, justify this claim.

e. Among those who had a negative test, what percentage delivered before 42 weeks?

f. Would you advise a pregnant woman to spend $90 for the test? Explain why or why not.

6.67 According to a study on partner abuse reported by the *Sacramento Bee* (July 14, 2000, p. A6, Associated Press), 25% of women with male partners had been assaulted by their current or a former partner, whereas 11% of women with female partners had been assaulted. What is the relative risk of assault for women with male partners compared to women with female partners? Write your answer in a sentence that would be understood by someone with no training in statistics.

**6.68** In Exercise 6.36, data were given for admissions to two graduate programs for men and women. The data are given again here for use in this exercise, with the combined data presented as well.

|  | Program A | | Program B | | Combined | |
|---|---|---|---|---|---|---|
|  | Admit | Deny | Admit | Deny | Admit | Deny |
| **Men** | 400 | 250 | 50 | 300 | 450 | 550 |
| **Women** | 50 | 25 | 125 | 300 | 175 | 325 |

**a.** Give the odds of being admitted versus being denied to program A separately for men and women. Then give the odds ratio for women compared to men. Write a sentence explaining the odds ratio in words.

**b.** Give the odds of being admitted versus being denied for the combined programs separately for men and women. Then give the odds ratio for women compared to men. Write a sentence explaining the odds ratio in words.

**c.** Using the results from parts (a) and (b), write one sentence that sounds as if the university favors women applicants and another sentence that sounds as if the university favors male applicants.

## Dataset Exercises

**Statistics⚫Now™** Datasets **are required** to solve these exercises and can be found at **http://1pass.thomson.com** or on your CD.

6.69 For this exercise, use the **UCDavis2** dataset. The variable **WtFeel** contains UC Davis student responses to the question "Do you think you are: Underweight? About Right? Overweight?"

a. Create a contingency table for the relationship between **sex** (female or male) and perception of weight (**WtFeel**).

b. For each sex category, calculate the conditional percentages for the perception of weight variable. Of the men, what percentage thinks they are underweight? Of the women, what percentage thinks they are underweight?

c. In Section 6.5, the chi-square test was used only for $2 \times 2$ tables, but it can be used for larger tables as well. Use software to do a chi-square test of statistical significance for the relationship between sex and perception of weight. What is the $p$-value of the test? On the basis of this $p$-value, what conclusion can be reached about the relationship?

6.70 For this exercise, use the **UCDavis2** dataset. The variable **Cheat** contains answers to a question about whether the respondent would tell the instructor if he or she saw another student cheating on an examination.

a. Create a two-way table for the relationship between **sex** (male or female) and the variable **Cheat.** Of all students, what percentage said that they would tell the instructor about witnessing cheating?

b. Of the females, what percentage said that they would tell the instructor about witnessing cheating? What percentage of the males said that they would do this?

c. Is the observed relationship between sex and response to the cheating question statistically significant? Justify your answer by reporting the results of a chi-square test.

d. For this question about cheating, what population do you think is represented by the students in the data set?

6.71 For this exercise, use the **GSS-02** dataset. The variable **gunlaw** has responses to a question about whether the respondent would favor a law requiring a police permit to be obtained before the purchase of any gun. The variable **owngun** indicates whether or not there are any guns in the respondent's home (including garages and sheds).

a. Write null and alternative hypotheses about the possible relationship between **owngun** and **gunlaw.**

b. Create a two-way table that summarizes the relationship between **owngun** and **gunlaw.** What percentage of all respondents had a gun in their home?

c. Among those with a gun in their home, what percentage favors a gun permit law? Among those with no gun in their home, what percentage favors a gun permit law? On the basis of the difference between these two percentages, would you say that there is a relationship between the two variables? Why or why not?

d. What is the $p$-value of a chi-square test of statistical significance for the relationship between **owngun** and *gunlaw*? Is the observed relationship statistically significant?

e. Explain what it means to say that an observed relationship is statistically significant.

6.72 For this exercise, use the **GSS-02** dataset. Write a paragraph that describes the relationship between *race* (respondent's race) and *cappun* (opinion about capital punishment). Include a two-way table of counts and a relevant set of conditional percentages.

**Statistics⬡Now**™ Preparing for an exam? Assess your progress by taking the post-test at **http://1pass.thomson.com.**

**Mentor**™
    Do you need a live tutor for homework problems? Access vMentor at **http://1pass.thomson.com** for one-on-one tutoring from a statistics expert.

# 7

*Mario Tama/Getty Images*

## *Does the shuffle function really mix the song order randomly?*

**See Case Study 7.2** *(p. 268)*

# Probability

We rarely stop to think about the precise meaning of the word *probability*. When we speak of the probability that we will win a lottery based on buying a single ticket, are we using the word in the same way that we do when we speak about the probability that we will buy a new house in the next five years?

Statistical methods are used to evaluate information in uncertain situations and probability plays a key role in that process. Remember our definition of statistics from Chapter 1: *Statistics is a collection of procedures and principles for gathering data and analyzing information in order to help people make decisions when faced with uncertainty.* Decisions like whether to buy a lottery ticket, whether to buy an extended warranty on a computer, or which of two courses to take are examples of decisions that you may have to make that involve uncertainty and the evaluation of probabilities.

Probability calculations are a key element of *statistical inference,* in which we use sample information to make conclusions about a larger population. For example, consider Case Study 1.6 in which 22,071 physicians were randomly assigned to take either aspirin or a placebo. In the aspirin group, there were 9.42 heart attacks per 1000 participating doctors, whereas in the placebo group there were 17.13 heart attacks per 1000 participants. It can be determined that the probability is only .00001 that the observed difference between heart attack rates would be so large if in truth there was no difference between taking aspirin and taking placebo. This is strong evidence that the observed difference did not occur just by chance. From this, we conclude that taking aspirin does reduce the risk of a heart attack. ∎

## 7.1 Random Circumstances

The next case study is hypothetical but contains the kinds of situations that people encounter every day involving probability. We will use the elements of this story throughout the chapter to illustrate the concepts and calculations necessary to understand the role probability plays in our lives.

Last week, Alicia went to her physician for a routine medical exam. This morning, her physician phoned to tell her that one of her tests came back positive, indicating that she may have a disease that we will simply call D. Thinking there must be some mistake, Alicia inquired about the accuracy of the test. The physician told her that the test is 95% accurate as to whether someone has disease D or not. In other words, when someone has D, the test detects it 95% of the time. When someone does not have D, the test is correctly negative 95% of the time.

Therefore, according to the physician, even though only 1 out of 1000 women of Alicia's age actually has D, the test is a pretty good indicator that Alicia may have the disease. Alicia doesn't know it yet, but her physician is wrong to imply that it's likely that Alicia has the disease. Actually, her chance of having the disease is small, even given the positive test result. Later in this chapter you will discover why this is true.

Alicia had planned to spend the morning studying for her afternoon statistics class. At the beginning of each class, her professor randomly selects three different students to answer questions about the material. There are 50 students in the class, so Alicia reasons that she is not likely to be selected. Rather than studying, she uses the time to search the Web for information about disease D.

At the statistics class that afternoon, the 50 student names are written on slips of paper and put into a bag. One name is drawn (without replacement) for each of the three questions. Alicia twice breathes a sigh of relief. She is not picked to answer either of the first two questions. But probability is not in Alicia's favor on this day: She gets picked to answer the third question.

## Random Circumstances in Alicia's Day

A **random circumstance** is one in which the outcome is unpredictable. In many cases, *the outcome is not determined until we observe it.* It was not predetermined that Alicia would be selected to answer one of the questions in class. This happened when the professor drew her name out of the bag. In other cases, *the outcome is already determined, but our knowledge of it is uncertain.* Alicia either has disease D or not, but she and her physician don't know which possibility is true.

One lesson of this chapter is that the probabilities associated with random circumstances sometimes depend on other random circumstances. Alicia's test results were positive. The probability that Alicia would have a positive test depends on another random circumstance, which is whether or not she actually has disease D. In her statistics class, Alicia was not selected to answer questions 1 or 2. So at the time of the drawing of a name for the third question, the probability that she would be selected was 1/48 because there were only 48 names left in the bag instead of the original 50. If she had been selected to answer either question 1 or question 2, her probability of being selected to answer question 3 would have been 0.

Here is a list of the random circumstances in Alicia's story and the possible outcomes for each of them:

*Random Circumstance 1:* Disease status

- Alicia has D.
- Alicia does not have D.

*Random Circumstance 2:* Test result

- Test is positive.
- Test is negative.

*Random Circumstance 3:* First student's name is drawn

- Alicia is selected.
- Alicia is not selected.

*Random Circumstance 4:* Second student's name is drawn

- Alicia is selected.
- Alicia is not selected.

*Random Circumstance 5:* Third student's name is drawn

- Alicia is selected.
- Alicia is not selected.

### Assigning Probabilities to the Outcomes of a Random Circumstance

We said that the probability was 1/48 that Alicia would be picked for the third question given that she had not been picked for either of the first two. Note that this probability is expressed as a fraction. A probability is a value between 0 and 1 and is written either as a fraction or as a decimal fraction. From a purely mathematical point of view, a **probability** simply is a number between 0 and 1 that is assigned to a possible outcome of a random circumstance. Additionally, for the complete set of distinct possible outcomes of a random circumstance, the total of the assigned probabilities must equal 1.

In practice, of course, we should assign probabilities to outcomes in a meaningful way. A probability should provide information about how likely it is that a particular outcome will be the result of a random circumstance. In the next section, we discuss two different ways to assign and interpret probabilities. One of these ways, the relative frequency definition of probability, forms the foundation for the statistical inference methods that we will examine in later chapters.

7.1 Exercises are on page 269.

---

*thought question 7.1* Based on your understanding of probability and random events, assign probabilities to the two possible outcomes for Random Circumstance 3.*

---

*thought question 7.2* At the beginning of Alicia's day, the outcomes of the five random circumstances listed were uncertain to her. Which of them were uncertain because *the outcome was not yet determined* and which were uncertain because of Alicia's *knowledge of the outcome?***

---

# 7.2 Interpretations of Probability

The word *probability* is so common that in all probability, you will encounter it today in everyday language. But we rarely stop to think about the precise meaning of the word. When we speak of the probability that we will win a lot-

---

*HINT: See Example 7.3 on page 233.
**HINT: Only two of them were already determined.

tery based on buying a single ticket, are we using the word in the same way that we do when we speak about the probability that we will buy a new house in the next five years? We can quantify the chance of winning the lottery exactly, but our assessment of the chance that we will buy a new house is based on our personal and subjective belief about how life will evolve for us.

The conceptual difference illustrated by these two different situations creates two different interpretations of what is meant by the word *probability*. In most situations in statistics, a probability is assigned to a possible outcome on the basis of what will or has happened over the long run of repeatedly observing a random circumstance. There are situations, however, in which a probability may be assigned based on the expert assessment of an individual.

## The Relative Frequency Interpretation of Probability

The relative frequency interpretation of probability applies to situations in which we can envision repeatedly observing the results of a random circumstance. For example, it is easy to imagine flipping a coin over and over again and counting the number of heads and tails. It makes sense to interpret the probability that the coin lands with heads up to be the relative frequency, over the long run, with which it lands with heads up. When we say that the probability of flipping heads is 1/2, we can interpret this to mean that about half of a large number of flips will result in heads.

Here are some other situations to which the relative frequency interpretation of probability can be applied:

- Buying lottery tickets regularly and observing how often you win.
- Drawing a student's name out of a hat and seeing how often a particular student is selected.
- Commuting to work daily and observing whether or not a certain traffic signal is red when we encounter it.
- Surveying many adults and determining what proportion smokes.
- Observing births and noting how often the baby is a female.

| *definition* | In situations that we can imagine repeating many times, we define the **probability** of a specific outcome as *the proportion of times it would occur over the long run.* This also is called the **relative frequency** of that particular outcome. |
|---|---|

Notice the emphasis on what happens *in the long run.* We cannot accurately assess the probability of a particular outcome by observing it only a few times. For example, consider a family with five children in which only one child is a boy. We should not take this as evidence that the probability of having a boy is only 1/5. To more accurately assess the probability that a baby is a boy, we have to observe many births.

**Example 7.1** | **Probability of Male Versus Female Births** According to the U.S. Centers for Disease Control, the long-run relative frequency of males born in the United

**Table 7.1** Relative Frequency of Male Births over Time

| Weeks of Watching | Total Births | Total Boys | Proportion of Boys |
|---|---|---|---|
| 1 | 30 | 19 | .633 |
| 4 | 116 | 68 | .586 |
| 13 | 317 | 172 | .543 |
| 26 | 623 | 383 | .615 |
| 39 | 919 | 483 | .526 |
| 52 | 1237 | 639 | .517 |

States is about .512. In other words, over the long run, 512 male babies and 488 female babies are born per 1000 births. (*Source:* http://www.cdc.gov/nchs/data/nvsr/nvsr53/nvsr53_20.pdf.)

Suppose we were to tally births and the relative frequency of male births in a certain city for the next year. Table 7.1, which was generated with a computer simulation, shows what we might observe. In the simulation, the chance was .512 that any individual birth was a boy, and the average number of births per week was 24. Notice how the *proportion* or *relative frequency* of male births is relatively far from .512 over the first few weeks but then settles down to around .512 in the long run. After 52 weeks, the *relative frequency* of boys was $639/1237 = .517$. If we used only the data from the first week to estimate the probability of a male birth, our estimate would have been far from the actual value. We need to look at a large number of observations to accurately estimate a probability. ■

### Determining the Relative Frequency Probability of an Outcome

There are two ways to determine a relative frequency probability. The first method involves *making an assumption about the physical world.* The second method involves *making a direct observation of how often something happens.*

***Method 1: Make an Assumption about the Physical World***   Sometimes, it is reasonable to assign probabilities to possible outcomes based on what we think about physical realities. We generally assume, for example, that coins are manufactured in such a way that they are equally likely to land with heads or tails up when flipped. Therefore, we conclude that the probability of a flipped coin showing heads up is 1/2.

---

**Example 7.2**

Statistics⬡Now™

Watch a video example at **http://1pass.thomson.com** or on your CD.

**A Simple Lottery**   A lottery game that is run by many states in the United States is one in which players choose a three-digit number between 000 and 999. A player wins if his or her three-digit number is chosen. If we assume that the physical mechanism that is used to draw the winning number gives each possibility an equal chance, we can determine the probability of winning. There are 1000 possible three-digit numbers (000, 001, 002, . . . , 999), so the probability that the state picks the player's number is 1/1000. In the long run, a player should win about 1 out of 1000 times. Note that this does not mean that a player will win *exactly* once in every thousand plays. ■

---

**Example 7.3**

**The Probability That Alicia Has to Answer a Question**   In Case Study 7.1, there were 50 student names in the bag when a student was selected to answer the first question. If we assume that the names have been well mixed in the bag, then each student is equally likely to be selected. Therefore, the probability that Alicia would be selected to answer the first question was 1/50. ∎

*Method 2: Observe the Relative Frequency*   Another way to determine the probability of a particular outcome is by *observing the relative frequency of the outcome over many, many repetitions of the situation.* We illustrated this method with Example 7.1 when we observed the relative frequency of male births in a given city over a year. If we consider many births, we can get a very accurate figure for the probability that a birth will be a male. As we mentioned, the relative frequency of male births in the United States has been consistently close to .512. In 2002, for example, there was a total of 4,021,726 live births in the United States, of which 2,057,979 were males. Assuming that the probability does not change in the future, the probability that a live birth will result in a male is about 2,057,979/4,021,726 = .5117. (*Source:* http://www.cdc.gov/nchs/data/nvsr/nvsr53/nvsr53_20.pdf.)

**Example 7.4**

**The Probability of Lost Luggage**   According to Krantz (1992, p. 33; original source: U.S. Department of Transportation), 1 in 176 passengers on U.S. airline carriers will temporarily lose their luggage. This number is based on data collected over the long run and is found by dividing the number of passengers who temporarily lost their luggage by the total number of passengers. Another way to state this fact is to say that the probability is 1/176, or about .006, that a randomly selected passenger on a U.S. carrier will temporarily lose luggage. ∎

### Proportions and Percentages as Probabilities

Relative frequency probabilities are often derived from the proportion of individuals who have a certain characteristic, or a certain outcome in a random circumstance. The relative frequency probabilities are therefore numbers between 0 and 1. It is commonplace, however, to be somewhat loose in how relative frequencies are expressed. For instance, here are four ways to express the relative frequency of lost luggage:

- The proportion of passengers who lose their luggage is 1/176 or about .006.
- About 0.6% of passengers lose their luggage.
- The probability that a randomly selected passenger will lose his or her luggage is about .006.
- The probability that you will lose your luggage is about .006.

This last statement is not exactly correct, because *your* probability of lost luggage depends on a number of factors, including whether or not you check any luggage and how late you arrive at the airport for your flight. The probability of .006 really applies only to the long-run relative frequency across all passengers, or the probability that a *randomly selected* passenger is one of the unlucky ones.

### Estimating Probabilities from Observed Categorical Data

When appropriate data are available, the long-run relative frequency interpretation of probability can be used to estimate probabilities of outcomes and combinations of outcomes for categorical variables. Assuming that the data are representative, the probability of a particular outcome is estimated to be the relative frequency (proportion) with which that outcome was observed. As we learned in Case Study 1.3 and in Chapter 3, such estimates are not precise, and an approximate margin of error for the estimated probability is calculated as $1/\sqrt{n}$. We will learn a more precise way to calculate the margin of error in Chapter 10.

**Example 7.5**  |  **Nightlights and Myopia Revisited**  In an example discussed in Chapters 2 and 6, we investigated the relationship between nighttime lighting in early childhood and the incidence of myopia later in childhood. The results are shown again in Table 7.2.

**Table 7.2**  Nighttime Lighting in Infancy and Eyesight

| Slept with: | No Myopia | Myopia | High Myopia | Total |
|---|---|---|---|---|
| **Darkness** | 155 (90%) | 15 (9%) | 2 (1%) | 172 |
| **Nightlight** | 153 (66%) | 72 (31%) | 7 (3%) | 232 |
| **Full Light** | 34 (45%) | 36 (48%) | 5 (7%) | 75 |
| **Total** | 342 (71%) | 123 (26%) | 14 (3%) | 479 |

Assuming that these data are representative of a larger population, what is the approximate probability that someone from that population who sleeps with a nightlight in early childhood will develop some degree of myopia? There were 232 infants who slept with a nightlight. Of those, 72 developed myopia and 7 developed high myopia, for a total of 79 who developed some degree of myopia. Therefore, the relative frequency of some myopia among nightlight users is $(72 + 7)/232 = 79/232 = .34$. From this, we can conclude that the approximate probability of developing some myopia, given that an infant slept with a nightlight, is .34. This estimate is based on a sample of 232 people, so there is a margin of error of about $1/\sqrt{232} = .066$.  ■

## The Personal Probability Interpretation

The relative frequency interpretation of probability clearly is limited to repeatable conditions. Yet uncertainty is a characteristic of most events, whether or not they are repeatable under similar conditions. We need an interpretation of probability that can be applied to situations even if they will never happen again.

If you decide to drive downtown this Saturday afternoon, will you be able to find a good parking space? Should a movie studio release its potential new hit movie before Christmas, when many others are released, or wait until January when it might have a better chance to be the top box-office attraction? If the United States enters into a trade alliance with a particular country, will that cause problems in relations with another country? Will you have a better time if you go to Cancun or to Florida for your next spring break?

These are unique situations, not likely to be repeated. They require people to make decisions based on an assessment of how the future will evolve. We could each assign a personal probability to these events, based on our own knowledge and experiences, and we could use that probability to help us with our decision. Different people may not agree on what the probability of an event happening is, but nobody would be considered to be wrong.

| definition | We define the **personal probability** of an event to be the degree to which a given individual believes that the event will happen. Sometimes, the term **subjective probability** is used because the degree of belief may be different for each individual. |
|---|---|

There are very few restrictions on personal probabilities. They must fall between 0 and 1 (or, if expressed as a percent, between 0 and 100%). They must also fit together in certain ways if they are to be **coherent.** By *coherent,* we mean that your personal probability of one event doesn't contradict your personal probability of another. For example, if you thought that the probability of finding a parking space downtown Saturday afternoon was .20, then to be coherent, you must also believe that the probability of not finding one is .80. We will explore some of these logical rules later in this chapter.

### How We Use Personal Probabilities

People routinely base decisions on personal probabilities. This is why committee decisions are often so difficult. For example, suppose a committee is trying to decide which candidate to hire for a job. Each member of the committee has a different assessment of the candidates, and they may disagree on the probability that a particular candidate would best fit the job. We are all familiar with the problem juries sometimes have when trying to agree on someone's guilt or innocence. Each member of the jury has his or her own personal probability of guilt and innocence. One of the benefits of committee or jury deliberations is that they may help members to reach some consensus in their personal probabilities.

Personal probabilities often take relative frequencies of similar events into account. For example, the late astronomer Carl Sagan believed that the probability of a major asteroid hitting the earth soon is high enough to be of concern. "The probability that the Earth will be hit by a civilization-threatening small world in the next century is a little less than one in a thousand" (Araf, 1994, p. 4). To arrive at that probability, Sagan obviously could not use the long-run frequency definition of probability. He would have to use his own knowledge of astronomy combined with past asteroid behavior.

Notice that unlike relative frequency probabilities, personal probabilities assigned to unique events are not equivalent to proportions or percentages. For instance, it does not make sense to say that major asteroids will hit approximately 1 out of 1000 earths in the next century.

## A Philosophical Issue About Probability

There is some debate about how to represent probability when an outcome has been determined but is still unknown, such as if you have flipped a coin but not yet looked at it. Technically, any particular outcome has either happened or not. If it has happened, its probability of happening is 1; if it hasn't, its probability of happening is 0.

For instance, in Case Study 7.1, Alicia either has the disease or not, but no one knows which is true. Because 1 out of 1000 women of her age have the disease, without any additional information, we might say that the probability that

Alicia has it is 1/1000. But one could argue that this probability statement is meaningless, because she either has the disease or not.

As another example, suppose that a woman is pregnant but no one knows the sex of the fetus. Is the probability that she will have a boy either 0 or 1, or is it .512? Suppose you flip a coin but cover it with your hand before anyone sees it. Is the probability of a head still 1/2 until it is revealed?

The answer to the dilemma in these kinds of situations is that technically, once an outcome has been determined, a relative frequency probability statement no longer makes sense. However, a personal probability statement about what outcome has occurred or may ultimately be revealed does make sense. This interpretation is often what is meant when we discuss such occurrences in everyday life, even in situations in which relative frequency information is used to determine the probability value, such as in the three examples just discussed.

Another example of this type of situation is the construction of a 95% confidence interval, introduced in Chapter 3, which we will study in detail in Chapters 10 and 11. Before the sample is chosen, a probability statement makes sense. The probability is .95 that a sample will be selected for which the computed 95% confidence interval covers the truth. After the sample has been chosen, "the die is cast." Either the computed confidence interval will cover the truth, or it won't. That's why we say that we have *95% confidence* that the interval is correct, rather than saying that *the probability* that it is correct is .95.

7.2 Exercises are on page 269.

---

*in summary*      Interpretations of Probability

The *relative frequency probability* of an outcome is the proportion of times the outcome would occur over the long run. Relative frequency probabilities can be determined by either of these methods:

- Making an assumption about the physical world and using it to define relative frequencies

- Observing relative frequencies of outcomes over many repetitions of the same situation or measuring a representative sample and observing relative frequencies of possible outcomes

The *personal probability* of an outcome is the degree to which a given individual believes it will happen. Sometimes data from similar events in the past and other knowledge are incorporated when determining personal probabilities.

---

*thought question 7.3*   You are about to enroll in a course for which you know that 20% of the students will receive a grade of A. Do you think that the probability that *you* will receive an A in the class is .20? Do you think the probability that a randomly selected student in the class will receive an A is .20? Explain the difference in these two probabilities, using the distinction between relative frequency probability and personal probability in your explanation.*

---

**\*HINT:** Look at the preceding "In Summary" box. Which of the methods described applies in each case?

# 7.3 Probability Definitions and Relationships

In this section, we cover the fundamental definitions for how outcomes of one or more random circumstances may be related to each other. We also give some elementary rules for calculating probabilities. In the next section, we will cover more detailed rules for calculating probabilities

The first set of definitions has to do with the possible outcomes of a random circumstance. A specific possible outcome is called a **simple event,** and the collection of all possible outcomes is called the **sample space** for the random circumstance. The term **event** is used to describe any collection of one or more possible outcomes.

*definition*

- The **sample space** is the collection of unique, nonoverlapping possible outcomes of a random circumstance.
- A **simple event** is one outcome in the sample space. In other words, a simple event is a possible outcome of a random circumstance.
- An **event** is a collection of one or more simple events in the sample space. Events are often written using capital letters A, B, C, and so on.

**Example 7.6** | **Days per Week of Drinking Alcohol** Suppose that we randomly sample college students and ask each student how many days he or she drinks alcohol in a typical week. The *simple events* in the *sample space* of possible outcomes are 0 days, 1 day, 2 days, and so on, up to 7 days. These simple events will not be equally likely. We may want to learn the probability of the *event* that a randomly selected student drinks alcohol on 4 or more days in a typical week. The *event* "4 or more" is comprised of the *simple events* {4 days, 5 days, 6 days, and 7 days}. ■

## Assigning Probabilities to Simple Events

The notation $P(A)$ is used to denote the probability that event A occurs. A probability value is assigned to each *simple event* (unique outcome) using either relative frequency or personal probabilities. The relative frequency method is used most often.

### Two Conditions for Valid Probabilities

The assigned probabilities for the possible outcomes of a random circumstance must meet the following two conditions:

1. Each probability is between 0 and 1.
2. The sum of the probabilities over all possible simple events is 1. In other words, the total probability for all possible outcomes of a random circumstance is equal to 1.

### Probabilities for Equally Likely Simple Events

If there are $k$ equally likely simple events in the sample space, then the probability is $1/k$ for each one. For instance, suppose that a fair six-sided die is rolled and the number that lands face up is observed. Then the sample space is {1, 2, 3, 4, 5, 6}, and the probability is 1/6 for each outcome. As another example, suppose that two fair coins are tossed and the outcome of interest is the combination of results for the first two tosses. There are four equally likely events, {HH, HT, TH, TT}, where H = heads and T = tails, so the probability is 1/4 for each outcome.

**Example 7.7**

**Probabilities for Some Lottery Events** Example 7.2 (p. 233) described a lottery game in which players choose a three-digit number between 000 and 999.

> *Random Circumstance:* A three-digit winning lottery number is selected.
> *Sample Space:* {000, 001, 002, 003, . . . , 997, 998, 999}. There are 1000 simple events.
> *Probabilities for Simple Event:* The probability that any specific three-digit number is a winner is 1/1000. We assume that all three-digit numbers are equally likely.

The assumption that all simple events are equally likely can be used to find probabilities for events that include multiple simple events.

- Let event A = the last digit is a 9. Notice that event A contains the subset of all simple events that end with a 9. We can write this as A = {009, 019, . . . , 999}. Because one out of every ten three-digit numbers is included in this set, $P(A) = 1/10$.

- Let event B = the three digits are all the same. This event includes ten simple events, which are {000, 111, 222, 333, 444, 555, 666, 777, 888, 999}. Because event B contains 10 of the 1000 equally likely simple events, $P(B) = 10/1000$, or 1/100.  ■

## Complementary Events

An event either happens or it doesn't. A randomly selected person either has blue eyes or does not have blue eyes. Alicia either has a disease or does not. "Opposite" events are called **complementary events.**

---

*definition*    One event is the **complement** of another event if the two events do not contain any of the same simple events *and* together they cover the entire sample space. For an event A, the notation $A^C$ represents the complement of A.

---

Because the total probability for a sample space must be equal to 1, the probabilities of complementary events must sum to 1. In symbols, $P(A) + P(A^C) = 1$. As a result, $P(A^C) = 1 - P(A)$. In words, the probability that an event does not happen is equal to one minus the probability that it does.

**Example 7.8**

**The Probability of Not Winning the Lottery** If A is the event that a player buying a single three-digit lottery ticket wins, then $A^C$ is the event that he or she does not win. As we have seen in Example 7.2, $P(A) = 1/1000$, so it makes sense that $P(A^C) = 1 - 1/1000 = 999/1000$.  ■

## Mutually Exclusive Events

We sometimes are interested in two events with distinctly separate simple events that may not cover all possible outcomes in the sample space. For example, we might want to know the probability that a randomly selected man is unusually short or tall; for instance, is either at least 6 inches shorter than the mean height or is at least 6 inches taller than the mean height. The events "at least 6 inches shorter than the mean" and "at least 6 inches taller than the mean" are separate events that cannot occur simultaneously, so they are called *mutually exclusive* or *disjoint events.*

| *definition* | Two events are **mutually exclusive** if they do not contain any of the same simple events (outcomes). Equivalent terminology is that the two events are **disjoint.** |
| --- | --- |

**Example 7.9** | **Mutually Exclusive Events for Lottery Numbers**  Here are two mutually exclusive (disjoint) events when a winning lottery number is drawn from the set 000 to 999.

> A = All three digits are the same.
> B = The first and last digits are different.

Notice that events A and B together do not cover all possible three-digit numbers, so they are not complementary events. ■

When A and B are mutually exclusive events, the probability that either event A or event B occurs is found by adding probabilities for the two events. In symbols, *P*(either A or B) = *P*(A) + *P*(B) for mutually exclusive events. For example, the probability that a randomly selected person has either blue eyes or green eyes is equal to the probability of having blue eyes plus the probability of having green eyes.

## Independent and Dependent Events

The next two definitions have to do with whether the fact that one event will take place (or has taken place) affects the probability that a second event could happen.

| *definition* | • Two events are **independent** of each other if knowing that one will occur (or has occurred) does not change the probability that the other occurs.<br>• Two events are **dependent** if knowing that one will occur (or has occurred) changes the probability that the other occurs. |
| --- | --- |

The definitions can apply *either* to events *within the same random circumstance* or to events *from two separate random circumstances.* As an example of two dependent events within the same random circumstance, suppose that a random digit from 0 to 9 is selected. If we know that the randomly selected digit is an *even number,* our assignment of a probability to the event that *the number is 4* changes from 1/10 to 1/5. Often, we examine questions about the dependence of the outcomes of two separate random circumstances. For instance, does the knowledge of whether a randomly selected person is male or female (random circumstance 1) affect the probability that he or she is opposed to abortion (random circumstance 2)? If the answer is yes, the outcomes of the two random circumstances are dependent events.

**Example 7.10**

Statistics △ Now™

Watch a video example at **http://1pass.thomson.com** or on your CD.

**Winning a Free Lunch**  Some restaurants display a glass bowl in which customers deposit their business cards, and a drawing is held once a week for a free lunch. Suppose you and your friend Vanessa each deposit a card in two consecutive weeks. Define

Event A = You win in week 1.
Event B = Vanessa wins in week 1.
Event C = Vanessa wins in week 2.

Events A and B are *dependent* events. If you win (event A), then Vanessa cannot win (event B cannot happen). In this case, P(B) = 0. If you do not win (A does not happen), it is possible that Vanessa could win, so P(B) > 0. Notice that events A and B refer to the same random circumstance: the drawing in week 1.

Events A and C are *independent*. Knowing whether you win in the first week (event A) gives no information about the probability that Vanessa wins in the second week (event C). Notice that events A and C refer to two *different* random circumstances: the drawings in separate weeks. ■

In Example 7.10, the events from two different random circumstances were independent. This is not always the case. The next example illustrates that events from two separate random circumstances may be dependent.

**Example 7.11**

**The Probability That Alicia Has to Answer a Question**  In Case Study 7.1 (p. 230), Alicia was worried that her statistics teacher might call on her to answer a question during class. Define

Event A = Alicia is selected to answer question 1.
Event B = Alicia is selected to answer question 2.

In Example 7.3, we determined that P(A) = 1/50. Are A and B independent? Notice that they refer to different random circumstances, but P(B) is affected by knowing whether or not A occurred. If Alicia answered question 1, her name is already out of the bag, so P(B) = 0. If event A does not occur, there are 49 names left in the bag including Alicia's, so P(B) = 1/49. Therefore, knowing whether or not A occurred changes P(B). ■

## Conditional Probabilities

Suppose that two events are dependent. From the definition of dependent events, it follows that knowing that one event will occur (or has occurred) *does* change the probability that the other occurs. Therefore, we need to distinguish between two probabilities:

- P(B), the unconditional probability that the event B occurs
- P(B|A), read "probability of B given A," the conditional probability that the event B occurs *given that we know A has occurred or will occur*

| *definition* | The **conditional probability** of the event B, given that the event A occurs, is the long-run relative frequency with which event B occurs when circumstances are such that A also occurs. This probability is written as $P(B|A)$. |
|---|---|

**Example 7.12**

**Probability That a Teenager Gambles Depends on Gender**   Based on a survey of nearly all of the ninth- and twelfth-graders in Minnesota public schools in 1998, a total of 78,564 students, Stinchfield (2001) reported proportions admitting that they had gambled at least once a week during the previous year. He found that these proportions differed considerably for males and females, as well as for ninth- and twelfth-graders. One result he found was that 22.9% of the ninth-grade boys but only 4.5% of the ninth-grade girls admitted gambling at least weekly. Assuming these figures are representative of all Minnesota ninth-graders, we can write these as conditional probabilities for ninth-grade Minnesota students:

$P$(student is weekly gambler|student is boy) = .229
$P$(student is weekly gambler|student is girl) = .045

The vertical line | denotes the words *given that*. The probability that the student is a weekly gambler, *given that* the student is a boy, is .229. The probability that the student is a weekly gambler, *given that* the student is a girl, is .045.

Notice the dependence between the outcomes for the random circumstance "weekly gambling habit" and the outcomes for the random circumstance "gender." Knowledge of a student's gender changes the probability that he or she is a weekly gambler. ■

7.3 Exercises are on page 270.

---

*thought question 7.4*   Remember that there were 50 students in Alicia's statistics class and that student names were not put back in the bag after being selected. Define the events A = Alicia is selected to answer question 1 and B = Alicia is selected to answer question 2. Describe each of these four conditional probabilities in words and determine a value for each.

$P(B|A)$,       $P(B^C|A)$,       $P(B|A^C)$,       $P(B^C|A^C)$

Now, based on your answers, can you formulate a rule about the value of $P(B|A) + P(B^C|A)$?*

---

*in summary*     Key Probability Definitions

- The **sample space** is the collection of unique, nonoverlapping possible outcomes of a random circumstance.
- A **simple event** is one outcome in the sample space. In other words, a simple event is a possible outcome of a random circumstance.

---

**\*HINT:** For $P(B|A)$ and $P(B|A^C)$, see Example 7.11.

The events are independent, so Rule 3b applies. The probability that *both* individuals share your birth month is the probability that A occurs *and* B occurs:

$$P(\text{A and B}) = P(\text{A})P(\text{B}) = \frac{1}{12} \times \frac{1}{12} = \frac{1}{144} = .007$$

The Extension of Rule 3b allows this result to be generalized. For instance, the probability that four unrelated strangers *all* share your birth month is $(1/12)^4$. ■

## Determining a Conditional Probability (Rule 4)

*definition*  |  **Rule 4 (conditional probability):** This rule is simply an algebraic restatement of Rule 3a, but it is sometimes useful to use the following form of that rule, which gives the probability that B occurs *given* that A occurs:

$$P(\text{B}|\text{A}) = \frac{P(\text{A and B})}{P(\text{A})}$$

The assignment of letters to events A and B is arbitrary, so it is also true that

$$P(\text{A}|\text{B}) = \frac{P(\text{A and B})}{P(\text{B})}$$

Often, a conditional probability is given to us directly. In Example 7.12, for instance, it was reported that the estimated conditional probability is .045 that a ninth-grader is a weekly gambler if the student is a female. In other cases, the physical situation can make it easy to determine a conditional probability. For instance, in drawing 2 cards from an ordinary deck of 52 cards, the probability that the second card drawn is red, *given* that the first card drawn was red, is 25/51 because 25 of the remaining 51 cards are red. Occasionally, it is useful to use Rule 4, an algebraic restatement of the multiplication rule (Rule 3a), to calculate a conditional probability.

**Example 7.18**  |  **Probability That Alicia Is Picked for the First Question Given That She Is Picked to Answer a Question**  This example continues Case Study 7.1 (p. 230). Here is an example for which it is easy to see the logical answer. If *we know* that Alicia is picked to answer one of the questions, what is the probability that it was the first question? Once we know that she had to answer one of the three questions, the conditional probability that it was the first one should be 1/3. All three questions would be equally likely to be the question for which Alicia was picked.

Let's use Rule 4 to verify this logic. Define

A = Alicia is called on to answer the first question; $P(\text{A}) = 1/50$.
B = Alicia is called on to answer any one of the questions; $P(\text{B}) = 3/50$, since there are 50 students and 3 students are chosen to answer questions.

Notice that the event that A and B *both* occur is synonymous with the event A occurring, since A is a subset of B. If A has occurred, then B also has occurred.

Therefore, $P(A \text{ and } B) = 1/50$. Applying Rule 4, we have

$$P(A|B) = \frac{P(A \text{ and } B)}{P(B)} = \frac{1/50}{3/50} = \frac{1}{3} \blacksquare$$

**Example 7.19**

**The Probability of Guilt and Innocence Given a DNA Match**  This example is modified from one given by Paulos (1995, p. 72). Suppose that there has been a crime and it is known that the criminal is a person within a population of 6,000,000. Further, suppose it is known that in this population only about one person in a million has a DNA type that matches the DNA found at the crime scene, so let's assume there are six people in the population with this DNA type. Someone in custody has this DNA type. *We know* the person's DNA matches, but what is the probability that he is actually innocent? In the absence of any other evidence beyond the DNA match, the probability of guilt is 1/6 because the criminal is one of the six individuals who have this DNA type in the population. Therefore, the probability of innocence is 5/6.

We now know the answer, but to illustrate Rule 4, we will look at how it can be used to determine this probability. Imagine randomly selecting someone from the population. Define

Event A = DNA of randomly chosen person matches DNA at the crime scene.
Event B = person selected is innocent of the crime.

We want to determine

$$P(B|A) = \frac{P(A \text{ and } B)}{P(A)}$$

The event "A and B" is the event that the selected person is innocent *and* the DNA matches. There are five such people in the population, so $P(A \text{ and } B) = 5/6,000,000$. The probability of a DNA match is $P(A) = 6/6,000,000$, and

$$P(B|A) = \frac{P(A \text{ and } B)}{P(A)} = \frac{5/6,000,000}{6/6,000,000} = \frac{5}{6}$$

If you were on the jury, it would be important to realize that without additional evidence, the probability that this person is *innocent* is 5/6, even though the DNA matches. The prosecutor surely would emphasize a different conditional probability—specifically, the very small probability that an innocent person's DNA type would match the crime scene DNA (remember, there are 5,999,999 innocent people). Do not confuse these two different conditional probabilities.

You may be thinking that it is preposterous that the police would just happen to arrest someone whose DNA matches the DNA found at the crime scene. But, just as with fingerprints, there are now data bases of DNA matches, so it is quite reasonable that the police would find someone to arrest who has a DNA match, even if that person had nothing to do with the crime. ∎

> ***thought question 7.6***  Explain why the tree diagram in Figure 7.1 (p. 256) displayed disease status first and test results second rather than the other way around. *

# 7.6  Using Simulation to Estimate Probabilities

Some probabilities are so difficult or time-consuming to calculate that it is easier to **simulate** the situation repeatedly by using a computer or calculator and observe the relative frequency of the event of interest. If you simulate the random circumstance $n$ times and the outcome of interest occurs in $x$ out of those $n$ times, then the estimated probability for the outcome of interest is $x/n$. This is an estimate of the long-run relative frequency with which the outcome would occur in real life.

**Example 7.28**

**Getting All the Prizes**  Refer to the situation in Example 7.21 (p. 252). Suppose cereal boxes each contain one of four prizes and any box is equally likely to contain any of the four prizes. You would like to collect all four prizes. If you buy six boxes of cereal, how likely is it that you will get all four prizes?

It would be quite time-consuming to solve this problem using the rules presented in the previous sections. It is much simpler to simulate the situation using a computer or calculator. Table 7.3 shows 50 simulations using the Minitab computer software. In each case, six digits were drawn, each equally likely to be 1, 2, 3 or 4. The outcomes that contain all four digits are highlighted in bold.

**Table 7.3**  50 Simulations of Prizes from Cereal Boxes

| | | | | |
|---|---|---|---|---|
| 112142 | 443222 | 323324 | 321223 | **332314** |
| **224123** | 342324 | **232413** | 121123 | 422244 |
| 244412 | **342431** | 121333 | **244132** | 434224 |
| 234323 | **144332** | 121142 | **432121** | 144313 |
| 443313 | 211141 | **421342** | 441211 | 111134 |
| **113432** | 433424 | **312314** | 241114 | 313411 |
| 214443 | 222422 | 144441 | **213141** | 312232 |
| 312341 | 411124 | 111422 | 312213 | **323314** |
| **143124** | 111131 | **441233** | 121223 | 424433 |
| 213341 | 114333 | **243311** | 442244 | **214133** |

Counting the outcomes in bold reveals that 19 out of 50, or .38, of these simulations satisfy the condition that all four prizes are collected. In fact, this simulation had particularly good results. There are $4^6 = 4096$ equally likely possible outcomes for listing six digits randomly chosen from 1, 2, 3, 4, and 1560 of those outcomes include all four digits. So the actual probability of getting all four prizes is 1560/4096 = .3809. ∎

In general, determining the accuracy of simulation for estimating a probability is similar to determining the margin of error when estimating a propor-

*\***HINT:**  See the paragraph just before Example 7.27 on page 257.

tion from a sample survey. The probability estimated by simulation will be no farther off than $1/\sqrt{n}$ most of the time, where $n$ is the number of simulated repetitions of the situation.

This information about the accuracy of the simulation approach for estimating probabilities should make it clear that you need to repeat the simulated situation a large number of times to obtain accurate results. Therefore, it is most useful if you can use a computer to generate the simulated outcomes and also to count how many of them satisfy the event of interest.

**Example 7.29**

**Statistics⌂Now™**

For software help, download your Minitab, Excel, TI-83, SPSS, R, and JMP manuals from **http://1pass .thomson.com**, or find them on your CD.

**Finding Gifted ESP Participants** An ESP test is conducted by randomly selecting one of five video clips and playing it in one building, while a participant in another building tries to describe what is playing. Later, the participant is shown the five video clips and is asked to determine which one best matches the description he or she had given. By chance, the participant would get this correct with probability 1/5. Individual participants are each tested eight times, with five new video clips each time. They are identified as "gifted" if they guess correctly at least five times out of the eight tries. Suppose people actually do have some ESP and can guess correctly with probability .30 (instead of the .20 expected by chance). What is the probability that a participant will be identified as "gifted"?

In Chapter 8, you will learn how to solve this kind of problem, but we can simulate the answer using a random number table or generator that produces the digits 0, 1, 2, . . . , 9 with equal likelihood. Table 3.2 (page 81) is an example of a random number table. Many calculators and computers will simulate these digits. Here are the steps needed for one "repetition":

- Each "guess" is simulated with a digit, equally likely to be 0 to 9.
- For each participant, we simulate eight "guesses" resulting in a string of eight digits.
- If a digit is 7, 8, or 9, we count that guess as "correct," so $P(\text{correct}) = 3/10 = .3$, as required in the problem. If the digit is 0 to 6, the guess is "incorrect." (There is nothing special about 7, 8, 9; we could have used any three digits.)
- If there are five or more "correct" guesses (digits 7, 8, 9), we count that as "gifted."

The entire process is repeated many times, and the proportion of times the result is a "gifted" participant is an estimate of the desired probability.

***Process and Results Using Minitab***

1. Generate 1000 rows of eight columns, using random integers from 0 to 9. Each row represents the eight guesses for one participant.
2. Use the "code" feature to recode all of the values, (7,8,9)→1, (0,1,2,3,4,5,6)→0.
3. Sum the eight columns; for each row (each participant) the sum is the number of "correct" guesses. This is equivalent to the number of 1s in the row after recoding.
4. The sums are now in a column with 1000 rows. Tally this column to find out how many of the participants had five or more "correct guesses."

5. The results of doing this were as follows:

| Number Correct | 0 | 1 | 2 | 3 | 4 | 5 | 6 | 7 | 8 |
|---|---|---|---|---|---|---|---|---|---|
| Frequency | 64 | 187 | 299 | 271 | 119 | 44 | 14 | 2 | 0 |

Notice that there were $44 + 14 + 2 + 0 = 60$ individuals who got five or more correct, so the probability of finding a "gifted" participant each time is about $60/1000 = .06$. In other words, if everyone is equally talented and each guess is correct with probability .3, then there will be five or more correct guesses out of eight tries with probability .06, or about 6% of the time. ■

*7.6 Exercises are on page 274.*

---

***thought question 7.7***   Does using simulation to estimate probabilities rely on the relative frequency or the personal probability interpretation of probability? Explain. Whichever one you chose, could simulation be used to find probabilities for the other interpretation? Explain.*

---

# 7.7   Flawed Intuitive Judgments About Probability

People have poor intuition about probability assessments. In Case Study 7.1, Alicia's physician made a common mistake by informing her that the medical test was a good indicator of her disease status because it was 95% accurate as to whether she had the disease or not. As we learned in Example 7.23, the probability that she actually has the disease is small, about 2%. In this section, we explore this phenomenon, called *confusion of the inverse,* and some other ways in which intuitive probability assessments can be seriously flawed.

## Confusion of the Inverse

Eddy (1982) posed the following scenario to 100 physicians:

> One of your patients has a lump in her breast. You are almost certain that it is benign, in fact you would say there is only a 1% chance that it is malignant. But just to be sure, you have the patient undergo a mammogram, a breast X-ray designed to detect cancer.
>
> You know from the medical literature that mammograms are 80% accurate for malignant lumps and 90% accurate for benign lumps. In other words, if the lump is truly malignant, the test results will say that it is malignant 80% of the time and will falsely say it is benign 20% of the time. If the lump is truly benign, the test results will say so 90% of the time and will falsely declare that it is malignant only 10% of the time.
>
> Sadly, the mammogram for your patient is returned with the news that the lump is malignant. What are the chances that it is truly malignant?

---

*****HINT:** Could you tell a computer what proportions to use if you and your friend do not agree on what to use?

Most of the physicians to whom Eddy posed this question thought the probability that the lump was truly malignant was about 75% or .75. In truth, given the probabilities described in the scenario, *the probability is only .075.* The physicians' estimates were ten times too high! When Eddy asked the physicians how they arrived at their answers, he realized they were confusing the answer to the actual question with the answer to a different question. "When asked about this, the erring physicians usually report that they assumed that the probability of cancer given that the patient has a positive X-ray was approximately equal to the probability of a positive X-ray in a patient with cancer" (1982, p. 254).

Robyn Dawes has called this phenomenon **confusion of the inverse** (Plous, 1993, p. 132). The physicians were confusing the conditional probability of cancer *given a positive X-ray* with the inverse, the conditional probability of a positive X-ray *given that the patient has cancer!*

It is not difficult to see that the correct answer is indeed .075. Let's construct a hypothetical table of 100,000 women who fit this scenario. In other words, these are women who would present themselves to the physician with a lump for which the probability that it was malignant seemed to be about 1%. Thus, of the 100,000 women, about 1%, or 1000 of them, would have a malignant lump. The remaining 99%, or 99,000, would have a benign lump.

Further, given that the test was 80% accurate for malignant lumps, it would show a malignancy for 800 of the 1000 women who actually had one. Given that it was 90% accurate for the 99,000 women with benign lumps, it would show benign for 90%, or 89,100 of them, and malignant for the remaining 10%, or 9900 of them. Table 7.4 shows how the 100,000 women would fall into these possible categories.

**Table 7.4**  Breakdown of Actual Status Versus Test Status for a Rare Disease

|  | Test Is Malignant | Test Is Benign | Total |
|---|---|---|---|
| **Actually Malignant** | 800 | 200 | 1000 |
| **Actually Benign** | 9900 | 89,100 | 99,000 |
| **Total** | 10,700 | 89,300 | 100,000 |

Let's return to the question of interest. Our patient has just received a positive test for malignancy. Given that her test showed malignancy, what is the actual probability that her lump is malignant? Of the 100,000 women, 10,700 of them would have an X-ray show malignancy. But of those 10,700 women, only 800 of them actually have a malignant lump! Thus, given that the test showed a malignancy, the probability of malignancy is just $800/10,700 = 8/107 = .075$.

Sadly, many physicians are guilty of confusion of the inverse. Remember, in a situation in which the *base rate* for a disease is very low and the test for the disease is less than perfect, there will be a relatively high probability that a positive test result is a false positive. If you ever find yourself in a situation similar to the one just described, you may wish to construct a table like the one above.

To determine the probability of a positive test result being accurate, you need only three pieces of information:

1. What the base rate or probability that you are likely to have the disease is, without any knowledge of your test results.
2. What the **sensitivity** of the test is, which is the proportion of people who correctly test positive when they actually have the disease.
3. What the **specificity** of the test is, which is the proportion of people who correctly test negative when they don't have the disease.

Notice that items 2 and 3 are measures of the accuracy of the test. They do not measure the probability that people have the disease when they test positive or the probability that they do not have the disease when they test negative. Those probabilities, which are obviously the ones of interest to the patient, can be computed by constructing a table similar to Table 7.4 or by the other methods shown earlier in this chapter.

## Specific People Versus Random Individuals

According to the Federal Aviation Administration, between 1987 and 1996, there were 39 fatal airline accidents for regularly scheduled flights on U.S. carriers. During that same time period, there were 76,557,000 flight departures. That means that the relative frequency of fatal accidents was about one accident per 2 million departures. Based on these kinds of statistics, you will sometimes hear statements like "The probability that you will be in a fatal plane crash is 1 in 2 million" or "The chance that your marriage will end in divorce is 50%."

Do these probability statements really apply to you personally? In an attempt to personalize the information, reporters express probability statements in terms of individuals when they actually apply to the aggregate. Obviously, if you never fly, the probability that you will be in a fatal plane crash is 0. Here are two equivalent, correct ways to restate the aggregate statistics about fatal plane crashes:

- In the long run, about 1 flight departure out of every 2 million ends in a fatal crash.
- The probability that a randomly selected flight departure ends in a fatal crash is about 1/2,000,000.

If you have had a terrific marriage for 30 years, it is not likely to end in divorce now. Your probability of a divorce surely is less than the 50% figure often reported in the media. Here are two correct ways to express aggregate statistics about divorce:

- In the long run, about 50% of marriages end in divorce.
- At the beginning of a randomly selected marriage, the probability that it will end in divorce is about .50.

Notice the emphasis on "random selection" in the second version of each statement. Sometimes the phrase *randomly selected* is omitted, but it should always be understood that randomness is part of the communication.

# Coincidences

When one of the authors of this book was in college in upstate New York, she visited Disney World in Florida during summer break. While there, she ran into three people she knew from her college, none of whom were there together. A few years later, she visited the top of the Empire State Building in New York City and ran into two friends (who were there together) and two additional unrelated acquaintances. Years later, when traveling from London to Stockholm, she ran into a friend at the airport in London while waiting for the flight. The friend not only turned out to be taking the same flight but had been assigned an adjacent seat.

These events are all examples of what would commonly be called coincidences. They are certainly surprising, but are they improbable? Most people think that coincidences have low probabilities of occurring, but we shall see that our intuition can be quite misleading regarding such phenomena. We will adopt the definition of **coincidence** proposed by Persi Diaconis and Fred Mosteller (1989, p. 853):

*definition* | A **coincidence** is a surprising concurrence of events, perceived as meaningfully related, with no apparent causal connection.

The mathematically sophisticated reader may wish to consult the article by Diaconis and Mosteller, in which they provide some instructions on how to compute probabilities for coincidences. For our purposes, we will need nothing more sophisticated than the simple probability rules we encountered earlier in this chapter.

Following are some examples of coincidences that at first glance seem highly improbable.

**Example 7.30** | **Two George D. Brysons** "My next-door neighbor, Mr. George D. Bryson, was making a business trip some years ago from St. Louis to New York. Since this involved weekend travel and he was in no hurry, . . . and since his train went through Louisville, he asked the conductor, after he had boarded the train, whether he might have a stopover in Louisville.

"This was possible, and on arrival at Louisville he inquired at the station for the leading hotel. He accordingly went to the Brown Hotel and registered. And then, just as a lark, he stepped up to the mail desk and asked if there was any mail for him. The girl calmly handed him a letter addressed to Mr. George D. Bryson, Room 307, that being the number of the room to which he had just been assigned. It turned out that the preceding resident of Room 307 was another George D. Bryson" (Weaver, 1963, pp. 282–283). ∎

**Example 7.31** | **Identical Cars and Matching Keys** Plous (1993, p. 154) reprinted an Associated Press news story describing a coincidence in which a man named Richard Baker and his wife were shopping on April Fool's Day at a Wisconsin shopping center. Mr. Baker went out to get their car, a 1978 maroon Concord, and drove it around to pick up his wife. After driving for a short while, they noticed items in the car that did not look familiar. They checked the license plate, and sure enough, they had someone else's car. When they drove back to the shopping center (to find the police waiting for them), they discovered that the owner of the car they were driving was a Mr. Thomas Baker, no relation to Richard Baker. Thus, both Mr. Bakers were at the same shopping center at the same time, with identical cars and with matching keys. The police estimated the odds as "a million to one." ■

**Example 7.32**

**Winning the Lottery Twice** Moore (1997, p. 330) reported on a *New York Times* story of February 14, 1986, about Evelyn Marie Adams who won the New Jersey lottery twice in a short time period. Her winnings were $3.9 million the first time and $1.5 million the second time. When Ms. Adams won for the second time, the *New York Times* claimed that the odds of one person winning the top prize twice were about 1 in 17 trillion. Then, in May 1988, Robert Humphries won a second Pennsylvania lottery, bringing his total winnings to $6.8 million. ■

Most people think that the events just described are exceedingly improbable, and they are. *What is not improbable is that someone, somewhere, someday will experience those events or something similar.*

When we examine the probability of what appears to be a startling coincidence, we ask the wrong question. For example, the figure quoted by the *New York Times* of 1 in 17 trillion is the probability that a *specific* individual who plays the New Jersey State Lottery exactly twice will win both times (Diaconis and Mosteller, 1989, p. 859.) However, millions of people play the lottery every day, and it is not surprising that someone, somewhere, someday would win twice.

In fact, Purdue professors Stephen Samuels and George McCabe (cited in Diaconis and Mosteller, 1989, p. 859) calculated those odds to be practically a sure thing. They calculated that there was at least a 1 in 30 chance of a double winner in a four-month period and better than even odds that there would be a double winner in a seven-year period somewhere in the United States. And they used conservative assumptions about how many tickets past winners had purchased.

When you experience a coincidence, remember that there are over 6 billion people in the world and over 280 million in the United States. If something has a 1 in a million probability of occurring to each individual on a given day, it will occur to an average of over 280 people in the United States *each day* and to over 6000 people in the world *each day*. Of course, probabilities of specific events depend on individual circumstances, but we hope you can see that it is not unlikely that something surprising will happen quite often.

**Example 7.33**

**Sharing the Same Birthday** Here is a famous example that you can use to test your intuition about surprising events. How many people would need to be gathered together to be at least 50% sure that two of them share the same birthday (month and day, not necessarily year)? Most people provide answers that are much higher than the correct one, which is only 23 people.

There are several reasons why people have trouble with this problem. If your answer was somewhere close to 183, or half the number of birthdays, then you may have confused the question with another one, such as the probability that someone in the group has *your* birthday or that two people have a specific date as their birthday.

It is not difficult to see how to calculate the appropriate probability, using our probability rules. Notice that the only way to avoid two people having the same birthday is if all 23 people have different birthdays. To find that probability, we simply use the rule that applies to the word *and* (Rule 3), thus multiplying probabilities. The probability that the first three people have different birthdays is the probability that the second person does not share a birthday with the first, which is 364/365 (ignoring February 29), and the third person does not share a birthday with either of the first two, which is 363/365. (Two dates were already taken.)

Continuing this line of reasoning, the probability that none of the 23 people share a birthday is

$$\frac{(364)(363)}{(365)(365)} \cdots \frac{(343)}{(365)} = .493$$

The probability that at least two people share a birthday is the probability of the complement, or $1 - .493 = .507$.

If you find it difficult to imagine that this could be correct, picture it this way. Imagine each of the 23 people shaking hands with the other 22 people and asking them about their birthday. There would be 253 handshakes and birthday conversations. Surely, there is a relatively high probability that at least one of those pairs would discover a common birthday.

By the way, the probability of a shared birthday in a group of ten people is already better than one in nine, at .117. (There would be 45 handshakes.) With only 50 people, it is almost certain, with a probability of .97. (There would be 1225 handshakes.) ■

The technique used in Example 7.33 can be used to illustrate why some coincidences are not at all unlikely. Consider the situation described in Example 7.30, in which two George D. Brysons occupied the same hotel room in succession. It is not at all unlikely that occasionally, two successive occupants of a hotel room somewhere, sometime will have the same name. To see that this is so, consider the probability of the complement—that no matter who occupies a hotel room, the next occupant will have a different name. Even ignoring common names, if we add these probabilities across many thousands of hotel rooms and many days, weeks, and years, it is extremely unlikely that successive occupants would always have different names.

**Example 7.34**    **Unusual Hands in Card Games**  As a final example of unlikely coincidences, consider a card game such as bridge, in which a standard 52-card deck is dealt to four players, so they each receive 13 cards. Any specific set of 13 cards is equally likely, each with a probability of about 1 in 635 billion. You would probably not be surprised to get a mixed hand, say, the 4, 7, and 10 of hearts; 3, 8, 9, and jack of spades; 2 and queen of diamonds; and 6, 10, jack, and ace of clubs. Yet that specific hand is just as unlikely as getting all 13 hearts! The point is that any very specific event, surprising or not, has extremely low probability, but there are many, many surprising events, and their combined probability is quite high.

Magicians sometimes exploit the fact that many small probabilities add up to one large probability by doing a trick in which they don't tell you what to expect in advance. They set it up so that *something* surprising is almost sure to happen. When it does, you are likely to focus on the probability of *that particular outcome,* rather than realizing that a multitude of other outcomes would have also been surprising and that one of them was likely to happen.

To summarize, most coincidences seem improbable only if we ask for the probability of that specific event occurring at that time, to us. If, instead, we ask the probability of it occurring some time, to someone, the probability can become quite large. Further, because of the multitude of experiences we each have every day, it is not surprising that some of them may appear to be improbable. That specific event is undoubtedly improbable. What is not improbable is that something "unusual" will happen to each of us once in awhile. ■

## The Gambler's Fallacy

Another common misperception about random events is that they should be self-correcting. Another way to state this is that people think the long-run frequency of an event should apply even in the short run. This misperception has classically been called the **gambler's fallacy.** Researchers William Gehring and Adrian Willoughby (2000) have discovered that this fallacy and related decision making while gambling are reflected in activity in the brain, in "a medial-frontal region in or near the anterior cingulate cortex" (p. 2279).

The gambler's fallacy can lead to poor decision making, especially if applied to gambling. For example, people tend to believe that a string of good luck will follow a string of bad luck in a casino. Unfortunately, independent chance events have no such memory. Making ten bad gambles in a row doesn't change the probability that the next gamble will also be bad.

Notice that the gambler's fallacy primarily applies to independent events. The gambler's fallacy may not apply to situations in which knowledge of one outcome affects probabilities of the next. For instance, in card games using a single deck, knowledge of what cards have already been played provides information about what cards are likely to be played next, and gamblers routinely make use of that information. If you normally receive lots of mail but have received none for two days, you would probably (correctly) assess that you are likely to receive more than usual the next day.

A somewhat different but related misconception is what Tversky and Kahneman (1982) call the belief in the **law of small numbers,** "according to which [people believe that] even small samples are highly representative of the populations from which they are drawn" (p. 7). They report that "in considering tosses of a coin for heads and tails, for example, people regard the sequence HTHTTH to be more likely than the sequence HHHTTT, which does not appear random, and also more likely than the sequence HHHHTH, which does not represent the fairness of the coin" (p. 7). Remember that any specific sequence of heads and tails is just as likely as any other sequence if the coin is fair, so the idea that the first sequence is more likely is a misperception.

*7.7 Exercises are on page 274.*

---

***thought question 7.8***    If you wanted to pretend to be psychic, you could do a "cold reading" on someone you do not know. Suppose you are doing this for a 25-year-old woman. You make statements such as the following:

- I see that you are thinking of two men, one with dark hair and the other one with slightly lighter hair or complexion. Do you know who I mean?
- I see a friend who is important to you but who has disappointed you recently.
- I see that there is some distance between you and your mother that bothers you.

Using the material in this section, explain why this would often work to convince people that you are psychic. *

---

## case study 7.2    Doin' the iPod Random Shuffle

The ability to play a collection of songs in a random order is a popular feature of portable digital music players. As an example, an Apple iPod Shuffle player with 512 megabytes of memory can store about 120 songs. Players with larger memory can store and randomly order thousands of songs. When the shuffle function is used, the stored songs are played in a random order.

We mention the iPod because there has been much grumbling, particularly on the Internet, that its shuffle might not be random. Some users complain that a song might be played within the first hour in two or three consecutive random shuffles. A similar complaint is that there are clusters of songs by the same group or musician within the first hour or so of play. *Newsweek* magazine's technology writer, Steven Levy (31 January 2005), wrote, "From the day I first loaded up my first Pod, it was as if the little devil liked to play favorites. It had a particular fondness for Steely Dan, whose songs always seemed to pop up two or three times in the first hour of play. Other songs seemed to be exiled to a forgotten corner" (p. 10). Conspiracy theorists even accuse Apple of playing favorites, giving certain musicians a better chance to have their songs played early in the shuffle.

One fundamental cause of the complaints may be a dislike of the consequences of randomness. People may be expecting a more systematic reordering than randomness often provides. For instance, there seems to be a belief that two songs by the same musician or group should not land near each other in the reordering. In a random shuffle, every possible reordering of the items is equally likely. (With a 120 song list, the number of possible shuffles is about 669 followed by 196 zeros.) Among all possible reorderings, some will have a few songs by the same musician being played nearly consecutively. This is analogous to occasionally getting several heads in a row when flipping coins or finding three aces within the first few cards of a randomly shuffled deck of playing cards.

For illustrative purposes, suppose that an iPod Shuffle user has stored ten albums with twelve songs per album. What is the probability that two or more songs from the same album will be among the first four songs played in a random shuffle? We can find this by first determining the probability that all of the first four songs of the shuffle are from different albums and then subtracting that probability from 1. Define

Event A = first song is anything.
Event B = second song is from different album than first.
Event C = third song is from different album than first two picked.
Event D = fourth song is from different album than first three picked.

The probability that all four are from different albums is

$$P(\text{A and B and C and D}) = \frac{120}{120} \times \frac{108}{119} \times \frac{96}{118} \times \frac{84}{117} = .53$$

Thus, the probability that all four songs are not from different albums, meaning that at least two songs are from the same album, is

$$P(\text{at least two of first four are from the same album}) = 1 - .53$$
$$= .47$$

At least two of the first four songs will be from the same album about one half of the time. If we extend the calculation to the first five songs in the random order, we can use the same method that was just used to learn that the probability is .671 that at least two songs will be from the same album. In fact, at least three of the first five songs in the random shuffle will be from the same album about 10% of the time.

We have set up a specific scenario to do these calculations, but the basic moral of the story remains the same for any situation in which we randomly select or order items. We too often imagine that we see patterns in random sequences and we tend to expect more structure then randomness provides. For example, we may expect a random shuffle to play one song by every musician before playing a second song by any musician. If this does not happen, we think there is a nonrandom pattern.

---

*****HINT:** Think about all of the possible ways in which these statements could be true for yourself.

# Key Terms

## Section 7.1
random circumstances, 230
probability, 231, 232

## Section 7.2
relative frequency, 232
personal probability, 236
subjective probability, 236
coherent probabilities, 236

## Section 7.3
simple event, 238
sample space, 238
event, 238
complementary events, 239
complement, 239

mutually exclusive events, 240
disjoint events, 240
independent events, 240
dependent events, 240
conditional probability, 242

## Section 7.4
probability rule 1: complements, 243
probability rule 2: addition, 244
addition rule, 244
probability rule 3: multiplication, 245
multiplication rule, 246
probability rule 4: conditional probability, 247
sample with replacement, 249–250
sample without replacement, 249–250

## Section 7.5
Bayes Rule, 252
tree diagram, 256

## Section 7.6
simulation, 259

## Section 7.7
confusion of the inverse, 262
sensitivity, 263
specificity, 263
coincidence, 264
gambler's fallacy, 267
law of small numbers, 267

# Exercises

● Denotes basic skills exercises

◆ Denotes dataset is available in StatisticsNow at **http://1pass.thomson.com** or on your CD but is **not required** to solve the exercise.

**Bold-numbered exercises** have answers in the back of the text and fully worked solutions in the Student Solutions Manual.

**Statistics⬛Now**™ Go to the StatisticsNow website at **http://1pass.thomson.com** to:
• Assess your understanding of this chapter
• Check your readiness for an exam by taking the Pre-Test quiz and exploring the resources in the Personalized Learning Plan

## Section 7.1

**7.1** ● According to a U.S. Department of Transportation website (http://www.dot.gov/airconsumer), 76.1% of domestic flights flown by the top ten U.S. airlines from June 1998 to May 1999 arrived on time. Represent this in terms of a random circumstance and an associated probability.

**7.2** ● Jan is a member of a class with 20 students. Each day for a week (Monday to Friday), a student in Jan's class is randomly selected to explain how to solve a homework problem. Once a student has been selected, he or she is not selected again that week. If Jan was not one of the four students selected earlier in the week, what is the probability that she will be picked on Friday? Explain how you found your answer.

**7.3** Identify three random circumstances in the following story, and give the possible outcomes for each of them:

It was Robin's birthday and she knew she was going to have a good day. She was driving to work, and when she turned on the radio, her favorite song was playing. Besides, the traffic light at the main intersection she crossed to get to work was green when she arrived, something that seemed to happen less than once a week. When she arrived at work, rather than having to search as she usually did, she found an empty parking space right in front of the building.

**7.4** Find information on a random circumstance in the news. Identify the circumstance and possible outcomes, and assign probabilities to the outcomes. Explain how you determined the probabilities.

## Section 7.2

**7.5** ● Suppose you live in a city that has 125,000 households with telephones and a polling organization randomly selects 1000 of them to phone for a survey. What is the probability that your household will be selected?

**7.6** ● Is each of the following values a legitimate probability value? Explain any "no" answers.
a. .50
b. .00
**c.** 1.00
**d.** 1.25
e. −.25

**7.7** ● Which interpretation of probability applies to each of the following situations? If it's the relative frequency interpretation, explain which of the methods listed in the "In Summary" box at the end of Section 7.2 applies.

a. If a spoon is tossed 10,000 times and lands with the rounded head face up 3,000 of those times, we would say that the probability of the rounded head landing face up for that spoon is about .30.

b. In a debate with you, a friend says that she thinks there is a 50–50 chance that God exists.

c. Based on data from the 1991 to 1993 General Social Survey, the probability in those years was about $612/1669 = .367$ that a randomly selected adult who had ever been married had divorced.

**7.8** ● A car dealer has noticed that one out of 25 new-car buyers returns the car for warranty work within the first month.

a. Write a sentence expressing this fact as a proportion.

b. Write a sentence expressing this fact as a percent.

c. Write a sentence expressing this fact as a probability.

**7.9** Explain which interpretation of probability (relative frequency or personal) applies to each of these statements and how you think the probability was determined.

**a.** According to Krantz (1992, p. 161), the probability that a randomly selected American will be injured by lightning in a given year is $1/685,000$.

**b.** According to my neighbor, the probability that the tomato plants she planted last month will actually survive to produce fruit is only $1/2$.

c. According to the nursery where my neighbor purchased her tomato plants, if a plant is properly cared for, the probability that it will produce tomatoes is .99.

d. The probability that a husband will outlive his wife (for U.S. couples) is $3/10$ (Krantz, 1992, p. 163).

**7.10** Refer to Exercise 7.3. Suppose Robin wants to find the probability associated with the outcomes in the random circumstances contained in the story. Identify one of the circumstances, and explain how she could determine the probabilities associated with its outcomes.

**7.11** Casino games often use a fair die that has six sides with 1 to 6 dots on them. When the die is tossed or rolled, each of the six sides is equally likely to come out on top. Using the physical assumption that the die is fair, determine the probability of each of the following outcomes for the number of dots showing on top after a single roll of a die:

a. Six dots.

b. One or two dots.

c. An even number of dots.

**7.12** A computer solitaire game uses a standard 52-card deck and randomly shuffles the cards for play. Theoretically, it should be possible to find optimal strategies for playing and then to compute the probability of winning based on the best strategy. Not only would this be an extremely complicated problem, but also

the probability of winning for any particular person depends on that person's skill. Explain how an individual could determine his or her probability of winning this game. (*Hint:* It might take a while to determine the probability!)

**7.13** Give an example of a situation for which a probability statement makes sense but for which the relative frequency interpretation could not apply, such as the probability given by Carl Sagan for an asteroid hitting the earth.

**7.14** Alicia's statistics class meets 50 times during the semester, and each time it meets, the probability that she will be called on to answer the first question is $1/50$. Does this mean that Alicia will be called on to answer the first question exactly once during the semester? Explain.

**7.15** Every day, John buys a lottery ticket with the number 777 for the lottery described in Example 7.2. He has played 999 times and has never won. He reasons that since tomorrow will be his 1000th time and the probability of winning is $1/1000$, he will have to win tomorrow. Explain whether or not John's reasoning is correct.

**7.16** Refer to Example 7.5 (p. 235). What is the probability that a randomly selected child who slept in darkness would develop some degree of myopia?

## Section 7.3

**7.17** ● A student wants to send a bouquet of roses to her mother for Mother's Day. She can afford to buy only two types of roses and decides to randomly pick two from the following four varieties: Blue Bell, Yellow Success, Sahara, and Aphrodite. Label the varieties B, Y, S, A.

a. Make a list of the six simple events in the sample space.

b. Assuming that all outcomes are equally likely, what is the probability that she will pick Sahara and Aphrodite?

**7.18** ● Remember that the event $A^C$ is the complement of the event A.

**a.** Are A and $A^C$ mutually exclusive? Explain.

**b.** Are A and $A^C$ independent? Explain.

**7.19** ● A penny and a nickel are tossed. Explain whether the outcomes for the two coins are:

a. Independent events.

b. Complementary events.

c. Mutually exclusive events.

**7.20** ● Suppose that events A and B are mutually exclusive with $P(A) = 1/2$ and $P(B) = 1/3$.

a. Are A and B independent events? Explain how you know.

b. Are A and B complementary events? Explain how you know.

**7.21** ● Suppose that A, B, and C are all disjoint possible outcomes for the same random circumstance. Explain whether each of the following sets of probabilities is possible.

**a.** $P(A) = 1/3$, $P(B) = 1/3$, $P(C) = 1/3$.

**b.** $P(A) = 1/2$, $P(B) = 1/2$, $P(C) = 1/4$.

c. $P(A) = 1/4$, $P(B) = 1/4$, $P(C) = 1/4$.

7.22 ● Jill and Laura have lunch together. They flip a coin to decide who pays for lunch and then flip a coin again to decide who pays the tip. Define a possible outcome to be who pays for lunch and who pays the tip, in order—for example "Jill, Jill."
a. List the simple events in the sample space.
b. Are the simple events equally likely? If not, why not?
c. What is the probability that Laura will have to pay for lunch and leave the tip?

7.23 Refer to Example 7.10 (p. 241), in which you and your friend Vanessa enter a drawing for a free lunch in week 1 and again in week 2. Events defined were A = you win in week 1, B = Vanessa wins in week 1, C = Vanessa wins in week 2.
a. Are events B and C independent? Explain.
b. Suppose that after week 1, the cards that were not drawn as the winning card are retained in the bowl for the drawing in week 2. Are events B and C independent? Explain.

7.24 When a fair die is tossed, each of the six sides (numbers 1 to 6) is equally likely to land face up. Two fair dice, one red and one green, are tossed. Explain whether the following pairs of events are disjoint:
**a.** A = red die is a 3; B = red die is a 6.
**b.** A = red die is a 3; B = green die is a 6.
c. A = red die and green die sum to 4; B = red die is a 3.
d. A = red die and green die sum to 4; B = red die is a 4.

**7.25** Refer to Exercise 7.24. For each part, explain whether the two events are independent.

7.26 When 190 students were asked to pick a number from 1 to 10, the number of students selecting each number were as follows:

| Number | 1 | 2 | 3 | 4 | 5 | 6 | 7 | 8 | 9 | 10 | Total |
|---|---|---|---|---|---|---|---|---|---|---|---|
| Frequency | 2 | 9 | 22 | 21 | 18 | 23 | 56 | 19 | 14 | 6 | 190 |

a. What is the approximate probability that someone asked to pick a number from 1 to 10 will pick the number 3?
b. What is the approximate probability that someone asked to pick a number from 1 to 10 will pick one of the two extremes, 1 or 10?
c. What is the approximate probability that someone asked to pick a number from 1 to 10 will pick an odd number?

7.27 According to Krantz (1992, p. 102), "Women between the ages of 20 and 24 have a 90 percent chance of being fertile while women between 40 and 44 have only a 37 percent chance of bearing children [i.e., being fertile]." Define appropriate events, and write these statements as conditional probabilities, using appropriate notation.

7.28 Refer to Exercise 7.27. Which method of finding probabilities do you think was used to find the "90 percent chance" and "37 percent chance"? Explain.

**7.29** Refer to Exercises 7.27 and 7.28. Suppose an American woman is randomly selected. Are her age and her fertility status independent? Explain.

7.30 Refer to Case Study 7.1. Define $C_1$, $C_2$, and $C_3$ to be the events that Alicia is called on to answer questions 1, 2, and 3, respectively.
a. Based on the physical situation used to select students, what is the (unconditional) probability of each of these events? Explain.
b. What is the *conditional* probability of $C_3$, *given* that $C_1$ occurred?
c. Are $C_1$ and $C_3$ independent events? Explain.

7.31 Use the information given in Case Study 7.1 and the "physical assumption" method of assigning probabilities to argue that on any given day, the probability that Alicia has to answer one of the three questions is 3/50.

## Section 7.4

7.32 ● A fair coin is tossed three times. The event "A = getting all heads" has probability = 1/8.
**a.** Describe in words what the event $A^C$ is.
**b.** What is the probability of $A^C$?

7.33 ● Two fair dice are rolled. The event "A = getting the same number on both dice" has probability = 1/6.
a. Describe in words what the event $A^C$ is.
b. What is the probability of $A^C$?

7.34 ● Two fair coins are tossed. Define

A = Getting a head on the first toss.

B = Getting a head on the second toss.

A and B = Getting a head on both the first and second tosses.

A or B = Getting a head on the first toss, or the second toss, or both tosses.

a. Find $P(A)$ = the probability of A.
b. Find $P(B)$ = the probability of B.
c. Using the multiplication rule (Rule 3b), find $P(A \text{ and } B)$.
d. Using the addition rule (Rule 2a), find $P(A \text{ or } B)$.

**7.35** ● Julie is taking English and history. Suppose that at the outset of the term, her probabilities for getting A's are

$P$(grade of A in English class) = .70.

$P$(grade of A in history class) = .60.

$P$(grade of A in both English and history classes) = .50.

**a.** Are the events "grade of A in English class" and "grade of A in history class" independent? Explain how you know.
**b.** Use the addition rule for two events (Rule 2) to find the probability that Julie will get at least one A between her English and history classes.

7.36 ● In a recent election, 55% of the voters were Republicans, and 45% were not. Of the Republicans, 80% voted for Candidate X, and of the non-Republicans, 10% voted for Candidate X. Consider a randomly selected voter. Define

A = Voter is Republican.

B = Voted for Candidate X.

a. Write values for $P(A)$, $P(A^C)$, $P(B|A)$, and $P(B|A^C)$.
b. Find $P(A$ and $B)$, and write in words what outcome it represents.
c. Find $P(A^C$ and $B)$, and write in words what outcome it represents.
d. Using the results in parts (b) and (c), find $P(B)$. (*Hint:* The events in parts (b) and (c) cover all of the ways in which B can happen.)
e. Use the result in part (d) to state what percent of the vote Candidate X received.

7.37 ● In each situation, explain whether the selection is made with replacement or without replacement.
   a. The three digits in the lottery in Example 7.2 (p. 233).
   b. The three students selected to answer questions in Case Study 7.1 (p. 230).
   c. The five people selected for extra security screening while boarding a particular flight.
   d. The two football teams selected to play in the Rose Bowl in a given year.
   e. The cars stopped by the police for speeding in five consecutive mornings on the same stretch of highway.

7.38 In Example 7.17 (p. 246), we found the probability that both of two unrelated strangers share your birth month. In this exercise, we find the probability that at least one of the two strangers shares your birth month. Assume that all 12 months are equally likely.
   **a.** What is the probability that the first stranger does *not* share your birth month?
   **b.** What is the probability that the second stranger does *not* share your birth month?
   **c.** What is the probability that *neither* of them shares your birth month?
   **d.** Use the result from part (c) to find the probability that at least one of them shares your birth month.

7.39 Refer to Exercise 7.38. In this exercise, another method is used for finding the probability that at least one of two unrelated strangers shares your birth month.
   a. What is the probability that the first stranger shares your birth month?
   b. What is the probability that the second stranger shares your birth month?
   c. What is the probability that both of them share your birth month?
   d. Use Rule 2a (for "either A or B") to find the probability that at least one of the strangers shares your birth month.

7.40 In Example 7.15, we found the probability that a woman with two children either has two girls or two boys is .5002. What is the probability that she has one child of each sex?

7.41 Harold and Maude plan to take a cruise together, but they live in separate cities. The cruise departs from Miami, and they each book a flight to arrive in Miami an hour before they need to be on the ship. Their travel planner explains that Harold's flight has an 80% chance of making it on time for him to get to the ship and that Maude's flight has a 90% chance of making it on time.
   **a.** How do you think the travel planner determined these probabilities?

b. Assuming that the probabilities quoted are long-run relative frequencies, do you think whether Harold's plane is on time is independent of whether Maude's plane is on time on the particular day they travel? Explain. (*Hint:* Consider reasons why planes are delayed.)
c. Whether realistic or not, assuming that the probabilities are independent, what is the probability that Harold and Maude will both arrive on time?
d. What is the probability that one of the two will be cruising alone?

7.42 A robbery has been committed in an isolated town. Witnesses all agree that the criminal was driving a red pickup truck and had blond hair. Evidence at the scene indicates that the criminal also smoked cigarettes. Police determine that 1/50 of the vehicles in town are red pickup trucks, 30% of the residents smoke, and 20% of the residents have blond hair. The next day they notice a red pickup truck whose driver has blond hair and is smoking. They arrest the driver for the robbery.
   a. Discuss whether or not you think it is reasonable to assume that whether someone drives a red pickup truck, smokes cigarettes, and has blond hair are all independent traits.
   b. Assuming that the traits listed are all independent, what proportion of the vehicle owners in town are red pickup truck owners who smoke and have blond hair?
   c. Continuing to assume the traits listed are independent, if there are 10,000 vehicle owners in town, how many of them fit the description of the criminal?
   d. Assuming that there is no other evidence, what is the probability that the driver arrested by the police is innocent of the robbery?
   e. The prosecuting attorney argues that the probability that someone would own a red pickup truck and smoke cigarettes and have blond hair is very small, so the person arrested by the police must be guilty. Do you agree with this reasoning? Explain.

7.43 A raffle is held in a club in which 10 of the 40 members are good friends with the president. The president draws two winners.
   **a.** If the two winners are drawn *with replacement*, what is the probability that a friend of the president wins each time?
   **b.** If the two winners are drawn *without replacement*, what is the probability that a friend of the president wins each time?
   **c.** If the two winners are drawn *with replacement*, what is the probability that neither winner is a friend of the president?
   **d.** If the two winners are drawn *without replacement*, what is the probability that neither winner is a friend of the president?

## Section 7.5

7.44 ● A public library carries 50 magazines that focus on either news or sports. Thirty of the magazines focus on news and the remaining 20 focus on sports. Among the 30 news magazines, 20 include international news and 10 include national, state, or local news only. Among the 20 sports magazines, 5 focus on international sports, and the remaining 15 focus on national, state, or local sports.

a. Create a table illustrating these numbers, with type of magazine (news, sports) as the row variable.

**b.** A customer randomly picks a magazine from the shelf and notices that it is a news magazine. Given that it is a news magazine, what is the probability that it includes international news?

**c.** What proportion of the magazines include international news or sports?

7.45 ● In a large general education class, 60% (.6) are science majors, and 40% (.4) are liberal arts majors. Twenty percent (.2) of the science majors are seniors, while 30% (.3) of the liberal arts majors are seniors.

a. If there are 100 students in the class, how many of them are science majors?

b. If there are 100 students in the class, how many of them are science majors and seniors?

c. Create a hypothetical table of 100 students with major (science, liberal arts) as the row variable and class (senior, nonsenior) as the column variable, illustrating the proportions given in the exercise.

d. Use the table created in part (c) to determine what percent of the class are seniors.

7.46 ● Refer to Exercise 7.45.

a. Create a tree diagram for this situation.

b. Use the tree diagram in part (a) to determine what percentage of the class are seniors.

**7.47** ● In a computer store, 30% (.3) of the computers in stock are laptops and 70% (.7) are desktops. Five percent (.05) of the laptops are on sale, while 10% (.1) of the desktops are on sale. Use a tree diagram to determine what percentage of the computers in the store are on sale.

7.48 ● In an Italian restaurant, a waitress has observed that 80% of her customers order coffee and 25% of her customers order both biscotti and coffee. Define

A = a randomly selected customer orders coffee.

B = a randomly selected customer orders biscotti.

a. Express the waitress's observations as probability statements involving events A and B.

b. What is the conditional probability that a randomly selected customer would order a biscotti, given that he or she orders coffee?

c. What is the conditional probability that a customer would not order a biscotti, given that he or she orders coffee?

7.49 ● Two students each use a random number generator to pick a number between 1 and 7. What is the probability that they pick the same number?

**7.50** A standard poker deck of cards contains 52 cards, of which four are aces. Suppose two cards are drawn sequentially, so that one random circumstance is the result of the first card drawn and the second random circumstance is the result of the second card drawn. Go through the five steps listed under "Specific Set of Steps for Finding Probabilities" to find the probability that the first card is an ace and the second card is not an ace.

7.51 Refer to the example toward the end of Section 7.4 of drawing two students without replacement from a class of 30 students in which 3 are left-handed. Draw a tree diagram to illustrate that the probability of selecting two left-handed students is 1/145.

7.52 Suppose that a magnet high school includes grades 11 and 12, with half of the students in each grade. Half of the senior class and 30% of the junior class are taking calculus. Create a hypothetical table of 100 students for this situation. Use grade (junior, senior) as the row variable.

**7.53** Refer to Exercise 7.52. Suppose a calculus student is randomly selected to accompany the math teachers to a conference. What is the probability that the student is a junior?

7.54 An airline has noticed that 40% of its customers who buy tickets don't take advantage of advance-purchase fares and the remaining 60% do. The no-show rate for those who don't have advance-purchase fares is 30%, while for those who do have them, it is only 5%.

a. Create a tree diagram for this situation.

b. Create a "hypothetical hundred thousand" table for this situation.

c. Using the tree diagram from part (a) or the table from part (b), find the percent of customers who are no-shows.

d. Given that a customer is a no-show, what is the probability that he or she had an advance-purchase fare?

e. Refer to the answer you found in part (d). Write a sentence presenting this result that someone with no training in statistics would understand.

7.55 In a test for ESP, a picture is randomly selected from four possible choices and hidden away. Participants are asked to describe the hidden picture, which is unknown to anyone in the environment. They are later shown the four possible choices and asked to identify the one they thought was the hidden target picture. Because the correct answer had been randomly selected from the four choices before the experiment began, the probability of guessing correctly by chance alone is 1/4 or .25. Each time the experiment is repeated, four new pictures are used.

a. Researchers are screening people for potentially gifted ESP participants for a later experiment. They will accept only people who get three correct answers in three repetitions of the screening experiment. What is the probability that someone who is just guessing will be accepted for the later experiment?

b. Refer to part (a). Suppose that an individual has some ESP ability and can correctly guess with probability .40 each time. What is the probability that such a person will be accepted for the later experiment?

c. Refer to parts (a) and (b). Suppose that half of the people in the population being tested are just guessing and have no ESP ability, while the other half can actually guess correctly with probability .40 each time. Given that someone is accepted for the upcoming experiment, what is the probability that the person actually has ESP ability?

## Section 7.6

**7.56** ● Five fair dice were tossed, and the sum of the resulting tosses was recorded. This process was repeated 10,000 times using a computer simulation. The number of times the sum of the five tosses equaled 27 was 45. What is the estimated probability that the sum of the five dice will be 27?

**7.57** ● The observed risk of an accident per month at a busy intersection without any stoplights was 1%. The potential benefit of adding a stoplight was studied by using a computer simulation modeling the typical traffic flow for a month at that intersection. In 10,000 repetitions, a total of 50 simulated accidents occurred.
   a. What is the estimated probability of an accident in a month at the intersection with a stoplight added?
   b. Would adding the stoplight be a good idea? Explain.

**7.58** ● Refer to Example 7.28 (p. 259) and use the results given in Table 7.3 to estimate probabilities of the following outcomes:
   **a.** Prize 4 is received at least three times.
   **b.** Prize 4 is received at least three times *given* that the full collection of all four prizes was not received.
   **c.** Prize 4 is received at least three times *and* the full collection of all four prizes was received.

**7.59** Refer to the simulation results given on page 261 for Example 7.29. What is the estimated probability that a participant would guess four or more correctly?

**7.60** Refer to Example 7.29 (p. 260). Explain how you would change the simulation procedure if the assumption was that everyone was randomly guessing, so that the probability of a correct guess was .20 each time.

**7.61** Refer to Example 7.29 (p. 260). Suppose the probability of a correct guess each time is .40.
   a. Explain how you would simulate this situation.
   b. Carry out the simulation, and estimate the probability that a participant will be identified as gifted.

**7.62** Suppose that in a lottery game players choose three digits, each from the set 0 to 9. A winning sum is then selected from the possibilities 0 (0 + 0 + 0) to 27 (9 + 9 + 9) by randomly choosing three digits and using their sum. For instance, if the three digits 483 are selected, then a winning ticket is any ticket in which the three digits sum to 4 + 8 + 3 = 15. Suppose the winning sum is 15.
   a. Using the physical assumption that all 1000 choices (000, 001, . . . , 999) are equally likely, find the probability that three digits chosen at random will sum to 15.
   b. Simulate playing this game 1000 times, randomly selecting the three digits each time. Based on the simulation, find the approximate probability of winning when the winning sum is 15.
   c. Based on information about how people respond when asked to pick random digits, do you think the

proportion of people who would actually pick three digits summing to 15 is close to the probabilities you found in parts (a) and (b)? Explain.

**7.63** Janice has noticed that on her drive to work, there are several things that can slow her down. First, she hits a red light with probability .3. If she hits the red light, she also has to stop for the commuter train with probability .4, but if she doesn't hit the red light, she has to stop with only probability .2.
   **a.** Find the probability that Janice has to stop for both the red light and the train.
   **b.** Find the probability that Janice has to stop for the red light *given* that she has to stop for the commuter train.
   c. Explain how you would conduct a simulation to estimate the probabilities in parts (a) and (b).
   d. Conduct the simulation described in part (c). Use at least 50 repetitions.
   e. Compare the answers that you computed in parts (a) and (b) to the estimated probabilities from your simulation in part (d).

## Section 7.7

**7.64** ● The Pap smear is a screening test to detect cervical cancer. Estimate the sensitivity and specificity of the test if a study of 200 women with cervical cancer resulted in 160 testing positive, and in another 200 women without cervical cancer, 4 tested positive.

**7.65** ● In tossing a fair coin ten times, if the first nine tosses resulted in all tails, will the chance be greater than .5 that the tenth toss will turn up heads? Explain.

**7.66** ● Which of the following sequences resulting from tossing a fair coin five times is most likely: HHHHH, HTHHT, or HHHTT? Explain your answer.

**7.67** ● Using material from Section 7.7, explain what is wrong with the following statement: "The probability that you will win the million dollar lottery is about the same as the probability that you will give birth to quintuplets."

**7.68** The University of California at Berkeley's *Wellness Encyclopedia* (1991) contains the following statement in its discussion of HIV testing: "In a high-risk population, virtually all people who test positive will truly be infected, but among people at low risk the false positives will outnumber the true positives. Thus, for every infected person correctly identified in a low-risk population, an estimated ten noncarriers [of HIV] will test positive" (p. 360). Suppose that you have a friend who is part of this low-risk population but who has just tested positive.
   a. Using the numbers in the statement above, what is the probability that your friend actually carries the virus?
   b. Understandably, your friend is upset and doesn't believe that the probability of being infected with HIV isn't really near 1. After all, the test is accurate, and it came out positive! Explain to your friend how the quoted statement can be true even though the test is very accurate both for people with HIV and

| Simple Event | BBB | BBG | BGB | GBB | BGG | GBG | GGB | GGG |
|---|---|---|---|---|---|---|---|---|
| X = No. of Girls | 0 | 1 | 1 | 1 | 2 | 2 | 2 | 3 |
| Probability | 1/8 | 1/8 | 1/8 | 1/8 | 1/8 | 1/8 | 1/8 | 1/8 |

We can find the probability for each value of $X$ by adding the probabilities of the simple events that have that value. The probabilities for the possible values of $X$ are as follows:

$P(X = 0) = 1/8$    (BBB is the only event with $X = 0$)

$P(X = 1) = 3/8$    (added probabilities for the three simple events with $X = 1$)

$P(X = 2) = 3/8$    (added probabilities for the simple events with $X = 2$)

$P(X = 3) = 1/8$    (GGG is the only event with $X = 3$)

This set of probabilities make up the probability distribution (pdf) for $X$ = number of girls. Notice that these probabilities sum to 1 over all possible values of $X$ and that all probabilities are between 0 and 1. ∎

## Graphing the Probability Distribution Function

It is often useful to represent a probability distribution function wih a picture similar to a histogram or bar graph. The possible values are placed on the horizontal axis, and their probabilities are placed on the vertical axis. A bar is drawn centered on each possible value, with the height of the bar equal to the probability for that value. Sometimes the bars are scaled so that their total area is 1. Then the probability for any combination of values can be found as the area of the bars (rectangles) for those values. This method will be displayed in the next example.

**Example 8.7**

**Graph of pdf for Number of Girls**  In Example 8.6, we determined the probability distribution function (pdf) for the number of girls in three births. Figure 8.1 displays the pdf. Notice that the bar over each value forms a rectangle that has a width of 1 and a height equal to the probability for that value. Therefore, the area of each bar is equivalent to the probability for that value. For instance, $P(X = 2) = 3/8$ coincides with the area of the rectangle centered on $X = 2$, a rectangle that has height = 3/8 and width = 1. Later in this chapter, you will learn that equating probability to an area is a crucial and necessary idea when working with continuous variables. For discrete variables, however, we generally can use more direct ways to find the probability that $X$ takes a partic-

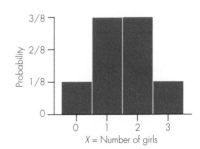

**Figure 8.1** ▪ The pdf for $X$ = Number of Girls in Three Births

ular value, so for discrete variables, we usually don't have to find the area of a rectangle to find a probability. ∎

# The Cumulative Distribution Function of a Discrete Random Variable

A **cumulative probability** is the probability that the value of a random variable is *less than or equal* to a specific value. Following are some examples:

- Probability of getting *6 or fewer* answers right when guessing at answers to ten true-false questions
- Probability that there are *2 or fewer* televisions in a randomly selected household
- Probability that *3 or fewer* individuals have type O+ blood in a random sample of $n = 10$ people

The notation for a cumulative probability is $P(X \leq k)$, where $X$ represents the random variable and $k$ is a specific value.

---

**definition** | The **cumulative distribution function (cdf)** for a random variable $X$ is a rule or table that provides the probabilities $P(X \leq k)$ for any real number $k$. Generally, the term **cumulative probability** refers to the probability that $X$ is less than or equal to a particular value.

---

For a discrete random variable, the cumulative probability $P(X \leq k)$ is the sum of probabilities for all values of $X$ less than or equal to $k$.

**Example 8.8** | **Cumulative Distribution for the Number of Girls**  In Example 8.6, we found that the pdf for $X$ = number of girls among three children in a family is as follows:

| $k$ | 0 | 1 | 2 | 3 |
|---|---|---|---|---|
| $P(X = k)$ | 1/8 | 3/8 | 3/8 | 1/8 |

The cumulative distribution function (cdf) for $X$ is as follows:

| $k$ | 0 | 1 | 2 | 3 |
|---|---|---|---|---|
| $P(X \leq k)$ | 1/8 | 4/8 | 7/8 | 1 |

For each specific value of $X$, the cumulative probability is the sum of probabilities for all values less than or equal to that value. Calculations are as follows:

$$P(X \leq 0) = 1/8$$
$$P(X \leq 1) = 1/8 + 3/8 = 4/8$$
$$P(X \leq 2) = 1/8 + 3/8 + 3/8 = 7/8$$

The cumulative probability for $X = 3$ must equal 1 because all possible values of $X$ are less than or equal to 3. ∎

# Finding Probabilities for Complex Events Based on Random Variables

The probability rules in Chapter 7 all apply to events defined for random variables and can be used to find probabilities of complicated events.

**Example 8.9**

**A Mixture of Children**  What is the probability that a family with three children will have at least one child of each sex? In Example 8.6, we defined the random variable $X$ = number of girls in three children. If there is at least one child of each sex, then either the family has one girl and two boys ($X = 1$) or two girls and one boy ($X = 2$). In Example 8.6, we found that the probability is 3/8 for each of these two possibilities. The probability can be found as

$$P(X = 1 \text{ or } X = 2) = P(X = 1) + P(X = 2) = 6/8 = 3/4$$

Notice that we can simply add these probabilities because the events $X = 1$ and $X = 2$ are mutually exclusive events. A family with three children cannot have both exactly one girl ($X = 1$) and exactly two girls ($X = 2$). ■

***thought question 8.2***   In Example 8.9, we added the probabilities of $X = 1$ and $X = 2$ because they were mutually exclusive events. For any discrete random variable $X$, is it always true that $X = k$ and $X = m$ are mutually exclusive events, where $k$ and $m$ represent two values that $X$ can have? Explain.*

**Example 8.10**

**Probabilities for Sum of Two Dice**  Many gambling games use dice in which the number of dots rolled is equally likely to be 1, 2, 3, 4, 5, or 6. For instance, craps is a game in which two dice are rolled and the sum of the dots on the two dice determines whether the gambler wins, loses, or continues rolling the dice.

Let $X$ = sum of two fair dice. Here is the sample space for the two dots and the corresponding values of the random variable $X$:

| Dots | X | Dots | X | Dots | X | Dots | X | Dots | X | Dots | X |
|------|---|------|---|------|---|------|---|------|---|------|---|
| 1,1 | 2 | 2,1 | 3 | 3,1 | 4 | 4,1 | 5 | 5,1 | 6 | 6,1 | 7 |
| 1,2 | 3 | 2,2 | 4 | 3,2 | 5 | 4,2 | 6 | 5,2 | 7 | 6,2 | 8 |
| 1,3 | 4 | 2,3 | 5 | 3,3 | 6 | 4,3 | 7 | 5,3 | 8 | 6,3 | 9 |
| 1,4 | 5 | 2,4 | 6 | 3,4 | 7 | 4,4 | 8 | 5,4 | 9 | 6,4 | 10 |
| 1,5 | 6 | 2,5 | 7 | 3,5 | 8 | 4,5 | 9 | 5,5 | 10 | 6,5 | 11 |
| 1,6 | 7 | 2,6 | 8 | 3,6 | 9 | 4,6 | 10 | 5,6 | 11 | 6,6 | 12 |

There are 36 equally likely simple events in the sample space, so each simple event has probability 1/36. To construct the probability distribution (pdf) for $X$, count the number of simple events resulting in each value of $X$ and divide by 36.

*****HINT:** See step 4 on page 284. Are there simple events for which $X = k$ and $X = m$ at the same time?

The probability distribution for the sum of two dice is as follows:

| k | 2 | 3 | 4 | 5 | 6 | 7 | 8 | 9 | 10 | 11 | 12 |
|---|---|---|---|---|---|---|---|---|----|----|----|
| **P(X = k)** | 1/36 | 2/36 | 3/36 | 4/36 | 5/36 | 6/36 | 5/36 | 4/36 | 3/36 | 2/36 | 1/36 |

Figure 8.2 displays the pdf with the rectangles for $X = 4$, 5, and 6 shaded. We can find $P(4 \leq X \leq 6)$ by adding the areas of the rectangles over the interval or by adding the probabilities for all individual values in that range. Here, $P(4 \leq X \leq 6) = 3/36 + 4/36 + 5/36 = 12/36 = 1/3$.

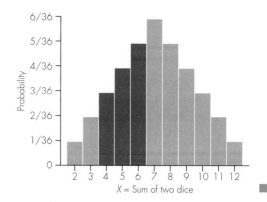

**Figure 8.2** ▍ Probability distribution function for sum of two dice showing $P(4 \leq X \leq 6)$

8.2 Exercises are on page 321.

*thought question* **8.3** Refer to the probability distribution for the sum of two dice. What is the value of $P(X \leq 4)$? What does this probability measure? Explain what is measured by the value of $1 - P(X \leq 4)$.*

# 8.3 Expectations for Random Variables

If we know probabilities for all possible values of a random variable, we can determine the mean outcome over the long run. This long-run average is called the *expected value* of the random variable. Synonymously, it is the **mean value of the random variable.**

As an example, suppose you play a lottery game once every week. For one play, the random variable that interests you is the amount that you win or lose. You won't be able to accurately predict the value of this random variable in advance for any given week, but as we show in this section, you will be able to determine the average amount that you can expect to win or lose per play over many weeks. This long-run average amount won or lost per play is the expected value, or mean value, of the random variable that describes possible outcomes for a single play.

*****HINT:** What possible sums of $k$ are included if $X \leq 4$?

| definition | The **expected value** of a random variable $X$ is the mean value of the variable in the sample space or population of possible outcomes. *Expected value* can also be interpreted as the mean value that would be obtained from an infinite number of observations of the random variable. |
| --- | --- |

## Calculating Expected Value

The calculation of expected value, or mean value, for a discrete random variable is

Expected value = Sum of "value × probability"

In this formula, *value* is a possible numerical outcome for the random variable, and *probability* = the probability of that outcome. We compute "*value × probability*" separately for each possible outcome and then add these quantities to find the expected value.

**Example 8.11**    **Gambling Losses**  Suppose that in a gambling game, the probability of winning $2 is .3 and the probability of losing $1 is .7. Let $X$ = amount a player "gains" on a single play. The following table gives the probability distribution for $X$ by listing possible values along with their probabilities. In the table, a loss of $1 is written as a negative "gain."

| **X = amount gained** | $2 | −$1 |
| --- | --- | --- |
| **Probability** | .3 | .7 |

What is the mean amount a player will "gain" per bet over many plays? The calculation is

Expected value = Sum of "value × probability"
$$= (\$2)(.3) + (-\$1)(.7)$$
$$= -\$0.10$$

Over a large number of plays, a player can expect to lose an *average* of 10 cents per play. Notice that the expected value in this example is not a possible outcome of a single play. ∎

### Notation and Formula for Expected Value

The notation $E(X)$ represents the mean or expected value of a random variable $X$. The Greek letter $\mu$ ("mu") can also be used to represent the mean or expected value. In other words, $\mu = E(X)$.

| formula | **Calculating Expected Value for a Discrete Random Variable** <br><br> If $X$ is a discrete random variable with possible values $x_1, x_2, x_3, \ldots$, occurring with probabilities $p_1, p_2, p_3, \ldots$, then the **expected value** of $X$ is calculated as <br><br> $$E(X) = \sum x_i p_i$$ <br> The formula says to calculate the sum of "value times probability." |
| --- | --- |

**Example 8.12**

Statistics△Now™

Watch a video example at **http://1pass.thomson.com** or on your CD.

**California Decco Lottery Game**  Decco is one of the many games played in the California lottery over the years. In this game, a player chooses four cards, one each from the four suits in a regular deck of playing cards. For example, the player might choose "4 of hearts," "3 of clubs," "10 of diamonds," and "jack of spades." The state draws a winning card from each suit. When one or more of the player's choices matches the winning cards drawn, the player wins a prize, the prize amount depending on the number of matches.

We will define the random variable $X$ to be a player's net gain for any single play. Table 8.1 lists the prizes, net gains, and probabilities for all possible numbers of matches, including 0 matches. It costs $1.00 to play the game, so the net gain for any prize is $1.00 less than the prize awarded. Notice that we count a free ticket as an even trade because it would cost $1.00 to buy a new one, but see the technical note following this example for discussion of whether a free ticket actually is an even trade.

**Table 8.1**  Probabilities for the Decco Game

| Number of Matches | Prize | $X$ = Net Gain | Probability |
|---|---|---|---|
| 4 | $5,000 | $4,999 | .000035 |
| 3 | $50 | $49 | .00168 |
| 2 | $5 | $4 | .0303 |
| 1 | Free ticket | 0 | .242 |
| 0 | None | −$1 | .726 |

How much would you win or lose per ticket in this game, over the long run? The expected value of $X$, denoted as $E(X)$, is the answer, and the calculation is

$$E(X) = (\$4{,}999 \times .000035) + (\$49 \times .00168) + (\$4 \times .0303)$$
$$+ (\$0 \times .242) + (-\$1 \times .726)$$
$$= -\$0.35$$

This tells us that over many repetitions of the game, players will *lose* an *average* of 35 cents per play. From the perspective of the Lottery Commission, this means that it pays out about 65 cents in prizes for each $1 ticket they sell for this game. ■

---

*technical note*  **Free Tickets Are Not Worth the Price**

The situation is actually worse than presented in Example 8.12. Many lottery games give a free ticket as a prize, and you may think the value of that ticket is equal to its cost, usually $1.00. It's not, unless you can sell the ticket to someone else for that price. The real value of the ticket is the expected win for one play of the game. In the computation in Example 8.12, for simplicity, we used a "net gain" of 0 when the prize was a free ticket, but because it is actually worth less than that, the average loss per play is even more than $0.35. We will continue to use an expected loss of $0.35 for simplicity in discussions of the Decco game.

Notice that the expected value in Example 8.12 is not one of the possible outcomes, so the phrase *expected value* is a bit misleading. In other words, the expected value may not be a value that is ever expected on a single random outcome. Instead, it is the average over the long run.

*thought question* **8.4**    Suppose the probability of winning in a gambling game is .001, and when a player wins, his or her net gain is $999. When a player loses, the net amount lost is $1 (the cost to play). Is this game fair? Why or why not? How would you define a fair game? Does the number of times the game is played affect your view of whether the game is fair? Explain. *

## Standard Deviation for a Discrete Random Variable

We first encountered the idea of standard deviation in Chapter 2, where we learned how to compute the standard deviation as a measure of spread for a quantitative variable and then used it in the Empirical Rule. Similarly, the **standard deviation of a discrete random variable** can be useful to quantify how spread out the possible values of a discrete random variable might be, weighted by how likely each value is to occur. As with the standard deviation of a set of measurements, the standard deviation of a random variable is essentially the average distance the random variable falls from its mean, or expected value, over the long run.

*formula*    **Variance and Standard Deviation of a Discrete Random Variable**

If $X$ is a random variable with possible values of $x_1, x_2, x_3, \ldots$, occurring with probabilities $p_1, p_2, p_3, \ldots$, and with expected value $E(X) = \mu$, then,

**Variance of $X$** $= V(X) = \sigma^2 = \sum (x_i - \mu)^2 p_i$

**Standard deviation of $X$** = square root of $V(X) = \sigma = \sqrt{\sum (x_i - \mu)^2 p_i}$

The sum is taken over all possible values of the random variable $X$.

**Example 8.13**    **Stability or Excitement—Same Mean, Different Standard Deviations** Suppose you decide to invest $100 in a scheme that you hope will make some money. You have two choices of investment plans, and you must decide which one to choose. For each $100 invested under the two plans, the possible net gains one year later and their probabilities are as shown:

| Plan 1 | | Plan 2 | |
|---|---|---|---|
| $X$ = Net Gain | Probability | $Y$ = Net Gain | Probability |
| $5,000 | .001 | $20 | .3 |
| $1,000 | .005 | $10 | .2 |
| $0 | .994 | $4 | .5 |

*****HINT:** A game is fair if the expected value of the gain is the same as the cost to play.

Examine the possible gains and associated probabilities. Which plan would you choose? Let's calculate the expected value for each plan. The expected value will tell us what the average long-run gain (loss?) per investment is for a particular plan.

For Plan 1,

$$E(X) = \$5,000 \times (.001) + \$1,000 \times (.005) + \$0 \times (.994) = \$10.00$$

For Plan 2,

$$E(Y) = \$20 \times (.3) + \$10 \times (.2) + \$4 \times (.5) = \$10.00$$

In other words, the expected net gain is the same for the two plans.

What is different about the two plans is the variability in the possible net gains. Under the first plan, you could make a lot of money but with a low probability. Most likely, you will make nothing at all. Under the second plan, you will be sure to make a small amount, but there is no possibility of a big gain. This difference in the variability in possible amounts under the two plans can be quantified by finding the standard deviations.

The calculations of the standard deviations are displayed in Table 8.2. Summing the third column under each plan provides the variances. Taking the square root of a variance provides the standard deviation of the net gain for a plan.

**Table 8.2** Standard Deviation Calculations for Example 8.13

| Plan 1 | | | Plan 2 | | |
|---|---|---|---|---|---|
| $(X - \mu)^2$ | $p$ | $(X - \mu)^2 p$ | $(Y - \mu)^2$ | $p$ | $(Y - \mu)^2 p$ |
| $(\$5,000 - \$10)^2 = \$24,900,100$ | .001 | \$24,900.1 | $(\$20 - \$10)^2 = \$100$ | .3 | \$30 |
| $(\$1,000 - \$10)^2 = \$980,100$ | .005 | \$4,900.5 | $(\$10 - \$10)^2 = 0$ | .2 | 0 |
| $(\$0 - \$10)^2 = 100$ | .994 | \$99.4 | $(\$4 - \$10)^2 = \$36$ | .5 | \$18 |

For Plan 1,

$$V(X) = \$29,900.00 \qquad \text{and} \qquad \sigma = \sqrt{29,900.00} = \$172.92$$

For Plan 2,

$$V(Y) = \$48.00 \qquad \text{and} \qquad \sigma = \sqrt{48} = \$6.93$$

These values demonstrate that the possible outcomes for Plan 1 are much more variable than the possible outcomes for Plan 2. If you wanted to invest cautiously, you would choose Plan 2, but if you wanted to have the chance to gain a large amount of money, you would choose Plan 1. ■

## Expected Value (Mean) and Standard Deviation for a Population

When we have a population of measurements, we can think of each individual measurement as the value of a random variable, and we can calculate the mean and standard deviation in either of two ways. We can create a probability distribution for the measurements and use the definitions of expected value and standard deviation in this section. Or we can find the mean and standard deviation by using the individual measurements directly in the formulas of Chapter 2. The two methods will give the same answers.

**Example 8.14**

**Mean Hours of Study for the Class Yesterday**   A class of $N = 200$ students is asked, "To the nearest hour, how many hours did you study yesterday?" The instructor is interested only in the responses of the students in his class, so the 200 students in the class constitute a population. The probability distribution of the responses is shown in Table 8.3.

**Table 8.3** Hours of Study Yesterday for a Population of College Students

| $X$ = Hours of Study | Number of Students | Probability |
|---|---|---|
| 0 | 16 | 16/200 = .08 |
| 1 | 28 | 28/200 = .14 |
| 2 | 50 | 50/200 = .25 |
| 3 | 36 | 36/200 = .18 |
| 4 | 26 | 26/200 = .13 |
| 5 | 22 | 22/200 = .11 |
| 6 | 14 | 14/200 = .07 |
| 7 | 6 | 6/200 = .03 |
| 8 | 2 | 2/200 = .01 |
| Total | 200 | 1.00 |

What is the mean number of hours studied the previous day in this population? The answer can be found by using the formula $E(X) = \Sigma\, x_i p_i$, and the calculation is

$$E(X) = (0 \times .08) + (1 \times .14) + (2 \times .25) + (3 \times .18) + (4 \times .13)$$
$$+ (5 \times .11) + (6 \times .07) + (7 \times .03) + (8 \times .01)$$
$$= 2.96 \text{ hours}$$

This mean can also be calculated in the usual way that we average 200 individual values, which is that we can total the 200 individual values and divide that total by $N = 200$. ■

8.3 Exercises are on page 322.

---

*formula*   **Mean and Standard Deviation for a Population**

Suppose a population has $N$ individuals and a measurement $X$ is of interest. Define:

$k_i$ = value of $X$ for individual $i$.

$x_1, x_2, x_3, \ldots$ as the distinct possible values for the measurement $X$.

$p_1, p_2, p_3, \ldots$ as the proportions of the population with the values $x_1, x_2, x_3, \ldots$.

Then, the **population mean and standard deviation** are

$$E(X) = \mu = \frac{1}{N}\Sigma\, k_i = \Sigma\, x_i p_i$$

$$\text{Standard deviation of } X = \sigma = \sqrt{\frac{\Sigma(k_i - \mu)^2}{N}} = \sqrt{\Sigma(x_i - \mu)^2 p_i}$$

# 8.4  Binomial Random Variables

In this section, we consider an important family of *discrete* random variables called *binomial random variables*. Certain conditions must be met for a variable to fall into this family, but the basic idea is that a binomial random variable is a count of how many times an event occurs (or does not occur) in a particular number of independent observations or trials that make up a random circumstance.

## Binomial Experiments and Binomial Random Variables

The number of heads in three tosses of a fair coin, the number of girls in six independent births, and the number of men who are six feet tall or taller in a random sample of ten adult men from a large population are all examples of binomial random variables. A **binomial random variable** is defined as $X$ = number of successes in the $n$ trials of a binomial experiment.

| | |
|---|---|
| *definition* | A **binomial experiment** is defined by the following conditions: <br><br> 1. There are $n$ "trials," where $n$ is specified in advance and is not a random value. <br> 2. There are two possible outcomes on each trial, called "success" and "failure" and denoted S and F. <br> 3. The outcomes are independent from one trial to the next. <br> 4. The probability of a "success" remains the same from one trial to the next, and this probability is denoted by $p$. The probability of a "failure" is $1 - p$ for every trial. |

Each row in Table 8.4 contains an example of a binomial experiment and a binomial random variable. For each example, go through the conditions for a binomial experiment and make sure you understand why those conditions apply to that example.

**Table 8.4** Examples of Binomial Experiments and Binomial Random Variables

| Random Circumstance | Random Variable | Success | Failure | $n$ | $p$ |
|---|---|---|---|---|---|
| (1) Toss three fair coins | $X$ = number of heads | Head | Tail | 3 | 1/2 |
| (2) Roll a die eight times | $X$ = number of 4s and 6s | 4, 6 | 1, 2, 3, 5 | 8 | 2/6 = 1/3 |
| (3) Randomly sample 1000 U.S. adults | $X$ = number who have seen a UFO | Seen UFO | Have not seen UFO | 1000 | Proportion of all adults who have seen a UFO |
| (4) Roll two dice once | $X$ = number of times sum is 7 | Sum is 7 | Sum not 7 | 1 | 6/36 = 1/6 |

### Additional Features of Binomial Random Variables

The examples in Table 8.4 illustrate a few subtle features of binomial random variables. Keep these points in mind when trying to determine whether a random variable fits the binomial description:

dard deviation, and $P(X = k)$ when $X$ is a binomial random variable. You won't have to know how to do these derivations, but the results are useful for later applications in this textbook, so remember where to find them.

| *formula* | **Mean, Standard Deviation, and Variance of a Binomial Random Variable** |
|---|---|

For a *binomial random variable X* based on $n$ trials with success probability $p$,

The **mean** is $\mu = E(X) = np$.

The **standard deviation** is $\sigma = \sqrt{np(1 - p)}$.

*Note:* For any random variable, the standard deviation is the square root of the **variance.** For a binomial random variable, the variance is $V(X) = np(1 - p)$, which leads to the formula for standard deviation.

**Example 8.18**

**Is There Extraterrestrial Life?**  Suppose that 50% of a large population would say "yes" if they were asked, "Do you believe there is extraterrestrial life?" A sample of $n = 100$ is taken from this population, and each person in the sample is asked the question about belief in extraterrestrial life. The random variable $X$ = number in the sample who say "yes" is approximately a binomial random variable with $n = 100$ and $p = .5$.

- The mean of $X$ is $\mu = np = 100(.5) = 50$.
- The standard deviation is $\sqrt{np(1 - p)} = \sqrt{100(.5)(1 - .5)} = 5$.

In repeated samples of size 100 from this population, on average 50 people would say "yes" to the question. The amount by which that number would differ from sample to sample is represented by the standard deviation of 5. We will discuss this kind of "sample variability" in much more detail in Chapter 9. ■

8.4 Exercises are on page 323.

---

> ***thought question 8.6***  In Example 8.17, we determined the probability that you could guess your way to a passing score on a quiz with 15 true-false questions. If you did guess at each of 15 true-false questions, what is the expected value of $X$ = number of correct answers? Is the expected value a possible score on the quiz? What exactly does the expected value tell us?[*]

---

## *case study* 8.1    Does Caffeine Enhance the Taste of Cola?

**Statistics⬙Now™** Read the original source on your CD.

Soft drink manufacturers claim that caffeine enhances the flavor of cola drinks and that this is why they make caffeinated colas. Researchers from

Johns Hopkins University disputed this claim in a study reported in the *Archives of Family Medicine* (Griffiths and Vernotica, 2000). The researchers conducted a taste test to find out if people can tell by tasting whether a cola contains caffeine or not.

*(continued)*

---

[*]**HINT:** Use the formula for $E(X)$ for a binomial random variable.

Each participant tried 20 times to identify which of two cola drinks was "sample A" and which was "sample B." Unknown to the participants, one sample (A or B) contained caffeine and one did not. The particular sample (A or B) with caffeine was the same for all 20 trials. On each trial, a participant tasted drinks from two unlabeled cups and guessed which was sample A. After every trial, the participant was told whether he or she was correct. To learn about the tastes of the two samples, participants did five practice trials before the 20 trials that counted in the results.

Suppose that a participant cannot ever detect the difference between the cola samples and randomly guesses on every trial. In that case, $X$ = number of correct guesses is a binomial random variable with $n = 20$ trials and $p = .5$ (the probability of correctly guessing just by luck).

The investigators called a participant's flavor detection performance "significant" when the individual made 15 or more correct identifications in 20 trials. The probability is only .0207 (about 1 in 50) that somebody who guesses every time could do this well. The calculation of this probability is

$$P(X \geq 15) = 1 - P(X \leq 14) = 1 - .9793 = .0207.$$

The value of $P(X \leq 14)$ could be found using Minitab, Excel, or a calculator such as a TI-83.

Twenty-five individuals participated, and only two made 15 or more correct identifications. This isn't much better than what would be expected if everybody randomly guessed for all trials. As we determined above, the probability of 15 or more correct when randomly guessing is .0207, which is about 1 in 50. If all 25 participants guessed, we would expect $25(1/50) = 0.5$, about 0 or 1, participants, to fall into the "significant" category.

The researchers interpreted their results to mean that people generally can't taste whether a cola contains caffeine. Soft drink manufacturers objected to several aspects of the study, including the sample sizes for trials and participants and the way in which the researchers prepared the caffeinated cola samples.

# 8.5 Continuous Random Variables

We learned in Section 8.1 that a *continuous random variable* is one for which the outcome can be any value in an interval or collection of intervals. In practice, all measurements are rounded to a specified number of decimal places, so we may not be able to accurately observe all possible outcomes of a continuous variable. For example, the limitations of weighing scales keep us from observing that a weight may actually be 128.3671345993 pounds. Generally, however, we call a random variable a continuous random variable if there are a large number of observable outcomes covering an interval or set of intervals.

For a discrete random variable, we can find the probability that the variable $X$ exactly equals a specified value. We can't do this for a continuous random variable. For a continuous random variable, we are only able to find the probability that $X$ falls between two values. In other words, unlike discrete random variables, continuous random variables do not have probability distribution functions specifying the exact probabilities of specified values. Instead, they have probability density functions, which are used to find probabilities that the random variable falls into a specified interval of values.

| definition | The **probability density function** for a continuous random variable $X$ is a curve such that the area under the curve over an interval equals the probability that $X$ is in that interval. In other words, the probability that $X$ is between the values $a$ and $b$ is the area under the density curve over the interval between the values $a$ and $b$. |
|---|---|

## Notation for Probability in an Interval

- The two endpoints of an interval are represented using the letters $a$ and $b$.
- The interval of values of $X$ that falls between $a$ and $b$, including the two endpoints, is written $a \leq X \leq b$.
- The probability that $X$ has a value between $a$ and $b$ is written $P(a \leq X \leq b)$.

For example, the probability that a randomly selected 3-year-old weighs between 31 and 33 pounds would be written $P(31 \leq X \leq 33)$. The random variable $X =$ weight of a 3-year-old, $a = 31$ and $b = 33$.

**Example 8.19**

Statistics⊖Now™

Watch a video example at **http://1pass.thomson.com** or on your CD.

**Time Spent Waiting for the Bus**    A bus arrives at a bus stop every 10 minutes. If a person arrives at the bus stop at a random time, how long will he or she have to wait for the next bus? Define the random variable $X =$ waiting time until the next bus arrives. The value of $X$ could be any value between 0 and 10 minutes, and $X$ is a continuous random variable. (In practice, the limitations of watches would force us to round off the exact time.) Figure 8.5 shows the probability density function for the waiting time. Possible waiting times are along the horizontal axis, and the vertical axis is a density scale. The height of the "curve" is .1 for all $X$ between 0 and 1, so the total area between 0 and 10 minutes is $(10)(.1) = 1$.

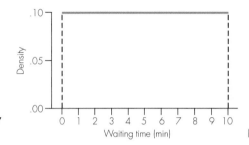

**Figure 8.5** ▮ A uniform probability density function for Example 8.19

The density function shown in Figure 8.5 is a flat line that covers the interval of possible values for $X$. There is a "uniformity" to this density curve in that every interval with the same width has the same probability. A random variable with this property is called a **uniform random variable** and is the simplest example of a continuous random variable. For this type of variable, the probability for an interval equals the area of a rectangle.

**Example 8.20**

Statistics⊖Now™

For software help, download your Minitab, Excel, TI-83, SPSS, R, and JMP manuals from **http://1pass.thomson.com**, or find them on your CD.

**Probability That the Waiting Time is Between 5 and 7 Minutes**    Suppose we want to find the probability that the waiting time $X$ was in the interval from 5 to 7 minutes. The general principle for any continuous random variable is that the probability $P(a \leq X \leq b)$ is the "area under the curve" over the interval from $a$ to $b$. In this example, the "area under the curve" is the area of a rectangle that has width $= 7 - 5 = 2$ minutes and height $= .1$. This area is $(2)(.1) = .2$, which is the probability that the waiting time is between 5 and 7 minutes. In Figure 8.6, the shaded area represents the desired probability.

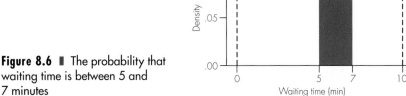

**Figure 8.6** ▮ The probability that waiting time is between 5 and 7 minutes

Theoretically, the use of calculus is needed to find an area under a density curve (unless it's a simple rectangle or other simple shape, as in Examples 8.19 and 8.20). In practice, however, tables of appropriate probabilities usually are available. Computer software and graphing calculators also can be used to "automate" the calculations for the most commonly encountered models.

8.5 Exercises are on page 325.

> ***thought question 8.7***   The total area under the probability density function over the entire range of values the random variable $X$ can possibly have is the same for all continuous random variables. What is that total area? What probability does it represent?
>
>     Suppose the random variable is the amount of rainfall in Davis, California, for a randomly selected year and can range from 0 inches to 50 inches. What is the area under the appropriate density curve over the range from 0 to 50 inches? What probability does it represent?*

## 8.6   Normal Random Variables

The most commonly encountered type of continuous random variable is the **normal random variable** which has a specific form of a bell-shaped probability density curve called a **normal curve.** A normal random variable is also said to have a **normal distribution.** Any normal random variable is completely characterized by specifying values for its mean $\mu$ and standard deviation $\sigma$.

    Nature provides numerous examples of measurements that follow a normal curve. The fact that so many different kinds of measurements follow a normal curve is not surprising. On many attributes, the majority of people are somewhat close to average, and as you move farther from the average, either above or below, there are fewer people with such values.

### Features of Normal Curves and Normal Random Variables

As with any continuous random variable, the probability that a normal random variable falls into a specified interval is equivalent to an area under its density curve. Also, $P(X = k) = 0$, meaning that the probability is 0 that a normal random variable $X$ exactly equals any specified value.

    Some features shared by all normal curves and normal random variables $(X)$ are as follows:

1. The normal curve is symmetric and bell-shaped (but not all symmetric bell-shaped density curves are normal curves).
2. $P(X \leq \mu) = P(X \geq \mu) = .5$, meaning that there are equal probabilities for a measurement being less than the mean and greater than the mean. This happens because the curve is symmetric about the mean.
3. $P(X \leq \mu - d) = P(X \geq \mu + d)$ for any positive number $d$. This means that the probability that $X$ is more than $d$ units below the mean equals the probability that $X$ is more than $d$ units above the mean.

---

***\*HINT:*** For $X$ = Davis rainfall, as described, what is $P(0 \leq X \leq 50)$?

4. The Empirical Rule holds:

- $P(\mu - \sigma \leq X \leq \mu + \sigma) \approx .68$
- $P(\mu - 2\sigma \leq X \leq \mu + 2\sigma) \approx .95$
- $P(\mu - 3\sigma \leq X \leq \mu + 3\sigma) \approx .997$

**Example 8.21**

**Statistics⌂Now™**

Watch a video example at **http://
1pass.thomson.com** or on your CD.

**College Women's Heights**  The distribution of the heights of college women can be described reasonably well by a normal curve with mean $\mu = 65$ inches (5 foot 5 inches) and standard deviation $\sigma = 2.7$ inches. A normal curve with these characteristics is shown in Figure 8.7. Heights are shown along the horizontal axis, with tick marks given at the mean and at 1, 2, and 3 standard deviations above and below the mean. Notice that half (.5) of the area is above the mean of 65.0 and half is below it.

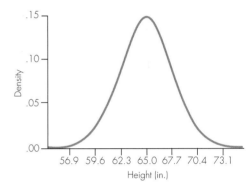

**Figure 8.7** ▌ Normal curve for heights of college women

Most of the normal curve is over the interval from 56.9 inches to 73.1 inches. This is consistent with the Empirical Rule that we encountered in Chapter 2. One part of the Empirical Rule is that for a bell-shaped curve, about 99.7% of the values fall within 3 standard deviations of the mean in either direction. Here, that interval is $65 \pm 3 \times 2.7$, which is $65 \pm 8.1$ inches, or 56.9 and 73.1 inches. The Empirical Rule is generally used as an approximate rule for describing sample data with a bell-shaped curve, but the features of that rule are exact characteristics of a normal curve model. ▓

***thought question 8.8***   Which of the following measurements do you think are likely to have a normal distribution: heights of college men, incomes of 40-year-old women, pulse rates of college athletes? Explain your reasoning for each variable. For those variables that are likely to be normally distributed, give approximate values for the mean and standard deviation.*

## Standardized Scores

We learned in Chapter 2 that a **standardized score,** also called a **z-score,** is the distance between a specified value and the mean, measured in number of

**\*HINT:**  Which of these variables might have large outliers or a skewed distribution?

standard deviations. We repeat the definition here using notation for random variables.

| formula | **Calculating a z-Score** |

The formula for converting any value $x$ to a **z-score** is

$$z = \frac{\text{Value} - \text{Mean}}{\text{Standard deviation}} = \frac{x - \mu}{\sigma}$$

A $z$-score measures the number of standard deviations that a value falls from the mean.

**Example 8.22**

Statistics⬡Now™

Watch a video example at **http://1pass.thomson.com** or on your CD.

**The z-Score for a Height of 62 Inches** Assume that for the population of college women, mean height is 65 inches, and the standard deviation of heights is 2.7 inches. The $z$-score corresponding to a height of 62 inches is

$$z = \frac{\text{Value} - \text{Mean}}{\text{Standard deviation}} = \frac{62 - 65}{2.7} = \frac{-3}{2.7} = -1.11$$

This $z$-score tells us that 62 inches is 1.11 standard deviations below the mean height for this population. ■

## Finding Probabilities for $z$-Scores

A normal random variable with mean $\mu = 0$ and standard deviation $\sigma = 1$ is said to be a **standard normal random variable** and to have a **standard normal distribution.** When we convert values for any normal random variable to $z$-scores, it is equivalent to converting the random variable of interest to a standard normal random variable. We use the letter Z to represent a standard normal random variable.

Tables and software packages are available that give probabilities for the standard normal distribution. Table A.1, "Standard Normal Probabilities," in this book's Appendix (and also inside the back cover) is a table of cumulative probabilities for a standard normal random variable. In other words, we can use Table A.1 to find $P(Z \le z^*)$, the probability that a standard normal random variable (Z) is less than a specific value (written as $z^*$).

### Reading Table A.1 in the Appendix

As an example, the following (incomplete) row from Table A.1 shows that $P(Z \le 1.82) = .9656$.

| z | .00 | .01 | .02 | .03 | .04 | .05 | .06 | .07 | .08 | .09 |
|---|-----|-----|-----|-----|-----|-----|-----|-----|-----|-----|
| **1.8** | ⟶ | | .9656 | | | | | | | |

The first column, with the heading "$z$," gives the digit before the decimal place and the first decimal digit of a value of $z$. The column headings give the second decimal digit for $z$. Cumulative probabilities are given in the body of the table.

As a second example of reading Table A.1, this row of the table gives the information that $P(Z \le -2.59) = .0048$.

| z | .00 | .01 | .02 | .03 | .04 | .05 | .06 | .07 | .08 | .09 |
|---|-----|-----|-----|-----|-----|-----|-----|-----|-----|-----|
| **−2.5** | $\longrightarrow$ | | | | | | | | | .0048 |

Here are two other examples that you should verify by looking in Table A.1 to be sure you understand how to read the table:

$$P(Z \le 1.31) = .9049$$
$$P(Z \le -2.00) = .0228$$

At the bottom of both pages of the table, there is a section titled "In the Extreme" where cumulative probabilities are given for selected "extreme" $z$-scores. For instance, from that section of the table, you can learn that $P(Z \le -4.75) = .000001$ (one in a million!).

***Technical Detail:*** $P(Z = z^*) = 0$, so it is equivalent to write $P(Z \le z^*)$ or $P(Z < z^*)$.

## How to Solve General Normal Curve Problems

Software such as Minitab and Excel and some calculators can be used to find probabilities for any general normal random variable. In using these tools, it is not necessary to convert values to $z$-scores. However, an important fact about normal random variables is that any probability problem about a normal random variable can be converted to a problem about a standard normal variable.

### Finding a Cumulative Probability $P(X \le a)$ for Any Normal Random Variable

Suppose $X$ is a normal random variable with mean $\mu$ and standard deviation $\sigma$. There are two steps in finding $P(X \le a)$, the probability that $X$ is less than or equal to the value $a$. These steps are not necessary if you have a computer or calculator that can determine $P(X \le a)$ directly.

**Step 1:** Calculate a $z$-score for the value $a$; call it $z^*$.

**Step 2:** Use a table (such as Table A.1), calculator, or computer to find $P(Z \le z^*)$, the cumulative probability for $z^*$. This equals the desired probability $P(X \le a)$.

Figure 8.8 illustrates the nature of $P(X \le a)$. It is the area to the left of the value $a$ under the appropriate normal curve. This is a cumulative probability for the value $a$.

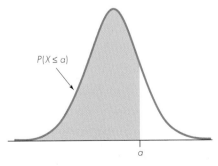

**Figure 8.8** ■ $P(X \le a)$, the probability that $X$ is less than or equal to the value $a$

**Example 8.23**

**Probability That Height Is Less Than 62 Inches** Assume that the heights of college women have a normal distribution with mean $\mu = 65$ inches and standard deviation $\sigma = 2.7$ inches. What is the probability that a randomly selected college woman is 62 inches or shorter? This is the same as asking what *proportion* of college women are 62 inches or shorter.

**Step 1:** $z^* = \dfrac{62 - 65}{2.7} = -1.11$

**Step 2:** $P(X \leq 62) = P(Z \leq -1.11) = .1335$ (found by using Table A.1)

In other words, about 13% of college women are 62 inches or shorter. Figure 8.9 illustrates the area of interest under the normal curve for college women's heights.

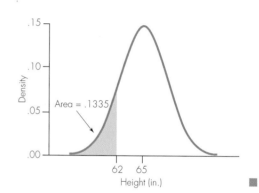

**Figure 8.9** ▌ College women's heights and P(height ≤ 62 inches)

## Useful Probability Relationships for Normal Random Variables

The following three rules are useful for finding probabilities for a normal random variable $X$:

**Rule 1:** $P(X > a) = 1 - P(X \leq a)$
**Rule 2:** $P(a < X < b) = P(X \leq b) - P(X \leq a)$
**Rule 3:** $P(X > \mu + d) = P(X < \mu - d)$

These relationships are illustrated in Figures 8.10, 8.11, and 8.12.

- Rule 1, shown in Figure 8.10, can be used to find the probability that $X$ is *greater than* the value $a$.

- Rule 2, shown in Figure 8.11, can be used to find the probability that $X$ is *between* the values $a$ and $b$.

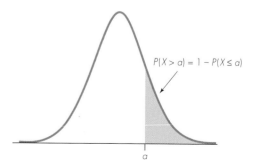

**Figure 8.10** ▌ P(X > a), the probability that X is greater than the value a

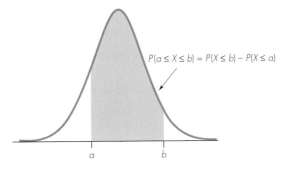

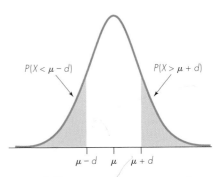

**Figure 8.11** ▮ $P(a \le X \le b)$, the probability that $X$ is between the values $a$ and $b$

**Figure 8.12** ▮ Symmetry property of a normal random variable

- Rule 3, shown in Figure 8.12, is a *symmetry* property. There are equal probabilities for $X$ being more than $d$ units above the mean and more than $d$ units below the mean.

Generally, in solving normal curve problems, it helps to draw a picture of the problem, highlight the desired area, and then figure out how the desired area relates to the rules just given.

Keep in mind that all questions about normal random variables can be converted to equivalent problems about $z$-scores. The rules given for general normal random variables also apply to $Z$, the standard normal random variable ($z$-scores). For $Z$, the mean is 0, so Rule 3 becomes $P(Z > +d) = P(Z < -d)$.

### Examples of "Greater Than" Problems

The next two examples demonstrate how to find the probability that a normal random variable has a value greater than a specified value.

**Example 8.24**

**Statistics△Now™**

For software help, download your Minitab, Excel, TI-83, SPSS, R, and JMP manuals from **http://1pass .thomson.com**, or find them on your CD.

**Probability That Z Is Greater Than 1.31**    There are two equivalent ways to find the probability that the standard normal random variable $Z$ is greater than 1.31. First, we can use Rule 1:

$$P(Z > 1.31) = 1 - P(Z \le 1.31)$$
$$= 1 - .9049 = .0951$$

We used Table A.1 in the Appendix to look up the probability $P(Z \le 1.31)$. Figure 8.13 illustrates the area of interest and the solution. Notice that the desired

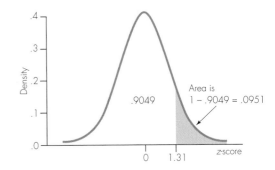

**Figure 8.13** ▮ Probability that $Z$ is greater than 1.31

probability equals the area to the right of 1.31 under the density curve. As an alternative method of solution, we could use the symmetry relationship (Rule 3):

$$P(Z > 1.31) = P(Z < -1.31) = .0951 \quad \blacksquare$$

**Example 8.25**

**Statistics⬡Now**™

Watch a video example at **http://1pass.thomson.com** or on your CD.

For software help, download your Minitab, Excel, TI-83, SPSS, R, and JMP manuals from **http://1pass.thomson.com**, or find them on your CD.

**Probability That Height Is Greater Than 68 Inches**   As in previous examples in this section, we assume that the heights of college women have a normal distribution with mean $\mu = 65$ inches and standard deviation $\sigma = 2.7$ inches. What is the probability that a randomly selected college woman is taller than 68 inches?

The simplest solution (from Rule 1) is to first find the probability that the woman is shorter than 68 inches and then subtract that probability from 1. The steps are as follows:

$$z = \frac{68 - 65}{2.7} = 1.11$$

$$P(Z \le 1.11) = .8665 \quad \text{(from Table A.1)}$$
$$P(Z > 1.11) = 1 - P(Z \le 1.11)$$
$$= 1 - .8665 = .1335$$

This is the probability that a randomly selected woman is taller than 68 inches. Figure 8.14 illustrates the problem and solution.

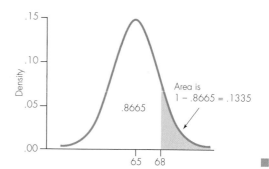

**Figure 8.14** ▌ Probability that a college woman is taller than 68 inches

In Example 8.25, we found that the probability that a randomly selected college woman is taller than 68 inches is .1335. In Example 8.23, we found the same value (.1335) for the probability that a college woman is shorter than 62 inches. This was not a coincidence. Both problems are about being more than 3 inches from the mean (65 inches). By symmetry (Rule 3), the answers must be the same.

### Examples for "Between a and b" Regions

The following two examples show how to find the probability that a normal random variable is between two specified values.

**Example 8.26**

**Probability That Z Is Between −2.59 and 1.31**   The probability that a standard normal variable Z is between −2.59 and 1.31 is found by taking the difference between the probabilities to the left of these two values. Table A.1 in the Appendix can be used to look up the cumulative probabilities. The solution is

$$P(-2.59 \le Z \le 1.31) = P(Z \le 1.31) - P(Z \le -2.59)$$
$$= .9049 - .0048 = .9001$$

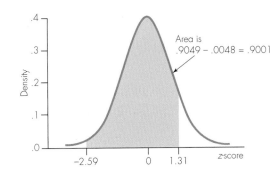

**Figure 8.15** ∎ Probability that
$Z$ is between $Z = -2.59$ and
$Z = 1.31$

Figure 8.15 shows the relevant area under the standard normal density curve
and the solution. ∎

**Example 8.27**

**Probability That a Vehicle Speed Is Between 60 and 70 mph** Suppose
that vehicle speeds ($X$) at a highway location have a normal distribution with
mean $\mu = 62$ mph and standard deviation $\sigma = 5$ mph. What is the probability
that the speed of a randomly selected vehicle at this location is between 60 and
70 mph? The desired probability is written as $P(60 \leq X \leq 70)$. To find this probability, we use the relationship

$$P(60 \leq X \leq 70) = P(X \leq 70) - P(X \leq 60)$$

We next calculate a $z$-score for each of the values 60 and 70 mph and determine
the cumulative probability for each $z$-score. The difference between these two
cumulative probabilities is the probability that a vehicle speed is between 60
and 70 mph.

$$\text{For } X = 70, z = \frac{70 - 62}{5} = 1.6; P(Z \leq 1.6) = .9452$$

$$\text{For } X = 60, z = \frac{60 - 62}{5} = -0.4; P(Z \leq -0.4) = .3446$$

$$P(60 \leq X \leq 70) = P(Z \leq 1.6) - P(Z \leq -0.4)$$
$$= .9452 - .3446 = .6006$$

Table A.1 was used to look up values for $P(Z \leq 1.6)$ and $P(Z \leq -0.4)$. Figure 8.16
illustrates the solution.

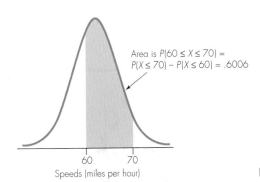

**Figure 8.16** ∎ Probability that a
vehicle speed is between 60 and
70 mph

| TI-84 *tip* | **Calculating Normal Curve Probabilities** |
|---|---|

- To begin, press 2nd VARS to reach the **Distribution** menu.

- Use **2:normalcdf(** to find the probability that a normal random variable has a value in a specified interval.

- Enter numerical values (separated by commas) to complete the expression **normalcdf(** *lower value, upper value, mean, standard deviation*). For example, **normalcdf(** 60, 70, 62, 5) gives the probability that a normal random variable with $\mu = 62$ and $\sigma = 5$ has a value between 60 and 70.

*Note:* To find a cumulative probability using **normalcdf(,** make *lower value* be an extreme negative value. To find the probability greater than a specified value, make *upper value* be an extreme positive value.

## Finding Percentiles

In some problems, we want to know what value of a variable has a given percentile ranking. For example, we may want to know what pulse rate is the 25th percentile of pulse rates for men. Notice that the word **percentile** refers to the *value* of a variable. The **percentile rank** corresponds to the *cumulative probability* (the area to the left under the density curve) for that value.

Suppose that the 25th percentile of pulse rates for adult males is 64 beats per minute. This means that 25% of men have a pulse rate below 64. The *percentile* is 64 beats per minute (a value of the variable), and the *percentile rank* is 25% or .25 (a cumulative probability). Figure 8.17 illustrates this concept.

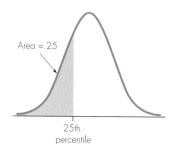

Area = .25

25th
percentile

**Figure 8.17** ▪ The 25th percentile for a random variable with a normal distribution

### Finding a Percentile or a Value with Specified Cumulative Probability

Remember that the goal is to find the value of the variable that has a given cumulative probability. You may be able to use software or a calculator to directly determine the solution.

Using Table A.1 to find a percentile for a normal random variable, there are two steps:

**Step 1:** Find the value $z^*$ that has the specified cumulative probability. This does not involve calculation; find the percentile rank in the body of the table and read the $z$-score from row and column headings.

**Step 2:** Compute $x = z^*\sigma + \mu$. This is the desired percentile.

**Example 8.28**

**The 75th Percentile of Systolic Blood Pressures** Suppose that the blood pressures of men aged 18 to 29 years old have a normal distribution with mean $\mu = 120$ and standard deviation $\sigma = 10$. What value of blood pressure is the 75th percentile for this population?

Figure 8.18 illustrates the problem. We want to find the blood pressure for which the cumulative probability (area to the left under the density curve) is .75. With Table A.1, the steps are as follows:

**Step 1:** Search for .75 among the cumulative probabilities in the body of Table A.1 and then look at the row and column labels to read $z^*$. The approximate $z$-score for the 75th percentile is $z^* = 0.67$.

**Step 2:** $x = z^*\sigma + \mu = (0.67)(10) + 120 = 126.7$

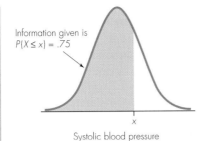

Information given is
$P(X \leq x) = .75$

x

Systolic blood pressure

**Figure 8.18** ▮ The 75th percentile for systolic blood pressure

8.6 Exercises are on page 325.

The 75th percentile of blood pressures for men aged 18 to 29 years old is 126.7, or approximately 127. ▮

---

**MINITAB tip**    **Calculating Normal Curve Probabilities and Percentiles**

- Use **Calc>Probability Distributions>Normal.**
- Specify the mean and standard deviation.
- To find $P(X \leq k)$, select "Cumulative Probability" and also specify the value of $k$ in the box labeled "Input Constant."
- To find a percentile, select "Inverse Cumulative" and also specify the cumulative probability for the percentile in the box labeled "Input Constant."

*Note:* It's not necessary to compute $z$-scores when using Minitab for determining normal curve probabilities.

---

# 8.7 Approximating Binomial Distribution Probabilities

One useful application of the normal distribution is that it can be used to approximate probabilities for some other types of random variables. In this section, we learn how to use the normal distribution to approximate cumulative probabilities for binomial random variables.

When $X$ has a binomial distribution with a large number of trials, the binomial probability formula is difficult to use because the factorial expressions in the formula become very large. The work required to find binomial probabilities when $n$ is large is enormous, and even some computer programs and calculators may not be able to do the task. Fortunately, the normal distribution can be used to approximate probabilities for a binomial random variable when this situation occurs.

<table>
<tr><td>*definition*</td><td>The **normal approximation to the binomial distribution** is based on the following result, derived from mathematics. If $X$ is a binomial random variable based on $n$ trials with success probability $p$, and $n$ is sufficiently large, then $X$ is also approximately a normal random variable. For this normal random variable,<br><br>$$\text{Mean} = \mu = np$$<br>$$\text{Standard deviation} = \sigma = \sqrt{np(1-p)}$$<br><br>**Conditions:** The approximation works well when both $np$ and $n(1-p)$ are at least 10.</td></tr>
</table>

**Example 8.29**

**The Number of Heads in 30 Flips of a Coin**  Figure 8.19 displays the probability distribution function of a binomial random variable based on $n = 30$ trials and a success probability $p = .5$. This distribution could, for example, describe $X =$ the number correct if you are just guessing on a 30-question true-false quiz. The bell-shaped pattern of the distribution is clear, and a normal curve could be used to approximate this distribution because both $np$ and $n(1 - p)$ are greater than 10.

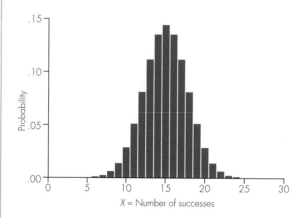

**Figure 8.19** ▌ Binomial distribution for $n = 30$ and $p = .5$

Recall that 99.7%, or nearly all, of a normal curve falls within 3 standard deviations of the mean. For the binomial random variable displayed in Figure 8.19, $\mu = (30)(.5) = 15$, and $\sigma = \sqrt{30 \times .5 \times (1 - .5)} = 2.74$. The interval $\mu \pm 3\sigma$ is approximately $15 \pm 8$, or 7 to 23. Notice how this interval fits with the binomial distribution shown in Figure 8.19. The probability is small that $X$ would take a value outside this interval. ▩

## Approximating Cumulative Probabilities for a Binomial Random Variable

The normal curve approximation can be used to find cumulative probabilities for binomial random variables, as follows:

$$P(X \le k) \approx P(Z \le z^*)$$

where

$$z^* = \frac{k - np}{\sqrt{np(1 - p)}}$$

In other words, to find a cumulative probability, do the following:

1. Calculate a *z*-score for the value of interest, *k*.
2. Use the standard normal curve to determine the cumulative probability for the *z*-score found in step 1.

More complex probabilities can be found as well, using the same relationships as for any normal random variable *X*.

It is important to note that you should not use the normal curve approximation if your statistical software or calculator can determine a binomial probability exactly. In this regard, the capabilities of different software programs and calculators vary. For the next example, for instance, Minitab versions 13 and higher can give an exact answer using the binomial distribution, but older versions cannot.

**Example 8.30** | **Political Woes** A politician is convinced that at least 50% of her constituents favor a woman's right to choose abortion but is concerned because the latest poll of 500 adults in her district found that slightly fewer than half, only 240, or 48% of the sample, supported this position. If indeed 50% of adults in the district support abortion, what is the probability that a random sample of 500 adults would find 240 or fewer of them holding this position?

If $X$ = number in sample who support abortion and if 50% of all adults in the district actually support it, then $X$ is a binomial random variable with $n = 500$ and $p = .5$. The desired probability is $P(X \leq 240)$. It would be tedious to find this probability by computing and adding binomial probabilities for $X = 0$ to 240! It is much easier to use the normal approximation, with

$$\mu = np = 250$$
$$\sigma = \sqrt{np(1 - p)} = 11.2$$

The solution is $P(X \leq 240) \approx P\left( Z \leq \frac{240 - 250}{11.2} \right) = P(Z \leq -0.89) = .1867.$

Thus, it is not unlikely that she could observe a minority of 48% or less of the sample favoring abortion, even if 50% of the district favors it. ■

**Example 8.31** | **Guessing and Passing a True-False Test** In Example 8.29 and Figure 8.19, we displayed probabilities for a binomial random variable with $n = 30$ and $p = .5$. Suppose you need 70%, or 21 questions, correct to pass a 30-question true-false test on which you are just guessing. Find $P(X \geq 21)$. Notice that the complement of this event is *not* $P(X \leq 21)$; it's $P(X \leq 20)$. Using the normal approximation to the binomial with $n = 30$ and $p = .5$, we have

$$P(X \geq 21) = 1 - P(X \leq 20) \approx 1 - P\left( Z \leq \frac{20 - np}{\sqrt{np(1 - p)}} \right) = 1 - P(Z \leq 1.83)$$

$$= 1 - .9664 = .0336.$$

Using Minitab, we find the exact probability that $X \geq 21$ is $1 - .9786 = .0214$.

8.7 Exercises are on page 326.

Figure 8.19 illustrates the exact binomial pdf for this situation. Notice that $P(X \geq 21)$ corresponds to the areas of the rectangles centered above 21, 22, ..., 30, of which only the ones above 21, 22, and 23 are visible. After that, the probabilities, and thus the heights of the rectangles, are essentially 0. For instance, $P(X = 24) = .0006.$ ∎

---

| *technical note* | **Continuity Correction** |
|---|---|

Notice in Figure 8.19 that the rectangle centered on $X = 21$ actually begins at 20.5, not at 21.0. But in the normal approximation to the binomial calculation, we found the area under a normal curve starting at 21, not 20.5. In doing so, we omitted half of the rectangle centered at 21. To make the approximation more accurate, sometimes a **continuity correction** is used by adding or subtracting .5, based on which rectangles are desired. For Example 8.29,

$$P(X \geq 21) \approx P\left( Z \geq \frac{20.5 - (30)(.5)}{\sqrt{(30)(.5)(1 - .5)}} \right) = P(Z \geq 2.01) = P(Z \leq -2.01)$$

$$= .0222$$

which is much closer to the exact probability of .0214 than was the approximation of .0336 that was found without the continuity correction.

---

# 8.8 Sums, Differences, and Combinations of Random Variables

There are many instances in which we want information about combinations of random variables. Suppose, for example, that your final score in a class is found by adding together 25% of each of two midterm exam scores (Exam1 and Exam2) and 50% of your final exam score (Exam3). Expressed as an equation, the final score is calculated as follows:

Final score = .25 × Exam1 + .25 × Exam2 + .50 × Exam3

If you know the mean for each exam, it is easy to calculate the mean final score. It probably makes sense to you that the mean final score can be found by combining the mean exam scores in the same way that an individual's scores are combined. If the means on the three exams are $\mu_1$, $\mu_2$, and $\mu_3$, the formula for the mean final score is

Mean of final scores = .25 × $\mu_1$ + .25 × $\mu_2$ + .50 × $\mu_3$

For instance, if the exam means are 72, 76, and 80, respectively, then the mean of the final scores is

Mean of final scores = (.25 × 72) + (.25 × 76) + (.50 × 80)
$$= 18 + 19 + 40 = 77$$

We are assuming the class constitutes the entire population of interest, so the means are represented as population means ($\mu$) instead of sample means ($\bar{x}$).

The formula for calculating the mean final score, however, would hold for sample means as well.

## Linear Combinations of Random Variables

The combination of exam scores that we just considered is an example of a *linear combination* of random variables. In a linear combination, we add (or subtract) variables, but some of the combined variables may be multiplied by a numerical value, as occurred in calculating the final score. For two random variables $X$ and $Y$, the most commonly encountered linear combinations are the **sum** $X + Y$ and the **difference** $X - Y$.

---

*definition*    A **linear combination** of random variables, $X, Y, \ldots$ is a combination of the form

$$L = aX + bY + \cdots$$

where $a$, $b$, etc. are numbers, which could be positive or negative. The most commonly encountered linear combinations of two variables are

Sum $= X + Y$

Difference $= X - Y$

---

### Mean and Standard Deviation for Linear Combinations

A number of rules tell us about the statistical properties of a linear combination of random variables. The most basic rule is for the mean. This rule applies to any set of random variables, whether they are discrete or continuous, independent of each other or not, and regardless of their distributions. The only assumption is that the variables that are combined all have finite means.

---

*formula*    **Mean of a Linear Combination of Random Variables (Including Sums and Differences)**

If $X, Y, \ldots$ are random variables, $a, b, \ldots$ are numbers, either positive or negative, and

$$L = aX + bY + \cdots$$

The mean of $L$ is

$$\text{Mean}(L) = a\,\text{Mean}(X) + b\,\text{Mean}(Y) + \cdots$$

In particular,

$$\text{Mean}(X + Y) = \text{Mean}(X) + \text{Mean}(Y)$$

$$\text{Mean}(X - Y) = \text{Mean}(X) - \text{Mean}(Y)$$

---

Rules for the variance and standard deviation of a linear combination are complicated, and we will only consider a specific situation, which is that the random variables are independent. Two random variables are statistically

**independent** if the probability for any event associated with one random variable is not altered by whether or not any event for the other random variable has happened. In a more practical sense, it is usually safe to say that *two random variables are independent if there is no physical connection between the two variables or if there is no apparent reason why the value of one variable should influence the value of the other.*

| formula | **Variance and Standard Deviation of a Linear Combination of Independent Random Variables** |
|---|---|

If $X$, $Y$, etc. are independent random variables, $a$, $b$, etc. are numbers, and

$$L = aX + bY + \cdots$$

then the *variance* and *standard deviation* of $L$ are

$$\text{Variance}(L) = a^2\,\text{Variance}(X) + b^2\,\text{Variance}(Y) + \cdots$$
$$\text{Standard deviation of } L = \sqrt{\text{Variance}(L)}$$

In particular,

$$\text{Variance}(X + Y) = \text{Variance}(X) + \text{Variance}(Y)$$
$$\text{Variance}(X - Y) = \text{Variance}(X) + \text{Variance}(Y)$$

Keep in mind that the standard deviation is the square root of the variance. Notice also that the variance of the *difference* is the *sum* of the variances because $b = -1$ and $b^2 = +1$.

Simply knowing the mean and standard deviation of a combination of random variables is of little practical use in most cases. To find probabilities associated with various outcomes, we also need to know the distribution.

## Combining Independent Normal Random Variables

Some families of random variables keep the same shape when linear combinations of variables within the class are formed. This occurs for normal curve distributions. Any linear combination of normally distributed variables also has a normal distribution, but we consider only independent variables and focus on rules for the *sum* and for the *difference* of two such variables.

| formula | **Linear Combinations of Independent Normal Random Variables** |
|---|---|

If $X$, $Y$, etc. are independent, normally distributed random variables and $a$, $b$, etc. are numbers, either positive or negative, then the random variable $L = aX + bY + \cdots$ is normally distributed. In particular,

$X + Y$ is normally distributed with mean $\mu_X + \mu_Y$ and standard deviation $\sqrt{\sigma_X^2 + \sigma_Y^2}$.

$X - Y$ is normally distributed with mean $\mu_X - \mu_Y$ and standard deviation $\sqrt{\sigma_X^2 + \sigma_Y^2}$.

## Section 8.7

normal approximation to the binomial distribution, 312–313

continuity correction, 314

## Section 8.8

linear combination of random variables, 315

sum of random variables, 315, 316

difference of random variables, 315, 316

independent random variables, 316

# Exercises

- ● Denotes basic skills exercises
- ◆ Denotes dataset is available in StatisticsNow at **http:// 1pass.thomson.com** or on your CD but is **not required** to solve the exercise.

**Bold-numbered exercises** have answers in the back of the text and fully worked solutions in the Student Solutions Manual.

**Statistics⌂Now**™  Go to the StatisticsNow website at **http://1pass.thomson.com** to:

- Assess your understanding of this chapter
- Check your readiness for an exam by taking the Pre-Test quiz and exploring the resources in the Personalized Learning Plan

## Section 8.1

8.1  ● A book is randomly chosen from a library shelf. For each of the following characteristics of the book, decide whether the characteristic is a continuous or a discrete random variable:
   a. Weight of the book (e.g., 2.3 pounds).
   b. Number of chapters in the book (e.g., 10 chapters).
   c. Width of the book (e.g., 8 inches).
   d. Type of book (0 = hardback or 1 = paperback).
   e. Number of typographical errors in the book (e.g., 4 errors).

8.2  ● Decide whether each of the following characteristics of a television news broadcast is a continuous or a discrete random variable:
   a. Number of commercials shown (e.g., five commercials).
   b. Length of the first commercial shown (e.g., 15 seconds).
   c. Whether there were any fatal car accidents reported (0 = there were no fatal accidents reported, 1 = there was at least one fatal accident reported).
   d. Whether or not rain was forecast for the next day (0 = no rain forecast, 1 = rain forecast).

8.3  ● For each characteristic, explain whether the random variable is continuous or discrete.
   a. The number of left-handed individuals in a sample of 100 people.
   b. Time taken to complete an exam for students in a class.
   c. Vehicle speeds at a highway location.
   d. The number of accidents reported last year at a highway location.

8.4  Suppose you regularly play a lottery game. Give an example of a discrete random variable in this context.

8.5  Refer to Example 8.1 (p. 280), the scheduling of an outdoor event. Give an example of another continuous random variable (in addition to temperature) and another discrete random variable (in addition to number of planes flying overhead) that would influence the enjoyment of the event. Give the sample space for those random variables.

## Section 8.2

8.6  ● What is the missing value, represented by "?" in each of the following probability distribution functions?
   a.

| k | 1 | 4 | 5 | 7 |
|---|---|---|---|---|
| $P(X = k)$ | 1/6 | 1/6 | 1/6 | ? |

   b.

| k | 100 | 200 | 300 | 400 | 500 |
|---|---|---|---|---|---|
| $P(X = k)$ | .1 | .2 | .3 | .3 | ? |

8.7  ● A professor gives a weekly quiz with varying numbers of questions and uses a randomization device to decide how many questions to include. Let the random variable $X$ = the number of questions on an upcoming quiz. The probability distribution for $X$ is given in the following table, but one probability is missing. What is the probability that $X = 30$?

| No. of Questions, X | 10 | 15 | 20 | 25 | 30 |
|---|---|---|---|---|---|
| Probability | .05 | .20 | .50 | .15 | ? |

8.8  ● Let the random variable $X$ = number of phone calls you will get in the next 24 hours. Suppose the possible values for $X$ are 0, 1, 2, or 3 and their probabilities are .1, .1, .3, and .5, respectively. For instance, the probability that you will receive no calls is .1.
   a. Verify that the "Conditions for Probabilities for Discrete Random Variables" given in Section 8.2 are met.

---

● Basic skills    ◆ Dataset available but not required    **Bold-numbered** exercises answered in the back

b. Draw a picture of the probability distribution function.

c. Find the cumulative distribution function.

**8.9** ● For a fair coin tossed three times, the eight possible simple events are HHH, HHT, HTH, THH, HTT, THT, TTH, TTT. Let $X$ = number of heads. Find the probability distribution for $X$ by writing a table showing each possible value of $X$ along with the probability that value occurs.

**8.10** ● Suppose the probability distribution for $X$ = number of jobs held during the past year for students at a school is as follows:

| No. of Jobs, $X$ | 0 | 1 | 2 | 3 | 4 |
|---|---|---|---|---|---|
| Probability | .14 | .37 | .29 | .15 | .05 |

a. Find $P(X \leq 2)$, the probability that a randomly selected student held two or fewer jobs during the past year. .80

b. Find the probability that a randomly selected student held either one or two jobs during the past year.

c. Find $P(X > 0)$, the probability that a randomly selected student held at least one job during the past year.

d. Write a table that lists the cumulative probability distribution function for $X$.

**8.11** A kindergarten class has three left-handed children and seven right-handed children. Two children are selected without replacement for a shoe-tying lesson. Let $X$ = the number who are left-handed.

a. Write the simple events in the sample space. For instance, one simple event is RL, indicating that the first child is right-handed and the second one is left-handed.

b. Find the probability for each of the simple events in the sample space. (*Hint:* A tree diagram may help you solve this.)

c. Find the probability distribution function for $X$.

d. Draw a picture of the probability distribution function for $X$.

**8.12** Refer to Example 8.10 (p. 287), in which the probability distribution is given for the sum of two fair dice. Use the distribution to find the probability that the sum is even.

**8.13** Consider three tosses of a fair coin. Write the sample space, and then find the probability distribution function for each of the following random variables:

a. $X$ = number of tails.

b. $Y$ = the difference between the number of heads and the number of tails.

c. $Z$ = the sum of the number of heads and the number of tails.

**8.14** A woman decides to have children until she has her first girl or until she has four children, whichever comes first. Let $X$ = number of children she has. For simplicity, assume that the probability of a girl is .5 for each birth.

a. Write the simple events in the sample space. Use B for boy and G for girl. For instance, one simple event is BG, because the woman quits once she has a girl.

b. Find the probability for each of the simple events in the sample space.

c. Find the probability distribution function for $X$.

d. Draw a picture of the probability distribution function for $X$.

**8.15** Refer to Example 8.10 (p. 287). Find the cumulative distribution function (cdf) for the sum of two fair dice.

**8.16** Explain which would be of more interest in each of the following situations, the probability distribution function or the cumulative probability distribution function for $X$. If you think both would be of interest, explain why.

a. A politician wants to know how her constituents feel about a proposed new law. She hires a survey firm to take a random sample of voters in her district, and $X$ = number in the sample who oppose the new law.

b. Each time someone is exposed to a virus, he or she could become infected with it. Let $X$ = number of exposures a person can sustain before becoming infected. For instance, if $X$ = 0, the person becomes infected on the first exposure; if $X$ = 1, the person becomes infected on the second exposure, and so on.

c. A pharmaceutical company wants to show that its new drug is effective for lowering blood pressure. The drug is administered to a sample of 800 patients with high blood pressure, and $X$ = the number for whom the drug reduces blood pressure.

d. A caterer who serves dinner at wedding receptions is told in advance the number of people who have said they will attend; $X$ is the number of people who actually attend the dinner.

**8.17** The following table gives the probability distribution for $X$ = classes skipped yesterday by students at a college.

| Classes skipped, $X$ | 0 | 1 | 2 | 3 | 4 |
|---|---|---|---|---|---|
| Probability | .73 | .16 | .06 | .03 | .02 |

a. What is the probability that a randomly selected student skipped either two or three classes yesterday?

b. What is the probability that a randomly selected student skipped at least one class yesterday?

c. Write a table that gives the cumulative distribution function for $X$.

## Section 8.3

**8.18** ● Burt pays $30 a year for towing insurance. He thinks that the probability that he will need to have his car towed once in the next year is 1/10 (.1) and the probability that he will have to have it towed more than once is zero. It will cost $100 if his car is towed if he doesn't have insurance but will cost nothing if he does have

1 2 3 4
.5 .5 .5 .5

insurance. Let $X$ = Burt's cost next year for towing and/or insurance.

**a.** If Burt buys the insurance, $X$ has only one possible value. What is it?

**b.** If Burt doesn't buy the insurance, $X$ has two possible values. List the two values and their probabilities.

c. Refer to parts (a) and (b). In each case, find $E(X)$. Using these two values for $E(X)$, explain whether Burt should buy the insurance.

8.19 ● The Brann family wants to have children. They are working with a financial advisor who informs them that on the basis of families with similar characteristics, the probability distribution for the random variable $X$ = number of children they might have is as follows:

| No. of Children | 0 | 1 | 2 | 3 |
|---|---|---|---|---|
| Probability | .05 | .60 | .30 | .05 |

a. What is the expected value of $X$?

b. Explain what "expected value" means in this example.

c. Is $E(X)$ a possible outcome for the number of children the Brann family will have?

**8.20** ● The probability that Mary will win a game is .01, so the probability that she will not win is .99. If Mary wins, she will be given $100; if she loses, she must pay $5. If $X$ = amount of money Mary wins (or loses), what is the expected value of $X$?

8.21 ● A random variable $X$ has the following probability distribution:

| Values of $X$ | −1 | 0 | 1 |
|---|---|---|---|
| Probability | .25 | .50 | .25 |

a. Calculate the mean of $X$.

b. Calculate the variance of $X$.

c. Calculate the standard deviation of $X$.

8.22 ● A random variable $X$ has the following probability distribution:

| Values of $X$ | −2 | 0 | 2 |
|---|---|---|---|
| Probability | .25 | .50 | .25 |

a. Calculate the mean of $X$.

b. Calculate the variance of $X$.

c. Calculate the standard deviation of $X$.

**8.23** Find the expected value for the number of girls in a family with three children, assuming that boys and girls are equally likely. Use the probability distribution function given in Example 8.6 (p. 284).

8.24 Find the expected value for the sum of two fair dice. The probability distribution function was found in Example 8.10 (p. 287).

8.25 Find the standard deviation for the sum of two fair dice.

**8.26** Find the standard deviation for the net gain in the California Decco lottery game in Example 8.12 (p. 290).

8.27 An insurance company expects 10% of its policyholders to collect claims of $500 this year and the remaining 90% to collect no claims. What is the expected value for the amount they will pay out in claims per person?

8.28 Refer to Exercise 8.27. If the insurance company wants to make a net profit of $10 per policyholder for the year, how much should it charge each person for insurance? Ignore administrative and other costs.

**8.29** In 1991, 72% of children in the United States were living with both parents, 22% were living with mother only, 3% were living with father only, and 3% were not living with either parent (*World Almanac*, 1993, p. 945).

a. What is the expected value for the *number* of parents a randomly selected child was living with in 1991?

b. Does the concept of expected value have a meaningful interpretation for this example? Explain.

8.30 Suppose the probability that you get an A in any class you take is .3 and the probability that you get a B is .7. To construct a grade point average, an A is worth 4.0 and a B is worth 3.0.

a. What is the expected value for your grade point average?

b. Would you expect to have this grade point average for each quarter or semester of your school career? Explain.

8.31 Suppose you have to cross a train track on your commute. The probability that you will have to wait for a train is 1/5, or .20. If you don't have to wait, the commute takes 15 minutes, but if you have to wait, it takes 20 minutes.

a. What is the expected value of the time it takes you to commute?

b. Is the expected value ever equal to the actual commute time? Explain.

## Section 8.4

**8.32** ● A fair coin is flipped 200 times. The random variable $X$ = number of heads out of the 200 tosses.

**a.** Is $X$ a binomial random variable? If so, specify $n$ and $p$. If not, explain why not.

**b.** What is the expected value of $X$?

8.33 ● For each of the examples below, decide if $X$ is a binomial random variable. If so, specify $n$ and $p$. If not, explain why not.

a. $X$ = number of heads from flipping the same coin ten times, where the probability of a head = 1/2.

b. $X$ = number of heads from flipping two coins five times each, where the probability of a head for one coin = 1/2 and the probability of a head for the other coin = 1/4.

c. $X$ = number of cities in which it will rain tomorrow among five neighboring cities located within ten miles of each other.

d. $X$ = number of children who will get the flu this winter in a kindergarten class with 20 children.

**8.34** ● For each of the following binomial random variables, specify $n$ and $p$:

a. A fair die is rolled 30 times. $X$ = number of times a 6 is rolled.

b. A company puts a game card in each box of cereal and 1/100 of them are winners. You buy ten boxes of cereal, and $X$ = number of times you win.

c. Jack likes to play computer solitaire and wins about 30% of the time. $X$ = number of games he wins out of his next 20 games.

**8.35** ● Refer to Exercise 8.34. Find $\mu$ in each case.

**8.36** ● Assuming that $X$ is a binomial random variable with $n = 10$ and $p = .20$, find the probability for each of the following values of $X$:

a. $X = 5$.

b. $X = 2$.

c. $X = 1$.

d. $X = 9$.

**8.37** ● Find the mean and standard deviation for a binomial random variable $X$ with:

a. $n = 10$, $p = .50$.

b. $n = 1$, $p = .40$.

c. $n = 100$, $p = .90$.

d. $n = 30$, $p = .01$.

**8.38** Explain which of the conditions for a binomial experiment is *not* met for each of the following random variables:

a. A football team plays 12 games in its regular season. $X$ = number of games the team wins.

b. A woman buys a lottery ticket every week for which the probability of winning anything at all is 1/10. She continues to buy them until she has won three times. $X$ = the number of tickets she buys.

c. A poker hand consists of 5 cards drawn from a standard deck of 52 cards. $X$ = the number of aces in the hand.

**8.39** A computer chess game and a human chess champion are evenly matched. They play ten games. Find probabilities for the following events.

a. They each win five games.

b. The computer wins seven games.

c. The human chess champion wins at least seven games.

**8.40** In an ESP test, a "participant" tries to draw a hidden "target" photograph that is unknown to anyone in the room. After the drawing attempt, the participant is shown four choices and asked to determine which one had been the real target. The real target is randomly selected from the four choices in advance, so the probability of a correct match by chance is 1/4. The test is repeated ten times, using four new photographs each time.

a. Go through the conditions for a binomial experiment, and explain how this situation fits each one of them, assuming that the participant is just guessing each time.

b. Let $X$ = number of correct choices in the ten tests. If the participant is just guessing, is $X$ a binomial ran-

dom variable? If not, explain why not. If so, specify $n$ and $p$.

c. If the participant is just guessing, find $P(X \geq 6)$.

d. Suppose the participant actually has some psychic ability and can get each answer correct with probability .5 instead of .25. Find $P(X \geq 6)$.

e. Compare the answers in parts (c) and (d). If the participant actually selects six of the ten answers correctly, would you believe that he or she was just guessing or that he or she was using some psychic ability? Explain your answer. (Note that there is no correct answer here; your reasoning is what counts.)

**8.41** For each of the following situations, assume that $X$ is a binomial random variable with the specified $n$ and $p$, and find the requested probability.

**a.** $n = 10$, $p = .5$, $P(X = 4)$.

**b.** $n = 10$, $p = .3$, $P(X \geq 4)$.

**c.** $n = 10$, $p = .3$, $P(X \leq 3)$.

**d.** $n = 5$, $p = .1$, $P(X = 0)$.

**e.** $n = 5$, $p = .1$, $P(X \geq 1)$.

**8.42** For each of the following scenarios, write the desired probability in a format such as $P(X \leq 10)$ and specify $n$ and $p$. Do not actually compute the desired probability. If you cannot specify a numerical value for $p$, write in words what it represents.

a. A random sample of 1000 adults is drawn from the United States. $X$ = the number in the sample who are living with a partner but are not married. The desired probability is the probability that at least 1/4 of the sample fits this description.

**b.** A pharmaceutical company claims that 20% of those taking its new allergy medication will experience drowsiness. To test this claim, it randomly assigns 500 people to take the new medication and measures $X$ = the number who experience drowsiness. The desired probability is the probability that 110 or more people in the sample experience drowsiness if the claim is true.

c. A student has not studied the material for a 20-question true-false test and simply guesses on each question. A passing grade is 70%; the desired probability is the probability that the student passes the test.

**8.43** Find the expected value and standard deviation for a binomial random variable with each of the following values of $n$ and $p$:

a. $n = 10$, $p = 1/2$.

b. $n = 100$, $p = 1/4$.

c. $n = 2500$, $p = 1/5$.

d. $n = 1$, $p = 1/10$.

e. $n = 30$, $p = .4$.

**8.44** Use the direct formulas for expected value and standard deviation to verify that the mean and standard deviation for a binomial random variable with $n = 2$ and $p = .5$ are $\mu = 1$ and $\sigma = 0.7071$, respectively. (*Hint:* The only possible values of $X$ are 0, 1, 2. Find their probabilities and use the formulas at the end of Section 8.3.)

## Section 8.5

**8.45** ● Suppose that the time students wait for a bus can be described by a uniform random variable $X$, where $X$ is between 0 minutes and 60 minutes.
   a. What is the probability that a student will wait between 0 and 30 minutes for the next bus?
   b. What is the probability that a student will have to wait at least 30 minutes for the next bus?

**8.46** A game is played with a spinner on a circle, like the minute hand on a clock. The circle is marked evenly from 0 to 100, so, for example, the 3:00 position corresponds to 25, the 6:00 position to 50, and so on. The player spins the spinner, and the resulting number is the number of seconds he or she is given to solve a word puzzle. Let $X$ = amount of time the player is given to solve the puzzle.
   **a.** Explain how you know that $X$ is a uniform random variable.
   **b.** Write down the density function for $X$. Be sure to specify the range of values for which it holds.
   **c.** Find $P(X \le 15$ seconds$)$.
   d. Find $P(X \ge 40$ seconds$)$.
   e. Draw a picture of the density function of $X$, and use it to illustrate the probabilities that you found in parts (c) and (d).
   f. What is the expected value of $X$? This was not covered in the text, so explain your reasoning.

**8.47** Give an example of a uniform random variable that might occur in your daily life.

**8.48** Draw the density "curve" corresponding to each of the following random variables, then shade the area corresponding to the desired probability. You *do not* need to compute the probability.
   a. $X$ is a uniform random variable from 10 to 20, $P(10 \le X \le 13)$.
   b. $X$ is a uniform random variable from 0 to 20, $P(X \le 5)$.
   c. $X$ is a uniform random variable from 0 to 8, $P(X > 4)$.

## Section 8.6

**8.49** ● In each situation below, calculate the standardized score (or z-score) for the value $x$:
   a. Mean $\mu = 0$, standard deviation $\sigma = 1$, value $x = 1.5$.
   **b.** Mean $\mu = 10$, standard deviation $\sigma = 6$, value $x = 4$.
   c. Mean $\mu = 10$, standard deviation $\sigma = 5$, value $x = 0$.
   d. Mean $\mu = -10$, standard deviation $\sigma = 15$, value $x = -25$.

**8.50** ● For each value of $z^*$, find the cumulative probability $P(Z \le z^*)$:
   a. $z^* = 0$.
   **b.** $z^* = -0.35$.
   **c.** $z^* = 0.35$.
   d. $z^* = 1.96$.
   e. $z^* = -2.33$.
   f. $z^* = 2.58$.
   g. $z^* = 1.65$.

**8.51** ● Weights $(X)$ of men in a certain age group have a normal distribution with mean $\mu = 180$ pounds and standard deviation $\sigma = 20$ pounds. Find each of the following probabilities:
   a. $P(X \le 200)$ = probability the weight of a randomly selected man is less than or equal to 200 pounds.
   b. $P(X \le 165)$ = probability the weight of a randomly selected man is less than or equal to 165 pounds.
   c. $P(X > 165)$ = probability the weight of a randomly selected man is more than 165 pounds.

**8.52** ● Draw the density "curve" corresponding to each of the following normal random variables, then shade the area corresponding to the desired probability. You *do not* need to compute the probability.
   a. $X$ is a normal random variable with $\mu = 75$ and $\sigma = 5$, $P(70 \le X \le 85)$.
   b. $X$ is a normal random variable with $\mu = 15$ and $\sigma = 10$, $P(5 \le X \le 20)$.
   c. $X$ is a normal random variable with $\mu = 100$ and $\sigma = 15$, $P(X > 115)$.

**8.53** ● Find the following probabilities for a standard normal random variable $Z$:
   **a.** $P(Z \le -1.4)$.
   b. $P(Z \le 1.4)$.
   c. $P(-1.4 \le Z \le 1.4)$.
   d. $P(Z \ge 1.4)$.

**8.54** ● Find the following probabilities for a standard normal random variable $Z$:
   a. $P(Z \le -3.72)$.
   b. $P(-3.72 \le Z \le 3.72)$.
   c. $P(Z \ge 15)$.

**8.55** Find the following probabilities for $X$ = pulse rates of women, for which the mean is 75 and the standard deviation is 8. Assume a normal distribution.
   **a.** $P(X \le 71)$.
   b. $P(X \ge 85)$.
   c. $P(59 \le X \le 95)$.

**8.56** Refer to Exercise 8.55 about the pulse rates of women. What is the 10th percentile of women's pulse rates?

**8.57** Find the value $z^*$ that satisfies each of the following probabilities for a standard normal random variable $Z$:
   **a.** $P(Z \le z^*) = .025$.
   b. $P(Z \le z^*) = .975$.
   c. $P(-z^* \le Z \le z^*) = .95$.

**8.58** Find the following probabilities for Verbal SAT test scores $X$, for which the mean is 500 and the standard deviation is 100. Assume that SAT scores are described by a normal curve.
   a. $P(X \le 500)$.
   b. $P(X \le 650)$.
   c. $P(X \ge 700)$.
   d. $P(500 \le X \le 700)$.

**8.59** Heights for college women are normally distributed with mean = 65 inches and standard deviation = 2.7 inches. Find the proportion of college women whose heights fall into the following ranges:
   **a.** Between 62 inches and 65 inches.
   b. Between 60 inches and 70 inches.

c. Less than 70 inches.

d. Greater than 60 inches.

e. Either less than 60 inches *or* greater than 70 inches.

**8.60** Refer to Exercise 8.59. Find the height such that about 25% of college women are shorter than that height. What is the percentile ranking for that height?

8.61 Refer to Exercises 8.59 and 8.60. Find the height such that about 10% of college women are taller than that height. What is the percentile ranking for that height?

8.62 Assume that right handspan measurements for college women are approximately normally distributed, with mean 20 cm and standard deviation 1.8 cm, and that for men, they are normally distributed, with mean 22.5 cm and standard deviation 1.5 cm. Measure your own right handspan.

a. Assuming that you fit with the population of college students of your sex, what is the *z*-score corresponding to your right handspan measurement? (Be sure to give your sex and right handspan measurement with your answer.)

b. What proportion of students of your sex have right handspan measurements smaller than yours?

**8.63** Suppose the yearly rainfall totals for a city in northern California follow a normal distribution, with mean of 18 inches and standard deviation of 6 inches. For a randomly selected year, what is the probability that total rainfall will be in each of the following intervals?

**a.** Less than 10 inches.

**b.** Greater than 30 inches.

**c.** Between 15 and 21 inches.

**d.** Greater than 35 inches.

**8.64** Refer to Exercise 8.63. Suppose that in a given year, the total rainfall is only 6 inches. You work for the local newspaper, and your editor has asked you to write a story about the terrible drought the town is suffering and how abnormal the situation is. Write a few sentences that you could use to explain the statistical facts to your readers (and your editor). Be sure to comment on whether or not you agree that the situation is terribly abnormal.

## Section 8.7

8.65 ● Suppose that a fair coin is flipped $n = 100$ times.

a. Calculate the mean and standard deviation of $X =$ number of heads. (*Hint:* The variable $X$ has a binomial distribution.)

b. Use the normal approximation to the binomial distribution to estimate the probability that the number of heads is greater than or equal to 60.

c. Repeat part (a) using the continuity correction.

**8.66** ● Suppose that $p = .512$ is the proportion of one-child families in which the child is a boy.

**a.** For a random sample of $n = 50$ one-child families, estimate the probability that there will be 20 or fewer boys. Use the normal approximation to the binomial distribution.

**b.** Repeat part (a) using the continuity correction.

8.67 Use the normal approximation to the binominal distribution to approximate the stated probability in each of the following scenarios:

a. A random sample of 1000 eligible voters is drawn, and $X =$ number who actually voted in the last election. It is known that 60% of all eligible voters did vote. Find (approximately) $P(X \le 620)$.

b. To be eligible for a certain job, women must be at least 62 inches tall, and 87% of women meet this criterion. In a random sample of 2000 women, $X =$ number who qualify for the job (based on height). Find (approximately) $P(X \le 1700)$.

8.68 In a test for extrasensory perception, the participant repeatedly guesses the suit of a card randomly sampled with replacement from an ordinary deck of cards. There are four suits, all equally likely. The participant guesses 100 times, and $X =$ number of correct guesses.

a. Explain why $X$ is a binomial random variable, and specify $n$ and $p$, assuming that the participant is just guessing.

b. Find the mean and standard deviation for $X$ if the participant is just guessing.

c. Suppose that the participant guesses correctly 33 times. Find the approximate probability of guessing this well or better by chance.

d. Suppose that the participant guesses correctly 50 times. Find the approximate probability of guessing this well or better by chance. In that circumstance, would you be convinced that the participant was doing something other than just guessing? Explain.

## Section 8.8

8.69 ● Suppose the heights of adult males in a population have a normal distribution with mean $\mu = 70$ inches and standard deviation $\sigma = 2.8$ inches. Two unrelated men will be randomly sampled. Let $X =$ height of the first man and $Y =$ height of the second man.

a. Consider $D = X - Y$, the difference between the heights of the two men. What type of distribution will the variable $D$ have?

b. What is the mean value for the distribution of $D$?

c. Assuming independence between the two men, find the standard deviation of $D$.

d. Determine the probability that the first man is more than 3 inches taller than the second man. That is, find $P(D > 3)$.

e. Find the probability that one of the men is at least 3 inches taller than the other. That is, find the probability that either $(D > 3)$ or $(D < -3)$.

8.70 ● Suppose the length of time a person takes to use an ATM machine is normally distributed with mean $\mu = 100$ seconds and standard deviation $\sigma = 10$ seconds. There are $n = 4$ people ahead of Jackson in a line of people waiting to use the machine. He is concerned about $T =$ total time the four people will take to use the machine.

**a.** What is the mean value of $T$ = total time for the four people ahead of Jackson?

**b.** Assuming that the times for the four people are independent of each other, determine the standard deviation of $T$.

**c.** Jackson hopes the total time $T$ is less than or equal to 360 seconds (6 minutes). Find $P(T < 360)$, the probability that the total waiting time is less than 360 seconds.

8.71 ● Ed and Taylor work together on a group quiz that has 15 multiple-choice questions, each with four choices for the possible answer. Unfortunately, neither had time to study, so they decide to randomly guess at all answers. Ed guesses answers for the first 7 questions, and Taylor guesses for the other 8 questions.

**a.** For any single question on the quiz, what is $p$ = chance of a correct guess?

**b.** Let $X$ = number of correct guesses that Ed makes. What are the values of $n$ and $p$ for the binomial distribution that describes $X$?

**c.** Let $Y$ = number of correct guesses that Taylor makes. What are the values of $n$ and $p$ for the binomial distribution that describes $Y$?

**d.** Consider $X + Y$ = total correct guesses that Ed and Taylor make. What are the values of $n$ and $p$ for the binomial distribution that describes $X + Y$?

8.72 ● The variable $X$ has a normal distribution with mean $\mu = 75$ and standard deviation $\sigma = 6$. The variable $Y$ has a normal distribution with mean $\mu = 70$ and standard deviation $\sigma = 8$. Variables $X$ and $Y$ are independent.

**a.** Find the mean and standard deviation of the sum $X + Y$.

**b.** Find the mean and standard deviation of the difference $X - Y$.

**8.73** Give the mean, variance, and standard deviation of the sum $X + Y$ in each of the following cases. If possible, name what distribution the sum has as well.

**a.** $X$ is a binomial random variable with $n = 10$, $p = .5$. $Y$ is a binomial random variable with $n = 20$, $p = .4$. $X$ and $Y$ are independent.

**b.** $X$ is a normal random variable with $\mu = 100$ and $\sigma = 15$.
$Y$ is a normal random variable with $\mu = 50$ and $\sigma = 10$.
$X$ and $Y$ are independent.

**c.** $X$ is a normal random variable with $\mu = 100$ and $\sigma = 15$.
$Y$ is a binomial random variable with $n = 200$, $p = .25$.
$X$ and $Y$ are independent.

8.74 Ethan has a midterm in his statistics class, which starts 10 minutes after the scheduled end of his biology class. The biology teacher rarely ends class on time though, and the amount of time he is overtime is approximately normally distributed, with a mean of 2 minutes and a standard deviation of 1/2 minute. The time it takes Ethan to get from one class to another is also normal, with a mean of 6 minutes and a standard deviation of 1 minute.

**a.** Is it reasonable to assume that the two times are independent? Explain.

**b.** Assuming that the two times are independent, what is the probability that Ethan will be late for his exam?

8.75 Joe performs remarkably well on remote viewing ESP tests, which require the "viewer" to draw a picture, then determine which of four possible photos was the intended "target." Because the target is randomly selected from among the four choices, the probability of a correct match by chance is 1/4, or .25. Joe participates in three experiments with $n$ = 10, 20, and 50 trials, respectively. In the three experiments, his numbers correct are 4, 8, and 20, respectively, for a total of 32 correct out of 80 trials.

**a.** For each experiment, what is the expected number correct if Joe is just guessing?

**b.** Over all 80 trials, what is the expected number correct? Explain how you found your answer.

**c.** For each experiment separately, find the approximate probability that someone would get as many correct as Joe did, or more, if the person was just guessing.

**d.** Out of the 80 trials overall, find the approximate probability that someone would get as many correct as Joe did (32), or more, if the person was just guessing.

**e.** Compare your answers in parts (c) and (d). If Joe were trying to convince someone of his ability, would it be better to show the three experiments separately, or would it be better to show the combined data? Explain.

**8.76** Annmarie is trying to decide whether to take the train or the bus into the city. The train takes longer but is more predictable than the bus because there are no traffic delays. Train times are approximately normally distributed, with mean 60 minutes and standard deviation 2 minutes, while bus times are approximately normally distributed, with mean 50 minutes and standard deviation 8 minutes. The bus and train times are independent of each other. Find the probability that the train is faster on any given day.

8.77 Charles, Julia, and Alex are in grades 4, 3, and 2, respectively, and are representing their school at a spelling bee. The school's team score is the sum of the number of words the individual students spell correctly out of 50 words each. Different words are given for each grade level. From practicing at school, it is known that the probability of spelling each word correctly is .9 for Charles and .8 for the younger two, Julia and Alex.

**a.** Find the mean and standard deviation for the number of correct words for each child.

**b.** Find the mean and standard deviation for the team score.

**c.** Assume that the individual scores are independent. Does the team score have a binomial distribution? Explain.

d. Although not obvious from the material in this chapter, the team score would be approximately normal with the mean and standard deviation you were asked to find in part (b). If last year's team score was 131, what is the approximate probability that this year's team scores as well or better?

## Chapter Exercises

**8.78** The vehicle speeds at a particular interstate location can be described by a normal curve. The mean speed is 67 mph, and the standard deviation is 6 mph. What proportion of vehicle speeds at this location are faster than 75 mph?

**8.79** Do you think that a normal curve would be a good approximation to the distribution of the ages of all individuals in the world? Briefly explain your answer.

**8.80** A histogram is drawn of the weights of all students in a class of 100 men and 100 women. Would this histogram have a bell shape? Briefly explain your answer.

**8.81** Suppose that 10% of a population is left-handed. What is the probability that in a sample of $n = 10$ individuals, 3 or more individuals are left-handed?

**8.82** New spark plugs have just been installed in a small airplane with a four-cylinder engine. There is one spark plug per cylinder, so four spark plugs have been installed. For each spark plug, the probability that it is defective and will fail during its first 20 minutes of flight is 1/10,000, independent of the other spark plugs.
  a. For each spark plug, what is the probability that it will not fail?
  b. If one spark plug fails, the plane will shake and not climb higher, but it can be landed safely. What is the probability that this happens?
  c. If two or more spark plugs fail, the plane will crash. What is the probability that this happens?

**8.83** In the casino game of roulette, a gambler can bet on which of 38 numbers will be the result when the roulette wheel is spun. On a $2 bet, a gambler gains $70 if he or she picks the right number but loses the $2 otherwise.
  **a.** Let $X$ = amount gained or lost on a $2 bet on a roulette number. Write out the probability distribution of $X$.
  **b.** Calculate $E(X)$, the mean value of $X$. What does this value indicate about the advantage that a casino has over roulette players?

**8.84** Suppose that a college determines the following distribution for $X$ = number of courses taken by full-time students this semester:

| k | 3 | 4 | 5 | 6 | 7 |
|---|---|---|---|---|---|
| P(X = k) | .07 | .14 | .52 | .25 | .02 |

  a. Write out the cumulative distribution function of $X$.

  b. What is the probability that a randomly selected full-time student is taking five or fewer courses this semester?
  c. What is the probability that a randomly selected full-time student is taking more than five courses?

**8.85** Refer to Exercise 8.84. What is the mean number of courses taken by full-time students?

**8.86** Explain which of the conditions for a binomial experiment is *not* met for each of the following random variables:
  **a.** A ten-question quiz has five true-false questions and five multiple-choice questions, each with four possible choices for the answer. A student randomly picks an answer for every question. $X$ = number of answers that are correct.
  **b.** Four students are randomly picked without replacement from a class of ten women and ten men. $X$ = number of women among the four selected students.

**8.87** The standard medical treatment for a certain disease is successful in 60% of all cases.
  a. The treatment is given to $n = 200$ patients. What is the probability that the treatment is successful for 70% or more of these 200 patients? (*Tip:* 70% of $n = 200$ is 140 patients.)
  b. The treatment is given to only $n = 20$ patients. What is the probability that the treatment is successful for 70% or more of these 20 patients?

**8.88** Kim and her sister Karen each plan to have four children. Assume that the probability of a girl is .50, independent across births.
  a. If $X$ = number of girls Kim will have, what is the distribution of $X$?
  b. If $Y$ = number of girls Karen will have, what is the distribution of $Y$?
  c. What is the distribution for $T$, the total number of girls Kim and Karen have?

**8.89** Shaun (3 years old) and Patrick (4 years old) are each allowed to pick one book for bedtime stories, and the parent on duty reads the two books sequentially. The lengths of time it takes to read the books from which they select are approximately normally distributed, with mean 5 minutes and standard deviation 2 minutes.
  **a.** Does it make sense to assume that the times for the two books are independent? Explain.
  **b.** On about what proportion of nights will storytime exceed 15 minutes, assuming that the times for the two books are independent?

**8.90** Use the rule referenced from Chapter 7 to verify that the relationship is correct for a standard normal random variable $Z$:
  a. Use Rule 1 of Chapter 7 to verify that $P(Z > a) = 1 - P(Z \le a)$.
  b. Use Rule 2b of Chapter 7 to verify that $P(Z \le b) = P(Z \le a) + P(a \le Z \le b)$. Use that result to show that $P(a \le Z \le b) = P(Z \le b) - P(Z \le a)$.

8.91 You will need a computer program that simulates a large number of binomial random variables at once for this exercise. Refer to Example 8.35 (p. 319), "Strategies for Studying When You Are Out of Time." Simulate what would happen if you used strategy 2, in which the total score was the sum of two binomial random variables, one with $n = 10$ and $p = .8$, the other with $n = 10$ and $p = .2$. Run 1000 simulations, and determine the proportion of them for which you would have received a passing grade (13 or more). Explain what you did.

# 9

Joos Mind /Stone/Getty Images

*Will her overeating lead to excessive weight gain during her freshman year?*

See Example 9.1 *(p. 334)*

# Understanding Sampling Distributions: Statistics as Random Variables

> This chapter introduces the reasoning that allows researchers to make conclusions about entire populations using relatively small samples of individuals. The secret to understanding how things work is to understand what kind of dissimilarity we should expect to see among different samples from the same population.

By this point in the book, you should have realized that one of the main purposes of statistical methods is to enable us to satisfy our curiosity about all kinds of things. Most of the time, the questions we want to answer extend beyond the data we are able to collect to questions about larger populations. Therefore, we need to know how to make use of sample data to answer questions about a bigger collection of individuals than the ones we have measured.

This chapter serves as an introduction to the reasoning that allows researchers to reach conclusions about entire populations on the basis of a relatively small sample of individuals. The basic idea is that we must work backward, from a sample to a population. We start with a question about a population such as "How many teenagers are infected with HIV?", "At what average age do left-handed people die?", or "What is the average income of all students at a large university?" We collect a sample from the population about which we have the question and measure the variable of interest. We can then answer the question of interest for the sample. Finally, on the basis of statistical theory, we will be able to determine how close our sample answer is to what we really want to know: the true answer for the population. ▌

## 9.1 Parameters, Statistics, and Statistical Inference

Most questions that we ask about large populations are translated into questions about specific summary characteristics of the group. For instance, suppose we collect data from a sample of students at our school on how many hours they study per week. We might characterize the data by finding the mean value for the sample. Then we might use that sample summary to estimate the mean weekly hours of study for the complete population of students.

A **parameter** is a number that is a summary characteristic of a population, a random situation, or a comparison of populations. Sometimes, the phrase **population parameter** is used to make it clear that a parameter is associated with a population instead of a sample. Examples of parameters are the proportion of adults in the world who are left-handed, the probability that any baby who sleeps with a nightlight will have myopia at age 20, and the difference in the mean incomes of all men and women in the same profession.

We assume that a parameter has a fixed, unchanging value. Usually, the value of a parameter is not known to us and will not be known to us because we will not be able to measure every unit in the population. Although we will not have the information necessary to find the numerical value of a population parameter, but we will be able to use statistical methods to make a good guess, as you will see in the remainder of this book.

A **statistic,** also called a **sample statistic,** is a number that is computed from a sample of values taken from a larger population. It is a summary characteristic of the sample data. The sample data may be collected in a sample survey, an observational study or an experiment. The term **sample estimate** or simply **estimate** is sometimes used for a sample statistic when the statistic is used to estimate the unknown value of a population parameter.

One of the big ideas in this chapter is that possible values of sample statistics are variable. If two different samples are taken from the same population, it is likely that the sample statistics will be different for those two samples. For instance, if we were to take two different random samples of 1000 adults and find the proportion in each sample that were left-handed, the proportions would probably not be exactly the same. In this scenario, the sample proportion is an example of a sample statistic. The corresponding population parameter is the proportion of the entire population that is left-handed. Notice that the population parameter remains fixed, although we do not know its numerical value. The value of a sample statistic may change from sample to sample, and we will know the value once we have measured a sample.

## Statistical Inference

The results in this chapter will be used for the remainder of the book to make conclusions about population parameters on the basis of sample statistics. The procedures we will use for making these conclusions are called **statistical inference** procedures. The two most common procedures are to find *confidence intervals* and to conduct *hypothesis tests*. Here, we give just a brief introduction to these two procedures, which will be developed throughout the rest of the book.

### Confidence Intervals

The first inference technique is to create a **confidence interval,** which is an interval of values that the researcher is fairly sure will cover the true, unknown value of the population parameter. In other words, we use a confidence interval to estimate the value of a population parameter. For instance, in Example 3.3 (p. 77), we used data from a Gallup poll of $n = 1003$ adult Americans to estimate that between 62% and 68% of U.S. adults would say that religion is very important in their lives. In this example, the population parameter of interest is the percent of all U.S. adults who would say that religion is very important in their lives. We don't know the true value of this parameter, but we estimate it to be in the interval 62% to 68%. We have already encountered confidence intervals in Case Study 1.3 and in Chapter 3, when we learned about the margin of error. We will explore confidence intervals further in Chapters 10 and 11.

### Hypothesis Testing

The second statistical inference technique is called **hypothesis testing** or **significance testing.** Hypothesis testing uses sample data to attempt to reject a hypothesis about the population. Usually researchers want to reject the notion that chance alone can explain the sample results. We encountered this idea in Chapter 6 when we learned how to determine whether or not the relationship between two categorical variables is statistically significant. The hypothesis that researchers set up to reject in that setting is that two categorical variables are unrelated to each other in the population, and that any observed relationship in the sample is simply due to chance. In most research settings, the *desired* conclusion is that the variables under scrutiny are related.

Hypothesis testing is applied to population parameters by specifying a **null value** for the parameter—a value that would indicate that nothing of interest is happening. For instance, a weight-loss clinic might postulate that the average weight loss for the population of clinic patrons is 0, so the null value would be 0. The clinic's goal, of course, would be to show that something more interesting is happening and that in fact the average weight loss is greater than 0.

Hypothesis testing proceeds by obtaining a sample, computing a sample statistic, and assessing how unlikely the sample statistic would be if the null parameter value were correct. In most cases, the researchers are trying to show that the null value is not correct. Achieving **statistical significance** is equivalent to rejecting the idea that the observed results are plausible if the null value is correct. For instance, if the weight-loss clinic could show that a sample of patrons had an average weight loss of 10 pounds, with minimal variability, then it would be hard to believe that a population weight loss of 0 yielded those results. We will explore hypothesis testing further in Chapters 12 and 13.

9.1 Exercises are on page 384.

*in summary*    Parameters, Statistics, and Statistical Inference

- A **parameter** is a numerical summary of a population. Its value is considered to be fixed and unchanging.
- A **statistic,** or **sample statistic,** is a numerical summary of a sample. Its value may be different for different samples. The term **sample estimate** or **estimate** is synonymous and reinforces the idea that a statistic often is used to estimate the corresponding population parameter.
- In **statistical inference,** information from a sample is used to make generalizations about a larger population. Sample statistics are used to make conclusions about population parameters.

# 9.2 From Curiosity to Questions About Parameters

One of the most important skills that a good researcher needs is the ability to translate curiosity about something into a question that can be answered. In statistics, this process usually requires us to formulate a question about a parameter. In this chapter and Chapters 10 to 13, we will learn how to formulate and answer questions about five different types of situations, represented by five different parameters. Fortunately, the statistical methods that we use in these five situations are almost identical. The details change, but the methods remain the same.

Let's see how curiosity about something can be translated into questions about parameters. Many students think that the appropriate statistical procedure to use to answer a question depends on what type of data is available. But in fact the process should work the other way around. The scientific question should be expressed as a question about a parameter, then data should be collected that are appropriate for making conclusions about the value of that parameter. Here is a diagram illustrating how the process of satisfying one's curiosity about something should work.

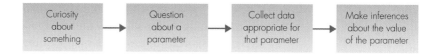

Example 9.1 illustrates how to turn curiosity into a question about a parameter. As you will see, we can ask different questions about the same issue, and each different question may lead to a different parameter of interest. To collect appropriate sample data, researchers must first decide which questions and parameters they wish to pursue.

**Example 9.1**

**The "Freshman 15"**   Do college students really gain weight during their freshman year? The lore is that they do, and this phenomenon has been called "the freshman 15" because of speculation that students typically gain as much as 15 pounds during the first year of college. How can we turn our curiosity about the freshman 15 into a question about a parameter? There are several ways in which we could investigate whether or not students gain weight during the first year. We might want to know what proportion of freshmen gain weight. A related question would be whether most students gain weight. Or we might want to know what the average weight gain is across all first-year students. We might want to know whether women gain more weight than men or vice versa. Here are two ideas for satisfying our curiosity, along with the standard notation that we use for the relevant parameters:

- Parameter = $p$ = proportion of the population of first-year college students who weigh more at the end of the year than they did at the beginning of the year.
- Parameter = $\mu_d$ = the mean (average) weight gain during the first year for the population of college students. The subscript $d$ indicates that the raw

all those who want to quit smoking would actually quit after using a nicotine patch for 8 weeks.

To collect data that will help us to answer questions about one proportion or probability, we measure *one categorical variable* for each person, unit, or trial in our survey or experiment. For instance, for each person in a smoking study, we would measure whether they quit smoking or not after 8 weeks of using a nicotine patch. For each live birth, we would measure whether the baby is male or female. In a survey, we categorize opinions for each person into one of two or more possible categories (such as favor, oppose, or no opinion), then ask what proportion have a certain response.

In each case, there is a single, usually unknown, population proportion or probability $p$, and we collect data to make inferences about it using a sample proportion, $\hat{p}$. Binomial experiments are the most common example of situations for which the parameter of interest is a population proportion or probability.

**Example 9.2**

**Opinions About Genetically Modified Food**    In Example 3.4, we learned that only 35% of a random sample of 1024 adults from the United States in June 2001 thought that genetically modified food was safe to eat. The *population* in this example is all adults in the United States in June 2001. The *population parameter* is

$p$ = the *proportion* of the population of all adults who would have responded that genetically modified food was safe to eat if they had been asked. We do not know the numerical value of $p$.

For the survey that was taken, the *sample statistic* was

$\hat{p} = .35$ = the *sample proportion* who responded that genetically modified food was safe to eat in the sample of 1024 adults who were asked.

Note that if a different sample of 1024 adults had been asked at that time, the population parameter would remain the same because it is not based on the sample, it is based on everyone in the population. The sample statistic probably would change because a different subset of the population would be chosen to respond. ■

**Example 9.3**

**Probability of Quitting with a Nicotine Patch**    In Case Study 4.3, we learned that 46% of volunteers who were randomly assigned to wear a nicotine patch had quit smoking after 8 weeks. The *population* in this example is all smokers who would volunteer for such a study. The *population parameter* is

$p$ = the *probability* that a randomly selected person from this population would quit smoking after 8 weeks of wearing a nicotine patch. We do not know the numerical value of $p$.

For the experiment that was conducted, the *sample statistic* was

$\hat{p} = .46$ = the proportion in the sample of 120 nicotine patch users who quit smoking after 8 weeks.

Notice that if a different sample of 120 people had been in the study, the population probability would remain the same but the sample statistic would probably change. ■

### Parameter 2: Difference in Two Population Proportions

Sometimes we want to compare a proportion or probability for two populations. We learned in Chapter 6 that one way to do this is by using the *relative risk*. For instance, in Case Study 1.6, we learned that for male volunteers in the study, taking aspirin daily reduced the chances of having a heart attack from 17.13 per thousand when taking a placebo to 9.42 per 1000 with aspirin. Thus, for this sample, the relative risk of a heart attack was 1.82, with a heart attack more likely after taking a placebo than after taking aspirin.

Another way to make a comparison in this situation is to examine the **difference in two population proportions.** This is appropriate when we have two populations or have two population groups formed by a categorical variable and we want to compare some feature of the two populations. In Case Study 1.6, the proportions having heart attacks for the placebo and aspirin samples were .01713 and .00942, respectively, for a difference of only .0077, or about 7.7 heart attacks per thousand people. We can use this information to estimate the difference in the probability of having a heart attack for the population if taking a placebo and if taking aspirin.

There are usually two questions of interest about a difference in two population proportions. First, we want to estimate the value of the difference. Second, often we want to test the hypothesis that the difference is 0, which would indicate that the two proportions are equal.

**Example 9.4**

**How Much More Likely Are Smokers to Quit with a Nicotine Patch?** In Case Study 4.3, we learned that 46% of smokers who wore a nicotine patch had quit after 8 weeks but only 20% of the volunteers who were randomly assigned to wear a placebo patch had quit. The corresponding proportions are .46 and .20, so the difference is .26. For the sample of volunteers in the experiment, an additional 26% quit with the nicotine patch compared to the placebo patch. For this example, we want to compare two population probabilities:

$p_1$ = the probability that a smoker who wears a placebo patch will quit smoking after 8 weeks.

$p_2$ = the probability that a smoker who wears a nicotine patch will quit smoking after 8 weeks.

We might be interested in estimating each of these probabilities, but the parameter of interest for this example is the difference:

$p_1 - p_2$ = the difference in population proportions who would quit if wearing a placebo patch compared to wearing a nicotine patch.

In particular, we would probably want to test the hypothesis that this difference is 0, which would indicate that the two patch types are equally effective. We can use the *sample statistic* to make this inference, which is

$\hat{p}_1 - \hat{p}_2 = .46 - .20 = .26$ = difference between the proportion of the samples who quit smoking after wearing a nicotine patch and after wearing a placebo patch.

Notice that if we were to repeat the experiment with a new set of volunteers, the *population* difference would remain the same but the *sample* difference would probably change. ∎

### Parameter 3: One Population Mean

With quantitative variables, one of the simplest summaries for the population is the average of the variable for everyone in the population, called the **population mean.** This can be estimated by using the *sample mean*, which we learned in Chapter 2 is just the arithmetic average for the sample.

**Example 9.5**   **Age at First Intercourse for Females**   In Example 3.18, we learned that in 1990, the average age at first intercourse for a sample of female teenagers was 17.2 years. We can use this to estimate what the mean age was for the entire population of teenagers in 1990. Therefore, the parameter of interest is

$\mu$ = mean age at first intercourse for all teenage girls in 1990.

For the sample that was available, the sample statistic was

$\bar{x}$ = 17.2 years

Notice that if we were to have taken a different sample at that time, the mean for the new *sample* would probably not be the same, but the *population* mean would not have changed.   ■

### Parameter 4: Population Mean for Paired Differences

We learned about *matched pairs* in Chapter 4. When we collect quantitative data on matched pairs we often are interested in the *difference* rather than in the two separate values. Data that are formed by taking the differences in matched pairs are called **paired differences.** Once we have taken differences, we usually are interested in the same things we would have been interested in had the differences been a single variable: location, shape, and spread.

The simplest summary measure is the **population mean for paired differences,** which is the mean that we would get if we took differences for the entire population of possible pairs. There are usually two questions of interest about a population mean for paired differences. First, we want to estimate the value of the mean difference. Second, often we want to test the hypothesis that the mean difference is 0, which would indicate that the population means for the two parts of the pair are equal.

Notice that we get the same numerical answer if we find the two means first and then subtract them as if we find the difference for each pair first and then find the mean. However, you should think of this situation as one in which you take the differences first and treat those as the original data. The reason for this is that we will need to use the standard deviation of the differences in our inference procedures, and that is not the same as the difference of the standard deviations.

**Example 9.6**   **Which Hand Is Bigger?**   In Example 5.11, we noticed that there was a strong correlation between the left and right handspan measurements for 190 college students. But does one hand have a wider span than the other, on average? For these students, the left hand had a slightly wider average span; the average was 21.34 cm for the left hand and 21.18 for the right hand, for a difference of 0.16 cm. (This held for both men and women, with a difference of about 0.145 cm for men and 0.168 cm for women.)

The population of interest is all college students, and the parameter of interest is

$\mu_d$ = population mean difference in left handspan and right handspan (left − right) for the population of college students.

The sample statistic for the 190 college students in our sample was

$$\bar{d} = 0.16 \text{ cm}$$

There are two questions of interest in this case. First, is there really a difference in the population, so that $\mu_d$ is different from 0? If so, what is the value of $\mu_d$? Notice that if we were to take a different sample of 190 students, the mean of the differences would probably change, but the population mean difference remains the same. ■

### Parameter 5: Difference in Two Population Means

Parameter 4 addressed the question of the mean for paired differences. The last parameter that we will study is the difference in two separate population means, measured with independent samples. With **independent samples,** the individuals in one sample are not coupled in any way with the individuals in the other sample. For instance, measuring opinions of brother and sister pairs would not produce independent samples, but measuring opinions of a random sample of girls and boys in a city would, even if a brother and sister both happened to be chosen. Independent samples also can be assumed when one sample is divided into two groups on the basis of a categorical grouping variable, such as male and female or smokers and nonsmokers.

As an example of this scenario, in Example 4.7, we learned that the average age at death for people in the study who were left-handed was 66 years, compared with an average age of 75 years for those who were right-handed. Does this represent a real difference in mean ages at death for the population of left- and right-handed people? (Recall that the study under discussion had some other problems, so if you are left-handed, don't worry!)

When we are interested in comparing the mean for the same quantitative variable in two different populations or two population groups formed by a categorical variable, the parameter of interest is the **difference in two population means.** There are usually two questions of interest about the difference in two population means. First, we want to estimate the value of the difference. Second, we often want to test the hypothesis that the difference is 0, which would indicate that the two population means are equal.

**Example 9.7**

**Do Girls and Boys Have First Intercourse at the Same Age on Average?** In Example 3.20, we also learned that in 1990, the average age at first intercourse for a sample of male teenagers was 16.5 years, while it was 17.2 years for females. Does this represent a true difference in the population of teenagers in 1990? We want to find the difference in the two parameters:

$\mu_1$ = age at first intercourse for the population of teenage boys in 1990.
$\mu_2$ = age at first intercourse for the population of teenage girls in 1990.

The parameter of interest is the difference in the two population means:

$\mu_1 - \mu_2$ = difference in population mean ages at first intercourse (boys − girls) in 1990.

The *sample statistic* based on the teenagers in the study is the difference in sample means:

$$\bar{x}_1 - \bar{x}_2 = 16.5 - 17.2 = -0.7 \text{ year (about eight and a half months).}$$

Notice that if we had asked a different sample of teenagers in 1990, the difference in sample means would probably not have been exactly the same, but the difference in population means would remain fixed. ■

It should be obvious to you by now that once we have conducted a survey, study, or experiment, the numerical values of the sample statistics are known. However, the numerical values of the population parameters can never be known in most cases. There are many reasons why this is the case. The population may be too large to measure in its entirety. The parameter may be associated with what would happen if everyone in a population were to follow a certain diet, take a certain drug, use a nicotine patch, and so on. To reiterate, one of our main goals is to estimate population parameters using sample statistics so we can answer questions of interest in situations in which the entire population cannot be measured.

9.2 Exercises are on page 385.

# THE ORGANIZATION OF CHAPTERS 9 TO 13

It is easier to learn something when you recognize how it relates to what you already know and when you understand recurring patterns. With that in mind, we have organized Chapters 9 to 13 to illustrate that the same basic approach is utilized in most of the statistical inference procedures covered in them. The chapters are organized around statistical inference procedures for the five parameters listed in Table 9.1 and described in Examples 9.2 to 9.7.

Before we learn how to find confidence intervals and test hypotheses for each parameter, we need to investigate the relationship between the parameter of interest in a problem and the corresponding sample statistic that estimates the parameter. This relationship will be formulated into something called a *sampling distribution* for the sample statistic, which is defined in the next section. Thus, for each of the five parameters, we will study three inference topics:

- The *sampling distribution* of the statistic that estimates the parameter
- A *confidence interval* procedure to estimate the parameter to within an interval of possibilities
- A *hypothesis test* procedure to test whether the parameter equals a specific value

Chapters 9 to 13 are organized and written to allow flexibility in the order of covering topics and situations. The coverage of each of the three topics begins with an *introductory module,* followed by a module for each of the big five parameters. Therefore, there are six modules per topic: the introductory topic and one for each of the five parameters.

Module locations are indicated by a color tab on the outer edge of the text page so that you can follow the three topics for a particular parameter through the chapters. The same color is used for the three topics covered for a parameter. You can study all six modules for one of the three inference topics (sampling distributions or confidence intervals or hypothesis tests) together, or you can pick one parameter of interest (such as a population mean) and learn all about methods related to it before moving on to learn about a different parameter. *Once you have studied the introductory module for a topic, you can then study the remaining ones in any order. Confidence interval and hypothesis-testing modules can be done in either order.*

Table 9.2 shows where to find the various modules. In the module designations, SD denotes the sampling distribution topic, CI denotes the confidence interval topic, and HT denotes the hypothesis testing topic. The five parameters are numbered from 1 to 5, and an introductory topic is numbered as parameter 0. Notice that all sampling distribution modules are in this chapter (Chapter 9), confidence interval modules are in Chapters 10 and 11, and hypothesis testing modules are in Chapters 12 and 13.

9.2 Exercises are on page 385.

**Table 9.2**   Organization of Chapters 9 to 13

| Parameter | Chapter 9: Sampling Distributions (SD) | Chapter 10: Confidence Intervals (CI) | Chapter 11: Confidence Intervals (CI) | Chapter 12: Hypothesis Tests (HT) | Chapter 13: Hypothesis Tests (HT) |
|---|---|---|---|---|---|
| 0. Introductory | SD Module 0 Overview of sampling distributions | CI Module 0 Overview of confidence intervals | | HT Module 0 Overview of hypothesis testing | |
| 1. Population Proportion ($p$) | SD Module 1 SD for one sample proportion | CI Module 1 CI for one population proportion | | HT Module 1 HT for one population proportion | |
| 2. Difference in two population proportions ($p_1 - p_2$) | SD Module 2 SD for difference in two sample proportions | CI Module 2 CI for difference in two population proportions | | HT Module 2 HT for difference in two population proportions | |
| 3. Population mean ($\mu$) | SD Module 3 SD for one sample mean | | CI Module 3 CI for one population mean | HT Module 3 HT for one population mean | |
| 4. Population mean of paired differences ($\mu_d$) | SD Module 4 SD for sample mean of paired differences | | CI Module 4 CI for population mean of paired differences | HT Module 4 HT for population mean of paired differences | |
| 5. Difference in two population means ($\mu_1 - \mu_2$) | SD Module 5 SD for difference in two sample means | | CI Module 5 CI for difference in two population means | HT Module 5 HT for difference in two population means | |

# 9.3  SD Module 0: An Overview of Sampling Distributions

You can probably guess by now that the relationship between sample statistics and population parameters will play a crucial role in statistical inference. It is this relationship that allows us to determine the accuracy of confidence intervals as estimates of population parameters. It also allows us to determine the extent to which sample results are surprising, given that we assume a potential parameter value to be true, as we will do in hypothesis tests.

Think of taking a random sample or conducting an experiment as one big random circumstance. A sample statistic is merely a number assigned to the outcome of that random circumstance. For instance, taking a random sample of 1000 adults and finding the proportion that are left-handed amounts to assigning a number (the sample proportion) to the outcome of a random circumstance (taking a random sample of 1000 adults and finding out whether they are left-handed). The number might be .12 for one sample (120 left-handers out of 1000), .09 for the next (90 left-handers), and so on. Recall from Chapter 8 that a *random variable* assigns a number to the outcome of a random circumstance. Thus, we can think of sample statistics as special cases of random variables because they fit the definition of random variables.

| *definition* | The **sampling distribution** for a statistic is the probability distribution of possible values of the statistic for repeated samples of the same size taken from the same population. |
|---|---|

In general, we can use a random variable's probability distribution to assess possibilities and accompanying probabilities for the different values of that variable. The sampling distribution for a statistic describes the possible values the statistic might have when random samples are taken from a population. For example, the sampling distribution would tell us the probability of getting a sample proportion of .12 left-handers in a sample of 1000, or .09 left-handers, and so on. This is important information because it can be used, among other things, to find the probability that a sample statistic will be within a specified distance of the unknown population parameter. That tells us how confident we can be that an interval extending that specified distance above and below the sample statistic will reach the population parameter. This reasoning will be explained in detail when we learn how to form confidence intervals.

Statisticians have worked out what the sampling distributions look like for the big five sample statistics. You will see that the general structure of the sampling distribution is the same for each of the five situations. In Section 9.10, we describe how to find sampling distributions for other situations.

**Example 9.8**

Statistics⬭Now™

Watch a video example at **http://1pass.thomson.com** or on your CD.

**Mean Hours of Sleep for College Students** The sample of Penn State students first discussed in Chapter 2 included the question "How many hours of sleep did you get last night?" The mean of the 190 responses was 7.1 hours. Suppose we were to ask another 190 students that same question. It is unlikely that the mean response would be exactly 7.1 hours again.

Figure 9.1 shows a histogram of possible sample means we might get from simulating this experiment 200 times. A normal curve is superimposed on top of the histogram to demonstrate that the sampling distribution of possible sample means is approximately a normal distribution. We will return to this idea later in this chapter.

To understand how this histogram was developed, think about a university that has 50,000 students. Suppose the distribution of hours of sleep students had the previous night actually has a mean of 7.1 hours and standard deviation of 2 hours. You and 199 friends (you are very popular) each go out and take a random sample of 190 of those students. You ask the students in your sample how many hours they slept last night and then calculate the mean of the 190 answers. Finally, you and your friends (200 of you) get together and draw a histo-

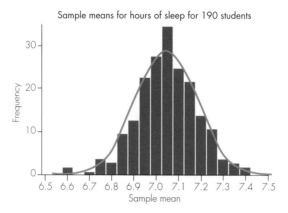

**Figure 9.1** ▌ The sampling distribution of mean hours of sleep for *n* = 190 students. The histogram is 200 simulated means; the smooth curve is the actual sampling distribution.

gram of your 200 different sample means. You would get a picture similar to Figure 9.1. Notice that most of you would get a sample mean around 7.0 or slightly more or less, but occasionally, someone would get an unusual sample with a mean as high as 7.4 hours or as low as 6.6 hours. ▪

In Chapter 8, we learned that a probability distribution specifies the probability of various outcomes for a random circumstance. For continuous random variables, knowing the details of the probability density function allows us to specify the probability that the value of the random variable falls into a specific interval. For instance, if we knew that test grades had a normal distribution with a mean of 70 and a standard deviation of 10, then we knew that a randomly selected test score from that population would be between 60 and 80 (mean ± one standard deviation) with probability .68, between 50 and 90 (mean ± two standard deviations) with probability .95, and so on.

A sampling distribution is the probability distribution for a sample statistic. It plays the same role as probability distributions do for other random variables. We can specify the probability that a sample statistic will fall into a specific interval if we know all of the details about its sampling distribution. For instance, in Example 9.8, if we knew all of the details of the smooth curve in Figure 9.1, we could answer questions such as "What is the probability that the mean hours of sleep (i.e., the sample mean) for a random sample of 190 students will be less than 6.9 hours?" or "What is the probability that the mean hours of sleep for a random sample of 190 students will be between 7 and 8 hours?" Notice that this is *not* the same question as "What is the probability that one randomly selected student slept between 7 and 8 hours?" Sampling distributions give us information about sample statistics, not about individual values for the population.

Notice that sampling distributions give us information about sample statistics, *given* that we know all about the population already. For instance, to get the sampling distribution in Figure 9.1, we assumed that we knew that the mean and standard deviation of the population of sleep times were 7.1 hours and 2 hours, respectively. In practice, knowing how to get information about possible sample statistics based on knowing values of the population parameter isn't very helpful. If we know about the population already, we don't need to find sample statistics. Therefore, why do we want to know about sampling distributions?

The answer is that we can turn the information around and use it for statistical inference. For instance, notice that in Figure 9.1, the possible sample means are all within about 0.5 hour of the population mean (7.1 hours in this case). If we know how close together the sample mean and the population mean are likely to be, then once we have a sample mean, we can make a reasonably good guess about the population mean. We will learn about sampling distributions in this chapter and then learn how to turn their information around to make inferences about population parameters in Chapters 10 to 13.

## A Common Format for the Five Sampling Distributions

The general format of the sampling distribution is the same for each of the five sample statistics under discussion—only the details change. *In each case, as long as certain conditions are met, the sampling distribution is approximately normal.* Remember that normal distributions are completely specified by knowing the mean and standard deviation. Therefore, to specify the sampling distribution for any one of our sample statistics, we need to know only how to determine the mean and standard deviation that are appropriate to that case. For each of the sampling distribution modules, we will give details for the mean and standard deviation of the sampling distribution and will list the conditions required for the approximate normality to hold.

There are some commonalities in the details of the mean and standard deviation. In each of the five cases, *the mean of the sampling distribution is the population parameter that corresponds to the sample statistic* (i.e., the parameter of interest estimated by the statistic). For instance, the mean of the sampling distribution of $\hat{p}$ (the sample proportion) is the population proportion $p$. As another example, the mean of the sampling distribution of $\bar{x}$ (the sample mean) is the population mean $\mu$. This type of result tells us that over all possible random samples from a population, the mean value of the sample statistic is equal to the population parameter.

*The standard deviation of a sampling distribution measures how the values of the sample statistic might vary across different samples from the same population.* The formula for the standard deviation of the sampling distribution differs for each of the five cases that we consider, but in each case, it depends on the sample size(s) and gets smaller for larger samples. In other words, as the sample size(s) get larger, the variability among possible values of the statistic from different samples gets smaller.

To differentiate the standard deviation of a sampling distribution from $\sigma$, the standard deviation of the individual population measurements, we need to add a qualifier to the terminology. We will do this by specifying the statistic as part of the terminology for the standard deviation. Thus, the standard deviation of the sampling distribution of a sample mean is called **the standard deviation of $\bar{x}$**. The standard deviation of the sampling distribution of a sample proportion is called **the standard deviation of $\hat{p}$,** and so on.

When we use these results for statistical inference, we often need to use sample information to estimate the standard deviation of the sampling distribution. To make it clear that the value is an estimate, we use the term **standard**

**error** to describe the estimated standard deviation for a sampling distribution. Thus, *the standard error of the mean* or *the standard error of* $\bar{x}$ are terms used for the estimate of *the standard deviation of* $\bar{x}$, and *the standard error of* $\hat{p}$ is the term used for the estimate of *the standard deviation of* $\hat{p}$.

9.3 Exercises are on page 386.

---

*in summary*     Sampling Distribution Definition and Features

- A **sampling distribution** is the probability distribution of a sample statistic. It describes how values of a sample statistic vary across all possible random samples of a specific size that can be taken from a population.

- The **mean value of a sampling distribution** is the mean value of a sample statistic over all possible random samples. For the big five scenarios, this mean equals the value of the population parameter.

- The **standard deviation of a sampling distribution** measures the variation among possible values of the sample statistic over all possible random samples. When referring to such a standard deviation, we include the name of the statistic being studied—for example, *the standard deviation of the mean.*

- The term **standard error** is used to describe the estimated value of the standard deviation of a statistic. Because the formula is different for each statistic, we include the name of the statistic—for example, *the standard error of the mean.*

---

# 9.4 SD Module 1: Sampling Distribution for One Sample Proportion

In this section, we cover sampling distributions for one sample proportion. However, the module includes substantial discussion and explanation that should help you to understand sampling distributions in general. The next module, for the difference in two proportions, is much shorter, giving the basic details and an example. This same pattern will be repeated in the modules for means. The module for one mean will provide substantial detail and explanation; the remaining two modules for mean differences and differences in means will provide the basics and an example only.

Suppose we conduct a binomial experiment with $n$ trials and get successes on $x$ of the trials. Or suppose we measure a categorical variable for a representative sample of $n$ individuals, and $x$ of them have responses in a certain category. In each case, we can compute the statistic $\hat{p} =$ the **sample proportion** $= x/n$, the proportion of trials resulting in success, or the proportion in the sample with responses in the specified category. If we repeated the binomial experiment or collected a new sample, we would probably get a different value for the sample proportion.

For instance, suppose we would like to know what proportion of a large population carries the gene for a certain disease. We sample 25 people and use the sample proportion from that sample to estimate the true answer. Suppose that in truth 40% of the population carries the gene, although this fact is unknown to us.

What will happen if we randomly sample 25 people from this population? Will we always find 10 people (40%) with the gene and 15 people (60%) without? Or will the number and proportion with the gene differ for different samples of $n = 25$? You may recognize that this situation is a binomial experiment and that if $X$ = number (out of 25) who carry the gene, then $X$ is a *binomial random variable* with $n = 25$ and $p = .4$.

## Many Possible Samples

Consider four different random samples of 25 people taken from this population. Remember that we are trying to estimate the proportion of the population with the gene, based on the sample *statistic*, which is the *sample proportion*. We do not know that, in truth, the population proportion (the parameter) is actually 40%. Here is what we would have concluded about the proportion of people who carry the gene, given four possible samples with $X$ as specified:

*Sample 1:*   $X = 12$, proportion with gene = $12/25 = .48 = 48\%$.
*Sample 2:*   $X = 9$, proportion with gene = $9/25 = .36 = 36\%$.
*Sample 3:*   $X = 10$, proportion with gene = $10/25 = .40 = 40\%$.
*Sample 4:*   $X = 7$, proportion with gene = $7/25 = .28 = 28\%$.

Notice that each one of these samples would have given a different answer and that the sample answer may or may not have matched the truth about the population, which is that 40% carry the gene.

In practice, when a researcher conducts a study similar to this one or a polling organization randomly samples a group of people to measure public opinion, only one sample is collected. There is no way to determine whether or not the sample is an accurate reflection of the population. However, statisticians have calculated what to expect for the vast majority of possible samples.

## The Normal Curve Approximation Rule for Sample Proportions

A result given in Section 8.7 is that with sufficiently large $n$, a binomial random variable is also approximately a normal random variable. A binomial random variable $X$ counts the number of times an event happens in $n$ trials, but the approximate normality also applies to the proportion, $\hat{p} = X/n$. Dividing each possible value of $X$ by the sample size $n$ does not change the shape of the distribution of possible values. In other words, the *sampling distribution for a sample proportion is approximately a normal distribution.*

We will call this result the **Normal Curve Approximation Rule for Sample Proportions.** Sometimes, we will refer to it simply as the *Rule for Sample Proportions.* This rule can be applied in two different situations in which we observe data:

*Situation 1:*   A random sample is taken from a relatively large actual population.
*Situation 2:*   A binomial experiment is repeated numerous times.

| |
|---|

*definition*    The **Normal Curve Approximation Rule for Sample Proportions** can be defined as follows:

> Let $p$ = population proportion of interest or binomial probability of success.
> Let $\hat{p}$ = corresponding sample proportion or proportion of successes.

If numerous samples or repetitions of the same size $n$ are taken, the distribution of possible values of $\hat{p}$ is approximately a normal curve distribution with

> Mean = $p$
>
> Standard deviation = s.d.$(\hat{p})$ = $\sqrt{\dfrac{p(1 - p)}{n}}$

This approximate normal distribution is called the **sampling distribution of $\hat{p}$.**

***Technical Note:*** The $n$ individuals in the sample or the repetitions must be independent, equivalent to Condition 3 in a binomial experiment.

### Conditions for Which the Normal Curve Approximation Rule for Sample Proportions Applies

In each of the two situations, three conditions must all be met for the Normal Curve Approximation Rule for Sample Proportions to apply:

**Condition 1: The Physical Situation**    In Situation 1, there is an actual population with a fixed proportion who have a certain trait, opinion, disease, and so on.

or

In Situation 2, there is a repeatable situation for which a certain outcome is likely to occur with a fixed relative frequency probability.

**Condition 2: Data Collection**    In Situation 2, a random sample is selected from the population, thus ensuring that the probability of observing the characteristic is the same for each sample unit.

or

In Situation 2, the situation is repeated numerous times, with the outcome each time independent of all other times.

**Condition 3: The Size of the Sample or Number of Trials**    The size of the sample or the number of repetitions is relatively large. The necessary size depends on the proportion or probability under investigation. It must be large enough that we are likely to see at least ten of each of the two possible responses or outcomes. In other words, $np$ and $n(1 - p)$ must each be at least 10.

### Examples of Scenarios for Which the Rule for Sample Proportions Applies

From your study of binomial random variables in Chapter 8, you should recognize that these are simply the conditions for a binomial random variable with large $n$. Here are some examples of scenarios that meet these conditions.

*Scenario 1: Election Polls*    A pollster wants to estimate the proportion of voters who favor a certain candidate. The voters are the population units, and favoring the candidate is the opinion of interest. A large sample is taken.

*Scenario 2: Television Ratings*   A television rating firm wants to estimate the proportion of households with television sets that are tuned to a certain television program. The collection of all households with television sets makes up the population, and being tuned to that particular program is the trait of interest.

*Scenario 3: Consumer Preferences*   A manufacturer of soft drinks wants to know what proportion of consumers prefer a new mixture of ingredients compared with the old recipe. The population consists of all consumers, and the response of interest is preference for the new formula over the old.

*Scenario 4: Testing ESP*   A researcher studying extrasensory perception wants to know the probability that people can successfully guess which of five symbols is on a hidden card. Each card is equally likely to contain each of the five symbols. There is no physical population. The repeatable situation of interest is a guess, and the response of interest is a successful guess. The researcher wants to determine whether the probability of a correct guess is higher than .20, the chance for a successful guess if there were no such thing as extrasensory perception. The number of guesses made by each person should be at least 50 for Condition 3 (number of trials) to hold. With random guessing, $p = 1/5 = .2$, so $n = 50$ trials would lead to $np = 10$ and $n(1 - p) = 40$.

---

**thought question 9.1**   Each of scenarios 1 to 4 can be thought of as a binomial experiment. Specify what constitutes a success in each case. Then construct your own example of a scenario for which the Rule for Sample Proportions would hold, and specify what constitutes a success for your example. *

---

**Example 9.9**

**Possible Sample Proportions Favoring a Candidate**  Suppose that of all voters in the United States, 40% are in favor of Candidate X for president. Pollsters take a sample of 2400 voters. What proportion of the sample would be expected to favor Candidate X? The rule tells us that the proportion of the sample who favor Candidate X is a random variable that has approximately a normal distribution. The mean and standard deviation for the distribution are

$$\text{Mean} = p = .4 \quad (40\% \text{ expressed as a proportion})$$

$$\text{s.d.}(\hat{p}) = \sqrt{\frac{p(1 - p)}{n}} = \sqrt{\frac{(.4)(1 - .4)}{2400}} = \sqrt{\frac{.24}{2400}} = \sqrt{.0001} = .01$$

Figure 9.2 shows a histogram of the sample proportions resulting from simulating this situation 400 times, with the appropriate normal curve superimposed on the histogram. Notice that as expected from the Empirical Rule in Chapter 2, the possible values cover the range $\mu \pm 3(\text{s.d.})$, in this case, $.4 \pm 3(.01)$.

---

**\*HINT:** A *success* is the characteristic or trait of interest. When constructing your own scenario, first look at Conditions 1 and 2 for the two different situations described.

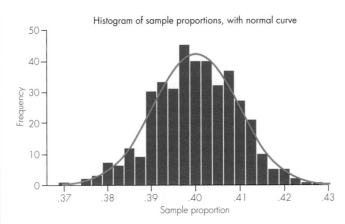

**Figure 9.2** ▍ Histogram of 400 sample proportions based on $n = 2400$, $p = .4$; normal curve with mean $= .4$ and standard deviation $= .01$

*thought question 9.2*   For Example 9.9, into what range of possible values should the sample proportion fall 95% of the time, according to the Empirical Rule? If the polling organization used a sample of only 600 voters instead, would this range of possible sample proportions be wider, more narrow, or the same? Explain your answer, and explain why it makes intuitive sense.*

Example 9.9 indirectly shows how a sampling distribution provides information about the accuracy of a sample statistic. In that example, we learned that with a sample size of $n = 2400$ voters, it is nearly certain that the proportion of voters favoring Candidate X in the sample will be within $\pm 3(.01) = \pm .03$ of the true population proportion. In Chapters 10 and 11, we will learn how to use that kind of information to form confidence intervals. In the next example, we will see how a sampling distribution is used to judge how an observed sample result fits with a potential value for a population proportion. In Chapters 12 and 13, we will learn how to use that kind of information to test hypotheses.

**Example 9.10**

Statistics◯Now™

Read the original source on your CD.

**Caffeinated or Not?**   Case Study 8.1 described a study in which 25 people each tried 20 times to guess whether a cola they taste is caffeinated or not caffeinated. For some of the 20 trials, the cola being tasted contained caffeine; in the others, it did not. Suppose we take the view that people cannot judge merely by taste whether caffeine is present. If this were so, then participants would actually be randomly guessing on each trial, so $p = .5$, where $p$ is the probability of correctly guessing whether caffeine is present or not on any trial. If this is the case for all individuals, then the overall experiment can be seen as a binomial experiment with $n = 500$ trials (25 people $\times$ 20 trials each) and $p = .5$.

We can use the Rule for Sample Proportions to approximate the sampling distribution of potential values of $\hat{p}$, the proportion of correct guesses in the

*HINT:** Use Example 9.10 for guidance.

sample of 500 trials. The sampling distribution will be approximately a normal curve with the following mean and standard deviation:

$$\text{Mean} = p = .5$$

$$\text{Standard deviation} = \text{s.d.}(\hat{p}) = \sqrt{\frac{.5(1-.5)}{500}} = .0224$$

Figure 9.3 shows this normal curve. On the basis of the Empirical Rule, we know that values fall in the interval $\mu \pm (2 \times \text{s.d.})$ about 95% of the time. So in this situation, there is about a 95% chance that the proportion of correct guesses in a sample of 500 trials will be in the interval $.5 \pm (2 \times .0224)$, or between about .455 and .545. In the actual experiment, 265 correct guesses were made, so the sample proportion was $\hat{p} = 265/500 = .53$. This is well within the interval of values that would occur for about 95% of samples of $n = 500$ trials in which a random guess is made on every trial. In other words, $\hat{p} = .53$ is not strong evidence against the hypothesis that in general, people cannot taste whether a cola contains caffeine or not.

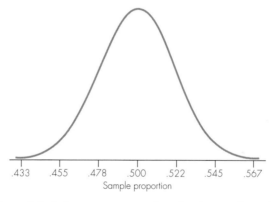

**Figure 9.3** ▌ Approximate sampling distribution of possible sample proportions when $n = 500$, $p = .5$

## Estimating the Population Proportion from a Single Sample Proportion

In Example 9.9 we learned about the range of possible values to expect in polls with $n = 2400$ and $p = .4$. In practice, when we actually take a random sample of 2400 voters in a political poll, we have only one sample proportion, and we don't know the true population proportion. However, we do know how far apart the sample proportion and the true population proportion are likely to be. That information is contained in the **standard deviation of $\hat{p}$,**

$$\text{s.d.}(\hat{p}) = \sqrt{\frac{p(1-p)}{n}}$$

In many situations, we will not know the value of $p$, the population proportion. Instead, we will use an observed sample proportion $(\hat{p})$ to estimate the unknown value of the parameter $p$. Thus, we estimate the standard deviation of $\hat{p}$ by using the sample value $\hat{p}$ in the formula. The estimated version is called the **standard error of $\hat{p}$,**

$$\text{s.e.}(\hat{p}) = \sqrt{\frac{\hat{p}(1 - \hat{p})}{n}}$$

For instance, if $\hat{p} = .39$ and $n = 2400$, the standard error is $\sqrt{.39(1 - .39)/2400} = .01$. This value estimates the theoretical standard deviation of the sampling distribution for sample proportions, based on a single sample. Because we know that the mean, which is the true proportion $p$, is almost surely within 3 standard deviations of the observed value $\hat{p}$, we know that $p$ is almost surely within the range $\hat{p} \pm 3(\text{s.e.}) = .39 \pm 3(.01) = .39 \pm .03$. So we know that the true proportion who support the candidate is almost surely between .36 and .42. Notice that the only numerical value we needed to determine this was the sample proportion $\hat{p}$, and of course, the sample size $n$. We will explore this idea in much more detail in Chapter 10.

9.4 Exercises are on page 386.

## 9.5   SD Module 2: Sampling Distribution for the Difference in Two Sample Proportions

Do the same proportions of men and women support the death penalty? If not, how do the two proportions differ? Has the proportion of adults who support same-sex marriage changed in the last decade? If so, by how much has it changed? By how much does getting a flu shot reduce the probability of getting the flu? These are all questions that can be answered by looking at the difference in two proportions.

To use the methods described here, we need to collect *independent samples* from the two groups or populations. Remember that with independent samples, the individuals in one sample are not coupled in any way with the individuals in the other sample. For instance, measuring the voting behavior of husbands and wives would not produce independent samples, but measuring voting behavior of randomly (and independently) selected samples of men and women in a neighborhood would.

Another method that allows us to assume that we have independent samples is to measure two variables for everyone in a random sample, where one variable is a categorical grouping variable. As an example, we might take a random sample of adults and, for each person, record the person's sex and whether he or she supports or opposes the death penalty. The men and women constitute the independent samples, and we can then examine the difference in the proportions favoring the death penalty in these two groups.

### Notation for the Difference in Two Proportions

In the notation scheme for two proportions, we will continue to use $p$ to denote a population proportion and $\hat{p}$ to denote a sample proportion, and we will use subscripts 1 and 2 to represent the group.

For the populations:

$p_1 = $ *population* proportion for the first population.
$p_2 = $ *population* proportion for the second population.

The parameter of interest is $p_1 - p_2 =$ the difference in *population* proportions. For the samples:

$\hat{p}_1 = sample$ proportion for the sample from the first population.
$\hat{p}_2 = sample$ proportion for the sample from the second population.

The sample statistic is $\hat{p}_1 - \hat{p}_2 =$ the difference in *sample* proportions.

## Conditions for the Sampling Distribution of $\hat{p}_1 - \hat{p}_2$ to Be Approximately Normal

The sampling distribution of the difference in two independent sample proportions is approximately normal when both of these conditions hold:

*Condition 1:*  Sample proportions are available for two independent samples, randomly selected from the two populations of interest.
*Condition 2:*  All of the quantities $n_1p_1$, $n_1(1 - p_1)$, $n_2p_2$, $n_2(1 - p_2)$ are at least 10. These quantities represent the expected numbers of successes and failures in each of the two samples.

## Mean, Standard Deviation, and Standard Error for the Sampling Distribution of $\hat{p}_1 - \hat{p}_2$

The **mean of the sampling distribution of $\hat{p}_1 - \hat{p}_2$** is

$p_1 - p_2 =$ the difference in *population* proportions.

The **standard deviation of $\hat{p}_1 - \hat{p}_2$** is

$$\text{s.d.}(\hat{p}_1 - \hat{p}_2) = \sqrt{\frac{p_1(1 - p_1)}{n_1} + \frac{p_2(1 - p_2)}{n_2}}$$

When we do not know the population proportions, we use sample proportions instead. Thus, the **standard error of $\hat{p}_1 - \hat{p}_2$** is

$$\text{s.e.}(\hat{p}_1 - \hat{p}_2) = \sqrt{\frac{\hat{p}_1(1 - \hat{p}_1)}{n_1} + \frac{\hat{p}_2(1 - \hat{p}_2)}{n_2}}$$

Remember that the mean of the sampling distribution for the situations we are considering is always the associated parameter. In this case, the parameter associated with the difference in two sample proportions is the difference in the corresponding population proportions.

---

*formula*

In general, the **standard error of the difference** between statistics from independent samples is determined as

$$\text{s.e.(difference)} = \sqrt{[\text{s.e.(statistic 1)}]^2 + [\text{s.e.(statistic 2)}]^2}$$

---

The same formula applies to standard deviations, with "s.d." replacing "s.e." on both sides of the equation. It follows from Section 8.8 in Chapter 8 in which the result is given that the variance of the difference between two independent random variables is the sum of their variances. This result applies to the differ-

*SD Module 2: $p_1 - p_2$*

ence $\hat{p}_1 - \hat{p}_2$ because the two sample proportions $\hat{p}_1$ and $\hat{p}_2$ are independent random variables. (This is the reason for the condition requiring independent samples.) The variance of any random variable is the standard deviation squared, and we learned in Module 1 that the standard deviation of $\hat{p}$ is $\sqrt{p(1 - p)/n}$. This formula applies to both $\hat{p}_1$ and $\hat{p}_2$. Putting all of these pieces together leads to the formulas given for the standard deviation and standard error of $\hat{p}_1 - \hat{p}_2$.

**Example 9.11**

**Men, Women, and the Death Penalty**   Are you in favor of the death penalty, or are you opposed to it? On the basis of surveys done in the past, women are more likely to oppose the death penalty than are men. In the 1993 General Social Survey (**GSS-93** on the CD for this book), the sample proportions opposing the death penalty were $\hat{p}_1 = 224/840 = .267$ for women and $\hat{p}_2 = 113/648 = .174$ for men, for a difference of $\hat{p}_1 - \hat{p}_2 = .267 - .174 = .093$.

If we were to repeat this survey with a new random sample of 840 women and 648 men, the difference $\hat{p}_1 - \hat{p}_2$ would probably not be exactly .093. What can we expect for the difference in repeated samples? To answer this question, we must pretend that we are omniscient and actually know the truth about the population proportions of men and women who oppose the death penalty. Let's suppose that, in truth, 27% of women and 17% of men oppose the death penalty, so $p_1 = .27$, $p_2 = .17$, and $p_1 - p_2 = .10$. Then the sampling distribution of $\hat{p}_1 - \hat{p}_2$ is approximately normal, with mean $p_1 - p_2 = .10$ and standard deviation

$$\text{s.d.}(\hat{p}_1 - \hat{p}_2) = \sqrt{\frac{p_1(1 - p_1)}{n_1} + \frac{p_2(1 - p_2)}{n_2}}$$

$$= \sqrt{\frac{.27(1 - .27)}{840} + \frac{.17(1 - .17)}{648}} = .021$$

Figure 9.4 illustrates this sampling distribution. It represents the possible differences in *sample* proportions of women and men (women − men) who oppose the death penalty, if in fact the difference in *population* proportions is .10, the true population proportions for women and men are .27 and .17, respectively, and the sample sizes are 840 and 648 for women and men, respec-

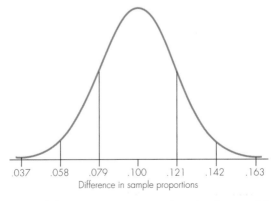

.037   .058   .079   .100   .121   .142   .163
Difference in sample proportions

**Figure 9.4** ▮ Approximate sampling distribution of $\hat{p}_1 - \hat{p}_2$ when $p_1 = .27$, $p_2 = .17$, $n_1 = 840$, and $n_2 = 648$

tively. Notice that the difference of .093 achieved in the 1993 General Social Survey would not have been surprising in this scenario. The figure shows a range spanning three standard deviations on either side of the mean. Lines showing one and two standard deviations on either side of the mean are illustrated. For example, using the Empirical Rule, we would expect the difference $\hat{p}_1 - \hat{p}_2$ to be between .079 and .121 in 68% of all surveys fitting this scenario. ■

9.5 Exercises are on page 388.

## 9.6  SD Module 3: Sampling Distribution for One Sample Mean

We now turn to the case in which the information of interest involves the mean or means of quantitative variables. For example, a company that sells oat products might want to know the mean cholesterol level people would have if everyone had a certain amount of oat bran in their diet. To help determine financial aid levels, a large university might want to know the mean income of all students on campus who work.

As with the modules for proportions, this module for one mean includes substantial discussion. The remaining two modules for means are much shorter, giving the basic details and an example.

Suppose a population consists of thousands or millions of individuals, and we are interested in estimating the mean of a quantitative variable. If we sample 25 people and compute the mean of the variable for that sample, how close will that *sample mean* be to the *population mean* we are trying to estimate? Each time we take a sample, we will get a different sample mean. Can we say anything about what we expect those means to be?

For example, suppose we are interested in estimating the average weight loss for everyone who attends a national weight-loss clinic for 10 weeks. Suppose that, unknown to us, the distribution of weight losses for everyone in this population is approximately normal with a mean of 8 pounds and a standard deviation of 5 pounds. Figure 9.5 shows a normal curve with these characteristics. If the weight losses are approximately normal, we know from the Empirical Rule in Chapter 2 that 95% of the individuals will fall within 2 standard deviations, or 10 pounds, of the mean of 8. In other words, 95% of the individual weight losses will fall between −2 (a gain of 2 pounds) and 18 pounds lost.

**SD Module 3: $\mu$**

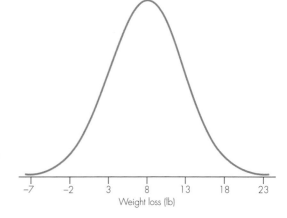

**Figure 9.5** ▌ Normal curve for individual weight losses with mean $\mu = 8$ pounds and standard deviation $\sigma = 5$ pounds

Weight loss (lb)

**Table 9.3** Four Potential Samples from a Population with $\mu = 8$, $\sigma = 5$

| | |
|---|---|
| **Sample 1:** | 1, 1, 2, 3, 4, 4, 4, 5, 6, 7, 7, 7, 8, 8, 9, 9, 11, 11, 13, 13, 14, 14, 15, 16, 16 |
| **Sample 2:** | −2, −2, 0, 0, 3, 4, 4, 4, 5, 5, 6, 6, 8, 8, 9, 9, 9, 9, 9, 10, 11, 12, 13, 13, 16 |
| **Sample 3:** | −4, −4, 2, 3, 4, 5, 7, 8, 8, 9, 9, 9, 9, 9, 10, 10, 11, 11, 11, 12, 12, 13, 14, 16, 18 |
| **Sample 4:** | −3, −3, −2, 0, 1, 2, 2, 4, 4, 5, 7, 7, 9, 9, 10, 10, 10, 11, 11, 12, 12, 14, 14, 14, 19 |

Table 9.3 lists some possible samples that could result from randomly sampling 25 people from this population; these were indeed the first four samples produced by a computer simulation of this situation. The weight losses have been put into increasing order for ease of reading. A negative value indicates a weight gain. (As an aside, notice that, as expected, $5/100 = 5\%$ of the values are outside of the range −2 to 18.) We are interested in the behavior of the possible sample means for samples such as these.

Here are the sample means and standard deviations for each of these four samples. You can see that the sample means, although all different, are relatively close to the population mean of 8. You can also see that the sample standard deviations are relatively close to the population standard deviation of 5. This latter fact is important, because although we are interested in using the sample mean to estimate the population mean, we will need to use the sample standard deviation to help find a standard error for this estimate.

*Sample 1*:    Mean = 8.32 pounds, standard deviation = 4.74 pounds.
*Sample 2*:    Mean = 6.76 pounds, standard deviation = 4.73 pounds.
*Sample 3*:    Mean = 8.48 pounds, standard deviation = 5.27 pounds.
*Sample 4*:    Mean = 7.16 pounds, standard deviation = 5.93 pounds.

## Conditions for the Sampling Distribution of the Mean to Be Approximately Normal

As with sample proportions, statisticians understand what to expect for the possible distribution of sample means in repeated sampling from the same population. Technically called the *sampling distribution of the sample mean,* we call this rule the **Normal Curve Approximation Rule for Sample Means,** or simply the *Rule for Sample Means* to convey what it says. Unlike the equivalent rule for proportions, it is not always necessary to have a large sample for this rule to work. If the population of measurements is bell-shaped, then the result holds for all sample sizes. The Rule for Sample Means applies in both of the following types of situations:

*Situation 1:*    The population of the measurements of interest is bell-shaped, and a random sample of any size is measured.

*Situation 2:*    The population of measurements of interest is not bell-shaped, but a *large* random sample is measured. Thirty is usually used as an arbitrary demarcation of "large," but if there are extreme outliers, it is better to have a larger sample.

There are actually only a limited number of situations for which the Rule for Sample Means does *not* apply. It does not apply at all if the sample is not random, and it does not apply for small random samples unless the original popu-

as *matched pairs* in which observations are taken on two halves of a pair. A matched pair can be two measurements from the same individual measured under two conditions (IQ after listening to Mozart and after sitting in silence), or at two different times (weight at the beginning and at the end of the first year of college), and so on. Or, a pair can be two separate individuals who form a natural pair, such as twins, couples (ages of husbands and wives), or individuals who are matched according to one or more important criteria.

Data that are collected as matched pairs are sometimes called **dependent samples** because the two observations are not statistically independent of each other. In the next module, we consider how to compare means for *independent* samples. In this module, we are concerned with the population mean difference for *dependent* observations, taken as matched pairs. The situation is analogous to the case for one mean, except that in that case, we observed a single measurement for each individual, and in this case, we start with two measurements but reduce them to a single measurement by taking the difference.

## Notation for the Mean of Paired Differences

The sample consists of $n$ pairs of observations. The response variable(s) is quantitative, so it makes sense to determine a mean. We are interested only in the difference between the two values within a pair of observations.

$d_i$ = the difference in the two measurements for individual $i$, where $i = 1, 2, \ldots, n$.

$\mu_d$ = the mean for the *population* of differences, if all possible pairs were to be measured.

$\sigma_d$ = the standard deviation for the *population* of differences.

$\bar{d}$ = the mean for the *sample* of differences.

$s_d$ = the standard deviation for the *sample* of differences.

---

*technical note*  **Mean of the Differences or Difference in Means?**

You might wonder why we make a distinction between the mean of the differences and the difference in the means, when they are in fact numerically equivalent. Why do we have to take the differences first and then find the sample statistics? Why not just find the average of the first number in each pair, then the average of the second number and subtract the two averages? For instance, to find the average difference in ages of husbands and wives, why don't we just find the average age for the husbands and the average age for the wives and then take the difference? The answer is that if we do not have the individual difference, then we can't find the standard deviation for the differences, and we need that to estimate $\mu_d$ and test hypotheses about it. Thus, while it is true that $\mu_d = \mu_1 - \mu_2$ (the population mean of the differences = the difference in the separate population means) and $\bar{d} = \bar{x}_1 - \bar{x}_2$ (the sample mean of the differences = the difference in the sample means), there is no way to find $s_d$ or $\sigma_d$ if all we have are the separate standard deviations $s_1$ and $s_2$ or $\sigma_1$ and $\sigma_2$. We need to know how spread out the differences are, not how spread out the values in each separate set are.

SD Module 4: $\mu_d$

# Conditions for the Sampling Distribution of $\bar{d}$ to Be Approximately Normal

Remember that the mean of paired differences is analogous to the case of one mean except that the measurements are differences. Therefore, exactly the same conditions that are required for the sampling distribution of $\bar{x}$ to be approximately normal also hold for the sampling distribution of $\bar{d}$. Either of the following two situations will work:

*Situation 1:* The population of differences is bell-shaped, and a random sample of any size is measured.

*Situation 2:* The population of differences is not bell-shaped, but a *large* random sample is measured. Thirty pairs are usually used as an arbitrary demarcation of "large," but if any of the differences are extreme outliers, it is better to have a larger sample.

As with other situations, it often is difficult to get a random sample, and researchers usually are willing to assume that these conditions hold if they can get a representative sample with no obvious sources of confounding or bias.

# Mean, Standard Deviation, and Standard Error for the Sampling Distribution of $\bar{d}$

The **mean of the sampling distribution of $\bar{d}$** is $\mu_d$, the population mean of the differences. The **standard deviation of $\bar{d}$** is

$$\textbf{s.d.}(\bar{d}) = \frac{\sigma_d}{\sqrt{n}}$$

The **standard error of $\bar{d}$** is

$$\textbf{s.e.}(\bar{d}) = \frac{s_d}{\sqrt{n}}$$

Remember that for the situations we are considering, the mean of the sampling distribution is always the associated parameter. Also remember that the standard error is used to estimate the standard deviation when $\sigma_d$ is not known.

**SD Module 4: $\mu_d$**

**Example 9.13**

**Suppose There Is No "Freshman 15"** In Example 9.1, we asked whether it was really true that students gain weight on average in the first year of college. We discussed a study that measured a sample of 60 students for 12 weeks and found that they gained an average of 4.2 pounds. Is it likely that the mean weight gain for a random sample of 60 students would be as high as 4.2 pounds *if* there is no average weight gain in the population of all students? To answer this question, we need to know what to expect the *sample* mean $\bar{d}$ to be for a sample of $n = 60$ measurements if the *population* mean $\mu_d$ is 0. The sampling distribution of $\bar{d}$ provides this information. To find the standard deviation for the sampling distribution, we need to know the standard deviation for the population of weight gains, $\sigma_d$. Let's be omniscient and suppose we know that it is 7 pounds. Then the sampling distribution of $\bar{d}$ is approximately normal with mean $= \mu_d = 0$ and standard deviation s.d.$(\bar{d}) = \sigma_d/\sqrt{n} = 7/\sqrt{60} = 0.904$, which we will round off to 0.9 pound.

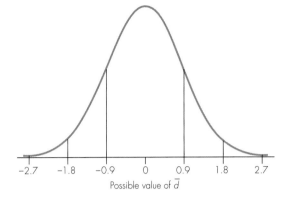

**Figure 9.8** ▌ Sampling distribution of $\bar{d}$ when $\mu_d = 0$, $\sigma_d = 7$, and $n = 60$

Possible value of $\bar{d}$

Figure 9.8 illustrates the sampling distribution for this situation; remember that the mean is $\mu_d = 0$ pounds and the standard deviation is s.d.$(\bar{d}) =$ 0.9 pound. Values that are 1, 2, and 3 standard deviations on either side of the mean of the distribution are marked. For instance, using the Empirical Rule, we would expect the sample mean difference to fall between −1.8 pounds and +1.8 pounds in 95% of all samples of size 60. This assumes that in fact the average weight gain in the population is 0 and the standard deviation of weight gains is 7 pounds.

The mean weight gain for the 60 students in the Cornell study was $\bar{d} =$ 4.2 pounds. We can see from Figure 9.8 that a sample average this large would be extremely unlikely under the assumed scenario that the true population mean weight gain is 0 and the standard deviation is 7 pounds. In Chapter 13, we will learn a more formal way to use the sample mean to determine whether it is unreasonable to assume that the population mean is a specific value—in this case, 0. ▌

9.7 Exercises are on page 390.

## 9.8  SD Module 5: Sampling Distribution for the Difference in Two Sample Means

Do women who smoke during pregnancy have babies with lower average birth weight than women who don't smoke? If so, what is the difference? Do people who practice yoga have lower blood pressure, on average, than people who do not? If so, by how much? Is the mean resting pulse rate for men lower than it is for women? Do men drive faster, on average, than women? If so, by how much? These are all questions that can be answered by looking at the difference in two means for independent samples.

To use the methods described here, we need to collect *independent samples* from the two groups or populations. Remember that with independent samples, the individuals in one sample are not coupled in any way with the individuals in the other sample. An equivalent procedure is to measure a categorical variable with two possible values for everyone in a random sample (along with the quantitative variable of interest) and use the categorical variable to form two groups. An example is to categorize mothers by whether or not they smoked during pregnancy and measure the birth weights of their newborn infants. The

SD Module 5: $\mu_1 - \mu_2$

parameter of interest would be the difference in mean birth weights for children of smokers and nonsmokers.

A common application of inference for the difference in two population means is in randomized experiments, in which the two groups are formed by randomly assigning one of two treatments to each individual in the study. Volunteers generally are used for these studies rather than random samples, and the results can be extended to the population of individuals who would volunteer if given the opportunity. The population means are conceptual rather than real because they are the means that would occur if everyone in the population were to be subjected to the treatment. For example, in a study of weight loss in obese men, volunteers were randomly assigned to exercise regularly or to follow a restricted-calorie diet. The population means are the mean weight losses that would occur for the population of men similar to the ones in the study *if* they were to follow these same regimens.

## Notation for the Difference in Two Means

The data consist of $n_1$ observations from the first population and $n_2$ observations from the second population. The response variable, measured in both samples, is quantitative.

$\mu_1 = $ *population* mean for the first population.
$\mu_2 = $ *population* mean for the second population.

The parameter of interest is $\mu_1 - \mu_2 = $ the difference in *population* means.

$\bar{x}_1 = $ *sample* mean for the sample from the first population.
$\bar{x}_2 = $ *sample* mean for the sample from the second population.

The sample statistic is $\bar{x}_1 - \bar{x}_2 = $ the difference in *sample* means.

$\sigma_1 = $ the *population* standard deviation for the first population.
$\sigma_2 = $ the *population* standard deviation for the second population.

$s_1 = $ the *sample* standard deviation for the sample from the first population.
$s_2 = $ the *sample* standard deviation for the sample from the second population.

## Conditions for the Sampling Distribution of $\bar{x}_1 - \bar{x}_2$ to Be Approximately Normal

An important condition in this situation is that the two samples must be *independent*.

Three ways to obtain independent samples are as follows:

- Take separate random samples from each of two populations such as men and women.

- Take a random sample from a population and divide the sample into two groups based on a categorical variable such as smoker and nonsmoker.

- Randomly assign participants in a randomized experiment to two treatment groups such as exercise or diet.

definition

The **Central Limit Theorem** states that if $n$ is sufficiently large, the sample means of random samples from a population with mean $\mu$ and finite standard deviation $\sigma$ are approximately normally distributed with mean $\mu$ and standard deviation $\sigma/\sqrt{n}$.

*Technical Note:* The mean and standard deviation given in the Central Limit Theorem hold for any sample size; it is only the "approximately normal" shape that requires $n$ to be sufficiently large.

You may notice that the Central Limit Theorem is nothing more than a re-statement of the Rule for Sample Means, except that in the latter, we specified the conditions that are needed to satisfy "if $n$ is sufficiently large." The Central Limit Theorem is often used in situations in which it isn't obvious that the statistic of interest is actually a sample mean. It may be obvious that the sampling distributions of $\bar{x}$ and $\bar{d}$ follow from the Central Limit Theorem, but it may not be obvious that the sampling distribution of $\hat{p}$, the Rule for Sample Proportions, does too.

The Rule for Sample Proportions follows from the Central Limit Theorem by defining each observation in the sample to be either 1 or 0. The observation is 1 if the individual sampled has the desired trait or opinion; the observation is 0 otherwise. The sample mean is simply the average of the 1s and 0s in the sample, which is the sample proportion $\hat{p}$. The population mean $= p$, the proportion of 1s in the population. The population standard deviation is the standard deviation of the 1s and 0s in the population, which can be shown to be $\sigma = \sqrt{p(1 - p)}$, so $\sigma/\sqrt{n} = \sqrt{p(1 - p)/n}$, which you should recognize as s.d.$(\hat{p})$.

*thought question 9.5*   Verify that if the raw data for each individual in a sample is 1 when the individual has a certain trait and 0 otherwise, then the sample mean is equivalent to the sample proportion with the trait. You can do this by using a formula, explaining it in words, or constructing a numerical example.*

Example 9.18

**California Decco Losses**   In Chapter 8 and in Example 9.17 of this section, we discussed the California Decco lottery game, for which the mean amount lost per ticket over millions of tickets sold is $\mu = \$0.35$. The standard deviation is $\sigma = \$29.67$, reflecting the large variability in possible amounts won or lost, ranging from a net win of \$4999 to a net loss of \$1. Suppose a store sells 100,000 tickets in a year. What are the possible values for the average loss per ticket? The Central Limit Theorem tells us the answer: Assuming that $n = 100,000$ is sufficiently large, the distribution of possible sample means is approximately normal, with a mean (loss) of \$0.35 and a standard deviation of $\sigma/\sqrt{n} = \$29.67/\sqrt{100,000} = \$0.09$. According to the Empirical Rule, the mean loss is almost surely in the range $\$0.35 \pm 3(\$0.09)$, or between 8 cents and 62 cents.

*HINT:* Write out a short list of 0s and 1s; then find the mean of the list, and compare that to the proportion of 1s in the list.

Remember that this is the average loss per ticket for a collection of 100,000 tickets. Notice that the *total* loss for the 100,000 is therefore likely to be between ($0.08 × 100,000) and ($0.62 × 100,000), or between $8000 and $62,000! Certainly, you can think of better ways to invest $100,000. ■

## Mean or Median as "Central"?

The Central Limit Theorem provides information about what to expect of the mean in large samples, but the median remains the statistic of choice for representing the "central" value in an extremely skewed set of numbers. For instance, consider two random samples of net winnings for ten Decco tickets:

| Net Amount Won ($n = 10$) | Mean | Median |
|---|---|---|
| −1, −1, −1, −1, −1, −1, −1, −1, 0, 0 | −$0.80 (80 cent loss) | −$1.0 ($1.00 loss) |
| −1, −1, −1, −1, −1, −1, −1, 0, 0, 4999 | $499.20 (won) | −$1.0 ($1.00 loss) |

Clearly, the median is a better measure of the "typical" net win in ten plays of this game. The problem is that for the occasional sample with a large outlier, the mean will be inflated, but this is not representative of the entire population. Remember that in this case, the population mean is 35 cents lost, and the population median is −$1.00. In fact, 72.6% of all tickets (as shown in Chapter 8) result in $1.00 lost or a −$1.00 net win.

---

***thought question 9.6*** The Central Limit Theorem does not specify what is meant by "a sufficiently large sample." What factor(s) about the population of values do you think determine how large is large enough for the approximate normal shape to hold? Consider the California Decco example. Do you think $n = 30$ would be large enough for the distribution of possible values for the average loss to be approximately normal? Why or why not?

Now consider the handspan measurements of females. Do you think $n = 30$ would be large enough for the approximate normal shape to hold? What is different about these two examples?[*]

---

## Sampling Distribution for Any Statistic

Every statistic has a sampling distribution, but the appropriate distribution may not always be normal or even be approximately bell-shaped. Statisticians have developed a theory that helps to determine the appropriate sampling distribution for many common statistics. It is also possible to construct an approximate sampling distribution for a statistic by actually taking repeated samples of the same size from a population and constructing a relative frequency histogram or table for the values of the statistic over the many samples. The next example illustrates this method.

[*]**HINT:** See Table 8.1 (p. 290) for the Decco game probability distribution. See Figure 2.4 (p. 29) in Section 2.5 for a histogram of female handspans.

**Example 9.19**

**Winning the Lottery by Betting on Birthdays**  In the Pennsylvania Cash 5 lottery game, players select five numbers from the integers 1 to 39, and the grand prize is won by anyone who selects the same five numbers as the ones randomly drawn by the lottery officials. Suppose that in such a game, someone bets numbers corresponding to the days of the month on which five family members were born, which is anecdotally a common strategy. Obviously, the highest integer that someone using that strategy could select is 31, so there is no chance of winning the grand prize at all if the highest of the winning draws is bigger than 31—that is, if it is 32 to 39. What is the probability of that happening?

The statistic in question is the *highest* of five integers randomly drawn without replacement from the integers 1 to 39—let's call it $H$. For instance, if the numbers selected are 3, 12, 22, 36, and 37, then $H = 37$. Determining the actual sampling distribution of $H$ is a complicated exercise in probability. However, we can approximate the sampling distribution of $H$ using lottery data supplied by the Pennsylvania Lottery Commission for repeated plays of the Cash 5 game.

The relative frequency histogram in Figure 9.12 shows the values of $H$ for the 1560 games between April 23, 1992, and May 18, 2000, obtained from the website www.palottery.com. (The game was played weekly at first, then daily.) Table 9.4 displays some of the possible values of $H$ and the frequency with which they occurred in the 1560 games. For instance, for one game, the highest of the five numbers was only 10, while for 70 out of the 1560 games (about 4.5% of the time), the high was 30. (Examples noted in this paragraph are in boldface in the table to help you find them.) Notice that the high was 31 or less for 444 of the games, which represents 28.46% of the games. In other words, all of the winning numbers are contained in the integers 1 to 31 in just over 28% of the games. For the remaining 72% of the games, the highest number is over 31. In fact, the most

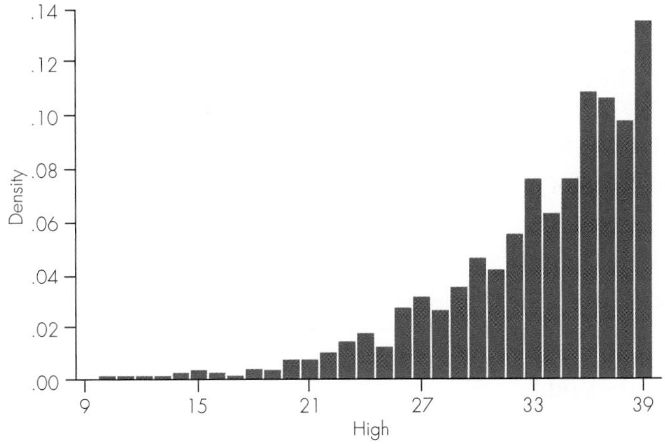

**Figure 9.12** ▌ Approximate sampling distribution for highest Cash 5 number

**Table 9.4** Partial Sampling Distribution for Highest Cash 5 Number (examples from text are in bold)

| $H =$ high | Count | Cumulative Count | Percentage | Cumulative Percentage |
|---|---|---|---|---|
| 10 | 1 | 1 | 0.06 | 0.06 |
| 11 | 1 | 2 | 0.06 | 0.13 |
| 12 | 1 | 3 | 0.06 | 0.19 |
| 13 | 1 | 4 | 0.06 | 0.26 |
| 14 | 2 | 6 | 0.13 | 0.38 |
| 15 | 3 | 9 | 0.19 | 0.58 |
| 16 | 2 | 11 | 0.13 | 0.71 |
| ⋮ | ⋮ | ⋮ | ⋮ | ⋮ |
| 29 | 55 | 309 | 3.53 | 19.81 |
| **30** | **70** | 379 | **4.49** | 24.29 |
| **31** | 65 | **444** | 4.17 | **28.46** |
| 32 | 85 | 529 | 5.45 | 33.91 |
| 33 | 118 | 647 | 7.56 | 41.47 |
| 34 | 98 | 745 | 6.28 | 47.76 |
| 35 | 118 | 863 | 7.56 | 55.32 |
| 36 | 169 | 1032 | 10.83 | 66.15 |
| 37 | 165 | 1197 | 10.58 | 76.73 |
| 38 | 152 | 1349 | 9.74 | 86.47 |
| **39** | **211** | 1560 | **13.53** | 100.00 |

common outcome is $H = 39$, the highest possible number, which occurred in 211/1560, or about 13.5% of all games played.

Notice that the shape of the probability distribution for $H$ is not symmetric and is skewed to the left. This example illustrates that not all sample statistics have bell-shaped sampling distributions. ■

***thought question 9.7*** Example 9.19 described a sample statistic, $H$ = highest number drawn, for the Cash 5 lottery game. Give another example of a sample statistic for the Cash 5 game, and describe what you think the shape of its sampling distribution would be. *

It is also possible to construct a sampling distribution for simple situations using the probability rules from Chapter 7. It requires listing all possible value of the sample statistic and finding the probability that each one will be the one achieved when a sample is taken. This can be done for discrete random variables only, because the rules of probability from Chapter 7 cannot readily be applied to continuous random variables.

**Example 9.20**

**Constructing a Simple Sampling Distribution for the Mean Movie Rating**
Sean is lazy about reading, so he does not read movie reviews. Instead, he relies on numerical ratings given by two movie critics, both of whom assign ratings of from one to four stars to the movies they review. Sean decides what movies to see on the basis of the mean of the two ratings. If the mean is at least 3, Sean will go see the movie. Unknown to Sean, the reviewers are lazy too, and they both randomly choose how many stars to give to each movie as follows: The probabilities of assigning one or two stars are .2 each, and the probabilities of assigning three or four stars are .3 each. What is the sampling distribution for the mean number of stars, assuming that the two critics make these assignments independently? What percentage of movies receive a mean rating of at least three stars? (Doesn't it sometimes seem that this is how critics assign their ratings?)

The sample statistic is the mean of the two ratings. There are a total of $4 \times 4 = 16$ combinations of ratings for the two critics. The probability of each combination is found by using the multiplication rule (Rule 3b, p. 246). For instance, the probability that they both assign one star to a movie is $(.2)(.2) = .04$. The sampling distribution for $\bar{x}$ can be found by first listing all of the possible samples of two ratings along with probabilities for the 16 possible samples, as shown in Table 9.5.

**Table 9.5** Possible Samples of Two Movie Ratings, Associated $\bar{x}$ and Probability

| Sample | 1, 1 | 1, 2 | 1, 3 | 1, 4 | 2, 1 | 2, 2 | 2, 3 | 2, 4 | 3, 1 | 3, 2 | 3, 3 | 3, 4 | 4, 1 | 4, 2 | 4, 3 | 4, 4 |
|---|---|---|---|---|---|---|---|---|---|---|---|---|---|---|---|---|
| $\bar{x}$ | 1 | 1.5 | 2 | 2.5 | 1.5 | 2 | 2.5 | 3 | 2 | 2.5 | 3 | 3.5 | 2.5 | 3 | 3.5 | 4 |
| Probability | .04 | .04 | .06 | .06 | .04 | .04 | .06 | .06 | .06 | .06 | .09 | .09 | .06 | .06 | .09 | .09 |

*HINT: Some possibilities are the lowest value, the median value, and the mean value.

The distinct values of $\bar{x}$ can now be listed, with their probabilities found by adding the probabilities for the samples that produced them. For example, two samples produce a mean of 1.5; they are (one star, two stars) and (two stars, one star). Their probabilities are each .04, so the probability that the mean rating is 1.5 stars is .04 + .04 = .08. The resulting sampling distribution for $\bar{x}$ is shown in Table 9.6.

**Table 9.6**   The Sampling Distribution for the Mean Movie Rating, $\bar{x}$

| Value of $\bar{x}$ | 1 | 1.5 | 2 | 2.5 | 3 | 3.5 | 4 |
|---|---|---|---|---|---|---|---|
| Probability | .04 | .08 | .16 | .24 | .21 | .18 | .09 |

9.10 Exercises are on page 394.

Notice that .21 + .18 + .09 = .48, or 48% of all movies will pass Sean's test of receiving a mean rating of at least 3.  ■

*skillbuilder applet*

## 9.11  Finding the Pattern in Sample Means

Statistics △ Now™

To explore this applet and work through this activity, go to Chapter 9 at **http://1pass.thomson.com** and click on Skillbuilder Applet, or view the applet on your CD.

The main idea for any sampling distribution is that it gives the pattern for how the potential value of a statistic may vary from sample to sample. The Rule for Sample Means tells us that in two common situations, a normal curve approximates the sampling distribution of the sample mean. The **SampleMeans** applet lets us see the pattern that emerges when we look at the means of many different random samples from the same population. Figure 9.13 illustrates the

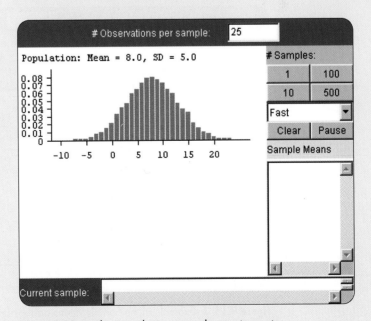

**Figure 9.13 ▮** The SampleMeans applet starting point

appearance of the applet when it is started. The histogram shown is a population of individual measurements with $\mu = 8$ and $\sigma = 5$, as for the weight-loss example in this chapter. Notice that the distribution of individual measurements is bell-shaped, so the Rule for Sample Means should hold for any sample size. At the start, the number of observations per sample is set at $n = 25$.

### What Happens

The applet simulates choosing simple random samples from the population. For each sample selected, the mean is calculated, displayed in the box under "Sample Means," and graphed in a histogram. The individual measurements in the most recently selected sample are shown at the bottom of the display. The buttons under "# Samples" can be used to select either 1, 10, 100, or 500 different samples at a time. Figure 9.14 illustrates an example in which two different samples of $n = 25$ per sample have been selected, while Figure 9.15 shows an example in which 500 different samples of $n = 25$ were selected. Notice that in Figure 9.14 the two distinct sample means are evident in the bottom histogram, while in Figure 9.15 the histogram of the 500 sample means (the bottom histogram) looks bell-shaped.

You can control the speed of the applet with the menu under the "# Samples" button. With slow and fast sampling, you see the histogram change as samples are added. With batch sampling, the histogram of sample means is not updated until all requested samples have been selected. The "# Observations per sample" box is used to change the sample size per sample. The "Clear" button clears all results and should be used whenever changing the sample size ("# Observations per sample").

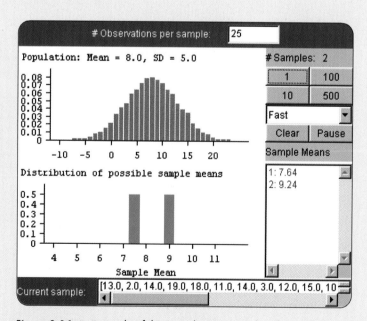

**Figure 9.14** ▌ Example of the SampleMeans applet after two samples of $n = 25$ have been selected

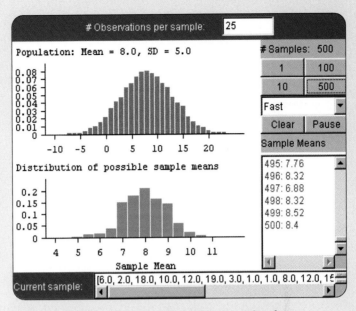

**Figure 9.15** ▌ Example of the SampleMeans applet after 500 samples of $n = 25$ have been selected

## What to Do

With the number of observations per sample set at 25, generate many different samples. Take note of the resulting shape of the histogram of sample means. Take note of the center and range of the distribution of sample means. Use the **Clear** button to clear the histogram, and then enter 100 as "# Observations per sample." Generate many different samples for samples of this size. Again, take note of the shape, center, and spread of the histogram of sample means. Verify that the range of the histogram of sample means is approximately $\mu \pm 3 \times (\sigma/\sqrt{n})$. Do the same things for "# Observations per sample" = 200. Experiment with the applet to formulate answers to the following questions. What feature of the distribution of sample means is affected by changing the sample size? What features are not affected by changing the sample size? How does changing the sample size affect the chance that a sample mean will fall in the interval 7.5 to 8.5 (within 0.5 pound of the true mean)?

## Lessons Learned

This applet lets you see that a normal curve approximates the distribution of sample means for different random samples of the same size from a population. This is evident from the shape of the histogram for sample means, assuming that you select many different samples of the same size. You can also see that increasing the number of observations in a sample decreases the variability among sample means. Notice that regardless of sample size, the histogram of sample means is approximately a normal curve centered at 8 pounds, the population mean. But increasing the sample size increases the likelihood that a sample mean will be close to 8 pounds, the population mean.

9.11 Exercises are on page 395.

## *case study 9.1*  Do Americans Really Vote When They Say They Do?

On November 8, 1994, a historic election took place in which the Republican Party won control of both houses of Congress for the first time since 1952. But how many people actually voted? On November 28, 1994, *Time* magazine (p. 20) reported that in a telephone poll of 800 adults taken during the two days following the election, 56% reported that they had voted. Considering that only about 68% of adults are registered to vote, that isn't a bad turnout.

But *Time* reported another disturbing fact along with the 56% reported vote. *Time* reported that, in fact, only 39% of American adults had voted, based on information from the Committee for the Study of the American Electorate.

Could it be that the results of the poll simply reflected a sample that, by chance, voted with greater frequency than the general population? The Rule for Sample Proportions can answer that question. Let's suppose that the truth about the population is, as reported by *Time*, that only 39% of American adults voted. Then the Rule for Sample Proportions tells us what kind of sample proportions we can expect in samples of 800 adults, the size used by the *Time* poll. The mean of the possibilities is .39, or 39%. The standard deviation is $\sqrt{(.39)(.61)/800} = .017$, or 1.7%.

Therefore, we are almost certain that the sample percentage based on a sample of 800 adults should fall within $3 \times 1.7\% = 5.1\%$ of the truth of 39%. In other words, if respondents were telling the truth, the sample percentage should be no higher than 44.1%—nowhere near the reported percentage of 56%.

In fact, if we combine the Rule for Sample Proportions with what we learned about normal curves in Chapter 8, we can say even more about how unlikely this sample result would be. If in truth only 39% (.39) of the population voted, the standardized score for the reported value of 56% (.56) is $z = (.56 - .39)/.017 = 10.0$. We know from Chapter 8 that it is virtually impossible to obtain a standardized score of 10.

Another example of the fact that *reported* voting tends to exceed *actual* voting occurred in the 1992 U.S. presidential election. According to the *World Almanac* (1995, p. 631), 61.3% of American adults reported voting in the 1992 election. In a footnote, the *Almanac* explains:

> Total reporting voting compares with 55.9 percent of population actually voting for president, as reported by Voter News Service. Differences between data may be the result of a variety of factors, including sample size, differences in the respondents' interpretation of the questions, and the respondents' inability or unwillingness to provide correct information or recall correct information.

Unfortunately, because figures are not provided for the size of the sample, we cannot assess whether or not the difference between the actual percentage of 55.9 and the reported percentage of 61.3 can be explained by the natural variability among possible sample proportions.

## *in summary*  Sampling Distributions

Assuming that appropriate conditions are met, the sampling distribution of each of the statistics in this summary table is approximately normal with mean = parameter and standard deviation as shown. The standard deviation is estimated using the standard error. The standardized statistic is always $z$ when the standard deviation is used in the denominator and is $z$ or $t$, as shown in the last column, when the standard deviation must be estimated and thus the standard error is used in the denominator.

| | Parameter | Statistic | Standard Deviation of the Statistic | Standard Error of the Statistic | Standardized Statistic with s.e. |
|---|---|---|---|---|---|
| **One Proportion** | $p$ | $\hat{p}$ | $\sqrt{\dfrac{p(1-p)}{n}}$ | $\sqrt{\dfrac{\hat{p}(1-\hat{p})}{n}}$ | $z$ |
| **Difference Between Proportions** | $p_1 - p_2$ | $\hat{p}_1 - \hat{p}_2$ | $\sqrt{\dfrac{p_1(1-p_1)}{n_1} + \dfrac{p_2(1-p_2)}{n_2}}$ | $\sqrt{\dfrac{\hat{p}_1(1-\hat{p}_1)}{n_1} + \dfrac{\hat{p}_2(1-\hat{p}_2)}{n_2}}$ | $z$ |

| | Parameter | Statistic | Standard Deviation of the Statistic | Standard Error of the Statistic | Standardized Statistic with s.e. |
|---|---|---|---|---|---|
| **One Mean** | $\mu$ | $\bar{x}$ | $\dfrac{\sigma}{\sqrt{n}}$ | $\dfrac{s}{\sqrt{n}}$ | $t$ |
| **Mean Difference, Paired Data** | $\mu_d$ | $\bar{d}$ | $\dfrac{\sigma_d}{\sqrt{n}}$ | $\dfrac{s_d}{\sqrt{n}}$ | $t$ |
| **Difference Between Means** | $\mu_1 - \mu_2$ | $\bar{x}_1 - \bar{x}_2$ | $\sqrt{\dfrac{\sigma_1^2}{n_1} + \dfrac{\sigma_2^2}{n_2}}$ | $\sqrt{\dfrac{s_1^2}{n_1} + \dfrac{s_2^2}{n_2}}$ | $t$ |

# Key Terms

# Exercises

● Denotes basic skills exercises

◆ Denotes dataset is available in StatisticsNow at **http:// 1pass.thomson.com** or on your CD but is **not required** to solve the exercise.

**Bold-numbered exercises** have answers in the back of the text and fully worked solutions in the Student Solutions Manual.

**Statistics⊖Now™** Go to the StatisticsNow website at **http://1pass.thomson.com** to:
- Assess your understanding of this chapter
- Check your readiness for an exam by taking the Pre-Test quiz and exploring the resources in the Personalized Learning Plan

## Section 9.1

**9.1** ● In each situation, explain whether the value given in bold print is a statistic or a parameter:
  **a.** A polling organization samples 1000 adults nationwide and finds that **72%** of those sampled favor tougher penalties for persons convicted of drunk driving.
  **b.** In the year 2000 census, the U.S. Census Bureau found that the median age of all American citizens was about **35** years.
  **c.** For a sample of 20 men and 25 women, there is a **14-centimeter** difference in the mean heights of the men and women.
  **d.** A writer wants to know how many typing mistakes there are in his manuscript, so he hires a proofreader, who reads the entire manuscript and finds **15** errors.

**9.2** ● A polling organization plans to sample 1000 adult Americans to estimate the proportion of Americans who think crime is a serious problem in this country. In the context of this poll, explain the difference between a statistic and a parameter. What is the parameter of interest to the polling organization? What will be the statistic?

**9.3** ● *Part (a) of this exercise is also Exercise 1.3.* Using Case Study 1.6 on pages 5–6 as an example:
  **a.** Explain the difference between a population and a sample.
  **b.** Use the proportion of placebo takers who had heart attacks, which was .017, to explain the difference between a population parameter and a sample statistic.

**9.4** ● As we learned in Chapter 3, a 95% confidence interval is an interval that we are 95% confident covers the truth. Does "the truth" in this definition refer to the truth about a sample statistic or about a population parameter? Or does it depend on the situation? Explain.

**9.5** A report in *Newsweek* about the relationship among diet, genes, and disease (Underwood and Adler, 2005), discussed the E4 allele for a protein known as Apo E and said "an estimated 15 to 30 percent of the population has at least one copy of the allele." People who have the allele are at higher risk for diabetes, heart disease in smokers, and Alzheimer's disease.
  **a.** What is the population parameter of interest in the quote?
  **b.** Would it make more sense to find a confidence interval for the population parameter you described in part (a) or to conduct a hypothesis test for a specific null value for it? Explain. If your answer is "a hypothesis test," specify the null value that you would test.
  **c.** The phrase "an estimated 15 to 30 percent" indicates that scientists don't know precisely what percentage of the population has the allele. Explain why they don't know.

**9.6** Some stockholders want to know whether the mean salary for male employees in a large company is higher than the mean salary for female employees. The company allows them access to salary information for a random sample of 100 male and 100 female employees, and the mean salaries are $41,000 for the males and $39,500 for the females.
  **a.** Are the mean salaries in this example statistics or parameters? Explain.
  **b.** Based on these means, can the shareholders determine that the mean salary for males in the company is higher than the mean salary for females? Explain.
  **c.** If they selected new random samples of 100 males and 100 females, would the average salaries of those samples be $41,000 and $39,500? Explain.

**9.7** Refer to Exercise 9.6. Suppose that the 100 men and 100 women in the sample were all of the company's employees. Now answer questions a and b, repeated here:
  **a.** Are the mean salaries in this example statistics or parameters? Explain.
  **b.** Based on these means, can the shareholders determine that the mean salary for males in the company is higher than the mean salary for females? Explain.

**9.8** In Section 9.1, read the paragraph that begins with "Hypothesis testing is applied to population parameters by . . ." on page 333. Now refer to Exercise 9.6, for which the population parameter of interest would be the difference in mean salaries for all males and all females in the company.
  **a.** What null value for the population parameter would indicate that there is no problem with equity in salaries?
  **b.** What numerical value based on the data from the random samples could be used as a sample statistic to estimate the population parameter of interest?
  **c.** Would statistical significance be more likely to be achieved if the sample statistic was higher than the value you specified in part (b), or if it was lower than that value? Explain.

## Section 9.2

**9.9** ● Explain whether $\hat{p}$ or $p$ is the correct statistical notation for each proportion described:
   **a.** The proportion that smokes in a randomly selected sample of $n = 300$ students in the eleventh and twelfth grades.
   **b.** The proportion that smokes among all students in the eleventh and twelfth grades in the United States.
   **c.** The proportion that is left-handed in a sample of $n = 250$ individuals.

**9.10** ● Explain whether $\bar{x}$ or $\mu$ is the correct statistical notation for each mean described.
   **a.** The mean hours of study per week was 15 hours for a sample of $n = 30$ students at a college.
   **b.** A university administrator determines that the mean age of all students at a college is 20.2 years.

**9.11** ● In a random sample of 300 parents in a school district, 104 supported a controversial new education program. The purpose of the survey was to estimate the proportion of all parents in the school district who support the new program.
   **a.** What is the research question of interest for this survey?
   **b.** What is the population parameter in this study? What is the appropriate statistical notation (symbol) for this parameter?
   **c.** What is the value of the sample estimate (statistic) in this study? What is the appropriate statistical notation (symbol) for this estimate?

**9.12** ● A large Internet provider conducted a survey of its customers. One question that it asked was how many e-mail messages the respondent had received the previous day. The mean number was 13.2.
   **a.** What is the research question of interest for this survey?
   **b.** What is the population parameter in this study? What is the appropriate statistical notation (symbol) for this parameter?
   **c.** What is the value of the sample estimate (statistic) in this study? What is the appropriate statistical notation (symbol) for this estimate?

**9.13** ● A medical researcher wants to estimate the difference in the proportions of women with high blood pressure for women who use oral contraceptives versus women who do not use oral contraceptives. In an observational study involving a sample of 900 women, the researcher finds that .15 (15%) of the 500 women who used oral contraceptives had high blood pressure, whereas only .10 (10%) of the 400 women who did not use oral contraceptives had high blood pressure.
   **a.** What is the research question of interest for this study?
   **b.** What is the population parameter in this study? What is the appropriate statistical notation for this parameter?
   **c.** What is the value of the sample estimate (statistic) in this study? What is the appropriate statistical notation for this estimate?

**9.14** ● A student doing a project for a statistics class wants to estimate the difference between the mean number of music CDs owned by male and female students. He surveys 50 male students and 45 female students about the number of music CDs they own. The mean number for the males is 110, and the mean number for females is 90.
   **a.** What is the research question of interest for this survey?
   **b.** What is the population parameter in this study? What is the appropriate statistical notation for this parameter?
   **c.** What is the value of the sample estimate (statistic) in this study? What is the appropriate statistical notation for this estimate?

**9.15** A study was done by randomly assigning 200 volunteers with sore throats to either drink a cup of herbal tea or use a throat lozenge to ease their pain. The percentages reporting relief for the two methods were compared. What are the research question, the population parameter, and the sample estimate for this study?

**9.16** A sample of 100 students at a university was asked how many hours a week they spent studying and how many they spent socializing. The difference was computed for each student. What are the research question, the population parameter, and the sample estimate for this study?

**9.17** Refer to Exercise 9.16. Does the situation described represent independent samples or paired data? Explain.

**9.18** For each of the following research questions, would it make more sense to collect independent samples or paired data? Explain.
   **a.** In the United States, on average, what is the age difference between husbands and wives?
   **b.** What is the difference in average ages at which teachers and plumbers retire?
   **c.** What is the difference in average salaries for high school graduates and college graduates?

**9.19** Give an example of a research question for which the population parameter of interest would be $p_1 - p_2$.

**9.20** Give an example of a research question for which the population parameter of interest would be $\mu_1 - \mu_2$.

**9.21** In Example 9.1 (p. 334), we discussed a few different parameters that might be of interest in exploring the question of whether first-year college students gain weight. In each of the following cases, explain what population(s) the sample(s) should be drawn from and what variable or variables would be measured for each individual in the sample(s).
   **a.** The parameter of interest is $p$ = proportion of the population of first-year college students who weigh more at the end of the year than they did at the beginning of the year.
   **b.** The parameter of interest is $\mu_d$ = the mean (average) weight gain during the first year for the population of college students.
   **c.** The parameter of interest is $\mu_1 - \mu_2$ = how much more weight college students gain, on average, during the year than do their contemporaries of the same age.

**9.22** For each of the following situations from earlier chapters, define the parameter of interest and the sample statistic that would be used to estimate that parameter, choosing from among the five described in Section 9.2. When defining the parameter, make sure you explain what constitutes the population.

a. In Case Study 1.3, we learned that 57% of a sample of teens who go out on dates say that they have been out with someone of another race or ethnic group.

b. In Example 2.2, we learned that 155 out of 172, or 90%, of babies who slept in darkness were free of myopia later in childhood.

c. In Example 2.2, we learned that while 90% of the babies in the sample who slept in darkness were free of myopia later in childhood, only 45% of the sample of babies who slept in full light were free of myopia later in childhood.

d. In Example 2.17, we learned that the mean height was 1602 millimeters for a sample of 199 married British women taken in 1980.

## Section 9.3

**9.23** ● Consider a situation in which a sample of 100 SAT scores is taken from the population of all people who took the test in a given year. Which would be more spread out: the sampling distribution of the sample mean for 100 SAT scores or a histogram of the 100 SAT scores reported in one sample? Explain.

**9.24** ● Explain what symbol ($\hat{p}$, $\mu$, etc.) would be used to represent each of the following:

a. The mean of the sampling distribution of $\hat{p}$.

b. One value from the sampling distribution of $\hat{p}$.

c. The mean of the sampling distribution of the sample mean.

d. One value from the sampling distribution of the sample mean.

**9.25** Consider a situation in which a random sample of 1000 adults is surveyed and the proportion that primarily buys organic vegetables is found. If a new random sample of 1000 adults is taken from the same population, explain whether each of the following would change:

a. The population proportion, $p$.

b. The sample proportion, $\hat{p}$.

c. The standard deviation of $\hat{p}$.

d. The standard error of $\hat{p}$.

e. The sampling distribution of $\hat{p}$, including its shape, mean, and standard deviation.

**9.26** Researchers would like to know the mean body temperature of runners at the completion of a marathon. They plan to take a random sample of 100 runners who complete a marathon and record their temperatures. Which of the following will the researchers know after they have taken the sample? Explain your answer in each part.

a. The value of the mean of the sampling distribution of the sample mean.

b. The approximate shape of the sampling distribution of the sample mean.

c. The standard deviation of the sampling distribution of the sample mean.

**9.27** Refer to Exercise 9.26. If a different sample of 100 marathon runners was taken from the same population, which of the following would be the same as it was for the first sample, and which would be different? Explain your answer in each part.

a. The value of the mean of the sampling distribution of the sample mean.

b. The approximate shape of the sampling distribution of the sample mean.

c. The standard deviation of the sampling distribution of the sample mean.

**9.28** Refer to Exercises 9.26 and 9.27. Using that context as an example, define in your own words what is meant by a sampling distribution of the sample mean.

**9.29** Refer to the smooth curve in Figure 9.1 (p. 346), illustrating the sampling distribution of mean hours of sleep for a sample of 190 college students, taken from a population of sleep hours with a mean of 7.1 hours and standard deviation of 2 hours. Explain whether each of the following would be likely if you were to take a sample of 190 students from the original population.

a. The mean hours of sleep for the 190 students would be between 8 and 9 hours.

b. The mean hours of sleep for the 190 students would be between 7 and 8 hours.

c. The number of hours of sleep reported by the first student in the sample would be between 8 and 9 hours.

d. The maximum number of hours of sleep reported by any student in the sample would be less than 8 hours.

**9.30** One result stated in Section 9.3 is that the standard deviation of the sampling distribution gets smaller as the sample size gets larger.

a. Using the context of Example 9.8 (p. 345), which described the sampling distribution of the mean of 190 sleep values, explain why this result makes sense.

b. Consider the extreme case in which everyone in the population is measured. In that case, what would be the standard deviation for the sampling distribution of the sample mean?

## Section 9.4

**9.31** ● Assuming that the Rule for Sample Proportions applies, calculate the mean and the standard deviation of the sampling distribution of possible sample proportions for each combination of sample size ($n$) and population proportion ($p$).

a. $n = 400$, $p = .5$.

b. $n = 1600$, $p = .5$.

c. $n = 64$, $p = .8$.

d. $n = 256$, $p = .8$.

**9.32** ● Refer to Exercise 9.31, and notice that for parts (a) and (b), the values of $n$ are different while $p$ is the same. This pattern also occurs in parts (c) and (d). Use the answers to Exercise 9.31 to describe how changing the sample size affects the mean and standard deviation of the distribution of possible sample proportions.

**9.33** ● A polling organization polls $n = 100$ randomly selected registered voters to estimate the proportion of a large population that intends to vote for Candidate Y in an upcoming election. Although it is not known by the polling organization, $p = .55$ is the actual proportion of the population that prefers Candidate Y.

a. Give the numerical value of the mean of the sampling distribution of $\hat{p}$.

b. Calculate the standard deviation of the sampling distribution of $\hat{p}$.

c. Use the Empirical Rule to find values that fill in the blanks in the following sentence. In about 99.7% of all randomly selected samples of $n = 100$ from this population, the sample proportion preferring Candidate Y will be between ___ and ___.

**9.34** ● In a random sample of $n = 200$ drivers, 50 individuals say that they never wear a seatbelt when driving.

a. Give a numerical value of $\hat{p} =$ sample proportion that never wears a seatbelt when driving.

b. Calculate the standard error of $\hat{p}$.

c. In a different survey of $n = 400$ drivers, 100 people say they never wear a seatbelt when driving. Give numerical values for $\hat{p}$ and the standard error of $\hat{p}$.

**9.35** ● In a random sample of $n = 500$ adults, 300 individuals say that they believe in love at first sight.

a. Calculate the value of $\hat{p} =$ sample proportion that believes in love at first sight.

b. Calculate the standard error of $\hat{p}$.

**9.36** Suppose the probability is $p = .2$ that a person who purchases an instant lottery ticket wins money, and this probability holds for every ticket purchased. Consider different random samples of $n = 64$ purchased tickets. Let $\hat{p} =$ sample proportion of winning tickets in a sample of 64 tickets.

a. Give the numerical value of the mean of the sampling distribution of $\hat{p}$.

b. Calculate the standard deviation of the sampling distribution of $\hat{p}$.

c. Use the Empirical Rule to find values that fill in the blanks in the following sentence: In about 68% of all randomly selected samples of $n = 64$ instant lottery tickets, the proportion of tickets that will be money winners will be between ___ and ___.

d. Use the Empirical Rule to find values that fill in the blanks in the following sentence: In about 95% of all randomly selected samples of $n = 64$ instant lottery tickets, the proportion of tickets that will be money winners will be between ___ and ___.

**9.37** Suppose medical researchers think that $p = .70$ is the proportion of all teenagers with high blood pressure whose blood pressure would decrease if they took calcium supplements. To test this theory, the researchers plan a clinical trial (experiment) in which $n = 200$ teenagers with high blood pressure will take regular calcium supplements.

a. Assume that $p = .70$ actually is the population proportion that would experience a decrease in blood pressure. What are the numerical values of the mean

and standard deviation of the sampling distribution of $\hat{p}$, the sample proportion, for samples of $n = 200$ teenagers?

b. Use the results of part (a) to calculate an interval that will contain the sample proportion $\hat{p}$ for about 99.7% of all samples of $n = 200$ teenagers.

c. In the clinical trial, 120 of the 200 teenagers taking calcium supplements experienced a decrease in blood pressure. What is the value of $\hat{p}$ for this sample? Is this value a parameter or a statistic?

d. Given the answers for parts (b) and (c), explain why the observed value of $\hat{p}$ could be used as evidence that the researchers might be wrong to think that $p = .70$.

**9.38** An automobile club comes to the aid of stranded motorists who are members. Over the long run, about 5% of the members utilize this service in any 12-month period. A small town has 400 members of the club. Consider the 400 members to be representative of the larger club membership. Let $\hat{p}$ be the proportion of them who will utilize the service in the coming 12-month period.

a. Explain why conditions 1 to 3 for the Normal Curve Approximation Rule for Sample Proportions are met in this scenario. State any assumptions that you need to make to do so.

b. What are the values of $n$ and $p$ in this situation?

c. Use the Rule for Sample Proportions to describe the sampling distribution of $\hat{p}$, including values for the mean and standard deviation.

d. The towing service in town contracts with the club to come to the aid of up to 28 members in the next 12-month period. What proportion is that of the 400 members in town? What is the approximate probability that the actual proportion requiring aid will exceed that value?

**9.39** Refer to Exercise 9.38, in which $\hat{p}$ is the sample proportion of 400 randomly selected club members who require aid in a 12-month period. Based on the Empirical Rule, into what interval should $\hat{p}$ fall about 95% of the time?

**9.40** Explain the difference between the *standard error of* $\hat{p}$ and the *standard deviation of the sampling distribution of* $\hat{p}$. Write down the formula for each one. Which one is more likely to be used in practice? Why?

**9.41** For each of the following, if the scenario fits the Rule for Sample Proportions, describe the sampling distribution of $\hat{p}$ and draw a picture of it. If the scenario doesn't fit, explain why not (in which case you do not have the tools to describe the sampling distribution of $\hat{p}$).

a. An auto insurance company has 1500 customers in a city, and over the long run, 6% of them will file a claim in any given year. Define $\hat{p}$ to be the proportion who file a claim in the next year.

b. A dean has found that over the years, about 15% of students who begin their college work in her program do not finish. This year, 90 students will begin her program; define $\hat{p}$ to be the proportion who will finish.

c. An essay contest has three winners, but the number of entrants varies each time it is offered, so the proportion of entrants who are winners also varies. Define $\hat{p}$ to be the proportion of entrants who will be winners the next time the contest is offered.

**9.42** Recent studies have shown that about 20% of American adults fit the medical definition of being obese. A large medical clinic would like to estimate what percentage of their patients are obese, so they take a random sample of 100 patients and find that 18 are obese. Suppose that in truth, the same percentage holds for the patients of the medical clinic as for the general population, 20%. Give a numerical value for each of the following:

**a.** The population proportion of obese patients in the medical clinic.

**b.** The proportion of obese patients for the sample of 100 patients.

**c.** The standard error of $\hat{p}$.

**d.** The mean of the sampling distribution of $\hat{p}$.

**e.** The standard deviation of the sampling distribution of $\hat{p}$.

**9.43** Refer to Exercise 9.42, in which 20% of the patients in a medical clinic are obese. If the clinic took repeated random samples of 100 observations and found the sample proportion who were obese, into what interval should those sample proportions fall about 95% of the time?

**9.44** A student commented, "Once a poll has been taken, $\hat{p}$ (the sample proportion with a certain opinion) is a known value, so it cannot have a distribution of possible values. Therefore, I don't understand what you mean by the sampling distribution of $\hat{p}$." Write an explanation for this student.

## Section 9.5

**9.45** ● If the population proportions are equal, what will be the mean of the sampling distribution of $\hat{p}_1 - \hat{p}_2$?

**9.46** ● For each of the following situations, specify the mean of the sampling distribution of the difference in sample proportions:

a. The populations are male and female registered voters in the United States, and it is of interest to compare the proportions of males and females who are registered with the Democratic Party. Suppose that, in truth, 30% of men and 36% of women are registered Democrats. Random samples of 500 will be taken from each population.

b. The populations are adults with college degrees and adults without college degrees. It is of interest to compare the proportions that support the legalization of marijuana. Suppose that, in truth, 30% of those with college degrees support it, and 20% of those without college degrees support it. Random samples of 100 will be taken from each population.

c. The populations are left-handed and right-handed junior high school students. It is of interest to compare the proportions that have learned to play a mu-

sical instrument. Suppose that, in truth, 40% of both populations have learned to play a musical instrument. A random sample of 800 right-handed students and 200 left-handed students will be taken.

**9.47** Refer to Exercise 9.46. In each part, specify the standard deviation of the sampling distribution of the difference in sample proportions.

**9.48** Refer to Exercises 9.46 and 9.47. The situations are repeated here. In each case, draw a picture of the sampling distribution of $\hat{p}_1 - \hat{p}_2$, similar to Figure 9.4 (p. 356).

a. The populations are male and female registered voters in the United States, and it is of interest to compare the proportions of males and females who are registered with the Democratic Party. Suppose that, in truth, 30% of men and 36% of women are registered Democrats. Random samples of 500 will be taken from each population.

b. The populations are adults with college degrees and adults without college degrees. It is of interest to compare the proportions that support the legalization of marijuana. Suppose that, in truth, 30% of those with college degrees support it, and 20% of those without college degrees support it. Random samples of 100 will be taken from each population.

c. The populations are left-handed and right-handed junior high school students. It is of interest to compare the proportions that have learned to play a musical instrument. Suppose that, in truth, 40% of both populations have learned to play a musical instrument. A random sample of 800 right-handed students and 200 left-handed students will be taken.

**9.49** For each of the following situations, explain whether the conditions hold for the sampling distribution of $\hat{p}_1 - \hat{p}_2$ to be approximately normal:

a. A polling organization wants to compare the proportions of men and women who favor a particular candidate in an upcoming election and plans to take random samples of 1000 men and 1000 women.

b. A researcher plans to take a random sample of 500 children and compare the proportion of them who live with both parents in the household to the proportion who live with just one parent in the household.

**c.** A company builds an expensive component for automobiles and wants to compare two methods of manufacturing them. The company plans to take a random sample of ten of the components manufactured by using each method and compare the proportion of them that fail from each of the two methods.

**9.50** There is some evidence that women tend to vote earlier in the day and men tend to vote later in the day. If men and women vote differently, then exit polls taken early in the day may not reflect the eventual results. Suppose that in a particular election, 51% of the voters who vote in the morning vote for Candidate A, while only 48% of those who vote later vote for Candidate A. An exit poll is based on 500 voters in the morning and

500 voters later in the day. Define $\hat{p}_1$ to be the proportion of the morning sample that voted for Candidate A and $\hat{p}_2$ to be the proportion of the later sample that voted for Candidate A.

a. Based on the morning voters only, describe the sampling distribution of $\hat{p}_1$. Draw a figure similar to Figure 9.3 (p. 353).

b. What is the mean of the sampling distribution of $\hat{p}_1 - \hat{p}_2$?

c. What is the standard deviation of the sampling distribution of $\hat{p}_1 - \hat{p}_2$?

d. Draw a picture of the sampling distribution of $\hat{p}_1 - \hat{p}_2$, similar to Figure 9.4 (p. 356).

e. Using the figure you drew in part (d), sketch the region of the figure that corresponds to having samples with $\hat{p}_1$ being *less* than $\hat{p}_2$.

f. Using your knowledge of how to find areas under normal curves from Chapter 8, find the probability that $\hat{p}_1 - \hat{p}_2 < 0$. In other words, find the probability that the exit poll shows a lower proportion voting for Candidate A in the morning ($\hat{p}_1$) than later in the day ($\hat{p}_2$).

9.51 Researchers would like to know whether the proportions of elementary school children who are obese differ in rural and urban areas. They plan to take random samples of 900 children from urban areas and 900 children from rural areas and compare the proportions that are obese. Define population 1 to be elementary school children in urban areas and population 2 to be elementary school children in rural areas.

a. Define in words what the parameter $p_1$ represents in this situation.

b. Define in words what $\hat{p}_1$ is in this situation.

c. Suppose that, unknown to the researchers, $p_1$ and $p_2$ are both .20 (20%). Describe the sampling distribution of $\hat{p}_1 - \hat{p}_2$, including its mean, standard deviation, and shape.

d. Draw a picture of the sampling distribution that you described in part (c), similar to Figure 9.4 (p. 356).

9.52 Refer to Exercise 9.51, including the information given in part (c). Which of the following could be specified before the samples are taken? Explain.

a. The value of the mean of the sampling distribution of $\hat{p}_1 - \hat{p}_2$.

b. The approximate shape of the sampling distribution of $\hat{p}_1 - \hat{p}_2$.

c. The standard deviation of the sampling distribution of $\hat{p}_1 - \hat{p}_2$.

d. The standard error of the sampling distribution of $\hat{p}_1 - \hat{p}_2$.

9.53 Refer to Exercises 9.51 and 9.52. Which of the following could be different if two different sets of samples of 900 children are taken? Explain.

a. The value of the mean of the sampling distribution of $\hat{p}_1 - \hat{p}_2$.

b. The approximate shape of the sampling distribution of $\hat{p}_1 - \hat{p}_2$.

c. The standard deviation of the sampling distribution of $\hat{p}_1 - \hat{p}_2$.

d. The standard error of the sampling distribution of $\hat{p}_1 - \hat{p}_2$.

9.54 When taking large random samples from two different populations and finding the sampling distribution of $\hat{p}_1 - \hat{p}_2$, explain which of the following would change if the sample sizes are changed.

a. The mean of the sampling distribution.

b. The approximate shape of the sampling distribution.

c. The standard deviation of the sampling distribution.

## Section 9.6

9.55 ● Explain whether the Rule for Sample Means can be used to describe the possible values of the sample mean in each scenario.

a. Mean normal body temperature will be determined for a randomly selected sample of 18 individuals. In the population of all humans, normal body temperature has approximately a normal distribution with mean $\mu = 98.2$ degrees Fahrenheit and standard deviation $\sigma = 0.5$.

b. Mean number of music CDs owned will be determined for a randomly selected sample of four college students. In the population of all college students, the distribution of number of CDs owned is skewed to the right.

c. Refer to part (b). The mean number of music CDs owned will be determined for a randomly selected sample of 900 college students.

9.56 ● The weights of men in a particular age group have mean $\mu = 170$ pounds and standard deviation $\sigma = 24$ pounds.

a. For randomly selected samples of $n = 16$ men, what is the standard deviation of the sampling distribution of possible sample means?

b. For randomly selected samples of $n = 64$ men, what is the standard deviation of the sampling distribution of possible sample means?

c. In general, how does increasing the sample size affect the standard deviation of the sampling distribution of possible sample means? Parts (a) and (b) provide a hint.

9.57 ● Example 9.8 (p. 345) concerned selecting random samples of $n = 190$ from a population of student responses to the question "How many hours of sleep did you get last night?" Suppose that in a large population of students, the mean amount of sleep the previous night was $\mu = 7.05$ hours and the standard deviation was $\sigma = 1.75$ hours. Consider randomly selected samples of $n = 190$ students.

a. What is the value of the mean of the sampling distribution of possible sample means?

b. Calculate the standard deviation of the sampling distribution of possible sample means.

c. Use the Empirical Rule to find values that fill in the blanks at the end of the following sentence: For 68% of all randomly selected samples of $n = 190$ students, the mean amount of sleep the previous night will be between ___ and ___ hours.

d. Use the Empirical Rule to fill in the blanks at the end of the following sentence. For 95% of all randomly selected samples of $n = 190$ students, the mean amount of sleep will be between ___ and ___ hours.

**9.58** ● ◆ The **cholest** dataset on the CD for this book includes cholesterol levels measured for 28 heart attack patients two days after their attacks. The sample mean for this sample is 253.9, and the standard deviation is 47.7.

a. Assuming that this was a sample from a larger population of heart attack patients, what symbols should be used to represent the values given for the mean and standard deviation?

b. Calculate the standard error of the sample mean.

**9.59** ● A randomly selected sample of $n = 60$ individuals over 65 years old takes a test of memorization skills. The sample mean is $\bar{x} = 53$, and the standard deviation is $s = 7.2$. Give the numerical value of the standard error of the mean.

**9.60** Vehicle speeds at a certain highway location are believed to have approximately a normal distribution with mean $\mu = 60$ mph and standard deviation $\sigma = 6$ mph. The speeds for a randomly selected sample of $n = 36$ vehicles will be recorded.

a. Give numerical values for the mean and standard deviation of the sampling distribution of possible sample means for randomly selected samples of $n = 36$ from the population of vehicle speeds.

b. Use the Empirical Rule to find values that fill in the blanks in the following sentence: For a random sample of $n = 36$ vehicles, there is about a 95% chance that mean vehicle speed in the sample will be between ___ and ___ mph.

c. Sample speeds for a random sample of 36 vehicles are measured at this location, and the sample mean is 66 mph. Given the answer to part (b), explain whether this result is consistent with the belief that the mean speed at this location is $\mu = 60$ mph.

**9.61** Small planes cannot fly well if the payload (people, luggage, and fuel) weighs too much. Suppose an airline runs a commuter flight that holds 40 people. The airline knows that the weights of passenger + luggage for typical customers on this flight is approximately normal with a mean of 210 pounds and a standard deviation of 25 pounds.

a. Draw a picture of this distribution.

b. Using the Rule for Sample Means, describe the sampling distribution of the mean weight of passenger + luggage for a random sample of 40 customers.

c. Draw a picture superimposing the distributions from parts (a) and (b). (Label it clearly, and remember that the total area under each curve must equal 1.)

d. Assume that customers on any particular flight are similar to a random sample. If the *total* weight of passengers and their luggage should not exceed 8800 pounds, what is the probability that a sold-out flight (40 passengers and their luggage) will exceed the weight limit? (*Hint:* Rewrite the desired limit as an average per passenger.)

**9.62** Explain the difference between the standard deviation of the sampling distribution of the mean and the standard error of the mean. Explain which one is more likely to be used in practice and why.

**9.63** Describe the sampling distribution for the statistic of interest in each of the following stories, including numerical values for the mean and standard deviation of the sampling distribution.

a. A fisherman takes tourists out fishing and has noticed that over the long run, the weights of a certain type of fish that is typically caught is approximately normal with a mean of 10 pounds and a standard deviation of 2 pounds. During a good day, the boatload of eight people catches the limit of 2 fish each, or a total of 16 fish. The statistic of interest is the mean weight of the 16 fish caught on a good day.

b. A fisherman hangs out at the local bar in the evening and hears tourists talking about the weights of the fish they caught that day. He notices that the weights mentioned seem to have a mean of about 20 pounds and a standard deviation of about 4 pounds. He records what he hears as the weights for the next 50 "fish stories" at the bar. The statistic of interest is the mean weight for the 50 stories. Assume that they represent a random sample of all such stories.

**9.64** A rental car company has noticed that the distribution of the number of miles customers put on rental cars per day is somewhat skewed to the right, with an occasional high outlier. The distribution has a mean of 80 miles and a standard deviation of 50 miles.

a. Draw a picture of a normal curve with this mean and standard deviation (80 miles and 50 miles), and use the information in the picture to verify that the distribution of number of miles must be skewed to the right and/or have occasional high outliers rather than being normally distributed.

b. What is the approximate distribution for the *mean* number of miles per day put on a typical one of the company's rental cars in a year (365 days)?

c. What is the approximate distribution for the *total* number of miles put on a typical one of the rental cars in a year (365 days)?

d. Do you think the necessary conditions for using the Rule for Sample Means to find the answer in part (b) are met in this situation? Explain.

## Section 9.7

**9.65** ● In each of the following situations, explain whether the comparison involves paired data or data from two independent samples:

a. Twenty sophomores are randomly selected from campus dormitories at a college, and 20 other sophomores are randomly selected from students who live off campus. The mean grade point averages of the two groups are compared.

b. Thirteen individuals each take a strength test both

**d.** Which value of $R$ is most likely to occur? Which value is least likely to occur?

9.106 A soft drink called Crash is so popular that vending machines allocate the top two selection buttons to it. Over the long run, of those who buy Crash, 60% push the top button and 40% push the lower button, and this behavior seems to be independent from one purchase to the next. Suppose a sample of $n = 10$ people buy Crash at a vending machine one morning and $X =$ the number who push the top button.

a. Is $X$ a statistic? Explain.

b. Describe the sampling distribution of $X$.

**9.107** In a large retirement community, 60% of households have no dogs, 30% have one dog, and 10% have two dogs. Following a news story about how the elderly have trouble walking their dogs in the winter weather, a magnanimous group of teenagers decides that it will randomly select two households in the retirement community each and call and offer to walk their dog(s). For each teen who participates, $T$ is the *total* number of dogs in the two households called.

**a.** List all possible values for $T$.

b. Find the sampling distribution for $T$. (*Hint:* This is simply a list of possible values for $T$ and their probabilities.)

c. If there are no dogs in the two households ($T = 0$), the teen starts over with a new sample of two households. If 1000 teens participate, for about how many of them will $T = 0$ for their first sample?

**9.108** The *I Ching* is an ancient Chinese system of asking for advice. In one version, three pennies are flipped, and if the number of heads is odd (1 or 3), a "broken line" is recorded, while if it is even (0 or 2), a "solid line" is recorded. This process is repeated six times, and then the results are used to identify one of the $2^6 = 64$ possible patterns of solid and broken lines and corresponding advice. Assume that the coins are fair and the flips are independent. Define $B$ to be the number of broken lines out of the six.

**a.** For each of the six lines, what is the probability that it is a "broken line"?

**b.** What are the possible values for $B$?

**c.** What is the probability that $B = 0$, so that all of the lines are solid?

**d.** Find the sampling distribution for $B$.

9.109 Are there situations for which the sampling distribution of a statistic consists of a single possible value, which has probability 1 of occurring each time? If not, explain why not. If so, explain how that could happen and give an example.

**9.110** Consider the population of net gains for tickets in the California Decco lottery game described in Examples 9.17 and 9.18. The possible values are $4999, $49, $4, 0, and −$1; the mean is −$0.35, and the standard deviation is $29.67. Suppose someone buys ten tickets each week.

**a.** Explain why the Rule for Sample Means (or the Central Limit Theorem) would *not* hold for the distribution of possible mean net gains for the weekly purchases of ten tickets.

**b.** Although the possible means for samples of ten tickets are *not* approximately normal, you can still specify numerical values for the mean and standard deviation of the sampling distribution of the mean in this situation. What are they?

9.111 Refer to the discussion of the Cash 5 lottery game in Example 9.19 and the accompanying Table 9.4 (p. 377).

a. In what proportion of the games is the highest number 35 or more?

b. What is the median value of $H$ for the 1560 Cash 5 games represented in Example 9.19?

c. In what proportion of the games is the highest number 15 or less?

**9.112** Define $L$ to be the *lowest* of the five numbers drawn in the Cash 5 lottery game in Example 9.19. Describe the shape of the sampling distribution of $L$, and explain how you know.

9.113 Use a computer to simulate the numbers drawn for the Cash 5 lottery game 1000 times, and draw a histogram for the values of $H =$ highest number drawn. Compare your histogram to the one shown in Figure 9.12 (p. 377).

9.114 Suppose that two fair dice are rolled and $H =$ higher of the two faces. If both dice show the same value, $H$ is that value. Construct the sampling distribution of $H$, listing all possible values and their probabilities.

**9.115** Suppose someone who takes a ten-question True-False test doesn't know any of the answers and randomly selects True or False for each question. Define the statistic $X =$ number correct.

**a.** Describe the sampling distribution of $X$. (*Hint:* This distribution was introduced in Chapter 8.)

**b.** What is the mean of this sampling distribution?

**c.** What is the standard deviation of this sampling distribution?

9.116 Refer to Exercise 9.115. Now define the statistic of interest to be the *proportion* correct, $X/10$. What is the *mean* of the sampling distribution of this statistic? Explain how you found your answer.

## Section 9.11: Skillbuilder Applet Exercises

**Statistics ◯Now™** Test yourself on these questions and explore the applets at **http://1pass.thomson.com** or on your CD.

*For Exercises 9.117 to 9.119, use the **SampleMeans** applet described in Section 9.11. It is on the CD for this book. The applet selects random samples from a bell-shaped population in which the mean $\mu = 8$ and the standard deviation $\sigma = 5$. For Exercises 9.120 and 9.121 use the related **TVMeans** applet.*

9.117 This exercise concerns possible sample means for random samples of $n = 36$.

a. Use the applet to generate 500 different random samples of $n = 36$ observations. What is the approximate shape of the resulting histogram of sample means?

b. Approximately, what were the smallest and largest values of the sample mean among the 500 samples you generated?

c. For random samples of $n = 36$, use the Empirical Rule to find the interval of values in which about 99.7% of all sample means will fall. Compare that interval to your answer for part (b).

9.118 This exercise concerns possible sample means for random samples of $n = 75$.
   a. Use the applet to generate 500 different random samples of $n = 75$ observations. What is the approximate shape of the resulting histogram of sample means?
   b. Approximately, what were the smallest and largest values among the sample means in the 500 samples you generated?
   c. For random samples of $n = 75$, use the Empirical Rule to find the interval of values into which the sample mean will fall about 99.7% of the time. Compare the interval to your answer for part (b).
   d. Describe any difference between the histograms created for part (a) of this exercise and part (a) of Exercise 9.117. (If you have not already generated the histogram for part (a) of Exercise 9.117, do so now.)

9.119 This exercise compares sample means for random samples of $n = 10$ and random samples of $n = 100$.
   a. Use the applet to generate ten different random samples of $n = 10$. List the ten sample means, and identify the lowest and highest values in the list.
   b. Use the applet to generate ten different random samples of $n = 100$. List the ten sample means, and identify the lowest and highest values in the list.
   c. Compare the results of parts (a) and (b). What is indicated about the effect of sample size on the variation among sample means for different samples?

9.120 Use the **TVMeans** applet, which selects random samples from a population modeled by using responses given by college students to a question asking how many hours they watch television in a typical week. The mean for the population is $\mu = 8.352$ hours and the standard deviation is $\sigma = 7.723$ hours.
   a. Generate 500 different random samples of $n = 5$ observations. Draw a sketch showing the shape of the histogram of sample means for random samples of $n = 5$.
   b. Generate 500 different random samples of $n = 49$ observations. Draw a sketch showing the shape of the histogram of sample means for random samples of $n = 49$.
   c. Refer to the two situations given in Section 9.6 for which the Normal Curve Approximation Rule for Sample Means holds. Explain which situation is present in the **TVMeans** applet. Also, explain how parts (a) and (b) illustrate that situation.

9.121 Refer to Exercise 9.120 about the **TVMeans** applet. Assume that the Normal Curve Approximation Rule for Sample Means holds for random samples of $n = 49$ observations.
   a. Use the Empirical Rule to find the interval of values into which the sample mean will fall about 95% of the time for random samples of $n = 49$.
   b. Use the Empirical Rule to find the interval of values into which the sample mean will fall about 68% of the time for random samples of $n = 49$.

## Chapter Exercises

9.122 Suppose you want to estimate the proportion of students at your college who are left-handed. You decide to collect a random sample of 200 students and ask them which hand is dominant. Go through the Conditions for Which the Rule for Sample Proportions Applies, and explain why the rule would apply to this situation.

9.123 Refer to Exercise 9.122. Suppose the truth is that .12, or 12%, of the students are left-handed and you take a random sample of 200 students. Use the Rule for Sample Proportions to draw a picture similar to Figure 9.3 (p. 353), showing the possible sample proportions for this situation.

9.124 According to the *Sacramento Bee* (2 April 1998, p. F5), "A 1997–98 survey of 1027 Americans conducted by the National Sleep Foundation found that 23% of adults say they have fallen asleep at the wheel in the last year."
   a. Conditions 2 and 3 needed to apply the Rule for Sample Proportions are met because this result is based on a large random sample of adults. Explain how condition 1 is also met.
   b. The article also said that based on the same survey) "37 percent of adults report being so sleepy during the day that it interferes with their daytime activities." If in truth, 40% of all adults have this problem, find the interval in which about 95% of all sample proportions should fall, based on samples of size 1027. Does the result of this survey fall into that interval?
   c. Suppose that a survey based on a random sample of 1027 college students was conducted and 25% reported being so sleepy during the day that it interferes with their daytime activities. Would it be reasonable to conclude that the population proportion of college students who have this problem differs from the proportion of all adults who have the problem (using .4 from part b)? Explain.

9.125 A recent Gallup poll found that of 800 randomly selected drivers surveyed, 70% thought that they were better-than-average drivers. In truth, in the population, at most only 50% of all drivers can be "better than average."
   a. Draw a picture of the possible sample proportions that would result from samples of 800 people from a population with a true proportion of .50.
   **b.** Would you be unlikely to see a sample proportion of .70, based on a sample of 800 people, from a population with a proportion of .50? Explain, using your picture from part (a).

9.126 Suppose that you are interested in estimating the average number of miles per gallon of gasoline your car can get. You calculate the miles per gallon for each of the next nine times you fill the tank. Suppose that in truth, the values for your car are bell-shaped, with a mean of 25 miles per gallon and a standard deviation of 1. Draw a picture of the possible sample means you are likely to get based on your sample of nine observations. Include the intervals into which 68%, 95%, and almost all of the potential sample means will fall, using the Empirical Rule from Chapter 2.

9.127 Refer to Exercise 9.126. Redraw the picture under the assumption that you will collect 100 measurements instead of only 9. Discuss how the picture differs from the one in the previous exercise.

9.128 Give an example of a scenario of interest to you for which the Rule for Sample Proportions would apply. Explain why the conditions allowing the rule to be applied are satisfied for your example.

**9.129** Suppose the population of IQ scores in the town or city where you live is bell-shaped, with a mean of 105 and a standard deviation of 15. Describe the distribution of possible sample means that would result from random samples of 100 IQ scores.

9.130 Suppose that 35% of the students at a university favor the semester system, 60% favor the quarter system, and 5% have no preference. Would a random sample of 100 students be large enough to provide convincing evidence that the quarter system is favored? Explain.

9.131 According to *USA Today* (April 20, 1998 Snapshot), a poll of 8709 adults taken in 1976 found that 9% believed in reincarnation, while a poll of 1000 adults taken in 1997 found that 25% held that belief.

a. Assuming that a proper random sample was used, verify that the sample proportion for the 1976 poll represents the population proportion to within about 1%.

b. Based on these results, would you conclude that the proportion of all adults who believe in reincarnation was higher in 1997 than it was in 1976? Explain.

**9.132** Suppose that 20% of all television viewers in the country watch a particular program.

**a.** For a random sample of 2500 households measured by a rating agency, describe the distribution of the possible sample proportions who watch the program.

**b.** The program will be cancelled if the ratings show less than 17% of households watching it in a random sample of households. Given that 2500 households are used for the ratings, is the program in danger of getting cancelled? Explain.

9.133 Use the Rule for Sample Means to explain why it is desirable to take as large a sample as possible when trying to estimate a population value.

9.134 According to the *Sacramento Bee* (2 April 1998, p. F5), Americans get an average of 6 hours and 57 minutes of sleep per night. A survey of a class of 190 statistics students at a large university found that they averaged 7.1 hours of sleep the previous night, with a standard deviation of 1.95 hours.

a. Assume that the population average for adults is 6 hours and 57 minutes, or 6.95 hours of sleep per night, with a standard deviation of 2 hours. Draw a picture illustrating how the Rule for Sample Means would apply to sample means for random samples of 190 adults.

b. Would the mean of 7.1 hours of sleep obtained from the statistics students be a reasonable value

to expect for the sample mean of a random sample of 190 adults? Explain.

c. Can the sample taken in the statistics class be considered a representative sample of all adults? Explain.

**9.135** Explain whether or not each of the following scenarios meets the conditions for which the Rule for Sample Proportions applies. If not, explain which condition is violated.

**a.** Unknown to the government, 10% of all cars in a certain city do not meet appropriate emissions standards. The government wants to estimate that percentage, so they take a random sample of 30 cars and compute the sample proportion that do not meet the standards.

**b.** The U.S. Census Bureau would like to estimate what proportion of households have someone at home between 7:00 P.M. and 7:30 P.M. on weeknights to determine whether or not that would be an efficient time to collect census data. They survey a random sample of 2000 households and visit them during that time to see whether or not someone is at home.

**c.** You want to know what proportion of days in typical years have rain or snow in the area where you live. For the months of January and February, you record whether or not there is rain or snow each day, and then you calculate the proportion.

**d.** A large company wants to determine what proportion of its employees are interested in on-site day care. For a random sample of 100 employees, the company calculates the sample proportion who are interested.

9.136 Explain whether or not you think the Rule for Sample Means applies to each of the following scenarios. If it does apply, specify the population of interest and the measurement of interest. If it does not apply, explain why not.

a. A researcher wants to know what the average cholesterol level would be if people restricted their fat intake to 30% of calories. He gets a group of patients who have had heart attacks to volunteer to participate, puts them on a restricted diet for a few months, then measures their cholesterol and calculates the mean value.

b. A large corporation would like to know the average income of its workers' spouses. Rather than going to the trouble to collect a random sample, they post someone at the exit of the building at 5:00 P.M. Everyone who leaves between 5:00 P.M. and 5:30 P.M. is asked to complete a short questionnaire on the issue. There are 70 responses.

c. A university wants to know the average income of its alumni. They select a random sample of 200 alumni and mail them a questionnaire. The university follows up with a phone call to those who do not respond within 30 days.

d. An automobile manufacturer wants to know the average price for which used cars of a particular model and year are selling in a certain state. It is able to obtain a list of buyers from the state motor vehicle division, from which a random sample of 20 buyers is selected. The manufacturer makes every effort to find out what those people paid for the cars and is successful in doing so.

9.137 In Case Study 9.1 (p. 382), we learned that about 56% of American adults actually voted in the presidential election of 1992, whereas about 61% of a random sample claimed that they had voted. The size of the sample was not specified, but suppose it was based on 1600 American adults, a common size for such studies.

a. Into what interval of values should the sample proportion fall 68%, 95%, and almost all of the time?

b. Do you think that the observed value of 61% is reasonable, based on your answer to part (a)?

c. Now suppose that the sample had been of only 400 people. Compute a standardized score to correspond to the reported percentage of 61%. Comment on whether or not you believe that people in the sample could all have been telling the truth, based on your result.

9.138 Suppose that the population of grade point averages (GPAs) for students at the end of their first year at a large university has a mean of 3.1 and a standard deviation of 0.5.

a. Draw a picture of the distribution of the sample mean GPA for a random sample of 100 students.

b. Find the probability that the sample mean will be 3.15 or greater.

**9.139** The administration of a large university wants to take a random sample to measure student opinion of a new food service on campus. It plans to use a continuous scale from 1 to 100, on which 1 is complete dissatisfaction and 100 is complete satisfaction. The administration knows from past experience with such questions that the standard deviation for the responses is going to be about 5, but it does not know what to expect for the mean. It wants to be almost sure that the sample mean is within plus or minus 1 point of the true population mean value. How large will the random sample have to be?

## Dataset Exercises

**Statistics⊕Now™** Datasets **are required** to solve these exercises and can be found at **http://1pass.thomson.com** or on your CD.

9.140 The data for this exercise are in the **GSS-93** dataset. The variable *gunlaw* is whether a respondent favors or opposes stronger gun control laws. Of the 1055 respondents, 870 are in favor. *For this exercise, assume these respondents represent a population.*

a. What is the population proportion $p$ who favor stronger gun control laws?

b. Suppose random samples of $n = 60$ are drawn from this population. Describe and draw the approximate sampling distribution for the proportion $\hat{p}$ who would favor stronger gun control laws.

c. Simulate 200 samples of size 60 from this population, and draw a histogram of the sample proportions who favor stronger gun control laws.

d. Compare your simulated distribution in part (c) with the sampling distribution that you described in part (b).

9.141 Use the **pennstate1** dataset for this exercise. The data for the variable *HrsSleep* are responses by $n = 190$ students to the question "How many hours did you sleep last night?" For this exercise, suppose they represent a population.

a. Simulate 200 samples of $n = 10$ responses for *HrsSleep.* Show the commands you used, or explain how you did the simulation.

b. Draw a histogram of the 200 sample means.

c. Describe the simulated sampling distribution you found in part (b).

d. Draw a histogram of the 200 sample medians for the samples you found in part (a).

e. Compare the sampling distributions for the sample mean and the sample median.

9.142 Use the data in the **GSS-93** dataset for this exercise. One of the questions asked was age in years. A five-number summary for age is 18, 32, 43, 58, and 89.

a. Draw a boxplot of the ages for the entire dataset.

b. Let $H$ = highest age in the sample for samples of size $n = 10$. Do any of the rules given in this chapter provide information about the sampling distribution of $H$? If so, explain.

c. Simulate 200 samples of size 10, and find $H$ for each one. Create a histogram for the 200 values of $H$.

d. Using the histogram in part (c) and any other numerical or graphical information you wish to provide, describe the sampling distribution of $H$.

9.143 Use the **Fantasy5** dataset for this exercise, with data from 2318 games of California's version of the same scheme as the Pennsylvania Cash 5 game.

a. Draw a picture of the sampling distribution of $H$, the highest number each day, and compare it to the equivalent picture for the Cash 5 game, shown in Figure 9.12 (p. 377). Do you expect them to be similar? Are they?

b. Consider each game to be a random sample of size 5, drawn without replacement from the integers 1 to 39. Which column in the dataset represents the median for each of these random samples?

c. Draw a picture of the sampling distribution of the median.

d. Find the mean of each of the 2318 samples of size 5. Draw a picture of the sampling distribution of the mean.

e. Compare the sampling distributions of the median and the mean. Which one is more spread out? Why do you think that is the case?

9.144 Use the **Cash 5** dataset for this exercise, with data from 1560 games of Pennsylvania's Cash 5 lottery game described in Section 9.10. The variables *High* and *Low* give the highest and lowest numbers picked in each game.

truth about the whole school? For instance, if we report that somewhere between 33% and 49% of all students at the school think that marijuana should be legalized, how confident can we be that we are correct?

The last of these questions is connected to the idea of a *confidence interval*, an interval of estimates computed from the sample that is likely to capture the population value.

We know from our discussion of sampling distributions in Chapter 9 that the observed value of 41% is only one of many possibilities. A different sample would probably yield a different number. In Chapter 9, we learned that if we had information about the population, we could describe the possible values we would get from taking different samples. In Chapters 10 and 11, we will work in the opposite direction. We will use sample information to say something about a fixed but unknown population value. We will do this by formulating an interval that we are fairly certain covers the true population value.

---

| *in summary* | Sampling Distributions and Confidence Intervals |

The basic idea of a **sampling distribution** is to use knowledge of the population to describe possible sample values. The basic idea of a **confidence interval** is to use knowledge of a sample to estimate a population value.

---

In this chapter, we start with the Confidence Interval (CI) Introductory Module (0), which is divided into two lessons. The first lesson will help you to understand confidence intervals, and the second lesson will show you how to construct them. In Module 1, we show how to construct a confidence interval for one proportion and go into detail about why the method works. In Module 2, we cover confidence intervals for a difference in two population proportions. The chapter concludes with a discussion of how to use confidence intervals to guide decisions about population parameters. Confidence intervals for the other three parameters, involving means, are covered in Chapter 11.

10.1 Exercises are on page 433.

# 10.2 CI Module 0: An Overview of Confidence Intervals

## Lesson 1: Understanding Confidence Intervals

To make our discussion of estimation clear, let's summarize and review some of the language and notation associated with the estimation problem.

- A **unit** is an individual person or object to be measured.
- The **population** (or **universe**) is the entire collection of units about which we would like information or the entire collection of measurements that we would have if we could measure the whole population. The popula-

tion might be imagined rather than actual, such as the population of out-comes if 100,000 coin flips were to be conducted.

- The **sample** is the collection of units that we will actually measure or the collection of measurements that we will actually obtain.

- The **sample size**, denoted by the letter $n$, is the number of units or measurements in the sample.

- A **population parameter** is a fixed number associated with a population, and in the context of confidence intervals, it is unknown and we would like to estimate it. An example is $p$, the proportion of a *population* with a particular characteristic.

- **Sample statistic, sample estimate,** and **point estimate** are all synonyms for a number computed from a sample that can be used to estimate the corresponding population parameter. An example is $\hat{p}$, the proportion of a *sample* with a particular characteristic, found as the number with the characteristic divided by the sample size $n$.

Finally, remember that we would ideally like to have a *randomly selected sample* from the population, but in Chapters 3 and 4, we learned that is not al-ways possible and that sometimes we can settle for less. Let's review the crucial criterion for statistical inference to be valid.

---

*definition* | The **Fundamental Rule for Using Data for Inference** is that available data can be used to make inferences about a much larger group *if the data can be con-sidered to be representative with regard to the question(s) of interest.*

---

*thought question 10.1* Each day, Maria gets dozens of e-mail messages. She keeps track of what proportion of the messages are spam or other junk and what proportion are interesting. Suppose she got 50 messages yesterday, and 20 of them were interest-ing. If the collection of messages on a single day is considered to be a sample of all e-mail messages she ever receives, explain the meaning of each of the following defi-nitions in the context of this example, and give numerical values where possible: *unit, population, sample, sample size, population parameter, sample statistic* (or *sample estimate*). Discuss whether you think the Fundamental Rule for Using Data for Inference would allow Maria to draw conclusions about the population proportion based on the sample proportion.*

Before learning the specifics of how to construct confidence intervals, it is important to understand why we use intervals to estimate a single number and what we mean by *confidence*. In this lesson, these ideas are covered; in the next lesson, we learn the details of computing confidence intervals for the big five parameters.

---

**\*HINT:** A "unit" is a single e-mail. For the last part, consider whether some days of the week are more likely than others to bring junk e-mail messages.

# The Concept of a Confidence Interval as an Interval Estimate

As we have already mentioned, one of the most common methods for using sample data to learn something about a population is to construct a *confidence interval* estimate for the unknown value of a population parameter.

| | |
|---|---|
| *definition* | A **confidence interval** is an interval of values computed from sample data that is likely to include the true population value. |

You might wonder why we need an interval of values when we are trying to estimate a single number: the true value of the population parameter. The answer is that we have incomplete and therefore possibly inaccurate information about the parameter's value unless we have measured the entire population. When all we have measured is a sample, which is usually the case in research, we must rely on the *sample estimate* as our best guess for the value of the population parameter. But we know that it isn't 100% accurate, so we must figure out how far off it might be in one direction or the other.

Remember that the term **point estimate** is sometimes used as a synonym for the *sample estimate* (also called the *sample statistic*). That's because it is a single number or *point* on the number line. In contrast, the term **interval estimate** is used as a synonym for *confidence interval*, particularly in discussing the numerical end result in a given situation. Even though it is an interval of values, an interval estimate is attempting to estimate one single, fixed value. Using an interval is important because we would like to be fairly confident that the true value is indeed covered by the possibilities given in the interval estimate. With a single number estimate, we would have no way to gauge how accurate it might be.

A confidence interval is always accompanied by a **confidence level,** which tells us how likely it is that the interval estimate actually captures the truth we are seeking. The most common confidence level used in research and in the media is 95%. Later in this chapter, we will learn how to determine an appropriate interval for any specified confidence level.

**Example 10.1**

**Teens and Interracial Dating: Case Study 1.3 Revisited**  In Case Study 1.3, we described a 1997 *USA Today*/Gallup poll of teenagers. One finding was that 57% of the 496 teens in the sample who date said that they had been out on a date with someone of another race or ethnic group. What can we say about this issue for the entire population of teenagers who date? In Case Study 1.3, we learned that the percentage of all teenagers who date in the United States who would say that they have dated interracially or dated someone of another ethnic group is likely to be between 52% and 62%. We formed this 95% confidence interval using the *sample estimate* of 57% and then adding and subtracting the *margin of error* of about 5%, which was reported with the poll. We will learn why this method works, as well as how to find the margin of error, in CI Module 1. ∎

## Interpreting the Confidence Level

Here is a typical confidence interval statement from a study of allergies in adult Americans that we will learn about in Example 10.2 in the next lesson: "Based on this sample, we are 95% confident that somewhere between 33% and 39% of all Americans suffer from allergies." Notice that this statement conveys a range of estimates of a population value as well as the likelihood that the interval does indeed capture the population value. The term *confidence interval* is used to describe the interval itself (33% to 39%), and the 95% used in the statement is called the *confidence level*.

The term **confidence level** is used to describe the chance that an interval actually contains the true population value in the following sense. Most of the time (quantified by the confidence level), intervals that are computed in this way will capture the truth about the population, but occasionally, they will not. In any given instance, the interval either captures the truth or it does not, but we will never know which is the case. Therefore, our confidence is in the *procedure*—it works most of the time—and the "confidence level" or "level of confidence" is the percentage of the time we expect it to work. For instance, a 95% confidence level means that confidence intervals computed by using the same procedure will include the true population value for 95% of all possible random samples from the population. In practice, we can't know for sure whether our confidence interval actually does capture the population value. We can feel fairly certain that it does because the procedure that we use will be successful most of the time.

---

*in summary*　　　Interpreting the Confidence Level

- For a confidence interval, the **confidence level** is the probability that the procedure that is used to determine the interval will provide an interval that includes the population parameter.
- The confidence level has a relative frequency probability interpretation: If we consider all possible randomly selected samples of the same size from a population, the *confidence level* is the fraction or percentage of those samples for which the confidence interval includes the population parameter.

*Note:* It is commonplace to express the confidence level as a percentage.

---

The most common confidence level that researchers use is 95%. In other words, researchers define "likely to include the true population value" to mean that they are 95% certain that the procedure will work. So they are willing to take a 5% risk that the interval does not actually include the true value. Sometimes researchers employ only 90% confidence, and occasionally, they use higher confidence levels like 98% or 99%.

It would be wonderful to be 100% confident that a confidence interval contains the population value. Unfortunately, 100% confidence is impossible unless either we sample the entire population or we provide an absurdly wide interval of estimates. For instance, if we said that the percentage of American college students who think marijuana should be legalized is somewhere between 0% and 100%, we would have 100% confidence that we're right. The cost of a high confidence level might be that the interval is so wide that it is not informative.

Be careful when giving information about a specific confidence interval computed from an observed sample. The confidence level expresses only how often the confidence interval procedure works in the long run. It does not tell us the probability that a specific interval includes the population value. For example, suppose that a 95% confidence interval estimate of the percentage of all Americans suffering from allergies is 33% to 39%. This interval either does or does not include the true unknown value of the population percentage, but we can never know which is the case because we can't know the population value.

> ***thought question* 10.2** Explain in your own words what it means to say that we have 95% confidence in the interval estimate. Then give an example of something you do in your life that illustrates the same concept: You follow the same procedure each time, and it either works (most of the time) or does not work to produce the desired result. What confidence level would you assign to the procedure in your example; that is, what percentage of the time do you think it produces your desired result?*

## Lesson 2: Computing Confidence Intervals for the Five Scenarios

The same general format applies to computing any confidence interval for any of the five parameters of interest. Remember that the five scenarios include estimating:

- One population proportion, $p$
- The difference in two population proportions for independent samples, $p_1 - p_2$
- One population mean, $\mu$
- The population mean of paired differences, $\mu_d$
- The difference in two population means for independent samples, $\mu_1 - \mu_2$

Confidence intervals for the first two parameters, involving proportions, are covered in this chapter. Confidence intervals for the remaining three parameters, involving means, are covered in Chapter 11.

In each case, the goal is to estimate the *population parameter* with an interval of values that we are fairly confident covers it. In each case, we have sample

---

**\*HINT:** What is a task you frequently perform that you can't always do successfully?

data from one sample (for $p$, $\mu$, or $\mu_d$) or two independent samples (for $p_1 - p_2$ or $\mu_1 - \mu_2$.)

Let's start with the end result: the general formula. We will include further details for each of the scenarios in the individual modules. For readers who want to know how to derive the formula, we will include details in the modules on one proportion and one mean. The details for the other three situations are similar but more tedious to write out. Here is the general format.

---

*definition*    A **confidence interval** or **interval estimate** for any of the five parameters can be expressed as

Sample estimate $\pm$ Multiplier $\times$ Standard error

The **multiplier** is a number based on the confidence level desired and determined from the standard normal distribution (for proportions) or Student's $t$-distribution (for means). Details are provided in the individual modules.

---

Let's look at each part of the formula:

- The *sample estimate,* also called the *sample statistic,* is the sample proportion, difference in two sample proportions, sample mean, and so on. For example, if the parameter of interest is a population proportion $p$, then the sample estimate is the sample proportion $\hat{p}$.

- The *multiplier* determines the amount of confidence we will have in the result. For instance, for a 95% confidence interval for a proportion, the appropriate multiplier is 1.96, sometimes rounded off to 2. For a 68% confidence interval for a proportion, the multiplier is 1.0. The reason for this will be explained in Module 1 on one proportion, in which you will also learn how to find the multiplier for any desired confidence level. You choose the level of confidence you want, which then determines the multiplier. The choice is up to you, but the higher the confidence level, the wider the interval will be.

- The *standard error* is the **standard error of the sample statistic,** which is an *estimate* of the standard deviation of the sample statistic. An example is the *standard error of $\hat{p}$,* which is $\sqrt{\hat{p}(1 - \hat{p})/n}$. The standard error formulas for the five situations are given in the summary table at the end of Chapter 9, on pages 382–383.

You can see that once we have sample data, we don't need all of the individual numbers to compute an interval estimate for the population parameter; we need only a few summary measures. The summary measures we need from the sample(s) are as follows:

- For situations involving proportions, the *sample size(s)* and the *sample proportion(s)*

- For situations involving means, the *sample size(s),* the *sample mean(s)* and the *sample standard deviation(s)*

### 95% Confidence Interval for One Proportion as an Illustration

To illustrate the general formula, here is how we would find a 95% confidence interval for one proportion:

- The sample estimate is the sample proportion, denoted by $\hat{p}$.
- The appropriate multiplier is 1.96, which we will round off to 2.
- The standard error of $\hat{p}$ (from Chapter 9) is $\sqrt{\hat{p}(1-\hat{p})/n}$.

So a 95% confidence interval for a population proportion can be written as follows:

Sample estimate $\pm$ Multiplier $\times$ Standard error

$$\hat{p} \pm 2 \times \sqrt{\frac{\hat{p}(1-\hat{p})}{n}}$$

Let's look at an example of computing an interval estimate using this formula.

**Example 10.2**

**Statistics ◯ Now™**

Watch a video example at **http:// 1pass.thomson.com** or on your CD.

**The Pollen Count Must be High Today** In April 1998, the Marist Institute for Public Opinion surveyed 883 randomly selected American adults about allergies. According to a report posted at the Institute's website (www.mipo .marist.edu), 36% of the sample answered "yes" to the question "Are you allergic to anything?" Therefore, the sample proportion who said yes is $\hat{p} = .36$. We will use the sample information to calculate a 95% confidence interval estimate of the population parameter $p$ = proportion of all American adults who are allergic to something.

Values for the parts of the formula Sample estimate $\pm$ Multiplier $\times$ Standard error are as follows:

- *Sample estimate* $= \hat{p} = .36$
- *Multiplier* $= 2$ (to achieve 95% confidence)
- *Standard error* $= \sqrt{\dfrac{\hat{p}(1-\hat{p})}{n}} = \sqrt{\dfrac{.36(1-.36)}{883}} = .016$

The 95% confidence interval is $.36 \pm 2 \times .016$, which is $.36 \pm .032$ or .328 to .392 (about 33% to 39%).

*Interpretation:* The confidence interval, .328 to .392, estimates the proportion of *all* American adults who have an allergy. In percentage terms, this is about 33% to 39%. The confidence level (95%) describes our confidence in the procedure used to determine the interval. In the long run, the procedure will be correct about 95% of the time. ■

### What Determines the Width of the Interval?

The width of a confidence interval is the difference between the lower and upper values of the interval. A relatively narrow confidence interval gives a more precise estimate of the population value than does a relatively wide confidence interval. The amount that we add and subtract to the sample estimate to create the interval determines its width. Notice from Example 10.2 that three factors affect the width of a confidence interval for a population proportion:

1. *The sample size, n.* This will always be part of the denominator of the standard error; therefore, in all situations, the larger the sample size(s), the narrower the interval will be. This makes sense because we have more information in a large sample than in a small one and therefore should be able to obtain a sample estimate that is closer to the truth for a large sample than for a small one.

2. *The sample proportion, $\hat{p}$.* Within this context, $\hat{p}$ is a measure of the natural variability among the units. If the proportion is close to either 1 or 0, most individuals have the same trait or opinion, so there is little natural variability, and the standard error is smaller. Notice that in the extreme, if $\hat{p}$ is 0 or 1, the standard error is 0, and the confidence interval has no width. As $\hat{p}$ gets closer to .5 from either direction, the standard error gets larger. It reaches its maximum when $\hat{p}$ is .5.

3. *The multiplier* 2. This is determined by the 95% confidence level. Later in the chapter, we will learn how to find other multipliers and will learn that the larger the multiplier, the more confidence we have that the interval covers the truth. The price of added confidence is that the interval will be wider.

The same three factors will influence the amount that is added and subtracted to the sample estimate to form a confidence interval for all five scenarios. The second factor, which in this case is the sample proportion, provides a measure of natural variability among the units. When the parameter involves means, this factor will be determined by the sample standard deviation(s).

10.2 Exercises are on page 433.

*in summary* | **Three Factors That Affect the Width of a Confidence Interval**

The three factors that determine the width of a confidence interval are as follows:

1. The sample size(s). Larger samples produce more narrow intervals.
2. The natural variability among units in the population(s). More natural variability results in wider intervals.
3. The level of confidence desired. The more confidence we want to have in the result, the wider the interval needs to be.

# 10.3 CI Module 1: Confidence Interval for a Population Proportion

The material in this module is divided into three lessons. If you are concerned only with how to compute a confidence interval for a proportion, you need only study Lesson 1, in which we explain how to implement the general formula to find a confidence interval for a population proportion for any confidence level. However, we hope that you will continue with Lesson 2, in which we explain why the formula works. In Lesson 3, we reconcile the formulas in this chapter with the confidence interval formula in Chapter 3, in which we added and subtracted a margin of error of $1/\sqrt{n}$ to the sample proportion to find an approximate 95% confidence interval for a population proportion. We also explain why the media reports the margin of error that was introduced in Chapter 3.

## Lesson 1:  Details of How to Compute a Confidence Interval for a Population Proportion

The confidence interval method described in this section is applicable to a surprisingly wide range of interesting research questions. The scenarios are similar to the ones that we encountered when we studied binomial random variables in Chapter 8. Here are two common settings in which we might want to estimate a population proportion or probability $p$:

1. *A population exists, and we are interested in knowing what proportion of it has a certain trait, opinion, characteristic, response to a treatment, and so on.* Some typical research questions for this setting are as follows:

   - What proportion of drivers are talking on a cell phone at any given moment?

   - What proportion of the population of married women have had an extramarital affair?

   - What proportion of a population would quit smoking if they were to wear a nicotine patch for eight weeks?

   - What proportion of new drivers have an accident within the first year of driving?

2. *A repeatable situation exists, and we are interested in the long-run probability of a specific outcome.* Some typical research questions for this setting are as follows:

   - What is the probability that a new fertility procedure will be successful for a randomly selected couple who tries it?

   - What is the probability that someone who has lived to be 80 will continue to live to be 90?

   - What is the probability that a randomly selected television of a certain model will fail before the warranty period is over?

   - What is the probability that a cash-filled wallet will be returned if it is found?

We can find a confidence interval with specified confidence level to answer each of these questions by collecting appropriate data from the population or repeatable situation.

CI Module 1: p

# How to Compute the Confidence Interval

In Module 0, we learned that a confidence interval for any of our five situations is given by the formula

Sample estimate ± Multiplier × Standard error

In the case of a confidence interval for one proportion:

*Sample estimate* = the sample proportion = $\hat{p}$

*Standard error* = s.e.$(\hat{p})$ = $\sqrt{\dfrac{\hat{p}(1 - \hat{p})}{n}}$

## Finding the Multiplier *z\**

The *multiplier* is denoted by *z\** and is found by using the standard normal distribution. Values of the multiplier for the most common confidence levels used by researchers are shown in Table 10.1. The multipliers are shown to three decimal places, but in practice they often are rounded to two or even fewer decimal places, especially when this rounding does not change the resulting confidence interval or confidence level by much. For example, the multiplier for a 95% confidence interval is 1.96, which is sometimes rounded to 2.0. The result is that the actual confidence level is 95.45% instead of 95%, a minor difference.

**Table 10.1** Confidence Intervals for a Population Proportion

| Confidence Level | Multiplier (*z\**) | Confidence Interval |
|---|---|---|
| 90 | 1.645 or 1.65 | $\hat{p}$ ± 1.65 standard errors |
| 95 | 1.960, sometimes rounded to 2 | $\hat{p}$ ± 2 standard errors |
| 98 | 2.326 or 2.33 | $\hat{p}$ ± 2.33 standard errors |
| 99 | 2.576 or 2.58 | $\hat{p}$ ± 2.58 standard errors |

**Example 10.3**

**Statistics Now™**

Watch videos of similar examples at **http://1pass.thomson.com** or on your CD.

For software help, download your Minitab, Excel, TI-83, SPSS, R, and JMP manuals from **http://1pass .thomson.com**, or find them on your CD.

**Is There Intelligent Life on Other Planets?** In a 1997 Marist Institute survey of 935 randomly selected Americans, 60% of the sample answered "yes" to the question "Do you think there is intelligent life on other planets?" (*Source:* www .mipo.marist.edu). Let's use this sample estimate to calculate a 90% confidence interval for the proportion of all Americans who believe that there is intelligent life on other planets.

Sample estimate = $\hat{p}$ = .60   (60% expressed as a proportion)

s.e.$(\hat{p})$ = $\sqrt{\dfrac{.6(1 - .6)}{935}}$ = .016

Multiplier = 1.65   (from Table 10.1 for .90 confidence level)

A 90% confidence interval for the population proportion is

Sample estimate ± Multiplier × Standard error

.60 ± 1.65 × .016

.60 ± .026

*Interpretation:* With 90% confidence, we can say that in 1997, the proportion of Americans who believed that there is intelligent life on other planets was in the range .60 ± .026 (or 60% ± 2.6%), or 57.4% to 62.6%.

We can increase our confidence that the interval covers the truth by increasing the confidence level. We pay a price, because the multiplier will be larger and the interval will be wider. For instance, for 98% confidence, the interval is *sample proportion ± 2.33 standard errors.* The calculation is .60 ± 2.33 × .016, which is .60 ± .037. When converted to percentages, the 98% confidence interval is 60% ± 3.7%. We can be fairly sure that in 1997, between 56.3% and 63.7% of Americans believed that there is life on other planets. Because the entire interval is above 50%, we can also state with high confidence that in 1997, a majority of Americans believed that there is intelligent life on other planets. ■

### Finding $z^*$ for Other Levels of Confidence

In general, the multiplier $z^*$ is the standardized score such that the area between $-z^*$ and $+z^*$ under the standard normal curve corresponds to the desired confidence level. Figure 10.2 illustrates this relationship. In practice, it would be unusual for anybody to use a confidence level other than the four shown in Table 10.1. Theoretically, however, any confidence level could be specified, and the standard normal curve would be used to determine the associated $z^*$ multiplier.

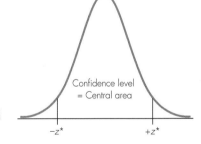

**Figure 10.2** ■ The relationship between the confidence level and the multiplier $z^*$. The confidence level is the probability between $-z^*$ and $z^*$ in the standard normal curve

**Example 10.4**

**50% Confidence Interval for Proportion Believing That Intelligent Life Exists Elsewhere**  Suppose we want a 50% confidence interval for the 1997 proportion that believed that intelligent life exists elsewhere. Figure 10.3 illustrates the problem that we have to solve to find the $z^*$ value for a 50% confidence level.

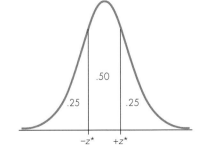

**Figure 10.3** ■ Finding $z^*$ for the 50% confidence level

To use the Standard Normal Curve table in Appendix A.1, we observe from Figure 10.3 that if the area between $-z^*$ and $+z^*$ is .50, the area to the right of $z^*$ is .25, and so the area to the left of $z^*$ is .75. From Appendix A.1, we learn that the $z$-value for which the cumulative probability is .75 is about $z^* = 0.67$. This means that a 50% confidence interval for a population proportion is *sample proportion* $\pm 0.67$ *standard errors,* which in this example is .60 $\pm 0.67 \times$ .016, or .60 $\pm$ .011. Notice that this interval is narrower than the 90% and 98% confidence intervals computed previously. Using a lower confidence level produces a narrower interval. ■

*in summary*  ## Formula for a Confidence Interval for a Population Proportion $p$

A formula for the **confidence interval for a population proportion** is

$$\hat{p} \pm z^* \sqrt{\frac{\hat{p}(1 - \hat{p})}{n}}$$

where

- $\hat{p}$ is the sample proportion.
- $z^*$ denotes the multiplier.
- $\sqrt{\dfrac{\hat{p}(1 - \hat{p})}{n}}$ is the **standard error of the sample proportion.**

Table 10.1 has $z^*$ multipliers for the most commonly used confidence levels. The multiplier $z^*$ is such that the area under the standard normal curve between $-z^*$ and $z^*$ corresponds to the confidence level.

## Conditions for Using the Formula

As has been mentioned, the research questions that can be addressed by using a confidence interval for a population proportion are ones that arise when data are collected in ways similar to a binomial experiment. A simple way to make sure the appropriate conditions are satisfied is to take a random sample from a large population. In addition, the sample size must be large enough that the normal curve approximation rule for proportions applies. This will become evident when we explain in Lesson 2 why the formula works. Therefore, the conditions that must be met to use the formula for a confidence interval for one population proportion are as follows:

1. The sample is a randomly selected sample from the population.
2. Both $n\hat{p}$ and $n(1 - \hat{p})$ should be at least 10 (although some authors say that these quantities need only be at least 5).

The first condition is an idealistic one that is sometimes difficult to achieve. A more lenient condition is given by the Fundamental Rule for Using Data for

Inference, which we repeat here: Available data can be used to make inferences about a much larger group *if the data can be considered to be representative with regard to the question(s) of interest.* In the context of this chapter, a confidence interval based on a sample proportion can be used to estimate the population proportion if the individuals in the sample can be considered to be representative of the individuals in the population for the trait, opinion, etc. of interest.

**Example 10.5** | **College Men and Ear Pierces**  During the 1990s, it became more common for American men, particularly young men, to wear an earring. In class surveys done between 1996 and 1998 in statistics classes at Penn State, 141 of 565 college men indicated that they had at least one ear pierce. The sample proportion with an ear pierce was $\hat{p} = 141/565 = .25$. The amount we add and subtract for a 95% confidence level is

$$\text{Multiplier} \times \text{standard error} = 2\sqrt{\frac{.25(1 - .25)}{565}} = .036$$

A 95% confidence interval for the population proportion is $.25 \pm .036$. This is .214 to .286, which in percentage terms is 21.4% to 28.6%.

There is an important question that should be asked about this sample: What population does it represent? Because all 565 men were from the same school, the sample may not represent all college men in the United States. It may not even represent all Penn State men because the surveyed classes were taken only by students in the liberal arts and social sciences. Perhaps the population represented is Penn State men in the liberal arts and social sciences.

Another difficulty here is that the sample was a convenience sample that simply included the men who happened to be in certain classes. In the theoretical development of a 95% confidence interval, the assumption is that we have a random sample. Rather than attempting to use the sample to estimate a population proportion, maybe we should only report that 25% of these 565 men had at least one ear pierce. It can be argued, however, that even though the sample was not randomly selected, it is likely to be representative of a larger group and therefore can be used to make inferences about that group. ∎

### Conditions When the Parameter Is a Long-Run Probability *p*

In settings for which the parameter *p* is a long-run probability instead of a proportion for an existing population, the first condition can be rewritten as follows:

*Alternative Condition 1:*  The sample proportion $\hat{p}$ is the proportion of times a specified outcome occurs in *n* repeated independent trials with fixed probability *p* that the specified outcome will occur.

The sample size requirement remains as stated in Condition 2; both $n\hat{p}$ and $n(1 - \hat{p})$ must be at least 10. Notice that this is equivalent to having at least 10 trials with the specified outcome and at least 10 trials without it. The next example illustrates this situation.

**Example 10.6** | **Would You Return a Lost Wallet?**  In a study that was repeated in various countries, staff from the magazine *Reader's Digest* planted "lost" wallets containing the equivalent of about $50, a name, a local address, a phone number, and family photos to find out whether people who found them would return them. For the study done in Canada, 120 wallets were "lost," and 77 of them were returned intact (Kiener, 2005), so the sample proportion of wallets returned is $\hat{p} = 77/120 = .64$. The sample size requirement is met, because there are at least 10 trials in which the wallet was returned and at least 10 in which it was not. The parameter of interest is the probability $p$ that a similarly "lost" wallet would be returned intact if lost in Canada. Let's find a 95% confidence interval estimate for $p$. The appropriate multiplier is 1.96, which we will round to 2.0, so the interval is

Sample estimate $\pm$ Multiplier $\times$ Standard error

$$\hat{p} \pm 2 \times \sqrt{\frac{\hat{p}(1-\hat{p})}{n}}$$

$$.64 \pm 2 \times \sqrt{\frac{(.64)(.36)}{120}}$$

$$.64 \pm 2 \times .044$$

$$.64 \pm .088$$

$$.55 \text{ to } .73$$

Therefore, we can be 95% confident that the probability of a lost wallet being returned under the conditions used in the experiment is between .55 and .73. ■

---

***thought question* 10.3** Suppose the legislature in a particular state wanted to know what proportion of students graduating from the state university last year were permanent residents of the state. The university had information for all students showing that 3900 of the 5000 graduates were state residents. Is a confidence interval appropriate for this situation? If so, compute the appropriate interval. If not, explain why not.*

---

**MINITAB *tip*** | **Calculating a Confidence Interval for a Proportion**

- To compute a confidence interval for a proportion, use **Stat>Basic Statistics>1 Proportion.** (This procedure is not in versions earlier than Version 12.)

- If the raw data are in a column of the worksheet, specify that column. If the data have already been summarized, click on "Summarized Data," and then specify the sample size and the count of how many observations have the characteristic of interest.

***Note:*** To calculate intervals in the manner described in this chapter, use the ***Options*** button, and click on "Use test and interval based on normal distribution." Note also that the confidence level can be changed by using the ***Options*** button.

---

**\*HINT:** Do the 5000 graduates constitute a sample or a population?

# Lesson 2: Understanding the Formula

## Developing the 95% Confidence Interval

Let's illustrate why the formula works by developing a 95% confidence interval and then generalizing to other confidence levels. Remember from the Empirical Rule introduced in Chapter 2 that about 95% of all values under a normal curve fall within two standard deviations of the mean.

The key to developing a 95% confidence interval for a proportion is the Normal Curve Approximation Rule for Sample Proportions described in Chapter 9. That rule describes the distribution of the sample proportions that would occur if we took many different random samples of the same size from a population. By this rule, if the sample size is sufficiently large, the distribution of possible values of $\hat{p}$ is approximately a normal curve. For this normal distribution,

- The mean is $p$, the population proportion.
- The standard deviation is $\sqrt{\dfrac{p(1-p)}{n}}$.

Because the distribution of possible sample proportions has a normal curve, we can use the Empirical Rule to make the following statement: *In about 95% of all samples, the sample proportion $\hat{p}$ will fall within 2 standard deviations of $p$, the true population proportion, so for about 95% of all samples,*

$$-2 \text{ standard deviations} < \hat{p} - p < 2 \text{ standard deviations}$$

This means that in about 95% of all samples, the difference between the sample proportion and the true population proportion is less than 2 standard deviations.

We have one difficulty. The standard deviation formula involves $p$, the true population proportion, a value that we don't know. The obvious solution is to put $\hat{p}$, the sample proportion, in the place of $p$ in the standard deviation formula. As we discussed in Chapter 9, when this is done, many researchers use the term **standard error** to describe the result. Let's review that terminology.

---

*definition*　　The **standard error of a sample proportion,** denoted by s.e.$(\hat{p})$, is

$$\text{s.e.}(\hat{p}) = \sqrt{\dfrac{\hat{p}(1-\hat{p})}{n}}$$

---

If we assume that the standard error of $\hat{p}$ is a reasonably close approximation to the standard deviation of $\hat{p}$, we can use the standard error in place of the standard deviation. So for approximately 95% of all random samples from a population, the difference between the sample proportion and the population proportion is described by the mathematical inequality

$$-2 \text{ standard errors} < \hat{p} - p < 2 \text{ standard errors}$$

This mathematical inequality says that in about 95% of all samples, the difference between the sample proportion and the population proportion is less than 2 standard errors. With two steps of algebra, this inequality can be changed to

$$\hat{p} - 2 \text{ standard errors} < p < \hat{p} + 2 \text{ standard errors}$$

This holds for about 95% of all samples and thus provides the formula for the 95% confidence interval.

## An Intuitive Explanation of 95% Confidence

You may recall that in Chapter 3, we learned how to compute a conservative 95% confidence interval for a sample proportion or percent. The formula that we used was

Sample proportion $\pm$ Margin of error

where the margin of error is $1/\sqrt{n}$ and often is reported by the media along with the results of a survey. In the next lesson, we will see how to reconcile this formula with the one we have developed for a 95% confidence interval, but for now, just assume that they are similar formulas and that the following comments apply to both. They are easier to write in terms of margin of error.

The margin of error (or 2 $\times$ standard error) has these characteristics:

- The difference between the sample proportion and the population proportion is *less* than the margin of error *about 95% of the time,* or for about 19 of every 20 sample estimates.

- The difference between the sample proportion and the population proportion is *more* than the margin of error *about 5% of the time,* or for about 1 of every 20 sample estimates.

In other words, for most sample estimates, about 95% of them, the actual error is quite likely to be smaller than the margin of error. Occasionally, about 5% of the time, the error might be larger than the margin of error. Unfortunately, we never know the actual amount of error in a particular estimate. We know only that most of the time, the actual error is less than the reported margin of error.

If we use the margin of error to form an interval around the sample estimate, there are two things that can happen:

- If, by the luck of the draw, we have one of the sample estimates for which the actual error is *smaller* than the margin of error, then the sample estimate is close enough to the true population value that the interval "sample estimate $\pm$ margin of error" will include the true value. Because at least 95% of all sample estimates have this characteristic, our method will work at least 95% of the time.

- If we happen to have one of the few sample estimates for which the actual error is *larger* than the margin of error, then the interval "sample estimate $\pm$ margin of error" will *not* include the true value. This will happen with no more than 5% of possible sample estimates.

## Developing the Interval for Other Levels of Confidence

There is nothing special about the 95% level of confidence except that the Empirical Rule provided us with the knowledge of what multiplier to use (2.0). As illustrated in Figure 10.2 on page 413, for any confidence level we can find the appropriate multiplier $z^*$ such that the desired percentage of the standard normal curve lies between $-z^*$ and $+z^*$. Call the confidence level $C\%$. To find $z^*$ notice that the standard normal curve can be partitioned into three sections: the area between $-z^*$ and $z^*$ is $C\%$, the area below $-z^*$ and the area above $z^*$ are each half of what's left, or $(100 - C\%)/2$. Therefore, $-z^*$ is the value that has area $(100 - C\%)/2$ below it, and can easily be read from Table A.1 in the Appendix. For instance, for 90% confidence level the area between $-z^*$ and $z^*$ is 90% and the area below $-z^*$ is 10%/2 or .05. Looking up .05 in the body of Table A.1, we find that $-z^*$ is between $-1.64$ (area is .0505) and $-1.65$ (area is .0495). Therefore, we use a multiplier of 1.645 for a 90% confidence interval.

Approximately $C\%$ of all samples would result in a value of $\hat{p}$ such that

$$-z^* \text{ standard errors} < \hat{p} - p < z^* \text{ standard errors}$$

Using algebra, we can restate the inequality by saying that approximately $C\%$ of all samples would result in the following being true:

$$\hat{p} - z^* \text{ standard errors} < p < \hat{p} + z^* \text{ standard errors}$$

This inequality provides the confidence interval formula. Notice that it says that for $C\%$ of all samples, $p$ will be contained in the interval from $(\hat{p} - z^*$ standard errors) to $(\hat{p} + z^*$ standard errors), which is the same as $\hat{p} \pm z^*$ standard errors, the general formula for a confidence interval for a proportion.

---

***thought question* 10.4**  As we noted toward the end of Section 9.4 s.e.$(\hat{p})$ is an estimate of s.d.$(\hat{p})$, and it is generally very close. Use the Empirical Rule from Chapters 2 and 8 and the Normal Curve Approximation Rule for Sample Proportions from Section 9.4 to draw a picture of possible sample proportions, using s.e.$(\hat{p})$ in place of s.d.$(\hat{p})$. Show what percentage of sample proportions you expect to fall into the following ranges:

Population proportion $\pm$ s.e.$(\hat{p})$
Population proportion $\pm$ 2 s.e.$(\hat{p})$
Population proportion $\pm$ 3 s.e.$(\hat{p})$

Then identify what connection margin of error has with the value "95%" in this picture.

---

## Lesson 3: Reconciling Margin of Error and 95% Confidence Intervals

In Module 0, we learned that a confidence interval for any of our five situations is given by the formula

Sample estimate $\pm$ Multiplier $\times$ Standard error

We also learned that in the case of a 95% confidence interval for one proportion,

$$\text{Multiplier} \times \text{standard error} = 2\sqrt{\frac{\hat{p}(1 - \hat{p})}{n}}$$

However, in Chapter 3, we learned that an approximate 95% confidence interval for a population proportion is given by the formula

Sample estimate $\pm$ Margin of error

---

**\*HINT:** Example 9.9 (p. 351) and the Thought Question accompanying it can be used for guidance.

where the margin of error was given as $1/\sqrt{n}$. How do we reconcile these different formulas?

First, the term **margin of error** is used to describe the Multiplier × Standard error in a 95% confidence interval for a population proportion. In this situation, it is equivalent to write "± Multiplier × Standard error" and "± Margin of error." But the margin of error is not always $1/\sqrt{n}$.

The margin of error that is often reported in the news is actually a conservative version of the real margin of error. It is an approximation that works best when the true population proportion $p$ is close to .5, and it is too large if $p$ is close to 0 or 1. Therefore, we call $1/\sqrt{n}$ the **conservative margin of error** for a 95% confidence interval. The formula given next describes a more accurate margin of error for estimating a proportion. Notice that if you use $\hat{p} = .5$ in the formula, you will get the conservative margin of error.

*formula*  **General Format for the Margin of Error**

For a 95% confidence level, a formula for the approximate **margin of error for a sample proportion** is

$$\text{Margin of error} \approx 2\sqrt{\frac{\hat{p}(1 - \hat{p})}{n}}$$

Recall that in Section 9.4, we learned that the standard error of a sample proportion is

$$\text{s.e.}(\hat{p}) = \sqrt{\frac{\hat{p}(1 - \hat{p})}{n}}$$

Thus, the "95% margin of error" is simply two standard errors, or 2 s.e.$(\hat{p})$. The margin of error is equivalent to Multiplier × Standard error, where the "Multiplier" is 2. When $p = .5$, this is equivalent to $1/\sqrt{n}$.

By now, you should realize that the formula for an approximate 95% confidence interval for a proportion can be written in several equivalent ways. When $\hat{p}$ is close to .5, the 95% margin of error is approximately $1/\sqrt{n}$, and all of the formulas are equivalent to the one learned in Chapter 3.

*in summary*  Various Forms of a 95% Confidence Interval for a Proportion

All of the following are equivalent ways of writing the formula for an **approximate 95% confidence interval for a proportion $p$:**

Sample estimate ± Margin of error

Sample estimate ± 2 standard errors

$$\hat{p} \pm 2\sqrt{\frac{\hat{p}(1 - \hat{p})}{n}}$$

and when $\hat{p}$ is close to .5, these formulas are essentially equivalent to

$$\hat{p} \pm 1/\sqrt{n}$$

Otherwise, this last formula produces a **conservative 95% confidence interval for $p$.**

**Example 10.7**

**Winning the Lottery and Quitting Work**  What would you do for the rest of your life if you won a lot of money in a lottery? In a 1997 poll conducted by the Gallup Organization, one of the questions was "If you won 10 million dollars in the lottery would you continue to work, or would you stop working?" The results were reported at the Gallup Organization's website (www.gallup.com).

Surprisingly, 59% of the 616 employed respondents answered that they would continue working, 40% said that they would stop working, and 1% had no opinion. The website article also gave this information about the poll:

> The current results are based on telephone interviews with a randomly selected sample of 1,014 adults, conducted August 22–25, 1997. Among this group, 616 are employed full-time or part-time. For results based on this sample of "workers," one can say with 95% confidence that the error attributable to sampling could be plus or minus 4 percentage points.

Gallup describes the margin of error with the phrase "could be plus or minus 4 percentage points," but the phrase "could be *as large as* plus or minus 4 percentage points" would have been more informative. The margin of error is a likely upper limit on the sampling error. Here, there is a 95% chance that the sampling error is actually smaller than 4 percentage points.

Let's find a 95% confidence interval for the proportion of the population who would quit working if they won the lottery (or at least who say they would). The report of the poll on the Gallup website said that the margin of error for the part of the poll based on the 616 employed people was 4% and that 40% of those respondents said that they would quit working. Because our formulas are based on proportions rather than percentages, we can translate these numbers into a *sample estimate* of $\hat{p} = .4$ and a margin of error of .04. Notice that the reported margin of error of .04 is based on the formula $1/\sqrt{n} = 1/\sqrt{616} = .0403$. That's the *conservative margin of error*. Let's find the more accurate margin of error.

The standard error is

$$\sqrt{\frac{\hat{p}(1-\hat{p})}{n}} = \sqrt{\frac{(.4)(.6)}{616}} = .01974$$

and therefore the more accurate margin of error is

$$2\sqrt{\frac{\hat{p}(1-\hat{p})}{n}} = 2(.01974) = .03948$$

You can see that the real margin of error of .03948 is very close to the conservative margin of error of .04 reported with the survey. This is because $\hat{p} = .4$ is close to .5, so the real margin of error and the conservative margin of error are very close. Therefore, we can find an approximate 95% confidence interval in all of these ways:

- *Sample estimate ± margin of error*, which is .4 ± .03948, essentially .36 to .44
- *Sample estimate ± 2 standard errors*, which is .4 ± 2(.01974), also .36 to .44
- $\hat{p} \pm 2\sqrt{\hat{p}(1-\hat{p})/n}$, which is just the previous bullet written in symbols, also .36 to .44
- $\hat{p} \pm 1/\sqrt{n}$ (the conservative version), which is .4 ± .04, also .36 to .44

Therefore, no matter which format we use, the answer is the same. An approximate 95% confidence interval for the proportion of working adults who would quit working if they won a large amount in the lottery is .36 to .44, or 36% to 44%.

Notice that the interval does not cover 50%; it resides completely below 50%. Therefore, it would be fair to conclude, with high confidence, that fewer than half of all working Americans think they would quit if they won the lottery (and more than half think they would continue working). ■

### Why Polling Organizations Report the Conservative Margin of Error

The formula $1/\sqrt{n}$ provides a **conservative estimate of the margin of error** in that it usually overestimates the actual size of the margin of error. This conservative number is used by polling organizations so that they can report a single margin of error for all questions on a survey. The more precise formula involves the sample proportion, so the exact margin of error might change from one question to the next within the same survey. The formula $1/\sqrt{n}$ provides a margin of error that works (conservatively) for all questions based on the same sample size, even if the sample proportions differ from one question to the next.

How do we reconcile the two different formulas for the margin of error? The two formulas for the margin of error are equivalent when the proportion used in the more complicated formula is .5. When we use .5 for $\hat{p}$, the margin of error is

$$2 \times \sqrt{\frac{.5(1 - .5)}{n}} = \frac{1}{\sqrt{n}}$$

An important part of this story is that when you use any proportion other than .5 in the more exact formula, the answer is smaller than what you get when you use .5. This means that the conservative formula will almost always overestimate the size of the margin of error.

Example 10.8 | **The Gallup Poll Margin of Error for *n* = 1000** The Gallup Organization often uses samples of about 1000 randomly selected Americans. For $n = 1000$, the conservative estimate of margin of error is $1/\sqrt{1000} = .032$, or about 3.2%. If the proportion for which the margin of error is used is close to .5, then the reported (conservative) margin of error of .032 will be a good approximation to the real, more accurate margin of error. However, if the proportion is much smaller or larger than .5, then using .032 as the margin of error will produce an interval that is wider than necessary for the desired 95% confidence. For instance, if $\hat{p}$ is .1 or .9, then the real margin of error is about $2\sqrt{(.1)(.9)/1000} = .019$. You can see that using $\hat{p} \pm .032$ instead of $\hat{p} \pm .019$ would give you an interval that is much wider than needed. ∎

The next example describes a situation in which an unnecessarily wide interval includes negative values, which are not possible for a population proportion.

**Example 10.9**

**Statistics⬭Now™**

Watch a video example at **http://1pass.thomson.com** or on your CD.

10.3 Exercises are on page 435.

**Allergies and Really Bad Allergies** In the Marist Institute allergy survey, about 3% said that they experienced "severe" allergy symptoms. The sample size is $n = 883$, so the conservative margin of error is $1/\sqrt{883} = .034$, or 3.4%. Using this conservative margin of error, a 95% confidence interval estimate of the percentage of all Americans who have severe allergies is $3\% \pm 3.4\%$, or $-0.4\%$ to 6.4%. Notice that the lower end of this interval is negative, an impossible value for a population percentage, so the interval would be reported as 0% to 6.4%.

Generally, when the sample proportion $\hat{p}$ is far from .5, as it is for the severe allergy symptoms group, the conservative margin of error is too conservative. Using $\hat{p} = .03$ in the more complicated exact formula produces .011, or 1.1%, as the margin of error value. The corresponding 95% confidence interval estimate of the population percentage with severe allergy symptoms is $3\% \pm 1.1\%$, which is 1.9% to 4.1%. ∎

---

**thought question 10.5**   Suppose that we collect data on "hours of sleep last night" from 64 students who are present at an 11:00 A.M. statistics class. The sample mean is 6.8 hours, the standard deviation of the data is 1.6 hours, and the standard error of the mean is 0.2 hour. What population, if any, is represented by this sample? Assuming that there is a population represented by this sample, explain why we cannot use the formula

$$\hat{p} \pm z^* \sqrt{\frac{\hat{p}(1 - \hat{p})}{n}}$$

to compute a 95% confidence interval for the population mean. What general formula in this chapter could be used to calculate the interval? Calculate the confidence interval.*

---

## 10.4  CI Module 2: Confidence Intervals for the Difference in Two Population Proportions

The objective in this section is to form a **confidence interval for the difference between two population proportions.** Again, we start with the same general format that we introduced in Module 0 for the five parameters of interest, which is

Sample estimate $\pm$ Multiplier $\times$ Standard error

**\*HINT:** Is the given formula appropriate for a mean? Find the definition box that provides a confidence interval for any of the five parameters. Table 10.1 can be used for the multiplier.

Let's review the situation that generates research questions for which the difference between two population proportions is of interest. We have two populations or groups from which independent samples are available. We are interested in comparing the two populations with respect to the proportion with a certain trait, opinion, or characteristic. For example, we might want to compare the proportions of males and females who support the death penalty. Or, we might want to compare the proportion of physicians that would have a heart attack if they were to take daily aspirin to the proportion that would have a heart attack if they were to take a daily placebo instead.

The notation is as follows:

|  | Population Proportion | Sample Size | Number with Trait | Sample Proportion |
|---|---|---|---|---|
| **Population 1** | $p_1$ | $n_1$ | $X_1$ | $\hat{p}_1 = \dfrac{X_1}{n_1}$ |
| **Population 2** | $p_2$ | $n_2$ | $X_2$ | $\hat{p}_2 = \dfrac{X_2}{n_2}$ |

The *sample estimate* is $\hat{p}_1 - \hat{p}_2$, the difference between the two sample proportions. The *standard error* is

$$\text{s.e.}(\hat{p}_1 - \hat{p}_2) = \sqrt{\frac{\hat{p}_1(1 - \hat{p}_1)}{n_1} + \frac{\hat{p}_2(1 - \hat{p}_2)}{n_2}}$$

The multiplier is denoted as $z^*$, and it is determined by using the standard normal distribution.

---

*formula*

**Confidence Interval for the Difference Between Two Population Proportions**

A **confidence interval** for $p_1 - p_2$ is

$$\hat{p}_1 - \hat{p}_2 \pm z^* \sqrt{\frac{\hat{p}_1(1 - \hat{p}_1)}{n_1} + \frac{\hat{p}_2(1 - \hat{p}_2)}{n_2}}$$

$z^*$ is the value of the standard normal variable with the desired confidence level as the area between $-z^*$ and $z^*$.

**Note:** Values of $z^*$ for common confidence levels can be found in Table 10.1, or in the last row of Table A.2 in the Appendix for reasons that are explained in CI Module 3 in Chapter 11.

---

## Conditions for a Confidence Interval for the Difference in Two Proportions

As is the case for all statistical inference procedures, there are conditions that should be present in order to use the confidence interval just described. For a confidence interval for the difference between two proportions, the principal conditions have to do with the sample sizes for the observed samples. Both conditions must hold.

*Condition 1:* Sample proportions are available based on independent, randomly selected samples from the two populations.

*Condition 2:* All of the quantities $n_1\hat{p}_1$, $n_1(1 - \hat{p}_1)$, $n_2\hat{p}_2$, and $n_2(1 - \hat{p}_2)$ are at least 10. These quantities represent the counts observed in the category of interest and not in that category, respectively, for the two samples.

Notice that these are the same conditions that are required for the sampling distribution of $\hat{p}_1 - \hat{p}_2$ to be approximately normal, except that the sample size condition now applies using the $\hat{p}$'s instead of the $p$'s.

**Example 10.10**

**Statistics☐Now™**

Watch videos of similar examples at **http://1pass.thomson.com** or on your CD.

For software help, download your Minitab, Excel, TI-83, SPSS, R, and JMP manuals from **http://1pass .thomson.com**, or find them on your CD.

**Snoring and Heart Attacks** P. G. Norton and E. V. Dunn conducted a study to determine whether there is a relationship between snoring and risk of heart disease (Norton and Dunn, 1985). They found that of the 1105 snorers they sampled, 86 had heart disease, while only 24 of the 1379 nonsnorers had heart disease. We will define population 1 to be snorers and population 2 to be nonsnorers. In each case, $p$ is the (unknown) population proportion with heart disease.

How much difference is there between the proportions with heart disease for the two populations? We can answer this question with a confidence interval. First recognize that the conditions for calculating this confidence interval are satisfied. Condition 1 is satisfied because the researchers collected independent samples from each population, and Condition 2 is satisfied because the relevant observed counts all are greater than 10. In the group that snores, the numbers with and without heart disease are 86 and 1019, respectively, while in the group that doesn't snore, the numbers with and without heart disease are 24 and 1355, respectively.

The sample proportions with heart disease in the two groups and the standard error of the difference between the two sample proportions are

$$\hat{p}_1 = \frac{86}{1105} = .0778, \qquad \hat{p}_2 = \frac{24}{1379} = .0174, \qquad \text{and} \qquad \hat{p}_1 - \hat{p}_2 = .0604$$

$$\text{s.e.}(\hat{p}_1 - \hat{p}_2) = \sqrt{\frac{\hat{p}_1(1 - \hat{p}_1)}{n_1} + \frac{\hat{p}_2(1 - \hat{p}_2)}{n_2}}$$

$$= \sqrt{\frac{.0778(.9222)}{1105} + \frac{.0174(.9826)}{1379}} = .0088$$

In general, the confidence interval estimate of the difference $\hat{p}_1 - \hat{p}_2$ is

$$\hat{p}_1 - \hat{p}_2 \pm z^* \text{ s.e.}(\hat{p}_1 - \hat{p}_2)$$
$$.0604 \pm z^*(.0088)$$

On the basis of this general statement, the following table includes confidence intervals for three different confidence levels. For each level, we show the appropriate $z^*$ multiplier and the completed answer to $.0604 \pm z^*(.0088)$. The last row of Table A.2 was used to determine the multipliers.

| Confidence Level | $z^*$ | Confidence Interval |
|---|---|---|
| 90% | 1.645 | .046 to .075 |
| 95% | 1.960 | .043 to .078 |
| 99% | 2.576 | .038 to .083 |

**CI Module 2: $p_1 - p_2$**

*As is always the case, the higher the level of confidence, the wider the interval.* From these intervals, it appears that the proportion of snorers with heart disease in the population is about 4% to 8% higher than the proportion of nonsnorers with heart disease.

As a final note about this example, we point out that it is also informative to compute the relative risk of heart disease for snorers compared to nonsnorers. This relative risk is the ratio of the sample proportions, which is $\hat{p}_1/\hat{p}_2 = .078/.0174 = 4.5$. In other words, we estimate that the risk of heart disease for snorers is about 4.5 times what the risk is for nonsnorers. Unfortunately, it is not straightforward to compute a confidence interval for relative risk in the population. Also, remember that on the basis of an observational study such as this, we can conclude only that there is a relationship between snoring and heart disease, not a cause-and-effect influence. ■

---

**MINITAB** *tip* | **Determining the Interval for the Difference in Proportions**

- To compute a confidence interval for the difference in two proportions, use **Stat>Basic Statistics>2 Proportions.** See the following note for information about inputting data.

- To change the confidence level, use the **Options** button. The default confidence level is 95%.

*Note:* The raw data for the response may be in one column, and the raw data for group categories (Subscripts) may be in a second column. Or the raw data for the two independent groups may be in two separate columns. Or the data may already be summarized. If so, click **Summarized Data,** then specify the sample size (trials) and the *number* of successes for each group.

---

**Example 10.11** | **Do You Always Buckle Up When Driving?** How often do you wear a seatbelt when driving a car? This was asked in the 2003 Youth Risk Behavior Surveillance System, a nationwide biennial survey of 9th- through 12th-graders in the United States. Possible responses to the question about seatbelt use were as follows: never, rarely, sometimes, most times, and always. Example 2.1 and Table 2.2 presented data on the connection between seatbelt use and course grades usually attained in school for 12th-graders who said that they drive. In this example, we compare female and male 12th-grade drivers. To do so, we find a 95% confidence interval for $p_1 - p_2 =$ difference between the population proportions of 12th-grade female and male drivers who would say they always wear a seatbelt when driving. The Minitab output for this confidence interval is on the next page. (The data are in the **YouthRisk03** dataset on the CD for this book.)

The sample proportions of females and males who said they always wear a seatbelt are given toward the top of the output under "Sample p." The values under "X" and "N" are the number who said they always wear a seatbelt and the sample size for each sex, respectively. Rounded to three decimal places, the sample proportion for females is $\hat{p}_1 = 915/1467 = .624$ and the sample propor-

Example 10.11

# Exercises

●   Denotes basic skills exercises

◆   Denotes dataset is available in StatisticsNow at **http:// 1pass.thomson.com** or on your CD but is **not required** to solve the exercise.

**Bold-numbered exercises** have answers in the back of the text and fully worked solutions in the Student Solutions Manual.

**Statistics⬡Now**™  Go to the StatisticsNow website at **http://1pass.thomson.com** to:
• Assess your understanding of this chapter
• Check your readiness for an exam by taking the Pre-Test quiz and exploring the resources in the Personalized Learning Plan

## Section 10.1

**10.1** ● In each part of this question, explain whether the proportion that is described is a sample proportion or a population proportion.

a. In the 1990 United States Census, it was found that about 1 in 9 Americans were at least 65 years old at that time.

b. A randomly selected group of 500 registered voters in your state is asked whom they intend to vote for in an upcoming election for governor. Based on the proportion choosing each candidate, the polling organization predicts the election winner.

**c.** In a clinical trial done to assess the effectiveness of a new medication for asthma, a satisfactory relief from symptoms was experienced by 55% of $n = 80$ participants.

**10.2** ● Fill in the blanks with the appropriate term.

a. A _____ gives information about a population value based on sample data.

b. A _____ gives information about possible sample values based on information about the population.

**10.3** A Gallup poll surveyed a random sample of 439 American teenagers. One question that they were asked was "How strict are your parents compared to most of your friends parents?" The choices were "more strict," "less strict," and "about the same." The choice "more strict" was selected by 171 of the respondents (Mazzuca, 2004).

a. What is the research question of interest?

b. What is the population parameter of interest?

c. What percentage of the sample responded with "more strict"?

d. Draw a figure like Figure 10.1 (p. 402) to illustrate this situation.

**10.4** Refer to Exercise 10.3 about a Gallup poll on how strict teenagers perceive their parents to be in comparison to other parents, based on a random sample of 439 teenagers. The choice "more strict" was selected by 171 of the respondents. Suppose that in fact the true population percentage of teenagers who think that their parents are more strict is 40%.

a. If a new sample of 439 teenagers from the same population were taken, which one of the following would be true? Explain your answer.

   i. The number selecting the choice "more strict" would be 171 again.

   ii. The number selecting the choice "more strict" could be 171 again.

   iii. The number selecting the choice "more strict" could not be 171 again.

b. If a new sample of 439 teenagers from the same population were taken, which one of the following would be true? Explain your answer.

   i. The true population percentage of teenagers who think that their parents are more strict would be 40%.

   ii. The true population percentage of teenagers who think that their parents are more strict could be 40%.

   iii. The true population percentage of teenagers who think that their parents are more strict could not be 40%.

**10.5** Would a confidence interval or a sampling distribution be more appropriate for answering each of the following questions?

a. Based on the fact that 65% of a random sample of 500 students at a large university said that they believe in life after death, what percentage of all students at the university believe in life after death?

b. Based on the fact that 65% of all students at a large university believe in life after death, what percentage of a random sample of 500 students at the university would say that they believe in life after death?

**10.6** Read the examples of questions based on a "parameter of interest" in the section on "Curiosity and Confidence Intervals." Create your own example of something that interests you, and translate it into a question about a population parameter.

## Section 10.2

*Exercises 10.7 to 10.18 correspond to the two Lessons in Section 10.2. Lesson 1 exercises are 10.7 to 10.12; Lesson 2 exercises are 10.13 to 10.18.*

**10.7** ● In a 1997 survey done by the Marist College Institute for Public Opinion, 36% of a randomly selected sample of $n = 935$ American adults said that they do not get enough sleep each night (*Source:* www.mipo.marist .edu). The margin of error was reported to be 3.5%.

a. Use the survey information to create a 95% confidence interval for the percentage that feels they don't get enough sleep each night. Write a sentence that interprets this interval. Specify the population.

**b.** Do you think this sample could be used to estimate the percentage of college students who think they don't get enough sleep each night? Why or why not?

**10.8** ● A CNN/Time Poll conducted in the United States October 23–24, 2002, asked, "Do you favor or oppose the legalization of marijuana?" In the nationwide poll of $n = 1007$ adults, 34% said that they favored legalization. The margin of error was given as 3.1% (*Source:* www.pollingreport.com).

a. Find a 95% confidence interval estimate of the percentage of American adults who favored the legalization of marijuana at the time of the poll.

b. Write a sentence that interprets the interval computed in part (a).

**10.9** Suppose 200 different researchers all randomly select samples of $n = 400$ individuals from a population. Each researcher uses his or her sample to compute a 95% confidence interval for the proportion that has blue eyes in the population. About how many of the confidence intervals will capture the population proportion? About how many of the intervals will not capture the population proportion? Briefly explain how you determined your answers.

**10.10** Taking into account the purpose of a confidence interval described in Section 10.2, explain what is wrong with the following statement: "Based on the survey data, a 95% confidence interval estimate of the sample proportion is .095 to .117."

**10.11** *Parade Magazine* reported that "nearly 3200 readers dialed a 900 number to respond to a survey in our Jan. 8 cover story on America's young people and violence" (Feb. 19, 1995, p. 20). Of those responding, "63.3% say they have been victims or personally know a victim of violent crime." Can the methods in this chapter legitimately be used to compute a 95% confidence interval for the proportion of Americans who fit that description? Explain why or why not.

**10.12** What is the probability that a 95% confidence interval will not cover the true population value? Explain.

**10.13** ● Explain whether the width of a confidence interval would increase, decrease, or remain the same as a result of each of the following changes:

a. Increase the confidence level from 95% to 99%.

b. Decrease the confidence level from 95% to 90%.

**10.14** ● For the following examples given in the introduction to this chapter, specify which of the five parameters stated on page 407 in Lesson 2 of Section 10.2 is appropriate. Give and define appropriate notation for the parameter of interest. If two symbols are involved, such as $\mu_1 - \mu_2$, define each symbol.

**a.** The parameter of interest is the proportion of the population that suffers from allergies.

**b.** The parameter of interest is the proportion of working adults who say that they would quit their jobs if they won a large amount in the lottery.

**c.** Does taking calcium help to reduce premenstrual symptoms? To investigate, we could devise a questionnaire that measured the severity of symptoms and then randomly assign women to take calcium or a placebo. The parameter of interest is the difference in what the population mean severity score would be if everyone took calcium versus the score if everyone took a placebo.

d. Do men and women feel the same away about dating someone who has a great personality without regard for looks? To investigate this, one parameter that we could use is the difference in population proportions of males and females who would answer "yes" if asked, "Would you date someone with a great personality even though you did not find them attractive?"

**10.15** Suppose that in a random sample of 300 employed Americans, there are 57 individuals who say that they would fire their boss if they could. Calculate a 95% confidence interval for the population proportion. Write a sentence or two that interprets this interval.

**10.16** *U.S. News and World Report* (Dec. 19, 1994, pp. 62–71) reported on a survey of 1000 American adults, conducted by telephone on December 2–4, 1994, designed to measure beliefs about apocalyptic predictions. One of the results reported was that 59% of the sample said that they believe the world will come to an end.

**a.** Calculate the standard error of the sample proportion who believe that the world will come to an end.

**b.** Calculate a 95% confidence interval for the population proportion. Interpret the interval in a way that could be understood by a statistically naïve reader.

**10.17** Refer to Exercise 10.3, which described a Gallup poll. The poll surveyed a random sample of 439 American teenagers, and one question that they were asked was "How strict are your parents compared to most of your friends' parents?" The choices were "more strict," "less strict," and "about the same." The choice "less strict" was selected by 30% of the respondents (Mazzuca, 2004).

a. What is the sample proportion that answered "less strict?" Use appropriate notation.

b. Calculate the standard error of the sample proportion.

c. Calculate a 95% confidence interval for the population proportion. Interpret the interval in a way that could be understood by a statistically naïve reader.

**10.18** For each of the following, explain whether it affects the width of a confidence interval, the center of a confidence interval, neither, or both.

a. The sample estimate, for the situations involving means.

b. The multiplier.

c. The standard error of the estimate.

d. The confidence level.

e. The sample standard deviation(s), for the situations involving means.

f. The sample proportion(s), for the situations involving proportions.

## Section 10.3

*Exercises 10.19 to 10.40 correspond to the three Lessons in Section 10.3. Lesson 1 exercises are 10.19 to 10.30; Lesson 2 exercises are 10.31 to 10.34; Lesson 3 exercises are 10.35 to 10.40.*

**10.19** ● Suppose a new treatment for a certain disease is given to a sample of 200 patients. The treatment was successful for 166 of the patients. Assume that these patients are representative of the population of individuals who have this disease.
**a.** Calculate the sample proportion successfully treated.
**b.** Determine a 95% confidence interval for the proportion of the population for whom the treatment would be successful. Write a sentence that interprets this interval.

**10.20** ● Refer to Exercise 10.19. Calculate a 98% confidence interval for the proportion successfully treated. Is this interval wider or narrower than the interval computed in part (b) of Exercise 10.19?

**10.21** ● ◆ In a survey of 190 college students, 134 students said that they believe there is extraterrestrial life (*Data source:* **pennstate3** dataset on the CD for this book).
**a.** Find $\hat{p}$ = sample proportion that believes that there is extraterrestrial life.
**b.** Find a 95% confidence interval estimate of the proportion of all college students who believe there is extraterrestrial life.

**10.22** ● In a poll done by Princeton Survey Research Associates in October 1998 for *Newsweek Magazine,* 56% of 753 randomly sampled American adults answered "yes" to the question "Do you personally believe that abortion is wrong?" (*Source:* www.pollingreport .com /abortion.htm.)
**a.** Calculate a 95% confidence interval for the October 1998 percentage of all Americans who believed that abortion is wrong.
**b.** On the basis of this confidence interval, is it reasonable to conclude that in October 1998 more than half of all Americans thought that abortion is wrong? Explain.

**10.23** ● For each confidence interval procedure, provide the confidence level.
**a.** Sample proportion $\pm$ 1.645 $\times$ standard error.
**b.** Sample proportion $\pm$ 2 $\times$ standard error.
**c.** Sample proportion $\pm$ 2.33 $\times$ standard error.
**d.** Sample proportion $\pm$ 2.58 $\times$ standard error.

**10.24** In a randomly selected sample of 400 registered voters in a community, 220 individuals say that they plan to vote for Candidate Y in the upcoming election.
**a.** Find the sample proportion planning to vote for Candidate Y.
**b.** Calculate the standard error of the sample proportion.
**c.** Find a 95% confidence interval for the proportion of the registered voter population who plan to vote for Candidate Y.
**d.** Find a 98% confidence interval for the proportion of the registered voter population who plan to vote for Candidate Y.

**10.25** In an ABCNews.com nationwide poll done in 2001, the proportion of respondents who thought that it should be illegal to use a handheld cellular telephone while driving a car was .69 (69%). The poll's sample size was 1027.
**a.** Find the value of the standard error of the sample proportion.
**b.** Find a 95% confidence interval estimate of the population proportion that thinks it should be illegal to use a cellular telephone while driving.
**c.** Find a 90% confidence interval estimate of the population proportion that thinks it should be illegal to use a cellular telephone while driving.

**10.26** Refer to Example 10.3 (p. 412), in which for a sample of 935 American adults, .60 was the proportion that thinks intelligent life exists elsewhere. Compute a 99% confidence interval for the population proportion.

**10.27** A randomly selected sample of 15 individuals is asked whether they are right-handed or left-handed. In the sample, only one person is left-handed. Explain why a confidence interval for the population proportion should not be computed by using the methods described in this chapter.

**10.28** Refer to Exercise 10.15. Calculate a 90% confidence interval for the population proportion who would fire their boss if they could.

**10.29** Determine the value of the $z^*$ multiplier that would be used to compute an 80% confidence interval for a population proportion.

**10.30** Find $z^*$ for each of the confidence levels in parts (a)–(c), and draw a picture similar to Figure 10.3 (p. 413). Then answer part (d).
**a.** 60%.
**b.** 86%.
**c.** 99.8%.
**d.** On the basis of the values of $z^*$ found in parts (a)–(c), explain what happens to the width of the confidence interval when the confidence level is increased.

**10.31** ● Refer to the derivation of a 95% confidence interval for $p$ on page 417 in Lesson 2 of Section 10.3. Use a similar argument to explain why a 68% confidence interval for $p$ is "Sample estimate $\pm$ Standard error."

**10.32** ● In constructing 99% confidence intervals for proportions, over the long run, about what percentage of the time would each of the following happen?
**a.** The difference between the sample proportion and the population proportion is less than the "Multiplier $\times$ Standard error."
**b.** The difference between the sample proportion and the population proportion is more than the "Multiplier $\times$ Standard error."

**10.33** ● Suppose that a polling organization reports that the margin of error is 4% for a sample survey. Explain what this indicates about the possible difference between a percentage determined from the survey data and the population value of the percentage.

**10.34** Suppose that a margin of error for a sample percentage is reported to be 3%. What is the probability that the difference between the sample percentage and the population percentage will be more than 3%? Interpret the meaning of this probability.

**10.35** ● For each combination of sample size and sample proportion, find the approximate margin of error for the 95% confidence level.
  **a.** $n = 100$, $\hat{p} = .56$.
  **b.** $n = 400$, $\hat{p} = .56$.
  c. $n = 400$, $\hat{p} = .20$.
  d. $n = 400$, $\hat{p} = .80$.
  e. $n = 1000$, $\hat{p} = .50$.

**10.36** ● Suppose a polling organization is conducting a survey to estimate the proportion of Americans who regularly attend religious services. The organization plans to gather data from a randomly selected sample of 550 individuals. On the basis of the "conservative" formula, what will be the margin of error for the survey?

**10.37** In a Gallup Youth Survey done in 2000, $n = 501$ randomly selected American teenagers were asked about how well they get along with their parents.
  **a.** According to the Gallup Organization, the margin of error for the poll was 5%. Verify that this figure is approximately correct.
  b. A survey result was that 54% of the sample said they get along "very well" with their parents. Using the reported margin of error, calculate a 95% confidence interval for the population proportion that gets along "very well" with their parents. Write a sentence that interprets this interval.
  c. Using the more exact formula for margin of error in Section 10.3, calculate a 95% confidence interval. Compare the answer to the answer in part (b).
  d. Another result in the same survey was that only 5% said that cheating on exams was "not at all serious." Using the reported margin of error, compute a 95% confidence interval for the population proportion that would say this. Also, compute the 95% confidence interval using the more exact formula for margin of error. Which interval is narrower?
  e. You should have found that the intervals in parts (b) and (c) were quite similar to each other but that the two intervals computed in part (d) were not. Explain why the two intervals agreed more closely in parts (b) and (c) than they did in part (d).

**10.38** In Example 10.9 (p. 423), the information given was that 3% of a nationwide randomly selected sample of $n = 883$ suffered from severe allergies.
  **a.** Determine the standard error of the sample proportion in this problem.
  **b.** Verify that the margin of error is about 1.1% for a 95% confidence interval for the percentage that suffers from severe allergies in the population represented by this sample.

**10.39** Suppose that a margin of error for a survey is reported to be 4%. Write a brief explanation of what this margin of error tells us about the accuracy of sample percentages from this survey.

**10.40** In a survey reported in a special edition of *Newsweek* (Spring/Summer 1999, *Health for Life*), only 3% of a

sample of 757 American women responded "Not at all satisfied" to the question "How satisfied are you with your overall physical appearance?"
  a. The margin of error for the poll was reported as 3.5%. Use the conservative formula for margin of error to verify that this is approximately correct.
  b. On the basis of the reported margin of error, calculate a 95% confidence interval for the percentage of all American women who are "not at all satisfied" with their physical appearance. Are all values in this interval valid estimates of the population percentage? Explain.
  c. Use the "exact" formula to calculate the margin of error.
  d. On the basis of the margin of error in part (c), calculate a 95% confidence interval for the population percentage. Write a sentence that interprets this confidence interval for somebody who doesn't know very much about statistics.

## Section 10.4

**10.41** ● In each situation, explain whether the method covered in Section 10.4 for finding a confidence interval for the difference in two proportions should be used or not:
  **a.** A survey is done to estimate the difference between the proportions of college students and high school students who smoke cigarettes. Data on cigarette smoking habits are gathered from randomly selected samples of 500 college and 500 high school students.
  **b.** Two new treatments for a serious disease are compared. Each treatment is used for five patients. The first treatment is successful for three of the five patients who used it. The second treatment is successful for only one of the five patients who used that treatment.
  **c.** An economist wants to estimate the difference between the mean annual income of college graduates and the mean annual income of high school graduates who did not go to college. Income data are collected from samples of 200 individuals in each educational degree group.

**10.42** ● ◆ In the General Social Survey, an ongoing nationwide survey done by the National Opinion Research Center at the University of Chicago, a question asked is whether a respondent favors or opposes capital punishment (the death penalty) for persons convicted of murder. The output for this exercise compares the proportions who said that they were opposed to the death penalty in the year 2000 and the year 1993. (*Sources:* **GSS-93** dataset on the CD for this book and http://csa.berkeley.edu:7502/archive.htm.)

| Sample | X | N | Sample p |
|--------|-----|------|----------|
| 2000 | 801 | 2565 | 0.312 |
| 1993 | 337 | 1488 | 0.226 |

Estimate for p(1) − p(2): 0.086
95% CI for p(1) − p(2): (0.058, 0.114)

a. What proportion of the year 2000 sample was opposed to the death penalty? What proportion of the year 1993 sample was opposed?

b. What is the estimated difference between the proportions opposed to the death penalty in the two years?

c. Write the 95% confidence interval given in the output. Then interpret this interval in the context of this situation.

d. Provide the formula that is used to calculate the interval, and substitute appropriate numerical values into the formula.

10.43 ● ◆ Refer to the output given for Exercise 10.42.

a. Find the value of s.e.$(\hat{p}_1 - \hat{p}_2)$ = standard error of the difference between the two sample proportions.

b. Find a 90% confidence interval for the difference between the population proportions opposed to the death penalty in the years 2000 and 1993. Use Example 10.10 (p. 425) for guidance.

10.44 In a study done in Maryland, investigators surveyed individuals by telephone about how often they get tension headaches (Schwartz et al., 1998). One response variable that was measured was whether or not the respondent had experienced an episodic tension-type headache (ETTH) in the prior year. A headache pattern was called "episodic" if the headaches occurred less often than 15 times a month; otherwise, the headaches were called "chronic." Of the 1600 women in the survey aged 18 to 29, 653 said that they had experienced episodic tension-type headaches in the last year. Of the 2122 women in the 30- to 39-year-old age group, the number having experienced episodic headaches was 995.

a. Estimate the proportion in each group that experienced an episodic headache in the prior year, and compute the difference in these two proportions.

b. Compute a 95% confidence interval for the difference between the proportions for these age groups in the population. Write a sentence that interprets this confidence interval.

**10.45** Refer to Exercise 10.44. In the sample, there were 4594 individuals with at least a college degree and $n = 7076$ individuals with at least a high school diploma (but not a college degree). Following is Minitab output with a 95% confidence interval for the difference in proportions experiencing episodic tension-type headaches in the two educational groups:

| Sample | X | N | Sample p |
|---|---|---|---|
| 1 | 2140 | 4594 | 0.466 |
| 2 | 2690 | 7076 | 0.380 |

Estimate for p(1) − p(2): 0.086
95% CI for p(1) − p(2): (0.068, 0.104)

a. Write two or three sentences that interpret the results.

b. Provide the formula that is used to calculate the interval, and substitute appropriate numerical values into the formula.

10.46 In a survey of college students, 70% (.70) of the 100 women surveyed said that they believe in love at first sight, whereas only 40% (.40) of the 80 men surveyed said that they believe in love at first sight.

a. Find the value of the difference between the sample proportions for men and women.

b. Find the standard error of the difference between the sample proportions.

c. Find an approximate 95% confidence interval for the difference between population proportions believing in love at first sight for men versus women.

**10.47** Suppose that a randomly selected sample of $n = 900$ registered voters is surveyed to estimate the proportions that will vote for the two candidates in an upcoming election. Fifty-five percent of those sampled say that they will vote for one candidate, while 45% indicate a preference for the other candidate. Explain why the method in Section 10.4 for computing a confidence interval for a difference in proportions should not be used to estimate the difference in the proportions planning to vote for the two candidates.

10.48 ◆ In Case Study 10.3 (p. 431), students in a statistics class at Penn State were asked, "Would you date someone with a great personality even though you did not find them attractive?" By gender, the results were that 61.1% of 131 women answered "yes" while 42.6% of 61 men answered "yes" (*Source:* **pennstate3** dataset on the CD for this book).

a. A 95% confidence interval for the difference in proportions of men and women who would say "yes" to the question if asked is given as .035 to .334. Write a sentence that interprets the interval.

b. What populations do you think are represented by these samples of men and women?

10.49 ◆ Refer to Exercise 10.48.

a. Compute a 99% confidence interval for the difference in proportions.

b. How does the width of the 99% confidence interval computed in part (a) of this problem compare to the width of the 95% confidence interval given in part (a) of Exercise 10.48?

c. Would a 90% confidence interval for the difference in proportions be wider or narrower than the 99% confidence interval computed in part (a)? Explain.

d. Does the interval computed in part (a) include the value 0? What does this tell us about whether there is a difference between college men and women for this question?

## Section 10.5

**10.50** ● A Gallup Organization poll of $n = 1004$ randomly selected American adults in July 2002 found that 55% of those surveyed felt that their weight was about right. The margin of error for the survey was given as 3% (*Source:* www.gallup.com).

**a.** Find a 95% confidence interval estimate of the percentage of American adults who think their weight is about right.

**b.** Based on the interval computed in part (a), explain whether it is reasonable to say that more than 50% of American adults think their weight is about right.

**10.51** ● Suppose that a health expert has claimed that 28% of college students smoke cigarettes. To investigate this claim, researchers survey a random sample of college students. Using this sample, a 95% confidence interval for the percentage of college students who smoke is found to be 19% to 24%. On the basis of this interval, explain whether the claim that 28% of all college students smoke is reasonable or not.

**10.52** ● A university is contemplating switching from the quarter system to a semester system. The administration conducts a survey of a random sample of 400 students and finds that 240 of them prefer to remain on the quarter system.

**a.** Construct a 95% confidence interval for the true proportion of all students who would prefer to remain on the quarter system.

**b.** Does the interval that you computed in part (a) provide convincing evidence that more than half of all students prefer to remain on the quarter system? Explain.

**10.53** Suppose that a 95% confidence interval for the proportion of men who experience sleep apnea (irregular breathing during sleep) is .11 to .17 and a 95% confidence interval for the proportion of women who experience sleep apnea is .04 to .08.

**a.** On the basis of these intervals, is it reasonable to conclude that the population proportions experiencing sleep apnea differ for men and women? Explain.

**b.** Find a 95% confidence interval for the difference in proportions of men and women in the population who experience sleep apnea.

**c.** On the basis of the interval you found in part (b), is it reasonable to conclude that the population proportions experiencing sleep apnea differ for men and women? Explain.

**10.54** As was described in Exercise 10.50, a July 2002 Gallup poll of $n = 1004$ randomly selected U.S. adults found that 55% felt that their weight was about right. In a similar poll in 1990, about 46% felt that their weight was about right. The margin of error for each poll was 3%.

**a.** Based on these results, is it reasonable to conclude that the population proportions thinking their weight is about right were different in 1990 and 2002?

**b.** Assuming that the sample size was 1004 in the 1990 poll as well, find a 95% confidence interval for the difference in the population proportions.

**c.** On the basis of the interval that you found in part (b), formulate a conclusion about whether the population proportion was different in 2002 than it was in 1990. Write your conclusion in words that

would be understood by someone with no training in statistics.

**10.55** *This exercise is also Exercise 1.12.* A random sample of 1001 University of California faculty members taken in December 1995 was asked, "Do you favor or oppose using race, religion, sex, color, ethnicity, or national origin as a criterion for admission to the University of California?" (Roper Center, 1996). Fifty-two percent responded "favor."

**a.** What is the population for this survey?

**b.** What is the approximate margin of error for the survey?

**c.** Based on the results of the survey, could it be concluded that a majority (over 50%) of *all* University of California faculty members favor using these criteria? Explain.

## Chapter Exercises

**10.56** In each situation, explain why you think the sample proportion should or should not be used to estimate the population proportion.

**a.** An Internet news organization asks visitors to its website to respond to the question "Are you satisfied with the president's job performance?" Of $n = 3500$ respondents, 61% say that they are not satisfied with the president's performance. On the basis of this survey, the organization writes an article saying that a majority of Americans are not satisfied with the president.

**b.** A convenience sample of $n = 400$ college students in two classes at the same university is used to estimate the proportion that is left-handed in the nationwide population of college students.

**10.57** A federal law that lowered the limit for a legal blood alcohol level for automobile drivers was signed by President Clinton in October 2000. At its website, the Gallup Organization reported the following survey results in an article dated October 26, 2000:

> 72% of Americans support lowering the drunk driving limit to 0.08% BAC. . . . A total 1,002 telephone interviews were conducted between July 20–Aug. 3, 2000, with a representative sample of the U.S. public age 16 and older. The findings are based on 930 respondents who identified themselves as licensed drivers. The margin of error equals plus or minus three percentage points.

**a.** The first sentence of the information from the Gallup Organization states, "72% of Americans support lowering the drunk driving limit to 0.08% BAC." Considering the difference between a sample and a population, explain why this sentence is not necessarily correct. Rewrite the sentence so that it is correct.

**b.** The article provides the information that "the margin of error equals plus or minus three percentage points." Write one or two sentences interpreting this value that could be understood by someone who does not know anything about statistics.

c. Using the information provided by the Gallup Organization, calculate a 95% confidence interval for the proportion that supports the lower legal blood alcohol limit. Write a sentence that interprets this interval. Be sure to specify the appropriate population.

d. On the basis of the interval computed in part (c), is it reasonable to conclude that more than half of American licensed drivers support the lower blood alcohol limit?

10.58 In Chapter 2, we saw data from a statistics class activity in which 190 students were asked to randomly pick one of the numbers 1, 2, 3, 4, 5, 6, 7, 8, 9, 10. The number 7 was picked by 56 students.

a. For the sample, calculate the proportion of students who picked 7.

b. Calculate the standard error for this sample proportion.

c. Calculate a 90% confidence interval for the population proportion.

d. Calculate a 95% confidence interval for the population proportion.

e. Calculate a 98% confidence interval for the population proportion.

f. What do the results of parts (c)–(e) indicate about the effect of confidence level on the width of a confidence interval?

g. On the basis of these confidence intervals, do you think that students choose numbers "randomly"? Explain.

**10.59** A study reported in the *Journal of Occupational and Environmental Medicine* (Franke et al., 1998, 40: 441–444) found that retired male Iowa policemen are more likely to have heart disease compared to other men with similar ages but different occupations. Data from the study were that about 32% of a sample of 232 retired policemen had heart disease but only 18% of a sample of 817 men with other occupations had heart disease.

**a.** In Chapter 4, we learned about the various types of studies. Is this an observational study or a randomized experiment? Explain.

**b.** Do you think that we can attribute the observed difference in heart disease percentages to job-related factors? Explain.

c. Graphically compare the policemen and the other men, using a bar chart.

d. Calculate a 95% confidence interval for the proportion of retired Iowa policemen who have heart disease.

e. Calculate a 95% confidence interval for the proportion of other retired men who have heart disease.

f. Why are the widths of the two confidence intervals different from each other?

10.60 Refer to Exercise 10.59. Find a 95% confidence interval for the difference in population proportions of male policemen and men of other occupations who have heart disease. On the basis of this confidence interval, is it safe to conclude that male police officers are more likely to experience heart disease than other men? Explain.

10.61 Use an example from this chapter to explain the difference between a sample proportion and a population proportion.

10.62 A professor wants to know whether her class of 60 students would prefer the final exam to be given as a take-home exam or an in-class exam. She surveys the class and learns that 45 of the 60 students would prefer a take-home. Explain why she should not use these data to compute a 95% confidence interval for the proportion preferring a take-home exam.

**10.63** The Gallup Organization used two different wordings of a question in a February 1999 poll about the death penalty. The two different questions were asked of two different random samples of Americans. Results reported at the Gallup website were

Question 1: Are you in favor of the death penalty for a person convicted of murder? (Based on 543 adults: margin of error plus or minus 5 percentage points)

| | For | Against | No Opinion |
|---|---|---|---|
| **99 Feb 8–9** | 71% | 22% | 7% |

Question 2: What do you think should be the penalty for murder—the death penalty or life imprisonment with absolutely no possibility of parole?

(Based on 511 adults: margin of error plus or minus 5 percentage points)

| | Death Penalty | Life Imprisonment | No Opinion |
|---|---|---|---|
| **99 Feb 8–9** | 56% | 38% | 6% |

a. Compare the responses to the two questions using a bar chart.

b. The margin of error for each question was reported as 5%. Verify that this is approximately true.

**c.** For the first question, calculate a 95% confidence interval for the percentage of all Americans who are "For" the death penalty. Use the margin of error reported by the Gallup Organization. Write a sentence that interprets this confidence interval.

**d.** For the second question, calculate a 95% confidence interval for the percentage of all Americans who would answer "Death Penalty" when asked whether the penalty for murder should be the death penalty or life imprisonment. Use the margin of error reported by the Gallup Organization. Write a sentence that interprets this interval.

10.64 Refer to Exercise 10.63, in which two different wordings of a question were used to estimate the proportion of Americans who are in favor of the death penalty.

a. Verify that the conditions are met to compute a confidence interval for the difference in two population proportions, where they are the proportions of the population who would support the death penalty if asked Question 1 and if asked Question 2, respectively.

b. Find a 95% confidence interval for the difference in the two population proportions.

c. On the basis of the confidence interval that you computed in part (b), formulate a conclusion about whether the wording of the question affects the proportion of the population who support the death penalty. Write a few sentences expressing your conclusion that would be understood by someone who has no training in statistics.

10.65 In 1998, two statistics classes for students in the liberal arts at a large northeastern university were asked, "Do you have a tattoo?" The responses were as follows:

|  | Yes | No |
|---|---|---|
| Women | 46 | 227 |
| Men | 32 | 175 |

a. What proportion of the women in this sample has a tattoo?

b. What proportion of the men in this sample has a tattoo?

c. Compare the men and women, using a bar graph.

d. What population do you think is represented by this sample? In other words, to whom do you think the results can be generalized?

e. For the college women, calculate a 95% confidence interval for the proportion of the population that has a tattoo. Write a sentence that interprets this interval.

f. For the college men, calculate a 95% confidence interval for the proportion of the population that has a tattoo. Write a sentence that interprets this interval.

g. Use the results of parts (e) and (f) to compare college men and women with regard to the proportion that has a tattoo.

10.66 Refer to Exercise 10.65, which reported results for the numbers of men and women who had a tattoo.

**a.** Find a 90% confidence interval for the difference in population proportions of men and women with a tattoo.

**b.** On the basis of the interval that you computed in part (a), formulate a conclusion about whether the population proportions of men and women with tattoos differ for the populations represented in this study. Explain.

10.67 Find the reported results of a poll in which a margin of error is also reported. Look in a weekly news magazine such as *Newsweek* or *Time,* in a newspaper such as the *New York Times,* or on the Internet.

a. Explain what question was asked in the poll and what margin of error was reported. Verify the margin of error with a calculation.

b. Present a 95% confidence interval for the results. Explain in words what the interval means for your example.

10.68 For Case Study 10.2 (nicotine patches and Zyban), verify the confidence intervals reported in Table 10.3.

**10.69** On January 30, 1995, *Time* magazine reported the results of a poll of adult Americans in which it asked, "Have you ever driven a car when you probably had too much alcohol to drive safely?" The exact results were not given, but from the information given, we can guess at what they were. Of the 300 men who answered, 189 (63%) said yes and 108 (36%) said no. The remaining 3 men weren't sure. Of the 300 women, 87 (29%) said yes and 210 (70%) said no; the remaining 3 women weren't sure.

**a.** Compute an approximate 95% confidence interval for the difference in proportions of men and women who would say yes to this question.

b. Write a sentence or two interpreting this interval.

10.70 Refer to Exercise 10.69. Calculate an approximate 95% confidence interval for the proportion of men who would say that they have ever driven a car when they probably have had too much to drink, and write a sentence or two that interprets the interval. Be specific about the population that the interval describes.

10.71 ◆ A sample of college students was asked whether they would return the money if they found a wallet on the street. Of the 93 women, 84 said "yes," and of the 75 men, 53 said "yes." Assume that these students represent all college students (*Source:* **UCDavis2** dataset on the CD for this book).

a. Find separate approximate 95% confidence intervals for the proportions of college women and college men who would say "yes" to this question.

b. Find an approximate 95% confidence interval for the difference in the proportions of college men and women who would say "yes" to this question.

c. Write a few sentences interpreting the intervals in parts (a) and (b).

**10.72** In the table below are data collected between 1996 and 1998 on ear pierces and tattoos for male Penn State students. Assume that these men represent a random sample of male Penn State students. The results were as follows:

|  | Tattoo? | | |
|---|---|---|---|
| **Ear Pierce?** | *No* | *Yes* | *Total* |
| **No** | 381 | 43 | 424 |
| **Yes** | 99 | 42 | 141 |
| **Total** | 480 | 85 | 565 |

a. For the population of men with no ear pierce, find a 90% confidence interval for the proportion with a tattoo.

b. For the population of men with an ear pierce, find a 90% confidence interval for the proportion with a tattoo.

**c.** Find a 99% confidence interval for the difference in proportions of men with a tattoo for the populations with and without an ear pierce. Write a sentence interpreting the interval.

10.73 Fatty acids that are present in fish oil may be useful for treating some psychiatric disorders. An article reported on September 3, 1998, at the Yahoo Health website described a randomized experiment done by a Harvard Medical School researcher in which 14 bipolar patients received fish oil daily, while 16 other patients received a placebo daily. After four months, 9 of the 14 patients who received fish oil had responded favorably, but only 3 of the 16 placebo patients had done so.

a. Calculate the difference between the proportions showing favorable response in the two groups, and also calculate a standard error for this difference.

b. Discuss whether the method described in Section 10.4 should be used to calculate a confidence interval for the difference. If you believe that it should, calculate the confidence interval.

c. In Chapter 4, blind and double-blind procedures were discussed. Explain how those concepts would be applied to this experiment.

## Dataset Exercises

10.74 For this exercise, use the **GSS-02** dataset. The dataset contains data from the 2002 General Social Survey, and the variable *polparty* indicates the political party preference of the respondent. Calculate a 95% confidence interval for the proportion of Americans who indicated that they prefer the Democratic party.

10.75 For this exercise, use the **GSS-02** dataset. The dataset contains data from the 2002 General Social Survey, and the variable *marijuan* indicates the respondent's opinion about the legalization of marijuana. Calculate a 95% confidence interval for the proportion of Americans who are opposed to the legalization of marijuana.

10.76 For this exercise, use the **pennstate3** dataset. The data are from a student survey in a statistics class at Penn State. The variable *atfirst* gives data for whether or not a student believes in love at first sight.

a. What is the sample proportion that believes in love at first sight?

b. Find a 95% confidence interval for the population proportion that believes in love at first sight.

c. Write a sentence that interprets the interval found in part (b).

10.77 Refer to Exercise 10.76 about the variable *atfirst* in the **pennstate3** dataset.

a. Create a two-way table that shows the relationship between the variables *atfirst* and *Sex*. (See Sections 2.3 and 6.1 to review two-way tables.)

b. What proportion of the females in the sample believe in love at first sight? What proportion of the males believe in love at first sight?

c. For each sex separately, find a 95% confidence interval for the population proportion that believes in love at first sight.

d. Find a 95% confidence interval for the difference in the population proportions. What does the confidence interval indicate about the difference between the proportions of males and females who believe in love at first sight?

10.78 Use the dataset **GSS-02.** The variable *owngun* indicates whether the respondent owns a gun or not, and the variable *polparty* contains the respondent's political party preference. Determine a 95% confidence interval for the difference in proportions of Republicans and Democrats who owned a gun that year.

# 11

*Geoffrey Clifford / The Image Bank / Getty Images*

*Do men lose more weight from dieting or from exercising?*

See Example 11.3 *(p. 449)*

# Estimating Means with Confidence

The parameters researchers estimate with confidence intervals are not limited to means and proportions. When more complicated procedures are needed, journal articles provide completed intervals and your job is to interpret them. Principles in this and the previous chapter can be used to understand any confidence interval.

In Chapter 10, we learned how to form and interpret a confidence interval for a population proportion and for the difference in two population proportions. Remember that a *confidence interval* is an interval of values computed from sample data that is likely to include the true population value. For example, based on a sample survey of 1000 Americans, we may be able to say with 95% confidence that the interval from .23 to .29 (or 23% to 29%) will cover the true proportion of all Americans who favor the legalization of marijuana. The *confidence level* for an interval is a proportion or percentage, such as 95%, that describes our confidence in the *procedure* that we used to determine the interval. We are confident that in the long run, the stated confidence level is the percentage of the confidence intervals computed with the procedure that will actually contain the true population value.

In this chapter, we expand the use of confidence intervals to parameters involving means of populations, including the population mean of one quantitative variable, the population mean of paired differences, and the difference in means for two populations. We may, for example, wish to know the mean number of hours per day that college students watch television. As an example of the mean of paired differences, we may want to know how much the mean IQ would change if everyone in the population were to listen to classical music for 30 minutes. As an example of the difference in two population means, we may want to know the difference in effectiveness of two diets, based on comparing the mean weight that would be lost if everyone in a population were to go on one diet to the mean weight that would be lost if they were to go on the other diet. In each of these types of situations, the objective is the same. We use sample information to form a confidence interval estimate of a population value. ▮

***Note to Readers:*** CI Module 0 in Chapter 10 must be covered before covering Chapter 11.

# 11.1 Introduction to Confidence Intervals for Means

Here, we will review material from Chapters 9 and 10 that is relevant to finding confidence intervals for parameters involving means. Because the material is scattered throughout those two chapters, it is useful to review it all in one place.

To begin, let's review the following two definitions, which are relevant to our discussion of the estimation of all population values:

- A **parameter** is a population characteristic. The numerical value of a parameter usually cannot be determined because we cannot measure all units in the population. We have to estimate the parameter using sample information.

- A **statistic,** or **estimate,** is a characteristic of a sample. A statistic estimates a parameter.

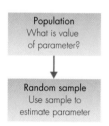

Figure 11.1 illustrates the statistical estimation problem. We wish to estimate a characteristic of a population. In more technical terms, we want to estimate a *parameter*. To do this, we take a random sample from the population and use the sample to calculate a sample *statistic* that estimates the parameter.

You will need to cover the CI Module 0 in Chapter 10 before the material in this chapter. From that introductory module, remember that the general format for a confidence interval for the five parameters that we are studying is

**Figure 11.1** ■ Sample information is used to estimate a population parameter.

$$\text{Sample estimate} \pm \text{Multiplier} \times \text{Standard error}$$

In this introduction, we will discuss each of the three components of this formula for the three situations involving means. After the introduction, there is a module for each of the three parameters, expanding on the material in this section.

Following are descriptions of the three research situations that can be analyzed by using a confidence interval estimate of a population parameter involving means. The parameters in these three situations are as follows:

1. $\mu$ = a population mean.
2. $\mu_d$ = the population mean of paired differences.
3. $\mu_1 - \mu_2$ = the difference between two population means.

Let's revisit the research situations for these parameters that were discussed in Chapter 9 and specify the *sample estimate* used in the confidence interval formula.

### Situation 1: Estimating the Mean of a Quantitative Variable

- *Example research questions:*

    What is the mean amount of time that college students spend watching TV per day?
    What is the mean pulse rate of women?

- *Population parameter:* $\mu$ (spelled *mu* and pronounced "mew") = population mean for the variable

- *Sample estimate:* $\bar{x}$ = the sample mean for the variable

## Situation 2: Estimating the Population Mean of Paired Differences for a Quantitative Variable

- *Example research questions:*

  What is the mean difference in weights for freshmen at the beginning and end of the first semester?

  What is the mean difference in age between husbands and wives in Britain?

- *Population parameter:* $\mu_d$ = population mean of the differences in the two measurements.

- *Sample estimate:* $\overline{d}$ = the mean of the differences for a sample of the two measurements.

## Situation 3: Estimating the Difference Between Two Populations with Regard to the Mean of a Quantitative Variable

- *Example research questions:*

  How much difference is there in average weight loss for those who diet compared to those who exercise to lose weight?

  How much difference is there between the mean foot lengths of men and women?

- *Population parameter:* $\mu_1 - \mu_2$, where $\mu_1$ and $\mu_2$ represent the means in population 1 and 2, respectively

- *Sample estimate:* $\overline{x}_1 - \overline{x}_2$, the difference between the two sample means

---

*thought question* **11.1**  Pick one of the example research questions provided in one of the three situations just described. Explain an appropriate sampling procedure, and then describe what the population parameter and the sample estimate would be for your example.*

---

# Paired Data Versus Independent Samples

It is important to understand the distinction between data collected as paired differences, for which the parameter of interest is $\mu_d$, and data collected as independent samples, for which the parameter of interest is $\mu_1 - \mu_2$.

## Paired Data

Suppose we measure the IQ score of each person in a sample after he or she listens to a relaxation tape and again after he or she listens to a Mozart tape. The difference between the two IQ measurements can be computed for each person, and we could use the sample differences to estimate the mean population difference. This would provide information about whether listening to Mozart produces higher IQ scores, on average, than listening to a relaxation tape does.

The term *paired data* (or *paired samples*) is used when pairs of variables are collected and used in this way. Generally, in such situations, we are interested

---

***HINT:*** Sampling methods are covered in Sections 3.3 and 3.4 of Chapter 3.

only in the population (and sample) of *differences,* not in the original data. Therefore, for paired quantitative variables, it is appropriate to use methods that are designed for inferences about a single population mean. The mean in these situations is the mean for the population of differences that would result if we measured both variables on all units in the population and then found the difference.

---

*definition*    The term **paired data** means that the data have been observed in natural pairs. Some ways in which paired data can occur are as follows:

- Each person is measured twice. The two measurements of the same characteristic or trait are made under different conditions.

- Similar individuals are paired before giving the treatments in an experiment. Each member of a pair then receives a different treatment. The same response variable is measured for all individuals.

- Two different variables are measured for each individual. There is interest in the amount of difference between the two variables.

---

### Independent Samples

Two samples are called **independent samples** when the measurements in one sample are not related to the measurements in the other sample. For example, a comparison of the mean pulse rates of men and women is a comparison of two *independent samples* as long as the individuals that are sampled are not related pairs, such as siblings or spouses. Independent samples are generated in a variety of ways. Here are the most common methods:

- Random samples are taken separately from two populations, and the same response variable is recorded for each individual.

- One random sample is taken, and a variable is recorded for each individual, but then units are categorized as belonging to one population or another, such as male and female.

- Participants are randomly assigned to one of two treatment conditions, such as diet or exercise, and the same response variable, such as weight loss, is recorded for each individual unit.

- Two random samples are taken from a population, and a separate variable is measured in each sample. For example, one sample may be asked about hours of Internet use, whereas a separate sample is asked about hours of television viewing. Then the results are compared. This method for collecting data is not used very often.

**Example 11.1**    **Pet Ownership and Stress** Many studies use a combination of paired data and independent samples. The independent samples are used to compare groups or treatment conditions, and the paired data are used to measure the change in a variable of interest for each person in both groups.

For example, suppose researchers were interested in knowing if owning a pet helps reduce blood pressure for seniors who live alone. They could design an experiment as follows: Recruit a representative group of volunteers who are willing to participate in the experiment. Randomly assign half of the group to receive a pet and the other half to serve as a control group. (To be fair, they could provide a pet to the control group at the end of the study.) Measure the initial blood pressure for all volunteers at the start of the experiment. Then measure their blood pressure again after six months. The data of interest is the change in blood pressure.

Notice that the *change in blood pressure* for each individual is an example of a *paired difference,* taken as (Initial blood pressure − Ending blood pressure) for each person. Collecting the data in this way is an example of the first method of obtaining paired data listed after the definition on page 446: "*Each person is measured twice. The two measurements of the same characteristic or trait are made under different conditions.*" In this example, blood pressure is measured under two different conditions: at the start of the experiment, and six months later, after owning a pet or not.

The *changes in blood pressure for the two groups,* those who are given a pet and the control group of those who are not given one, constitute *independent samples.* Collecting data in this way is an example of the third method of obtaining independent samples, listed on page 446: "*Participants are randomly assigned to one of two treatment conditions, and the same response variable is recorded for each individual unit.*" In this case, the two treatment conditions are being given a pet or not, and the response variable is the *change* in blood pressure from the beginning to the end of the experiment.

The parameters of interest in this experiment would be

$\mu_d$ = mean change in blood pressure for the *population* of all seniors who live alone, if they were to own a pet for six months.

$\mu_1 - \mu_2$ = difference in the mean change in blood pressure for the population of all seniors who live alone, *if* they were to own a pet for six months compared with *if* they were not to own a pet for six months. ∎

## Standard Errors for Sample Statistics Involving Means

Recall from Chapter 10 that the standard error of the sample statistic is a key component for computing confidence intervals for the situations that we are discussing here. Let's review what we have learned about standard errors in general and then revisit what we need to compute standard errors for statistics involving means.

| | |
|---|---|
| *definition* | *Rough Definition:* The **standard error** of a sample statistic measures, roughly, the average difference between the statistic and the population parameter. This "average difference" is over all possible random samples of a given size that can be taken from the population.<br><br>*Technical Definition:* The **standard error** of a sample statistic is the estimated standard deviation of the sampling distribution for the statistic. |

### Standard Error for a Sample Mean or the Mean of Paired Differences

The standard error for the mean of one quantitative variable or the mean of paired differences are found in exactly the same way, except that in the latter case, the data consist of differences in two data values instead of original data values. Therefore, the formula is the same, but the notation for the paired differences includes a subscript of "$d$" to remind us that it is computed using differences.

*formula*

**Standard Error of a Sample Mean**

$$\text{s.e.}(\bar{x}) = \frac{s}{\sqrt{n}}, \qquad s = \text{sample standard deviation}$$

**Standard Error of a Sample Mean of Paired Differences**

$$\text{s.e.}(\bar{d}) = \frac{s_d}{\sqrt{n}}, \qquad s_d = \text{sample standard deviation of the paired differences}$$

**Example 11.2**

Statistics⊘Now™

Watch a video example at **http://1pass.thomson.com** or on your CD.

**Mean Hours per Day That Penn State Students Watch TV**  In a class survey at Penn State, a question was "In a typical day, about how much time do you spend watching television?" Here is a numerical summary, generated by Minitab, of the responses (in hours):

| Variable | N | Mean | Median | TrMean | StDev | SE Mean |
|----------|-----|------|--------|--------|-------|---------|
| TV | 175 | 2.09 | 2.000 | 1.950 | 1.644 | 0.124 |

On the basis of this information, what is our estimate of the mean hours of television per day in the *population* represented by this sample? The sample mean, $\bar{x} = 2.09$ hours per day, is the statistic that estimates $\mu$, the population mean. From the Minitab output, we see that the standard error of the sample mean is $\text{s.e.}(\bar{x}) = 0.124$ (look under "SE Mean"). The calculation of this value is

$$\text{s.e.}(\bar{x}) = \frac{s}{\sqrt{n}} = \frac{1.644}{\sqrt{175}} = 0.124$$

In Section 11.2 (CI Module 3), we will use this information to compute a confidence interval estimate for the mean number of television hours watched by the population of Penn State students. ∎

The standard error in Example 11.2 measures, roughly, the average difference between the unknown value of the population mean $\mu$ and the sample mean $\bar{x}$ for samples of $n = 175$ from this population. Technically, the standard error estimates the standard deviation of the sampling distribution of possible sample means. The sample mean $\bar{x} = 2.09$ hours that is observed in this sample is just one of a plethora of possibilities for the sample mean. The distribution of these possibilities for $\bar{x}$ is centered on the unknown value $\mu$ hours, and the estimated standard deviation of the distribution of sample means is 0.124 hour. Thus, for instance, using the Empirical Rule from Chapter 2 and the approximate normality provided by the Normal Approximation Rule for Means in Chapter 9, we know that $\bar{x}$ should be within 0.124 hour of the true population mean $\mu$ in about 68% of all samples of size 175 from this population.

### Standard Error of the Difference Between Two Sample Means

Recall from Chapter 9 that the **standard error of the difference** between statistics from independent samples is determined as follows:

$$\text{s.e.(difference)} = \sqrt{[\text{s.e.(statistic 1)}]^2 + [\text{s.e.(statistic 2)}]^2}$$

In other words, even though we are *subtracting* the two statistics, we *add* the squared values of their standard errors. This is because the uncertainty in both estimates contributes to the overall uncertainty about the difference in them.

This formula leads to the following formula for the standard error of the difference between two sample means.

---

*formula* | **Standard Error of the Difference Between Two Sample Means**

The **standard error of the difference between two sample means** is

$$\text{s.e.}(\bar{x}_1 - \bar{x}_2) = \sqrt{\frac{s_1^2}{n_1} + \frac{s_2^2}{n_2}}$$

where $s_1$ and $s_2$ are the sample standard deviations for the two samples and $n_1$ and $n_2$ are the corresponding sample sizes.

---

**Example 11.3**

Statistics⬙Now™

Watch a video example at **http://1pass.thomson.com** or on your CD.

**Do Men Lose More Weight by Diet or by Exercise?** In a study done by Wood et al. (1988), 42 sedentary men were placed on a diet and 47 previously sedentary men were put on an exercise routine (study also reported by Iman, 1994, p. 258). The group on a diet lost an average of 7.2 kg, with a standard deviation of 3.7 kg. The men who exercised lost an average of 4.0 kg, with a standard deviation of 3.9 kg.

The difference between the sample means is an estimate of $\mu_1 - \mu_2$, the difference in the means of two hypothetical populations defined by the two weight-loss strategies in the study. The first hypothetical population is the population of weight losses for all sedentary men like the ones in this study if they were to be placed on this diet for a year. The second hypothetical population is the population of weight losses for those same men if they were to follow the exercise routine for a year.

The difference between the two means is $\bar{x}_1 - \bar{x}_2 = 7.2 - 4.0 = 3.2$ kg (about 7 pounds), with the dieters losing more weight than the exercisers. The standard error of the statistic $\bar{x}_1 - \bar{x}_2$ is

$$\text{s.e.}(\bar{x}_1 - \bar{x}_2) = \sqrt{\frac{s_1^2}{n_1} + \frac{s_2^2}{n_2}}$$

$$= \sqrt{\frac{3.7^2}{42} + \frac{3.9^2}{47}} = 0.81$$

In this example, the standard error measures, roughly, the average difference between the statistic $\bar{x}_1 - \bar{x}_2$ and the population parameter $\mu_1 - \mu_2$. Technically, the standard error estimates the standard deviation of the sampling distribution of possible sample differences in means, $\bar{x}_1 - \bar{x}_2$, for samples of sizes 42 and 47 from the populations. The observed sample difference of 3.2 kg is just one of many possibilities for the sample difference. For the distribution of the many possible values of the difference in sample means, the estimated standard deviation is s.e.$(\bar{x}_1 - \bar{x}_2) = 0.81$ kg.

In Section 11.4 (CI Module 5), we will use this information to compute a confidence interval estimate for the difference in what the mean weight loss would be in the population if all such men exercised versus if they dieted. ∎

---

***thought question* 11.2** Notice that all of the standard error formulas in this section have the sample size(s) in the denominator. This tells us that if the sample size is increased, the standard error will decrease (assuming that the sample statistics remain about the same). Refer to the rough definition of standard error, and explain why this relationship between sample size and standard error makes sense, based on that definition.*

---

## Using Student's *t*-Distribution to Determine the Multiplier

The main difference between finding confidence intervals for proportions and confidence intervals for means is the way the appropriate *multiplier* is found. In Chapter 10, we learned that the appropriate multiplier for confidence intervals involving proportions is found from a standard normal, or *z*-distribution. The notation for the multiplier in that case is $z^*$. In contrast, in finding confidence intervals for parameters involving means, the appropriate multiplier is found from **Student's *t*-distribution,** introduced in Chapter 9, and the notation for it is $t^*$.

Remember that theoretically, the *t*-distribution arises when a standardized statistic is calculated by using the sample standard deviation in place of the population standard deviation. This almost always is what happens in estimating means, because the population standard deviation $\sigma$ is rarely known. The rationale for using the *t*-distribution instead of the *z*-distribution will be given in CI Module 3, when we learn how to find a confidence interval for one mean.

A *t*-distribution has a bell shape, it is centered at 0, and it is more spread out than the standard normal curve in that there is more probability in the extreme areas than there is for the standard normal curve. This additional variability results from using *s*, which differs for each sample from the same population, instead of $\sigma$, which is a fixed constant for a given population of measurements.

A parameter called **degrees of freedom,** abbreviated as **df,** is associated with any *t*-distribution. For problems involving a *single mean* or the *mean of paired differences*, df $= n - 1$, where *n* is the sample size. We will learn more about what to use for the degrees of freedom for the difference in two means in CI Module 5, where that situation is covered.

---

***HINT:** Roughly, a standard error measures the average difference between a sample estimate and a parameter, so it measures how closely the sample estimate approximates the parameter.*

| *definition* | For a confidence interval for a population mean, the **$t^*$ multiplier** is the value in a $t$-distribution with df = $n - 1$ such that the area between $-t^*$ and $+t^*$ equals the desired confidence level. |
| --- | --- |

### Finding $t^*$ by Using Table A.2

Most statistical software packages will compute a confidence interval for a mean and for the other parameters involving means. If you are forced to do confidence interval calculations "by hand," it is relatively simple to complete the calculations. To find the multiplier for the desired confidence level, all you need is a table of the $t$-distribution (or the right calculator). Table A.2 in the Appendix shows $t^*$ multipliers for seven different confidence levels and varying degrees of freedom. To use this table, compute the degrees of freedom, choose a confidence level, and then look in the row and column for these values in the table. Figure 11.2 illustrates the relationship between the confidence level and the numerical value $t^*$.

As df gets large, the $t$-distribution gets closer to the standard normal curve because the sample standard deviation $s$ becomes a more accurate estimate of the population standard deviation $\sigma$. The $t$-distribution with df = infinity is identical to the standard normal curve. The last row of Table A.2 shows the values for that situation and can be used to find the multiplier for confidence intervals for proportions, where $z^*$ is used instead of $t^*$.

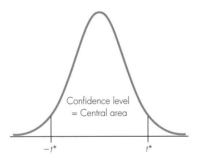

**Figure 11.2** ▮ The relationship between the confidence level and the $t^*$ multiplier

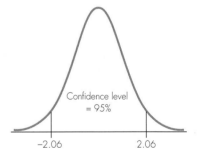

**Figure 11.3** ▮ The $t^*$ multiplier for a 95% confidence interval with df = 24

**Example 11.4**

**Finding the $t^*$ Values for 24 Degrees of Freedom and 95% or 99% Confidence Intervals** Suppose you want to find a 95% confidence interval and also a 99% confidence interval for the mean of a population based on a sample of $n = 25$ values. For the 95% interval, the appropriate multiplier is the value $t^*$ such that the area between $-t^*$ and $t^*$ is .95, or 95%, for a Student's $t$-distribution with degrees of freedom $n - 1 = 24$. The appropriate value, found in Table A.2, is 2.06. Figure 11.3 illustrates the relationship between 95% confidence and $t^* = 2.06$ for this example. For a 99% confidence interval, the appropriate value is $t^* = 2.80$, found in the "0.99" column of Table A.2. Notice that the multiplier for 99% confidence is greater than it is for 95% confidence, which will result in a wider interval. ▮

EXCEL *tip*

> The command TINV(p,d) provides the value $t^*$ for which the area *between* $-t^*$ and $t^*$ is $(1 - p)$ for a Student's $t$-distribution with degrees of freedom $d$. For example, to find $t^*$ for a 95% confidence interval with df = 24, use TINV(.05, 24). The result is 2.06, meaning that the area between $-2.06$ and 2.06 is .95 (or 95%) for a $t$-distribution with df = 24.

11.1 Exercises are on page 484.

*CI Module 3: μ*

# 11.2 CI Module 3: Confidence Intervals for One Population Mean

Make sure you read CI Module 0 (Section 10.2) and Section 11.1, "Introduction to Confidence Intervals for Means," before studying this section. The relevant background material is provided in those sections.

In Lesson 1 of this module, we describe how to find a confidence interval for a population mean using a sample of any size and with any level of confidence desired. In Lesson 2, we describe the special and simpler case of finding an approximate 95% confidence interval when the sample is large. This is a very common situation and therefore is worth a separate discussion.

## Lesson 1: Finding a Confidence Interval for a Mean for Any Sample Size and Any Confidence Level

Researchers often are interested in a very simple question about a population: If we were to measure a quantitative variable for every unit in the population, what would be the mean or average of those values? For example, what is the mean age of death for the population of left-handed adults? What is the mean human body temperature for healthy adults? What is the mean per capita income for all adults in your state? What was the mean price of a gallon of gasoline last week in the United States? What is the mean amount of sleep college students get per night?

All of these questions can be answered by taking an appropriate sample from the population of interest and computing a **confidence interval estimate of the mean** of the population, which is an interval of values computed from the sample data that we can be fairly confident covers the true population mean. How certain we can be is determined by the confidence level that we choose to use.

### How to Compute the Confidence Interval Estimate

By now, you should be familiar with the generic format for a confidence interval for any of the five parameters under discussion:

Sample estimate $\pm$ Multiplier $\times$ Standard error

To find a **confidence interval for a population mean,** the ingredients of the formula are as follows:

*Sample estimate* $= \bar{x} =$ the mean of all of the numbers in the sample. We learned how to compute the sample mean in Chapter 2.

*Multiplier* = $t^*$ = the value of Student's $t$-distribution with df = $n - 1$ and the desired confidence level as the area between $-t^*$ and $t^*$. We discussed how to find $t^*$ in Section 11.1.

*Standard error* = $s/\sqrt{n}$, where $s$ is the sample standard deviation (available by using computer software or found by using the formula on page 51 of Chapter 2) and $n$ is the number of individuals in the sample.

The result is sometimes called a *t-interval* because the multiplier is from Student's $t$-distribution.

Table 11.1 gives some examples of $t$ multipliers, found from Table A.2. If you don't have access to a calculator or computer to compute confidence intervals, make sure you understand how to find $t^*$ values. If the appropriate degrees of freedom are not in the table, find the closest values that are in Table A.2, then either average or use the larger of the two $t^*$ values. For example, for a sample size of 150, df = 149, which is not in Table A.2. The closest choices are df = 100 and df = 1000. For a 95% confidence interval, the respective $t^*$ values are 1.98 and 1.96. Therefore, you could use the average of 1.97 or be conservative and use the larger $t^*$ value of 1.98.

**Table 11.1**  Some $t^*$ Values from Table A.2 and Resulting Confidence Interval Formulas

| Sample Size $n$ | df = $n - 1$ | Confidence Level | Multiplier = $t^*$ | C.I. Formula |
|---|---|---|---|---|
| 25 | 24 | 90% | 1.71 | $\bar{x} \pm 1.71 \dfrac{s}{\sqrt{n}}$ |
| 25 | 24 | 95% | 2.06 | $\bar{x} \pm 2.06 \dfrac{s}{\sqrt{n}}$ |
| 25 | 24 | 99% | 2.80 | $\bar{x} \pm 2.80 \dfrac{s}{\sqrt{n}}$ |
| 101 | 100 | 95% | 1.98 | $\bar{x} \pm 1.98 \dfrac{s}{\sqrt{n}}$ |

Notice these features of $t^*$ values, evident from Table 11.1:

- As the confidence level *increases* (for constant df), the $t^*$ value *increases* as well. This result makes sense because to increase our confidence that the interval covers the true parameter value, we need to increase the width of the interval. Having a large multiplier accomplishes that.

- As the degrees of freedom *increase* (for constant confidence level), the value of $t^*$ *decreases*. This is because the uncertainty surrounding our sample estimate decreases when we have more data. A smaller value of $t^*$ results in a more precise interval.

## Conditions Required for Using the $t$ Confidence Interval

The methods that are presented in this section are derived mathematically by assuming that the sample has been randomly selected from a population in which the response variable has a normal distribution. The $t$-interval procedure, however, is a *robust* procedure because it works well over a wide range of

CI Module 3: $\mu$

situations. Put another way, the stated confidence level for a *t*-interval is approximately correct in many situations in which the assumption about a normally distributed response variable is not correct.

For large sample sizes, a *t*-interval is a valid estimate of the population mean, even in the presence of skewness. For small sample sizes, a *t*-interval can be used if the data are not skewed and contain no outliers.

### Two Situations for Which a *t* Confidence Interval for One Mean Is Valid

*Situation 1:*  The population of the measurements is bell-shaped, and a random sample of any size is measured. In practice, for small samples, the data should show no extreme skewness and should not contain any outliers.

*Situation 2:*  The population of measurements is not bell-shaped, but a *large* random sample is measured. A somewhat arbitrary definition of a "large" sample is $n \geq 30$, but if there are extreme outliers, it is better to have a larger sample.

Before calculating a confidence interval for a mean, first check that one of the situations just described holds. To look for outliers or skewness, plot the data using either a histogram, boxplot, dotplot, or stemplot. The sample mean and median can also be compared to each other. Differences between the mean and the median usually occur if the data are skewed — that is, contain outliers in one direction. If $n \geq 30$, the procedure in this section usually produces a valid confidence interval estimate of the population mean, but if the data are obviously skewed, you should question whether the median might be a better measure of location than the mean.

**Are Your Sleeves Too Short? The Mean Forearm Length of Men**  People are always interested in comparing themselves to others. If you are male, how does the length of your forearm, from elbow to wrist, compare to the average for the population of men? Suppose that the forearm lengths (in centimeters) for a randomly selected sample of $n = 9$ men are as follows:

$$25.5, 24.0, 26.5, 25.5, 28.0, 27.0, 23.0, 25.0, 25.0$$

Figure 11.4 shows a dotplot of these values. There is no obvious skewness, and there are no outliers. Assuming that the sample can be considered a random sample, situation 1 holds, and the necessary conditions for computing a confidence interval for the mean are present.

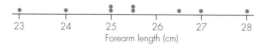

**Figure 11.4** ▮ Dotplot of the forearm lengths for Example 11.5

The sample mean is $\bar{x} = 25.5$ cm and the sample standard deviation is $s = 1.52$ cm. These values can be found using methods covered in Chapter 2. Values for two of the necessary ingredients for calculating a confidence interval for the population mean are as follows.

# Understanding Why the Confidence Interval Formula Works

In Chapter 9, we learned that the standardized statistic for a sample mean has a Student's $t$-distribution with df $= n - 1$. Understanding how that result is derived is beyond the scope of this text, but the result can be used to find the formula for a confidence interval for a population mean. We used a similar technique in CI Module 1 in Chapter 10 to derive the formula for a confidence interval for a population proportion.

Recall from Chapter 9 that

$$t = \frac{\bar{x} - \mu}{\text{s.e.}(\bar{x})}$$

has a $t$-distribution with df $= n - 1$. Suppose that we want a confidence interval with confidence level $C$; typically, $C$ is .95, .90, or .99. Define the value $t^*$ to be such that the area between $-t^*$ and $t^*$ is $C$ for a $t$-distribution with df $= n - 1$. For example, if we want a 95% confidence level, then $C = .95$, and we find $t^*$ so that the area between $-t^*$ and $t^*$ is .95.

Then we can write the following probability statement:

$$P\left(-t^* \leq \frac{\bar{x} - \mu}{\text{s.e.}(\bar{x})} \leq t^*\right) = C$$

$$= P(-t^* \times \text{s.e.}(\bar{x}) \leq \bar{x} - \mu \leq t^* \times \text{s.e.}(\bar{x}))$$

$$= P(-\bar{x} - t^* \times \text{s.e.}(\bar{x}) \leq -\mu \leq -\bar{x} + t^* \times \text{s.e.}(\bar{x}))$$

Multiply everything by $-1$, which reverses the inequality signs, to get

$$P(\bar{x} - t^* \times \text{s.e.}(\bar{x}) \leq \mu \leq \bar{x} + t^* \times \text{s.e.}(\bar{x})) = C$$

In other words, out of all the times we compute the interval $\bar{x} \pm t^* \times \text{s.e.}(\bar{x})$, the long-run probability that the result covers $\mu$ is given by the confidence level we chose, $C$.

Once we have data and compute a numerical interval, we can no longer apply the probability statement to the result. This is similar to other probability situations, such as flipping a coin. Before we determine the outcome (heads or tails), the probability of a head can be said to be .5. But after we have the outcome, the probability of getting a head is either 0 (if we got a tail) or 1 (if we got a head). It no longer makes sense to talk about the probability as .5. What we can continue to say is that the long-run probability of a head is .5.

Similarly, once we have observed data and computed a confidence interval, either it covers the true $\mu$ or it doesn't. Unlike the coin situation, we don't know whether it does or not, but it still doesn't make sense to talk about the probabilities of various outcomes when we have one observed outcome. All we can say is that we have a high degree of confidence in our procedure, determined by $C$, and that we are aware that there is a small chance that the confidence interval estimate will not cover the true parameter.

**CI Module 3: $\mu$**

> **MINITAB *tip*** | **Computing a Confidence Interval for a Mean**
>
> - To compute a confidence interval for a mean, use **Stat>Basic Statistics> 1-sample t.** Specify the column that contains the raw data. To change the confidence level, use the **Options** button. The default confidence level is 95%.
>
> - If the raw data are not available but the data are summarized so that you know values of the sample mean, standard deviation, and sample size, enter those values in the "Summarized data" area of the dialog box.
>
> *Note:* Use the **Graphs** button to create visual displays of the data.

## Lesson 2: Special Case: Approximate 95% Confidence Intervals for Large Samples

When we learned about Student's *t*-distribution, we learned that for large degrees of freedom, the distribution is very close to the standard normal distribution. Therefore, when the sample is large and thus the degrees of freedom are large, the $t^*$ multiplier will be close to the corresponding $z^*$ multiplier. You can see this empirically in Table A.2. For 95% confidence, the $z^*$ multiplier is 1.96 but is often rounded off to 2. The $t^*$ multiplier for df = 1000 is 1.96 as well; for df = 100, it is 1.98; and for df = 70, 80, or 90, $t^*$ is 1.99, which is still close to the exact $z^*$ value of 1.96. Even with as few as 30 df, the $t^*$ multiplier is 2.04, not far from the exact $z^*$ of 1.96 and even closer to the rounded-off value of 2.0.

In checking conditions for which the formula for a confidence interval for a mean applies, we see that it always applies in situation 2, when a *large* random sample is measured, regardless of the shape of the distribution. Therefore, when we have a large sample, we can simplify matters and find an *approximate* 95% confidence interval estimate using a multiplier of 2.0 instead of $t^*$. The argument works for the other parameters that are covered in this chapter and in Chapter 10 as well, so let's write the result to encompass them too.

> *definition* | For sufficiently large samples, the interval
>
> **Sample estimate ± 2 × Standard error**
>
> is an **approximate 95% confidence interval** for a population parameter. This is so for a proportion, a mean, the difference between two proportions, and the difference between two means.
>
> *Note:* The 95% confidence level describes how often the procedure provides an interval that includes the population value. For about 95% of all random samples of a specific size from a population, the confidence interval captures the population parameter.

**Example 11.8** | **Approximate 95% Confidence Interval for TV Time** In Example 11.6, we found that a 95% confidence interval for mean number of hours Penn State students watch TV was 2.09 ± (1.98)(0.124) or 2.09 ± 0.25, or 1.84 hours to

*CI Module 3: μ*

Statistics⌂Now™

Watch a video example at **http://1pass.thomson.com** or on your CD.

2.34 hours. The sample mean, based on a sample $n = 175$ students, was $\bar{x} = 2.09$ hours per day. The standard error is s.e. $(\bar{x}) = 0.124$, and an approximate 95% confidence interval estimate of $\mu$, the population mean, is

$$\text{Sample estimate} \pm 2 \times \text{Standard error}$$
$$\bar{x} \pm 2 \times \text{s.e.}(\bar{x})$$
$$2.09 \pm 2 \times 0.124$$
$$2.09 \pm 0.248$$
$$1.842 \text{ to } 2.338 \text{ hours}$$

You can see that the approximate 95% confidence interval is essentially the same as the one that was computed by using the multiplier $t^*$. It's obvious that this should be the case, since the $t^*$ multiplier was 1.98, very close to the multiplier of 2 used for the approximate interval.

In a research article, this confidence interval might be described as follows: *We are 95% confident that the mean time that Penn State students spend watching television per day is somewhere between 1.842 and 2.338 hours.* Actually, the report would probably round off the numbers and declare that, with 95% confidence, the mean is between 1.8 and 2.3 hours. Remember what is meant technically by that statement: We are 95% confident that the interval from 1.842 to 2.338 hours is correct in that it captures the mean television viewing hours for Penn State students. ■

11.2 Exercises are on page 486.

---

***thought question 11.3*** • What population do you think is represented by the sample of 175 students in Example 11.8? Do you think the Fundamental Rule for Using Data for Inference (reviewed on p. 404 of Chapter 10) holds in this case?

• Does the confidence interval in Example 11.8 tell us that *95% of the students* watch television between 1.842 and 2.338 hours per day? If not, what exactly does the interval tell us?*

---

## 11.3 CI Module 4: Confidence Interval for the Population Mean of Paired Differences

Remember that an important special case of a single mean of a population occurs when two quantitative variables are collected in pairs and we desire information about the *difference* between the two variables. For example, we may want to know the average difference between Verbal SAT scores for individuals before and after taking a training course. Or we may want to know the average difference in manual dexterity for the dominant and nondominant hands. In both of these examples, we would collect two measurements from each individual in the sample.

**\*HINT:** Refer to the statement in Example 11.8 that shows how the result might be described in a research article.

To analyze paired data, we begin by calculating the difference in the two measurements for each pair in the sample. As notation for an analysis of the differences, we could simply use the notation that was already developed for a single mean, but it is preferable to use notation emphasizing that the data consist of differences. Here is notation that can be used for paired data:

## Notation for Paired Differences

*Data:*    two variables for each of $n$ individuals or pairs; use the difference $d = x_1 - x_2$.

*Population parameter:*    $\mu_d$ = mean of differences for the population = $\mu_1 - \mu_2$.

*Sample estimate:*    $\bar{d}$ = sample mean of the differences = $\bar{x}_1 - \bar{x}_2$.

*Standard deviation and standard error:*    $s_d$ = standard deviation of the sample of differences; s.e.$(\bar{d}) = s_d/\sqrt{n}$.

*Confidence interval for $\mu_d$:*    $\bar{d} \pm t^* \times$ s.e.$(\bar{d})$, where df $= n - 1$ for the multiplier $t^*$.

## Conditions Required for Using the *t* Confidence Interval for the Mean of Paired Differences

Once the differences have been computed, the conditions that are required are the same as they are for a *t*-interval for one population mean. The distinction is that the conditions must hold for the data set of differences. Let's rewrite the two situations for which the *t*-interval formula is appropriate, using the terminology for the differences:

*Situation 1:*    The population of *differences* is bell-shaped, and a random sample of any size is measured. In practice, for small samples, the differences in the sample should show no extreme skewness and should not contain any outliers.

*Situation 2:*    The population of *differences* is not bell-shaped, but a *large* random sample is measured. A somewhat arbitrary definition of a "large" sample is $n \geq 30$ pairs, but if there are extreme outliers in the sample of differences, it is better to have a larger sample.

Before calculating a **confidence interval for the population mean of paired differences,** check that there are enough pairs to satisfy situation 2. If not, examine a plot of the differences—a histogram, dotplot, boxplot or stem-and-leaf plot—to make sure there are no excessive outliers or extreme skewness.

Notice that it is not equivalent to look at plots of the two original variables before taking the differences. It is quite possible to have extreme outliers or skewness in the original data but not in the differences. For example, suppose the data represent IQ after listening to a relaxation tape and after listening to Mozart for a sample of randomly selected adults. It is quite possible that there will be someone with an abnormally high or low IQ in the sample. For each of the two original IQ variables, that person's data may look like an outlier. But it is unlikely that the person's *difference* in IQ for the two conditions would stand out from the others.

**Example 11.9**

Statistics⬡Now™

Watch a video example at **http://
1pass.thomson.com** or on your CD.

For software help, download your
Minitab, Excel, TI-83, SPSS, R, and
JMP manuals from **http://1pass
.thomson.com**, or find them on
your CD.

**Screen Time—Computer Versus TV** The 25 students in a liberal arts course
in statistical literacy were given a survey that included questions on how many
hours per week they watched television and how many hours a week they used
a computer. The responses are shown in Table 11.2, along with the difference $d$
for each student. From these data, let's construct a 90% confidence interval for
$\mu_d$, the average *difference* in hours spent using the computer versus watching
television (Computer use − TV) for the population of students represented by
this sample.

**Table 11.2** Data on Weekly Hours of Computer Use and TV Viewing

| Student | Computer | TV | Difference | Student | Computer | TV | Difference |
|---------|----------|------|------------|---------|----------|------|------------|
| 1 | 30 | 2.0 | 28.0 | 14 | 5 | 6.0 | −1.0 |
| 2 | 20 | 1.5 | 18.5 | 15 | 8 | 20.0 | −12.0 |
| 3 | 10 | 14.0 | −4.0 | 16 | 30 | 20.0 | 10.0 |
| 4 | 10 | 2.0 | 8.0 | 17 | 40 | 35.0 | 5.0 |
| 5 | 10 | 6.0 | 4.0 | 18 | 15 | 15.0 | 0.0 |
| 6 | 0 | 20.0 | −20.0 | 19 | 40 | 5.0 | 35.0 |
| 7 | 35 | 14.0 | 21.0 | 20 | 3 | 13.5 | −10.5 |
| 8 | 20 | 1.0 | 19.0 | 21 | 21 | 35.0 | −14.0 |
| 9 | 2 | 14.0 | −12.0 | 22 | 2 | 1.0 | 1.0 |
| 10 | 5 | 10.0 | −5.0 | 23 | 9 | 4.0 | 5.0 |
| 11 | 10 | 15.0 | −5.0 | 24 | 14 | 0.0 | 14.0 |
| 12 | 4 | 2.0 | 2.0 | 25 | 21 | 14.0 | 7.0 |
| 13 | 50 | 10.0 | 40.0 | | | | |

First, check that the conditions for a confidence interval for one mean hold
for the differences. The sample size is under 30, so it is important to make sure
that there are no major outliers or extreme skewness. The sample mean and
median are 5.36 and 4.0 hours, respectively. In data ranging from −20 to +40,
these statistics are close enough to rule out extreme skewness. The boxplot of
the differences in Figure 11.7 provides further evidence that the appropriate
conditions are satisfied.

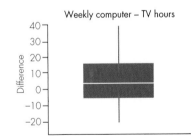

**Figure 11.7** ▮ Boxplot for difference between
computer and TV hours per week, $n = 25$

The standard deviation of the differences is 15.24 hours, and with $n = 25$,
the standard error is $15.24/\sqrt{25} = 3.05$. The df $= 25 − 1 = 24$, so from Table A.2,
for a 90% confidence interval, $t^* = 1.71$. The computation for a 90% confidence
interval is

$$\bar{d} \pm t^* \times \text{s.e.}(\bar{d})$$
$$5.36 \pm (1.71)(3.05)$$
$$5.36 \pm 5.22 \text{ or } 0.14 \text{ to } 10.58 \text{ hours}$$

CI Module 4: $\mu_d$

*Interpretation:* We are 90% confident that the average difference between computer usage and television viewing for students represented by this sample is covered by the interval from 0.14 to 10.58 hours per week, with more hours spent on computer usage than on television viewing. ■

## Interpreting a Confidence Interval for the Mean of Paired Differences

In addition to the usual interpretation of a confidence interval, it is often of interest to know whether the confidence interval for the mean of paired differences covers 0. If it does cover 0, then it is possible that the population mean is 0, indicating that the population means for the two measurements could be the same. If the interval does not cover 0, then we can be fairly certain that the population means for the two variables are different. How certain we can be depends on the confidence level that is used for the computation.

In Example 11.9, we found that a 90% confidence interval for the mean difference between hours of computer use and hours of television watching is 0.14 hour to 10.58 hours per week. The interval does not cover 0, so we can be fairly certain that for the population represented by this sample, students really do spend more time on computer usage than on watching television, on average. What conclusion do you think we would have reached if we had used a 95% confidence level? (See Exercise 11.36.)

> **thought question 11.4** For a fixed sample, explain why it is logical that a 95% confidence interval covers a wider range of values than a 90% confidence interval. Explain this in terms of our confidence that the procedure works in any given case. *

---

*in summary*   **Calculating a Confidence Interval for the Population Mean of Paired Differences**

A confidence interval for the population mean of paired differences $\mu_d$ (also called a **paired *t*-interval**) is

$$\bar{d} \pm t^* \times \text{s.e.}(\bar{d}), \quad \text{which is } \bar{d} \pm t^* \times \frac{s_d}{\sqrt{n}}$$

To determine this interval:

1. Make sure the appropriate conditions apply by checking sample size and/or a shape picture of the differences.
2. Choose a confidence level.
3. Compute the differences for the $n$ pairs in the sample, then find the mean $\bar{d}$ and standard deviation $s_d$ for those differences.

---

**\*HINT:** Consider the extreme case. How could you construct an interval that you are 100% confident covers the truth? Would it be wider or narrower than a 90% confidence interval?

4. Calculate the standard error of $\bar{d}$: s.e.$(\bar{d}) = \dfrac{s_d}{\sqrt{n}}$.

5. Calculate df $= n - 1$.

6. Use Table A.2 (or statistical software) to find the multiplier $t^*$.

Sometimes you will be provided with the summarized data rather than the original differences, as the next example illustrates. In that case, you won't have to carry out step 3 of the summary.

**Example 11.10**

Statistics⬛Now™

Read the original source on your CD.

**Meditation and Anxiety** Davidson et al. (2003) conducted a study in which volunteers were randomly assigned to a meditation group or a control group. The meditation group took an eight-week training course during which they were to practice meditation daily. (The control group was offered the training at the conclusion of the experiment.) The participants were given a standard test for level of anxiety, called the Spielberger State-Trait Anxiety Inventory. The "trait" part of the test, used in this experiment, is supposed to measure "trait anxiety," which is how the person generally feels. It is a characteristic that remains relatively constant. (The "state" anxiety part of the test measures how anxious someone is at the point in time when they are taking it and is used for studies of things such as test anxiety.) The test consists of 20 questions for which respondents rate themselves on a scale from 1 to 4. An example of a question is "I lack self-confidence," with choices ranging from 1 = "almost never" to 4 = "almost always." The maximum possible score on the test is 80, and the minimum is 20, but the majority of scores tend to be in the 30s and 40s.

The participants were given the test at the beginning of the study and again after eight weeks of meditation training. The *difference* in the scores were found for each of the $n = 21$ people in the meditation group who completed the test both times, with the goal of determining whether meditation reduces anxiety. The *parameter* of interest is the mean change in the anxiety score for the population if everyone were to be given the eight-week meditation training.

The article did not provide the original data, so the results that are presented here are approximated from information that was provided. Because the original data weren't provided, there is no way to check for outliers or skewness in the differences. We will proceed under the assumption that situation 1 is satisfied. Let's find a 95% confidence interval. The sample mean for the 21 differences was $\bar{d} = 4.5$, and the sample standard deviation was $s_d = 7.2$, so the standard error is

$$\text{s.e.}(\bar{d}) = \frac{s_d}{\sqrt{n}} = \frac{7.2}{\sqrt{21}} = 1.57$$

The df $= 21 - 1 = 20$, so from Table A.2, $t^* = 2.09$. Therefore, a 95% confidence interval estimate for the population mean reduction in anxiety if everyone were to be trained in meditation is

$$4.5 \pm 2.09 \times 1.57 \quad or \quad 4.5 \pm 3.3 \quad or \quad 1.2 \text{ to } 7.8$$

Notice that the entire interval is above 0. Therefore, we can say with high confidence that the mean anxiety level would go down in the population if everyone were to take the meditation training course. The control group scores ac-

**CI Module 4: $\mu_d$**

11.3 Exercises are on page 487.

tually went up slightly during the eight-week period, with a mean increase of about 1 point. ∎

---

| MINITAB *tip* | **Computing a Confidence Interval for the Mean of Paired Differences** |

These instructions are to be used if you have the raw data for the original two variables stored in two columns, with one pair in each row. There are two methods that can be used:

- First, calculate a column of differences using **Calc>Calculator,** then use **Stat>Basic Statistics>1-sample t.** Specify the column that contains the differences. To change the confidence level, use the **Options** button. The default confidence level is 95%.

- Alternatively, use **Stat>Basic Statistics>Paired t.** Specify the two columns that contain the raw data for the pair of measurements.

- If the raw data are not available but you know values of the mean and standard deviation of the sample of differences as well as the sample size, enter those values in the "Summarized data" area of the dialog box.

*Note:* Use the **Graphs** menu to create visual displays of the data to check for outliers and skewness.

---

| TI-84 *tip* | **Confidence Interval for a Mean or the Mean of Paired Differences** |

- Press [STAT] and scroll horizontally to **TESTS.** Scroll vertically to **8:Tinterval** and press [ENTER].

- If the individual data values are stored in a list (for example, L1) select **Data** as the input method. Enter the data list name (say, L1) and a confidence level. The entry for *Freq:* should be 1, the default value. Scroll to *Calculate* and press [ENTER].

- For paired data, enter the first value for each pair in list L1 and the second value for each pair in list L2. With a clean home screen use the keystrokes [2nd] [1] [−] [2nd] [2] [STO>] [2nd] [3] [ENTER] to create and store the differences in list L3. Then proceed as above, using L3 as the data list.

- If summary statistics are already known, select **Stats** as the input method. Enter values for the sample mean, standard deviation, sample size, and a confidence level. Scroll to *Calculate* and press [ENTER].

---

# 11.4  CI Module 5: Confidence Interval for the Difference in Two Population Means

## Lesson 1: The General (Unpooled) Case

Let's review the situation that generates research questions for which the difference between two population means is of interest. We have two populations or groups from which independent samples are available, or one population for

**Example 11.12**

**Parental Alcohol Problems and Child Hangover Symptoms**  Slutske, Pia-secki, and Hunt-Carter (2003) conducted a study on hangover symptoms in college students, using a sample of 1227 students who were enrolled in introductory psychology classes and who acknowledged that they currently use alcohol. The participants rated the severity of their experience with 13 hangover symptoms during the previous year, with possible ratings of 0 (never experience) to 4 (always experience). For the analysis in this example, the researchers simply counted how many of the symptoms received a nonzero rating for each student, so the score could range from 0 to 13. The mean and standard deviation for all of the students were 5.2 and 3.4, respectively. Thus, the average number of symptoms experienced at all during the previous year for this sample of students was 5.2 out of the 13 possible.

One concern that the researchers explored was that children whose parents had problems with alcohol might have problems of their own and therefore might experience more hangover symptoms. They asked the respondents the question "To your knowledge, have your biological parents had any of the following problems related to their use of alcohol: fulfilling obligations at school or work, physical health or emotional problems, problems with family or friends, legal problems (e.g., DUI, DWI)? (Slutske et al., 2003, p. 1444).

Twenty-three percent of the 1227 respondents (282) answered yes to the question, and 945 answered no. Let's find a 95% confidence interval estimate for the difference in mean hangover symptoms for the *populations* of students who would answer yes (population 1) and no (population 2). The sample statistics were as follows:

| Group | Mean | Standard Deviation |
|---|---|---|
| Parental alcohol problems ($n_1 = 282$) | $\bar{x}_1 = 5.9$ | $s_1 = 3.6$ |
| No parental alcohol problems ($n_2 = 945$) | $\bar{x}_2 = 4.9$ | $s_2 = 3.4$ |

So $\bar{x}_1 - \bar{x}_2 = 5.9 - 4.9 = 1$, and

$$\text{s.e.}(\bar{x}_1 - \bar{x}_2) = \sqrt{\frac{s_1^2}{n_1} + \frac{s_2^2}{n_2}} = \sqrt{\frac{3.6^2}{282} + \frac{3.4^2}{945}} = 0.24$$

This tells us that for this sample, the students with a parental history of alcohol problems experienced an average of one more hangover symptom in the past year than did students without that history. We would like to know what we can say about the difference in the mean number of hangover symptoms for the corresponding populations.

The sample sizes are large enough to proceed without checking the shape of the data. Using the conservative df $= 282 - 1 = 281$, we find from Table A.2 that $t^*$ is between 1.98 and 1.96. We will average them and use $t^* = 1.97$. The resulting confidence interval estimate for $\mu_1 - \mu_2$ is:

$$\bar{x}_1 - \bar{x}_2 \pm t^* \times \text{s.e.}(\bar{x}_1 - \bar{x}_2)$$
$$1 \pm (1.97)(0.24)$$
$$1 \pm 0.47$$
$$0.53 \text{ to } 1.47 \text{ symptoms}$$

CI Module 5: $\mu_1 - \mu_2$

Therefore, we can conclude that in the population of students similar to these, students with a biological parent who experienced alcohol problems are likely to have an average number of hangover symptoms that is from 0.53 to 1.47 higher than the average number of symptoms of students without that parental history. Notice that the entire interval is above 0, so we can be fairly certain that there really is a difference in the mean number of hangover symptoms in the populations of students with and without this parental history. In other words, we can be fairly certain that $\mu_1$ really is higher than $\mu_2$. ∎

## Approximate 95% Confidence Interval Using Multiplier 2

As with the other types of confidence intervals in Chapters 10 and 11, in certain circumstances, we can find an approximate 95% confidence interval for the difference in two means by using the multiplier 2 instead of using $t^*$. In finding a confidence interval for the difference in two means, the approximation will work well as long as the number of degrees of freedom is large. Remember that as degrees of freedom increase, the $t^*$ multiplier gets closer and closer to the corresponding $z^*$ multiplier. For 95% confidence, the $z^*$ multiplier is 1.96 but is often rounded up to 2.0.

When both sample sizes are at least 30 the approximation works well. Notice that in that case, the conservative degrees of freedom value is the minimum sample size $- 1 = 30 - 1 = 29$, and from Table A.2, we see that the $t^*$ multiplier for df = 29 is 2.05. But the df value of 29 is conservative, and in fact in this situation, the actual degrees of freedom (found by using Welch's approximation) would be quite a bit larger. Therefore, we recommend using the approximation as long as both sample sizes are at least 30.

---

*definition* | When both sample sizes are at least 30, the interval

**Difference in sample means ± 2 × Standard error**

is an **approximate 95% confidence interval for the difference in population means.**

---

**Example 11.13**

Statistics△Now™

Watch a video example at **http:// 1pass.thomson.com** or on your CD.

**Confidence Interval for Difference in Mean Weight Losses by Diet or Exercise** In Example 11.3, we learned about a study comparing the weight loss for men who dieted (group 1) and men who exercised (group 2). The sample mean weight losses were 7.2 kg and 4.0 kg, for a difference of 3.2 kg. In Example 11.3, we found that the standard error was 0.81. Therefore, an approximate 95% confidence interval for the difference in population means is

Sample difference $\pm 2 \times$ Standard error

$3.2 \pm 2 \times 0.81$

$3.2 \pm 1.62$

1.58 to 4.82 kg ∎

*thought question 11.5* In Section 11.3, we learned how to find a confidence interval for the mean of paired differences, which we used in Example 11.9 to estimate the mean difference in weekly computer and TV hours for a population of liberal arts students.

- Explain why it would not have been appropriate to use the methods in this section, for two independent samples, to estimate that mean difference, even though in either case the sample estimate is the difference in sample means, 5.36 hours.
- If the methods in this section had been erroneously used by treating computer usage and TV viewing hours as independent samples, do you think the standard error of the sample estimate would have been larger or smaller than it was in Example 11.9? Explain your answer using common sense, not formulas. Think about how much natural variability there would be in the data for two independent samples, compared with measuring both sets of hours on the same individuals.*

## Lesson 2: The Equal Variance Assumption and the Pooled Standard Error

In estimating the difference between two population means, it may sometimes be reasonable to assume that the populations have equal standard deviations. The term *variance* describes the squared standard deviation, so the assumption of equal standard deviations for the two populations is the same as assuming that the variances are equal. Using statistical notation, we can express the assumption of equal population variances as $\sigma_1^2 = \sigma_2^2 = \sigma^2$, where $\sigma^2$ represents the common value of the variance. With this **equal variance assumption,** information from both groups is combined in order to estimate the value of $\sigma^2$. The estimate of variance based on the combined or "pooled" data is called the **pooled variance.** The square root of the pooled variance is called the **pooled standard deviation,** and it is computed as

$$\text{Pooled standard deviation} = s_p = \sqrt{\frac{(n_1 - 1)s_1^2 + (n_2 - 1)s_2^2}{n_1 + n_2 - 2}}$$

*technical note* Notice that the pooled variance is a "weighted" average of the individual sample variances. The weights are $(n_1 - 1)/(n_1 + n_2 - 2)$ and $(n_2 - 1)/(n_1 + n_2 - 2)$, giving the variance of the larger sample more weight. If $n_1 = n_2$, then the weights are equal, and $s_p^2 = (s_1^2 + s_2^2)/2$.

Replacing the individual standard deviations $s_1$ and $s_2$ with the pooled version $s_p$ in the formula for standard error leads to the **pooled standard error for**

***HINT:*** Remember that the standard error measures how much the sample estimate is likely to vary from sample to sample. Do you think the paired sample estimate $\bar{d}$ would be likely to vary more than or less than the difference in the two separate means?

**the difference between two means:**

$$\text{Pooled s.e.}(\bar{x}_1 - \bar{x}_2) = \sqrt{\frac{s_p^2}{n_1} + \frac{s_p^2}{n_2}} = \sqrt{s_p^2\left(\frac{1}{n_1} + \frac{1}{n_2}\right)} = s_p\sqrt{\frac{1}{n_1} + \frac{1}{n_2}}$$

This all may seem quite complicated, but if the assumption of equal population variances is correct, this complication actually provides a cleaner mathematical solution for the determination of the multiplier $t^*$. In this case, the degrees of freedom are simply df $= n_1 + n_2 - 2$.

---

*formula*

**Pooled Confidence Interval for the Difference Between Two Population Means (Independent Samples)**

If we assume that the population variances are the same, then the confidence interval for $\mu_1 - \mu_2$, the difference between the population means, is

$$\bar{x}_1 - \bar{x}_2 \pm t^* \sqrt{s_p^2\left(\frac{1}{n_1} + \frac{1}{n_2}\right)}$$

$t^*$ is found by using a $t$-distribution with df $= n_1 + n_2 - 2$, and $s_p$ is the pooled standard deviation.

---

**MINITAB** *tip*    In Minitab, the default option for the two-sample $t$ procedure is the "unpooled" version in which the population variances are not assumed to be equal. To get the pooled version, check the dialog box item that says "Assume Equal Variances."

---

**Example 11.14**

Statistics⬡Now™

Watch a video example at **http://1pass.thomson.com** or on your CD.

For software help, download your Minitab, Excel, TI-83, SPSS, R, and JMP manuals from **http://1pass.thomson.com**, or find them on your CD.

CI Module 5: $\mu_1 - \mu_2$

**Pooled $t$-Interval for Difference Between Mean Female and Male Sleep Times**    Students in an introductory statistics class filled out a survey on a variety of issues, including how much sleep they had the previous night. Let's use the data to estimate how much more or less male students sleep than females students, on average. The class included 83 females and 65 males. Let's assume that these students are equivalent to a random sample of all students who take introductory statistics. How much difference is there between how long female and male students represented by this sample slept the previous night? To answer that question, we'll find a 95% confidence interval for $\mu_1 - \mu_2$ = difference in mean sleep hours for females versus males in the population. Here is the Minitab output for the pooled procedure in which we assume equal population variances for males and females

Two-sample T for sleep [with "Assume Equal Variance" option]

| Sex | N | Mean | StDev | SE Mean |
|---|---|---|---|---|
| Female | 83 | 7.02 | 1.75 | 0.19 |
| Male | 65 | 6.55 | 1.68 | 0.21 |

Difference = mu (Female) − mu (Male)
Estimate for difference: 0.461
95% CI for difference: (−0.103, 1.025)
T-Test of difference = 0 (vs not =): T-Value = 1.62     P-Value = 0.108     DF = 146
Both use Pooled StDev = 1.72

Notice that the sample standard deviations are very similar (1.75 for females and 1.68 for males), so it may be reasonable to assume that the population variances are similar as well. The sample mean given for females is $\bar{x}_1 = 7.02$ hours, the sample mean given for males is 6.55 hours, and the difference between these two sample means is reported on the output as "Estimate for difference: 0.461 hour." Notice that the sample means were rounded to two decimal places in the output which causes an inconsistency because the difference between 7.02 and 6.55 is 0.47. The value 0.461 is, however, the correct difference between the sample means.

The 95% confidence interval for the difference in mean hours of sleep for the populations of female and male students is **−0.103 to 1.025** hours (in bold and underlined in the output display). Because the interval covers 0, we can't rule out the possibility that the population means are equal for men and women, although for this sample, the women slept an average of about half an hour more than the men did.

The pooled standard deviation given at the bottom of the output is $s_p = 1.72$. Although not given in the output, the pooled standard error is 0.284. These two values can be found as follows:

$$s_p = \sqrt{\frac{(83 - 1)(1.75)^2 + (65 - 1)(1.68)^2}{83 + 65 - 2}} = \sqrt{2.957} = 1.72$$

$$\text{Pooled s.e.}(\bar{x}_1 - \bar{x}_2) = \sqrt{\frac{(1.72)^2}{84} + \frac{(1.72)^2}{65}} = 1.72\sqrt{\frac{1}{84} + \frac{1}{65}} = 0.284$$

In the pooled procedure, the degrees of freedom are found as df $= n_1 + n_2 - 2$, which in this example equals $83 + 65 - 2 = 146$. If Table A.2 is used, a conservative estimate of the multiplier $t^*$ is 1.98 (for df $= 100$ and 0.95 confidence). The confidence limits found by Minitab can be verified by calculating $0.461 \pm (1.98)(0.284)$. ∎

## Pooled or Unpooled?

In Example 11.14, the sample standard deviations for females and males had about the same values, so it was reasonable to use the assumption that the population standard deviations were equal. The confidence interval for the difference in means, however, would have been about the same even if the assumption of equal standard deviations had not been made. With the unpooled procedure, the 95% confidence interval for the difference in population means is −0.10 to 1.03 hours, quite close to the pooled version. One advantage for the pooled version is that finding the value of the degrees of freedom is simpler.

Sample standard deviations in two independent samples will almost never be identical in practice. So how do we know when it might be reasonable to use the pooled version of a confidence interval for a difference between two population means? And what is the risk of using the pooled procedure when, in truth, the population standard deviations differ? We will explore that question in detail when we consider hypothesis testing for the difference between two means in Chapter 13, but here we give some preliminary guidance.

CI Module 5: $\mu_1 - \mu_2$

If the larger value of the two sample standard deviations is from the group with the larger sample size, the pooled procedure will tend to give a wider confidence interval than the unpooled version and so would be a conservative estimate of the true difference, as the next example illustrates. Similar to when we used the conservative margin of error in a confidence interval for one proportion, it is acceptable to use the more conservative pooled procedure. But it is not good practice to knowingly create an interval that is wider than necessary. On the other hand, if the smaller of the two sample standard deviations is from the group with the larger sample size, the pooled version of the procedure may produce a misleading narrow interval. Generally, it is best to use the unpooled procedure unless the sample standard deviations are quite similar.

**Example 11.15**

**Statistics⊝Now™**

Watch a video example at **http://1pass.thomson.com** or on your CD.

**Sleep Time with and Without the Equal Variance Assumption** In Example 11.7, we compared confidence intervals for a single mean for the average amount of sleep for two types of statistics students. We can also find a confidence interval for the difference in mean hours of sleep for the two types of students. Here is the Minitab output for doing so, with and without the "Assume Equal Variance" option:

---

Two-Sample T-Test and Confidence Interval

Two-sample T for sleep [without "Assume Equal Variance" option]

| class | N | Mean | StDev | SE Mean |
|-------|-----|------|-------|---------|
| 10 | 25 | 7.66 | 1.34 | 0.27 |
| 13 | 148 | 6.81 | 1.73 | 0.14 |

95% CI for mu (10) − mu (13): (0.23, 1.46)
T-Test mu (10) = mu (13) (vs not =): T = 2.79    P = 0.0083    DF = 38

Two-sample T for sleep [with "Assume Equal Variance" option]

| class | N | Mean | StDev | SE Mean |
|-------|-----|------|-------|---------|
| 10 | 25 | 7.66 | 1.34 | 0.27 |
| 13 | 148 | 6.81 | 1.73 | 0.14 |

95% CI for mu (10) − mu (13): (0.13, 1.57)
T-Test mu (10) = mu (13) (vs not =): T = 2.33    P = 0.021    DF = 171
Both used Pooled StDev = 1.68

---

For this example, it is not reasonable to assume equal variances. Notice that doing so increases the width of the confidence interval. Without the assumption, the interval is from 0.23 to 1.46, but with the assumption, it is 0.13 to 1.57. A comparison of the standard error of the difference for the two situations illustrates why the width increased. Using the pooled estimate increases the standard error of the mean for the sample with $n = 25$.

$$\text{Unpooled s.e.}(\bar{x}_1 - \bar{x}_2) = \sqrt{\frac{(1.34)^2}{25} + \frac{(1.73)^2}{148}} = \sqrt{0.07 + 0.02} = 0.30$$

$$\text{Pooled s.e.}(\bar{x}_1 - \bar{x}_2) = \sqrt{\frac{(1.68)^2}{25} + \frac{(1.68)^2}{148}} = \sqrt{0.11 + 0.02} = 0.36$$

This example illustrates that it is conservative to use the pooled procedure if the two sample sizes are decidedly different and the larger sample standard deviation accompanies the larger sample size. ∎

11.4 Exercises are on page 488.

**CI Module 5: $\mu_1 - \mu_2$**

| MINITAB *tip* | **Computing a Confidence Interval for the Difference Between Two Means for Independent Samples** |
|---|---|

- To compute a confidence interval for the difference in two means, use **Stat>Basic Statistics>2 sample t.** Specify the location of the data. The raw data for the response (*Samples*) may be in one column, and the raw data for group categories (*Subscripts*) may be in a second column. Or the raw data for the two independent groups may be in two separate columns.

- If the raw data are not available but the data are summarized so that you know the sample mean, standard deviation, and sample size for each group, enter those values in the "Summarized data" area of the dialog box.

- To change the confidence level, use the **Options** button.

- To use the *pooled* standard error, click on **Assume Equal Variances.**

*Note:* Use the **Graphs** button to create a comparative dotplot or a comparative boxplot.

*in summary*

# Confidence Interval for the Difference in Two Population Means

A confidence interval for the difference in two population means $\mu_1 - \mu_2$ (also called a **two-sample *t*-interval**) is

$$\bar{x}_1 - \bar{x}_2 \pm t^* \times \text{s.e.}(\bar{x}_1 - \bar{x}_2)$$

where the degrees of freedom for $t^*$ and the standard error depend on whether the assumption of equal variances is reasonable. To determine this interval:

1. Make sure that the appropriate conditions apply by examining both sample sizes and/or a shape picture for each sample.
2. Choose a confidence level.
3. Calculate the sample means, $\bar{x}_1$ and $\bar{x}_2$, and the sample standard deviations, $s_1$ and $s_2$.
4. Determine whether the sample standard deviations are similar enough to assume that the population standard deviations are about equal. If so, the pooled procedure can be used.
5. Calculate the standard error s.e.$(\bar{x}_1 - \bar{x}_2)$, which is

$$\sqrt{\frac{s_1^2}{n_1} + \frac{s_2^2}{n_2}} \ (unpooled) \quad \text{or} \quad s_p\sqrt{\frac{1}{n_1} + \frac{1}{n_2}} \ (pooled)$$

   where $s_p$ is the pooled standard deviation (see p. 473).
6. Calculate the degrees of freedom. For the unpooled case, use Welch's approximation (see p. 468) or, conservatively, the smaller of $n_1 - 1$ and $n_2 - 1$. For the pooled case, use $n_1 + n_2 - 2$.
7. Use Table A.2 (or statistical software) to find the multiplier $t^*$.

CI Module 5: $\mu_1 - \mu_2$

# 11.5   Understanding Any Confidence Interval

The statistics and parameters that researchers examine are not limited to proportions, means, differences in means, or differences in proportions. Throughout this book, we have seen several other statistics, such as the median, the lower and upper quartiles, the correlation coefficient, and relative risk. For each of these statistics, there are formulas and procedures that allow us to compute confidence intervals for the corresponding population parameter. Some of those formulas and procedures, however, are quite complex.

In cases in which a complicated procedure is needed to compute a confidence interval, authors of journal articles usually provide the completed intervals. Your job as a reader is to be able to interpret those intervals. The principles that you have learned for understanding confidence intervals for means and proportions are directly applicable to understanding any confidence interval. As an example, let's consider the confidence intervals that were reported in one of our earlier case studies.

11.5 Exercises are on page 489.

---

*case study* **11.1**   Confidence Interval for Relative Risk: Case Study 4.4 Revisited

In Case Study 4.4, we discussed a study relating heart disease to baldness in men. A parameter of interest in that study was the relative risk of heart disease based on the degree of baldness. The investigators focused on the relative risk (RR) of myocardial infarction (a heart attack) for men with baldness compared to men without any baldness. Here is how they reported some of the results:

> For mild or moderate vertex baldness, the age-adjusted RR estimates were approximately 1.3, while for extreme baldness the estimate was 3.4 (95% CI, 1.7 to 7.0). . . . For any vertex baldness (i.e., mild, moderate, and severe combined), the age-adjusted RR was 1.4 (95% CI, 1.2 to 1.9). (Lesko et al., 1993, p. 1000)

The 95% confidence intervals for age-adjusted relative risk are not simple to compute, and they don't take the form "sample estimate $\pm 2 \times$ standard error." This is evidenced by the fact that the intervals are not symmetric about the sample relative risks given. However, these intervals have the same interpretation as any other confidence interval. For instance, with 95% confidence, we can say that the ratio of risks of a heart attack in the populations of men with extreme baldness and men with no baldness is somewhere between 1.7 and 7.0. In other words, men with extreme baldness are probably somewhere between 1.7 to 7 times as likely to experience a heart attack as men of the same age who have no baldness. Of course, these results assume that the men in the study are representative of the larger population.

---

**thought question 11.6**   Part of the quote in Case Study 11.1 said, "For any vertex baldness (i.e., mild, moderate, and severe combined), the age-adjusted RR was 1.4 (95% CI, 1.2 to 1.9)." Explain what is wrong with the following interpretation of this result, and write a correct interpretation:

*Incorrect Interpretation:* There is a .95 probability that the age-adjusted relative risk of a heart attack (for men with any vertex baldness compared to men without any) is between 1.2 and 1.9.*

---

**\*HINT:** The interpretation is confusing the long-run results for the procedure with the results for a specific interval. See the discussion in Chapter 10, just before Thought Question 10.2 (p. 407).

*in summary*    ## Confidence Interval Procedures

The basic structure of intervals for the parameters in the following table is

$$\text{Sample estimate} \pm \text{Multiplier} \times \text{Standard error}$$

The table describes the specific details for each type of parameter. The *Sample estimate* in the formula is the *Statistic* in the Table.

| | Parameter | Statistic | Standard Error | Multiplier | Example |
|---|---|---|---|---|---|
| **One Proportion** | $p$ | $\hat{p}$ | $\sqrt{\dfrac{\hat{p}(1-\hat{p})}{n}}$ | $z^*$ (see note 4) | 10.3 |
| **Difference Between Proportions** | $p_1 - p_2$ | $\hat{p}_1 - \hat{p}_2$ | $\sqrt{\dfrac{\hat{p}_1(1-\hat{p}_1)}{n_1} + \dfrac{\hat{p}_2(1-\hat{p}_2)}{n_2}}$ | $z^*$ (see note 4) | 10.10 |
| **One Mean** | $\mu$ | $\bar{x}$ | $\dfrac{s}{\sqrt{n}}$ | $t^*$ (see note 1) | 11.5 |
| **Mean Difference, Paired Data** | $\mu_d$ | $\bar{d}$ | $\dfrac{s_d}{\sqrt{n}}$ | $t^*$ (see note 1) | 11.9 |
| **Difference Between Means (unpooled)** | $\mu_1 - \mu_2$ | $\bar{x}_1 - \bar{x}_2$ | $\sqrt{\dfrac{s_1^2}{n_1} + \dfrac{s_2^2}{n_2}}$ | $t^*$ (see note 2) | 11.11 |
| **Difference Between Means (pooled)** | $\mu_1 - \mu_2$ | $\bar{x}_1 - \bar{x}_2$ | $s_p\sqrt{\dfrac{1}{n_1} + \dfrac{1}{n_2}}$ | $t^*$ (see note 3) | 11.14 |

[1] Use Table A.2, df $= n - 1$.
[2] Use Table A.2,

$$\text{df} \approx \frac{\left(\dfrac{s_1^2}{n_1} + \dfrac{s_2^2}{n_2}\right)^2}{\dfrac{1}{n_1 - 1}\left(\dfrac{s_1^2}{n_1}\right)^2 + \dfrac{1}{n_2 - 1}\left(\dfrac{s_2^2}{n_2}\right)^2}$$

or use df $= \min(n_1 - 1, n_2 - 1)$.
[3] Use Table A.2, df $= n_1 + n_2 - 2$; see Section 11.4, Lesson 2 for the formula for $s_p$.
[4] Use last row of Table A.2, or the $z^*$ multipliers:

| Conf. Level | $z^*$ |
|---|---|
| 0.90 | 1.645 or 1.65 |
| 0.95 | 1.960 or 2.00 |
| 0.98 | 2.326 or 2.33 |
| 0.99 | 2.576 or 2.58 |

# Key Terms

### Section 11.1
parameter, 444
statistic, 444
estimate, 444
paired data, 445–446
independent samples, 446
standard error, rough definition, 447
standard error, technical definition, 447

standard error of a sample mean, 448
standard error of a sample mean of paired differences, 448
standard error of the difference, 449
standard error of the difference between two sample means, 449
Student's *t*-distribution, 450
degrees of freedom (df), 450
*t*\* multiplier, 451

### Section 11.2
confidence interval estimate of the mean, 452
confidence interval for a population mean, 452
*t*-interval, 453, 455
approximate 95% confidence interval (for large samples), 460

## Section 11.3

## Section 11.4

# Exercises

- ● Denotes basic skills exercises
- ◆ Denotes dataset is available in StatisticsNow at **http://
   1pass.thomson.com** or on your CD but is **not required** to
   solve the exercise.

**Bold-numbered exercises** have answers in the back of the text and
   fully worked solutions in the Student Solutions Manual.

**Statistics⊘Now**™ Go to the StatisticsNow website at
**http://1pass.thomson.com** to:
- Assess your understanding of this chapter
- Check your readiness for an exam by taking the Pre-Test quiz and
   exploring the resources in the Personalized Learning Plan

## Section 11.1

11.1 ● Would it be appropriate to use the ages of death of
First Ladies given in Chapter 2 (Table 2.5, p. 28) to find
a 95% confidence interval for the mean? If your answer
is yes, what is the parameter of interest? If your answer
is no, explain why not.

**11.2** ● Suppose you are told that the mean IQ score for children in a particular school is 105.3. Explain what additional information you would need in order to know
whether the mean of 105.3 is a statistic or a parameter.

11.3 ● You have been told that for university classes, students are supposed to spend two hours working outside of class for each hour in class. You are curious
to know the average amount of time students at your
school spend working outside of class. You contact
what you think is a representative sample of 100 students at your school and ask them, "How many hours
of schoolwork do you do outside of class for each hour
in class?" The mean for the 100 responses is 1.2 hours.
- a. What is the population of interest, and what is the
   research question you want to answer about it?
- b. Explain in words what the population parameter is
   in this study, and identify the appropriate symbol
   for it.
- c. What is the value of the sample estimate in this
   study, and what symbol is used for it?

**11.4** A multinational company plans to open new offices in
one of two countries where it has not operated before
and is trying to decide which country to choose. One
feature the company is interested in is how much

vacation time it will need to offer employees in the
two countries to be competitive with companies that
are already there. The company takes a representative sample of 50 workers in similar industries in each
country and asks them how many days of vacation
they are given per year. In Country A, the average is
22.5 days; in Country B, the average is 16.3 days.
- a. Of the three parameters that we considered in Section 11.1 ($\mu$, $\mu_d$, and $\mu_1 - \mu_2$), two of them would be
   of interest in this situation. Identify which two, and
   explain what they would represent in this situation.
   Note that one of them is represented twice.
- b. What is the sample estimate for the parameter
   $\mu_1 - \mu_2$ in this situation? Use appropriate notation,
   and give a numerical answer.

11.5 Give an example of a research question (not covered in
this chapter) for which the appropriate parameter is
- a. $\mu$
- b. $\mu_d$
- c. $\mu_1 - \mu_2$

11.6 Researchers at the MIND Institute at the University
of California–Davis analyzed the blood of 70 autistic
and 35 nonautistic children between the ages of 4
and 6 (Lau, 2005). One marker that they examined was
B cells, which is a type of immune cell that produces
antibodies in the blood. They found that the mean
level for the autistic children was substantially higher
than the mean level for the nonautistic children in the
study.
- a. What is the research question of interest in this part
   of the study? Write the question in words, not in
   terms of statistical notation.
- b. What is the population parameter of interest? Use
   appropriate notation, and explain in words what it
   means.
- c. Write the research question of interest in terms of
   the population parameter you defined in part (b).

11.7 In each part, identify whether it would be appropriate
to use a procedure for paired data or a procedure for
two independent samples to estimate the difference of
interest.
- **a.** A researcher wants to estimate the difference between the mean scores on a memorization test for
   people over 60 years old compared to people under
   40 years old.

**b.** Women give their actual and desired weights in a survey. An estimate of the mean difference is desired.

**c.** Two treatments for reducing cholesterol are compared. Forty people with high cholesterol use the first treatment, and 40 other people with high cholesterol use the second treatment. An estimate of the difference between the mean decrease in cholesterol for treatment 1 and the mean decrease in cholesterol for treatment 2 is desired.

11.8 For each of the following research questions, would it make more sense to collect paired data or independent samples? Explain.

a. What is the difference in the mean price of gasoline for all gas stations in the United States last week and this week?

b. What is the mean difference in height for the male and female in fraternal twin pairs in which there is one of each sex?

c. What is the difference in mean heights for males and females in the adult population?

d. How does the mean weight of adults in college towns compare with the mean weight of adults in other towns of similar size?

11.9 In the definition of paired data on page 446, three ways in which paired data can occur were given. They are repeated in the parts of this exercise. In each case, given an example of a research question for which paired data would be collected using the method described.

a. Each person is measured twice. The two measurements of the same characteristic or trait are made under different conditions.

b. Similar individuals are paired before giving the treatments in an experiment. Each member of a pair then receives a different treatment. The same response variable is measured for all individuals.

c. Two different variables are measured for each individual. There is interest in the amount of difference between the two variables.

**11.10** In the description of independent samples on page 446, four common methods were given for generating independent samples. They are repeated in the parts of this exercise. In each case, give an example of a research question for which independent samples would be collected using the method described.

**a.** Random samples are taken separately from two populations, and the same response variable is recorded for each individual.

b. One random sample is taken, and a variable is recorded for each individual, but then units are categorized as belonging to one population or another, such as male and female.

c. Participants are randomly assigned to one of two treatment conditions, such as diet or exercise, and the same response variable, such as weight loss, is recorded for each individual unit.

d. Two random samples are taken from a population, and a separate variable is measured in each sample. For example, one sample may be asked about hours of Internet use, whereas a separate sample is asked about hours of television viewing. Then the results are compared.

11.11 In each part, use the information given to calculate the standard error of the mean.

a. Mean height for a sample of $n = 81$ women is $\bar{x} = 64.2$ inches, and the standard deviation is $s = 2.7$ inches.

b. Mean systolic blood pressure for a sample of $n = 100$ men is $\bar{x} = 123.5$, and the standard deviation is $s = 9$.

c. Mean systolic blood pressure for a sample of $n = 324$ men is $\bar{x} = 123.5$, and the standard deviation is $s = 9$.

**11.12** Suppose a randomly selected sample of $n = 64$ men has a mean foot length of $\bar{x} = 27.5$ cm, and the standard deviation of the sample is 2 cm. Calculate the standard error of the sample mean. Write one or two sentences that interpret this value.

11.13 Refer to Exercise 11.12 about the foot lengths of men. A randomly selected sample of $n = 100$ women has a mean foot length of $\bar{x} = 24.0$ cm, and the standard deviation of the sample is 2 cm.

a. Calculate the standard error of the mean. Explain why the value is smaller than the standard error for the men in Exercise 11.12.

b. Now consider the difference between the mean foot lengths of men and women. Compute the standard error for the difference between sample means for the men and women.

**11.14** The pulse rates of $n = 50$ people are measured. These people then march in place for one minute, and the pulse rates are measured again. We are interested in estimating the mean difference between the two pulse rates for the population represented by this sample. Explain how you would compute the standard error for the sample mean difference. Be specific about what formula you would use.

11.15 If each of the following is decreased but everything else remains the same, will a confidence interval become wider, will it become narrower, or will the width remain the same? Explain.

a. The sample estimate.

b. The standard error.

c. The confidence level.

**11.16** Notice that the standard error formulas for means all involve sample standard deviations and sample sizes. With this knowledge, what can a researcher do, when designing a study, to try to make sure the standard error isn't too large to produce a meaningful confidence interval?

11.17 Suppose that a teacher asks her class how many hours they studied for the midterm exam. For the 40 students who are enrolled in the class, the mean was 5.5 hours, and the standard deviation was 2 hours. Explain why the teacher should not calculate a standard error for the mean and why she should not calculate a confidence interval to estimate the population mean.

**11.18** Refer to Figure 11.3 (p. 451), in which the relationship between $t^*$ and the 95% confidence level is illustrated for df = 24. Draw a similar picture for df = 24 and 90% confidence.

**11.19** Find the value of $t^*$ for each of the following situations.
  **a.** df = 15, 95% confidence level.
  b. df = 30, 90% confidence level.
  c. $n = 10$, 95% confidence interval for one mean.

**11.20** Explain what happens to the value of $t^*$ in each of the following cases.
  a. The confidence level is increased from 90% to 95%.
  b. The sample size is increased.
  c. The degrees of freedom are increased.
  d. The degrees of freedom are essentially infinite.
  e. The standard error of the mean is decreased because the standard deviation is decreased.

## Section 11.2

*Exercises 11.21 to 11.34 correspond to the two Lessons in Section 11.2. Lesson 1 exercises are 11.21 to 11.30; Lesson 2 exercises are 11.31 to 11.34.*

**11.21** ● In each case, use the information given to compute a confidence interval for the population mean $\mu$. Assume that all necessary conditions are present for using the method described in Section 11.2.
  **a.** $\bar{x} = 76$, $s = 6$, $n = 9$, 95% confidence level.
  b. $\bar{x} = 76$, $s = 6$, $n = 9$, 90% confidence level.
  c. $\bar{x} = 76$, $s = 6$, $n = 16$, 90% confidence level.
  d. $\bar{x} = 76$, $s = 8$, $n = 16$, 95% confidence level.
  e. $\bar{x} = 100$, $s = 8$, $n = 16$, 95% confidence level.

**11.22** ● What three factors affect the width of a confidence interval for a population mean? For each factor, indicate how an increase in the numerical value of the factor affects the interval width.

**11.23** ● Use Table A.2 to find the multiplier $t^*$ for calculating a confidence interval for a single population mean in each of the following situations:
  **a.** $n = 22$; confidence level = .98.
  **b.** $n = 5$; confidence level = .90.
  c. $n = 5$; confidence level = .99.
  d. $n = 25$; confidence level = .95.
  e. $n = 81$; confidence level = .95.

**11.24** ● A random sample of $n = 9$ men between 30 and 39 years old is asked to do as many situps as they can in one minute. The mean number is $\bar{x} = 26.2$, and the standard deviation is $s = 6$.
  a. Find the value of the standard error of the sample mean. Write a sentence that interprets this value. (Refer back to Section 11.1 for guidance).
  b. Find a 95% confidence interval for the population mean.
  c. Write a sentence that interprets the confidence interval in the context of this situation.

**11.25** Volunteers who had developed a cold within the previous 24 hours were randomized to take either zinc or placebo lozenges every 2 to 3 hours until their cold symptoms were gone (Prasad et al., 2000). Twenty-five participants took zinc lozenges, and 23 participants took placebo lozenges. The mean overall du-

ration of symptoms for the zinc lozenge group was 4.5 days, and the standard deviation of overall duration of symptoms was 1.6 days. For the placebo group, the mean overall duration of symptoms was 8.1 days, and the standard deviation was 1.8 days.
  a. Calculate a 95% confidence interval for the mean overall duration of symptoms in a population of individuals like those who used the zinc lozenges.
  b. Calculate a 95% confidence interval for the mean overall duration of symptoms in a population of individuals like those who used the placebo lozenges.
  c. On the basis of the intervals computed in parts (a) and (b), is it reasonable to conclude generally that taking zinc lozenges reduces the overall duration of cold symptoms? Explain why you think this is or is not an appropriate conclusion.
  d. In their paper, the researchers say that they checked whether it was reasonable to assume that the data were sampled from a normal curve population and decided that it was. How is this relevant to the calculations done in parts (a) and (b)?

**11.26** Refer to Exercise 11.25. Compute a 99% confidence interval for the mean duration of symptoms for individuals using the placebo lozenges. Write one or two sentences that interpret this interval.

**11.27** A randomly selected sample of $n = 12$ students at a university is asked, "How much did you spend for textbooks this semester?" The responses, in dollars, are

| 200 | 175 | 450 | 300 | 350 | 250 | 150 | 200 |
| 320 | 370 | 404 | 250 | | | | |

Computer output with a 95% confidence interval for the population mean is as follows:

| Variable | N | Mean | StDev | SE Mean | 95.0% CI |
|----------|---|------|-------|---------|----------|
| spent | 12 | 284.9 | 96.1 | 27.7 | (223.9, 346.0) |

  a. Draw either a dotplot or a boxplot of the data. Briefly discuss whether the necessary conditions for doing a confidence interval for the population mean are present or not.
  b. The standard error of the mean given in the output is 27.7 (under "SE Mean"). Verify this value by giving the correct formula and substituting appropriate values into this formula.
  c. What are the degrees of freedom for the $t^*$ multiplier? What is the value of the $t^*$ multiplier for the given confidence interval?
  d. Write a sentence that interprets the confidence interval. Be specific about the population that is being described.
  e. Calculate a 90% confidence interval for the population mean.

**11.28** Refer to Exercise 11.27. Use the Empirical Rule to create an interval that estimates the textbook expenses of 95% of the individual students at the university.

**11.29** Suppose that a highway safety researcher is studying the design of a highway sign and is interested in the mean maximum distance at which drivers are able to read the sign. The maximum distances (in feet) at which $n = 16$ drivers can read the sign are as follows:

| | | | | | | | |
|---|---|---|---|---|---|---|---|
| 440 | 490 | 600 | 540 | 540 | 600 | 240 | 440 |
| 360 | 600 | 490 | 400 | 490 | 540 | 440 | 490 |

a. Are the necessary conditions met for computing a confidence interval for the population mean distance? Assume that the sample can be considered to be a random sample from the population of interest. See page 454 for guidance.
b. Estimate the mean maximum distance at which drivers can read the sign. If conditions are not met, make appropriate adjustments. Use a 95% confidence interval, and write a sentence that interprets this interval.

11.30 In Example 11.6 (p. 455), we found that a 99% confidence interval for the mean number of hours Penn State students watch television in a day is 1.76 to 2.42 hours. Explain whether each of the following statements is a correct interpretation of this result.
a. We can be fairly confident that 99% of all Penn State students watch between 1.76 hours and 2.42 hours of television a day.
b. We can be fairly confident that the mean number of hours of television Penn State students watch per day is under 3 hours.
c. The probability is .99 that the mean number of hours of TV Penn State students watch per day is between 1.76 hours and 2.42 hours.

**11.31** ● The following scenarios were presented in Exercise 11.11, which asked for the standard error of the mean. Find the standard error of the mean, and then find an approximate 95% confidence interval for the population mean, using the approximation that is appropriate for large samples.
**a.** Mean height for a sample of $n = 81$ women is $\bar{x} = 64.2$ inches, and the standard deviation is $s = 2.7$ inches.
b. Mean systolic blood pressure for a sample of $n = 100$ men is $\bar{x} = 123.5$, and the standard deviation is $s = 9$.
c. Mean systolic blood pressure for a sample of $n = 324$ men is $\bar{x} = 123.5$, and the standard deviation is $s = 9$.

11.32 Example 11.3 (p. 449) described a study that compared mean weight loss for a diet plan to mean weight loss for an exercise plan. In the study, $n = 42$ men were put on a diet. The men who dieted lost an average of 7.2 kg, with a standard deviation of 3.7 kg.
a. Compute the standard error of the mean for the men who dieted. Write a sentence or two to explain what is measured by this value.
b. Compute an approximate 95% confidence interval for the mean weight loss for the men who dieted. Write a sentence that interprets this interval.

**11.33** This scenario was presented in Exercise 11.12, which asked for the standard error of the mean:

Suppose a randomly selected sample of 64 men has a mean foot length of 27.5 cm, and the standard deviation of the sample is 2 cm.

Calculate an approximate 95% confidence interval for the mean foot length of men. Write a sentence that interprets this interval.

11.34 Explain why the approximation for a 95% confidence interval for a mean that uses the value 2 as multiplier should not be used when the sample size is very small.

## Section 11.3

11.35 ● In Example 11.10 (p. 465), we found a 95% confidence interval for the mean reduction in trait anxiety if everyone in the population were to take the meditation training offered in the study. For the 21 people in the study, the mean reduction in anxiety was 4.5, and the standard deviation was 7.2.
a. Find a 99% confidence interval for the population mean.
b. On the basis of the interval you found in part (a), what can you conclude about the population mean reduction in anxiety?
c. Refer to the three ways paired data can occur, listed following the definition on page 446. Which of the three best describes the situation in this Exercise? Explain.

**11.36** ● In Example 11.9 (p. 463), we compared the number of hours of computer time with the number of hours watching TV by finding a 90% confidence interval.
**a.** Find a 95% confidence interval for this situation, and write a sentence interpreting it.
**b.** On the basis of the 95% confidence interval, can we conclude that for the population represented by this sample, students really do spend more time on computer usage than on watching television, on average? Explain.
**c.** Refer to the three ways paired data can occur, listed following the definition on page 446. Which of the three best describes the situation in this exercise? Explain.

11.37 A random sample of five college women was asked for their own heights and their mothers' heights. The researchers wanted to know whether college women are taller on average than their mothers. The results (in inches) were as follows:

| Pair | 1 | 2 | 3 | 4 | 5 |
|---|---|---|---|---|---|
| Daughter | 66 | 64 | 64 | 69 | 66 |
| Mother | 66 | 62 | 65 | 66 | 63 |

a. Define the parameter of interest in this situation.
b. Find a 95% confidence interval for the parameter you defined in part (a).
c. Using the interval in part (b), write a sentence or two about the relationship between women students' heights and their mothers' heights for the population. Your explanation should be written to be understood by someone with no training in statistics.

d. Explain two different things that the researchers could have done to obtain a narrower interval than the one you found in part (b).

**11.38** If a confidence interval for $\mu_d$ covers 0, can we conclude that the population means for the two measurements are identical? Explain.

11.39 A researcher was interested in knowing whether the mean weight of the second baby is higher, lower, or about the same as the mean weight of the first baby for women who have at least two children. She selected a representative sample of 40 women who had at least two children and asked them for the weights of the oldest two at birth. She found that the mean of the differences (first − second) was 5 ounces, with a standard deviation of 7 ounces.
  a. Why was the researcher's study design better than taking independent samples of mothers and asking the first sample about the weight of their first baby and the second sample about the weight of their second baby?
  b. Define the parameter of interest. Use appropriate notation.
  c. Write the mean of 5 ounces and the standard deviation of 7 ounces using appropriate statistical notation.
  d. Find a 90% confidence interval for the parameter that you defined in part (b). Write a sentence or two interpreting the interval.

11.40 For a sample of $n = 20$ women aged 18 to 29, responses to the question "How tall would you like to be?" are recorded along with actual heights. In the sample, the mean desired height is 66.7 inches, the mean actual height is 64.9 inches, and the mean difference is $\bar{d} = 1.8$ inches. The standard deviation of the differences is $s_d = 2.1$ inches.
  a. Compute a 95% confidence interval for the mean difference between desired and actual height, and write a sentence that interprets this interval. Be specific about what population is described by the interval.
  b. What are the necessary conditions for the validity of the confidence interval computed in part (a)?
  c. Explain why the following statement is not correct: "On the basis of the confidence interval computed in part (a), we can conclude that all women aged 18 to 29 would like to be taller."

## Section 11.4

*Exercises 11.41 to 11.51 correspond to the two Lessons in Section 11.4. Lesson 1 exercises are 11.41 to 11.47; Lesson 2 exercises are 11.48 to 11.51.*

**11.41** ● Suppose you were given a 95% confidence interval for the difference in two population means. What could you conclude about the population means if
  **a.** The confidence interval did *not* cover zero?
  **b.** The confidence interval *did* cover zero?

**11.42** ● Data on the testosterone levels (ng/dl in saliva) of men in different professions were given in a paper published in the *Journal of Personality and Social Psychology* (Dabbs et al., 1990). The researchers suggest that there are occupational differences in mean testosterone level. Medical doctors and university professors were two of the occupational groups for which means and standard deviations were given along with a dotplot of the raw data. Computer output for a 95% confidence interval for the difference in mean testosterone levels of medical doctors and university professors is as follows:

```
Two-sample T for MD vs Profs

          N      Mean    StDev    SE Mean
MD        16     11.60   3.39     0.85
Prof      10     10.70   2.59     0.82

Difference = mu MD − mu Profs
Estimate for difference: 0.90
95% CI for difference: (−1.54, 3.34)
DF = 22
```

  **a.** The 95% confidence interval is given as −1.54 to 3.34. The unpooled procedure was used to compute this interval. Show how the interval was computed by giving the formula and identifying numerical values for all elements of the formula. Note the degrees of freedom are given as "DF = 22."
  b. Interpret the confidence interval for the difference in means. Be specific about the populations that are described. What assumptions are you making in this interpretation?
  c. On the basis of the 95% confidence interval given, what can be concluded generally about the difference in mean testosterone levels of medical doctors and professors?

11.43 ● ◆ Each of 63 students in a statistics class used their nondominant hand to print as many letters of the alphabet, in order, as they could in 15 seconds. The following output for this exercise gives results for a 95% confidence interval for the difference in population means for females and males. The "unpooled" procedure was used. (*Data source:* **letters** dataset on the CD for this book).

```
Two-sample T for Exercise 11.43

Sex       N      Mean    StDev    SE Mean
Female    29     12.55   4.01     0.74
Male      34     13.65   4.46     0.77

Difference = mu (Female) − mu (Male)
Estimate for difference: −1.10
95% CI for difference: (−3.23, 1.04)
T-Test of difference = 0 (vs not =):
T-Value = −1.03    P-Value = 0.309    DF = 60
```

a. What was the mean number of letters printed by females in the sample? What was the mean number printed by males? What is the difference between the sample means for males and females?
b. What is the 95% confidence interval given in the output for the difference between population means? Write a sentence that interprets this interval.

c. Refer to part (b). On the basis of the 95% confidence interval, what conclusion can be made about whether the population means for males and females differ or not? Explain.

d. Find the value of the (unpooled) standard error of the difference between sample means.

e. Note that the value for degrees of freedom for this problem are given at the end of the output as "DF=60." Use Table A.2 to find the value of $t^*$ that was used in finding the interval.

**11.44** Refer to Exercise 11.25 about the effect of zinc lozenges on the duration of cold symptoms. For $n = 25$ in the zinc lozenge group, the mean overall duration of symptoms was 4.5 days, and the standard deviation was 1.6 days. For $n = 23$ in the placebo group, the mean overall duration of symptoms was 8.1 days and the standard deviation was 1.8 days.

a. Calculate $\bar{x}_1 - \bar{x}_2$ = difference in sample means and also compute the unpooled s.e.$(\bar{x}_1 - \bar{x}_2)$ = standard error of the difference in means.

b. Compute a 95% confidence interval for the difference in mean days of overall symptoms for the placebo and zinc lozenge treatments, and write a sentence interpreting the interval. Use the unpooled standard error and use the smaller of $n_1 - 1$ and $n_2 - 1$ as a conservative estimate of degrees of freedom.

c. Is the interval computed in part (b) evidence that the population means are different? Explain.

**11.45** For a sample of 36 men, the mean head circumference is 57.5 cm with a standard deviation equal to 2.4 cm. For a sample of 36 women, the mean head circumference is 55.3 cm with a standard deviation equal to 1.8 cm.

**a.** Find the value of the difference between the sample means for men and women.

**b.** Find the standard error of the difference between the sample means in this context.

**c.** Find an approximate 95% confidence interval for the difference between population mean head circumferences for men versus women.

**11.46** Refer to Exercises 11.12 and 11.13, in which the foot lengths of a random sample of 64 men had a mean of 27.5 cm and a standard deviation of 2 cm, while a random sample of 100 women had a mean foot length of 24.0 cm and a standard deviation of 2 cm. Compute an approximate 95% confidence interval for the difference between the mean foot lengths of men and women. Write a sentence that interprets this interval.

**11.47** In Example 11.12 (p. 471), we learned about a study of hangover symptoms in college students (Slutske et al., 2003, on the CD for this book). The students answered questions about alcohol use and hangovers, including a count of how many out of a list of 13 possible hangover symptoms they had experienced in the past year. For the 470 men, the mean number of symptoms was 5.3; for the 755 women, it was 5.1. The standard deviation was 3.4 for each of the two samples.

a. Define the parameter of interest using appropriate notation.

b. Find the standard error of the difference in means.

c. Find an approximate 95% confidence interval for the difference in population means.

d. On the basis of your interval in part (c), do you think that there is a difference in the population mean number of hangover symptoms experienced by college men and women? Explain.

**11.48** ● What assumption is necessary in order to use the pooled procedure for finding a confidence interval for the difference between two population means?

**11.49** ● Refer to the output for Exercise 11.43 about letters printed with the nondominant hand.

a. Find the value of $s_p$ = pooled standard deviation for these data.

b. For the pooled two-sample procedure, find the (pooled) standard error of the difference in sample means.

c. What would be the degrees of freedom for the pooled procedure?

**11.50** Refer to Exercises 11.25 and 11.44. Use the pooled procedure to compute a 95% confidence interval for the difference in mean days of symptoms for the zinc and placebo treatments.

**11.51** Following are data for 20 individuals on resting pulse rate and whether the individual regularly exercises or not.

| Person | Pulse | Regularly Exercises |
|--------|-------|---------------------|
| 1 | 72 | No |
| 2 | 62 | Yes |
| 3 | 72 | Yes |
| 4 | 84 | No |
| 5 | 60 | Yes |
| 6 | 63 | Yes |
| 7 | 66 | No |
| 8 | 72 | No |
| 9 | 75 | Yes |
| 10 | 64 | Yes |
| 11 | 62 | No |
| 12 | 84 | No |
| 13 | 76 | No |
| 14 | 60 | Yes |
| 15 | 52 | Yes |
| 16 | 60 | No |
| 17 | 64 | Yes |
| 18 | 80 | Yes |
| 19 | 68 | Yes |
| 20 | 64 | Yes |

a. Calculate a 95% confidence interval for the difference in means for the two groups defined by whether the individual regularly exercises or not. Be sure to verify necessary conditions and state any assumptions. Use the unpooled procedure.

b. Repeat part (a) using the pooled procedure.

c. Compare the unpooled and pooled results, and discuss which procedure is more appropriate.

## Section 11.5 and Case Studies 11.1 and 11.2

**11.52** ● ◆ Responses from $n = 204$ college students to the question "About how many CDs do you own?" are used to determine 95% confidence intervals for the median and the mean. (*Note:* The data are in the dataset **pennstate2** on the CD for this book.)

a. A 95% confidence interval for the population median is 50 to 60. Write two or three sentences interpreting this result. Be specific about the population that the interval describes, and also describe any implied assumptions about the sample.

b. A 95% confidence interval for the mean $\mu$ is 62.7 to 82.9. Compare this result to the result in part (a), and explain why this comparison suggests that the data for music CDs owned may be skewed.

c. Sample statistics summarizing the data are $\bar{x} = 72.8$, $s = 72.2$, $n = 204$. Use this information to verify the 95% confidence interval for the mean given in part (b).

11.53 ● Suppose you were given a 95% confidence interval for the relative risk of disease under two different conditions. What could you conclude about the risk of disease under the two conditions if

a. The confidence interval did not cover 1.0?

b. The confidence interval did cover 1.0?

**11.54** ● In a twelve-year study done at ten medical centers, Cole et al. (2000) investigated whether the decrease in heart rate over the first two minutes after stopping treadmill exercise was a useful predictor of subsequent death during the study period. Based on observations of $n = 5234$ participants, the investigators concluded: "Abnormal heart rate recovery predicted death (relative risk 2.58 [CI, 2.06 to 3.20]). After adjustment for standard risk factors, fitness, and resting and exercise heart rates, abnormal heart rate recovery remained predictive (adjusted relative risk, 1.55 [CI, 1.22 to 1.98])." Explain why the given confidence intervals are evidence that abnormal heart rate recovery after treadmill exercise is a predictor of subsequent death during the study period.

11.55 Refer to Case Study 11.2 (p. 479). Use the baseline data for the placebo group to find an approximate 95% confidence interval for the mean PMS complex score for the population of all women like the ones in this study. Write a sentence interpreting the interval.

11.56 Refer to Case Study 11.2 (p. 479).

a. Use the baseline data to find an approximate 95% confidence interval for the difference in means for the populations represented by the placebo and calcium groups at the start of the study.

b. Explain why the researchers would want the interval computed in part (a) to cover zero.

## Section 11.6 Skillbuilder Applet Exercises

**Statistics ⌂ Now™** Test yourself on these questions and explore the applet at **http://1pass.thomson.com** or on your CD.

*For these exercises, use the* **ConfidenceLevel** *applet described in Section 11.6. It is on the CD for this book.*

11.57 Set the confidence level at 68%.

a. Use the *sample!* button to generate a new random sample. Write the confidence interval displayed,

and indicate whether it captures the population mean (170) or not.

b. Generate 150 different samples. How many times did the confidence interval capture the population mean? What percentage of the samples gave a confidence interval that captured the population mean?

c. Refer to part (b). Compare the number and percentage of times that the confidence interval captured the population mean to what might be expected for the 68% confidence level.

11.58 Change the confidence level to 95%. Repeat the parts of Exercise 11.57 using the 95% confidence level.

11.59 Change the confidence level to 99%. Repeat the parts of Exercise 11.57 using the 99% confidence level.

11.60 For the same random sample, give the 90%, 95%, and 99% confidence intervals for the population mean. (This can be done by observing what occurs when the scrollbar is used to change the confidence level.) What is indicated about the relationship between the width of the confidence interval and the confidence level?

11.61 Write a definition of the confidence level for a confidence interval procedure.

## Chapter Exercises

*Be sure to include a check of the appropriate conditions as part of your work when you compute a confidence interval.*

11.62 Find an exact 95% confidence interval (use $t^*$) for the mean weight loss after a year of exercise for the population of men represented by those in the study in Example 11.3 (p. 449). There were 47 men in the exercise group in the sample, with mean weight loss of 4.0 kg and standard deviation of 3.9 kg.

**11.63** In computing a confidence interval for a population mean $\mu$, explain whether the interval would be wider, more narrow, or neither as a result of each of the following changes. Assume that features that are not mentioned (confidence level, mean, standard deviation, sample size) remain the same.

**a.** The level of confidence is changed from 90% to 95%.

**b.** The sample size is doubled.

**c.** A new sample of the same size is taken, and $\bar{x}$ increases by 10.

11.64 Example 2.17 (p. 49) reported that a random sample of 199 married British women had a mean height of 1602 mm, with standard deviation of 62.4 mm.

a. Find a 99% confidence interval for the mean height of all women represented by this sample. Write a sentence interpreting the interval.

b. Give a 99% confidence interval for the mean height *in inches* of all women represented by those in this sample (1 mm = 0.03937 in.).

11.65 Refer to Exercise 11.64. The sample included the heights of the women's husbands as well. For these men, the mean height was 1732.5 mm, with standard deviation 68.8 mm.

a. Find a 90% confidence interval for the mean height of the population of men represented by this sample.

b. Given the information provided in this and the previous exercise, is it possible to compute an approximate 95% confidence interval for the average difference in height between the husband and wife in British couples? If so, compute the interval. If not, explain what additional numerical information you would need.

**11.66** Example 2.16 (p. 48) gave the weights in pounds of the 18 men on the crew teams at Oxford and Cambridge universities in 1991–92. The data were as follows:

*Cambridge:* 188.5, 183.0, 194.5, 185.0, 214.0, 203.5, 186.0, 178.5, 109.0

*Oxford:* 186.0, 184.5, 204.0, 184.5, 195.5, 202.5, 174.0, 183.0, 109.5

**a.** Explain why it is not appropriate to use the methods in this chapter to compute a confidence interval for the mean weight of college crew team members, even if we assume these 18 men represent a random sample of all college crew team members.

b. The first eight values listed for Cambridge represent the weights of the eight rowers on the team. Assuming these eight men represent a random sample of all Cambridge crew team rowers over the years, compute a 90% confidence interval for the mean of that population.

c. Make the same assumption as in part (b) about the weights of the first eight men in the Oxford sample. Compute a 90% confidence interval for the difference in mean weights of Cambridge and Oxford rowers. Explain whether you used the pooled or unpooled standard error, and why.

d. Write a sentence interpreting the interval in part (c).

**11.67** In Stroop's Word Color Test, 100 words that are color names are shown in colors different from the word. For example, the word *red* might be displayed in blue. The task is to correctly identify the display color of each word; in the example just given the correct response would be blue. Gustafson and Kallmen (1990) recorded the time needed to complete this test for $n = 16$ individuals after they had consumed alcohol and for $n = 16$ other individuals after they had consumed a placebo drink flavored to taste as if it contained alcohol. Each group included eight men and eight women. In the alcohol group, the mean completion time was 113.75 seconds and the standard deviation was 22.64 seconds. In the placebo group, the mean completion time was 99.87 seconds and the standard deviation was 12.04 seconds.

a. Calculate the unpooled standard error for the difference in the two sample means.

b. Calculate a 95% confidence interval for the difference in population means. For the unpooled procedure, the approximate degrees of freedom are df = 22.

c. On the basis of the confidence interval computed in part (b), can we conclude that the population means for the two groups are different? Why or why not?

d. Can we verify the necessary conditions for computing the confidence interval in part (b)? If so, verify the conditions. If not, explain why it is not possible.

**11.68** ◆ An experiment was done by 15 students in a statistics class at the University of California at Davis to determine whether manual dexterity was better for the dominant hand compared to the nondominant hand (left or right). Each student measured the number of beans they could place into a cup in 15 seconds, once with the dominant hand and once with the nondominant hand. The order in which the two hands were measured was randomized for each student.

a. Explain whether a comparison of beans placed using the dominant and nondominant hands is a comparison of paired data or two independent samples.

b. Referring to material earlier in the book, explain why the order of the two hands was randomized rather than, for instance, having each student test the dominant hand first.

c. The data are presented in the table below and are given also in the **beans** dataset on the CD for this book. Assuming that these students represent a random sample from the population, compute a 90% confidence interval for the mean difference in the number of beans that can be placed into a cup in 15 seconds by the dominant and nondominant hands.

| Student | 1 | 2 | 3 | 4 | 5 | 6 | 7 | 8 | 9 | 10 | 11 | 12 | 13 | 14 | 15 |
|---|---|---|---|---|---|---|---|---|---|---|---|---|---|---|---|
| Dominant | 22 | 19 | 18 | 17 | 15 | 16 | 16 | 20 | 17 | 15 | 17 | 17 | 14 | 20 | 26 |
| Nondom. | 18 | 15 | 13 | 16 | 17 | 16 | 14 | 16 | 20 | 15 | 17 | 17 | 18 | 18 | 25 |
| Difference | 4 | 4 | 5 | 1 | −2 | 0 | 2 | 4 | −3 | 0 | 0 | 0 | −2 | 2 | 1 |

d. Write a sentence or two using the interval in part (c) to address the question of whether manual dexterity is better, on average, for the dominant hand.

**11.69** In a random sample of 199 married British couples (described in Chapter 2), there were 170 for which both ages were reported. The average difference in husband's age and wife's age was 2.24 years, with standard deviation of 4.1 years.

a. Find an approximate 95% confidence interval for the average age difference for all married British couples. Write a sentence interpreting the interval.

**b.** Were the samples of husband ages and wife ages collected as paired data or as two independent samples? Briefly explain.

**11.70** A study was conducted on pregnant women and the subsequent development of their children (Olds et al., 1994). One of the questions of interest was whether the IQ of children would differ for mothers who smoked at least 10 cigarettes a day during pregnancy and those who did not smoke at all. The mean IQ at age 4 for the children of the 66 nonsmokers was 113.28 points, while for the children of the 47 smokers, it was 103.12 points. The standard deviations were not reported, but from other information that was provided, the pooled standard deviation was about 13.5 points.

a. Assuming that these women and children are a representative sample, find a 95% confidence interval (use $t^*$) for the difference in mean IQ scores at age 4 for children of mothers who do not smoke during pregnancy and those whose mothers smoke at least 10 cigarettes a day during pregnancy. Write a sentence interpreting the interval.

b. One of the statements in the article was "after control for confounding background variables . . . the average difference observed at 36 and 48 months was reduced to 4.35 points (95% CI: 0.02, 8.68) [Olds et al., 1994, p. 224]." Explain the purpose of this statement, and interpret the result given.

**11.71** Examples 11.2, 11.6, and 11.8 in Sections 11.1 and 11.2 used a sample of $n = 175$ students to estimate the mean daily hours that Penn State students watch television. The following summary shows results for daily hours of television by gender:

| Sex | N | Mean | St Dev | SE Mean |
|-----|-----|------|--------|---------|
| Female | 116 | 1.95 | 1.51 | 0.14 |
| Male | 59 | 2.37 | 1.87 | 0.24 |

a. Calculate the difference between the mean values for the females and males. Is this a statistic or a parameter? Explain.

b. For each gender, verify the standard error of the mean ("SE Mean" in the output) by substituting appropriate values into the correct formula.

c. An approximate 95% confidence interval for the difference between means is $-0.14$ to $+0.97$ hours. On the basis of this interval, what can we conclude, if anything, about the difference in the mean daily television watching times of men and women? Justify your answer.

d. What formula was used to compute the interval given in part (c)? Substitute appropriate values into this formula

**11.72** In Example 11.11 (p. 468), we compared the time in seconds to cross an intersection when someone was or was not staring at the driver. We did not assume equal population variances. Following is the Minitab (Version 12) output that resulted from clicking "Assume Equal Variance":

Two-sample T for CrossTime

| Group | N | Mean | StDev | SE Mean |
|-------|-----|------|-------|---------|
| NoStare | 14 | 6.63 | 1.36 | 0.36 |
| Stare | 13 | 5.592 | 0.822 | 0.23 |

95% CI for mu (NoStare) − mu (Stare): (0.14, 1.94)
T-Test mu (NoStare) = mu (Stare) (vs >): T = 2.37
P = 0.013     DF = 25
Both use Pooled StDev = 1.14

a. What is the value of $s_p$?

b. Compare the 95% confidence interval with the one in Example 11.11, and discuss whether it is rea-

sonable to use the "equal variance" assumption for this study.

**11.73** Suppose 100 researchers each plan to independently gather data and construct a 90% confidence interval for a population mean. If $X =$ the number of those intervals that actually cover the population means, then $X$ is a binomial random variable.

**a.** What is a "success" for this binomial random variable?

**b.** What is the numerical value of $p$, the probability of a success?

**c.** What is the expected number of intervals that will cover their population means?

**d.** Will each researcher know whether he or she has a "successful" interval? Explain.

**e.** What is the probability that all 100 researchers actually get intervals that cover their population means of interest?

## Dataset Exercises

**Statistics⌂Now**™ Datasets **are required** to solve these exercises and can be found at **http://1pass.thomson.com** or on your CD.

**11.74** For this exercise, use the dataset **pennstate1**. Case Study 1.1 presented data given by college students in response to the question, "What is the fastest you have ever driven a car? ___ mph" The variable *fastest* contains the responses to this question. Determine a 95% confidence interval for the difference in the mean response of college men and college women. Interpret the interval, state any assumptions about the sample, and verify necessary conditions.

**11.75** For this exercise, use the dataset **pennstate1**. The variable *RtSpan* has the raw data for measurements of the stretched right handspans of $n = 190$ Penn State students.

a. Calculate a 95% confidence interval for the mean stretched handspan in the population of men represented by this sample. Write a sentence that interprets the interval.

b. Calculate a 95% confidence interval for the mean stretched handspan in the population of women represented by this sample. Write a sentence that interprets the interval.

c. Compute a 95% confidence interval to estimate the difference in the mean handspans of college men and women. Write a sentence that interprets the interval.

**11.76** For this exercise, use the **UCDavis2** dataset. In the survey, students reported ideal height (*IdealHt*) and actual height (*height*).

a. For the men in the sample, calculate a 90% confidence interval to estimate the mean difference between ideal and actual height. Be certain to graphically analyze the data in order to verify necessary conditions.

b. Repeat part (a) for the women in the sample.

c. Based on the intervals computed in parts (a) and (b), is there a difference between men and women with regard to mean difference between ideal and actual height? Explain.

d. Define $\mu_1$ to be the population mean difference (ideal height − actual height) for men, and define $\mu_2$ to be the same for women. Find a 95% confidence interval for $\mu_1 - \mu_2$. What does the result say about how men and women differ on the issue?

11.77 For this exercise, use the **cholest** dataset. The variable *control* contains cholesterol levels for individuals who have not had a heart attack, while the variable *2-day* contains cholesterol levels for heart attack patients two days after the attack.

a. Calculate a 98% confidence interval for the mean cholesterol level of individuals who have not had a heart attack. Interpret the interval, state any assumptions, and verify necessary conditions.

b. Calculate a 98% confidence interval for the mean cholesterol level of heart attack patients two days after the attack. Interpret the interval, state any assumptions, and verify necessary conditions.

c. Calculate a 98% confidence interval for the mean difference in the cholesterol levels of the populations represented by the two samples described in this problem. Interpret the interval, state any assumptions, and verify necessary conditions.

11.78 For this exercise, use the **deprived** dataset. Students in a statistics class were asked whether they generally felt sleep deprived (variable name = **Deprived**) and also were asked how many hours they usually slept per night (variable name = **SleepHrs**).

a. For the sample, find the mean hours of sleep per night for students who said that they are sleep deprived and for students who said that they are not sleep deprived. What is the difference in the two sample means?

b. Find a 95% confidence interval for the difference in population mean hours of sleep for students who would say that they are not sleep deprived versus students who would say that they are sleep deprived. Interpret the interval in the context of this situation.

**Statistics ⊖ Now™** Preparing for an exam? Assess your progress by taking the post-test at **http://1pass.thomson.com.**

**Mentor™**
Do you need a live tutor for homework problems? Access vMentor at **http://1pass.thomson.com** for one-on-one tutoring from a statistics expert.

# 12

## *Is it their personalities or their looks?*

**See Example 12.15** *(p. 522)*

Laurence Mouton/PhotoAlto/Getty Images

# Rejecting the Null Hypothesis

The phrase **statistically significant** is used to describe the results when the researcher has decided that the $p$-value is small enough to decide in favor of the alternative hypothesis. The **level of significance,** also called the **$\alpha$ level** ("alpha" level), is the borderline for deciding that the $p$-value is small enough to justify choosing the alternative hypothesis. A result is statistically significant when the $p$-value is less than the chosen level of significance.

Most researchers use the convention that we can reject the null hypothesis when the $p$-value is less than .05 (or 5%). In other words, the *level of significance* is usually set at $\alpha = .05$, and the result is *statistically significant* when the $p$-value is less than .05. In some research disciplines, the phrase *highly significant* is used to describe the result when the $p$-value is less than .01 (1%), and the phrase *marginally significant* is used when the $p$-value is between .05 and .10. $p$-Values greater than .10 are rarely considered sufficient evidence to reject the null hypothesis. Researchers frequently criticize these somewhat arbitrary borderlines, but they are used by many scientific journals.

*in summary*     ## Interpreting a $p$-Value

In any statistical hypothesis test, the smaller the $p$-value is, the stronger is the evidence against the null hypothesis. It is common research practice to call a result *statistically significant* when the $p$-value is less than .05.

# Stating the Two Possible Conclusions

The two possible conclusions of a hypothesis test are as follows:

- When the $p$-value is small, we *reject the null hypothesis* or, equivalently, we *accept the alternative hypothesis.* "Small" is defined as a $p$-value $\leq \alpha$, where $\alpha =$ level of significance (usually .05).

- When the $p$-value is not small, we conclude that we *cannot reject the null hypothesis* or, equivalently, that *there is not enough evidence to reject the null hypothesis.* "Not small" is defined as a $p$-value $> \alpha$, where $\alpha =$ level of significance (usually .05).

The conclusion that we "cannot reject the null hypothesis" may seem wordy to you, and you may be tempted to say that you accept the null hypothesis. Resist that temptation because *it is almost never correct to say,* "I accept the null hypothesis." The problem with "accepting the null" is that this wording makes it seem that we are convinced that the null hypothesis is true. The strategy of hypothesis testing is to reject the null hypothesis when the evidence is convincingly against it. If we collect only a small sample amount of data, we may not see convincing evidence for the alternative hypothesis because the sampling error is so large. That is certainly *not* equivalent to accepting that the null hypothesis is true. For instance, if we observe three births and two are boys, we would certainly not be willing to *accept* the hypothesis "2/3 of all births are boys" even though the observed data don't provide evidence that it is false.

---

***thought question* 12.3** Suppose an ESP test is conducted by having someone guess whether each of $n$ coin flips will result in heads or tails. The null hypothesis is that $p = .5$, and the alternative is that $p > .5$, where $p$ = probability of guessing correctly. Suppose one participant guesses $n = 10$ times and gets 6 right, while another guesses $n = 100$ times and gets 60 right. In each case, the percent correct was 60%. Do you think the $p$-value would be lower in one case than in the other, or would it be the same? Explain.*

---

**Example 12.5**

**A Jury Trial** If you are on a jury in the U.S. judicial system, you must presume that the defendant is innocent unless there is enough evidence to conclude that he or she is guilty. Therefore the two hypotheses are as follows:

*Null hypothesis:* The defendant is innocent.
*Alternative hypothesis:* The defendant is guilty.

The prosecution collects evidence in much the way that researchers collect data, in the hope that the jurors will be convinced that such evidence would be extremely unlikely if the assumption of innocence were true. Consistent with our thinking in hypothesis testing, in many cases, we would not *accept* the hypothesis that the defendant is innocent. We would simply conclude that the evidence was not strong enough to rule out the possibility of innocence. In fact, in the United States, the two conclusions juries are instructed to choose from are "guilty" and "not guilty." A jury would never conclude that "the defendant is innocent." ■

## Lesson 3: What Can Go Wrong: The Two Types of Errors and Their Probabilities

Whenever we use a sample of data to make a decision about a larger population, we may make a mistake. In testing statistical hypotheses, there are two potential decisions, and each decision brings the possibility of an error. Decision makers and researchers should consider the consequences of these possible errors when they create and apply their decision-making rules.

**Example 12.6**

**Errors in the Courtroom** Let's use the courtroom analogy as an illustration. Viewed from within the hypothesis testing framework, the two hypotheses are these:

*Null hypothesis:* The accused is *innocent*.
*Alternative hypothesis:* The accused is *guilty*.

---

**\*HINT:** The $p$-value is the conditional probability of observing 60% or more correct, given that the participant is just guessing. Would it be harder to guess at least 6 correct in 10 tries or at least 60 correct in 100 tries?

And the two possible decisions for a judge or jury are these:

*Do not reject the null hypothesis:* The accused is *not guilty.*
*Reject the null hypothesis:* The accused is *guilty.*

For trials in general, here are the possible errors and the consequences that accompany those errors:

- *Possible error 1:* A "guilty" verdict for a person who is really innocent.
  *Consequence:* An innocent person is falsely convicted. The guilty party remains free.

- *Possible error 2:* A "not guilty" verdict for a person who committed the crime.
  *Consequence:* A criminal is not punished.

In the U.S. court system, a false conviction (possible error 1) is generally viewed as the more serious error. Not only is an innocent person punished, but also a guilty one remains free. Courtroom rules and rules affecting pretrial investigations tend to reflect society's concern about incorrectly punishing an innocent person. ■

**Example 12.7**

Statistics⬙Now™

Watch a video example at **http://
1pass.thomson.com** or on your CD.

**Errors in Medical Tests**  Imagine that you are tested to determine whether you have a disease. The lab technician or physician who evaluates your results must make a choice between two hypotheses:

*Null hypothesis:* You do not have the disease.
*Alternative hypothesis:* You have the disease.

Unfortunately, many laboratory tests for diseases are not 100% accurate. There is a chance that the result is wrong. Consider the two possible errors and their consequences:

- *Possible error 1:* You are told you have the disease, but you actually don't. The test result was a **false positive.**
  *Consequence:* You will be unnecessarily concerned about your health, and you may receive unnecessary treatment, possible suffering adverse side effects.

- *Possible error 2:* You are told that you do not have the disease, but you actually do. The test result was a **false negative.**
  *Consequence:* You do not receive treatment for a disease that you have. If this is a contagious disease, you may infect others.

Which error is more serious? In most medical situations, the second possible error, a false negative, is more serious, but this could depend on the disease and the follow-up actions that are taken. For instance, in a screening test for cancer, a false negative could lead to a fatal delay in treatment. Initial test results that are "positive" for cancer are usually followed up with a retest so that a false positive may be discovered quickly. ■

# Type 1 and Type 2 Errors in Statistical Hypothesis Testing

The situation that is encountered in statistical hypothesis testing is about the same as in the courtroom and medical analogies. One notable difference is that in statistical hypothesis testing, random sampling error causes the errors defined in this section. When a sample is used, the "luck of the draw" could be such that the sample statistic does not properly represent the population value. This is particularly likely when the sample size is small and the corresponding standard error is large.

In the terminology of statistical hypothesis testing, the two possible errors are called **type 1** and **type 2** errors.

*definition*

A **type 1 error** can occur only when the null hypothesis is actually true. The error occurs by concluding that the alternative hypothesis is true.

A **type 2 error** can occur only when the alternative hypothesis is actually true. The error occurs by concluding that the null hypothesis cannot be rejected.

The numbering of the error types indicates which of the two hypotheses, (1) null or (2) alternative, is actually true. For example, a type 1 error is the error that occurs when the first hypothesis, the null, is really true but we decide in favor of the alternative.

**Example 12.8**

**Calcium and the Relief of Premenstrual Symptoms** In Case Study 11.2, we learned about a randomized "blind" study in which 466 women who suffered from premenstrual symptoms (PMS) were randomly assigned to take either calcium supplements or a placebo. The research group believed that the use of calcium supplements would reduce the severity of the symptoms. The hypotheses of interest to the researchers are as follows:

$H_0$: For reducing PMS symptoms, calcium is not better than placebo.
$H_a$: For reducing PMS symptoms, calcium is better than placebo.

A type 1 error would occur if the researchers decided that calcium is better than placebo when, in fact, it is not (the null hypothesis is true). A type 2 error would occur if the researchers decided that calcium is not better than placebo when, in fact, it is (the alternative hypothesis is true).

In the actual study, published in the *American Journal of Obstetrics and Gynecology* (Thys-Jacobs et al. 1998), the authors concluded that there was a statistically significant difference between the effectiveness of calcium supplements and placebo. Ideally, this conclusion is correct and no error was made. If future research and experience prove that calcium supplements are actually not effective, these researchers will have made a type 1 error by choosing the alternative when the null hypothesis is true. If a type 1 error was made in this situation, women may be taking calcium to reduce their symptoms when it does not actually work. On the other hand, if the researchers had not found a difference but calcium really is effective (a type 2 error) then women would not know about a treatment that is actually effective. ■

**Example 12.9**

**Medical Tests Revisited** In Example 12.7, we saw the two possible errors that might occur when you are tested for the presence of disease. Let's rephrase the possible errors in statistical terminology.

- A type 1 error occurs when the diagnosis is that you have the disease when actually you do not. In other words, *a type 1 error is a false positive.*

- A type 2 error occurs when the diagnosis is that you do not have the disease when actually you do. In other words, *a type 2 error is a false negative.* ∎

## Probabilities and Consequences of the Two Types of Errors

There always is a trade-off between the chances of making the two types of errors. For instance, in the U.S. judicial system, the procedures that are designed to minimize the probability of a false conviction carry with them an increased risk of erroneous not guilty verdicts. But if court rules were made more lenient to decrease the risk of incorrect acquittals, the probability of false convictions would increase.

In any given situation, it is important to understand the consequences of the two types of errors and to find the right balance between making the more serious and the less serious of the two types of errors. To know how to do that, we need to know how to determine the probability of making each type of error.

The first thing to realize about the probabilities of making type 1 or type 2 errors is that they are *conditional probabilities.* If the null hypothesis is true, you can make a type 1 error (by rejecting it), but you cannot make a type 2 error. If the alternative hypothesis is true, you can make a type 2 error (by failing to reject the null hypothesis), but you cannot make a type 1 error. Therefore, it makes sense to talk about the probability of making a type 1 error *given* that the null hypothesis is true and the probability of making a type 2 error *given* that the alternative hypothesis is true.

Figure 12.1 provides a tree diagram illustrating the possible errors and their probabilities. Notice that the first set of branches illustrate the truth, which is either the null hypothesis $H_0$ or the alternative hypothesis $H_a$. Because one or the

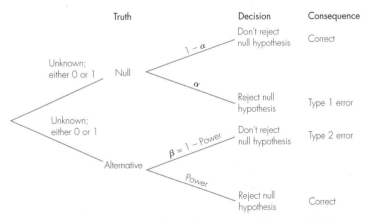

**Figure 12.1** ∎ Tree diagram of possible errors and their probabilities

other of these is definitely true, the appropriate probabilities for those branches are 1 (for the one that is true) and 0 (for the one that is not true). Unfortunately, we don't know which one is true, so we cannot specify which branch should be labeled 0 and which should be labeled 1. Now let's investigate the probabilities assigned to the second set of branches, which represent the decision made.

## Probability of a Type 1 Error and the Level of Significance

When we defined a *p-value* in Lesson 2 we learned two things:

- The *p*-value is the *conditional probability of observing a test statistic as extreme as that observed or more so, given that the null hypothesis is true.*
- We reject the null hypothesis when the *p*-value is less than the *level of significance,* denoted by the Greek letter $\alpha$ ("alpha").

From these two facts it follows that the probability of making a type 1 error, given that the null hypothesis is true, is the probability that the *p*-value is less than $\alpha$. As will be explained shortly, that probability is in fact $\alpha$ itself! Therefore, the level of significance $\alpha$ that we use for a hypothesis test also provides the conditional **probability of a type 1 error.**

---

*definition*   When the null hypothesis is true, the **probability of a type 1 error,** the level of significance, and the $\alpha$-level are all equivalent. When the null hypothesis is not true, a type 1 error cannot be made.

---

To see the correspondence between the level of significance and the probability of a type 1 error, suppose the null value is the correct population parameter value. In that case, the distribution of the test statistic ($z$ or $t$) is known and is either the standard normal or the Student's $t$-distribution. The *p*-value is the tail area at and beyond the observed test statistic in the tail(s) specified by the alternative hypothesis. Suppose the .05 level of significance is being used. By chance alone, the test statistic will fall into the most extreme 5% of the distribution just 5% of the time. When that happens, the *p*-value will be less than .05, the null hypothesis will be rejected, and a type 1 error will have been made. The same holds for any specified value for $\alpha$.

This correspondence means that with the standard level of significance of .05 or 5%, about 5% of all hypothesis tests done *when the null hypothesis is really true* will result in a false claim of statistical significance. Notice that this is not the same thing as saying that 5% of all rejected null hypotheses should not have been rejected. In fact, some statisticians and researchers argue that in the "real world," the null hypothesis is rarely true. If this argument is correct, the overall number of type 1 errors in practice may be quite small.

## Type 2 Errors and Power

The probability of making a type 1 error is easy to specify because it involves the assumption that the population parameter is equal to a specific null value. With that assumption, we know precisely what distribution the test statistic has and can thus find probabilities associated with possible outcomes.

It is not so simple to find the **probability of a type 2 error.** Remember that a type 2 error is made when the alternative hypothesis is true. But the alternative hypothesis covers a wide range of possibilities. If the truth is close to the null value, it may be difficult for sample data to provide adequate evidence to reject the null hypothesis, and the probability of making a type 2 error will be high. If the truth is far from the null value, then the sample data will likely provide enough evidence to reject the null value, and the probability of a type 2 error will be low.

There are three factors that affect the probability of making a type 2 error:

1. The sample size; larger $n$ reduces the probability of a type 2 error without affecting the probability of a type 1 error.
2. The level of significance used for the test; larger $\alpha$ reduces the probability of a type 2 error by increasing the probability of a type 1 error.
3. The actual value of the population parameter, which is not in the researcher's control. The farther the truth falls from the null value (as long as it is in the alternative hypothesis), the lower is the probability of a type 2 error.

## Power of a Hypothesis Test

Rather than focusing on the risk of making a mistake, many investigators prefer to focus on the chance that their sample will provide the evidence necessary to make the right choice. A common question when a study is planned is "If something is really going on in the population, what is the probability that we will be able to detect it?" This question leads to the statistical meaning of the word *power.*

| | |
|---|---|
| *definition* | When the alternative hypothesis is true, the probability of making the correct decision is called the **power** of a test. |

Power is simply the flip side of the risk of a type 2 error, and the same factors that influence the risk of a type 2 error also affect the power.

The concept of power is the same for all statistical hypothesis tests, but the technical formulas and details differ depending on the situation. The "by hand" calculations can be difficult; fortunately, some statistical software programs (including Minitab) can be used to easily determine the power of a test under specified conditions. We will discuss power for specific examples in the modules on one proportion (HT Module 1) and one mean (HT Module 3).

There is no notation for *power,* but the Greek letter $\beta$ is sometimes used to indicate the probability of making a type 2 error. Therefore, we can write power as $1 - \beta$, as illustrated on the appropriate branch of the tree diagram in Figure 12.1.

There are two features of power that apply to all hypothesis tests and that researchers should keep in mind when they plan a study:

- *The power increases when the sample size is increased.* This makes sense because when the sample size is increased, the standard error is decreased, leading to larger values of the test statistic. Also, the sample statistic is a more accurate estimate of the population value, making it easier to detect a difference between the true population value and the null value.

- *The power increases when the difference between the true population value and the null hypothesis value increases.* This makes sense because the probability of detecting a large difference is higher than the probability of detecting a small difference. However, remember that the truth about the population is not something that the researcher can control or change.

Researchers should evaluate power before they collect data that will be used to do hypothesis tests to make sure they have sufficient power to make the study worthwhile. Occasionally, news reports, especially in science magazines, will insightfully note that a study may have failed to find a relationship between two variables because the test had low power. This is a common consequence of conducting research with samples that are too small, but it is one that is often overlooked in media reports.

## The Trade-Off in Probability for the Two Types of Error

If we wanted to be sure that we rarely made a type 1 error, we could simply set the level of significance at a ridiculously low value, making it almost impossible to reject the null hypothesis when it was true. However, if we did this, we would obviously be at great risk of making a type 2 error, because our stringent criterion would make it difficult to reject the null hypothesis even when it was not true.

You can see in Figure 12.1 (p. 507), that you cannot find the probability of one of the two types of errors just by knowing the other one. You can also see that the researcher can specify the probability of a type 1 error directly—it is the level of significance $\alpha$. For a fixed $\alpha$, increasing the sample size(s) will reduce the risk of a type 2 error. Here are some guidelines for choosing $\alpha$, based on the seriousness of the two types of errors:

- If the consequences of a type 1 error are very serious, then the level of significance $\alpha$ should be chosen to be small, perhaps .01 or less. Remember that *if* the null hypothesis is true, the mistake that can be made is to reject it, and the probability of doing so is given by $\alpha$. In this situation, we want to make it more difficult to reject the null hypothesis.

- If the consequences of a type 2 error are very serious, it is desirable to have high power and low probability of making a type 2 error. Remember that a type 2 error occurs when the alternative hypothesis is true but the null hypothesis is not rejected. Therefore, to lower the probability of making a type 2 error, the level of significance $\alpha$ should be high, perhaps .10, to make it easier to reject the null hypotheses. The trade-off is that if the null hypothesis is true, the probability of a type 1 error is higher than it would be if a smaller $\alpha$ were used.

- The only way to reduce the probability of a type 2 error for any specified probability of a type 1 error is to take a larger sample.

12.2 Exercises are on page 539.

# 12.3 HT Module 1: Testing Hypotheses About a Population Proportion

In this section, we illustrate the five steps to hypothesis testing in the context of testing a specified value for a population proportion. We encountered this situation in Chapter 8 when we learned about binomial random variables. In that case we focused on $X$ = number of successes, whereas in this section we focus on the sample proportion, $X/n$. However, the scenarios in which the methods apply are equivalent, and the population parameter $p$ in this section is equivalent to the probability of success $p$ in a binomial setting.

***Steps in any Hypothesis Test***

> **Step 1:** *Determine the* null *and* alternative *hypotheses.*
>
> **Step 2:** *Verify necessary data conditions, and if they are met, summarize the data into an appropriate* test statistic.
>
> **Step 3:** *Assuming that the null hypothesis is true, find the p-value.*
>
> **Step 4:** *Decide whether or not the result is* statistically significant *based on the p-value.*
>
> **Step 5:** *Report the conclusion in the context of the situation.*

Examples 12.1 and 12.2 were illustrations of how to write the null and alternative hypotheses in words and in formulas when the parameter of interest is a population proportion. For instance, in Example 12.2 about the proportion favoring a lower legal blood alcohol limit, the null hypothesis was $H_0$: $p \leq .5$ and the alternative was $H_a$: $p > .5$. The value specified in the hypotheses for the population proportion of interest is called the **null value** and is denoted by the symbol $p_0$. In Example 12.2 just described, the value of $p_0$ is .5.

The possible null and alternative hypotheses are one of these three choices, depending on the research question:

1. $H_0$: $p = p_0$  versus  $H_a$: $p \neq p_0$  (two-sided)
2. $H_0$: $p \geq p_0$  versus  $H_a$: $p < p_0$  (one-sided)
3. $H_0$: $p \leq p_0$  versus  $H_a$: $p > p_0$  (one-sided)

Often, the null hypothesis for a one-sided test is written as $H_0$: $p = p_0$ instead of including the values above or below $p_0$ as well. We will write it both ways in this chapter to give you practice with the two ways in which you may encounter the null hypothesis in journal articles. Remember that a *p*-value is computed by assuming that the null hypothesis is true, and the specific null value $p_0$ is what is assumed to be the truth about the population for that computation.

## The z-Test for a Proportion

When we have a sufficiently large random sample, we can use a **z-test** to examine hypotheses about a population proportion. The test is called a "*z*-test" because the **test statistic** is a standardized score (*z*-score) for measuring the difference between the sample proportion and the null hypothesis value of the population proportion. The logic of the test is as follows:

- Determine the sampling distribution of possible sample proportions when the true population proportion is $p_0$ (called the *null value*), the value specified in $H_0$.
- Using properties of this sampling distribution, calculate a standardized score ($z$-score) for the observed sample proportion $\hat{p}$.
- If the standardized score has a large magnitude, conclude that the sample proportion $\hat{p}$ would be unlikely if the null value $p_0$ is true, and reject the null hypothesis.

The principal idea is that we determine how likely it would have been for the sample proportion to have occurred if the null hypothesis is true. For example, if the null value is $p_0 = .5$ and the observed sample proportion is $\hat{p} = .65$, we would determine how likely it is that we could observe a sample proportion as large as .65 (or larger) if the population proportion was .5.

## Conditions for Conducting the $z$-Test

Whenever we conduct a hypothesis test, we must make sure that the data meet certain conditions. These conditions are the assumptions that were made when the theory was derived for the test statistic that we are using to test our hypotheses. For testing proportions with the procedure described in this section, there are two conditions that should be true for the sample:

1. The sample should be a random sample from the population or the data should come from a binomial experiment with independent trials.
2. The quantities $np_0$ and $n(1 - p_0)$ should both be at least 10.

The first condition is not always practical, and most researchers are willing to use the test procedure as long as the sample is considered to be representative of the population for the question of interest. For instance, a college statistics class may be representative of all college students on questions such as which of two soft drinks is preferred or whether their right foot is longer than their left foot.

The second condition is a sample size requirement. To test the null hypothesis that $p = .5$, for example, we would need a sample of at least $n = 20$ to meet this condition. If the null hypothesis is that $p = .1$, the sample size would have to be at least $n = 100$. Our standard may be somewhat conservative; some authors suggest that $np_0$ and $n(1 - p_0)$ only need to be larger than 5 (e.g., Ott, 1998, p. 370).

Let's go through the five steps of hypothesis testing for an example, and then formalize the details.

**Example 12.10**

**Statistics⬡Now™**

For software help, download your Minitab, Excel, TI-83, SPSS, R, and JMP manuals from **http://1pass .thomson.com**, or find them on your CD.

**The Importance of Order in Voting** This example was first introduced in Chapter 2. In a student survey, a statistics teacher asked his students to "randomly pick a letter from these two choices—S or Q." For about half of the students, the order of the letters S and Q was reversed so that the end of the instruction read "Q or S." The purpose of the activity was to determine whether there might be a preference for choosing the first letter. A tendency to pick the first choice offered has been noted in several settings, including in elections.

**Step 1:** *Determine the null and alternative hypotheses.*

Use the letter $p$ to represent the proportion of the population that would pick the first letter. The null hypothesis is a statement of "nothing happening." If there is no general preference for either the first or second letter, $p = .5$ (because there are two choices). So the "null value" for $p$ is $p_0 = .5$.

The alternative hypothesis usually states the researcher's belief or speculation. The purpose of the activity was to find out whether there is a general preference for picking the first choice, and a preference for the first letter would mean that $p$ is greater than .5. The null and alternative hypotheses can be summarized as follows:

$H_0$: $p = .5$
$H_a$: $p > .5$

- These hypotheses are statements about the larger population that this sample represents. Hypothesis tests are always used to say something about the population from which the sample has been taken.

- A more literal statement of the null hypothesis would be $H_0$: $p \leq .5$ because it is possible that less than half of the population would choose the first letter. Even if the null hypothesis is modified in this way, the value .5 is used as the assumed null value in the calculations.

**Step 2:** *Verify necessary data conditions, and if they are met, summarize the data into an appropriate test statistic.*

Before computing the test statistic, we verify that our sample meets the necessary conditions for using the $z$-statistic. With $n = 190$ and $p_0 = .5$, both $np_0$ and $n(1 - p_0)$ equal 95, a quantity that is larger than 10, so the sample size condition is met. The sample, however, is not really a random sample—it is a convenience sample of students who were enrolled in this class. It does not seem that this will bias the results for this question, so we will behave as though the sample was a random sample.

Among 92 students who saw the order "*S* or *Q*," 61 picked *S*, the first choice. Among the 98 students who saw the order "*Q* or *S*," 53 picked *Q*, the first choice. In all, 114 of 190 students picked the first choice of letter. Expressed as a proportion, this is $114/190 = .60$.

Data analysts nearly always use statistical software to do the calculations for a hypothesis test. Figure 12.2 displays output for this problem from the Minitab

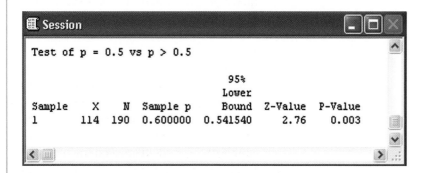

**Figure 12.2** ▌ Minitab output for *S* or *Q* choice (Version 14, Normal Distribution Option)

statistical program. We first work through the remainder of the hypothesis test utilizing this output. Following that, we will show the detailed calculations for a hypothesis test for a population proportion.

The output items labeled "X," "N," and "Sample p" summarize the calculation of the sample proportion. The sample proportion, $\hat{p} = .60$, is the primary information for the test statistic, but we need to determine the corresponding $z$-statistic to find the $p$-value. From the output, we see under the label "Z-Value" that the value of the $z$-statistic is 2.76. The $z$-value is a standardized score for measuring the difference between the sample proportion, $\hat{p} = .60$, and the null hypothesis value, $p_0 = .50$.

**Step 3:** *Assuming that the null hypothesis is true, find the p-value.*

The $p$-value for this hypothesis test is the answer to this question: *If the true p is .5, what is the probability that, for a sample of 190 people, the sample proportion could be as large as .60 (or larger)?*

The answer to this question is also the answer to a corresponding question about the $z$-statistic: *If the null hypothesis is true, what is the probability that the z-statistic could be as large as 2.76 (or larger)?*

We can either use computer output to find this value or find it from Table A.1, the Standard Normal Table in the Appendix. Details of using the table will be given later in this chapter. In the Minitab output, we can locate the answer, "0.003," at the far right under the heading "P-Value." If the population proportion is really .5, the probability is only .003, about 3 in 1000, that we could observe a sample proportion as large as .60 or larger.

*Note:* In the specific context of hypothesis tests for proportions, it is unfortunate that the letter "$p$" could have two different meanings. Do not confuse the population proportion "$p$" with the $p$-value for the test, which is sometimes abbreviated to just "$p$" in research articles and computer output. You should be able to sort out which is which based on the context.

**Step 4:** *Decide whether or not the result is statistically significant based on the p-value.*

The convention used by most researchers is to declare statistical significance when the $p$-value is smaller than .05. The $p$-value in this example is .003, so we can reject the null hypothesis.

**Step 5:** *Report the conclusion in the context of the situation.*

The statistical conclusion was to reject the null hypothesis that $p = .50$. In this situation, the conclusion is that *there is statistically significant evidence that the first letter presented is preferred.* This conclusion is a generalization that applies to the population represented by this sample. In other words, we conclude that for the entire population of individuals represented by these students, there would be a preference for choosing the first letter presented.

Conclusions are strengthened when another study replicates the results. The instructor repeated this activity in another class of 327 students. In that class, 58% of the students picked the first letter, which is very close to the 60% found in the original study. ▶

# Details for Calculating the $z$-Statistic

Our fundamental concern is the difference between the sample proportion and the null hypothesis value for the population proportion. If the sample proportion is sufficiently far from the null value, we can reject the null hypothesis. For example, if the null hypothesis is that $p = .5$, a sample proportion of .9 based on a relatively large random sample would be strong evidence against the null hypothesis. A sample proportion equal to .52 would be far less convincing evidence against the statement that the true $p = .5$.

We can measure the difference between the sample proportion and the null hypothesis value with a standardized score. From the Rule for Sample Proportions in Chapter 9, we know that the possible values of the sample proportion $\hat{p}$, based on a sample of size $n$, are approximately normal with

- mean = $p$ = true population proportion
- standard deviation = $\sqrt{\dfrac{p(1 - p)}{n}}$

When we assume that the null hypothesis is true, so that $p = p_0$, we assign special names to the mean and standard deviation. In particular, the term *null value* is used for $p_0$, and the term *null standard error* is used for the standard deviation when $p_0$ is substituted for $p$.

Technically the term *null standard deviation* should be used because we reserve the use of the term *standard error* for situations in which sample estimates are used in the formula for a standard deviation. However, we use *null standard error* to be consistent with the terminology for hypotheses testing used in the other situations discussed in Chapters 12 and 13. Under the null hypothesis, the sampling distribution of $\hat{p}$ is approximately normal with

- mean = null value = $p_0$
- standard deviation = null standard error = $\sqrt{\dfrac{p_0(1 - p_0)}{n}}$

The $z$-statistic for the significance test is

$$z = \frac{\text{Sample estimate} - \text{Null value}}{\text{Null standard error}} = \frac{\hat{p} - p_0}{\text{Null standard error}} = \frac{\hat{p} - p_0}{\sqrt{\dfrac{p_0(1 - p_0)}{n}}}$$

- $\hat{p}$ represents the sample estimate of the proportion
- $p_0$ represents the specific value in the null hypothesis
- $n$ is the sample size

**Example 12.10** *(cont.)*

**Calculating the $z$-Statistic for the $S$ or $Q$ Problem** For the "random" choice between $S$ and $Q$, the null hypothesis was $p = .5$, and the proportion of the sample of $n = 190$ students that picked the first letter was $\hat{p} = .6$. In Figure 12.2 (p. 513) we saw that the $z$-statistic for the hypothesis test is 2.76.

**HT Module 1: p**

**Example 12.11**

The calculation is

$$z = \frac{\text{Sample estimate} - \text{Null value}}{\text{Null standard error}} = \frac{.6 - .5}{\sqrt{\dfrac{.5(1 - .5)}{190}}} = \frac{.10}{.0362} = 2.76$$

**Do Fewer Than 20% Experience Medication Side Effects?**  Suppose that a pharmaceutical company wants to claim that side effects will be experienced by fewer than 20% of the patients who use a particular medication. In a clinical trial with $n = 400$ patients, they find that 68 patients experienced side effects.

**Step 1:**  *The null and alternative hypotheses are*

$H_0$: $p \geq .20$     (company's claim is not true)
$H_a$: $p < .20$     (company's claim is true)

Notice that the pharmaceutical company's claim is used as the alternative hypothesis. Also notice that the null hypothesis is a region of values. To calculate the $z$-statistic, it is standard practice to use the value that separates the null and alternative hypothesis regions, in this case $p_0 = .20$.

**Step 2:**  *Verify necessary data conditions, and if they are met, summarize the data into an appropriate test statistic.*

There were 400 patients, so both $np_0$ and $n(1 - p_0)$ are large enough to proceed with a test based on the $z$-statistic. Although most clinical trials use volunteer patients, presumably the company would choose volunteers who are representative of the larger population of potential users of the medication, so for practical purposes we will accept that the necessary conditions are met. Out of 400 patients, 68 experienced side effects. The sample proportion that experienced side effects is $\hat{p} = 68/400 = .17$. The $z$-statistic for the hypothesis test is

$$z = \frac{\text{Sample estimate} - \text{Null value}}{\text{Null standard error}} = \frac{.17 - .20}{\sqrt{\dfrac{.20(1 - .20)}{400}}} = \frac{-.03}{.02} = -1.5$$

Notice that the $z$-statistic is negative. This occurs because the sample proportion is less than the null value, the result that the pharmaceutical company wanted. We will complete the remaining three steps for this example after we present the general format for finding a $p$-value. ▶

## Computing the *p*-Value for the *z*-Test

Recall that a $p$-value is computed by assuming the null hypothesis is true and then determining the probability of a result as extreme as, or more extreme than, the observed test statistic in the direction of the alternative hypothesis. For the $z$-test of hypotheses about a population proportion, the $p$-value probability is found by using the normal curve. This is a consequence of the Normal Curve Approximation Rule for Sample Proportions introduced in Chapter 9. The details of "determining the probability of a result as extreme as (or more extreme than) the observed test statistic" are consistent with the direction specified in the alternative hypothesis:

- For a *greater than* alternative hypothesis, find the probability that the test statistic $z$ could have been *equal to or greater than* what it is.
- For a *less than* alternative, find the probability that the test statistic $z$ could have been *equal to or less than* what it is.
- For a *two-sided alternative hypothesis,* the *p*-value includes the probability areas in both extremes of the distribution of the test statistic $z$.

The correspondence between the type of alternative hypothesis and the *p*-value area is summarized in Table 12.1.

**Table 12.1** Alternative Hypothesis Regions and *p*-Value Areas

| Statement of H$_a$ | | *p*-Value Area | Normal Curve Region |
|---|---|---|---|
| $p < p_0$ | (less than) | Area to the left of $z$ (even if $z > 0$) | |
| $p > p_0$ | (greater than) | Area to the right of $z$ (even if $z < 0$) | |
| $p \neq p_0$ | (not equal) | $2 \times$ area to the right of $|z|$ | |

The reasoning behind including both extremes for the two-sided alternative hypothesis is as follows. Suppose we want to test whether the true percentage of heads is 50% when we spin a coin on its edge and watch how it lands. We spin the coin on its edge 100 times, and it lands heads up 60 times. Remember that "*the p-value is computed by assuming the null hypothesis is true and then determining the probability of a result as extreme (or more extreme) as the observed test statistic in the direction of the alternative hypothesis.*" In this case, "in the direction of the alternative hypothesis" includes both above and below 50%. So 40% heads is as extremely far into the alternative hypothesis region (as far from the null value of 50%) as is 60% heads. Both extremes provide equivalent support for the alternative hypothesis.

If we use Minitab to do the hypothesis test, the program will automatically compute and report the *p*-value. To calculate the *p*-value "by hand," we can use Table A.1, the Standard Normal Table, or we could use a spreadsheet program like Excel (the command NORMSDIST($z$) finds the area to *left* of $z$).

**Example 12.11** *(cont.)*

**Statistics** ⟳ **Now**™

Watch a video example at **http:// 1pass.thomson.com** or on your CD.

**The Medication Side Effects Problem**  We have already done steps 1 and 2 for this example. The null and alternative hypotheses are H$_0$: $p \geq .20$ and H$_a$: $p < .20$, and the $z$-statistic is $z = -1.5$. We now complete the final three steps.

**Step 3:** *Assuming that the null hypothesis is true, find the p-value.*

The alternative hypothesis is that the *proportion is less than* a specified value, so the *p*-value is the *area to the left* of the observed $z$-statistic. We can use Table A.1, the Standard Normal Table, to determine this probability, or we can

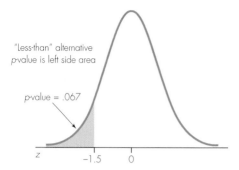

**Figure 12.3** ▮ *p*-Value for medication side effects example

use the Excel instruction NORMSDIST($-1.5$). For $z = -1.5$, the area to the left is about .067, and this is the *p*-value. It is the answer to this question: *If the true p is .2, what is the probability that, for a sample of 400 people, the sample proportion could be as small as .17 (or smaller)?*

The relationship between the *p*-value and the *z*-statistic for this problem is illustrated in Figure 12.3.

**Step 4:** *Decide whether or not the result is statistically significant based on the p-value.*

If the usual .05 standard for statistical significance is used, the *p*-value of .067 is not quite small enough to reject the null hypothesis.

**Step 5:** *Report the conclusion in the context of the situation.*

The sample evidence was in the desired direction for the pharmaceutical company, but it was not strong enough to conclusively reject the null hypothesis. The company cannot reject the idea that the proportion of the population who would experience side effects is .20 (or more). If this is an important issue for the company, it should consider gathering additional data. As we will learn in the next section, sample size affects statistical significance. This example also illustrates why it is not a good idea to "accept the null hypothesis." We would not be convinced that 20% or more of the population would experience side effects when only 17% of the sample did so. ▮

**Example 12.10** *(cont.)*

**Statistics⬢Now™**

For software help, download your Minitab, Excel, TI-83, SPSS, R, and JMP manuals from **http://1pass .thomson.com**, or find them on your CD.

**The *p*-Value for the "Greater Than" Alternative in the *S* or *Q* Example**
For the "choose *S* or *Q*" problem, the alternative hypothesis is $H_a: p > .5$. When an alternative hypothesis is that a *proportion is greater than* a specified value, the *p*-value is the *area to the right* of the observed *z*-statistic. We determined before that $z = 2.76$ and the corresponding *p*-value is .003. The *p*-value, which is the area to the right of 2.76, is illustrated in Figure 12.4.

To determine this *p*-value "by hand," we can use Table A.1 in the Appendix to learn that for $z = 2.76$, the area to the left is roughly .997. This is the probability that we could get a *z*-score smaller than 2.76, so the probability we could get a *z*-score greater than 2.76 is $1 - .997 = .003$. To calculate this *p*-value with Excel, enter = $1 -$ @NORMSDIST(2.76) into a cell of the spreadsheet. (The @ symbol tells Excel to invoke that function.)

### Step 2: Verify Necessary Data Conditions, and If They Are Met, Summarize the Data into an Appropriate Test Statistic

***Conditions Necessary for a z-Test of the Difference in Two Proportions to Be Valid***

- Independent samples are available from the two populations.
- The number with the trait or response of interest and the number without the trait or response of interest is at least 10 in each sample.

Make sure the data collection was done so as to ensure independent samples, then simply make sure there are at least 10 with and without the trait or response of interest in each of the two samples.

### Continuing Step 2: Computing the Test Statistic for a z-Test for Two Proportions

The sample statistic $\hat{p}_1 - \hat{p}_2$ (difference between sample proportions) estimates the parameter $p_1 - p_2$ (difference between population proportions). To find the null standard error, assume that the null hypothesis is true, so that $p_1 = p_2 = p$. Estimate the common population proportion $p$ using all of the data:

$$\hat{p} = \frac{n_1\hat{p}_1 + n_2\hat{p}_2}{n_1 + n_2} = \frac{X_1 + X_2}{n_1 + n_2}$$

This combined estimate is simply the sample proportion for both samples combined. It is used instead of separate estimates of $p_1$ and $p_2$ to find the null standard error. The rationale for this substitution is that the test is based on the distribution of the test statistic *assuming that the null hypothesis is true.* If $p_1 = p_2$, then $\hat{p}$ is the best estimate for each of the population proportions because it makes use of all available sample data. The standardized test statistic is

$$z = \frac{\text{Sample statistic} - \text{Null value}}{\text{Null standard error}} = \frac{\hat{p}_1 - \hat{p}_2 - 0}{\sqrt{\dfrac{\hat{p}(1 - \hat{p})}{n_1} + \dfrac{\hat{p}(1 - \hat{p})}{n_2}}}$$

$$= \frac{\hat{p}_1 - \hat{p}_2 - 0}{\sqrt{\hat{p}(1 - \hat{p})\left(\dfrac{1}{n_1} + \dfrac{1}{n_2}\right)}}$$

If the null hypothesis is true, the sampling distribution of this $z$-statistic is approximately the standard normal curve.

### Step 3: Assuming That the Null Hypothesis Is True, Find the *p*-Value

The standard normal curve is used to find the *p*-value:

- For $H_a$: $p_1 - p_2 > 0$, the *p*-value is the area above $z$ (a one-tailed test), even if $z$ is negative.
- For $H_a$: $p_1 - p_2 < 0$, the *p*-value is the area below $z$ (a one-tailed test), even if $z$ is positive.
- For $H_a$: $p_1 - p_2 \neq 0$, the *p*-value is $2 \times$ area above $|z|$ (a two-tailed test).

See Table 12.1 (p. 517) for guidance. Do not confuse the *p-value* with the population or sample *proportions,* which also use the letter *p*.

HT Module 2: $p_1 - p_2$

### Steps 4 and 5: Decide Whether or Not the Result Is Statistically Significant Based on the *p*-Value and Make a Conclusion in the Context of the Situation

Choose a significance level ("alpha"); standard practice is $\alpha = .05$. The result is statistically significant if the *p*-value $\leq \alpha$. Interpret the conclusion in the context of the situation. Whenever two populations are compared, the manner in which the data were collected should also be considered. Remember from Chapter 4 that an experiment is generally a stronger proof of causation than an observational study is.

**Example 12.17** *(cont.)*

**Statistics⟁Now**™

For software help, download your Minitab, Excel, TI-83, SPSS, R, and JMP manuals from **http://1pass .thomson.com**, or find them on your CD.

**HT Module 2: $p_1 - p_2$**

### The Prevention of Ear Infections

**Step 1:** *Determine the null and alternative hypotheses.* The researchers hoped to show that xylitol reduces ear infections. The parameter $p_1$ is the proportion who would get an ear infection in the population of children similar to those in the study if taking the placebo. The parameter $p_2$ is the proportion who would get an ear infection in that population if taking the xylitol. Therefore, the null and alternative hypotheses of interest are as follows:

$$H_0: p_1 - p_2 = 0 \quad \text{(or } p_1 = p_2\text{)}$$
$$H_a: p_1 - p_2 > 0 \quad \text{(or } p_1 > p_2\text{)}$$

A one-sided alternative is used because the researchers want to show that the proportion getting an ear infection is significantly lower in the xylitol group than in the placebo group.

**Step 2:** *Verify necessary data conditions, and if they are met, summarize the data into an appropriate test statistic.* There are at least ten children in each sample who did and did not get ear infections, so the conditions are met.

- $\hat{p}_1 = \dfrac{68}{165} = .412$ and $\hat{p}_2 = \dfrac{46}{159} = .289$

- The sample statistic is $\hat{p}_1 - \hat{p}_2 = .412 - .289 = .123$.

- The combined proportion is $\hat{p} = \dfrac{68 + 46}{165 + 159} = \dfrac{114}{324} = .35$.

- null s.e.$(\hat{p}_1 - \hat{p}_2) = \sqrt{\hat{p}(1 - \hat{p})\left(\dfrac{1}{n_1} + \dfrac{1}{n_2}\right)}$

$$= \sqrt{(.35)(.65)\left(\dfrac{1}{165} + \dfrac{1}{159}\right)} = .053$$

- $z = \dfrac{\text{Sample statistic} - \text{Null value}}{\text{Null standard error}} = \dfrac{.123}{.053} = 2.32$

**Steps 3, 4, and 5:** Because the alternative is $H_a: p_1 - p_2 > 0$, the *p*-value is the area above $z = 2.32$. Figure 12.8 illustrates this probability. From the normal probability table (Appendix Table A.1), we learn that the probability is .9898 that $z$ is less than 2.32, so *p*-value $= 1 - .9898 = .0102$. We can reject the null hypothesis and attach the label "statistically signifi-

The $z$-value for the hypothesis test is

$$z = \frac{\text{Sample estimate } - \text{ Null value}}{\text{Null standard error}}$$

For both of our hypothesis tests, the sample proportion is .55 and the null hypothesis value is .5, so the top portion of the calculation is the same for both cases. The $z$-value changes because the sample size affects the standard error.

- When $n = 60$, the null standard error $= \sqrt{\dfrac{.5(1 - .5)}{60}} \approx .065$.

- When $n = 960$, the null standard error $= \sqrt{\dfrac{.5(1 - .5)}{960}} \approx .016$.

Notice that increasing the sample size decreases the null standard error. Generally, a standard error roughly measures the "typical difference" between a statistic and a population parameter. If the null hypothesis that $p = .5$ were actually true, the consequence of taking a larger sample is that the sample proportion would be likely to fall closer to the null value (.5) than it would for a smaller sample. As a result, the difference between the sample proportion of .55 and the null value of .5 is relatively greater when the sample size is large than when it is small.

One more comment is in order about the example just presented. If in fact the null hypothesis were true, and $p = .50$, then a sample proportion as large as .55 would probably not have occurred in the larger sample. It is quite feasible for a sample of $n = 60$ to produce a sample proportion that differs from the truth by as much as .55 differs from .50. But it is quite unlikely for a sample of $n = 960$ to produce a sample proportion this far from the true proportion.

In summary, it is the size of the standard error that changes based on the size of the sample, and increasing the sample size gives a smaller standard error. As a result, a particular absolute difference between the sample proportion and null value is more significant with a large sample than with a small sample. No matter how large the sample, the sample proportion is likely to fall within two or so standard errors of the true population proportion. If the difference between the sample proportion and the null value is more than this, it is significant.

*thought question* **12.6**  If the null hypothesis is true, the correct conclusion is "cannot reject the null hypothesis." If the alternative hypothesis is true, the correct conclusion is "reject the null hypothesis." Increasing the sample size increases the probability of making the correct conclusion in one of these cases, but not in the other. Your job is to figure out which is which. Here is a hint: The conclusion "cannot reject the null hypothesis" is made as long as the sample proportion falls within a reasonable number of standard errors of the null value. When the null hypothesis is true, does the likelihood of this happening change as the sample size increases? Explain, and use this explanation to determine which conclusion has increased probability of being made correctly with increasing sample size.*

*****HINT:** Think about the probability that the sample proportion falls within 2 standard errors of the null value $p_0$ when it is the truth. Does that probability depend on the sample size?

## Real Importance Versus Statistical Significance

The phrase *statistically significant* means only that the data are strong enough to reject a null hypothesis. A result that is significant in the statistical meaning of the word is not necessarily significant in the more common meaning of the word. One reason for this, of course, is that the research study might not address an important real-world issue. Even when the research issue is important, however, statistical significance does not guarantee practical significance.

A *p*-value provides information about the conclusiveness of the evidence against the null hypothesis, but it does *not* provide information about the *magnitude* of the effect. In some instances, the magnitude of a statistically significant effect can be so small that the practical effect is not important. If the sample size is large enough, almost any null hypothesis can be rejected because there is almost always at least a slight relationship between two variables, a slight difference between two groups, or a slight deviation from the status quo.

**Example 12.19**

**Birth Month and Height** The headline of a Reuters news article posted at the Yahoo Health News website on February 18, 1998, was "Spring Birthday Confers Height Advantage." The article describes an Austrian study of the heights of 507,125 military recruits. In an article in the journal *Nature*, the researchers reported their finding that men born in the spring were, on average, about 0.6 centimeter taller than men born in the fall (Weber et al., *Nature*, 1998, 391: 754–755). This is a small difference; 0.6 centimeter is only about 1/4 of an inch. The sample size for the study is so large that even a very small difference will earn the title *statistically significant*. Do you think the practical import of this difference warranted the headline? ■

12.5 Exercises are on page 545.

---

*case study* **12.1**    The Internet and Loneliness:
Case Study 1.7 Revisited

We encountered the concept that a statistically significant result may not have practical importance when we read Case Study 1.7. That case study examined research done at Carnegie Mellon University in which the principal finding was that Internet usage may lead to feelings of loneliness and depression. This finding received a lot of media attention, but a close look at the study shows that the actual effects were quite small. For example, according to the *New York Times*, "one hour a week on the Internet was associated, on average, with an increase of 0.03, or 1 percent on the depression scale" (Harman, 30 August 1998, p. A3).

---

*thought question* **12.7**    In this section, we noted that a research finding of "statistical significance" does not necessarily indicate that the finding is of practical importance. In situations in which the hypothesis test is about a specified value for a population parameter, explain why it would be helpful to present a confidence interval for the parameter along with the finding of statistical significance. *

***HINT:*** Think about what information a confidence interval for a proportion provides.

## case study 12.2   An Interpretation of a *p*-Value Not Fit to Print

One of the most common mistakes that researchers and journalists make has to do with the interpretation of the *p*-value. We learned that a *p*-value does not tell us the probability that a hypothesis is true. When a *p*-value is erroneously interpreted in this manner, the resulting statements can be completely misleading.

In an article entitled, "Probability Experts May Decide Pennsylvania Vote," the *New York Times* (April 11, 1994, p. A15) reported on the use of statistics to try to decide whether there was fraud in a special election held in Philadelphia. When a state senator from Pennsylvania's Second Senate District died in 1993, a special election was held to fill the seat. The Democratic candidate, William Stinson, defeated the Republican candidate, Bruce Marks, by the narrow margin of 20,518 votes to 20,057 votes, a difference of only 461 votes.

The Republicans were disturbed when they saw the comparison of voting booth results and absentee ballot results, shown in Table 12.3. The Republican Marks beat Stinson in the voting booths, but Stinson, the election winner, had a huge advantage in the absentee ballots. The Republicans charged that the election was fraudulent and asked the courts to disallow the absentee ballot votes on the basis of suspicion of fraud. In February 1994, three months after the election, Philadelphia Federal District Court Judge Clarence Newcomer disqualified all absentee ballots and ruled that the Republican, Marks, should be seated. The Democrats appealed this ruling, and both sides hired statisticians to help sort out what might have happened.

**Table 12.3** Election Results by Type of Vote

|  | Democrat Stinson | Republican Marks | Difference |
|---|---|---|---|
| **Voting Booth** | 19,127 | 19,691 | −564 |
| **Absentee** | 1391 | 366 | +1025 |
| **Total** | 20,518 | 20,057 | +461 |

One statistical expert, Orley Ashenfelter, examined 22 previous senatorial elections in Philadelphia to determine the relationship between votes cast in the voting booth and those cast by absentee ballot. Using the previous data, he calculated a regression equation to predict the difference in absentee ballot votes for the two parties based on a given amount of difference between the parties in the voting booth.

In the voting booth, there was a difference of 564 votes in favor of the Republicans. Using his equation, Ashenfelter estimated that when there was this much difference in favor of the Republicans in the voting booths, there would be a difference of 133 votes in favor of the Republicans in the absentee ballots. Instead, the difference for absentee ballots in the disputed election was 1025 votes in favor of the *Democrats*.

Of course, everyone knows that chance events play a role in determining what happens in any given election. A hypothesis test can be used to evaluate the possible effects of randomness. Ashenfelter considered the null hypothesis that there was no fraud and calculated that the *p*-value was 6%. In other words, he determined that *if there was no fraud,* the probability was 6% that the Democratic advantage in absentee votes would be as large as 1025 votes when the Republican advantage in the voting booths was 565 votes.

Unfortunately, the author of the *New York Times* article misinterpreted this probability in a way that leads readers to believe that the election was probably fraudulent. When you read the following quote from the article, see whether you can detect the mistake in interpretation:

> More to the point, there is some larger probability that chance alone would lead to a sufficiently large Democratic edge on the absentee ballots to overcome the Republican margin on the machine balloting. And the probability of such a swing . . . Professor Ashenfelter calculates, was about 6 percent. Putting it another way, if past elections are a reliable guide to current voting behavior, *there is a 94 percent chance that irregularities in the absentee ballots, not chance alone,* swung the election to the Democrat, Professor Ashenfelter concludes (Passell, 1994, p. A15, italics added).

The author's statement that there is a 94% chance that irregularities swung the election is wrong. He mistakenly interpreted the *p*-value to be the probability that the null hypothesis is true, so he reported what he thought to be the probability that the election was fraudulent. In this case, the *p*-value tells us only that if we assume that there was no fraud, there is a 6% chance that we would see results like these. The *p*-value cannot be used to find the probability that the election was fraudulent.

The case went through two appeals, but the original decision made by Judge Newcomer to disallow the absentee ballots was upheld each time. The statistical evidence was not a major factor in these decisions. The appeal judges found that there was enough evidence of irregular absentee voting procedures to justify Judge Newcomer's decision.

# Key Terms

**Section 12.1**

hypothesis test, steps in any, 496

**Section 12.2**

null hypothesis ($H_0$), 497

alternative hypothesis ($H_a$), 497

one-sided hypothesis test, 498

one-tailed hypothesis test, 498

two-sided hypothesis test, 498

two-tailed hypothesis test, 498

null value, 498, 501

null standard error, 501

test statistic, 501

$p$-value, 501, 502

level of significance, 501, 503

Bayesian statistics, 502

statistically significant, 503

$\alpha$ level, 503

false positive, 505

false negative, 505

type 1 error, 506

type 2 error, 506

probability of a type 1 error, 508

probability of a type 2 error, 509

power, 509

**Section 12.3**

null value $p_0$, 511

$z$-test for a proportion, 511

$z$-test statistic for proportion, 515

$p$-value, test for proportion, 516–517

exact $p$-value, test for proportion, 521–522

rejection region, 523

power and sample size, 524–525

**Section 12.4**

hypothesis test for comparing two population proportions, 528

$z$-test (hypothesis test) for difference in two proportions, 529

conditions for $z$-test for difference in two proportions, 529

**Section 12.5**

sample size and statistical significance, 533

statistical significance vs. importance, 536

# Exercises

- Denotes basic skills exercises

- Denotes dataset is available in StatisticsNow at **http://1pass.thomson.com** or on your CD but is **not required** to solve the exercise.

**Bold-numbered exercises** have answers in the back of the text and fully worked solutions in the Student Solutions Manual.

**Statistics Now™** Go to the StatisticsNow website at **http://1pass.thomson.com** to:

- Assess your understanding of this chapter
- Check your readiness for an exam by taking the Pre-Test quiz and exploring the resources in the Personalized Learning Plan

## Section 12.1

**12.1** ● One example of a possible hypothesis of interest given in Section 12.1 is based on the fact that full moons occur on average every 29.53 days, and we might classify babies as to whether or not they were born during the 24-hour period surrounding a full moon. The hypothesis of interest is that the proportion of babies born during a full moon is 1/29.53, as it would be if births were uniformly distributed across days.
**a.** What is the population of interest in this situation?
**b.** Specify the population parameter of interest.
**c.** Write the hypothesis of interest using appropriate notation (introduced in Chapter 9).

**12.2** ● One example of a possible hypothesis of interest given in Section 12.1 is whether the proportion favoring the death penalty is the same for teenagers as it is for adults.
**a.** What are the populations of interest in this situation?
**b.** Specify the population parameter of interest.
**c.** Write the hypothesis of interest using appropriate notation (introduced in Chapter 9).

**12.3** ● Retired professional tennis players Martina Navratilova, Monica Seles, John McEnroe, and Jimmy Connors are all left-handed. Define the parameter $p$ to be the proportion of professional tennis players (current and retired) who are left-handed. Researchers are interested in the hypothesis that $p > .10$. (About 10% of the general population is left-handed.)
**a.** What is the population of interest in this situation?
**b.** Write the hypothesis in words, being careful to specify the population and the parameter of interest, by completing this sentence: "Researchers are interested in the hypothesis that. . . ."

**12.4** ● Define the parameter $p_1 - p_2$ to be the difference in the proportions of 21-year-old men and women in the United States who have a high school diploma. Researchers are interested in the hypothesis that $p_1 - p_2 = 0$.
**a.** What are the populations of interest in this situation?
**b.** Write the hypothesis in words, being careful to specify the populations and the parameter of inter-

---

● Basic skills ◆ Dataset available but not required **Bold-numbered** exercises answered in the back

est, by completing this sentence: "Researchers are interested in the hypothesis that . . . ."

12.5 Refer to the five steps in any hypothesis test listed in Section 12.1. Only one of the steps can be performed before the data are collected. Which step is it?

12.6 Give an example of a possible hypothesis of interest about a population proportion. Be sure to specify the population and the population parameter the hypothesis is about.

**12.7** Give an example of a possible hypothesis of interest about the difference between two population proportions. Be sure to specify the populations and the population parameter the hypothesis is about.

## Section 12.2

*Exercises 12.8 to 12.40 correspond to the three Lessons in Section 12.2. Lesson 1 exercises are 12.8 to 12.18; Lesson 2 exercises are 12.19 to 12.28; Lesson 3 exercises are 12.29 to 12.40.*

12.8 ● Determine whether each of these statements is an example of a null hypothesis or an example of an alternative hypothesis.
   a. The average weight of Canadian geese is the same as the average weight of Canadian warblers.
   b. The proportion of books in the local public library that are novels is higher than the proportion of books in the university library that are novels.
   c. The average price of wool jackets in New York City is lower in the summer than in the winter.
   d. The proportion of students who receive A grades from Professor Harrington is the same as or higher than the proportion of students who receive A grades from Professor Cantor.

12.9 ● For each of the following situations, write the alternative hypothesis.
   a. The null hypothesis is $H_0$: $p = .30$, and it is a two-sided hypothesis test.
   b. The null hypothesis is $H_0$: $p \leq .45$.
   c. The null hypothesis is $H_0$: $p \geq .60$.

**12.10** ● Is each of the following statements a valid null hypothesis? If not, explain why not.
   **a.** In a sample of students, the mean pulse rate for the men is equal to the mean pulse rate for the women.
   **b.** The average weight of newborn boys is the same as the average weight of newborn girls.
   **c.** The proportion of cars in California that are white is higher than the proportion of cars in Maine that are white.

12.11 ● State the null and alternative hypotheses for each of the following potential research questions. In each situation, also indicate whether the hypothesis test will be one-sided or two-sided.
   a. Do female college students study more, on average, than male college students do?
   b. Compared to men who are not bald, are bald men more likely to have heart disease?
   c. Is there a correlation between height and head circumference?

   d. Will an increase in the speed limits on interstate highways lead to an increase in the highway fatality rate?
   e. When a coin is spun on its edge, is the probability that it will land heads up equal to .5?

**12.12** Suppose that a statistics teacher asks his students to each randomly pick one of the numbers 0, 1, 2, 3, 4. His general theory is that the proportion who pick the number 0 will be less than what would be expected from random selection. Use $p$ to represent the population proportion that would pick the number 0.
   **a.** What is the specific value of $p$ that corresponds to true random selection?
   **b.** In terms of $p$, write null and alternative hypotheses for this situation.

12.13 Suppose the present success rate in the treatment of a particular psychiatric disorder is .65 (65%). A research group hopes to demonstrate that the success rate of a new treatment will be better than this standard. Use the letter $p$ to represent the success rate of the new treatment. Write null and alternative hypotheses for $p$.

**12.14** Do you think researchers should determine whether to use a one-sided or two-sided hypothesis test before they look at the sample data or after they look at it? Explain.

12.15 For each of the following situations, write the null and alternative hypotheses in words (as in Example 12.1, p. 497) and in symbols (as in Example 12.2, p. 498).
   a. In the last census, taken five years ago, it was determined that 6% of school-aged children in a certain state lived with their grandparent(s). To support a bill on tax breaks for seniors, a congressional member plans to take a random sample of school-aged children to determine if that percentage has increased.
   b. An anthropologist is trying to determine whether the people in a certain region are descended from the same ancestors as those in another region she has studied. She knows that in the region she has studied, 15% of the people have a certain unique genetic trait. She plans to take a random sample of people in the new region and test them for that trait.
   c. Paul likes a certain candy that comes in a bag with mixed colors, each with its own flavor. The candy company's website claims that 30% of these candies are red, which is Paul's favorite. He doubts their claim and thinks that fewer than 30% are red. He is willing to assume that the candy is randomly placed into bags, and he plans to buy several bags of candy and count the proportion of red pieces.

**12.16** Refer to Exercise 12.15. In each situation, the hypotheses are about a population proportion $p$. Explain what the population is and what the proportion of interest is in each case.

12.17 For the following two situations, write the null hypothesis in symbols. Remember from Chapter 9 that the symbol for one population means is $\mu$ and for the difference in two population means is $\mu_1 - \mu_2$.

a. The mean weight for the population of newborn babies is the same in the United States as it is in England.

b. The sticker on a new car states that the car gets 32 miles per gallon for highway driving. Marisa wants to test that hypothesis.

**12.18** In the following two situations from Exercise 12.17, what is the null value?

a. The mean weight for the population of newborn babies is the same in the United States as it is in England.

**b.** The sticker on a new car states that the car gets 32 miles per gallon for highway driving. Marisa wants to test that hypothesis.

**12.19** ● Consider this quote: "In a recent survey, 61 out of 100 consumers reported that they preferred plastic bags instead of paper bags for their groceries. If there is no difference in preference in the population, the chance of such extreme results in a sample of this size is about .03. Because .03 is less than .05, we can conclude that there is a statistically significant difference in preference." Give a numerical value for each of the following.

a. The $p$-value.

b. The level of significance, $\alpha$.

c. The sample proportion.

d. The sample size.

e. The null value.

**12.20** ● State the conclusion that would be made in each of the following situations.

**a.** Level of significance = .05, $p$-value = .10.

**b.** Level of significance = .05, $p$-value = .01.

c. Level of significance = .01, $p$-value = .99.

d. Level of significance = .01, $p$-value = .002.

**12.21** ● Suppose that a woman thinks she might be pregnant, so she takes a pregnancy test. Considering this situation to be analogous to hypothesis testing, write each of the following in words.

a. The null hypothesis. (*Hint:* Remember that the null hypothesis generally states that there is nothing going on.)

b. The alternative hypothesis.

c. The conclusion if there is not enough evidence to reject the null hypothesis.

d. The conclusion if the null hypothesis is rejected.

**12.22** ● Consider testing the null hypothesis that there is no relationship between smoking and getting a certain disease versus the alternative hypothesis that smokers are more likely than nonsmokers to get the disease. Explain in words what would be concluded about the relationship in the population if

**a.** The $p$-value is .33, and the level of significance is .05.

b. The $p$-value is .03, and the level of significance is .05.

**12.23** ● Refer to part (a) of Exercise 12.22. Use the scenario to explain why it is not appropriate to *accept* the null hypothesis.

**12.24** About 10% of the human population is left-handed. Suppose that a researcher speculates that artists are more likely to be left-handed than are other people in the general population. The researcher surveys 150 artists and finds that 18 of them are left-handed.

a. Define the parameter of interest and give the null value.

b. State the researcher's null and alternative hypotheses.

c. What proportion of the sample of artists is left-handed?

d. To calculate a $p$-value for the hypothesis test, what probability should the researcher calculate? Make your answer specific to this situation.

e. Write the formula for the test statistic in words and plug in the appropriate values in the numerator.

**12.25** An article in *USA Today* (Elias, June 3, 1999, p. D1) describes a study done by Georgia State University psychologist James Dabbs in which he found that women who are trial lawyers (litigators) are more likely to have male children than are women lawyers who are not trial lawyers. According to the article, "58% of litigators' kids were boys vs. 44% [boys] for the others. The odds of this happening by chance are less than 5%, he says."

a. Write the formula for the test statistic in words and plug in the appropriate values in the numerator.

b. Consider the second sentence in the quote. What statistical term describes this result?

c. The quote from the article tells us that a $p$-value is less than 5%. Write one sentence that describes exactly what the $p$-value measures in this setting.

d. Dabbs believes that the result can be attributed to the women's testosterone levels because in a previous study, he found that, on average, women litigators have higher testosterone levels than women lawyers who aren't litigators. Do you think this conclusion is justified? Explain why or why not.

**12.26** Given the convention of declaring that a result is "statistically significant" if the $p$-value is .05 or less, what decision would be made concerning the null and alternative hypotheses in each of the following cases? Be explicit about the wording of the decision.

**a.** $p$-value = .35.

**b.** $p$-value = .001.

**c.** $p$-value = .04.

**12.27** Two researchers are testing the null hypothesis that a population proportion $p$ is .25 and the alternative hypothesis that $p > .25$. Both take a sample of 100 observations. Researcher A finds a sample proportion of .29, and Researcher B finds a sample proportion of .33. For which researcher will the $p$-value of the test be smaller? Explain without actually doing any computations.

**12.28** A physician claims that as soon as his patients have a positive pregnancy test, he is generally able to predict the sex of the baby and that his probability of being right is greater than the one-half what would be expected if he were just guessing. A skeptic challenges his claim, and the physician decides to collect data for the next ten pregnancies to try to support his claim.

a. From the perspective of the physician, what are the null and alternative hypotheses for this situation?

b. Out of the ten pregnancies, he is correct six times. Write in words what probability needs to be computed to find the *p*-value.

c. The *p*-value for this test is .377. Does that prove that the skeptic is correct? Explain.

d. State the appropriate conclusion, using $\alpha = .05$.

e. Assuming that the physician really does have some ability to make these predictions, what would you recommend that he do to increase the chance of proving his claim?

**12.29** ● Explain whether each of the following statements is true or false.

**a.** The *p*-value is the probability that the null hypothesis is true.

**b.** If the null hypothesis is true, then the level of significance is the probability of making a type 1 error.

**c.** A type 2 error can only occur when the null hypothesis is true.

**d.** The probability of making a type 1 error plus the probability of making a type 2 error is 1.

12.30 ● Can the two types of error both be made in the same hypothesis test? Explain.

**12.31** ● Define events as follows: A = Null hypothesis is true. $A^C$ = Alternative hypothesis is true. B = Null hypothesis is not rejected. $B^C$ = Null hypothesis is rejected. For each of the following outcomes, explain whether a type 1 error, a type 2 error, or a correct decision has been made.

**a.** A and B.

**b.** $A^C$ and B.

c. A and $B^C$.

d. $A^C$ and $B^C$.

12.32 ●A medical insurance company wants to know whether the proportion of its customers requiring a hospital stay during a year will decrease if it provides coverage for certain types of alternative medicine. The company conducts a one-year study in which it gives insurance coverage for alternative medicine to 5000 randomly selected customers. Using the data from this study, the company tests the following hypotheses about the effect of offering the alternative medicine coverage to its customers:

> $H_0$: proportion requiring a hospital stay will not decrease
>
> $H_a$: proportion requiring a hospital stay will decrease

If the null hypothesis is rejected, the company will offer this coverage to all of its customers in the future.

a. Explain what a type 1 error would be in this situation.

b. Explain what a type 2 error would be in this situation.

c. Explain which type of error would be more serious for the insurance company.

d. Explain which type of error would be more serious for the customers.

12.33 Medical researchers now believe there may be a link between baldness and heart attacks in men.

a. State the null hypothesis and the alternative hypothesis for a study used to investigate whether or not there is such a relationship.

b. Discuss what would constitute a type 1 error in this study.

c. Discuss what would constitute a type 2 error in this study.

**12.34** Consider medical tests, in which the null hypothesis is that the patient does not have the disease and the alternative hypothesis is that they do.

**a.** Give an example of a medical situation in which a type 1 error would be more serious.

**b.** Give an example of a medical situation in which a type 2 error would be more serious.

12.35 Explain which type of error, type 1 or type 2, could be made in each of the following cases.

a. The null hypothesis is true.

b. The alternative hypothesis is true.

c. The null hypothesis is not rejected.

d. The null hypothesis is rejected.

**12.36** A politician is trying to decide whether to vote for a new tax bill that calls for substantial reforms. A random sample of voters in his district led him to believe the alternative hypothesis, $H_a$: $p > .5$, where *p* is the proportion of all voters in his district who support the bill. As a consequence, he decides to vote for the bill.

a. What would a type 1 error be in this situation, and what would be the consequences for the politician?

b. What would a type 2 error be in this situation, and what would be the consequences for the politician?

**c.** Explain which error would be more serious, if either, for the politician in this example.

**d.** Given the situation described, could the politician have made a type 1 error or a type 2 error?

12.37 Refer to Exercise 12.21, in which a woman thinks that she might be pregnant so she takes a pregnancy test. This situation is analogous to hypothesis testing.

a. Explain in words what a type 1 error would be in this situation.

b. Explain in words what a type 2 error would be in this situation.

c. Which type of error do you think is more serious in this situation? Explain.

12.38 A researcher is deciding whether to use a sample size of 100 or whether to increase the sample size to 200. Explain how this choice will affect the power of any hypothesis tests done using data from the resulting sample.

12.39 A researcher is deciding whether to use a level of significance equal to .05 or a level of significance equal to .01. Explain how this choice will affect the power of the hypothesis test.

**12.40** Refer to Figure 12.1 on page 507. Explain what probabilities can be assigned to each of the following, and whether they are conditional probabilities or not.
a. The null hypothesis is true.
b. The alternative hypothesis is true.
c. The null hypothesis is rejected.
d. The null hypothesis is not rejected.

## Section 12.3

**12.41** ● Find the $p$-value for each of these situations, taking into account whether the test is one-sided or two-sided.
**a.** $z$-statistic = 2.10, $H_0$: $p$ = .10, $H_a$: $p \neq .10$.
**b.** $z$-statistic = 2.00, $H_0$: $p$ = .6, $H_a$: $p < .6$.
**c.** $z$-statistic = $-1.09$, $H_0$: $p$ = .5, $H_a$: $p < .5$.
**d.** $z$-statistic = 4.25, $H_0$: $p$ = .25, $H_a$: $p > .25$.

**12.42** ● Refer to Exercise 12.41. For each of parts (a) to (d), specify the rejection region for $\alpha$ = .05, then reach a conclusion for the test using the rejection region rule.

**12.43** ● In each of the following, determine whether the conditions for conducting a $z$-test for a proportion are met. If not, explain why not.
**a.** Twenty students are randomly selected from the list of all sorority and fraternity members at a university, to determine if a majority of sorority and fraternity students favor a new policy on alcohol on campus. The hypotheses are as follows:

$H_0$: $p$ = .50
$H_a$: $p > .50$

**b.** Twenty employees of a large company are randomly selected to determine whether the proportion of company employees who are left-handed exceeds the national proportion of 10% who are left-handed. The hypotheses are as follows:

$H_0$: $p$ = .10
$H_a$: $p > .10$

**c.** A company employs 500 stockbrokers. They are all surveyed to find out whether a majority believes the market will go up in the next year.

$H_0$: $p$ = .50
$H_a$: $p > .50$

**d.** A market research firm wants to know whether more than 30% of the people who visit a mall actually buy something. A researcher stands by the exit door starting at noon and asks 50 people as they are leaving whether they bought anything.

$H_0$: $p$ = .30
$H_a$: $p > .30$

**12.44** ● Refer to Exercise 12.43. In each of the parts, define in words the population parameter $p$.

**12.45** ● For each of the following calculate the $z$-statistic:
**a.** Sample size $n$ = 30; sample proportion $\hat{p}$ = .60.

$H_0$: $p$ = .50
$H_a$: $p \neq .50$

**b.** Sample size $n$ = 60; sample proportion $\hat{p}$ = .10.

$H_0$: $p$ = .25
$H_a$: $p < .25$

**c.** Sample size $n$ = 500; sample proportion $\hat{p}$ = .30.

$H_0$: $p$ = .20
$H_a$: $p > .20$

**d.** Sample size $n$ = 200; sample proportion $\hat{p}$ = .50.

$H_0$: $p$ = .80
$H_a$: $p < .80$

**12.46** ● Refer to Exercise 12.45. In each case, calculate the $p$-value for the test.

**12.47** Explain the difference between the *null standard error* used in the denominator of a $z$-statistic for a test for a proportion and the *standard error* used in finding a confidence interval for a proportion.

**12.48** *Time Magazine* reported that in a 1994 survey of 507 randomly selected adult American Catholics, 59% answered yes to the question "Do you favor allowing women to be priests?" (*Time,* 26 December–2 January 1995, pp. 74–76).
a. Set up the null and alternative hypotheses for deciding whether more than half of American Catholics favor allowing women to be priests.
b. What conditions are necessary for using a $z$-statistic to test the hypotheses in part (a)? Are those conditions met here? Explain.
c. Compute the test statistic for this situation.
d. Calculate the $p$-value for the test.
e. On the basis of the $p$-value, make a conclusion for this situation. Write it in both statistical language and in words that someone with no training in statistics would understand.
f. Calculate a 95% confidence interval for the proportion of American Catholics who favor allowing women to be priests. Does this confidence interval confirm the conclusion that a majority of American Catholics favor allowing women priests? Explain.

**12.49** In Exercise 12.24, we described a survey done to determine whether artists are more likely to be left-handed than others in the general population. If we use $p$ to represent the proportion of all artists who are left-handed, the hypotheses are $H_0$: $p$ = .10 and $H_a$: $p > .10$. The sample result was that 18 artists of 150 surveyed (or 12%) are left-handed.
a. From the given information, do we know if the conditions necessary to use a $z$-statistic for this test are met? Explain.

**12.91** For a test of $H_0$: $p = .25$ versus $H_a$: $p > .25$, for what range of values of the sample proportion $\hat{p}$ will the $p$-value for the test be *greater* than .5? Explain your answer.

**12.92** For a one-sided hypothesis test for a proportion in which the alternative hypothesis is $H_a$: $p < p_0$, for what values of the test statistic $z$ will the $p$-value be *greater* than .5? Explain your answer.

**12.93** ◆ In Chapter 2, we saw data from a statistics class activity in which students were asked to "randomly" pick a number (integer) from 1 to 10. Of the 190 students, 56 picked the number 7. Carry out the five steps of hypothesis testing to determine whether people similar to these students are more likely to pick the number 7 than would be expected by chance if all numbers were equally likely. (*Source:* **pennstate1** dataset on the CD for this book.)

**12.94** A professor planned to give an examination in a large class on the Monday before Thanksgiving vacation. Some students asked whether he could change the date because so many of their classmates had at least one other exam on that date. They speculated that at least 40% of the class had this problem. The professor agreed to poll the class, and if there was convincing evidence that the proportion with at least one other exam on that date was greater than .40, he would change the date. Of the 250 students in the class, 109 reported that they had another exam on that date.
a. What proportion of the class reported that they had another exam on that date?
**b.** Is the proportion you found in part (a) a sample proportion or a population proportion?
c. The professor conducted a $z$-test of $H_0$: $p = .40$ versus $H_a$: $p > .40$ and found $z = 1.16$ and $p$-value = .125. He said that he would not move the exam because the null hypothesis cannot be rejected, and there is not convincing evidence that the population proportion is greater than .40. What is wrong with his reasoning? [*Hint:* Refer to part (b).]

**12.95** According to *USA Today* (Snapshot, April 20, 1998, also referenced in Exercise 9.131), a random sample of 8709 adults taken in 1976 found that 9% believed in reincarnation. A poll of 1000 adults in 1997 found that 25% believed in reincarnation. Test the hypothesis that the proportion of adults believing in reincarnation in 1997 was higher than the proportion in 1976.

**12.96** In a survey of students at the University of California, Davis, students were asked which of two popular soft drinks they preferred—let's call them brand C and brand P. Of the 159 respondents, 80 preferred brand P and 79 preferred brand C. Assuming that these students represent all students at UCD, test whether there is a preference for one drink over the other for students at UCD.

**12.97** Refer to Exercise 12.96. Some students were asked if they preferred C or P, while other students were asked if they preferred P or C. Of the 159 respondents, 86 responded with the first drink presented. Test the hypothesis that the first drink presented is more likely to be selected than the second drink presented.

**12.98** Refer to Example 12.10 (p. 512), in which convincing evidence was found for the importance of order in a "random" selection between two letters. Suppose a student in your class ran for president of the campus student association and lost. After reading Example 12.10, she realized that because her name is at the end of the alphabet, she was listed after all of her opponents on the ballot and the election was unfair to her. She demands a reelection with names listed in random order.
a. Which type of error, type 1 or type 2, could have been made in Example 12.10?
b. If the type of error that you specified in part (a) was made, is the reelection justified? Explain.
c. Now suppose the other type of error had been made, and the student read the results of the study in that case instead. Explain what the consequences would be for the student and the student government.
d. Which type of error do you think is more serious for the student who ran for president? Explain.

**12.99** Refer to Case Study 1.6, comparing heart attack rates for men who had taken aspirin or placebo. Suppose the observed proportions of .017 and .0094 are actually the correct population proportions who would have heart attacks with placebo and with aspirin. The following Minitab output shows the power of a one-sided test for two proportions for this situation for three sample sizes. The samples are the number of participants in *each* group (aspirin and placebo). Suppose you are the statistician advising a research team about conducting a new study to confirm the results of the old study. The researchers comment that samples of size 500 in each condition should be sufficient, since the effect is obviously so strong, based on the small $p$-value for the previous study. What would you advise? Explain.

```
Testing proportion 1 = proportion 2 (versus >)
Calculating power for proportion 1 = 0.017 and proportion 2 = 0.0094
Alpha = 0.05  Difference = 0.0076

Sample
Size       Power
 500       0.2768
1000       0.4380
3000       0.8250
```

**12.100** A Gallup poll taken on a random sample of Canadian adults in February 2000 asked the question "Do you favour or oppose marriages between people of the same sex?" A similar poll was taken in April 1999 (Edwards and Mazzuca, 2000). The Minitab output for the "two proportions" procedure (with the "pooled estimate for the test") is as follows, where Sample 1 is the February 2000 poll and Sample 2 is the April 1999 poll. X is the number who answered "favour":

| Sample | X | N | Sample p |
|--------|-----|------|----------|
| 1 | 431 | 1003 | 0.429711 |
| 2 | 360 | 1000 | 0.360000 |

Estimate for p(1) − p(2): 0.0697109
95% CI for p(1) − p(2): (0.0270067, 0.112415)
Test for p(1) − p(2) = 0 (vs not = 0): Z = 3.19
P-Value = 0.001

a. Define appropriate notation and write the null and alternative hypotheses to test whether the proportion of adult Canadians who favor marriages between people of the same sex was different in April 1999 and February 2000.

b. Using the Minitab output, go through the remaining four steps of hypothesis testing to test the hypotheses that you defined in part (a).

c. Use the confidence interval given in the Minitab output to test the hypotheses in part (a).

12.101 Refer to Example 12.17 (pp. 528 and 530), comparing the proportion of children with ear infections while taking xylitol or a placebo. Read the discussion of type 1 and type 2 errors, and explain the consequences of each type of error if it were to have been made in this study.

**12.102** Case Study 1.6 presented data on 22,071 physicians who were randomly assigned to take aspirin or a placebo every other day for five years. Of the 11,037 taking aspirin, 104 had a heart attack, while of the 11,034 taking placebo, 189 had a heart attack.

a. Test whether the proportion having heart attacks in the population of men similar to the ones in this study would be smaller if taking aspirin than if taking a placebo.

b. What would the consequences of type 1 and type 2 errors be for this study? Which do you think would be more serious?

12.103 Refer to Exercise 12.102. The test showed that the difference in proportions that had heart attacks after taking aspirin and after taking a placebo was highly statistically significant. Suppose the study had 2,200 participants instead of over 22,000, but that the proportions having heart attacks in the two groups were about the same as they were in the original study. In other words, suppose 10 out of 1100 in the aspirin group had a heart attack and 19 out of 1100 in the placebo group had a heart attack. For a two-sided test of the hypothesis that the proportions likely to have a heart attack are equal after taking aspirin and taking placebo, the test statistic is $z = 1.68$.

a. Using $\alpha = .05$, what would you conclude?

b. Discuss this example in the context of the material presented in Section 12.5.

**12.104** Perry et al. (1999) wanted to test the folklore that women who have not been given information about the sex of their unborn child can guess it at better than chance levels. They asked a sample of 104 pregnant women to guess the sex of their babies, and 57

guessed correctly. Assuming chance guessing would result in 50% correct guesses, test the hypothesis that women can guess at a better than chance level. Carry out the test using $\alpha = .05$, and make sure to state a conclusion.

12.105 Refer to Exercise 12.104. The authors also wanted to test whether women with different levels of education would differ in their ability to guess the sex of their babies correctly. Of the $n = 45$ women with more than 12 years of education, 32 guessed correctly. Of the $n = 57$ women with 12 or fewer years of education, 24 guessed correctly. (Two women did not report their level of education.) Is there sufficient evidence to conclude that women in the populations with these two different levels of education differ in their ability to guess the sex of their baby? Carry out the test using $\alpha = .05$, and make sure to state a conclusion.

12.106 A Gallup poll taken in August 2000 (Chambers, 2000) asked U.S. adults in a random sample of $n = 1019$ about their satisfaction with K-12 education. One question was: "Overall, how satisfied are you with the quality of education students receive in grades kindergarten through grade twelve in the U.S. today—would you say: completely satisfied, somewhat satisfied, somewhat dissatisfied or completely dissatisfied?" There were 622 respondents who said that they were somewhat or completely dissatisfied. Is this sufficient evidence to conclude that a majority of U.S. adults were dissatisfied in August 2000? Carry out the test using $\alpha = .05$, and make sure to state a conclusion.

**12.107** Refer to Exercise 12.106. A Gallup poll the previous August (1999) asked $n = 1028$ U.S. adults the same question, and 524 responded that they were somewhat or completely dissatisfied.

a. Repeat the test in Exercise 12.106 and make a conclusion for August 1999.

b. Referring to the data presented in the previous exercise and in this one, and assuming the samples were taken independently, is there sufficient evidence to conclude that the proportion of U.S. adults who were dissatisfied changed from August 1999 to August 2000? Justify your answer.

## Dataset Exercises

**Statistics⌂Now™**  Datasets **are required** to solve these exercises and can be found at **http://1pass.thomson.com** or on your CD.

12.108 For this exercise, use the dataset **UCDavis2**. The variable **Friends** is a response to the question "Who do you find it easiest to make friends with? People of the (circle one) Same Sex or Opposite Sex?" The responses are S = same or O = opposite.

a. Is there sufficient evidence that students find it easier to make friends with the opposite sex than with the same sex? What assumption(s) did you have to make about the sample in order for the testing procedure to be valid?

b. Conduct the analysis separately for men and for women, using the variable *Sex* to identify males (M) and females (F).

c. Refer to part (b). Compare the results for males and females by writing a summary of the results in words that would be understood by someone who has not studied statistics.

d. Compare your results in parts (a) and (b). Explain how they are similar and how they are different, if at all. Explain any discrepancies you found.

e. Is there sufficient statistical evidence to conclude that the population proportions of men and women who would answer "Opposite Sex" differ? Carry out the five steps of the appropriate hypothesis test.

12.109 For this exercise, use the dataset **UCDavis2.** The variable *Seat* is a response to the question "Where do you typically sit in a classroom?" Possible responses were F = Front, M = Middle, B = Back.

a. Test the hypothesis that a majority of students prefer to sit in the middle of the room.

b. Specify the population of interest, and comment on whether this sample is an appropriate representation of it.

12.110 The dataset **GSS-02** includes information on *sex* and on whether respondents think marijuana should be legal (*marijuan*). Use the dataset to test the hypothesis that the proportions of males and females who thought marijuana should be legal differed for the 2002 population.

**Statistics ⬯ Now™** Preparing for an exam? Assess your progress by taking the post-test at **http://1pass.thomson.com.**

**⬯ Mentor™**
 Do you need a live tutor for homework problems? Access vMentor at **http://1pass.thomson.com** for one-on-one tutoring from a statistics expert.

# 13

*Do they watch more or less television than the men in their lives?*

See Example 13.4 *(p. 568)*

Photodisc/Getty Images

# Testing Hypotheses About Means

To evaluate the quality of a research study, consider such features as sampling procedure, methods used to measure individuals, wordings of questions, and possible lurking variables. When conclusions are based on hypothesis tests, additional issues and cautions discussed in this chapter should be considered.

In this chapter, we continue our discussion of hypothesis testing, initiated in Chapter 12. Make sure you are familiar with HT Module 0, "An Overview of Hypothesis Testing" (Section 12.2) before covering this chapter.

Before we delve into the details, it is worth reviewing three cautions we have already encountered concerning the use of data to make inferences beyond the sample:

1. Inference is only valid if the sample is representative of the population for the question of interest.
2. Hypotheses and conclusions apply to the larger population(s) represented by the sample(s).
3. If the distribution of a quantitative variable is highly skewed, we should consider analyzing the median rather than the mean. Methods for testing hypotheses about medians are a special case of **nonparametric methods,** covered as Supplemental Topic 2 on the CD and briefly in Section 16.3. ▪

## 13.1 Introduction to Hypothesis Tests for Means

In Chapter 12, we learned the logic and basic steps of hypothesis testing and applied them to $z$-tests for one proportion and the difference in two proportions. In this chapter, we will look at the details of hypothesis tests for the same three situations for which we developed confidence intervals in Chapter 11, all of which involve population means:

- Hypotheses about one population mean, $\mu$. For example, we might wish to determine whether or not the mean body temperature for humans is 98.6 degrees Fahrenheit.

*Note to Readers:* HT Module 0 in Chapter 12 must be covered before covering Chapter 13.

- Hypotheses about the population mean difference for paired data, $\mu_d$. For example, we might ask whether the mean difference in height between college men and their fathers is 0. The data for each pair is "son's height − father's height," and we are interested in the mean of those differences for the population of father-son pairs in which "son" is a college student.

- Hypotheses about the difference between the means of two populations, $\mu_1 - \mu_2$. For example, we might ask whether the mean time of television watching per week is the same for men and women.

Tests to answer these kinds of questions are sometimes called **significance tests.**

The terms **hypothesis testing** and **significance testing** are synonymous. The term *significance testing* arises because the conclusion about whether to reject a null hypothesis is based on whether a sample statistic is "significantly" far from what would be expected if the null hypothesis were true in the population. Equivalently, we declare **statistical significance** and reject the null hypothesis if there is a relatively small probability that an observed difference or relationship in the sample would have occurred if the null hypothesis holds in the population. One major principle for all significance tests is that we declare "statistical significance" when the *p*-value of the test is "small."

The significance tests for the three situations that we examine in this chapter all have the same general format because the test statistic is a standardized score that measures the difference between an observed statistic and the null value of a parameter. This was also the case in Chapter 12 where we used a *z*-statistic to test hypotheses about a population proportion and the difference in two population proportions. The basic format that we followed in Chapter 12 and will also follow in this chapter is as follows:

- *The null hypothesis defines a specific value of a population parameter, called the null value.* In Example 13.1, for instance, the null hypothesis is that the mean normal body temperature of humans is $\mu = 98.6$ degrees Fahrenheit. Sometimes the null value is defined for the difference between two population parameters. In Example 13.4, the null hypothesis is that the difference between the mean amounts of time college men and women watch television in a typical day is $\mu_1 - \mu_2 = 0$.

- *A relevant statistic is calculated from sample information and summarized into a "test statistic."* In each case, we begin by determining the sample equivalent of the parameter of interest. For instance, we use $\bar{x}$ = mean of a sample of temperature measurements to estimate $\mu$ = population mean temperature. We then measure the difference between the sample statistic and the null value using the **standardized statistic,** which for hypotheses involving means is:

$$t = \frac{\text{Sample statistic } - \text{ Null value}}{\text{Null standard error}}$$

For hypotheses about means, the *standardized statistic* is called a ***t*-statistic,** and the *t*-distribution is used to find the *p*-value.

For hypotheses about proportions, the *standardized statistic* is called a *z*-statistic, and the standard normal distribution is used to find the *p*-value.

- *A p-value is computed on the basis of the standardized "test statistic."* The *p*-value is calculated by temporarily assuming the null hypothesis to be true and then calculating the probability that the test statistic could be as large in magnitude as it is (or larger) in the direction(s) specified by the alternative hypothesis.

- *On the basis of the p-value, we either reject or fail to reject the null hypothesis.* The most commonly used criterion (level of significance) is that we reject the null hypothesis when the *p*-value is less than .05. In many research articles, *p*-values are simply reported, and readers are left to draw their own conclusions. Remember that a *p*-value measures the strength of the evidence against the null hypothesis, and the smaller the *p*-value, the stronger the evidence against the null (and for the alternative).

These general ideas are consistent with the five steps outlined in Chapters 6 and 12 that are used for *any* hypothesis test. Let's review those five steps:

**Step 1:** Determine the *null* and *alternative* hypotheses.

**Step 2:** Verify necessary data conditions, and if they are met, summarize the data into an appropriate *test statistic*.

**Step 3:** Assuming that the null hypothesis is true, find the *p-value*.

**Step 4:** Decide whether or not the result is *statistically significant* based on the *p-value*.

13.1 Exercises are on page 588.

**Step 5:** Report the conclusion in the context of the situation.

## 13.2  HT Module 3: Testing Hypotheses About One Population Mean

For questions about **the mean** of a quantitative variable **for one population,** the null hypothesis typically has the form

$$H_0: \mu = \mu_0 \quad \text{(mean is a specified value)}$$

The alternative may either be one-sided ($H_a: \mu < \mu_0$ or $H_a: \mu > \mu_0$) or two-sided ($H_a: \mu \neq \mu_0$), depending on the research question of interest. The usual procedure for testing these hypotheses is called a **one-sample *t*-test** because the *t*-distribution is used to determine the *p*-value. Let's look at an example of a situation in which a one-sample *t*-test would be used.

**Example 13.1**

Statistics⬡Now™

Watch a video example at **http://1pass.thomson.com** or on your CD.

For software help, download your Minitab, Excel, TI-83, SPSS, R, and JMP manuals from **http://1pass.thomson.com**, or find them on your CD.

**Normal Human Body Temperature**  What is normal body temperature? A paper published in the *Journal of the American Medical Association* presented evidence that normal body temperature may be less than 98.6 degrees Fahrenheit, the long-held standard (Mackowiak et al., 1992). The value 98.6 degrees seems to have come from determining the mean in degrees Celsius, rounding up to the nearest whole degree (37 degrees), and then converting that number to Fahrenheit using $32 + (1.8)(37) = 98.6$. Rounding up may have produced a result higher than the actual average, which may therefore be lower than 98.6 degrees. To test this, the null hypothesis is $\mu = 98.6$, and we are only interested in rejecting in favor of lower values, so the alternative is $\mu < 98.6$. We can write these hypotheses as follows:

$$H_0: \mu = 98.6$$
$$H_a: \mu < 98.6$$

Suppose that a random sample of $n = 18$ normal body temperatures is as follows:

| 98.2 | 97.8 | 99.0 | 98.6 | 98.2 | 97.8 | 98.4 | 99.7 | 98.2 |
| 97.4 | 97.6 | 98.4 | 98.0 | 99.2 | 98.6 | 97.1 | 97.2 | 98.5 |

For these data, the sample mean is $\bar{x} = 98.217$. This is lower than 98.6, but to determine whether this is a statistically significant difference from 98.6, the question we must answer is: What is the probability that the sample mean could be 98.217 or less if the population mean is actually 98.6 (the null value)?

The answer is the *p*-value that we will use to decide between the two hypotheses. We will return to this example after we provide some details for the five steps of a one-sample *t*-test, but we won't keep you in suspense. The *p*-value for this example is .015, so the data support the alternative hypothesis. ▸

For a one-sample *t*-test, details of the five steps in the hypothesis test are as follows:

### Step 1: Determine the Null and Alternative Hypotheses

Remember that in general, the null hypothesis states the status quo and the alternative hypothesis states the research question of interest. For a one-sample *t*-test, the possible null and alternative hypotheses are one of these three choices, depending on the research question:

$(1)\ H_0: \mu = \mu_0 \quad \text{versus} \quad H_a: \mu \neq \mu_0 \quad \text{(two-sided)}$

$(2)\ H_0: \mu = \mu_0 \quad \text{versus} \quad H_a: \mu < \mu_0 \quad \text{(one-sided)}$

$(3)\ H_0: \mu = \mu_0 \quad \text{versus} \quad H_a: \mu > \mu_0 \quad \text{(one-sided)}$

Sometimes, when the alternative hypothesis is one-sided, the null hypothesis is written to include values in the other direction from the alternative hypotheses as well as equality, for instance, $H_0: \mu \leq \mu_0$, but the "null value" used when computing the test statistic is always $\mu_0$.

### Step 2: Verify Necessary Data Conditions, and If They Are Met, Summarize the Data into an Appropriate Test Statistic

Recall from the previous three chapters that whenever we carry out a statistical inference procedure, we must check to see whether certain data conditions are met. These conditions are connected to the assumptions that were made when the theory was derived for the particular situation. Carrying out a *t*-test for one population mean requires that one of the same two situations holds as for constructing a confidence interval for a mean, given in Chapter 11. Here are those situations, slightly modified from the way in which they were stated in Chapter 11.

### Situations for Which a One-Sample t-Test Is Valid

*Situation 1:* The population of the measurements of interest is approximately normal, and a random sample of any size is measured. In practice, the method is used as long as there is no evidence that the shape is notably skewed or that there are extreme outliers.

*Situation 2:* The population of measurements of interest is not approximately normal, but a *large* random sample is measured. Thirty is usually used as an arbitrary demarcation of "large," but if there are extreme outliers or extreme skewness, it is better to have a larger sample.

For small or moderate sample sizes, use a boxplot, histogram, dotplot, or stem-and-leaf plot to check for notable skewness or extreme outliers. Checking the relative sizes of the sample mean and sample median might also be useful. Skewness and outliers both cause these two statistics to differ from each other. If either skewness or outliers are present, a one-sample *t*-test should not be used. In this event, a statistical test called the *sign test* can be used to analyze hypotheses about the *median*. We provide the details of the sign test in Supplemental Topic 2 on the CD for this book. The procedure is available in most statistical software, including Minitab.

For a large sample size, the *t*-test will work well for testing hypotheses about the mean even if there is some skewness. However, when the data are obviously skewed, you should question whether the mean is the appropriate parameter to analyze. For skewed data, the median may be a better measure of location, and test procedures for analyzing the median may be more appropriate than the *t*-test.

### Continuing Step 2: The Test Statistic for a One-Sample t-Test

The next part of step 2 is to compute a test statistic. The statistic $\bar{x}$, the sample mean, estimates the population mean $\mu$. The "null standard error" for the denominator is the usual standard error of $\bar{x}$, s.e.$(\bar{x}) = s/\sqrt{n}$. Because this standard error does not depend on the null value $\mu_0$, the word *null* can be omitted from the description of the denominator standard error. The test statistic, called the *t*-statistic, is a standardized score for measuring the difference between the sample mean and the null hypothesis value of the population mean and is written as

$$t = \frac{\text{Sample mean} - \text{Null value}}{\text{Standard error}} = \frac{\bar{x} - \mu_0}{s/\sqrt{n}}$$

This particular *t*-statistic has approximately a *t*-distribution with df $= n - 1$.

### Step 3: Assuming That the Null Hypothesis Is True, Find the p-Value

Using the *t*-distribution with df $= n - 1$, the *p*-value is the area in the tail(s) beyond the test statistic *t*, as follows:

- For $H_a$: $\mu < \mu_0$ (a one-sided test), the *p*-value is the area below *t*, even if *t* is positive.

- For $H_a$: $\mu > \mu_0$ (a one-sided test), the *p*-value is the area above *t*, even if *t* is negative.

- For $H_a$: $\mu \neq \mu_0$ (a two-sided test), the *p*-value is $2 \times$ area above $|t|$.

HT Module 3: $\mu$

**Table 13.1** Alternative Hypothesis Regions and *p*-Value Areas

| Statement of H$_a$ | | *p*-Value Area | *t*-Curve Region |
|---|---|---|---|
| $\mu < \mu_0$ | (less than) | Area to the left of *t* (even if *t* > 0) | |
| $\mu > \mu_0$ | (greater than) | Area to the right of *t* (even if *t* < 0) | |
| $\mu \neq \mu_0$ | (not equal) | 2 × area to the right of \|*t*\| | |

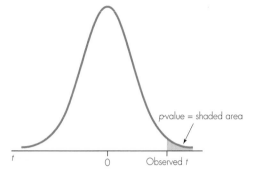

**Figure 13.1** ▍ The *p*-value for a one-tailed *t*-test with H$_a$: $\mu > \mu_0$

Table 13.1 summarizes and illustrates this information. The *p*-value is found in exactly the same way for the other hypothesis-testing situations considered in this chapter, including tests for the population mean difference (paired data) and the difference in population means (independent samples).

Most software programs, including Minitab and Excel, will do the work of finding the *p*-value for you. You can also find a range for the *p*-value using tables of the Student *t*-distribution. Table A.3 in the Appendix gives the areas to the right of eight different values of *t* for various values of degrees of freedom. The type of area given in Table A.3 is shown in Figure 13.1, which shows the area comprising the *p*-value for a one-tailed *t*-test with H$_a$: $\mu > \mu_0$. For instance, in Example 13.2 later in this chapter (p. 562), the *p*-value is the area to the right of *t* = 2.68, and the degrees of freedom are df = 9. In Table A.3, the *t*-value *t* = 2.68 is between the column headings 2.5 and 3.0, so the *p*-value is somewhere between .015 and .007. A *t*-distribution is symmetric, so Table A.3 can also be used to estimate *p*-values for a "less than" alternative hypothesis. This will be demonstrated when we complete Example 13.1.

### Steps 4 and 5: Decide Whether or Not the Result Is Statistically Significant Based on the *p*-Value and Report the Conclusion in the Context of the Situation

These two steps remain the same for all of the hypothesis tests considered in this book. Choose a level of significance $\alpha$, and reject the null hypothesis if the *p*-value is less than $\alpha$. Otherwise, conclude that there is not enough evidence to support the alternative hypothesis. It is standard to use $\alpha = .05$.

**Example 13.1** *(cont.)*

Statistics⊘Now™

For software help, download your Minitab, Excel, TI-83, SPSS, R, and JMP manuals from **http://1pass .thomson.com**, or find them on your CD.

### The Five Steps for Testing the Hypotheses About Normal Body Temperature

**Step 1:** The appropriate null and alternative hypotheses for this situation are as follows:

H$_0$: $\mu = 98.6$ degrees

H$_a$: $\mu < 98.6$ degrees

The parameter $\mu$ is the mean body temperature in the human population. The research question is whether $\mu$ is smaller than the long-held standard, so the alternative hypothesis displays a one-sided test.

HT Module 3: $\mu$

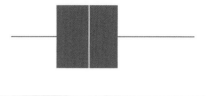

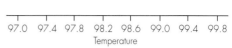

**Figure 13.2** ▮ Boxplot of the sample for Example 13.1

97.0   97.4   97.8   98.2   98.6   99.0   99.4   99.8
Temperature

**Step 2:** Verify necessary data conditions, and if they are met, summarize the data into an appropriate test statistic. The boxplot in Figure 13.2 illustrates that there are no outliers, and the shape does not appear to be notably skewed. Although not shown on the boxplot, the sample mean = $\bar{x} = 98.217$ is almost identical to the median of 98.2. This comparison provides additional evidence that skewness and outliers are not a problem, so the one-sample $t$-test can be used.

We used Minitab to carry out the test, and we can identify the elements necessary for constructing the $t$-statistic using the following Minitab output:

Test of mu = 98.600 vs mu < 98.600

| Variable | N | Mean | StDev | SE Mean | T | P |
|----------|---|------|-------|---------|---|---|
| Temperature | 18 | 98.217 | 0.684 | 0.161 | -2.38 | 0.015 |

The key elements are the following:

- The sample statistic is $\bar{x} = 98.217$ (under "Mean").
- The standard error is s.e. $(\bar{x}) = \dfrac{s}{\sqrt{n}} = \dfrac{0.684}{\sqrt{18}} = 0.161$ (under "SE Mean").
- $t = \dfrac{\text{Sample statistic} - \text{Null value}}{\text{Standard error}} = \dfrac{98.217 - 98.6}{0.161} = \dfrac{-0.383}{0.161}$
  $= -2.38$ (under "T")

**Step 3:** The $p$-value is given as .015 in the last column of output. It was calculated as the area to the left of $t = -2.38$ in a $t$-distribution with $n - 1 = 18 - 1 = 17$ degrees of freedom, and Figure 13.3 illustrates this area. If we use Table A.3, we can determine that the $p$-value is between

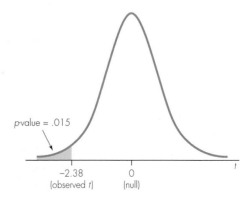

$p$-value = .015

−2.38            0
(observed $t$)    (null)

$t$

**Figure 13.3** ▮ $p$-Value region for Example 13.1

HT Module 3: $\mu$

.016 and .010. Because the $t$-distribution is symmetric, the area to the left of $t = -2.38$ equals the area to the right of $t = +2.38$. The value $t = 2.38$ is between the column headings 2.33 and 2.58 in the table, and for 17 degrees of freedom, the corresponding one-sided $p$-values shown in the table are .016 and .010.

**Step 4:** If we use $\alpha = .05$ as the level of significance criterion, the results are statistically significant because .015, the $p$-value of the test, is less than .05. In other words, we can reject the null hypothesis.

**Step 5:** We can conclude, on the basis of these data, that the mean temperature in the human population is actually less than 98.6 degrees. ▸

---

**EXCEL *tip***    **Finding the *p*-Value for a *t*-Test**

In Example 13.1, the $p$-value can also be found in Excel by typing @TDIST(2.38,17,1) in any cell. Notice that Excel requires you to give the absolute value of the test statistic. The "1" indicates that you want the area in one tail only, since this is a one-tailed test.

*Caution about using Excel for one-sided alternative hypothesis:* If the sample mean happens to be in the direction opposite to that specified by the alternative hypothesis, then $p$-value $> .5$ and Excel gives $1 - p$-value. For instance, in Example 13.1, if the sample mean had been above 98.6 and had resulted in $t = +2.38$, then the appropriate $p$-value would *still* be the area to the *left* of $t$, or to the *left* of 2.38, which is $1 - .015 = .985$. Excel gives only areas to the *right* of the specified value of $t$ for one-tailed tests and allows only positive values of $t$ to be specified.

---

*in summary*    ## The Steps for Testing a Single Mean with a "One-Sample *t*-Test"

**Step 1:** Determine the *null* and *alternative* hypotheses.

*Null hypothesis:*        $H_0: \mu = \mu_0$

*Alternative hypothesis:*    $H_a: \mu \neq \mu_0$  *or*  $H_a: \mu > \mu_0$  *or*  $H_a: \mu < \mu_0$

where the format of the alternative hypothesis depends on the research question of interest.

**Step 2:** Verify necessary data conditions, and if they are met, summarize the data into an appropriate *test statistic.*

If $n$ is large, or if there are no extreme outliers or skewness, compute

$$t = \frac{\text{Sample mean} - \text{Null value}}{\text{Standard error}} = \frac{\bar{x} - \mu_0}{s/\sqrt{n}}$$

**Step 3:** Assuming that the null hypothesis is true, find the *p-value.*

Using the $t$-distribution with df $= n - 1$, the $p$-value is the area in the tail(s) beyond the test statistic $t$, as follows:

**Step 1:** This is a paired-data design. Let $\mu_d$ = population mean difference between no alcohol and alcohol measurements if all pilots were to take these tests. Null and alternative hypotheses about $\mu_d$ are

$H_0: \mu_d = 0$ seconds

$H_a: \mu_d > 0$ seconds        (i.e., no alcohol > alcohol)

The alternative hypothesis is one-sided because we hope to show that if performance does change, there is longer useful performance in the "No Alcohol" condition.

**Step 2:** Figure 13.4 displays a boxplot of the differences for the ten participants. A check of the necessary conditions for doing a $t$-test reveals that while the sample size is small and the dataset of differences for the sample does have some skewness, outliers and extreme skewness do not appear to be serious problems. So a paired $t$-test will be used to examine this question.

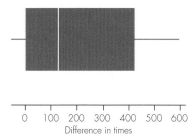

**Figure 13.4** ▌ Boxplot of differences for Example 13.2

Minitab output for analyzing the mean difference follows. Notice that the analysis involves only the sample of $n = 10$ differences and does not otherwise utilize the original performance scores.

| Test of mu = 0.0 vs mu > 0.0 | | | | | | |
| Variable | N | Mean | StDev | SE Mean | T | P |
| --- | --- | --- | --- | --- | --- | --- |
| Diff | 10 | 195.6 | 230.5 | 72.9 | 2.68 | 0.013 |

The $t$-statistic shown in the output is $t = 2.68$ (under "T" in the output). The calculation of the test statistic is as follows:

- The sample statistic is the observed mean difference, $\bar{d} = 195.6$.

- The standard error is s.e.$(\bar{d}) = \dfrac{s_d}{\sqrt{n}} = \dfrac{230.5}{\sqrt{10}} = 72.9$.

- $t = \dfrac{\text{Sample statistic} - \text{Null value}}{\text{Standard error}} = \dfrac{195.6 - 0}{72.9} = 2.68$

**Step 3:** At the right side of the output, we see that the $p$-value is .013. Because the alternative hypothesis was "greater than," this $p$-value was computed as the area to the right of 2.68 in a $t$-distribution with df = 10 − 1 = 9, an area that is illustrated in Figure 13.5. If we had used Table A.3, we would have found that .007 < $p$-value < .015. Finding the exact $p$-value of .013 requires a computer or calculator that gives areas for the $t$-distribution. For instance, in Excel, if you type @TDIST(2.68,9,1) into any cell, the result will be .013 (rounded to three decimal places.)

HT Module 4: $\mu_d$

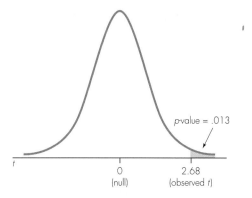

**Figure 13.5** ▮ *p*-Value region for Example 13.2

**Steps 4 and 5:** The *p*-value is .013, so we can reject the null hypothesis using the standard significance level of .05. Even with such a small experiment, we can declare that alcohol has a statistically significant effect and decreases useful performance time.

**Substitute Steps 3 and 4:** To find the critical value, use Table A.2 with df = 9. Because the alternative hypothesis is $H_a$: $\mu_d > 0$, for $\alpha = .05$ use the one-tailed $\alpha = .05$ column. The critical value is 1.83 and the rejection region is $t \geq 1.83$. Because the test statistic $t = 2.68$ is in the rejection region, we reject the null hypothesis. This is the same conclusion we reached in step 4 on the basis of the *p*-value. ▮

13.3 Exercises are on page 590.

---

**MINITAB** *tip*    **Computing a Paired *t*-Test**

- Use **Stat>Basic Statistics>Paired t.** Specify the two columns that contain the raw data for the pair of measurements.

- If the raw data are not available, but you know values of the mean and standard deviation of the sample of differences, and the sample size, enter those values in the "Summarized data" area of the dialog box.

- Alternatively, you can first calculate a column of differences using **Calc> Calculator,** and then use the *1-sample t* procedure.

*Note:* Use the **Graphs** button to create visual displays of the data.

---

*thought question* **13.2**    Suppose that in Example 13.2, the purpose of the study was to determine whether pilots should be allowed to consume alcohol the evening prior to their flights and the alcohol consumption occurred 12 hours before the measurement of time of useful performance. Refer to the discussion of type 1 and type 2 errors in Chapter 12. Explain what the consequences of each type of error would be in this example. Which would be more serious? Given the data and results of the study, which type of error could have been made?*

---

***HINT:*** Remember that a type 1 error can be made only when the null hypothesis is true. In this case, the null hypothesis is that, on average, alcohol has no effect on pilots' performance.

# 13.4 HT Module 5: Testing Hypotheses About the Difference in Two Population Means

## Lesson 1: The General (Unpooled) Case

It is often of interest to determine whether the means of populations represented by two independent samples of a quantitative variable differ. The two populations may be represented by two categories of a categorical variable, such as males and females, or they may be two hypothetical populations represented by different treatment groups in an experiment. In most cases, when comparing two means, the null hypothesis is that they are equal:

$$H_0: \mu_1 - \mu_2 = 0 \qquad (\text{or } \mu_1 = \mu_2)$$

The procedure for testing this null hypothesis is called the **two-sample *t*-test** or ***t*-test for the difference in two means.** Let's look at an example of a situation where a two-sample *t*-test would be used.

**Example 13.3**

Statistics◯Now™

For software help, download your Minitab, Excel, TI-83, SPSS, R, and JMP manuals from **http://1pass .thomson.com**, or find them on your CD.

**The Effect of a Stare on Driving Behavior** In Example 11.11, we discussed an experiment done by social psychologists at the University of California at Berkeley. The researchers either did not stare or did stare at automobile drivers stopped at a campus stop sign. In the experiment, the response variable was the time (in seconds) it took the drivers to drive from the stop sign to a mark on the other side of the intersection. The two populations represented by the observed times are the hypothetical ones that would consist of the times drivers like these would take to move through a similar intersection, either under normal conditions (no stare) or the experimental condition (stare). The hypothesis the researchers wished to test was that the stare would speed up the crossing times, so the mean crossing time would be greater (slower) for those who did not experience the stare than it would be for those who did. So the null and alternative hypotheses are

$$H_0: \mu_1 - \mu_2 = 0 \qquad (\text{or } \mu_1 = \mu_2)$$
$$H_a: \mu_1 - \mu_2 > 0 \qquad (\text{or } \mu_1 > \mu_2)$$

In these hypotheses, the subscript 1 is used to denote the No Stare population, and the subscript 2 is used to denote the Stare population. For the data given in Chapter 11, the mean crossing time was $\bar{x}_1 = 6.63$ seconds for $n = 14$ drivers who crossed under normal conditions (No Stare) and $\bar{x}_2 = 5.59$ seconds for $n = 13$ who crossed in the experimental condition (Stare). The difference between the sample means is 1.04 seconds. Is this difference large enough to be statistically significant evidence against the null hypothesis? We will answer this question after describing the details of the five hypothesis-testing steps for a two-sample *t*-test. ▶

### Step 1: Determine the Null and Alternative Hypotheses

The *null* hypothesis is usually that the difference in means is 0:

$$H_0: \mu_1 - \mu_2 = 0 \qquad (\text{or } \mu_1 = \mu_2)$$

The *alternative* hypothesis may either be one-sided or two-sided:

$$H_a: \mu_1 - \mu_2 > 0 \quad or \quad H_a: \mu_1 - \mu_2 < 0 \quad or \quad H_a: \mu_1 - \mu_2 \neq 0$$

In one-sided tests, the null hypothesis may be written to include an inequality in the opposite direction from the alternative hypothesis.

### Step 2: Verify Necessary Data Conditions, and If They Are Met, Summarize the Data into an Appropriate Test Statistic

The situations in which the $t$-test for the difference in two means is valid are similar to those given in Chapter 11 for constructing a confidence interval for the difference in two means.

#### Situations for Which a t-Test for the Difference in Two Population Means Is Valid

*Situation 1:*   Both populations represented by the measurements of interest are approximately normal, and random samples of any size are measured. In practice, the method is used as long as there is no evidence that the shapes are notably skewed or that there are extreme outliers.

*Situation 2:*   The populations of measurements of interest are not approximately normal, but a *large* random sample is measured from each one. Thirty is usually used as an arbitrary demarcation of "large," but if there are extreme outliers or extreme skewness, it is better to have larger samples.

*Independence:*   In either situation, the samples must be independent. In other words, they must not have been measured as paired or blocked data. A detailed description of methods for generating independent samples is given on page 446 in Section 11.1.

To verify the necessary conditions, examine histograms, stem-and-leaf plots, boxplots, or dotplots for each sample. Within each sample, a comparison of the mean and median can also provide information about possible skewness and outliers. If both of the two independent samples are large, we are in "Situation 2," and the two-sample $t$-test generally is valid for comparing the means unless there are remarkably extreme outliers. Keep in mind, however, that for skewed data, the median may be the better measure of location.

### Continuing Step 2: The Test Statistic for a Two-Sample t-Test

The relevant sample statistic is $\bar{x}_1 - \bar{x}_2 =$ the difference between sample means. As with a one-sample $t$-test, the standard error of $\bar{x}_1 - \bar{x}_2$ does not depend on whether the null hypothesis is true, so the term *null* is dropped from the denominator standard error. The "standardized" test statistic is

$$t = \frac{\text{Sample statistic} - \text{Null value}}{\text{Standard error}} = \frac{\bar{x}_1 - \bar{x}_2 - 0}{\sqrt{\dfrac{s_1^2}{n_1} + \dfrac{s_2^2}{n_2}}}$$

### Step 3: Assuming That the Null Hypothesis Is True, Find the p-Value

To find the $p$-value, use the $t$-distribution with appropriate degrees of freedom:

- For $H_a: \mu_1 - \mu_2 < 0$, the $p$-value is the area below $t$, even if $t$ is positive.
- For $H_a: \mu_1 - \mu_2 > 0$, the $p$-value is the area above $t$, even if $t$ is negative.
- For $H_a: \mu_1 - \mu_2 \neq 0$, the $p$-value is $2 \times$ area above $|t|$.

As discussed in Chapter 11, the $t$-distribution is only an approximation for the distribution of the $t$-statistic, and appropriate degrees of freedom are found by a complicated formula called *Welch's approximation*. The formula is given in Section 11.4 and will not be repeated here. In practice, if computer software such as Minitab is used, it will provide the numerical value of degrees of freedom found from the approximation. If software is not available, a conservative approach is to use the smaller of $n_1 - 1$ and $n_2 - 1$ as the degrees of freedom.

---

*technical note*  **Equal Variances and Pooled Standard Error**

If we are willing to make the assumption that the two populations we are comparing have the same (or similar) variance, then there is a procedure for which the $t$-distribution *is* the correct distribution of the $t$-statistic when the null hypothesis is true. It involves pooling both samples to estimate the common variance. The procedure is called a *pooled two-sample t-test*, and is discussed in Lesson 2 of this module.

---

### Steps 4 and 5: Decide Whether or Not the Result Is Statistically Significant Based on the *p*-Value and Report the Conclusion in the Context of the Situation

These steps proceed as with any hypothesis test. Choose a significance level ("alpha"), usually $\alpha = .05$, and declare the difference in means to be statistically significant if the $p$-value $\leq \alpha$. Report a conclusion appropriate for the situation. In drawing a conclusion about differences between two populations, it also is important to consider the way in which the data were gathered. Remember from Chapter 4 that data collected in an experiment can provide stronger evidence of causation than data from an observational study.

**Example 13.3** *(cont.)*

Statistics⚠Now™

For software help, download your Minitab, Excel, TI-83, SPSS, R, and JMP manuals from **http://1pass .thomson.com**, or find them on your CD.

#### The Effect of a Stare on Driving Behavior

**Step 1:** *State the hypotheses.* The subscript 1 is used to represent the No Stare population, and the subscript 2 is used to represent the Stare population. As was stated above, the appropriate null and alternative hypotheses are

$$H_0: \mu_1 - \mu_2 = 0 \qquad (\text{or } \mu_1 = \mu_2)$$
$$H_a: \mu_1 - \mu_2 > 0 \qquad (\text{or } \mu_1 > \mu_2)$$

The alternative hypothesis reflects the fact that the researchers thought staring would cause drivers to speed up, so the mean time should be slower (larger values) for the No Stare group and faster (smaller values) for the Stare group.

**Step 2:** *Verify necessary data conditions, and if they are met, determine the test statistic.* The data conditions were checked in Example 11.11. We saw that the conditions for conducting the $t$-test appeared to be valid, as there were no extreme outliers and no extreme skewness in the data. There were $n_1 = 14$ observations in the No Stare group and $n_2 = 13$ observations in the Stare group. The following Minitab (Release 12) output provides a confidence interval for the difference between the population means and results for a one-tailed test:

HT Module 5: $\mu_1 - \mu_2$

```
Two-sample T for CrossTime

Group      N      Mean      StDev      SE Mean
NoStare    14     6.63      1.36       0.36
Stare      13     5.59      0.822      0.23

95% CI for mu (NoStare) − mu (Stare ): (0.14, 1.93)
T-Test mu (NoStare) = mu (Stare ) (vs >): T = 2.41      P = 0.013      DF = 21
```

The last line of output shows us that the *t*-statistic is T = 2.41. The elements of the calculation are as follows:

- The sample statistic is $\bar{x}_1 - \bar{x}_2 = 6.63 - 5.59 = 1.04$ seconds.

- $\text{s.e.}(\bar{x}_1 - \bar{x}_2) = \sqrt{\dfrac{s_1^2}{n_1} + \dfrac{s_2^2}{n_2}} = \sqrt{\dfrac{1.36^2}{14} + \dfrac{0.822^2}{13}} = 0.43$

- $t = \dfrac{\text{Sample statistic} - \text{Null value}}{\text{Standard error}} = \dfrac{\bar{x}_1 - \bar{x}_2 - 0}{\text{s.e.}(\bar{x}_1 - \bar{x}_2)} = \dfrac{1.04 - 0}{0.43}$
  $= 2.41$

**Steps 3, 4, and 5:** *Determine the p-value and form a conclusion in context.* Note from the output that the *p*-value of the test is .013, so we can reject the null hypothesis and attach the label "statistically significant" to the results. The *p*-value is the probability that a sample difference could be as far or farther from 0 as 1.04 seconds is *in the positive direction,* if the population difference is actually 0. This probability is determined by using a *t*-distribution with df = 21 (the approximate degrees of freedom computed by Minitab using the Welch approximation formula described in Chapter 11) and finding the area to the right of the test statistic value, *t* = 2.41. From Table A.3, we would learn that the *p*-value is between .009 and .015. We can conclude that if all drivers were stared at, the mean crossing times at an intersection would be faster than under normal conditions. It can also be noted that these data were collected in a randomized experiment, so it is reasonable to conclude that the staring or not staring caused the difference between the mean crossing times.

**Substitute Steps 3 and 4:** As is always the case, the same conclusion is reached by using the *rejection region* approach. To find the *critical value* for this example, use Table A.2 with df = 21. (If Minitab had not provided the degrees of freedom, we would use the conservative value of smaller sample size − 1, which is 13 − 1 = 12, but we will assume that we know df = 21.) Because the alternative hypothesis is one-sided, $H_a: \mu_1 - \mu_2 > 0$, for $\alpha = .05$ use the one-tailed $\alpha = .05$ column. The critical value is 1.72, and the rejection region is $t \geq 1.72$. Because the test statistic $t = 2.41$ is in the rejection region, we reject the null hypothesis. This is the same conclusion that we reached on the basis of the *p*-value of .013, and because the *p*-value is more informative than the rejection region approach, it is always better to report it if it is known. ■

**Example 13.4**

Statistics⬡Now™

Watch a video example at **http:// 1pass.thomson.com** or on your CD.

**A Two-Tailed Test of Television Watching for Men and Women** In Example 11.2 we used information from *n* = 175 Penn State students to estimate the mean amount of time that college students watch television in a typical day. Do you think that the mean time watching television is *different* for the population of college men than it is for the population of college women?

**Step 1:** A two-sided alternative hypothesis is appropriate because there is no prior knowledge that one specific sex might have a higher mean. Therefore, if $\mu_1$ and $\mu_2$ are the mean population daily television-viewing hours for men and women, respectively, for the populations represented by these students, then the appropriate hypotheses are

$$H_0: \mu_1 - \mu_2 = 0$$
$$H_a: \mu_1 - \mu_2 \neq 0$$

**Step 2:** The conditions are met for a two-sample $t$-test to be valid because television viewing hours are available for large independent samples of men and women. Minitab provides the test statistic and furnishes the information necessary to show how to compute it by hand:

| Sex | N | Mean | StDev | SE Mean |
|-----|-----|------|-------|---------|
| Male | 59 | 2.37 | 1.87 | 0.24 |
| Female | 116 | 1.95 | 1.51 | 0.14 |

95% CI for mu (Male) − mu (Female): (−0.14, 0.98)
T-Test of mu (Male) − mu (Female) (vs not =): T = 1.49     P = 0.140     DF = 97

In the last line of output, we see that the test statistic is T = 1.49. The calculation is as follows:

- $\bar{x}_1 - \bar{x}_2 = 2.37 - 1.95 = 0.42$ hour

- $\text{s.e.}(\bar{x}_1 - \bar{x}_2) = \sqrt{\dfrac{s_1^2}{n_1} + \dfrac{s_2^2}{n_2}} = \sqrt{\dfrac{1.87^2}{59} + \dfrac{1.51^2}{116}} = 0.281$

- $t = \dfrac{\text{Sample statistic} - \text{Null value}}{\text{Standard error}} = \dfrac{0.42 - 0}{0.281} = 1.49$

**Steps 3, 4, and 5:** *p-value and conclusion.* The $p$-value is the probability that a sample mean difference could be as far or farther from 0 as 0.42 hour is, *in either direction,* if the population mean difference is actually 0. This is determined by finding the probability that the value of a $t$-distribution is above 1.49 or below $-1.49$. For this test, use approximate df = 97 (see the Minitab output above) or, if necessary, use df = $59 - 1 = 58$ (minimum sample size $-$ 1). In either case, the $p$-value is 2 times the area to the right of 1.49 in a $t$-distribution with the stated degrees of freedom (97 or 58), which is .14 for either degrees of freedom. Figure 13.6 (see next page) illustrates this area and demonstrates that it is actually the area to the left of $-1.49$ plus the area to the right of 1.49. It is simpler computationally to find 2 $\times$ the area to the right of 1.49.

Notice that although the sample mean of the daily viewing hours is higher for the men, the $p$-value of .14 is greater than .05, so we cannot reject the null hypothesis that $\mu_1 - \mu_2 = 0$. On the basis of these *samples,* there is insufficient evidence to conclude that the mean *population* television viewing hours for college men and women are different.

**Substitute Steps 3 and 4:** As is always the case, the same conclusion is reached by using the *rejection region* approach. The conclusion in this case is that we cannot reject the null hypothesis. To find the critical value for this example, use Table A.2 with df = 58. (If Minitab had not provided

For two-sided alternative,
p-value is 2 times the area
past the observed t.

−1.49    0    1.49
(observed t)

**Figure 13.6** ∎ p-Value region for Example 13.4

the degrees of freedom, we would have to use the conservative value of smaller sample size − 1, which is 59 − 1 = 58, so we will use it here.) Because the alternative hypothesis is two-sided, $H_a: \mu_1 - \mu_2 \neq 0$, for $\alpha = .05$ use the two-tailed $\alpha = .05$ column. The critical value falls between 2.01 and 2.00; we will use the more conservative 2.01 as the critical value. The rejection region is $t \leq -2.01$ and $t \geq 2.01$. Because the test statistic $t = 1.49$ is *not* in the rejection region, we *cannot* reject the null hypothesis. This is the same conclusion we reached based on the p-value of .14. ∎

---

*in summary*   The Steps for a Two-Sample *t*-Test (Unpooled)

**Step 1:** Determine the *null* and *alternative* hypotheses.

*Null hypothesis:*          $H_0: \mu_1 - \mu_2 = 0$
*Alternative hypothesis:*   $H_a: \mu_1 - \mu_2 \neq 0$   *or*   $H_a: \mu_1 - \mu_2 > 0$
                            *or*   $H_a: \mu_1 - \mu_2 < 0$

where the format of the alternative hypothesis depends on the research question of interest and the order in which population 1 and population 2 are defined.

**Step 2:** Verify necessary data conditions, and if they are met, summarize the data into an appropriate *test statistic.*

Verify that both *n*'s are large *or* that there are no extreme outliers or skewness in either sample. The samples must also be independent. Compute the test statistic:

$$t = \frac{\text{Sample statistic} - \text{Null value}}{\text{Standard error}} = \frac{\bar{x}_1 - \bar{x}_2 - 0}{\sqrt{\dfrac{s_1^2}{n_1} + \dfrac{s_2^2}{n_2}}}$$

**Step 3, 4, and 5:** Proceed exactly as instructed in the In Summary box, "The Steps for Testing a Single Mean with a 'One-Sample *t*-Test'" on pages 558–559, except in the instructions for step 3, use df = (smaller of $n_1 - 1$, $n_2 - 1$) and replace $\mu$ with $(\mu_1 - \mu_2)$ and $\mu_0$ with 0.

HT Module 5: $\mu_1 - \mu_2$

# Lesson 2: The Pooled Two-Sample $t$-Test

The use of the $t$-distribution for finding $p$-values for the two-sample $t$-test is an approximation. However, if the population standard deviations are equal, then a more precise method is available. The **pooled two-sample $t$-test** is based on the assumption that $\sigma_1 = \sigma_2 = \sigma$, the common standard deviation for both populations. The sample variances are combined to provide a **pooled sample variance, $s_p^2$,** and **pooled standard deviation, $s_p$.** The formula for the pooled sample variance is

$$\text{Pooled sample variance} = s_p^2 = \frac{(n_1 - 1)s_1^2 + (n_2 - 1)s_2^2}{n_1 + n_2 - 2}$$

The computations for the $t$-statistic and $p$-value remain the same as with the regular two-sample $t$-test except for two small changes. First, the pooled sample variance is used in place of the individual sample variances in computing the standard error. Second, the degrees of freedom are df $= n_1 + n_2 - 2$.

*in summary*    Pooled Two-Sample $t$-Test

The test statistic is

$$Pooled\ t = \frac{\text{Sample statistic} - \text{Null value}}{\text{Pooled standard error}} = \frac{\bar{x}_1 - \bar{x}_2 - 0}{\text{Pooled standard error}}$$

where

$$\text{Pooled standard error} = \text{Pooled s.e.}(\bar{x}_1 - \bar{x}_2) = \sqrt{\frac{s_p^2}{n_1} + \frac{s_p^2}{n_2}} = s_p\sqrt{\frac{1}{n_1} + \frac{1}{n_2}}$$

Under the null hypothesis, the pooled $t$-statistic has a $t$-distribution with df $= n_1 + n_2 - 2$. All other steps are the same as for the unpooled two-sample $t$-test.

Although the pooled $t$-test provides an exact solution and the unpooled version provides only an approximate solution, we do not recommend that the pooled $t$-test be used except in very specific circumstances. It is better to have an approximate solution that is known to work well than to use an exact solution based on an assumption that cannot be verified.

## Guidelines for Using the Pooled Two-Sample $t$-Test

- When $n_1 = n_2$, the pooled and unpooled standard errors are equal, so the $t$-statistic also is the same for both the unpooled and pooled procedures. The pooled procedure, however, is not exact unless the population standard deviations are equal, so it is generally preferable to use Welch's approximate df. If this value is not available, when the two sample sizes are equal, it is an acceptable substitute to use df $= n_1 + n_2 - 2$. This recommendation can be extended to include situations in which the sample sizes are close.

AT Module 5: $\mu_1 - \mu_2$

- When $n_1$ and $n_2$ are very different, the pooled test can be quite misleading unless the sample standard deviations are similar. Some books recommend using the pooled test as long as the ratio of sample standard deviations (larger $s$/smaller $s$) is $\leq 2$, but this cutoff is completely arbitrary, and we do not recommend it.

- It can be shown by using algebra that

$$\text{Pooled s.e.}(\bar{x}_1 - \bar{x}_2) \approx \sqrt{\frac{s_1^2}{n_2} + \frac{s_2^2}{n_1}}$$

  where the symbol "$\approx$" means "is approximately equal to." Therefore, using the pooled standard error is essentially equivalent to reversing the roles of $n_1$ and $n_2$ in the standard error. If the larger sample produces the smaller sample standard deviation, the standard error will tend to be underestimated, the $t$-statistic will tend to be too large, and the null hypothesis will be rejected too easily. Therefore, if the sample sizes are very different and the smaller standard deviation accompanies the larger sample size, we do not recommend using the pooled procedure.

- If the sample sizes are very different, the standard deviations are similar, and the larger sample size produced the larger standard deviation, the pooled $t$-test is acceptable because it will be conservative.

**Example 13.5**

**Misleading Pooled $t$-Test for Television Watching for Men and Women**  In Example 13.4 (p. 568), we compared television viewing hours for male and female college students and could not reject the hypothesis that they had the same population means. The sample size for males was only about half of what it was for females, but the standard deviation was larger for males. Here are the summary statistics:

|        | Sample Size | Mean | Standard Deviation | Variance |
|--------|-------------|------|--------------------|----------|
| Male   | $n_1 = 59$  | 2.37 | 1.87               | 3.50     |
| Female | $n_2 = 116$ | 1.95 | 1.51               | 2.28     |

The difference in means is 0.42 hour, the unpooled standard error is 0.281, the test statistic is 1.49, and the resulting $p$-value is .14. Here are the computations for the *pooled t-test:*

$$s_p^2 = \frac{(n_1 - 1)s_1^2 + (n_2 - 1)s_2^2}{n_1 + n_2 - 2} = 2.69, \qquad s_p = 1.64$$

$$\text{Pooled s.e.}(\bar{x}_1 - \bar{x}_2) = \sqrt{\frac{s_p^2}{n_1} + \frac{s_p^2}{n_2}} = s_p \sqrt{\frac{1}{n_1} + \frac{1}{n_2}} = 1.64 \sqrt{\frac{1}{59} + \frac{1}{116}} = 0.262$$

$$t = \frac{\text{Sample statistic} - \text{Null value}}{\text{Pooled standard error}} = 0.42/0.262 = 1.60$$

$$\text{df} = 59 + 116 - 2 = 173; \qquad p\text{-value} = .11$$

Although the *p*-value for the pooled test still would not lead to rejection of the null hypothesis, it is noticeably smaller than the *p*-value for the unpooled test. In similar situations, in which sample sizes are very different and the larger sample has the smaller variance, the pooled *t*-test and inflated test statistic could provide a misleading conclusion to reject the null hypothesis. ■

**Example 13.6**

**Statistics⬭Now™**

For software help, download your Minitab, Excel, TI-83, SPSS, R, and JMP manuals from **http://1pass .thomson.com**, or find them on your CD.

**Legitimate Pooled *t*-Test for Comparing Male and Female Sleep Time** In Example 11.14 (p. 475) we compared the mean sleep time for male and female college students by finding a confidence interval for the difference. The data were collected in statistics classes at the University of California, Davis, at the beginning of the spring 2000 quarter. Here is a summary of the results from Minitab, displaying both the unpooled and pooled versions:

Two Sample T-Test and Confidence Interval

Two sample T for sleep [without "Assume Equal Variance" option]

| sex | N | Mean | StDev | SE Mean |
|-----|-----|------|-------|---------|
| f | 83 | 7.02 | 1.75 | 0.19 |
| m | 65 | 6.55 | 1.68 | 0.21 |

95% CI for mu (f) − mu (m): (−0.10, 1.02)
T-Test mu (f) = mu (m) (vs not =): T = 1.62     P = 0.11     DF = 140

Two sample T for sleep [with "Assume Equal Variance" option]

| sex | N | Mean | StDev | SE Mean |
|-----|-----|------|-------|---------|
| f | 83 | 7.02 | 1.75 | 0.19 |
| m | 65 | 6.55 | 1.68 | 0.21 |

95% CI for mu (f) − mu (m): (−0.10, 1.03)
T-Test mu (f) = mu (m) (vs not =): T = 1.62     P = 0.11     DF = 146
Both use Pooled StDev = 1.72

13.4 Exercises are on page 591.

Notice that in this case, the sample sizes are similar, as are the sample standard deviations. Consequently, the pooled standard deviation of 1.72 is similar to the separate standard deviations of 1.75 and 1.68. The test statistic of 1.62 is identical under both procedures, and the *p*-value of .11 is the same. The use of the pooled procedure was warranted in this case. ■

---

**MINITAB *tip*** | **Computing a Two-sample *t*-Test to Compare Means**

- To carry out a two-sample *t*-test, use **Stat>Basic Statistics>2-sample t.** Specify the location of the data. The raw data for the response (*Samples*) may be in one column, and the raw data for group categories (*Subscripts*) may be in a second column. Or, the raw data for responses in the two independent groups may be in two separate columns.

- If the raw data are not available, but the data are summarized so that you know the sample mean, standard deviation, and sample size for each group, enter those values in the "Summarized data" area of the dialog box.

- To use the pooled standard error, click on ***Assume Equal Variances.***

- To specify the alternative hypothesis, use the ***Options*** button.

| SPSS *tip* | **Computing a Two-sample *t*-Test** |
|---|---|

- To compare two population means using data from two independent samples, use **Analyze>Compare Means>Independent-Samples T Test.** Results are given for both the unpooled and pooled procedures.

- The *p*-values given are for a two-sided alternative hypothesis. For a one-sided alternative, divide a given *p*-value by 2 (assuming that the ordering of the sample means is consistent with the alternative hypothesis).

*thought question* **13.3**   The paired *t*-test introduced in Section 13.3 and the two-sample *t*-test introduced in this section are both used to compare two sets of measurements, and the null hypothesis in both cases is usually that the mean population difference is 0. Explain the difference in the situations for which they are used. Suppose researchers wanted to know if college students spend more time watching TV or exercising. Explain how they could collect data appropriate for a paired *t*-test and how they could collect data appropriate for a two-sample *t*-test.*

*Note to Readers:* The remainder of this chapter applies to all of the material covered in Chapters 10 to 13, not just to hypothesis tests for means.

# 13.5   The Relationship Between Significance Tests and Confidence Intervals

In the situations covered in Chapters 10–13, you may have noticed a direct correspondence between the values covered by a confidence interval and the results of two-sided hypothesis tests. The correspondence is precise for tests involving one or two population means and a two-sided alternative.

## Confidence Intervals and Tests with Two-Sided Alternatives

In testing one population mean or the difference in two population means with the hypotheses

$H_0$: parameter = null value   *and*   $H_a$: parameter ≠ null value

- If the null value is covered by a $(1 - \alpha)100\%$ confidence interval, the null hypothesis is not rejected, and the test is not statistically significant at level $\alpha$.

- If the null value is not covered by a $(1 - \alpha)100\%$ confidence interval, the null hypothesis is rejected, and the test is statistically significant at level $\alpha$.

Don't be confused by the cumbersome notation $(1 - \alpha)100\%$. The correspondence is between a 95% confidence interval and a significance level of $\alpha = .05$, between a 99% confidence interval and a significance level of $\alpha = .01$, and so on.

---

*****HINT:** If you were in the study and they asked you (and other participants) about both time watching TV and time exercising, would the data be paired or independent samples?

When you are first learning to distinguish among the situations, it helps to have some clues. Here are tips for helping you to determine which parameter to use, assuming the choices are from the five covered in Chapters 10–13.

- *Are two different sample sizes given, or is it clear that two independent samples were taken?* If so, the parameter of interest must be either $p_1 - p_2$ or $\mu_1 - \mu_2$.

- *Is one or more sample means given?* If so, the parameter is probably a mean or difference in two means.

- *Are sample counts or proportions given?* If so, the parameter is probably a proportion or a difference in two proportions.

After you determine the parameter of interest and whether to conduct a hypothesis test and/or construct a confidence interval, you can use the summary tables at the end of this chapter and at the end of Chapter 11 to review the computational details of the analysis.

## Examples of Choosing an Inference Procedure

The following examples utilize studies from other chapters in this textbook. In each case, think about what type of inference is appropriate.

**Example 13.9**

**Statistics ⬡Now™**

Read the original source on your CD.

**Kids and Weight Lifting**   In Case Study 4.2 (p. 124), a randomized experiment was presented in which children were randomly assigned to one of two weight-lifting conditions or a control condition. Muscular strength and endurance were measured after eight weeks. One of the results reported in Case Study 4.2 was "leg extension strength significantly increased in both exercise groups compared with that in the control subjects" (Faigenbaum et al., 1999, p. 85).

Notice that the researchers are comparing each of the two weight-lifting conditions to the control condition. Let's focus on comparing the heavy-lifting group to the control group for the reported response variable, leg extension strength.

*Decision 1: Confidence interval, hypothesis test, or both?*   Based on the statement about the results, the researchers were clearly interested in knowing *if* there is a significant difference in leg extension strength after heavy lifting compared with no weight lifting. They were also most likely interested in the *magnitude* of the difference. Therefore, both inference procedures would be appropriate.

*Decision 2: What is the appropriate parameter to investigate?*   Let's investigate the three questions that are used to determine which parameter is appropriate.

- *Is there one sample or two? If there are two, are they independent samples, or are they paired?* It is clear that there are two samples and that they are independent. One group of children did heavy lifting, and the other group did no weight training.

- *For one sample, is the variable that is measured on each unit quantitative or categorical?* This question is not relevant because there are two samples.

> • *For two independent samples, one variable identifies the two samples and is categorical. Is the other variable also categorical?* The training (heavy or none) identifies the two samples. The other variable that is measured is the leg extension strength, which is not categorical. Presumably, a quantitative measurement was made for each child.

On the basis of the answers to these questions, the appropriate column of Table 13.3 is "Two Independent Samples" and the appropriate row is "Quantitative (Means)." Therefore, the parameter of interest is $\mu_1 - \mu_2$, the difference in population means for leg extension strength if all similar children were to participate in the heavy-lifting or control conditions. ∎

**Example 13.10** **Loss of Cognitive Functioning** Example 4.4 (p. 137) reported on an observational study of 6000 older people, which found that 70% of the participants did not lose functioning over time. What inference procedure can be used for this result?

> *Decision 1: Confidence interval, hypothesis test, or both?* There is no information provided about the study that would lead to a natural hypothesis testing question. The reported result, that 70% of the *sample* lost cognitive functioning over time, leads to a consideration of the margin of error and confidence interval estimate for the proportion of the *population* in this age group that loses cognitive functioning over time. Therefore, hypothesis testing is not appropriate, but a confidence interval is.
>
> *Decision 2: What is the appropriate parameter to investigate?* The sample statistic reported is a sample proportion for a single sample. People in the study were categorized by whether or not they lost cognitive functioning. Therefore, the parameter of interest is $p$ = proportion that loses cognitive functioning over time for the population of older people represented by this sample. ∎

13.6 Exercises are on page 594.

# 13.7 Effect Size

In most of the hypothesis-testing situations in Chapters 12 and 13, we are interested in comparing a population mean or proportion to a specific null value, or to another population mean or proportion. In many research situations, we would like to know something about the magnitude of the comparison. The test statistic and $p$-value for a test are not useful for this purpose because they depend on the size of the sample. A confidence interval may not be useful because it depends on the units of measurement, and we want a measure that is independent of the units of measurement. This is particularly true if we want to compare research results across studies, in which different measurement units may have been used or slightly different treatments may have been assigned.

In general, the **effect size** for a research question is a measure of how much the truth differs from chance or from a control condition. The most common effect-size measure used for a single mean is

$$d = \frac{\mu_1 - \mu_0}{\sigma}$$

where $\mu_1$ is the true population mean, $\mu_0$ is the null value, and $\sigma$ is the population standard deviation. Notice that $d$ measures the distance between the true mean and the null mean value in terms of number of standard deviations.

As an example, suppose it is known that the mean IQ in a certain population under usual conditions is 110, with a standard deviation of 15. Researchers speculate that listening to classical music temporarily boosts IQ. If indeed the mean IQ after listening to classical music is 115, then the effect size is $(115 - 110)/15 = 1/3$. In other words, listening to classical music boosts the mean IQ by one third of a standard deviation.

The most common effect-size measure for comparing two means is

$$d = \frac{\mu_1 - \mu_2}{\sigma}$$

where $\mu_1$ and $\mu_2$ are the two population means and $\sigma$ is the population standard deviation, assumed to be the same for both populations. If this assumption is not reasonable and if one population corresponds to a control group condition, then the standard deviation for the control condition is used. If that is not possible either, then a pooled standard deviation is used. Notice that $d$ measures the distance between $\mu_1$ and $\mu_2$ in terms of number of standard deviations, using the standard deviation for the individual populations as the yardstick. In most situations, the order of the difference is not important to the magnitude of the effect, so the absolute value of $d$ is reported.

As an example, suppose that resting pulse rates for those who don't exercise have a mean of 76 and standard deviation of 8 beats per minute, while resting pulse rates for those who do exercise have a mean of 68 with the same standard deviation. Then the effect size for the difference is $(76 - 68)/8 = 1$. In other words, average resting pulse rates for those who exercise and those who don't exercise differ by 1 standard deviation.

There are many uses for effect sizes, some of which require $d$ to be estimated from data. There are various suggestions for estimating $d$ based on complicated statistical principles (see, e.g., Hedges and Olkin, 1985), but the simplest method is to substitute sample values for the population parameters in $d$.

*formula*    **Estimating Effect Sizes for Means for One and Two Samples**

For a single sample, the estimated effect size is $\hat{d} = \dfrac{\bar{x} - \mu_0}{s}$, where $\bar{x}$ and $s$ are the sample mean and standard deviation, respectively.

For two samples, the estimated effect size is $\hat{d} = \dfrac{\bar{x}_1 - \bar{x}_2}{s}$, where $\bar{x}_1$ and $\bar{x}_2$ are the two sample means and $s$ is either the standard deviation of the control condition measurements (if there is one) or the pooled sample standard deviation.

Notice that there is a relationship between the test statistic $t$ and the estimated effect size. The formula for the estimated effect size includes everything in the formula for the $t$ statistic *except* the influence of the sample size(s).

> *formula* | **Relationship Between Test Statistic *t* and Estimated Effect Size**
>
> For one sample: $\quad t = \dfrac{\bar{x} - \mu_0}{\frac{s}{\sqrt{n}}} = \dfrac{\sqrt{n}(\bar{x} - \mu_0)}{s} = \sqrt{n}\,\hat{d}$, so $\hat{d} = \dfrac{t}{\sqrt{n}}$
>
> For two samples: $\quad$ Pooled $t = \dfrac{\bar{x}_1 - \bar{x}_2}{s\sqrt{\dfrac{1}{n_1} + \dfrac{1}{n_2}}} = \dfrac{\hat{d}}{\sqrt{\dfrac{1}{n_1} + \dfrac{1}{n_2}}}$

Although we will not cover other effect-size measures, the relationship for *t*-tests holds in most hypothesis-testing situations for which effect sizes are used. Rosenthal (1991) provides a summary of many of these procedures, and notes that in most cases,

Test statistic = Size of effect × Size of study

as we have seen in the tests for means.

## Interpreting an Effect Size

Effect sizes provide information about how strong a difference or effect is in the population, relative to another population or a hypothesized value. The interpretation is similar to the interpretation of a standardized score for a specific individual value. In fact, you should notice that the effect size *d* for a single mean of a normal population is simply the standardized score for $\mu_1$ relative to the normal distribution with mean $\mu_0$ and standard deviation $\sigma$.

Cohen (1988) defined small, medium, and large effect sizes for $|d|$ somewhat arbitrarily to be 0.2, 0.5, and 0.8, respectively. In other words, if the true mean $\mu_1$ is half of a standard deviation away from the null mean value ($|d| = 0.5$), Cohen called this a medium effect. If the true mean $\mu_1$ is only 1/5 of a standard deviation away from the null mean ($|d| = 0.2$), Cohen called it a small effect; similarly for 0.8 as a large effect. These are arbitrary, but he said that a small effect should be detectable only through statistics, a medium effect should be obvious to a careful observer, and a large effect should be obvious to any observer.

**Example 13.11** | **Could Aliens Tell That Women Are Shorter?** Suppose an alien were to visit Earth and observe humans. Would it be obvious that men are taller, on average, than women? The mean heights of adult men and women are about 70 inches and 65 inches, respectively, with standard deviations of about 2.5 inches within each population. Therefore, the effect size for the difference in heights is approximately $d = (70 - 65)/2.5 = 2.0$. Such a large difference should be immediately obvious to anyone.

However, suppose the difference in men's and women's heights was smaller. The following table shows what differences correspond to small, medium, and large effect sizes. Think about whether you would notice the difference in mean heights for two groups whose means differed by these amounts.

If $p$-value > significance level, do not reject $H_0$.
If $p$-value ≤ significance level, reject $H_0$, accept $H_a$, and conclude result is statistically significant.

| | Parameter | Statistic | Null Value | Null Standard Error | t or z; df for t | Example |
|---|---|---|---|---|---|---|
| **One Proportion** | $p$ | $\hat{p}$ | $p_0$ | $\sqrt{\dfrac{p_0(1-p_0)}{n}}$ | $z$ | 12.12 |
| **Difference Between Proportions** | $p_1 - p_2$ | $\hat{p}_1 - \hat{p}_2$ | Usually 0 | $\sqrt{\hat{p}(1-\hat{p})\left(\dfrac{1}{n_1}+\dfrac{1}{n_2}\right)}$ | $z$ | 12.17 |
| **One Mean** | $\mu$ | $\bar{x}$ | $\mu_0$ | $\dfrac{s}{\sqrt{n}}$ | $t$ <br> df $= n - 1$ | 13.1 |
| **Mean Difference, Paired Data** | $\mu_d$ | $\bar{d}$ | Usually 0 | $\dfrac{s_d}{\sqrt{n}}$ | $t$ <br> df $= n - 1$ | 13.2 |
| **Difference Between Means (unpooled)** | $\mu_1 - \mu_2$ | $\bar{x}_1 - \bar{x}_2$ | Usually 0 | $\sqrt{\dfrac{s_1^2}{n_1}+\dfrac{s_2^2}{n_2}}$ | $t$ <br> df = Welch's approximation or smaller of $n_1 - 1, n_2 - 1$ | 13.3 |
| **Difference Between Means (pooled)** | $\mu_1 - \mu_2$ | $\bar{x}_1 - \bar{x}_2$ | Usually 0 | $s_p\sqrt{\dfrac{1}{n_1}+\dfrac{1}{n_2}}$ | $t$ <br> df $= n_1 + n_2 - 2$ | 13.6 |

# Key Terms

● Basic skills ◆ Dataset available but not required **Bold-numbered** exercises answered in the back

# Exercises

● Denotes basic skills exercises

◆ Denotes dataset is available in StatisticsNow at **http:// 1pass.thomson.com** or on your CD but is **not required** to solve the exercise.

**Bold-numbered exercises** have answers in the back of the text and fully worked solutions in the Student Solutions Manual.

**Statistics ⬡ Now**™ Go to the StatisticsNow website at **http://1pass.thomson.com** to:
• Assess your understanding of this chapter
• Check your readiness for an exam by taking the Pre-Test quiz and exploring the resources in the Personalized Learning Plan

*For all exercises requiring hypothesis tests, make sure you check conditions, define all parameters, and state your conclusion in words in the context of the problem.*

## Sections 13.1

**13.1** ● Explain whether or not each of the following statements is true.
  a. Hypotheses and conclusions from hypothesis testing apply only to the samples on which they are based.
  b. The *p*-value is calculated with the assumption that the null hypothesis is true.
  c. One of the two possible conclusions in hypothesis testing is to accept the null hypothesis.
  d. The statements "reject the null hypothesis" and "accept the alternative hypothesis" are equivalent.

**13.2** ● For each of the following research questions, specify whether the parameter of interest is one population mean, the population mean of paired differences, or the difference between the means of two populations.
  **a.** Nutrition trends have changed over the years, and this may affect growth. Researchers want to know whether the mean height of 25-year-old women is the same as the mean height of 45-year-old women.
  b. You plan to fly from New York to Chicago and have a choice of two flights that leave at about the same time. You are able to find out how many minutes late each flight was for a random sample of 25 days over the past few years. (You have data for both flights on the same 25 days.) You want to know whether one flight has a higher population mean for the number of minutes the flight is late.
  c. Residents of a neighborhood have been complaining about speeding cars. The speed limit in the area is 25 miles per hour. The local police monitor the situation by recording the speed of 100 randomly selected cars. They want to know whether the mean speed of all cars that drive through the area is higher than 25 miles per hour.
  d. Refer to part (c). After collecting the data described in part (c), the police decide to install an electronic roadside sign that shows cars how fast they are driv-

ing. They then record the driving speed of another 100 randomly selected cars. They want to know whether the mean speed is lower after installation of the sign than it was beforehand.

**13.3** ● Refer to Exercise 13.2. For each part, write and define the notation that would be used for the population parameter of interest. Make sure you specify the population(s) to which the parameter applies.

**13.4** ● Refer to Exercises 13.2 and 13.3. Specify the null value that the researchers are interested in testing, and write the null hypothesis using the appropriate symbol(s) for the parameter of interest. If you haven't already done so in Exercise 13.3, make sure you specify the population(s) to which the parameter applies.

**13.5** Explain why a "standardized statistic" is used in hypothesis testing, instead of simply using the difference between the sample statistic and the null value as the test statistic.

**13.6** Explain why the null hypothesis for a significance test is rejected when the *p*-value is small rather than when it is large.

**13.7** Explain why each of the following terms is used for the procedure outlined in Section 13.1.
  a. Hypothesis testing.
  b. Significance testing.

## Section 13.2

**13.8** ● If you were given a set of data for which the question of interest was about the population mean, explain how you would determine whether it would be valid to do a one-sample *t*-test using that set of data.

**13.9** ● Review the five steps for any hypothesis test. Can any of the five steps be done *before* collecting the data? Explain.

**13.10** ● Define the parameter of interest, then use it to write the null and alternative hypotheses (in symbols) for each of the following research questions.
  a. Many cars have a recommended tire pressure of 32 psi (pounds per square inch). At a roadside vehicle safety checkpoint, officials plan to randomly select 50 cars for which this is the recommended tire pressure and measure the actual tire pressure in the front left tire. They want to know whether drivers on average have too little pressure in their tires.
  b. The box of Yvette's favorite cereal states that the net contents weigh 12 ounces. Yvette is suspicious of this claim because the package never seems full to her. She plans to measure the weight of the contents of the next 20 boxes she buys and find out whether she is being short-changed.
  c. Nonprofit hospitals must provide a certain amount of charity care to maintain nonprofit, tax-exempt status. Although the amount differs by area, let's suppose that all such hospitals are required to provide charity services equivalent to 4% of net pa-

tient revenue. A legislator is concerned that there is inadequate enforcement of this regulation. He plans to audit a random sample of 30 nonprofit hospitals and assess the percentage of net patient revenue they spend on charity care. He will then test whether the population mean is at least 4%.

**13.11** ● Give the value of the test statistic $t$ in each of the following situations.

a. $H_0: \mu = 50$, $\bar{x} = 60$, $s = 90$, $n = 100$.

b. Null value $= 100$, sample mean $= 98$, $s = 15$, sample size $= 40$.

c. $H_0: \mu = 250$, $\bar{x} = 270$, standard error $= 5$, $n = 100$.

**13.12** ● Find the $p$-value and draw a sketch showing the $p$-value area for each of the following situations in which the value of $t$ is the test statistic for the hypotheses given:

a. $H_0: \mu = \mu_0$, $H_a: \mu > \mu_0$, $n = 28$, $t = 2.00$.

b. $H_0: \mu = \mu_0$, $H_a: \mu > \mu_0$, $n = 28$, $t = -2.00$.

c. $H_0: \mu = \mu_0$, $H_a: \mu \neq \mu_0$, $n = 81$, $t = 2.00$.

d. $H_0: \mu = \mu_0$, $H_a: \mu \neq \mu_0$, $n = 81$, $t = -2.00$.

**13.13** ● Use Table A.2 to find the critical value and rejection region in each of the following situations. Then determine whether the null hypothesis would be rejected. In each case the null hypothesis is $H_0: \mu = 100$.

a. $H_a: \mu > 100$, $n = 21$, $\alpha = .05$, test statistic $t = 2.30$.

b. $H_a: \mu > 100$, $n = 21$, $\alpha = .01$, test statistic $t = 2.30$.

c. $H_a: \mu \neq 100$, $n = 21$, $\alpha = .05$, test statistic $t = 2.30$.

d. $H_a: \mu \neq 100$, $n = 21$, $\alpha = .01$, test statistic $t = 2.30$.

e. $H_a: \mu > 100$, $n = 10$, $\alpha = .05$, test statistic $t = 1.95$.

f. $H_a: \mu < 100$, $n = 10$, $\alpha = .05$, test statistic $t = -1.95$.

g. $H_a: \mu < 100$, $n = 10$, $\alpha = .05$, test statistic $t = 1.95$.

h. $H_a: \mu \neq 100$, $n = 10$, $\alpha = .05$, test statistic $t = 1.95$.

**13.14** ◆ The dataset **cholest** on the CD for this book includes cholesterol levels for heart attack patients and for a group of control patients. It is recommended that people try to keep their cholesterol level below 200. The following Minitab output is for the control patients:

Test of mu = 200.00 vs mu < 200.00

| Variable | N | Mean | StDev | SE Mean | T | P |
|---|---|---|---|---|---|---|
| control | 30 | 193.13 | 22.30 | 4.07 | −1.69 | 0.051 |

a. What are the null and alternative hypotheses being tested? Write them in symbols.

b. What is the mean cholesterol level for the sample of control patients?

c. How many patients were in the sample?

d. Use the formula for the standard error of the mean to show how to compute the value of 4.07 reported by Minitab.

e. What values does Minitab report for the test statistic and the $p$-value?

f. Identify the numbers that were used to compute the $t$-statistic, and verify that the reported value is correct.

g. What conclusion would be made in this situation, using a .05 level of significance?

**13.15** Suppose a study is done to test the null hypothesis $H_0: \mu = 100$. A random sample of $n = 50$ observations results in $\bar{x} = 102$ and $s = 15$.

a. What is the null standard error in this case?

b. Plug numbers into the formula

$$\frac{\text{Sample statistic} - \text{Null value}}{\text{Null standard error}}$$

c. On the basis of the information given, can the $p$-value for this test be found? If so, find it. If not, explain what additional information would be needed.

**13.16** It has been hypothesized that the mean pulse rate for college students is about 72 beats per minute. A sample of Penn State students recorded their sexes and pulse rates. Assume that the samples are representative of all Penn State men and women for pulse rate measurements. The summary statistics were as follows:

| Sex | n | Mean | StDev |
|---|---|---|---|
| Female | 35 | 76.9 | 11.6 |
| Male | 57 | 70.42 | 9.95 |

a. Test whether the pulse rates of all Penn State men have a mean of 72.

b. Test whether the pulse rates of all Penn State women have a mean of 72.

c. Write a sentence or two summarizing the results of parts (a) and (b) in words that would be understood by someone with no training in statistics.

**13.17** Refer to Exercise 13.10(a), which posed the following research question: "Many cars have a recommended tire pressure of 32 psi (pounds per square inch). At a roadside vehicle safety checkpoint, officials plan to randomly select 50 cars for which this is the recommended tire pressure and measure the actual tire pressure in the front left tire. They want to know whether drivers on average have too little pressure in their tires." Suppose the experiment is conducted, and the mean and standard deviation for the 50 cars tested are 30.1 psi and 3 psi, respectively. Carry out the five steps to test the appropriate hypotheses.

**13.18** A cell phone company knows that the mean length of calls for all of its customers in a certain city is 9.2 minutes. The company is thinking about offering a senior discount to attract new customers but first wants to know whether the mean length of calls for current customers who are seniors (65 and over) is the same as it is for the general customer pool. The only way to identify seniors is to conduct a survey and ask people whether they are over age 65. Using this method, the company contacts a random sample of 200 seniors and records the length of their last call. The sample mean and standard deviation for the 200 calls are 8 minutes and 10 minutes, respectively.

a. Do you think the data collected on the 200 seniors are approximately bell-shaped? Explain.

---

● Basic skills    ◆ Dataset available but not required    **Bold-numbered** exercises answered in the back

b. Is it valid to conduct a one-sample *t*-test in this situation? Explain.

c. In spite of how you may have answered part (b), carry out the five steps to test the hypotheses of interest in this situation.

13.19 A university is concerned that it is taking students too long to complete their requirements and graduate; the average time for all students is 4.7 years. The dean of the campus honors program claims that students who participated in that program in their first year have had a lower mean time to graduation. Unfortunately, there is no automatic way to pull the records of all of the thousands of students who have participated in the program; they must be pulled individually and checked. A random sample of 30 students who had participated is taken, and the mean and standard deviation for the time to completion for those students are 4.5 years and 0.5 year, respectively. Carry out the five steps to test the hypotheses of interest in this situation.

13.20  ◆ The survey in the **UCDavis2** data set on the CD accompanying this book asked students if they preferred to sit in the front, middle, or back of the class and also asked them their heights. The following data are the heights for 15 female students who said they prefer to sit in the back of the class.

68, 62, 65, 69, 68, 69, 64, 66, 69, 68, 62, 64, 67, 68, 65

The mean height for the population of college females is 65 inches. Carry out the five steps to test the claim that the mean height for females who prefer to sit in the back of the room is higher than it is for the general population. In other words, test whether females who prefer to sit in the back of the room are taller than average.

**13.21** ◆ Refer to Exercise 13.20. The following data are the heights for the 38 females who said they prefer to sit in the front of the classroom

66, 63, 63, 66, 65.5, 63, 60, 64, 63, 68, 68, 66, 62.5, 65, 64, 63, 66, 63, 63, 67, 66, 66, 62, 65, 63.5, 60, 61, 62, 63, 60, 65, 62, 63, 63, 62, 65, 63, 66

The mean height for the population of college females is 65 inches. Carry out the five steps to test the claim that the mean height for females who prefer to sit in the front of the room is lower than it is for the general population. In other words, test whether females who prefer to sit in the front of the room are shorter than average.

## Section 13.3

13.22 ● Explain how a paired *t*-test and a one-sample *t*-test are different and how they are the same.

13.23 ● If you were given a data set consisting of pairs of observations for which the question of interest was if the population mean of the differences was 0, explain the steps you would take to determine whether it is valid to use a paired *t*-test.

**13.24** ● Give the value of the test statistic *t* in each of the following situations, then find the *p*-value or *p*-value range for a two-tailed test.

a. $H_0: \mu_d = 0, \overline{d} = 4, s_d = 15, n = 50$.
**b.** $H_0: \mu_d = 0, \overline{d} = -4, s_d = 15, n = 50$.
c. $H_0: \mu_d = 0, \overline{d} = 0, s_d = 15, n = 50$.

**13.25** Most people complain that they gain weight during the December holidays, and Yanovski et al. (2000) wanted to determine whether that was the case. They sampled the weights of 195 adults in mid-November and again in early to mid-January. The mean weight change for the sample was a gain of 0.37 kg, with a standard deviation of 1.52 kg. State and test the appropriate hypotheses. Be sure to carefully define the population parameter(s) you are testing.

13.26 ◆ In Exercise 11.68 a study was reported in which students were asked to place as many dried beans into a cup as possible in 15 seconds with their dominant hand, and again with their nondominant hand (in randomized order). The differences in number of beans (dominant hand–nondominant hand) for 15 students were as follows:

4, 4, 5, 1, −2, 0, 2, 4, −3, 0, 0, 0, −2, 2, 1

The data also are given in the dataset **beans** on the CD for this book.

a. The research question was whether students have better manual dexterity with their dominant hand than with their nondominant hand. Write the null and alternative hypotheses.

b. Check the necessary conditions for doing a one sample *t*-test.

c. Carry out the test using $\alpha = .05$.

d. Carry out the test using $\alpha = .10$.

e. Write a conclusion about this situation that would be understood by other students of statistics.

13.27 ◆ Data from the dataset **UCDavis1** on the CD for this book included information on height (***height***) and mother's height (***momheight***) for 93 female students. Here is the output from the Minitab paired *t* procedure comparing these heights:

| Paired T for height − momheight | | | | |
|---|---|---|---|---|
| | N | Mean | StDev | SE Mean |
| height | 93 | 64.4495 | 2.5226 | 0.2616 |
| momheight | 93 | 63.1645 | 2.6284 | 0.2726 |
| Difference | 93 | 1.28495 | 2.64719 | 0.27450 |

95% lower bound for mean difference: 0.82884
T-Test of mean difference = 0 (vs > 0): T-Value = 4.68    P-Value = 0.000

a. It has been hypothesized that college students are taller than they were a generation ago and therefore that college women should be significantly taller than their mothers. State the null and alternative hypotheses to test this claim. Be sure to define any parameters you use.

b. Using the information in the Minitab output, the test statistic is $t = 4.68$. Identify the numbers that were used to compute the *t*-statistic, and verify that the stated value is correct.

c. What are the degrees of freedom for the test statistic?

d. Carry out the remaining steps of the hypothesis test.

e. Draw a sketch that illustrates the connection between the $t$-statistic and the $p$-value in this problem.

13.28 In Case Study 5.1 (p. 179), results were presented for a sample of 63 men who were asked to report their actual weight and their ideal weight. The mean difference between actual and ideal weight was 2.48 pounds, and the standard deviation of the differences was 13.77 pounds. Is there sufficient evidence to conclude that for the population of men represented by this sample the actual and ideal weights differ, on average? Justify your answer by showing all steps of a hypothesis test.

**13.29** Although we have not emphasized it, the paired $t$-test can be used to test hypotheses in which the null value is something other than 0. For example, suppose that the proponents of a diet plan claim that the mean amount of weight lost in the first three weeks of following the plan is 10 pounds. A consumer advocacy group is skeptical and measures the beginning and ending weights for a random sample of 20 people who follow the plan for three weeks. The mean and standard deviation for the difference in weight at the two times are 8 pounds and 4 pounds, respectively.

a. What is the parameter of interest? Be sure to specify the appropriate population.

b. What are the null and alternative hypotheses?

**c.** What is the value of the test statistic?

**d.** What is the $p$-value for the test?

e. What conclusion can the consumer advocacy group make?

13.30 A company manufactures a homeopathic drug that it claims can reduce the time it takes to overcome jet lag after long-distance flights. A researcher would like to test that claim. She recruits nine people who take frequent trips from San Francisco to London and assigns them to take a placebo for one of their trips and the drug for the other trip, in random order. She then asks them how many days it took to recover from jet lag under each condition. The results are as follows:

| | | | | | Person | | | | |
|---|---|---|---|---|---|---|---|---|---|
| | *1* | *2* | *3* | *4* | *5* | *6* | *7* | *8* | *9* |
| **Placebo** | 7 | 8 | 5 | 6 | 5 | 3 | 7 | 8 | 4 |
| **Drug** | 4 | 4 | 4 | 6 | 6 | 2 | 8 | 6 | 2 |

Carry out the five steps to test the appropriate hypotheses.

13.31 Many people have high anxiety about visiting the dentist. Researchers want to know if this affects blood pressure in such a way that the mean blood pressure while waiting to see the dentist is higher than it is an hour after the visit. Ten individuals have their systolic blood pressures measured while they are in the dentist's waiting room and again an hour after the conclusion of the visit to the dentist. The data are as follows:

| | | | | | Person | | | | | |
|---|---|---|---|---|---|---|---|---|---|---|
| | *1* | *2* | *3* | *4* | *5* | *6* | *7* | *8* | *9* | *10* |
| **B.P. Before** | 132 | 135 | 149 | 133 | 119 | 121 | 128 | 132 | 119 | 110 |
| **B.P. After** | 118 | 137 | 140 | 139 | 107 | 116 | 122 | 124 | 115 | 103 |

a. Write the parameter of interest in this situation.

b. Write the null and alternative hypotheses of interest.

c. Carry out the remaining steps to test the hypotheses you specified in part (b).

## Section 13.4

*Exercises 13.32 to 13.43 correspond to the two Lessons in Section 13.4. Lesson 1 exercises are 13.32 to 13.37; Lesson 2 exercises are 13.38 to 13.43.*

13.32 ● In each of the following situations, determine whether the alternative hypothesis was $H_a: \mu_1 - \mu_2 > 0$, $H_a: \mu_1 - \mu_2 < 0$, or $H_a: \mu_1 - \mu_2 \neq 0$.

a. $H_0: \mu_1 - \mu_2 = 0$, $t = 2.33$, df = 8, $p$-value = 0.048.

**b.** $H_0: \mu_1 - \mu_2 = 0$, $t = -2.33$, df = 8, $p$-value = 0.024.

**c.** $H_0: \mu_1 - \mu_2 = 0$, $t = 2.33$, df = 8, $p$-value = 0.976.

d. $H_0: \mu_1 - \mu_2 = 0$, $t = -2.33$, df = 8, $p$-value = 0.976.

13.33 ● For each of the following situations, identify whether a paired $t$-test or a two-sample $t$-test is appropriate:

a. The weights of a sample of 15 marathon runners were taken before and after a training run to test whether marathon runners lose dangerous levels of fluids during a run.

b. Random samples of 200 new freshmen and 200 new transfer students at a university were given a 50-question test on current events to test whether the level of knowledge of current events differs for new freshmen and transfer students.

c. Sixty students were matched by initial pulse rate, with the two with the highest pulse forming a pair, and so on. Within each pair, one student was randomly chosen to drink a caffeinated beverage, while the other one drank an equivalent amount of water. Their pulse rates were measured 10 minutes later, to test whether caffeine consumption elevates pulse rates.

13.34 ● Calculate the value of the unpooled test statistic $t$ in each of the following situations. In each case, assume the null hypothesis is $H_0: \mu_1 - \mu_2 = 0$.

**a.** $\bar{x}_1 = 35$, $s_1 = 10$, $n_1 = 100$; $\bar{x}_2 = 33$, $s_2 = 9$, $n_2 = 81$.

b. The difference in sample means is 48, s.e.$(\bar{x}_1 - \bar{x}_2) = 22$.

c. Minitab output:

| | N | Mean | StDev | SE Mean |
|---|---|---|---|---|
| Sample 1 | 68 | 80.58 | 4.22 | 0.51 |
| Sample 2 | 68 | 78.55 | 3.31 | 0.40 |

13.35 ● Do hardcover and softcover books likely to be found on a professor's shelf have the same average number of pages? Data on the number of pages for eight hardcover and seven softcover books from a professor's shelf were presented in Example 4.2 (p. 121) and are in the file **ProfBooks** on the CD for this book. The Minitab output from the "2-sample t" procedure follows. Assume that the books are equivalent to a random sample.

```
Cover   N    Mean    StDev   SE Mean
Hard    8    307     134     47
Soft    7    429.4   80.9    31

95% CI for mu (Hard) – mu (Soft): (–246, 2)
T-Test mu (Hard) = mu (Soft)(vs not =): T = –2.16
P = 0.054    DF = 11
```

a. Give the null and alternative hypotheses using symbols.
b. What is the value of the test statistic *t*?
c. Identify the numbers that were used to compute the *t*-statistic, and verify that the reported value is correct.
d. What conclusion would be made using a .05 level of significance? Write the conclusion in statistical terms and in the context of the problem.

**13.36** Example 2.14 (p. 43) gave the weights of eight rowers on each of the Cambridge and Oxford crew teams. The weights are shown again here. Assuming that these men represent appropriate random samples, test the hypothesis that the mean weight of rowers on the Cambridge and Oxford crew teams are equal versus the alternative that they are not equal.

*Cambridge:*  188.5, 183.0, 194.5, 185.0, 214.0, 203.5, 186.0, 178.5

*Oxford:*  186.0, 184.5, 204.0, 184.5, 195.5, 202.5, 174.0, 183.0

**13.37** Case Study 1.1 presented data given in response to the question "What is the fastest you have ever driven a car? ____ mph." The summary statistics are:

*Females:*  $n = 102$, mean = 88.4, standard deviation =14.4

*Males:*  $n = 87$, mean = 107.4, standard deviation = 17.4

Assuming that these students represent a random sample of college students, test whether the mean fastest speed driven by college men and college women is equal versus the alternative that it is higher for men.

**13.38** Example 11.7 (p. 457) presented results for the number of hours slept the previous night from a survey given in two statistics classes. One class was a liberal arts class; the other class was a general introductory class. The survey was given following a Sunday night after classes had started. For simplicity, let's assume that these classes represent a random sample of sleep hours for college students in liberal arts and non-liberal arts majors. The data are as follows:

|          | n   | Mean | St. Dev. |
|----------|-----|------|----------|
| Lib. arts | 25 | 7.66 | 1.34 |
| Non-Lib. arts | 148 | 6.81 | 1.73 |

a. Test the hypothesis that the mean number of hours of sleep for the two populations of students are equal versus the alternative that they are not equal. Use the unpooled *t*-test. (*Note:* The approximate df = 38 for the unpooled test.)
b. The figure below displays a dotplot of the data. Briefly explain what is indicated about the necessary conditions for doing a two-sample *t*-test.
c. Repeat the hypothesis test using the pooled procedure. Compare the results to those in part (a), and discuss which procedure you think is more appropriate in this situation.

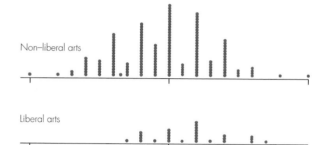

**13.39** Example 11.3 (p. 449) presented data from a study in which sedentary men were randomly assigned to be placed on a diet or exercise for a year to lose weight. Forty-two men were placed on a diet, while the remaining 47 were put on an exercise routine. The group on a diet lost an average of 7.2 kg, with a standard deviation of 3.7 kg. The men who exercised lost an average of 4.0 kg, with a standard deviation of 3.9 kg.
a. State and test appropriate null and alternative hypotheses to determine whether the mean weight loss would be different under the two routines for the population of men similar to those in this study.
b. Explain how you decided whether to do a pooled or unpooled test in part (a).

**13.40** Do students sleep more in Pennsylvania or in California? Data from surveys in elementary statistics classes at Penn State University and the University of California at Davis resulted in the following summary statistics for the number of hours students sleep:

|          | n   | Mean | St. Dev. | S.E. Mean |
|----------|-----|------|----------|-----------|
| UC Davis | 173 | 6.93 | 1.71 | 0.13 |
| Penn State | 190 | 7.11 | 1.95 | 0.14 |

Assume that these students are representative of all students at those two schools. Is there sufficient evidence to conclude that the mean hours of sleep are different at the two schools? Carry out all steps of the hypothesis test, and define all parameters.

13.41 Students in a statistics class at Penn State were asked, "About how many minutes do you typically exercise in a week?" Responses from the *women* in the class were as follows:

60, 240, 0, 360, 450, 200, 100, 70, 240, 0, 60, 360, 180, 300, 0, 270

Responses from the *men* in the class were as follows:

180, 300, 60, 480, 0, 90, 300, 14, 600, 360, 120, 0, 240

a. Draw appropriate graphs to check whether the conditions for conducting a two-sample *t*-test are met. Discuss the results of your graphs.
b. What additional assumption or condition is required if conclusions are to be made about amount of exercise for the population of all Penn State students on the basis of these sample results?
c. Assume that the conditions are met, and conduct a test to determine whether the mean amount of exercise differs for men and women.

**13.42** Researchers speculate that drivers who do not wear a seatbelt are more likely to speed than drivers who do wear one. The following data were collected on a random sample of 20 drivers who were clocked to see how fast they were driving (miles per hour), and then were stopped to see whether they were wearing a seat belt (Y = yes, N = no).

| Driver | | | | | | | | | | | | | | | | | | | |
|---|---|---|---|---|---|---|---|---|---|---|---|---|---|---|---|---|---|---|---|
| *1* | *2* | *3* | *4* | *5* | *6* | *7* | *8* | *9* | *10* | *11* | *12* | *13* | *14* | *15* | *16* | *17* | *18* | *19* | *20* |
| **Speed** | | | | | | | | | | | | | | | | | | | |
| 62 | 60 | 72 | 85 | 68 | 64 | 72 | 72 | 75 | 63 | 62 | 84 | 76 | 60 | 66 | 63 | 64 | 80 | 52 | 64 |
| **Seatbelt** | | | | | | | | | | | | | | | | | | | |
| Y | Y | N | N | Y | Y | Y | N | Y | Y | N | N | N | Y | N | N | Y | Y | Y | Y |

Do these results support the claim that the mean speed is higher for the population of drivers who do not wear seatbelts than for the population of drivers who do?
a. Carry out the five steps of hypotheses testing using the unpooled procedure.
b. Repeat part (a) using the pooled procedure.
c. Compare the unpooled and pooled results and discuss which procedure is more appropriate.
d. Carry out the unpooled test using the rejection region approach with $\alpha = .05$.

13.43 In Example 11.12 (p. 471) a study by Slutske, Piasecki, and Hunt-Carter (2003; and on the CD for this book) was presented, in which the mean number of hangover symptoms was compared for students whose parents have alcohol problems and students whose parents do not. Researchers are interested in knowing if the mean number of hangover symptoms is higher for the population of students whose parents have alcohol problems than for the population whose parents do not. The sample statistics are as follows:

| Group | Mean | Standard deviation |
|---|---|---|
| **Parental alcohol problems ($n_1 = 282$)** | $\bar{x}_1 = 5.9$ | $s_1 = 3.6$ |
| **No parental alcohol problems ($n_2 = 945$)** | $\bar{x}_2 = 4.9$ | $s_2 = 3.4$ |

a. Carry out the five steps of hypothesis testing using the unpooled procedure.
b. Repeat part (a) using the pooled procedure.
c. Compare the unpooled and pooled results and discuss which procedure is more appropriate.

## Section 13.5

**13.44** ● In each of the following cases, explain whether the null hypothesis $H_0$: $\mu = 25$ can be rejected. Use $\alpha = .05$.
**a.** 95% confidence interval for $\mu$ is (10 to 30), $H_a$: $\mu \neq 25$.
b. 95% confidence interval for $\mu$ is (26 to 50), $H_a$: $\mu \neq 25$.
**c.** 90% confidence interval for $\mu$ is (10 to 30), $H_a$: $\mu > 25$.
d. 90% confidence interval for $\mu$ is (10 to 30), $H_a$: $\mu < 25$.
**e.** 90% confidence interval for $\mu$ is (26 to 50), $H_a$: $\mu > 25$.
f. 90% confidence interval for $\mu$ is (26 to 50), $H_a$: $\mu < 25$.

13.45 ● Refer to the rules for the relationship between confidence intervals and two-sided alternatives, given in the two bullets on page 574.
a. Rewrite the rules specifically for $\alpha = .05$.
b. Rewrite the rules specifically for $\alpha = .01$.

13.46 ● Refer to the rules for the relationship between confidence intervals and one-sided tests given in the three bullets and sentence preceding them on page 575.
a. Rewrite the rules specifically for $\alpha = .05$.
b. Rewrite the rules specifically for $\alpha = .01$.

13.47 Each of the following presents a two-sided 95% confidence interval and the alternative hypothesis of a corresponding hypothesis test. In each case, state a conclusion for the test, including the level of significance you are using.
**a.** C.I. for $\mu$ is (101 to 105), $H_a$: $\mu \neq 100$.
**b.** C.I. for $p$ is (.12 to .28), $H_a$: $p < .10$.
c. C.I. for $\mu_1 - \mu_2$ is (3 to 15), $H_a$: $\mu_1 - \mu_2 > 0$.
d. C.I. for $p_1 - p_2$ is ($-.15$ to .07), $H_a$: $p_1 - p_2 \neq 0$.

13.48 As was stated in Section 13.5, "a confidence interval can be used as an alternative way to conduct a two-sided significance test." If a test were conducted by using this method, would the p-value for the test be available? Explain.

**13.49** For each of the following situations, can you conclude whether a 90% confidence interval for $\mu$ would include the value 10? If so, make the conclusion. If not, explain why you can't tell.
**a.** $H_0$: $\mu = 10$, $H_a$: $\mu < 10$, do not reject the null hypothesis for $\alpha = .05$.

b. $H_0$: $\mu = 10$, $H_a$: $\mu < 10$, reject the null hypothesis for $\alpha = .05$.

c. $H_0$: $\mu = 10$, $H_a$: $\mu \neq 10$, do not reject the null hypothesis for $\alpha = .10$.

d. $H_0$: $\mu = 10$, $H_a$: $\mu \neq 10$, reject the null hypothesis for $\alpha = .10$.

## Section 13.6

**13.50** ● Refer to the five parameters given in Table 13.3 (p. 578). For each of the following situations, identify which of the five is or are appropriate. Give the parameter(s) in symbols, and define what the symbols represent in the context of the situation.

**a.** Researchers want to compare the mean running times for the 50-yard dash for first-grade boys and girls. They select a random sample of 35 boys and 32 girls in first grade and time them running the 50-yard dash.

b. Researchers want to know what proportion of a certain type of tree growing in a national forest suffers from a disease. They test a representative sample of 200 of the trees from around the forest and find that 15 of them have the disease.

c. Researchers want to know what percentage of adults have a fear of going to the dentist. They also want to know the average number of visits made to a dentist in the past 10 years for adults who have that fear. They ask a random sample of adults whether or not they fear going to the dentist and also how many times they have gone in the past 10 years.

d. Refer to part (c). Researchers also want to compare the average number of visits made by adults who fear going to the dentist with the average number of visits for those who don't have the fear.

**13.51** ● Refer to Exercise 13.50. In each case, explain whether a confidence interval, a hypothesis test, or both would be more appropriate.

**13.52** Give an example of a situation for which the appropriate inference procedure would be each of the following:

a. A hypothesis test for one proportion.

b. A hypothesis test and confidence interval for a paired difference mean.

c. A confidence interval for one mean.

d. A confidence interval for the difference in two means for independent samples.

**13.53** Refer to each of the following scenarios from exercises in various chapters of this book. In each case, determine the most appropriate inference procedure(s), including the appropriate parameter. If you think inference about more than one parameter may be of interest, answer the question for all parameters of interest. Explain your choices.

**a.** A sample of college students was asked whether they would return the money if they found a wallet on the street. Of the 93 women, 84 said "yes," and of the 75 men, 53 said "yes." Assume these students represent all college students.

b. A study was conducted on pregnant women and subsequent development of their children (Olds et al., 1994). One of the questions of interest was whether the IQ of children would differ for mothers who smoked at least 10 cigarettes a day during pregnancy and those who did not smoke at all.

c. Max likes to keep track of birthdays of people he meets. He has 170 birthdays on his birthday calendar. One cold January night, he comes up with the theory that people are more likely to be born in October than they would be if all 365 days were equally likely. He consults his birthday calendar and finds that 22 of the 170 birthdays are in October.

d. The dataset **cholest** reports cholesterol levels of heart attack patients 2, 4, and 14 days after the heart attack. Data for days 2 and 4 were available for 28 patients. The mean difference (2 day − 4 day) was 23.29, and the standard deviation of the differences was 38.28.

## Section 13.7

**13.54** ● Compute the effect size for each of the following situations, and state whether it would be considered closer to a small, medium, or large effect:

a. In a one-sample test with $n = 100$, the test statistic is $t = 2.24$.

b. In a one-sample test with $n = 50$, the test statistic is $t = -2.83$.

c. In a paired-difference test with $n = 30$ pairs, the test statistic is $t = 1.48$.

d. In a test for the difference in two means with independent samples, with $n_1 = 40$ and $n_2 = 50$, the test statistic is $t = -2.33$.

**13.55** ● Refer to Table 13.4 (p. 585), which presents power for a one-sided, one-sample $t$-test. In a test of $H_0$: $\mu_d = 0$ versus $H_a$: $\mu_d > 0$, suppose the truth is that the population of differences is a normal distribution with mean $\mu_d = 2$ and standard deviation $\sigma = 4$.

**a.** Recalling the Empirical Rule from Chapter 2, draw a picture of this distribution, showing the ranges into which 68% and 95% of the differences fall.

b. On your picture, indicate where the null value of 0 falls.

**c.** If a sample of size 20 is taken, what is the power for the test, assuming that a .05 level of significance will be used?

d. Explain in words what probability the power of the test represents.

**13.56** ● In Table 13.4 (p. 585), it is shown that the power is .40 for a one-sided, one-sample $t$-test with .05 level of significance, $n = 50$, and true effect size of 0.2. Would the power be higher or lower for each of the following changes?

a. The true effect size is 0.4.

b. The sample size used is $n = 75$.

c. The level of significance used is .01.

**13.57** (Computer software required.) Find the power for the following one-sample $t$-test situations. In each case, assume a .05 level of significance will be used.

**a.** Effect size = 0.3, sample size = 45, $H_a: \mu > \mu_0$.

b. Sample size = 30, $H_a: \mu \neq 10$, true mean = 13, $\sigma = 4$.

c. Effect size = $-1.0$, sample size = 15, $H_a: \mu < \mu_0$.

13.58 (Computer software required.) In parts (a) to (d), find the sample size necessary to achieve power of .80 for a one-sample $t$-test with $H_a: \mu > \mu_0$ and level of significance of .05 for each of the following effect sizes:

a. 0.2

b. 0.4

c. 0.6

d. 0.8

e. Make a scatterplot of the sample size (vertical axis) versus the effect size for the effect sizes in parts (a) to (d). Does the relationship between the effect size and the sample size required to achieve 80% power appear to be linear? If not, what is the nature of the relationship?

13.59 For a $z$-test for one proportion, a possible effect size measure is $(p_1 - p_0)/\sqrt{p_0(1 - p_0)}$ where $p_0$ is the null value and $p_1$ is the true population proportion.

a. What is the relationship between this effect size and the $z$-test statistic for this situation?

b. Does the effect size fit the relationship "Test statistic = Size of effect × Size of study"? If not, explain why not. If so, show how it fits.

c. What would be a reasonable way to estimate this effect size?

**13.60** Refer to the effect-size measure in Exercise 13.59. For parts (a) to (c), compute the effect size.

**a.** $p_1 = .35, p_0 = .25$.

b. $p_1 = .15, p_0 = .05$.

c. $p_1 = .95, p_0 = .85$.

**d.** On the basis of the results in parts (a) to (c), does this effect size stay the same when $p_1 - p_0$ stays the same? Explain.

e. In statistical software for computing power for a test for a single proportion, unlike for a single mean, both $p_1$ and $p_0$ must be specified, rather than just the difference between them. Explain why it is not enough to specify the difference, using the results of the previous parts of this exercise.

13.61 Explain why it is more useful to compare effect sizes than $p$-values in trying to determine whether many studies about the same topic have found similar results.

## Section 13.8

**13.62** ● Refer to the list of items in Section 13.8. Explain which ones should be of concern if the sample size(s) for a test are large.

13.63 ● Refer to the list of items in Section 13.8. Explain which ones should be of concern if the sample size(s) for a test are small.

13.64 Refer to item 6 in the list of concerns in Section 13.8. Is that statement the same thing as saying that the null hypothesis is likely to be true in about 1 out of 20 tests that have achieved statistical significance? Explain.

13.65 Refer to the list of items in Section 13.8.

a. For which of the concerns would the $p$-value for a test be useful to have? Explain why in each case.

b. For which of the concerns would a confidence interval estimate for the parameter be useful to have? Explain why in each case.

c. For which of the concerns would the sample size(s) be useful to know? Explain why in each case.

## Chapter Exercises

**13.66** For each of the following research questions, specify the parameter and the value that constitute the null hypothesis of "parameter = null value." In other words, define the population parameter of interest and specify the null value that is being tested.

**a.** Do a majority of Americans between the ages of 18 and 30 think that the use of marijuana should be legalized?

b. Is the mean of the Math SAT scores in California in a given year different from the target mean of 500 set by the test developers?

**c.** Is the mean age of death for left-handed people lower than that for right-handed people?

d. Is there a difference in the proportions of male and female college students who smoke cigarettes?

13.67 Refer to Exercise 13.66. In each case, specify whether the alternative hypothesis would be one-sided or two-sided.

13.68 Suppose a one-sample $t$-test of $H_0: \mu = 0$ versus $H_a: \mu \neq 0$ results in a test statistic of $t = 0.65$ with df = 14. Suppose a new study is done with $n = 150$, and the sample mean and standard deviation turn out to be exactly the same as in the first study.

a. What conclusion would you reach in the original study, using $\alpha = .05$?

b. What conclusion would you reach in the new study, using $\alpha = .05$?

c. Compare your results in parts (a) and (b) and comment.

13.69 A study is conducted to find out whether results of an IQ test are significantly higher after listening to Mozart than after sitting in silence. Explain what has happened in each of the following scenarios:

a. A type 1 error was committed.

b. A type 2 error was committed.

c. The power of the test was too low to detect the difference that actually exists.

d. The power of the test was so high that a very small difference resulted in a statistically significant finding.

**13.70** In a random sample of 170 married British couples, the difference between the husband's and wife's ages had a mean of 2.24 years and a standard deviation of 4.1 years.

a. Test the hypothesis that British men are significantly older than their wives, on average.

b. Explain what is meant by the use of the term *significant* in part (a), and discuss how it compares with the everyday use of the word.

**13.71** Suppose a highway safety researcher makes modifications to the design of a highway sign. The researcher believes that the modifications will make the mean maximum distance at which drivers are able to read the sign greater than 450 feet. The maximum distances (in feet) at which $n = 16$ drivers can read the sign are as follows:

| 440 | 490 | 600 | 540 | 540 | 600 | 240 | 440 |
| 360 | 600 | 490 | 400 | 490 | 540 | 440 | 490 |

Use this sample to determine whether the mean maximum sign-reading distance at which drivers can read the sign is greater than 450. Show all five steps of a hypothesis test, and be sure to state a conclusion. If conditions are not met, make appropriate adjustments. Also, describe any assumptions that you make.

**13.72** In Exercise 11.70, a study is described in which the mean IQs at age 4 for children of smokers and nonsmokers were compared. The mean for the children of the 66 nonsmokers was 113.28 points, while for the children of the 47 smokers it was 103.12 points. Assume that the pooled standard deviation for the two samples is 13.5 points and that a pooled $t$-test is appropriate. Test the hypothesis that the population mean IQ at age 4 is the same for children of smokers and of nonsmokers versus the alternative that it is higher for children of nonsmokers.

**13.73** Refer to Exercise 13.72. One of the statements made in the research article and reported in Exercise 11.70 was "After control for confounding background variables . . . the average difference observed at 36 and 48 months was reduced to 4.35 points (95% CI: 0.02, 8.68)." Use this statement to test the null hypothesis that the difference in the mean IQ scores for the two populations is 0 versus the alternative that it is greater than 0. (Notice that the results have now been adjusted for confounding variables such as parents' IQ, whereas the data given in Exercise 13.72 had not.)

**13.74** It is believed that regular physical exercise leads to a lower resting pulse rate. Following are data for $n = 20$ individuals on resting pulse rate and whether the individual regularly exercises or not. Assuming that this is a random sample from a larger population, use this sample to determine whether the mean pulse is lower for those who exercise. Clearly show all five steps of the hypothesis test.

| Person | Pulse | Regularly Exercises | Person | Pulse | Regularly Exercises |
|--------|-------|---------------------|--------|-------|---------------------|
| 1 | 72 | No | 11 | 62 | No |
| 2 | 62 | Yes | 12 | 84 | No |
| 3 | 72 | Yes | 13 | 76 | No |
| 4 | 84 | No | 14 | 60 | Yes |
| 5 | 60 | Yes | 15 | 52 | Yes |
| 6 | 63 | Yes | 16 | 60 | No |
| 7 | 66 | No | 17 | 64 | Yes |
| 8 | 72 | No | 18 | 80 | Yes |
| 9 | 75 | Yes | 19 | 68 | Yes |
| 10 | 64 | Yes | 20 | 64 | Yes |

**13.75** In an experiment conducted by one of the authors, ten students in a graduate-level statistics course were given this question about the population of Canada: "The population of the U.S. is about 270 million. To the nearest million, what do you think is the population of Canada?" (The population of Canada at the time was slightly over 30 million.) The responses were as follows:

20, 90, 1.5, 100, 132, 150, 130, 40, 200, 20

Eleven other students in the same class were given the same question with different introductory information: "The population of Australia is about 18 million. To the nearest million, what do you think is the population of Canada?" The responses were as follows:

12, 20, 10, 81, 15, 20, 30, 20, 9, 10, 20

The experiment was done to demonstrate the *anchoring effect*, which is that responses to a survey question may be "anchored" to information provided to introduce the question. In this experiment, the research hypothesis was that the individuals who saw the U.S. population figure would generally give higher estimates of Canada's population than would the individuals who saw the Australia population figure.

a. Write null and alternative hypotheses for this experiment. Use proper notation.

b. Test the hypotheses stated in part (a). Be sure to state a conclusion in the context of the experiment.

c. As a step in part (b), you should have created a graphical summary to verify necessary conditions. Do you think that any possible violations of the necessary conditions have affected the results of part (b) in a way that has produced a misleading conclusion? Explain why or why not.

**13.76** ◆ The dataset **cholest** on the CD for this book reports cholesterol levels of heart attack patients 2, 4, and 14 days after the heart attack. Data for days 2 and 4 were available for 28 patients. The mean difference (2 day − 4 day) was 23.29, and the standard deviation of the differences was 38.28. A histogram of the differences was approximately bell-shaped. Is there sufficient evidence to indicate that the mean cholesterol level of heart attack patients decreases from the second to fourth days after the heart attack?

**13.77** Refer to Exercise 13.76. Suppose physicians will use the answer to that question to decide whether to retest patients' cholesterol levels on day 4. If there is no conclusive evidence that the cholesterol level goes down, they will use the day 2 level to decide whether to prescribe drugs for high cholesterol. If there is evidence that cholesterol goes down between days 2 and 4, patients will all be retested on day 4, and the prescription drug decision will be made then.

a. What are the consequences of type 1 and type 2 errors for this setting?

b. Which type of error do you think is more serious?

## Dataset Exercises

**Statistics ⬡ Now™** Datasets **are required** to solve these exercises and can be found at **http://1pass.thomson.com** or on your CD.

13.78 Refer to Exercises 13.76 and 13.77. Using the dataset **cholest,** determine whether there is sufficient evidence to conclude that the cholesterol level drops, on average, from day 2 to day 14 after a heart attack.

13.79 The dataset **UCDavis2** includes information on *Sex, Height,* and *Dadheight.* Use the data to test the hypothesis that college men are taller, on average, than their fathers. Assume that the male students in the survey represent a random sample of college men.

13.80 The dataset **UCDavis2** includes grade point average *GPA* and answers to the question: "Where do you typically sit in a classroom (circle one): Front, Middle, Back." The answer to this question is coded as F, M, B for the variable *Seat.* Assuming these students represent all college students, test whether there is a difference in mean GPA for students who sit in the front versus the back of the classroom.

13.81 The dataset **deprived** includes information on self-reported amount of sleep per night and whether a person feels sleep-deprived for $n = 86$ college students (*Source:* Laura Simon, Pennsylvania State University). Assume the students represent a random sample of college students. Is the mean number of hours of sleep per night lower for the population of students who feel sleep-deprived than it is for the population of students who do not? Use an unpooled test.

**Statistics ⬡ Now™** Preparing for an exam? Assess your progress by taking the post-test at **http://1pass.thomson.com.**

**Mentor™**
Do you need a live tutor for homework problems? Access vMentor at **http://1pass.thomson.com** for one-on-one tutoring from a statistics expert.

# 14

*How is her handspan related to her height?*

See Figure 14.1 *(p. 600)*

Royalty-free/Corbis

- $E(Y)$ represents the mean or expected value of $y$ for individuals in the population who all have the same particular value of $x$. Note that $\hat{y}$ is an estimate of $E(Y)$.

- $\beta_0$ is the **intercept** of the straight line in the **population.**

- $\beta_1$ is the **slope** of the straight line in the **population.** Note that if the slope $\beta_1 = 0$, there is no linear relationship in the population.

Unless we measure the entire population, we cannot know the numerical values of $\beta_0$ and $\beta_1$. These are population parameters that we estimate using the corresponding sample statistics. In the handspan and height example, $b_1 = 0.35$ is a sample statistic that estimates the population parameter, $\beta_1$, and $b_0 = -3$ is a sample statistic that estimates the population parameter $\beta_0$.

---

**technical note** | **Multiple Regression**

In **multiple regression,** the mean of the response variable is a function of two or more explanatory variables. Put another way, in multiple regression we use the values of more than one explanatory (predictor) variable to predict the value of a response variable. For example, a college admissions committee might predict college GPA for an applicant based on using Verbal SAT, Math SAT, high school GPA, and class rank. The general structure of an equation for doing this might be

$$\text{College GPA} = \beta_0 + \beta_1 \text{ Verbal SAT} + \beta_2 \text{ Math SAT} + \beta_3 \text{ HS GPA} + \beta_4 \text{ Class Rank}$$

As in simple regression, numerical estimates of the parameters $\beta_0$, $\beta_1$, $\beta_2$, $\beta_3$, and $\beta_4$ would be determined from a sample. On the CD for this book, multiple regression is covered in Supplemental Topic 3.

---

## Assumptions About Deviations from the Regression Line in the Population

To make statistical inferences about the population, two assumptions about how the $y$ values vary from the population regression line are necessary. First, we assume that there is a **constant variance,** meaning that the general size of the deviation of $y$ values from the line is the same for all values of the explanatory variable ($x$). This assumption may or may not be correct in any particular situation, and a scatterplot should be examined to see whether it is reasonable or not. In Figure 14.1 (p. 600), the constant variance assumption looks reasonable because the magnitude of the deviation from the line appears to be about the same across the range of observed heights.

The second assumption about the population is that for any specific value of $x$, the distribution of $y$ values is a normal distribution. Equivalently, this assumption is that deviations from the population regression line have a normal curve distribution. Figure 14.3 (see next page) illustrates this assumption along with the other elements of the population regression model for a linear relationship. The line $E(Y) = \beta_0 + \beta_1 x$ describes the mean of $y$, and the normal curves describe deviations from the mean.

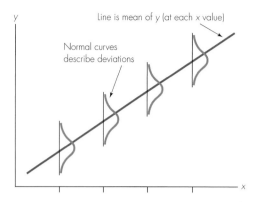

**Figure 14.3** ▌ Regression model for population

## The Simple Regression Model for a Population

A useful format for expressing the components of the population regression model is

$$y = \text{Mean} + \text{Deviation}$$

This conceptual equation states that for any individual, the value of the response variable ($y$) can be constructed by combining two components:

1. The *mean*, which in the population is the line $E(Y) = \beta_0 + \beta_1 x$ if the relationship is linear. There are other possible relationships, such as curvilinear, a special case of which is a quadratic relationship, $E(Y) = \beta_0 + \beta_1 x + \beta_2 x^2$. Relationships that are not linear will not be discussed further in this book.
2. The individual's *deviation* = $y - mean$, which is what is left unexplained after accounting for the mean $y$ value at that individual's $x$ value.

This format also applies to the sample, although technically, we should use the term *estimated mean* when referring to the sample regression line.

**Example 14.2**   **Mean and Deviation for Height and Handspan Regression** Recall from Example 14.1 that the sample regression line for handspans ($y$) and heights ($x$) is $\hat{y} = -3 + 0.35x$. Although it is not likely to be true, let's assume for convenience that this equation also holds in the population. If your height is $x = 70$ in. and your handspan is $y = 23$ cm, then

$$\text{Mean} = -3 + 0.35(70) = 21.5$$
$$\text{Deviation} = y - \text{Mean} = 23 - 21.5 = 1.5$$
$$y = 23 = \text{Mean} + \text{Deviation} = 21.5 + 1.5$$

In other words, the mean handspan for people with your height is 21.5 cm, and your handspan is 1.5 cm above that mean. ■

In the theoretical development of procedures for making statistical inferences for a regression model, the collection of all *deviations* in the population is assumed to have a normal distribution with mean 0 and standard deviation $\sigma$

(so the variance is $\sigma^2$). The value of the standard deviation $\sigma$ is an unknown population parameter that is estimated by using the sample. This standard deviation can be interpreted in the usual way in which we interpret a standard deviation. It is, roughly, the average distance between individual values of $y$ and the mean of $y$ as described by the regression line. In other words, it is roughly the size of the average deviation across all individuals in the range of $x$ values.

Keeping the regression notation straight for populations and samples can be confusing. Although we have not yet introduced all relevant notation, a summary at this stage will help you to keep it straight.

14.1 Exercises are on page 626.

---

*in summary*    ## Simple Linear Regression Model: Population and Sample Versions

For $(x_1, y_1), (x_2, y_2), \ldots, (x_n, y_n)$, a sample of $n$ observations of the explanatory variable $x$ and the response variable $y$ from a large population, the **simple linear regression model** describing the relationship is as follows.

**Population Version**

> *Mean:* $E(Y) = \beta_0 + \beta_1 x$
>
> *Individual:* $y_i = \beta_0 + \beta_1 x_i + \varepsilon_i = E(Y_i) + \varepsilon_i$

The deviations $\varepsilon_i$ are assumed to follow a normal distribution with mean 0 and standard deviation $\sigma$.

**Sample Version**

> *Mean:* $\hat{y} = b_0 + b_1 x$
>
> *Individual:* $y_i = b_0 + b_1 x_i + e_i = \hat{y}_i + e_i$

where $e_i$ is the *residual* for individual $i$. The sample statistics $b_0$ and $b_1$ *estimate* the population parameters $\beta_0$ and $\beta_1$. The mean of the residuals is 0, and the residuals can be used to estimate the population standard deviation $\sigma$.

---

# 14.2 Estimating the Standard Deviation for Regression

Recall that the standard deviation in the regression model measures, roughly, the average deviation of $y$ values from the mean (the regression line). Expressed another way, the **standard deviation for regression** measures the general size of the residuals. This is an important and useful statistic for describing individual variation in a regression problem, and it also provides information about how accurately the regression equation might predict $y$ values for individuals. A relatively small standard deviation from the regression line indicates that individual data points generally fall close to the line, so predictions based on the line will be close to the actual values.

The calculation of the estimate of standard deviation is based on the sum of the squared residuals for the sample. This quantity is called the **sum of squared errors** and is denoted by **SSE.** Synonyms for "sum of squared errors" are **residual sum of squares** or **sum of squared residuals.** To find the SSE, residuals are calculated for all observations, then the residuals are squared and summed. The *standard deviation from the regression line* is

$$s = \sqrt{\frac{\text{Sum of squared residuals}}{n-2}} = \sqrt{\frac{\text{SSE}}{n-2}}$$

and this sample statistic estimates the population standard deviation $\sigma$.

---

*formula*

**Estimating the Standard Deviation for a Simple Regression Model**

The formula for estimating the standard deviation for a simple regression model is

$$\text{SSE} = \sum(y_i - \hat{y}_i)^2 = \sum e_i^2$$

$$s = \sqrt{\frac{\text{SSE}}{n-2}} = \sqrt{\frac{\sum(y_i - \hat{y}_i)^2}{n-2}}$$

The statistic $s$ is an estimate of the population standard deviation $\sigma$.

***Technical Note:*** Notice the difference between the estimate of $\sigma$ in the regression situation compared to what it would be if we simply had a random sample of the $y_i$'s without information about the $x_i$'s:

*Sample of y's only:*   $s = \sqrt{\dfrac{\sum(y_i - \bar{y})^2}{n-1}}$

*Sample of $(x, y)$ pairs, linear regression:*   $s = \sqrt{\dfrac{\sum(y_i - \hat{y}_i)^2}{n-2}}$

Remember that in the regression context, $\sigma$ is the standard deviation of the $y$ values at *each x,* not the standard deviation of the whole population of $y$ values.

---

**Example 14.3**

**Relationship Between Height and Weight for College Men** Figure 14.4 displays regression results from the Minitab program and a scatterplot for the relationship between $y$ = weight (pounds) and $x$ = height (inches) in a

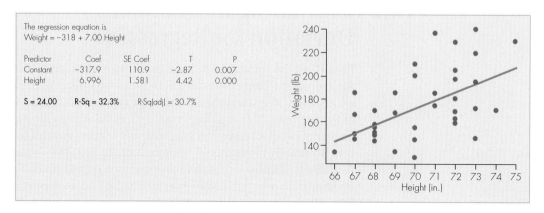

**Figure 14.4** ▌ The relationship between weight and height for $n = 43$ college men

Statistics⬠Now™

Watch a video example at **http://1pass.thomson.com** or on your CD.

sample of $n = 43$ men in a statistics class. The regression line for the sample is $\hat{y} = -318 + 7x$, and this line is drawn onto the plot. We see from the plot that there is considerable variation from the line at any given height. The standard deviation, shown in the last row of the computer output to the left of the plot, is "S = 24.00." This value roughly measures, for any given height, the general size of the deviations of individual weights from the mean weight for the height.

The standard deviation from the regression line can be interpreted in conjunction with the Empirical Rule for bell-shaped data stated in Section 2.7. Recall, for instance, that about 95% of individuals will fall within 2 standard deviations of the mean. As an example, consider men who are 72 inches tall. For men with this height, the estimated average weight determined from the regression equation is $-318 + 7.00(72) = 186$ pounds. The estimated standard deviation from the regression line is $s = 24$ pounds, so we can estimate that about 95% of men 72 inches tall have weights within $2 \times 24 = 48$ pounds of 186 pounds, which is $186 \pm 48$, or 138 to 234 pounds. Think about whether this makes sense for all the men you know who are 72 inches (6 feet) tall. ■

---

*in summary*   **Interpreting the Standard Deviation for Regression**

The standard deviation for regression estimates the standard deviation of the differences between values of $y$ and the regression equation that relates the mean value of $y$ to $x$. In other words, it measures the general size of the differences between actual and predicted values of $y$.

---

*thought question 14.1*   Regression equations can be used to predict the value of a response variable for an individual. What is the connection between the accuracy of predictions based on a particular regression line and the value of the standard deviation from the line? If you were deciding between two different regression models for predicting the same response variable, how would your decision be affected by the relative values of the standard deviations for the two models?[*]

## The Proportion of Variation Explained by $x$

In Chapter 5, we learned that the squared correlation $r^2$ is a useful statistic. It is used to measure how well the explanatory variable explains the variation in the response variable. This statistic is also denoted as $R^2$ (rather than $r^2$), and the value is commonly expressed as a percentage. Researchers typically use the phrase **"proportion of variation explained by $x$"** in conjunction with the value of $r^2$. For example, if $r^2 = .60$ (or 60%), the researcher may write that the explanatory variable explains 60% of the variation in the response variable.

---

[*]**HINT:** Read the first paragraph of this section (p. 605).

The formula for $r^2$ presented in Chapter 5 was

$$r^2 = \frac{\text{SSTO} - \text{SSE}}{\text{SSTO}}$$

The quantity **SSTO** is the sum of squared differences between observed $y$ values and the sample mean $\bar{y}$. It measures the size of the deviations of the $y$ values from the overall mean of $y$, whereas **SSE** measures the deviations of the $y$ values from the predicted values of $y$.

**Example 14.4**

**$R^2$ for Heights and Weights of College Men**  In Figure 14.4 for Example 14.3 (p. 606), we can find the information "R-sq = 32.3%" for the relationship between weight and height. A researcher might write "the variable height explains 32.3% of the variation in the weights of college men." This isn't a particularly impressive statistic. As we noted before, there is substantial deviation of individual weights from the regression line, so a prediction of a college man's weight based on height may not be particularly accurate.  ■

*thought question* **14.2**  Look at the formula for SSE, and explain in words under what condition SSE = 0. Now explain what happens to $r^2$ when SSE = 0, and explain whether that makes sense according to the definition of $r^2$ as "proportion of variation in $y$ explained by $x$."*

**Example 14.5**

Statistics⏁Now™

Watch a video example at **http://1pass.thomson.com** or on your CD.

For software help, download your Minitab, Excel, TI-83, SPSS, R, and JMP manuals from **http://1pass .thomson.com**, or find them on your CD.

**Driver Age and Highway Sign-Reading Distance**  In Example 5.2 (p. 153), we examined data for the relationship between $y$ = maximum distance (feet) at which a driver can read a highway sign and $x$ = the age of the driver. There were $n$ = 30 observations in the dataset. Figure 14.5 displays Minitab regression

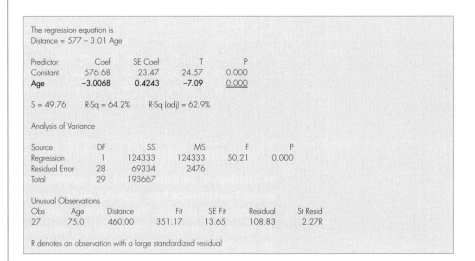

**Figure 14.5**  ▮ Minitab output: Sign-reading distance and driver age

---

***HINT:**  Remember that SSE stands for "sum of squared errors." The formula for $r^2$ is given just before Example 14.4.

output for these data. The equation describing the linear relationship in the sample is

$$\text{Average distance} = 577 - 3.01 \times \text{Age}$$

From the output, we learn that the standard deviation from the regression line is $s = 49.76$ and $R$-sq = 64.2%. Roughly, the average deviation from the regression line is about 50 feet, and the proportion of variation in sign-reading distances explained by age is .642 or 64.2%.

The analysis of variance table provides the pieces needed to compute $r^2$ and $s$:

$$\text{SSE} = 69{,}334$$

$$s = \sqrt{\frac{\text{SSE}}{n-2}} = \sqrt{\frac{69{,}334}{28}} = 49.76$$

$$\text{SSTO} = 193{,}667$$

$$\text{SSTO} - \text{SSE} = 193{,}667 - 69{,}334 = 124{,}333$$

$$r^2 = \frac{\text{SSTO} - \text{SSE}}{\text{SSTO}} = \frac{124{,}333}{193{,}667} = .642, \text{ or } 64.2\% \quad \blacksquare$$

14.2 Exercises are on page 626.

# 14.3  Inference About the Slope of a Linear Regression

In this section, we will learn how to carry out a hypotheses test to determine whether we can infer that two variables are linearly related in the larger population represented by a sample. We will also learn how to use sample regression results to calculate a confidence interval estimate of a population slope.

## Hypothesis Test for a Population Slope

The **statistical significance of a linear relationship** can be evaluated by testing whether or not the population slope is 0. If the slope is 0 in a simple linear regression model, the two variables are not related because changes in the $x$ variable will not lead to changes in the $y$ variable. The usual null hypothesis and alternative hypothesis about $\beta_1$, the slope of the population regression line $E(Y) = \beta_0 + \beta_1 x$, are

$H_0: \beta_1 = 0$     (the population slope is 0, so $y$ and $x$ are *not linearly related*)
$H_a: \beta_1 \neq 0$     (the population slope is not 0, so $y$ and $x$ *are linearly related*)

The alternative hypothesis may be one-sided or two-sided, although most statistical software uses the two-sided alternative.

The test statistic used to do the hypothesis test is a $t$-statistic with the same general format that we used in Chapters 12 and 13. That format, and its application to this situation, is

$$t = \frac{\text{Sample statistic} - \text{Null value}}{\text{Standard error}} = \frac{b_1 - 0}{\text{s.e.}(b_1)}$$

This is a standardized statistic for the difference between the sample slope and 0, the null value. Notice that a large value of the sample slope (either positive or negative) relative to its standard error will give a large value of $t$. If the mathematical assumptions about the population model described in Section 14.1 are correct, the statistic has a $t$-distribution with $n - 2$ degrees of freedom. The $p$-value for the test is determined using that distribution.

It is important to be sure that the necessary conditions are met when using any statistical inference procedure. The necessary conditions for using this test, and how to check them, will be discussed in Section 14.5.

"By hand" calculations of the sample slope and its standard error are cumbersome. Fortunately, the regression analysis of most statistical software includes a $t$-statistic and a $p$-value for this significance test.

*formula*

**Formula for the Sample Slope and Its Standard Error**

In case you ever need to compute the values by hand, here are the formulas for the sample slope and its standard error:

$$b_1 = r\frac{s_y}{s_x}$$

$$\text{s.e.}(b_1) = \frac{s}{\sqrt{\sum(x_1 - \bar{x})^2}} \qquad \text{where } s = \sqrt{\frac{\text{SSE}}{n - 2}}$$

In the formula for the sample slope, $s_x$ and $s_y$ are the sample standard deviations of the $x$ and $y$ values, respectively, and $r$ is the correlation between $x$ and $y$.

**Example 14.6**

Statistics ◁Now™

For software help, download your Minitab, Excel, TI-83, SPSS, R, and JMP manuals from **http://1pass .thomson.com**, or find them on your CD.

**Hypothesis Test for Driver Age and Sign-Reading Distance** Figure 14.5 (p. 608) for Example 14.5 presents Minitab output for the regression of sign-reading distance ($y$) and driver age. The part of the output that is used to test the statistical significance of the observed relationship is shown in bold. This line of output gives values for the sample slope, the standard error of the sample slope, the $t$-statistic, and the $p$-value for the test of

$H_0: \beta_1 = 0$      (the population slope is 0, so $y$ and $x$ are *not linearly related*)
$H_a: \beta_1 \neq 0$      (the population slope is not 0, so $y$ and $x$ *are linearly related*)

The test statistic is

$$t = \frac{\text{Sample statistic} - \text{Null value}}{\text{Standard error}} = \frac{b_1 - 0}{\text{s.e.}(b_1)} = \frac{-3.0068 - 0}{0.4243} = -7.09$$

The $p$-value (underlined in the output) is given to three decimal places as .000. This means that the probability is virtually 0 that the sample slope could be as far from 0 or farther than it is if the population slope really is 0. Because the $p$-value is so small, we can reject the null hypothesis and infer that the linear relationship observed between the two variables in the sample represents a real relationship in the population. ■

*technical note*

Most statistical software reports a $p$-value for a two-sided alternative hypothesis when doing a test for whether the slope in the population is 0. It may sometimes make sense to use a one-sided alternative hypothesis instead. In that case, the $p$-value for the one-sided alternative is (reported $p/2$) if the sign of $b_1$ is consistent with $H_a$, but is $1 - $ (reported $p/2$) if it is not.

# Confidence Interval for the Population Slope

The significance test of whether or not the population slope is 0 tells us only whether we can declare the relationship to be statistically significant. If we decide that the true slope is not 0, we might ask, "What is the value of the slope?" We can answer this question with a **confidence interval for $\beta_1$, the population slope.**

The format for this confidence interval is the same as the general format used in Chapters 10 and 11, which is

Sample statistic $\pm$ Multiplier $\times$ Standard error

The sample statistic is $b_1$, the slope of the least-squares regression line for the sample. As has been shown already, the standard error formula is complicated, and we will usually rely on statistical software to determine this value. The "multiplier" will be labeled $t^*$ and is determined by using a $t$-distribution with df $= n - 2$. Table A.2 can be used to find the multiplier for the desired confidence level.

---

*formula* | **Formula for Confidence Interval for $\beta_1$, the Population Slope**

A confidence interval for $\beta_1$ is

$$b_1 \pm t^* \text{ s.e.}(b_1)$$

The multiplier $t^*$ is found by using a $t$-distribution with $n - 2$ degrees of freedom and is such that the probability between $-t^*$ and $+t^*$ equals the confidence level for the interval.

---

**Example 14.7** | **95% Confidence Interval for Slope Between Age and Sign-Reading Distance** In Figure 14.5 (p. 608), we see that the sample slope is $b_1 = -3.01$ and s.e.$(b_1) = 0.4243$. There are $n = 30$ observations, so df $= 28$ for finding $t^*$. For a 95% confidence level, $t^* = 2.05$ (see Table A.2). The 95% confidence interval for the population slope is

$$-3.01 \pm 2.05 \times 0.4243$$
$$-3.01 \pm 0.87$$
$$-3.88 \text{ to } -2.14$$

With 95% confidence, we can estimate that in the population of drivers represented by this sample, the *mean* sign-reading distance decreases somewhere between 2.14 and 3.88 feet for each one-year increase in age. ■

---

*thought question* **14.3** In previous chapters, we learned that a confidence interval can be used to determine whether a hypothesized value for a parameter can be rejected. How would you use a confidence interval for the population slope to determine whether there is a statistically significant relationship between x and y? For example, why is the interval that we just computed for the sign-reading example evidence that sign-reading distance and age are related?[*]

---

[*]**HINT:** What is the null value for the slope? Section 13.5 discusses the connection between confidence intervals and significance tests.

> **SSPS *tip***    **Calculating a 95% Confidence Interval for the Slope**
>
> - Use **Analyze>Regression>Linear Regression.** Specify the *y* variable in the *Dependent* box and specify the *x* variable in the *Independent(s)* box.
> - Click **Statistics** and then select "Confidence intervals" under "Regression Coefficients."

## Testing Hypotheses About the Correlation Coefficient

In Chapter 5, we learned that the correlation coefficient is 0 when the regression line is horizontal. In other words, if the slope of the regression line is 0, the correlation is 0. This means that the results of a hypothesis test for the population slope can also be interpreted as applying to equivalent hypotheses about the correlation between *x* and *y* in the population.

We use different notation to distinguish between a correlation computed for a sample and a correlation within a population. It is commonplace to use the Greek letter $\rho$ (pronounced "rho") to represent the correlation between two variables within a population. Using this notation, null and alternative hypotheses of interest are as follows:

$H_0: \rho = 0$    (*x* and *y* are not correlated)
$H_a: \rho \neq 0$    (*x* and *y* are correlated)

The results of the hypothesis test described before for the population slope $\beta_1$ can be used for these hypotheses as well. If we reject $H_0: \beta_1 = 0$, we also reject $H_0: \rho = 0$. If we decide in favor of $H_a: \beta_1 \neq 0$, we also decide in favor of $H_a: \rho \neq 0$.

Many statistical software programs, including Minitab, will give a *p*-value for testing whether the population correlation is 0 or not. This *p*-value will be the same as the *p*-value given for testing whether the population slope is 0 or not.

The following Minitab output is for the relationship between pulse rate and weight in a sample of 35 college women. Notice that .292 is given as the *p*-value for testing that the slope is 0 (look under P in the regression results) and for testing that the correlation is 0. Because this is not a small *p*-value, we cannot reject the null hypotheses for the slope and the correlation.

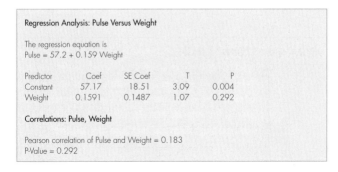

Regression Analysis: Pulse Versus Weight

The regression equation is
Pulse = 57.2 + 0.159 Weight

| Predictor | Coef | SE Coef | T | P |
|---|---|---|---|---|
| Constant | 57.17 | 18.51 | 3.09 | 0.004 |
| Weight | 0.1591 | 0.1487 | 1.07 | 0.292 |

Correlations: Pulse, Weight

Pearson correlation of Pulse and Weight = 0.183
P-Value = 0.292

## The Effect of Sample Size on Significance

The size of a sample always affects whether a specific observed result achieves statistical significance. For example, $r = 0.183$ is not a statistically significant correlation for a sample size of $n = 35$, as in the pulse and weight example, but it would be statistically significant if $n = 1000$. With very large sample sizes, weak relationships with low correlation values can be statistically significant. The "moral of the story" here is that with a large sample size, it may not be saying much to say that two variables are significantly related. This means only that we think that the correlation is not precisely 0. To assess the practical significance of the result, we should carefully examine the observed strength of the relationship.

14.3 Exercises are on page 627.

**technical note**  The usual $t$-statistic for testing whether the population slope is 0 in a linear regression could also be found by using a formula that involves only $n =$ sample size and $r =$ correlation between $x$ and $y$. The algebraic equivalence is

$$t = \frac{b_1}{\text{s.e.}(b_1)} = \sqrt{n-2}\frac{r}{\sqrt{1-r^2}}$$

In the output for the pulse rate and body weight example just given, notice that the $t$-statistic for testing whether the slope $\beta_1 = 0$ is $t = 1.07$. This was calculated as

$$t = \frac{b_1}{\text{s.e.}(b_1)} = \frac{0.1591}{0.1487} = 1.07$$

The sample size is $n = 35$, and the correlation is $r = 0.183$, so an equivalent calculation of the $t$-statistic is

$$t = \sqrt{n-2}\frac{r}{\sqrt{1-r^2}} = \sqrt{35-2}\frac{0.183}{\sqrt{1-0.183^2}} = 1.07$$

This second method for calculating the $t$-statistic illustrates two ideas. First, there is a direct link between the correlation value and the $t$-statistic that is used to test whether the slope is 0. Second, notice that for any fixed value of $r$, increasing the sample size $n$ will increase the size of the $t$-statistic. And the larger the value of the $t$-statistic, the stronger is the evidence against the null hypothesis.

## 14.4 Predicting $y$ and Estimating Mean $y$ at a Specific $x$

In this section, we cover two different types of intervals that are used to make inferences about the response variable ($y$). The first type of interval *predicts the value of y* for an individual with a specific value of $x$. For example, we may want to predict the freshman year GPA of a college applicant who has a 3.6 high school GPA. The second type of interval *estimates the mean value of y* for a population of individuals who all have the same specific value of *x*. As an example, we may want to estimate the mean (average) freshman year GPA of all college applicants who have a 3.6 high school GPA.

## Predicting the Value of *y* for an Individual

An important use of a regression equation is to estimate or predict the unknown value of a response variable for an *individual* with a known specific value of the explanatory variable. Using the data described in Example 14.5 (p. 608) for instance, we can predict the maximum distance at which an individual can read a highway sign by substituting his or her age for *x* in the sample regression equation. Consider a person who is 21 years old. The predicted distance for such a person is approximately $\hat{y} = 577 - 3.01(21) = 513.79$, or about 514 feet.

There will be variation among 21-year-olds with regard to the sign-reading distance, so the predicted distance of 513.79 feet is not likely to be the exact distance for the next 21-year-old who views the sign. Rather than predicting that the distance will be exactly 513.79 feet, we should instead predict that the distance will be within a particular interval of values.

A **95% prediction interval** describes the values of the response variable (*y*) for 95% of all individuals with a particular value of *x*. This interval can be interpreted in two equivalent ways:

1. The 95% prediction interval estimates the central 95% of the values of *y* for members of the population with a specified value of *x*.
2. The probability is .95 that a randomly selected individual from the population with a specified value of *x* falls into the corresponding 95% prediction interval.

We don't always have to use a 95% prediction interval. A prediction interval for the value of the response variable (*y*) can be found for any specified central percentage of a population with a specified value of *x*. For example, a 75% prediction interval describes the central 75% of a population of individuals with a particular value of the explanatory variable (*x*).

*definition*    A **prediction interval** estimates the value of *y* for an individual with a particular value of *x*, or equivalently, the range of values of the response variable for a specified central percentage of a population with a particular value of *x*.

Notice that a prediction interval differs conceptually from a confidence interval. A confidence interval estimates an unknown population parameter, which is a numerical characteristic or summary of the population. An example in this chapter is a confidence interval for the slope of the population line. A prediction interval, however, does not estimate a parameter; instead, it estimates the potential data value for an individual. Equivalently, it describes an interval into which a specified percentage of the population may fall.

Condition 4, which is that deviations from the regression line are normally distributed, is difficult to verify, but it is also the least important of the conditions because the inference procedures for regression are *robust*. This means that if there are no major outliers or extreme skewness, the inference procedures work well even if the distribution of *y* values is not a normal distribution. In Chapters 11 and 13, we learned that confidence intervals and hypothesis tests for a mean or a difference between two means also were robust.

To examine the distribution of the deviations from the line, a *histogram of the residuals* is useful, although for small samples, a histogram may not be informative. A more advanced plot called a *normal probability plot* can also be used to check whether the residuals are normally distributed, but we do not provide the details in this text. Figure 14.9 displays a histogram of the residuals for Example 14.3 (p. 606). It appears that the residuals are approximately normally distributed, so Condition 4 is met.

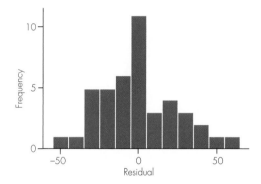

**Figure 14.9** ▌ Histogram of residuals for Example 14.3

Condition 5 follows from the data collection process. It is met as long as the units are measured independently. It would not be met if the same individuals were measured across the range of *x* values, such as if *x* = average speed and *y* = gas mileage were to be measured for multiple tanks of gas on the same cars. More complicated models are needed for *dependent* observations; those models will not be discussed in this book.

## Corrections When Conditions Are Not Met

There are some steps that can be taken if Condition 1, 2, or 3 is not met. If Condition 1 is not met, more complicated models can be used. For instance, Figure 14.10 (see next page) shows a typical plot of residuals that occurs when a straight-line model is used to describe data that are curvilinear. It may help to think of the residuals as prediction errors that would occur if we used the regression line to predict the value of *y* for the individuals in the sample. In the plot shown in Figure 14.10, the "prediction errors" are all negative in the central region of *x* and nearly all positive for outer values of *x*. This occurs because the wrong model is being used to make the predictions. A curvilinear model, such as the quadratic model discussed earlier, may be more appropriate.

Condition 2, that there are no influential outliers, can be checked graphically with the scatterplot of *y* versus *x* and the plot of residuals versus *x*. The

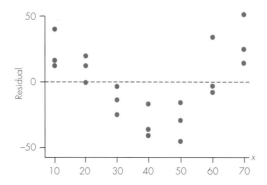

**Figure 14.10** ▌ A residual plot indicating the wrong model has been used

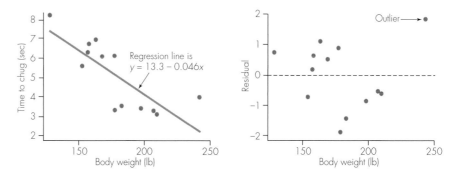

**Figure 14.11** ▌ Scatterplot and corresponding residual plot with an outlier

appropriate correction if there are outliers depends on the reason for the outliers. The same considerations and corrective action that we discussed in Chapter 2 would be taken, depending on the cause of the outlier. For instance, Figure 14.11 shows a scatterplot and a residual plot for the data of Exercise 5.73. A potential outlier is seen in both plots.

In this example, the $x$ variable is weight, and the $y$ variable is time to chug a beverage. The outlier probably represents a legitimate data value. The relationship appears to be linear for weights ranging up to about 210 pounds, but then it appears to change. It could either become quadratic or level off. We do not have enough data to determine what happens for higher weights. The solution in this case would be to remove the outlier and use the linear regression relationship only for body weights under about 210 pounds. Determining the relationship for higher body weights would require a larger sample of individuals in that range.

If either Condition 1 or Condition 3 is not met, a **transformation** may be required. This is equivalent to using a different model. Fortunately, often the same transformation will correct problems with Conditions 1, 3, and 4. For instance, when the response variable is monetary, such as salaries, it is often more appropriate to use the relationship

$$\ln(y) = b_0 + b_1 x + e$$

In other words, we assume that there is a linear relationship between the natural log of $y$ and the $x$ values. This is called a **log transformation on the $y$'s.** We will not pursue transformations further in this book.

14.5 Exercises are on page 629.

---

| MINITAB *tip* | **Plotting and Storing Residuals** |
| --- | --- |

- Use **Stat>Regression>Regression.** The *Graphs* button of the dialog box offers options for several possible residual plots for the regression.
- To store residuals in a column of the worksheet, use the *Storage* button of the dialog box for regression.

---

## *case study* 14.1   A Contested Election

Case Study 12.2 (p. 537), discussed allegations of fraud in a 1993 Philadelphia special election for state senator. The Democratic candidate, William Stinson, beat the Republican candidate, Bruce Marks, by a difference of 461 votes, but Judge Clarence Newcomer overturned the results in a decision issued on April 24, 1994. Judge Newcomer wrote, "The court will order the certification of Bruce Marks as the winner of the 1993 Special Election in the Second Senatorial District, because this court finds that the record of evidence overwhelmingly supports the finding that Bruce Marks would have won the election but for wrongdoing" (Newcomer, 1994).

Judge Newcomer based his decision partially on statistical evidence presented by expert witnesses. Other evidence suggested that a number of absentee ballots had been filed illegally, and a statistical analysis indicated that the absentee count was indeed unusual when compared to the machine vote count.

The data used for the statistical analysis consisted of two variables for each of the 21 senatorial elections held in the previous 11 years (1982–1992) in the Philadelphia area. The variables were

$y$ = *Diff absentee* = (Democrat candidate − Republican candidate) votes by absentee ballot

$x$ = *Diff machine* = (Democrat candidate − Republican candidate) votes by machine ballot

The data for the 21 prior elections were given by economist and expert witness Orley Ashenfelter, in a report to Judge Newcomer (Ashenfelter, 1994), and they are presented in Table 14.1 along with the contested 1993 results.

Figure 14.12 (see next page) shows a plot of the data, with the contested election indicated. In the contested election, the absentee difference (Democrat − Republican) appears well above what might be expected

**Table 14.1** Differences in Democratic and Republican Votes in Philadelphia Elections

| Year | District | Diff Absentee | Diff Machine |
| --- | --- | --- | --- |
| 82 | 2 | 346 | 26427 |
| 82 | 4 | 282 | 15904 |
| 82 | 8 | 223 | 42448 |
| 84 | 1 | 593 | 19444 |
| 84 | 3 | 572 | 71797 |
| 84 | 5 | −229 | −1017 |
| 84 | 7 | 671 | 63406 |
| 86 | 2 | 293 | 15671 |
| 86 | 4 | 360 | 36276 |
| 86 | 8 | 306 | 36710 |
| 88 | 1 | 401 | 21848 |
| 88 | 3 | 378 | 65862 |
| 88 | 5 | −829 | −13194 |
| 88 | 7 | 394 | 56100 |
| 90 | 2 | 151 | 700 |
| 90 | 4 | −349 | 11529 |
| 90 | 8 | 160 | 26047 |
| 92 | 1 | 1329 | 44425 |
| 92 | 3 | 368 | 45512 |
| 92 | 5 | −434 | −5700 |
| 92 | 7 | 391 | 51206 |
| 93 (Contested) | 2 | 1025 | −564 |

based on the overall connection between absentee and machine votes. Notice also that there is an additional outlier—for the election held in District 1 in 1992. No explanation or discussion of that outlier occurred in the court records.

*(continued)*

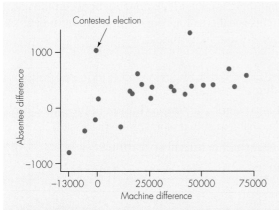

**Figure 14.12** ▌ Plot of absentee difference versus voting machine difference

Is the combination of absentee and machine differences in the contested election at all reasonable, if all other factors remain similar to circumstances of the other elections? The output shown at the bottom of this column is for a simple linear regression computed using data from the 21 prior elections. Included in the results are 95% confidence and prediction intervals for the difference in absentee ballots when the machine ballot difference is −564, the value observed in the contested election. In particular, notice that the 95% prediction interval for the absentee ballot difference for this machine ballot difference is −854.7 to 588.6. The actual absentee difference in the contested election was +1025, a difference far outside this range. Figure 14.13 displays this result graphically, giving "prediction limits" for the entire range of possible values of machine differences covered by the data.

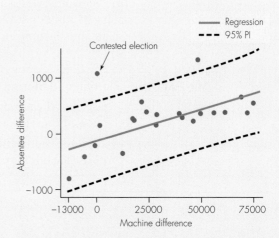

**Figure 14.13** ▌ Prediction limits for Diff Absentee, given machine difference

The analysis reported in the *New York Times* and quoted in Case Study 12.2 used the linear relationship that we have just used. However, an examination of the plot of the data in Figure 14.12 indicates that the relationship may actually be curvilinear. A plot of the residuals versus *x* (not shown) confirms this effect. Therefore, another analysis was done by including a "quadratic" term, $(\text{Diff Machine})^2$, in the regression equation. Results are shown in the output below, including 95% confidence and prediction intervals for the contested election. Notice that the prediction interval is even more disparate with the observed result of +1025 than it was with the simple linear regression equation.

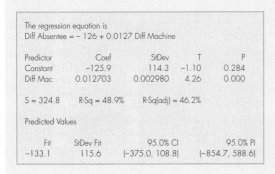

The regression equation is
Diff Absentee = − 126 + 0.0127 Diff Machine

| Predictor | Coef | StDev | T | P |
|---|---|---|---|---|
| Constant | −125.9 | 114.3 | −1.10 | 0.284 |
| Diff Mac | 0.012703 | 0.002980 | 4.26 | 0.000 |

S = 324.8     R-Sq = 48.9%     R-Sq(adj) = 46.2%

Predicted Values

| Fit | StDev Fit | 95.0% CI | 95.0% PI |
|---|---|---|---|
| −133.1 | 115.6 | (−375.0, 108.8) | (−854.7, 588.6) |

The regression equation is
Diff Absentee = − 219 + 0.0297 Diff Machine −0.000000 Quadratic

| Predictor | Coef | StDev | T | P |
|---|---|---|---|---|
| Constant | −219.0 | 105.4 | −2.08 | 0.052 |
| Diff Mac | 0.029709 | 0.006887 | 4.31 | 0.000 |
| Quadrati | −0.00000028 | 0.00000011 | −2.67 | 0.016 |

S = 282.6     R-Sq = 63.3%     R-Sq (adj) = 59.3%

Predicted Values

| Fit | StDev Fit | 95.0% CI | 95.0% PI |
|---|---|---|---|
| −235.9 | 107.7 | (−462.1, −9.6) | (−871.2, 399.5) |

The *p*-value reported for the "quadratic term" is .016, and it is for a test of the null hypothesis that the quadratic term is not needed. Therefore, it is clear that the new model, including the quadratic term, is better than the simple linear model. Figure 14.14 shows the prediction limits based on the quadratic model over the range of machine differences for the model including the quadratic term.

The regression analysis and accompanying figures cast doubt on the results of the contested election, assuming that all other factors were the same as in the past. There may have been other factors that led more people who cast votes for the Democratic candidate to do so by absentee ballot in this election. That issue cannot be addressed using the statistical evidence available, but Judge Newcomer was convinced that other potential differences could not account for the statistical anomaly displayed by the regression analysis.

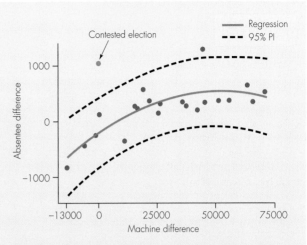

**Figure 14.14** ▌ Prediction limits for quadratic model

# Key Terms

# Exercises

● Denotes basic skills exercises

◆ Denotes dataset is available in StatisticsNow at **http://1pass.thomson.com** or on your CD but is **not required** to solve the exercise.

**Bold-numbered exercises** have answers in the back of the text and fully worked solutions in the Student Solutions Manual.

**Statistics ⬡ Now™** Go to the StatisticsNow website at **http://1pass.thomson.com** to:
• Assess your understanding of this chapter
• Check your readiness for an exam by taking the Pre-Test quiz and exploring the resources in the Personalized Learning Plan

## Section 14.1

**14.1** ● ◆ The **temperature** dataset on the CD for this book gives $y$ = mean April temperature (Fahrenheit) and $x$ = geographic latitude for 20 U.S. cities. The simple linear regression equation for the sample is $\hat{y} = 119 - 1.64\,x$.
   **a.** The value of latitude for Pittsburgh is 40. What is the predicted April temperature for Pittsburgh?
   **b.** The mean April temperature for Pittsburgh is 50. What is the residual for Pittsburgh?

**14.2** ● The population model for simple linear regression can be defined as

$$y_i = \beta_0 + \beta_1 x_i + \varepsilon_i$$

   **a.** Explain what the parameters $\beta_0$ and $\beta_1$ represent.
   b. Write the equation that represents how the mean of the $y$ variable relates to the $x$ variable.

**14.3** ● Ages of husbands and wives are recorded for a randomly selected sample of $n = 200$ married couples in the United States. The data will be used to do a simple linear regression with $y$ = husband's age and $x$ = wife's age.
   **a.** Explain the distinction between the regression line for the sample and the regression line for the population in this situation. Include descriptions of what is the sample and what is the population.
   b. Write the population model for the simple linear regression that describes how the age of a husband is related to the age of a wife. Write any necessary assumptions about deviations from the line.

**14.4** ● Suppose that medical researchers use sample data to find a regression line describing the relationship between $y$ = systolic blood pressure and $x$ = age for men between 40 and 60 years old and that the sample regression line is

   Mean systolic blood pressure = 85 + 0.9 (Age)

   **a.** Give the value of slope, and write a sentence that interprets it in the context of this situation.
   b. Estimate the mean systolic blood pressure of men who are 50 years old.

   c. Suppose that the systolic blood pressure of a 50-year-old man is 125. What is the value of the residual for this man?

**14.5** The linear relationship between number of bedrooms a house has and the selling price of the house is analyzed for a sample of $n = 250$ recently sold houses in a large metropolitan area. The sample regression line is

   Mean selling price = 72000 + 55000 (Bedrooms)

   Give the value of the slope of the regression line, and write a sentence that interprets it in the context of this situation.

**14.6** Heights and answers to the question "How much would you like to weigh?" were recorded for $n = 126$ women in a statistics class. A regression equation for $y$ = desired weight (in pounds) and $x$ = height (in inches) is

   Mean desired weight = $-65.4 + 2.90$ (Height)

   (*Data source:* William Harkness.)
   **a.** What is the value of the slope of this regression equation? Write a sentence that explains what this value shows about the linear relationship between desired weight and height for these women.
   **b.** Explain whether the proper notation for the slope of this equation is $b_1$ or $\beta_1$.
   **c.** What is the estimated mean desired weight for college women who are 65 inches tall?

**14.7** In the simple linear regression model for a population, the relationship between the $x$ variable and the mean of the $y$ variable is $E(Y) = \beta_0 + \beta_1 x$.
   **a.** Suppose $\beta_1 = 0$. What would be the appearance of this line? Draw a sketch illustrating your answer.
   b. Explain how the answer to part (a) illustrates the lack of a linear relationship between $x$ and $y$.

## Section 14.2

**14.8** ● An example in Sections 14.1 and 14.2 described the relationship between $y$ = handspan (cm) and $x$ = height (in.). The sample consisted of measurements of both variables for $n = 167$ college students. Computer output for this example is as follows:

```
The regression equation is
HandSpan = -3.00 + 0.351 Height

Predictor      Coef      SE Coef       T        P
Constant     -3.002       1.694     -1.77    0.078
Height        0.35057     0.02484   14.11    0.000

S = 1.301     R-Sq = 54.7%
```

   **a.** What is the value of the standard deviation from the regression line? Write a sentence that interprets this value.

b. What is the value of $R^2$ given in the output? Write a sentence that interprets this value.

**14.9** ● Find $r^2$ in each of the following situations:
   **a.** SSTO = 500, SSE = 300.
   **b.** SSTO = 200, SSE = 40.
   **c.** SSTO = 80, SSE = 0.
   **d.** SSTO = 100, SSE = 95.

**14.10** ● Example 14.5 (p. 608) gave the regression equation $y = 577 - 3.01x$ for the relationship between $\hat{y}$ = maximum distance at which a driver can read a highway sign and $x$ = driver age.
   **a.** What is the predicted value of the maximum sign-reading distance for a driver who is 21 years old?
   **b.** Compute the residual for a 21-year-old driver who can read the sign at a maximum distance of 525 feet.
   **c.** The standard deviation from the regression line in this example is approximately $s = 50$ feet. Use the Empirical Rule to give an interval that describes the maximum sign-reading distances for about 95% of all drivers who are 21 years old.
   **d.** Would it be unusual for a 21-year-old driver to be able to read the sign from 650 feet? Justify your answer.

**14.11** Refer to Exercise 14.1 about April temperatures and geographic latitudes for U.S. cities. Minitab output for the linear regression is as follows:

```
The regression equation is
AprTemp = 119 − 1.64 latitude

Predictor      Coef      SE Coef        T          P
Constant    118.776      4.467      26.59      0.000
latitude     −1.6436     0.1165     −14.11     0.000

S = 2.837      R-Sq = 91.7%
```

   **a.** In the output, what value is given for the standard deviation from the regression line? Write a sentence that interprets this value.
   **b.** In the output, the value of $R^2$ is given as 91.7%. Write a sentence that interprets this value.

**14.12** Sketch sample data for which the standard deviation from the regression line is $s = 0$.

**14.13** The least-squares regression line is $\hat{y} = 9 + 2x$ for these six observations of $x$ and $y$:

| x | 1 | 1 | 3 | 3 | 5 | 5 |
|---|---|---|---|---|---|---|
| y | 10 | 12 | 13 | 17 | 17 | 21 |

   **a.** For each observation, calculate the residual. Verify that the sum of the residuals is 0.
   **b.** Compute SSE, the sum of squared errors (residuals).
   **c.** For the sample, calculate the standard deviation from the regression line.

## Section 14.3

**14.14** ● ◆ The **UCDavis1** dataset on the CD for this book includes grade point averages and self-reported hours per week of watching television for students in a statistics class. In a simple linear regression for the relationship between $y$ = grade point average and $x$ = hours per week of television watching, the $p$-value is .552 for testing $H_0: \beta_1 = 0$ versus $H_a: \beta_1 \neq 0$.
   **a.** Is the observed relationship statistically significant or not? Explain.
   **b.** What would be the $p$-value for testing the null hypothesis that correlation between the variables is 0 in the population represented by the sample? Assume a two-sided alternative hypothesis.

**14.15** ● For men over the age of 40, a linear regression is done to examine the relationship between $y$ = systolic blood pressure and $x$ = age. A 95% confidence interval for the population slope $\beta_1$ is 0.3 to 0.7. In the context of this situation, write a sentence that interprets this confidence interval.

**14.16** ● Refer to the output given in Exercise 14.11 for the simple linear regression with $y$ = average April temperature and $x$ = geographic latitude for 20 U.S. cities.
   **a.** Write null and alternative hypotheses for testing whether there is a linear relationship between mean April temperature and latitude. Use proper statistical notation.
   **b.** Use information given in the output to test the hypotheses written in part (a). State a conclusion and give a reason for the conclusion.
   **c.** Using values given in the output, show how to calculate the $t$-statistic for testing the hypotheses stated in part (a).

**14.17** ◆ This output is for a linear regression relating $y$ = GPA and $x$ = hours per week spent using the computer for $n = 162$ college students. (*Note:* The data are in the **UCDavis1** dataset on the CD for this book.)

```
The regression equation is
GPA = 3.01 − 0.00555 computer

162 cases used 11 cases contain missing values

Predictor      Coef       SE Coef       T         P
Constant     3.00588      0.07335     40.98     0.000
computer    −0.005548     0.003979    −1.39     0.165
```

   **a.** Is there a statistically significant relationship between the two variables? Justify your answer using information from the output. Be specific about the null and alternative hypotheses that you are evaluating.
   **b.** Suppose that $\rho$ denotes the correlation between the two variables in the population represented by the sample. What is the $p$-value for testing $H_0: \rho = 0$ versus $H_a: \rho \neq 0$?

**14.18** The relationship between $y =$ hours of watching television in a typical day and $x =$ age was examined in Example 5.13. The data were gathered in the 2002 General Social Survey done by the National Opinion Research Center at the University of Chicago, and there were $n = 898$ observations. The correlation between the variables is $r = 0.15$, a value that is indicative of a weak connection between hours of watching television and age.

```
The regression equation is
tvhours = 2.08 + 0.0196 age

Predictor        Coef       SE Coef        T        P
Constant       2.0790       0.2177       9.55    0.000
age            0.019636     0.004400     4.46    0.000

S = 2.34365    R-Sq = 2.2%    R-Sq(adj) = 2.1%
```

a. What is the value of the slope of the sample regression equation? Write a sentence that interprets this value in the context of this situation.
b. Using proper statistical notation, write null and alternative hypotheses for testing the statistical significance of the relationship between the two variables.
c. Use information in the output to determine if there is a statistically significant relationship between age and daily hours of watching television. Include relevant information in your answer.
d. Explain whether or not you think the observed relationship has practical significance.

**14.19** In Example 14.3 (p. 606), a sample of $n = 43$ college men was used to determine the regression equation, Weight $= -318 + 7$(Height). Weights were measured in pounds and heights were measured in inches.
a. The slope of the line is 7. Explain whether the proper notation for this slope is $b_1$ or $\beta_1$.
b. The standard error for the slope was given as 1.581 in Figure 14.4. Calculate an approximate 95% confidence interval for the slope in the population. Write a sentence that interprets this interval.
c. It has been stated that for each 1-inch increase in height, average weight increases by 5 pounds. Explain whether the interval computed in part (b) supports this statement or not.

**14.20** A sample of 837 ninth-grade girls was asked how much time they spent watching music videos each week. The girls also were asked to rate how concerned they were about their weight on a scale of 0–100 (100 = extremely concerned) as well as to rate their perception of how important appearance is on a scale of 1–6 (6 = very important). The researchers (Borzekowski et al., 2000) wrote the following about the results: "When media use was separated into distinct media genres, only hours of watching music videos was related to perceived importance of appearance and weight concerns ($r = .12$, $p < 0.001$, and $r = .08$, $p < 0.05$, respectively)."

a. What are the null and alternative hypotheses for which $p$-values are given in the researchers' statement? What conclusions have the researchers made about these hypotheses?
b. Do the correlation values given by the researchers indicate weak or strong relationships between the variables? Support your answer with a sketch illustrating the appearance of a scatterplot for the given correlation values.

## Section 14.4

**14.21** ● For each part of this question, explain whether a confidence interval for the mean or a prediction interval for a value of $y$ should be used.
a. Estimate the first-year college GPA for a student whose high school GPA was 3.5.
b. Estimate the mean first-year college GPA for students whose high school GPA was 3.5.
c. Estimate the average income for individuals who have 20 years of experience within a particular occupation.
d. Estimate the adult height of a daughter of a woman who is 64 inches tall.

**14.22** ● Suppose a regression model is used to analyze the relationship between $y =$ grade point average and $x =$ number of classes missed in a typical week for college students.
a. In the context of this situation, explain what would be predicted by a prediction interval for $y$ when $x = 2$ classes missed per week.
b. In the context of this situation, explain what would be estimated by a confidence interval for $E(Y)$ when $x = 2$ classes missed per week.

**14.23** ● A college finds that among students who had a total SAT score of approximately 1200, about 90% have a first-year GPA in the range 2.7 to 3.7. Explain whether this interval is a prediction interval or a confidence interval for the mean.

**14.24** ● ◆ Forty students measure their resting pulse rates, then march in place for one minute and measure their pulse rates after the marching. The regression line for the sample is $\hat{y} = 17.79 + 0.894x$, where $y =$ pulse after marching and $x =$ pulse before marching. The following Minitab output gives a confidence interval and a prediction interval for pulse after marching when pulse before marching is 70. (*Data source:* **pulsemarch** dataset on the CD for this book.)

```
Fit      SE Fit      95.0% CI          95.0% PI
80.35    1.15      (78.02, 82.67)    (65.50, 95.19)
```

a. Write the 95% confidence interval for $E(Y)$. In the context of this situation, write a sentence that interprets this interval.
b. Write the 95% prediction interval for $y$. In the context of this situation, write a sentence that interprets this interval.

**14.25** Example 2.17 (p. 49) described the heights of a sample of married British women. The ages of the women and their husbands were available in the same dataset for $n = 170$ couples. For the sample, the linear regression line relating $y$ = husband's age and $x$ = wife's age is $\hat{y} = 3.59 + 0.967x$. The following results computed with Minitab give a confidence interval and prediction interval for husband's age when wife's age is 40:

| Fit | SE Fit | 95.0% CI | 95.0% PI |
|-----|--------|----------|----------|
| 42.258 | 0.313 | (41.641, 42.875) | (34.200, 50.316) |

a. Verify that the "Fit" is consistent with the predicted value that would be given by the regression equation.
b. Interpret the "95% CI" given by Minitab. Be specific about what the interval estimates.
c. Interpret the "95% PI" given by Minitab. Be specific about what the interval estimates.
d. Explain why the "95% PI" is much wider than the "95% CI."

**14.26** Refer to Exercise 14.25. Using the general format for a 95% confidence interval, verify the confidence interval for the mean given by Minitab. Notice that the standard error of the "Fit" is given.

**14.27** Suppose a linear regression analysis of the relationship between $y$ = systolic blood pressure and $x$ = age is done for women between 40 and 60 years old. For women who are 45 years old, a 90% confidence interval for $E(Y)$ is determined to be 128.2 to 131.3. Explain why it is incorrect to conclude that about 90% of women who are 45 years old have a systolic blood pressure in this range. Write a sentence that correctly interprets the interval.

## Section 14.5

**14.28** ● There are five conditions listed at the beginning of Section 14.5 that should be at least approximately true for linear regression. Which of the conditions can be checked by using each of the following methods? In each case, list all of them that can be checked.
a. Drawing a histogram of the residuals.
b. Drawing a scatterplot of the residuals versus the $x$ values.
c. Learning how the data were collected.
d. Drawing a scatterplot of the raw data, $y$ versus $x$.

**14.29** ● ◆ Refer to Exercise 14.24 about a linear regression for $y$ = pulse after marching in place and $x$ = pulse before marching in place. The figure for this exercise is a plot of residuals versus the pulse before marching for a sample of 40 students. Discuss what the plot indicates about Conditions 1, 2, and 3 for linear regression listed at the beginning of Section 14.5.

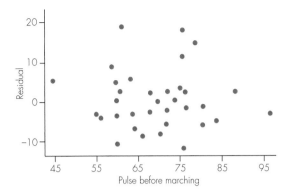

**Figure for Exercise 14.29**

**14.30** ● ◆ The figure for this exercise is a histogram of the residuals for a linear regression relating $y$ = height (in.) and $x$ = foot length (cm) for a sample of college men. Discuss what the histogram indicates about Conditions 2 and 4 for linear regression listed at the beginning of Section 14.5. (*Data source:* **heightfoot** dataset on the CD for this book.)

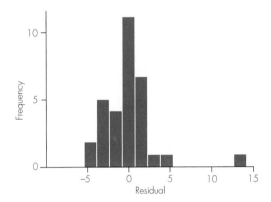

**14.31** The figure accompanying this exercise is a histogram of the residuals for a simple linear regression. What does this plot indicate about the necessary conditions for doing a linear regression? Be specific about which of the five necessary conditions are verified in this figure.

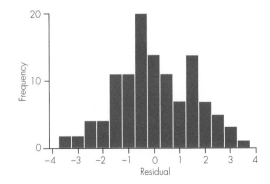

14.32 Plot residuals versus $x$ for the data given in Exercise 14.13.

**14.33** ◆ Observed data along with the sample regression line for the relationship between body weight (lb.) and neck girth (in.) in 19 female bears of various ages are shown in the figure below. (*Note:* The data are in the dataset **bears-female** on the CD for this book.)

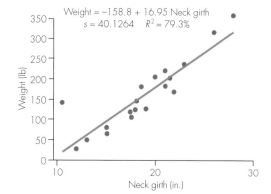

a. Which of the necessary conditions for linear regression appears to be violated in this dataset?
b. What corrective actions would you consider in order to properly estimate the relationship between body weight and neck girth?
c. Sketch a histogram that illustrates the pattern of the distribution of the residuals for this problem. Your sketch does not have to be accurate in the numerical details but should correctly show the pattern.

**14.34** The figure for this exercise shows data for the relationship between the average stopping distance (ft) of a car when the brakes are applied and vehicle speed (mph). The regression line for these data is also shown on the plot. (*Note:* The raw data were given in Exercise 5.7.)

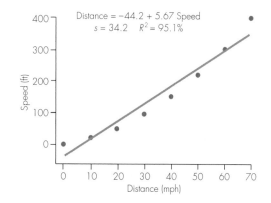

a. Which of the five necessary conditions for linear regression appears to be violated in this situation?
b. What corrective action would you take to correctly estimate the connection between stopping distance and vehicle speed?
c. Make a sketch that illustrates the pattern of a plot of residuals versus speed for the data shown in the figure in this exercise. Your sketch does not need to be numerically accurate but should correctly show the pattern of the plot.

## Chapter Exercises

**14.35** Data for $y$ = hours of sleep the previous day and $x$ = hours of studying the previous day for $n = 116$ college students were shown in Figure 5.14 (p. 168) and described in Example 5.14. Some regression results for those data are as follows:

The regression equation is
Sleep = 7.56 − 0.269 Study

| Predictor | Coef | SE Coef | T | P |
|---|---|---|---|---|
| Constant | 7.5555 | 0.2239 | 33.74 | 0.000 |
| study | −0.26917 | 0.06616 | —— | 0.000 |

S = 1.509    R-Sq = 12.7%    R-Sq(adj) = 11.9%

a. What is the estimated mean decrease in hours of sleep per 1-hour increase in hours of studying? What notation is used for this value?
b. Calculate an approximate 95% confidence interval for $\beta_1$. Write a sentence that interprets this interval.
c. Using proper statistical notation, write null and alternative hypotheses for assessing the statistical significance of the relationship.
d. We omitted from the output the $t$-statistic for testing the hypotheses of part (c). Compute the value of this $t$-statistic using other information shown in the output. What are the degrees of freedom for this $t$-statistic?
e. Is there a statistically significant relationship between hours of sleep and hours of studying? Justify your answer on the basis of information shown in the output.

**14.36** Refer to Exercise 14.35 about hours of sleep and hours of study.
a. What is the value of the standard deviation from the regression line? Write a sentence that interprets this value.
b. Calculate the predicted value of hours of sleep the previous day for a student who studied 4 hours the previous day.
c. Using the Empirical Rule, determine an interval that describes hours of sleep for approximately 95% of students who studied 4 hours the previous day.

d. The value of $R^2$ is given as 12.7%. Write a sentence that interprets this value.

**14.37** Refer to Exercises 14.35 and 14.36 about hours of sleep and hours spent studying. What is the intercept of the regression line? Does this value have a useful interpretation in the context of this problem? If so, what is the interpretation? If not, why not?

**14.38** Regression results for the relationship between $y =$ hours of sleep the previous day and $x =$ hours spent studying the previous day were given in Exercise 14.35. The figure for this exercise is a plot of residuals versus hours spent studying. What does the plot indicate about the necessary conditions for doing a linear regression? Be specific about which of the five conditions given in Section 14.5 are verified by this plot.

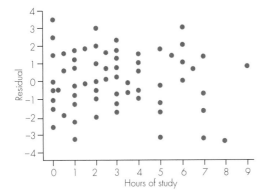

**14.39** Exercise 14.35 gave linear regression results for the relationship between $y =$ hours of sleep the previous day and $x =$ hours spent studying the previous day. Following is Minitab output showing a confidence interval and a prediction interval for hours of sleep when hours of studying = 3 hours:

| Fit | SE Fit | 95.0% CI | 95.0% PI |
|---|---|---|---|
| 6.748 | 0.142 | (6.466, 7.029) | (3.746, 9.750) |

Write down the interval given for predicting the hours of sleep when hours of studying is 3 hours. Give two different interpretations of this interval.

**14.40** Refer to the output given in Exercise 14.8 about hand-span and height. Compute a 90% confidence interval for the population slope. Write a sentence that interprets this interval.

**14.41** The following data are $x =$ average on five quizzes before the midterm exam and $y =$ score on the midterm exam for $n = 11$ students randomly selected from a multiple-section statistics class of about 950 students:

| $x =$ Quizzes | 80 | 68 | 94 | 72 | 74 | 83 | 56 | 68 | 65 | 75 | 88 |
|---|---|---|---|---|---|---|---|---|---|---|---|
| $y =$ Exam | 72 | 71 | 96 | 77 | 82 | 72 | 58 | 83 | 78 | 80 | 92 |

a. Plot the data, and describe the important features of this plot.

b. Using statistical software, calculate the regression line for this sample.

c. What is the predicted midterm exam score for a student with a quiz average equal to 75?

d. With statistical software, determine a 50% prediction interval for the midterm exam score of a student whose quiz average was 75.

**14.42** ◆ This exercise refers to the following Minitab output, relating $y =$ son's height to $x =$ father's height for a sample of $n = 76$ college males. (*Note:* The data are in the dataset **UCDavis1** on the CD for this book.)

```
The regression equation is
Height = 30.0 + 0.576 dadheight

76 cases used 3 cases contain missing values

Predictor        Coef      SE Coef       T         P
Constant       29.981        5.129      5.85    0.000
dadheigh       0.57568      0.07445     7.73    0.000

S = 2.657      R-Sq = 44.7%

Predicted Values [for dad's heights of 65, 70, and 74]

Fit        SE Fit         95.0% CI              95.0% PI
67.400      0.415      (66.574, 68.226)      (62.041, 72.759)
70.279      0.318      (69.645, 70.913)      (64.946, 75.612)
75.581      0.494      (71.596, 73.566)      (67.195, 77.967)
```

a. What is the equation for the regression line?

b. Identify the value of the $t$-statistic for testing whether or not the slope is 0. Verify that the value is correct using the formula for the $t$-statistic and the information provided by Minitab for the parts that go into the formula.

c. State and test the hypotheses about whether or not the population slope is 0. Use relevant information provided in the output.

d. Compute a 95% confidence interval for $\beta_1$, the slope of the relationship in the population. Write a sentence that interprets this interval.

**14.43** ◆ Refer to Exercise 14.42.

a. What is the value of $R^2$ for the observed linear relationship between height and father's height? Write a sentence that interprets this value.

b. What is the value of the correlation coefficient $r$?

**14.44** ◆ Refer to Exercises 14.42 and 14.43. The output provides prediction intervals and confidence intervals for father's heights of 65, 70, and 74 inches.

a. Verify that the "Fit" given by Minitab for father's height of 65 inches is consistent with the predicted height that would be given by the regression equation.

b. Write down the interval Minitab provided for predicting an individual son's height if his father's height is 70 inches. Provide two different interpretations for the interval.

c. Write a sentence that interprets the "95% CI" given for a son's height when the father's height is 74 inches.

d. Explain why the prediction interval is much wider than the corresponding confidence interval for each father's height provided.

e. The accompanying figure is a plot of the residuals versus father's height. Which of the five necessary conditions for regression listed in Section 14.5 appears to be violated? What corrective actions should you consider?

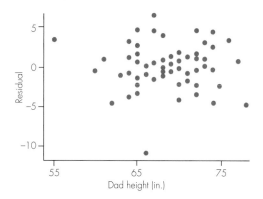
Dad height (in.)

14.45 The five steps for hypothesis testing were given in Chapters 12 and 13. Describe those steps as they apply to testing whether there is a relationship between two variables in the simple regression model.

14.46 Explain why rejecting $H_0: \beta_1 = 0$ in a simple linear regression model does not prove that the relationship is linear. To answer this question, you might find it helpful to consider the figure in Exercise 14.34, which shows stopping distance and vehicle speed for automobiles.

## Dataset Exercises

**Statistics⊜Now**™ Datasets **are required** to solve these exercises and can be found at **http://1pass.thomson.com** or on your CD.

14.47 Use the dataset **letters** for this exercise. A sample of 63 students wrote as many letters of the alphabet in order, as capital letters, as they could in 15 seconds using their dominant hand and then repeated this task using their nondominant hand. The variables **dom** and **nondom** contain the raw data for the results.

a. Plot $y = $ **dom** versus $x = $ **nondom.** Describe the important features of the plot.

b. Compute the simple regression equation for the relationship. What is the equation?

c. What are the values of the standard deviation from the regression line and $R^2$? Interpret these values in the context of this problem.

d. Determine a 95% confidence interval for the population slope. Write a sentence that interprets this interval.

e. Consider the statement, "On average, a student can write about 23 more letters in 15 seconds with the dominant hand than with the nondominant hand." What regression equation would accompany this statement? Based on your answers to parts (b) and (d), explain why this statement is reasonable.

14.48 Refer to the previous exercise about letters written with the dominant ($y$) and nondominant ($x$) hands.

a. Plot residuals versus $x = $ **nondom.** What does this plot indicate about conditions for using the linear regression model?

b. Create a histogram of the residuals. What does this plot indicate about the conditions for using the linear regression model?

**14.49** Use the **bears-female** dataset for this exercise. Weights (lb) and chest girths (in.) are given for $n = 19$ female wild bears. The corresponding variable names are **Weight** and **Chest.**

**a.** Plot $y = $ **Weight** versus $x = $ **Chest.** Describe the important features of the plot.

**b.** Compute a simple linear regression equation for $y = $ **Weight** and $x = $ **Chest.**

**c.** What is the value of $R^2$ for this relationship? Write a sentence interpreting this value.

**d.** What is the predicted weight for a bear with a chest girth of 40 inches?

**e.** Compute a 95% prediction interval for the weight of a bear with a chest girth of 40 inches. Write a sentence interpreting this interval.

**f.** Compute a 95% confidence interval for the mean weight of bears with a chest girth of 40 inches. Write a sentence interpreting this interval.

14.50 Use the dataset **heightfoot** for this exercise. Heights (in.) and foot lengths (cm) are given for 33 men.

a. Plot $y = $ height versus $x = $ foot length. What important features are evident in the plot?

b. Omit any outliers evident in the plot in part (a), and compute the linear regression line for predicting $y = $ height from $x = $ foot length.

c. Determine a 90% prediction interval for the height of a man whose foot length is 28 centimeters.

d. Discuss whether height can be accurately predicted from foot length. Use regression results to justify your answer.

e. For the data used for part (b), plot residuals versus $x = $ foot length. What does this plot indicate about the necessary conditions for doing a linear regression?

14.51 Refer to Exercise 14.50 about the relationship between height and foot length.

a. Do not omit any outliers. Use the complete dataset to determine a 90% prediction interval for the height of a man whose foot length is 28 centimeters.

b. Explain why the interval computed in part (a) is wider than the interval computed for part (c) of the previous exercise.

c. For the regression based on all data points, plot residuals versus $x$ = foot length. What does this plot indicate about the necessary conditions for doing a linear regression?

**Statistics⊖Now**™ Preparing for an exam? Assess your progress by taking the post-test at **http://1pass.thomson.com.**

**⟨vMentor** Do you need a live tutor for homework problems? Access vMentor at **http://1pass.thomson.com** for one-on-one tutoring from a statistics expert.

# 15

*John Giustina/Iconica/Getty Images*

### *With whom is it easiest to make friends?*

See Example 15.2 *(p. 637)*

# More About Inference for Categorical Variables

> Are the digits that are drawn in lotteries actually equally likely, as claimed? Are belief and performance on ESP tests related? Do males and females have the same opinions about whether it's easier to make friends with the same sex or the opposite sex? Chi-square tests can be used to answer these kinds of questions about categorical variables.

n Chapter 6, we learned techniques for analyzing a fundamental question that researchers often ask about two categorical variables, which is

> Is there a relationship between the variables so that the chance that an individual falls into a particular category for one variable depends upon the particular category they fall into for the other variable?

For instance, in Example 6.1, we saw that the likelihood of divorce is greater for smokers than it is for nonsmokers. The chi-square test is a procedure for assessing the statistical significance of a relationship between categorical variables. We introduced the basic ideas of this statistical method in Chapter 6; in this chapter, we examine the chi-square test more fully. We will use the chi-square procedure to test hypotheses about two-way tables, and we will also learn how the chi-square procedure is used to test hypotheses about a single categorical variable. ▮

## 15.1 The Chi-Square Test for Two-Way Tables

In Chapter 6, we learned these terms:

- The raw data from a **categorical variable** consist of group or category names that don't necessarily have any logical ordering. Your sex (male or female) and your handedness (left or right), for instance, are both categorical variables.

- A **two-way table** displays the counts of how many individuals fall into each possible combination of categories of two categorical variables. A two-way table may also be called a **contingency table.**

- Each combination of a row and a column is referred to as a **cell** of the contingency table.

- **Row percentages** are the percentages across a row of a contingency table. They are based on the total number of observations in the row.
- **Column percentages** are the percentages down a column of a contingency table. They are based on the total number of observations in the column.

When we consider two categorical variables, a question of interest is whether a relationship that is observed in a sample can be used to infer that there is a relationship in the population(s) from which the sample(s) was drawn. This question can be answered by using a procedure called a chi-square test. Section 6.4 gave a brief introduction to the chi-square test. In this section and the next, we expand that discussion. As usual, five steps are required to assess statistical significance.

## Step 1: The Null and Alternative Hypotheses

The **null** and **alternative hypotheses** about the categorical variables that form a two-way table are as follows:

$H_0$: The two variables are not related.
$H_a$: The two variables are related.

It is commonplace to use the word *associated* in place of the word *related*. Within a specific context, we may be able to write more informative versions of the general statements about the relationship. The next two examples illustrate different variations on how we might express null and alternative hypotheses for a two-way table.

**Example 15.1**

Statistics⌂Now™

Watch a video example at **http://1pass.thomson.com** or on your CD.

Read the original source on your CD.

**Ear Infections and Xylitol Sweetener**  Xylitol is a food sweetener that may also have antibacterial properties. In an experiment conducted in Finland, researchers investigated whether the regular use of chewing gum containing xylitol could reduce the risk of a middle ear infection for children in daycare centers (Uhari et al., 1998). The investigators randomly divided 533 children in daycare centers into three groups. One group regularly chewed gum that contained xylitol, another group regularly took xylitol lozenges, and the third group regularly chewed gum that did not contain xylitol. The experiment lasted for three months, and for each child, the researchers recorded whether the child had an ear infection during that period. (This experiment was also discussed in Example 12.17 (p. 528), where the data concerned xylitol administered in a syrup form rather than in gum.)

Table 15.1 displays the observed counts along with row percentages. The explanatory variable, which is the type of gum or lozenge each child took, is

**Table 15.1** Ear Infections and the Use of Xylitol

| Group | No | Yes | Total |
|---|---|---|---|
| | Ear Infection in Three Months? | | |
| Placebo Gum | 129 (72.5%) | 49 (27.5%) | 178 |
| Xylitol Gum | 150 (83.8%) | 29 (16.2%) | 179 |
| Xylitol Lozenge | 137 (77.8%) | 39 (22.2%) | 176 |
| Total | 416 (78%) | 117 (22%) | 533 |

displayed as rows. The response variable, which is whether the child had an ear infection, is displayed as columns. Notice that only 16.2% of the children in the xylitol gum group got an ear infection during the experiment, compared to 22.2% of the xylitol lozenge group and 27.5% of the placebo gum group.

The purpose of the experiment was to compare the three groups with regard to the proportion of children who experienced an ear infection. For this purpose, there are three relevant population parameters:

$p_1$ = proportion who would get an ear infection in a population given placebo gum

$p_2$ = proportion who would get an ear infection in a population given xylitol gum

$p_3$ = proportion who would get an ear infection in a population given xylitol lozenges

If the chance of falling into the ear infection category is not related to treatment method, the risk of an ear infection would be the same for the three treatments. So null and alternative hypotheses about the parameters are as follows:

$H_0$: $p_1 = p_2 = p_3$    (No relationship between treatment and outcome.)
$H_a$: $p_1$, $p_2$, and $p_3$ are not all the same.    (There is a relationship.)

Notice that the alternative hypothesis simply states that the three proportions are not all the same. A limitation of chi-square tests is that no particular direction for the relationship can be stated in the alternative hypothesis. We will revisit this ear infection study in several examples as we learn how to do a chi-square test. ■

**Example 15.2**

Statistics⊘Now™

Watch a video example at **http://1pass.thomson.com** or on your CD.

**With Whom Do You Find It Easiest to Make Friends?**   Students in a statistics class were asked, "With whom do you find it easiest to make friends?" Possible responses were "opposite sex," "same sex," and "no difference." Students were also categorized as male or female. We would like to determine whether there is a relationship between the sex of the respondent and response. To emphasize the comparison of men and women, we might express the null and alternative hypotheses as follows:

$H_0$: There would be no difference in the distribution of responses of men and women if the populations of them were asked this question. (No relationship between gender and response.)

$H_a$: There would be a difference in the distribution of responses of men and women if the populations of them were asked this question. (There is a relationship between gender and response.)

As is always the case in hypothesis testing, these hypotheses are statements about the larger populations represented by the samples of men and women. Again, notice that the alternative hypothesis does not specify any particular way in which the men and women might differ.

Table 15.2 (see next page) compares the responses of males and females, and Figure 15.1 displays the relevant conditional percentage distributions. We see that there appears to be a relationship. For instance, a decidedly higher percentage of females than males responded "opposite sex." We will continue the analysis of these data in Example 15.8 (p. 644). ■

**Table 15.2** Data For Example 15.2

| | With Whom Is It Easiest to Make Friends? | | | |
| --- | --- | --- | --- | --- |
| | Opposite Sex | Same Sex | No Difference | Total |
| **Females** | 58 (42%) | 16 (12%) | 63 (46%) | 137 |
| **Males** | 15 (22%) | 13 (19%) | 40 (59%) | 68 |
| **Total** | 73 (36%) | 29 (14%) | 103 (50%) | 205 |

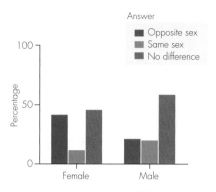

**Figure 15.1** ▌ Conditional distributions by gender for "With whom is it easiest to make friends?"

---

**technical note**  **Homogeneity and Independence**

It is generally sufficient to state the null and alternative hypotheses as $H_0$: No relationship and $H_a$: There is a relationship. There are two specific variations of these general statements, and the method of sampling dictates which variation is appropriate in a particular situation. If samples have been taken from separate populations, the null hypothesis statement is a statement of **homogeneity** (sameness) among the populations. Example 15.1 is such an instance. If a sample has been taken from a single population and two categorical variables are measured for each individual, the statement of no relationship is a statement of **independence** between the two variables. For instance, a null hypothesis may be that hair color and eye color are independent characteristics. (***Note:*** Independence was discussed in Chapter 7.) For sufficiently large samples, the same chi-square test is appropriate for both situations, so we will not be concerned about whether the null hypothesis is about homogeneity or about independence.

---

## Step 2: The Chi-Square Statistic for Two-Way Tables and Necessary Conditions

The basics of computing the chi-square statistic used for two-way tables are as follows.

- The counts in the cells of a two-way table of the sample data are called the **observed counts.**
- The chi-square statistic measures the difference between the observed counts and corresponding **expected counts.** The *expected counts* are hypothetical counts that would occur if the null hypothesis were true. The expected counts can be calculated as

$$\text{Expected} = \frac{\text{Row total} \times \text{Column total}}{\text{Total } n}$$

- The **chi-square statistic** is

$$\chi^2 = \sum_{\text{all cells}} \frac{(\text{Observed} - \text{Expected})^2}{\text{Expected}}$$

Scanning to the right, we see that 6.69 falls between 5.99 and 7.38. The corresponding column headings are .05 and .02. This means that the $p$-value is less than .05 but greater than .02. ∎

**Example 15.6**

**A Moderate $p$-Value** Suppose that a two-way table has three rows and three columns and the chi-square statistic is $\chi^2 = 8.12$. The degrees of freedom are df $= (3 - 1)(3 - 1) = 4$. To find the $p$-value, we scan the df $= 4$ row in Table A.5 and learn that 8.12 is between the entries 7.78 ($p = .10$) and 8.50 ($p = .075$). Thus, the $p$-value is between .075 and .10. We would write this as .075 $< p$-value $< .10$. Often, it is written simply as .075 $< p < .10$. In Excel, the command CHIDIST (8.12,4) tells us that the $p$-value is .087. ∎

**Example 15.7**

**A Tiny $p$-Value** Suppose that a two-way table has four rows and two columns and the chi-square statistic is $\chi^2 = 17.67$. The degrees of freedom are $(4 - 1)(2 - 1) = 3$. In Table A.5, we see that 17.67 is larger than the rightmost entry in the df $= 3$ row, so the $p$-value is less than .001 ($p < .001$). In Excel, the command CHIDIST (17.67,3) tells us that the $p$-value is .00051. ∎

> **thought question 15.1** Consider Example 15.2 (pp. 637–638) about gender and the question about with whom it's easiest to make friends. What are the degrees of freedom for the chi-square statistic for these data? To be statistically significant at the .05 level, how large would the calculated chi-square have to be?*

## Steps 4 and 5: Making a Decision and Reporting a Conclusion

An approximate $p$-value, or even a range for the $p$-value, is enough information to make a decision about the null hypothesis. If we use the .05 significance level, we simply have to determine whether the $p$-value is less than .05. In Example 15.6, for instance, we learned from Table A.5 that the $p$-value is between .075 and .10. Thus, we know that it is not less than .05, and we would fail to reject the null hypothesis.

The value in the .05 column of Table A.5 is referred to as the **critical value** for the .05 significance level. The region of chi-square values greater than the critical value is called the **rejection region.** The $p$-value will be less than .05 when the chi-square statistic is greater than this critical value, in other words, if the chi-square statistic is in the rejection region. Consequently, if the level of significance is $\alpha = .05$, two equivalent rules for declaring statistical significance based on a chi-square statistic are

- Reject the null hypothesis when the $p$-value is less than .05.

- Reject the null hypothesis when the chi-square statistic is greater than the entry in the .05 column of Table A.5 (the critical value), in other words, when the chi-square statistic is in the rejection region.

*HINT: You should use Table A.5 to determine this.

### Reporting a Conclusion in Context

Remember that the two possible conclusions for hypothesis tests are "Do not reject the null hypothesis" or "Reject the null hypothesis." The latter conclusion is equivalent to "accept the alternative hypothesis" or "conclude that the results are statistically significant."

In chi-square tests for two-way tables, it is important to write these conclusions in context. Suppose that we were testing whether there is a relationship between smoking (yes or no) and drinking alcohol (never, occasionally, often). Here are various ways to write the possible conclusions.

*Ways to write the conclusion "do not reject the null hypothesis"*

- The relationship between smoking and drinking alcohol is not statistically significant.

- The proportions of smokers who never drink, drink occasionally, and drink often are not significantly different from the proportions of nonsmokers who do so.

- There is insufficient evidence to conclude that there is a relationship in the population between smoking and drinking alcohol.

*Ways to write the conclusion "reject the null hypothesis"*

- There is a statistically significant relationship between smoking and drinking alcohol.

- The proportions of smokers in the population who never drink, drink occasionally, and drink often are not the same as the proportions of nonsmokers who do so.

- Smokers have significantly different drinking behavior than nonsmokers.

**Example 15.8**

**Making Friends**   Figure 15.4 shows partial Minitab output for a chi-square test of the relationship between sex and response to the question "With whom is it easiest to make friends?" (opposite sex, same sex, no difference). The data were first given in Table 15.2 (p. 638) as part of Example 15.2. The null hypothesis is that the response to the question is not related to the sex of the respondent. The alternative hypothesis is that these two variables are related.

In the last line of output, we see that the chi-square statistic is 8.515. Although Minitab provides the $p$-value, this value has been omitted to provide practice using Table A.5, the chi-square distribution table. The contingency table has two rows and three columns, so df $= (2 - 1)(3 - 1) = 2$. In the df $=$

Expected counts are printed below observed counts

|        | Opposite | Same  | No Diff | Total |
|--------|----------|-------|---------|-------|
| Female | 58       | 16    | 63      | 137   |
|        | 48.79    | 19.38 | 68.83   |       |
| Male   | 15       | 13    | 40      | 68    |
|        | 24.21    | 9.62  | 34.17   |       |
| Total  | 73       | 29    | 103     | 205   |

Chi-Sq = 1.740 + 0.590 + 0.494 +
         3.507 + 1.188 + 0.996 = 8.515

**Figure 15.4** ▮ Minitab output for Example 15.8

2 row of Table A.5, the computed chi-square statistic falls between the entries in the .025 column (7.38) and the .01 column (9.21). The $p$-value is between .01 and .025, so the result is statistically significant if we use the standard .05 significance level.

Equivalently, we can compare the value of the chi-square statistic to a critical value from Table A.5. Using the df = 2 row and the .05 column of Table A.5, we find that the critical value is 5.99. Reject the null hypothesis because the computed chi-square statistic of 8.515 is greater than the critical value of 5.99.

In the context of the situation, the conclusion is that there is a relationship between sex and response to the question asked. This conclusion applies to the larger population represented by the sample. Equivalently, conclude that there would be a difference in the distribution of responses of men and women if the populations of them were asked this question. Using the bar chart of row percentages shown in Figure 15.1 (p. 638), we can describe the nature of the difference between the females and males. Compared to men, women are more likely to say "opposite sex" and less likely to say "same sex." These conclusions are about the populations of men and women represented by the sample data. ■

## Supporting Analyses

We don't learn anything about the specific nature of a relationship from a chi-square test. The results of a statistically significant chi-square test should be accompanied by some other descriptive or inferential statistics that elaborate on the connection between the variables. Some or all of the following procedures may be useful for this purpose:

- Description of row (or column) percentages.
- Bar chart of counts or percentages.
- Examination of $\dfrac{(\text{Observed} - \text{Expected})^2}{\text{Expected}}$ within each cell. These values are called "contributions to chi-square." The cells with the largest values have contributed the most to the statistical significance of the relationship, so they deserve attention in any description of the relationship.
- Confidence intervals for important proportions or for differences between proportions. For instance, confidence intervals for the proportions getting an ear infection in the three treatment groups of the xylitol experiment would be informative.

15.1 Exercises are on page 658.

---

*in summary*      ## The Five Steps in a Chi-Square Test of a Relationship Between Two Variables

**Step 1:**  *Determine the null and alternative hypotheses.*

*Null hypothesis*          $H_0$: The two variables are not related.
*Alternative hypothesis:*  $H_a$: The two variables are related.

(*continued*)

**Step 2:** *Verify necessary data conditions, and if they are met, summarize the data into an appropriate test statistic.* If at least 80% of the expected counts are greater than 5 and none are less than 1, compute

$$\chi^2 = \sum_{\text{all cells}} \frac{(\text{Observed} - \text{Expected})^2}{\text{Expected}}$$

The *expected count* for a cell is $\dfrac{\text{Row total} \times \text{Column total}}{\text{Total } n}$.

**Step 3:** *Assuming that the null hypothesis is true, find the p-value.* Using the $\chi^2$ distribution with df $= (r - 1)(c - 1)$, where $r$ = number of rows and $c$ = number of columns in the two-way table, the *p*-value is the area in the tail to the right of the calculated test statistic $\chi^2$.

**Step 4:** *Decide whether or not the result is statistically significant based on the p-value.* Choose a significance level ("alpha"); the standard is $\alpha = .05$. The result is statistically significant if the *p*-value $\leq \alpha$.

**Step 5:** *Report the conclusion in the context of the situation.*

---

**MINITAB *tip*** | **Computing a Chi-Square Test for a Two-Way Table**

- If the raw data are stored in columns of the worksheet, use **Stat>Tables> Cross Tabulation and Chi-Square.** In the dialog box, specify a variable in the "For rows" box and a variable in the "For columns" box. Click **Chi-Square,** and select "Chi-Square analysis" along with any other desired options such as "Expected cell counts."

- If the data are already summarized into counts, enter the table of counts into columns of the worksheet, and then use **Stat>Tables>Chi-Square Test (Table in Worksheet).** In the dialog box, specify the worksheet columns that contain the counts.

# 15.2 Analyzing 2 × 2 Tables

There are a number of special circumstances that apply to analyzing $2 \times 2$ tables. First, there is a handy computational shortcut for computing the chi-square statistic for a $2 \times 2$ table that does not require computing expected counts. It can be accomplished with a calculator, without needing to write down any intermediate results.

### Shortcut Formula for a 2 × 2 Table

Suppose we label a $2 \times 2$ table as follows:

|        | Column 1 | Column 2 | Total |
|--------|----------|----------|-------|
| **Row 1** | $A$ | $B$ | $R_1$ |
| **Row 2** | $C$ | $D$ | $R_2$ |
| **Total** | $C_1$ | $C_2$ | $N$ |

The first modification of the chi-square procedure necessary for the goodness-of-fit situation involves the calculation of the expected counts. This calculation is straightforward and intuitively obvious. Suppose we hypothesize, for example, that 45% of a population has brown eyes, 35% has blue eyes, and 20% has an eye color other than brown or blue. If we randomly select $n =$ 200 people from this population, it probably makes sense to you that we would expect about 45% of 200 = $.45 \times 200 = 90$ people to have brown eyes if our hypothesis is correct. We also would expect about $.35 \times 200 = 70$ people to have blue eyes and $.20 \times 200 = 40$ people to have an eye color other than brown or blue.

The other modification of the chi-square procedure concerns the degrees of freedom. For a goodness-of-fit test, the degrees of freedom are computed as df $= k - 1$, where $k$ is the number of categories for the variable of interest. In the eye color example of the previous paragraph, there were three eye color categories. Consequently, the degrees of freedom for a chi-square goodness-of-fit test would be df $= 3 - 1 = 2$.

*thought question* **15.4**  Remember that the "degrees of freedom" for the chi-square test for a two-way table represents the largest number of cells for which you were "free" to find expected counts. The remaining expected counts were determined because the row and column totals had to be the same as they were for the observed counts. Explain how the same principle applies in specifying the degrees of freedom for a chi-square goodness-of-fit test, which are $k - 1$ when there are $k$ categories. *

*in summary*    ## The Five Steps in a Chi-Square Goodness-of-Fit Test

**Step 1:** *Determine the null and alternative hypotheses.*

> *Null hypothesis:*    $H_0$: The probabilities for the $k$ categories of a categorical variable are given by $p_1, p_2, \ldots, p_k$.
>
> *Alternative hypothesis:*    $H_a$: Not all probabilities specified in $H_0$ are correct.

***Note:*** The probabilities specified in the null hypothesis must sum to 1.

**Step 2:** *Verify necessary data conditions, and if they are met, summarize the data into an appropriate test statistic.* If at least 80% of the expected counts are greater than 5 and none are less than 1, compute

$$\chi^2 = \sum_{\text{all categories}} \frac{(\text{Observed} - \text{Expected})^2}{\text{Expected}}$$

where the expected count for the $i$th category is computed as $np_i$.

(*continued*)

---

**\*HINT:** The total expected count must equal the total observed count. If you know $k - 1$ expected counts, what must the remaining one be?

**Step 3:** *Assuming that the null hypothesis is true, find the p-value.* Using the $\chi^2$ distribution with df $= k - 1$, the *p*-value is the area to the right of the test statistic $\chi^2$.

**Step 4:** *Decide whether or not the result is statistically significant based on the p-value.* Choose a significance level ("alpha"); the standard is $\alpha = .05$. The result is statistically significant if the *p*-value $\le \alpha$. Equivalently, the result is statistically significant if the test statistic $\chi^2$ is at least as large as the value in the $\alpha$ column and df $= k - 1$ row of Table A.5.

**Step 5:** *Report the conclusion in the context of the situation.*

**Example 15.13**

Statistics⬭Now™

Watch a video example at **http://1pass.thomson.com** or on your CD.

For software help, download your Minitab, Excel, TI-83, SPSS, R, and JMP manuals from **http://1pass.thomson.com**, or find them on your CD.

### The Pennsylvania Daily Number

The Pennsylvania Daily Number is a state lottery game in which the state constructs a three-digit number by drawing a digit between 0 and 9 from each of three different containers. If the digits that were drawn, in order, were 3, 6, and 3, for example, then the daily number would be 363. In this example, we focus only on draws from the first container. If numbers are randomly selected, each value between 0 and 9 would be equally likely to occur. This leads to the following null hypothesis:

$$H_0: p = \frac{1}{10} \text{ for each of the 10 possible digits in first container}$$

Simply stated, the alternative hypothesis is that the null hypothesis is false. In this setting, that would mean that the probability of selection for some digits is different from 1/10.

Figure 15.6 shows the frequency distribution of draws from the first container for the $n = 500$ days between July 19, 1999, and November 29, 2000. The observed data were downloaded from the Pennsylvania lottery website. The expected count for each of the 10 possible outcomes is $1/10 \times 500 = 50$, and

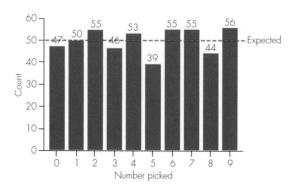

**Figure 15.6** ▌ Observed and expected counts for first digit drawn in the Pennsylvania Daily Number Game, July 19, 1999, through November 29, 2000

the red dashed line in the figure marks this value. The greatest difference between observed and expected occurs for the number 5, which was drawn from the first container only 39 times over these 500 days.

The chi-square goodness-of-fit statistic for these data is

$$\chi^2 = \sum_{\text{categories}} \frac{(\text{Observed} - \text{Expected})^2}{\text{Expected}}$$

$$= \frac{(47-50)^2}{50} + \frac{(50-50)^2}{50} + \frac{(55-50)^2}{50} + \frac{(46-50)^2}{50} + \frac{(53-50)^2}{50}$$

$$+ \frac{(39-50)^2}{50} + \frac{(55-50)^2}{50} + \frac{(55-50)^2}{50} + \frac{(44-50)^2}{50} + \frac{(56-50)^2}{50} = 6.04$$

There are 10 possible outcomes for the number drawn, so df $= 10 - 1 = 9$. To find the $p$-value, we find the probability that a chi-square statistic would be as large as or larger than 6.04 using a chi-square distribution with df $= 9$. From Table A.5, the chi-square distribution table, we learn that the $p$-value is greater than .50 because 6.04 is smaller than (to the left of) the entry of 8.34 under .50 in the row for df $= 9$. The Excel command CHIDIST(6.04,9) provides the information that the $p$-value is .736. This $p$-value is illustrated in Figure 15.7. The result is not statistically significant; the null hypothesis is not rejected.

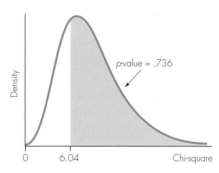

**Figure 15.7** ∎ $p$-Value for Example 15.13 is area to right of 6.04 in chi-square distribution with df $= 9$

---

*technical note* | **Mathematical Notation for Chi-Square Goodness-of-Fit Statistic**

The following notation can be used for the chi-square goodness-of-fit test:

$O_i$ = observed count in category $i$

$p_i$ = hypothesized probability for category $i$

$E_i = np_i$ = expected count in category $i$

$$\chi^2 = \sum_i \frac{(O_i - E_i)^2}{E_i}$$

$k$ = number of categories; df $= k - 1$

## Sample Size and the Goodness-of-Fit Test

The null hypothesis often is the desired hypothesis for the investigators when they do a goodness-of-fit test. In Example 15.13 about the Pennsylvania Daily Number, lottery officials would hope to "prove" that all numbers have an equal chance to be selected. This creates a technical and logical difficulty because with a small sample, a null hypothesis might be retained simply because there are not enough data to reject it. This was the basis for the advice given in Chapter 12 to use the phrase "cannot reject the null" rather than "accept the null." If the goal is to not reject the null hypothesis, the most scientifically valid approach is to use as large a sample as possible. This prevents the possible criticism that the null hypothesis was retained only because the sample was not large enough to provide conclusive evidence. On the other hand, the risk of a large sample size is that small, relatively unimportant departures from the null hypothesis can achieve statistical significance. It is often useful to include confidence intervals for the parameters if possible, to determine the magnitude of departures from the null hypothesis.

15.3 Exercises are on page 662.

---

| MINITAB *tip* | **Calculating the Chi-Square Statistic for Goodness-of-Fit** |
|---|---|

Minitab added a chi-square goodness-of-fit procedure in Version 14.2. The menu sequence is **Stat>Tables>Chi-Square Goodness-of-Fit.** Users of earlier versions can use the following steps:

- Enter the observed counts into a column. Suppose that it is C1.
- Enter the null hypothesis proportions into a second column. Suppose it is C2.
- Use **Cal>Calculator.** Create a column of expected counts by multiplying C2 by the total sample size. Suppose it is C3.
- Again use **Calc>Calculator.** Create (by computation) a column (say, C4) that contains a chi-square value. Given the column designations of the previous steps, the expression is SUM((C1−C3)**2/C3). Look carefully at the parenthesis locations. You can also compute (C1−C3)**2/C3 and store the results in C4, then sum C4. That will allow you to determine which categories are contributing the most to the value of the test statistic.

## case study 15.1    Do You Mind If I Eat the Blue Ones?

You'll finds six different colors of M&Ms in a bag of milk chocolate M&Ms: brown, red, yellow, blue, orange, and green. If you're like some people, you may have a color preference (or dislike) when it comes to eating M&Ms and might have fleeting thoughts about what the color distribution is or should be. At the M&Ms website, the distribution of colors for milk chocolate M&Ms was given as follows in 2000: brown 30%, red 20%, yellow 20%, blue 10%, orange 10%, and green 10%. (The color distribution has been changed since then.)

Students in a University of California at Davis statistics class (fall 2000 term) counted the frequency of each color among 6918 milk chocolate M&Ms from about 130 small (1.69-ounce) bags. All bags were purchased at the same store in California. The observed distribution of colors in this sample was as follows:

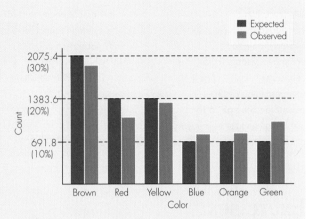

**Figure 15.8** ▌ Observed and expected counts for a sample of 6918 plain M&Ms

|  | **Color** | | | | | |
|---|---|---|---|---|---|---|
|  | **Brown** | **Red** | **Yellow** | **Blue** | **Orange** | **Green** |
| **Count** | 1911 | 1072 | 1308 | 804 | 821 | 1002 |
| **Percent** | 27.6% | 15.5% | 18.9% | 11.6% | 11.9% | 14.5% |
| **Expected %** | 30% | 20% | 20% | 10% | 10% | 10% |

Generally, the observed percentages in the UC Davis data are not far from the percentages advertised by the manufacturer, but some differences are evident. Most notably (especially for those who like red M&Ms), there were fewer red and more green candies than expected.

We can use a chi-square goodness-of-fit test to determine whether the difference between the expected and observed distributions is statistically significant. If we assign numbers to colors using the order in which the colors are listed above (1 = brown, 2 = red, and so on), the null hypothesis of interest is

$$H_0: p_1 = .3, \quad p_2 = .2, \quad p_3 = .2, \quad p_4 = .1, \quad p_5 = .1, \quad p_6 = .1$$

The first step in computing the chi-square statistic is to calculate the expected counts for each color. To do so, multiply the null hypothesis proportion for a color by the total sample size, $n = 6918$. For example, the expected count for brown is $.3 \times 6918 = 2075.4$. Figure 15.8 displays expected counts and observed counts for each color. As we noted before, the largest differences between observed and expected counts occur for red and green.

The chi-square goodness-of-fit statistic for these data is $\chi^2 = 268.75$, with degrees of freedom = number of categories − 1 = 6 − 1 = 5. In

Table A.5, the entry under .001 for df = 5 is 20.51. The computed chi-square statistic, 268.75, is so much larger than 20.51 that it is reasonable to say that the *p*-value is essentially 0. The result is highly significant in that it would be nearly impossible for a random sample of 6918 M&Ms to have a color distribution that differed from the null hypothesis distribution by as much (or more) as the observed sample does.

After obtaining these results, one of the authors posted a message about them to an Internet e-mail list for teachers of Advanced Placement statistics courses. She received several replies from other instructors who had done this activity. Nearly all described statistically significant departures from the color distribution given at the M&Ms website, but the nature of the difference varied considerably. In some cases, for example, the number of red candies was much more than expected; while in other cases, the number of red candies was much less than expected.

Why do so many different samples provide statistically significant evidence against the distribution given by the manufacturer? The answer probably has to do with the manufacturing and distribution process. A bag of M&Ms, or a group of bags purchased at the same store, may not be a *random sample* from a larger population. The color composition of the bags produced in a manufacturing plant on a given day could be affected by the available quantities of the various colors. Also, the process that is used to put candies into the bags may not produce a well-mixed random sampling of the colors. In other words, students counting M&M colors most likely are not using random or representative samples of the larger population.

# Key Terms

**Section 15.1**

categorical variable, 635
two-way table, 635
contingency table, 635
cell, 635
row percentage, 636
column percentage, 636
null hypothesis for two-way tables, 636
alternative hypothesis for two-way tables, 636

homogeneity, test for, 638
independence, test for, 638
observed counts for chi-square test, 638, 640
expected counts for chi-square test, 638, 639, 640
chi-square test statistic, 638
conditions for chi-square test, 640
p-value, 641, 642
degrees of freedom, 641

chi-square distribution, 641
critical value, 643
rejection region, 643

**Section 15.2**

Fisher's Exact Test, 650

**Section 15.3**

chi-square goodness-of-fit test, 652, 653–654

# Exercises

● Denotes basic skills exercises

◆ Denotes dataset is available in StatisticsNow at **http://1pass.thomson.com** or on your CD but is **not required** to solve the exercise.

**Bold-numbered exercises** have answers in the back of the text and fully worked solutions in the Student Solutions Manual.

**Statistics⬡Now**™ Go to the StatisticsNow website at **http://1pass.thomson.com** to:
• Assess your understanding of this chapter
• Check your readiness for an exam by taking the Pre-Test quiz and exploring the resources in the Personalized Learning Plan

## Section 15.1

**15.1** ● For each pair of variables, indicate whether or not a two-way table would be appropriate for summarizing the relationship. In each case, briefly explain why or why not.
    **a.** Satisfaction with quality of local K–12 schools (satisfied or not satisfied) and political party (Republican, Democrat, etc.).
    **b.** Height (centimeters) and foot length (centimeters).
    c. Gender and amount willing to spend on a home theater system.
    d. Age group (under 20, 21–29, etc.) and handedness (right-handed, left-handed, or ambidextrous).

15.2 ● In a nationwide survey, college students are asked how important religion is in their life (very, fairly, or not very) and whether they have ever cheated on a college exam (no or yes).
    a. Write null and alternative hypotheses about the two variables in this situation. Make your hypotheses specific to the context of this situation.
    b. Suppose that the p-value is .127 for a chi-square test of the hypotheses you wrote for part (a). Using a .05

level of significance, write a conclusion about the two variables in this exercise.
    c. What is the value of the degrees of freedom for the chi-square statistic in this situation?

15.3 ● Gender (female or male) and handedness (right-handed or left-handed) are recorded for a randomly selected sample of adults. Of the 100 women in the sample, 92 women are right-handed. Of the 80 men in the sample, 70 men are right-handed.
    a. Write a two-way table of observed counts.
    b. Determine expected counts for all combinations of gender and handedness.
    c. Calculate the value of the chi-square test statistic.

**15.4** ● In each of the following situations, give the p-value for the given chi-square statistic. Either use the information in Table A.5 to estimate a range for the p-value or use software to determine an exact value.
    **a.** $\chi^2 = 3.84$, df = 1.
    **b.** $\chi^2 = 6.7$ for a table with 3 rows and 3 columns.
    c. $\chi^2 = 26.23$ for a table with 2 rows and 3 columns.
    d. $\chi^2 = 2.28$, df = 9.

**15.5** ● Recall that the critical value for a chi-square test is the chi-square value for which the area to its right equals the level of significance. An observed relationship is statistically significant if the calculated chi-square statistic is greater than or equal to the critical value. Use Table A.5 to determine the critical value in each of the following situations.
    a. Level of significance is $\alpha = .05$; df = 1.
    **b.** Level of significance is $\alpha = .01$; table has 3 rows and 4 columns.
    c. Level of significance is $\alpha = .05$; df = 6.

15.6 ● For each of the following situations determine whether the result is statistically significant at the .05 level of significance.
    a. $\chi^2 = 2.89$, df = 1.
    b. $\chi^2 = 5.00$, df = 1.

c. $\chi^2 = 23.60$, df $= 4$.

d. $\chi^2 = 23.60$, df $= 15$.

**15.7** ● Refer to Exercise 15.6. For each part, determine whether the result is statistically significant at the .01 level of significance.

**15.8** Suppose investigators conduct a study on the relationship between birth order (first or only child, not first or only child) and activity preference (indoor or outdoor).

a. Write null and alternative hypotheses for the two variables in this situation.

b. Suppose the *p*-value of the study is not small enough to reject the null hypothesis. Write this conclusion in the context of the situation.

c. Now suppose that the *p*-value of the study is small enough to reject the null hypothesis. In the context of the situation, express the conclusion in two different ways.

**15.9** ◆ Students from two different statistics classes at UC Davis reported their gender (male or female) and recorded which class they were taking. One class is for liberal arts students, and the other is for non-liberal-arts students. The results for the 173 students are given below. In each cell, the observed counts are printed first, with the expected counts below them. Assume that these students are a representative sample of all UC Davis students who would take either class. (Raw data are in **UCDavis1** dataset on the CD for this book.)

|        | NonLib | LibArt | All    |
|--------|--------|--------|--------|
| Female | 83     | 11     | 94     |
|        | 80.42  | 13.58  | 94.00  |
| Male   | 65     | 14     | 79     |
|        | 67.58  | 11.42  | 79.00  |
| All    | 148    | 25     | 173    |
|        | 148.00 | 25.00  | 173.00 |

**a.** State the null and alternative hypotheses that could be tested from these data.

**b.** What are the degrees of freedom for the chi-square test statistic?

**c.** Compute the value of the chi-square test statistic.

**d.** Find the *p*-value or *p*-value range for the chi-square statistic found in part (c).

**e.** Make a conclusion and write it in the context of the situation. Be sure to state the level of significance you are using.

**15.10** The data for this exercise are from a sample of twelfth-graders, collected as part of the 2001 Youth Risk Behavior Surveillance System. The students were asked how often they wear a seat belt while driving, and for this exercise, we combine the responses for "never" and "rarely" and for "most times" and "always" and ignore the "sometimes" responses. A contingency table of the results for males and females follows.

|        | Seat Belt Use | | |
|--------|---------------|---------------|-------|
|        | Most Times or Always | Rarely or Never | Total |
| Female | 964 | 97 | 1061 |
| Male   | 924 | 254 | 1178 |
| Total  | 1888 | 351 | 2239 |

*Data source:* http://www.cdc.gov/nccdphp/dash/yrbs/.

a. Calculate the row percentages for females and the row percentages for males. Compare them.

b. Specify the null and alternative hypotheses that could be tested by using a chi-square statistic in this situation.

c. Calculate the expected count for the cell "Male, Most times or Always."

d. Using the expected count you calculated in part (c), obtain the expected counts for the remaining cells by subtraction. Remember that row and column totals are the same for expected counts as they are for observed counts.

e. Compute the chi-square statistic.

f. Use the chi-square statistic computed in part (e) to find a *p*-value or *p*-value range.

g. Form a conclusion, and write it in the context of the situation. Be sure to state the level of significance you are using.

**15.11** In the 2002 General Social Survey conducted by the National Opinion Research Center at the University of Chicago, participants were asked:

> *Do you favor or oppose the death penalty for persons convicted of murder?*
> *Do you think the use of marijuana should be made legal or not?*

A two-way table of counts for the responses to these two questions is as follows:

|               |        | Marijuana? | | |
|---------------|--------|-------|-----------|-------|
|               |        | Legal | Not Legal | Total |
|               | Favor  | 191   | 341       | 532   |
| Death Penalty? | Oppose | 104   | 171       | 275   |
|               | Total  | 295   | 512       | 807   |

*Data source:* http://csa.berkeley.edu:7502/archive.htm.

a. Is there an obvious choice for which variable should be the explanatory variable in this situation? Explain.

b. Calculate row and/or column percentages for the table. Use these percentages to describe the relationship (or absence of a relationship).

c. State null and alternative hypotheses about the two variables that have been used to create the contingency table.

d. Do a chi-square test of the hypotheses stated in part (c). State a conclusion, and support your conclusion with a *p*-value.

**15.12** For the expected counts shown in Table 15.3 (p. 640) for xylitol and ear infection data, verify that the null hypothesis "expected" proportion getting an ear infection is the same for the three treatment groups.

**15.13** An article at the Gallup website entitled "SOCIAL AUDIT, Gambling in America" included the following comparison of the responses of teenagers and adults to the question, "Generally speaking, do you approve or disapprove of legal gambling or betting?" The survey was conducted in April 1999.

| Opinion of Legal Gambling | | | | |
|---|---|---|---|---|
| | Approve | Disapprove | No Opinion | Total |
| **Adults** | 959 (63%) | 488 (32%) | 76 (5%) | 1523 |
| **Teens** | 261 (52%) | 235 (47%) | 5 (1%) | 501 |
| **Total** | 1220 | 723 | 81 | 2024 |

Source: Copyright © 1999 The Gallup Organization

a. State null and alternative hypotheses for this contingency table.
b. Here is Minitab output for a chi-square test. What conclusion can we reach about the hypotheses stated in part (a)? Justify this conclusion using information from the output.

```
Expected Counts are shown below observed counts

Rows: Age Group    Columns: Opinion

             Approve    Disappro    No Opini       All
Adults           959         488          76      1523
              918.01      544.04       60.95   1523.00

Teens            261         235           5       501
              301.99      178.96       20.05    501.00

All             1220         723          81      2024
             1220.00      723.00       81.00   2024.00

Chi-Square = 45.723,    DF = 2,    P-Value = 0.000
```

c. What is the expected count for the "Teens, Approve" cell? Show how to calculate this value.

**15.14** In Exercise 2.22, data were presented on seat belt use and grades for a sample of 2530 twelfth-graders, collected as part of the 2001 Youth Risk Behavior Surveillance System. Students were asked how often they wear a seat belt while driving; possible choices were never, rarely, sometimes, most times, and always. For this exercise, we combine the responses for "never" and "rarely" and for "most times" and "always" and ignore the "sometimes" responses. The following contingency table summarizes the results.

| Seat Belt Use | | | |
|---|---|---|---|
| Typical Grades | Most Times or Always | Rarely or Never | Total |
| **A or B** | 1354 | 180 | 1534 |
| **C** | 428 | 125 | 553 |
| **D or F** | 65 | 40 | 105 |
| **Total** | 1847 | 345 | 2192 |

Data source: http://www.cdc.gov/nccdphp/dash/yrbs/.

Carry out a chi-square test for the relationship between seat belt use and typical grades for twelfth-graders. Be sure to cover all five steps.

**15.15** The following drawing illustrates the water-level task. Several developmental psychologists have investigated performance on this task. The figure on the left shows the water level in a glass of water that is half full (or is it half empty?). The figure on the right shows the same glass tipped to the right, but the water level is not shown. The task is to correctly draw the water line in this glass.

Water line is shown      Draw the water line

A University of South Carolina psychologist examined the water-level task success rate of 50 students from each of five different colleges within the university (Kalichman, 1986). Within each college, the sample included 25 men and 25 women. The investigators showed each participant eight different drawings of tipped glasses. If the participant correctly drew the water line on four or more of the drawings, he or she was given a "passing" result. Here is a contingency table of the results:

| | Result on Task | | |
|---|---|---|---|
| College | Pass | Fail | Total |
| **Business** | 33 | 17 | 50 |
| **Language Arts** | 32 | 18 | 50 |
| **Social Sciences** | 25 | 25 | 50 |
| **Natural Sciences** | 38 | 12 | 50 |
| **Engineering** | 43 | 7 | 50 |
| **Total** | 171 | 79 | 250 |

a. Compare the colleges using row percentages. What differences are apparent from these percentages?
b. Write null and alternative hypotheses for this table.
c. What are the degrees of freedom for a chi-square test?
d. The chi-square statistic for this table is 16.915. Determine a $p$-value, and state a conclusion about the null and alternative hypotheses.
e. For each college, what are the expected counts for "Pass" and "Fail" outcomes?
f. (Obviously optional) What is the correct water line in the glass on the right?

## Section 15.2

**15.16** ● Refer to the data given for Exercise 15.9 about gender and type of statistics class students were taking.

Use the shortcut formula to calculate the value of the chi-square statistic for those data.

**15.17** ● Refer to the data given for Exercise 15.11 about opinion on the death penalty and opinion on the legalization of marijuana. Use the shortcut formula to calculate the value of the chi-square statistic for those data.

**15.18** The data in the table below first appeared in Exercise 6.11. The variables are height (short or not) and whether or not the student had ever been bullied in school for 209 secondary school students in England. The researchers gathered the data to test their hypothesis that short students are more likely to be bullied in school than other students are.

**Height and Bullying in School**

| Height | Ever Bullied | | |
|---|---|---|---|
| | Yes | No | Total |
| Short | 42 | 50 | 92 |
| Not Short | 30 | 87 | 117 |
| Total | 72 | 137 | 209 |

**a.** Should these data be analyzed by using a chi-square test or a z-test for comparing two proportions? Explain.

**b.** Explain why it makes sense to designate height as the explanatory variable.

**c.** Carry out an appropriate hypothesis test. Be sure to state null and alternative hypotheses, and clearly indicate your conclusion about these hypotheses.

**d.** Calculate separate 95% confidence intervals for the proportion ever bullied for short students and for students who are not short. Compare these two intervals. What does this comparison indicate about the relationship between height and the risk of being bullied?

**15.19** ● Refer to the data given for Exercise 15.18 about height and the risk of having been bullied for secondary school students. Use the shortcut formula to calculate the value of the chi-square statistic for those data.

**15.20** Refer to Exercise 15.10, in which the relationship between sex and seat belt use was examined.

a. Compute the chi-square statistic using the shortcut formula given in Section 15.2. Show your work.

b. What are the degrees of freedom for this situation?

c. Find the p-value or p-value range for the chi-square statistic found in part (a).

d. Form a conclusion, and write it in the context of the situation. Be sure to state the level of significance you are using.

**15.21** Refer to Exercise 15.13 about age and opinion on gambling. Would a z-test for the difference in two proportions have been appropriate instead of a chi-square test? If your answer is yes, explain whether the results would have been identical to the results of the chi-square test. If your answer is no, explain why not.

**15.22** In each of the following situations, explain which test would be most appropriate: a chi-square test, a one-sided z-test for the difference in two proportions, or a Fisher's Exact Test.

a. The manufacturer of a safety seal that is used in cars wants to know whether the safety seals perform as well in extreme cold temperatures as they do in normal temperatures. Testing the seals is expensive, so the manufacturer tests only six seals at an extreme cold temperature and six other seals at a normal temperature. Two of the six seals tested at the extreme cold fail and none of the six seals tested at a normal temperature fails.

b. In June, before an upcoming November presidential election, a random sample of 1000 voters was asked which candidate they would vote for if the election were held then. In September, a new random sample of 1000 voters is asked the same question. The polling agency wants to know if the population proportions planning to vote for each candidate are different in September than they were in June.

c. Frequent computer use may be a cause of carpal tunnel syndrome, a painful condition that affects people's hands and wrists. Researchers hypothesized that people who took typing classes in school would be less likely to suffer from carpal tunnel syndrome than those who did not take typing classes. To examine this hypothesis, the researchers asked a random sample of 3000 people who frequently used computers in their jobs whether they had ever taken typing classes and whether they had ever suffered from carpal tunnel syndrome.

**15.23** Refer to Exercise 15.22. In each case, specify the null and alternative hypotheses.

**15.24** The use of magnets has been proposed as a cure for various illnesses. Suppose that researchers conduct a study with 10 participants to determine whether using magnets as therapy reduces pain from migraine headaches. Five participants are randomly assigned to receive the magnet treatment, and of those, two report that their pain is reduced. The remaining five participants are given an indistinguishable sham treatment, and of those, none report that their pain is reduced. Fisher's Exact Test would be appropriate in this situation.

**a.** Write in words the question that the p-value would answer in this situation.

**b.** (Appropriate software required.) Carry out Fisher's Exact Test.

**15.25** Weindling et al. (1986; also in Hand et al., 1994, p. 15) were interested in the health of juvenile delinquents. They classified 16 boys who failed a vision test by whether or not they wore glasses and whether or not they were a juvenile delinquent. They were interested in knowing whether the likelihood of wearing glasses differed for juvenile delinquents compared to nondelinquents. The results are shown in the table at the top of the next page.

**Data for Exercise 15.25**

|  | Wears Glasses | Does Not Wear Glasses | Total |
|---|---|---|---|
| Delinquent | 1 | 8 | 9 |
| Nondelinquent | 5 | 2 | 7 |
| Total | 6 | 10 | 16 |

a. Specify the appropriate null and alternative hypotheses.

b. Explain why a chi-square test is not appropriate in this situation.

c. Find statistical software or a website that will conduct a Fisher Exact Test, and carry out the test.

d. Explain in words what the $p$-value for this test represents.

## Section 15.3

**15.26** ● In the following situations, give the expected count for each of the $k$ categories.

**a.** $k = 3$, $H_0$: $p_1 = p_2 = p_3 = 1/3$, and $n = 300$.

**b.** $k = 3$, $H_0$: $p_1 = 1/4$, $p_2 = 1/4$, $p_3 = 1/2$, and $n = 1000$.

**c.** $k = 4$, $H_0$: $p_1 = .2$, $p_2 = .4$, $p_3 = .1$, $p_4 = .3$, and $n = 2000$.

**d.** $k = 5$, $H_0$: all $p_i$ are the same, and $n = 2500$.

**15.27** ● In a chi-square goodness-of-fit test, is it possible for all of the expected counts to be larger than the corresponding observed counts? Explain.

**15.28** ● Explain whether each of these is possible in a chi-square goodness-of-fit test.

**a.** The chi-square statistic is negative.

**b.** The chi-square statistic is 0.

**c.** The expected counts are not whole numbers.

**d.** The observed counts are not whole numbers.

**e.** The probabilities specified in the null hypothesis sum to less than 1.

**f.** The degrees of freedom are larger than the number of categories.

**15.29** ● ◆ The following table shows the results from 190 students "randomly" choosing among the numbers from 1 to 10. For example, there were two students who chose the digit 1 and nine students who chose the digit 2. (Raw data are in the **pennstate1** dataset on the CD for this book.)

| Digit | 1 | 2 | 3 | 4 | 5 | 6 | 7 | 8 | 9 | 10 |
|---|---|---|---|---|---|---|---|---|---|---|
| Frequency | 2 | 9 | 22 | 21 | 18 | 23 | 56 | 19 | 14 | 6 |

Assuming that the students were actually choosing digits at random, the expected number of students choosing any digit is 19 ($19 = 190/10$), so all expected counts are 19.

a. State the null and alternative hypotheses. (*Hint*: Refer to Example 15.13 (p. 654).)

b. Calculate the chi-square goodness-of-fit test statistic.

c. What are the degrees of freedom for this test statistic?

d. Find the $p$-value or $p$-value range.

e. Form a conclusion, and write it in the context of the situation. Be sure to state the level of significance you are using.

**15.30** ● Suppose that on a typical day, the proportion of students who drive to campus is .3 (30%), the proportion who bike is .6 (60%), and the remaining .1 (10%) come to campus in some other way (e.g., walk, take the bus, get a ride). The campus sponsors a "spare the air" day to encourage people not to drive to campus on that day. They want to know whether the proportions using each mode of transportation on that day differ from the norm. To test this hypothesis, a random sample of 300 students that day was asked how they got to campus, with the following results:

| Method of Transportation | Drive | Bike | Other | Total |
|---|---|---|---|---|
| Frequency | 80 | 200 | 20 | 300 |

**a.** State the null and alternative hypotheses.

**b.** Find the expected counts for the three modes of transportation.

**c.** Calculate the chi-square goodness-of-fit statistic.

**d.** Find the $p$-value or $p$-value range.

**e.** Form a conclusion, and write it in the context of the situation. Be sure to state the level of significance you are using.

**15.31** In a class survey done in a statistics class, students were asked, "Suppose you are buying a new car and the model you are buying is available in three colors: silver, blue, or green. Which color would you pick?" Of the $n = 111$ students who responded, 59 picked silver, 27 picked green, and 25 picked blue. Is there sufficient evidence to conclude that the colors are not equally preferred? Carry out a significance test. Be sure to state the null hypothesis, and specify the population to which your conclusion applies.

**15.32** Refer to Exercise 15.31. Suppose a car manufacturer had hypothesized that 50% would prefer silver, 30% would prefer blue, and 20% would prefer green. Test the manufacturer's hypothesis.

**15.33** The following table contains the observed distribution of the last digit of the forecasted high temperature on December 10, 2000, for $n = 150$ U.S. and international cities. (*Source: New York Times*, Dec. 10, 2000, p. 47.)

| Last Digit | 0 | 1 | 2 | 3 | 4 | 5 | 6 | 7 | 8 | 9 |
|---|---|---|---|---|---|---|---|---|---|---|
| Count | 11 | 21 | 11 | 23 | 10 | 17 | 11 | 15 | 13 | 18 |

a. Compute expected counts for the null hypothesis that all digits 0, 1, . . . , 9 are equally likely to be the last digit of the forecasted high temperature.

b. Calculate the chi-square goodness-of-fit statistic for these data. What are the degrees of freedom for this statistic?

c. State a conclusion about the null hypothesis that all digits are equally likely to be the last digit selected. Explain.

**15.34** Suppose that a statistics teacher assigns her class of 60 students the homework task of flipping a coin two times, counting the number of heads, and submitting that number to her during the next class period.

**a.** List the possible sequences of results (in order) for two flips of a coin. Using this list, determine the probability distribution for $X$ = number of heads in two flips if the coin is fair.

**b.** Using proper notation, express the probabilities determined in part (a) as a null hypothesis about the number of heads that may occur in two flips of a fair coin.

**c.** The distribution of the number of heads reported by the 60 students in the class was as follows:

| Number of Heads | 0 | 1 | 2 |
|---|---|---|---|
| Number of Students | 8 | 40 | 12 |

Using these data, do a chi-square goodness-of-fit test to test the null hypothesis stated in part (b). Clearly show all steps, and state a conclusion about the null hypothesis.

**d.** If you were the teacher, what would you conclude about how the class may have done this assignment? Briefly justify your answer.

15.35 The California Daily 3 lottery game is identical to the Pennsylvania Daily Number game described in Example 15.13 (p. 654). The following table contains the observed distribution of all digits drawn in the California Daily 3 on the 200 days between May 14, 2000, and November 29, 2000. Three digits are drawn each day, so the sample size is $n = 600$ for this table.

| Digit | 0 | 1 | 2 | 3 | 4 | 5 | 6 | 7 | 8 | 9 |
|---|---|---|---|---|---|---|---|---|---|---|
| Count | 49 | 61 | 64 | 62 | 50 | 64 | 59 | 65 | 63 | 63 |

Carry out a significance test of the null hypothesis that all digits 0, 1, . . . , 9 are equally likely to be drawn.

**15.36** One of the authors of this book purchased four 1-pound bags of plain M&Ms at different stores in Pennsylvania to compare the color distribution to the one stated at the manufacturer's website in 2000. The observed results for the combined bags and the proportions alleged by the manufacturer are given in the following table. State and test the appropriate hypotheses to determine whether it is reasonable to assume that the observed colors are a random sample from a population with the manufacturer's alleged proportions.

| | Brown | Red | Yellow | Blue | Orange | Green | Total |
|---|---|---|---|---|---|---|---|
| **Observed** | 602 | 396 | 379 | 227 | 242 | 235 | 2081 |
| | (.289) | (.190) | (.182) | (.109) | (.116) | (.113) | |
| **Alleged Proportions** | .30 | .20 | .20 | .10 | .10 | .10 | |

## Chapter Exercises

15.37 In a survey reported in a special issue of *Newsweek* magazine (Special Edition: Health for Life, Spring/Summer 1999), $n = 747$ randomly selected women

were asked, "How satisfied are you with your overall appearance?" There were four possible responses to this question, and the following table shows the distribution of counts for the possible responses for each of three age groups. (*Note:* The counts were estimated from percentages given in *Newsweek*.)

| How Satisfied Are You with Your Overall Appearance? | | | | | |
|---|---|---|---|---|---|
| Age | Very | Somewhat | Not Too | Not at All | Total |
| **Under 30** | 45 | 82 | 10 | 4 | 141 |
| **30–49** | 73 | 168 | 47 | 6 | 294 |
| **Over 50** | 106 | 153 | 41 | 12 | 312 |
| **Total** | 224 | 403 | 98 | 22 | 747 |

**a.** Determine conditional percentages within each age group. Summarize these data using a bar chart, and describe what is shown about the relationship between age and satisfaction with appearance.

**b.** Minitab output for a chi-square test of association between age group and satisfaction with appearance follows. Verify that the degrees of freedom shown are correct.

**c.** Is there a statistically significant association between the two variables for this problem? Justify your answer with information from the output.

**d.** Refer to the output for part (b). Draw a sketch that illustrates the $p$-value for this problem, and write a sentence interpreting the $p$-value.

**e.** Refer to the output for part (b). What is the expected count for the "50+, Not Too Satisfied" cell? Show how to calculate this expected count.

```
Expected counts are printed below observed counts

            Very    Somewhat   Not Too   Not at All   Total
18–29        45        82         10         4         141
            42.28     76.07      18.50      4.15

30–49        73       168         47         6         294
            88.16    158.61      38.57      8.66

501         106       153         41        12         312
            93.56    168.32      40.93      9.19

Total       224       403         98        22         747

Chi-Sq = 0.175 + 0.463 + 3.904 + 0.006 +
         2.607 + 0.556 + 1.842 + 0.816 +
         1.655 + 1.395 + 0.000 + 0.860 = 14.278
DF = 6,    P-Value = 0.027
1 cells with expected counts less than 5.0
```

15.38 Refer to Exercise 15.37. Notice that the "contributions to chi-square" are given in the output for all cells. For instance, the "contribution to chi-square" for the row "18–29" and the column "Very" is 0.175, and for the row "18–29" and the column "Not at All" it is 0.006.

**a.** Identify the two cells with the highest "contributions to chi-square." Specify the numerical value of the "contribution" and the row and column categories for each of the two cells.

b. For each of the two cells identified in part (a), determine whether the expected count is higher or lower than the observed count.

c. Using the information in parts (a) and (b), explain how the women in those category combinations contribute to the overall conclusion for this study.

**15.39** Exercise 6.55 gave data on ear pierces and tattoos for a sample of 1375 college women. The data are presented again in the following table. The ear-pierce response is the total number of ear pierces for a woman, and this has been categorized.

**Ear Pierces and Tattoos, 1375 College Women**

| Pierces | No Tattoo | Have Tattoo |
|---|---|---|
| 2 or less | 498 | 40 |
| 3 or 4 | 374 | 58 |
| 5 or 6 | 202 | 77 |
| 7 or more | 73 | 53 |

*Data source: One of the authors.*

a. Describe the relationship with relevant percentages and an appropriate graph.

**b.** Write null and alternative hypotheses for the relationship between ear pierces and tattoos.

**c.** Test the null and alternative hypotheses written for part (b). Use $\alpha = .05$ for the level of significance. State a conclusion.

**15.40** In a study reported in the *Annals of Internal Medicine* (Lotufo et al., 2000), the investigators examined the possible relationship in men between baldness and the risk of coronary heart disease. Other researchers have reported a possible link between these two variables (see Case Study 4.4, for example). In this study, 19,112 men aged 40 to 84 years old were followed for 11 years. All men were free of coronary heart disease at the beginning of the study. During the study, the men were asked about any baldness pattern they may have had at the age of 45. The following table shows counts for the cross-classification of hair loss pattern at age 45 and whether a participant developed coronary heart disease during the study period or not. (*Note:* Vertex baldness occurs on top in the area of the crown or peak of the head.)

**Hair Loss Pattern and Coronary Heart Disease**

| Hair Pattern | Coronary Heart Disease | | |
|---|---|---|---|
| | Yes | No | Total |
| No Baldness | 548 | 7611 | 8159 |
| Baldness | | | |
| Frontal | 333 | 4075 | 4408 |
| Mild Vertex | 275 | 3148 | 3423 |
| Moderate Vertex | 163 | 1608 | 1771 |
| Severe Vertex | 127 | 1224 | 1351 |
| Total | 1446 | 17666 | 19112 |

a. Analyze the relationship between hair loss pattern and the risk of coronary heart disease. Write a short report in which you provide relevant de-

scriptive statistics as well the results of a test for statistical significance. If there is an association, describe the nature of the association.

b. Is this an observational study or an experiment? Briefly explain how the answer affects the interpretation of any observed link between hair loss pattern and coronary heart disease in men.

**15.41** Example 15.11 (p. 649) described an experiment in which students were classified as "sheep" who believe in ESP or as "goats" who do not. Each student then guessed the results of ten coin tosses. We classify students as "Stars" if they guessed five or more correctly and as "Duds" if they did not. The results are shown in the following table:

| | Stars | Duds | Total |
|---|---|---|---|
| **Sheep** | 79 | 33 | 112 |
| **Goats** | 48 | 32 | 80 |
| **Total** | 127 | 65 | 192 |

a. Describe the relationship between belief and performance with relevant percentages and an appropriate graph.

b. Write null and alternative hypotheses for the relationship between belief and performance. Be sure to state the population to which your hypotheses apply.

c. Test the null and alternative hypotheses written for part (b). Use $\alpha = .05$ for the level of significance. State a conclusion.

**15.42** Refer to Exercise 15.41, in which each student guessed the results of ten coin flips. If all students are just guessing, and if the coins are fair, then the number of correct guesses for each student should follow a binomial distribution.

**a.** What are the parameters $n$ and $p$ for the binomial distribution, assuming that the coins are fair and students were just guessing?

b. Specify the probabilities of getting 0 correct, 1 correct, . . . , 10 correct for this experiment if students were just guessing. (*Hint:* These are the probabilities in the pdf for a binomial distribution with parameters specified in part (a).)

**c.** The following table shows how many students got $\leq 2$ right, 3 right, 4 right, and so on, separately, for students classified as sheep (believe in ESP) and classified as goats (don't believe in ESP). Using your results from part (b), fill in the null probabilities that correspond to the hypothesis that students are just guessing.

| Number Correct | Sheep | Goats | Null Probabilities |
|---|---|---|---|
| $\leq 2$ | 6 | 5 | |
| 3 | 11 | 12 | |
| 4 | 16 | 15 | |
| 5 | 29 | 19 | |
| 6 | 28 | 16 | |
| 7 | 14 | 10 | |
| $\geq 8$ | 8 | 3 | |
| Total | 112 | 80 | |

d. Test the hypothesis that the population of sheep represented by these students guess according to the null probabilities specified in part (c). Give a *p*-value, and state a conclusion.

e. Repeat part (d) for the population of goats represented by these students.

15.43 Household sizes for the households participating in the 1996 General Social Survey conducted by the National Opinion Research Center at the University of Chicago are shown in the following table. The table also shows the proportion of households of each size in the survey and the corresponding proportions of all U.S. households of each size.

| Household Size | Number of Cases | Sample Proportion | U.S. Proportion |
|---|---|---|---|
| 1 | 744 | 0.256 | 0.257 |
| 2 | 988 | 0.340 | 0.322 |
| 3 | 454 | 0.156 | 0.169 |
| 4 | 453 | 0.156 | 0.149 |
| 5 or more | 265 | 0.081 | 0.103 |
| Total | 2904 | | |

*Sources: Sample data: SDA data archive, http://csa.berkeley.edu:7502; Population data: www.census.gov/populatio/socdemo/hh-fam/98ppla.txt (Table A. Households by Type and Selected Characteristics, 1998).*

a. Carefully state the null and alternative hypotheses for testing whether the sizes of households in the General Social Survey reflect the size of households for the United States as a whole.

b. Carry out the appropriate test for your hypotheses in part (a). State your conclusion in the context of the situation.

**15.44** The data in this exercise were first presented in Exercise 6.7. In the 1996 General Social Survey, religious preference and opinion about when premarital sex is wrong were among the measured variables. The contingency table of counts for these variables is shown in Exercise 6.7 and again here:

**Religious Preference and Opinion About Premarital Sex**

| | **When Is Premarital Sex Wrong?** | | | | |
|---|---|---|---|---|---|
| **Religion** | *Always* | *Almost Always* | *Sometimes* | *Never* | *Total* |
| **Protestant** | 355 | 117 | 227 | 384 | 1083 |
| **Catholic** | 62 | 37 | 120 | 226 | 445 |
| **Jewish** | 0 | 3 | 14 | 34 | 51 |
| **None** | 20 | 13 | 45 | 147 | 225 |
| **Other** | 15 | 13 | 23 | 40 | 91 |
| **Total** | 452 | 183 | 429 | 831 | 1895 |

*Source: SDA archive at http://csa.berkeley.edu:7502.*

**a.** Are the conditions necessary for carrying out a chi-square test met? Explain.

**b.** Test whether or not there is a statistically significant relationship between these two variables. Show all five steps for the hypothesis test.

15.45 Case Study 10.3 (p. 431) described a survey in which students in a statistics class were asked, "Would you date someone with a great personality even though you did not find them attractive?" The results were

that 80 of 131 women answered "yes" and 26 of 61 men answered "yes."

a. Construct a contingency table for this situation, being careful to identify the explanatory and response variables.

b. State the appropriate hypotheses for this situation.

c. Given the hypotheses that you specified in part (b), would it be more appropriate to conduct a chi-square test or a *z*-test, or does it not matter?

d. Conduct the appropriate test of the hypotheses you specified in part (b). Carry out all five steps.

15.46 Refer to Exercise 15.45. Use the shortcut formula for $2 \times 2$ tables to compute the chi-square statistic. Show the formula with all numbers entered.

**15.47** Wilding and Cook (2000) asked 352 males and 376 females to listen to a male voice and a female voice. One week later, they attempted to identify the voices they had heard in lineups of six male and six female voices. The results are shown in the following table. Using appropriate subsets of the data, conduct a test of the null hypotheses in parts (a) to (d).

| | **Identified Male Voice?** | | **Identified Female Voice?** | |
|---|---|---|---|---|
| **Listener's Sex** | *Yes* | *No* | *Yes* | *No* |
| **Male (n = 352)** | 145 | 207 | 132 | 220 |
| **Female (n = 376)** | 162 | 214 | 191 | 185 |
| **Total (n = 728)** | 307 | 421 | 323 | 405 |

**a.** $H_0$: Sex of listener and ability to identify a male voice are not related.

**b.** $H_0$: Sex of listener and ability to identify a female voice are not related.

**c.** $H_0$: Sex of listener and ability to identify a voice of the same sex are not related.

**d.** $H_0$: Sex of listener and ability to identify a voice of the opposite sex are not related.

e. Using the results from parts (a) to (d), write a short paragraph about the relationship between listener's sex, speaker's sex, and ability to subsequently identify a voice.

15.48 In a class survey, statistics students were asked, "Which one of these choices describes your perception of your weight: about right, overweight, underweight?" The table in Exercise 6.10 displayed the results by sex. Displayed below are those results along with the Minitab output for a chi-square test. (*Source: The authors.*)

Expected counts are printed below observed counts

| | About right | Overwt | Underwt | Total |
|---|---|---|---|---|
| F | 87 | 39 | 3 | 129 |
| | 91.88 | 25.56 | 11.56 | |
| M | 64 | 3 | 16 | 83 |
| | 59.12 | 16.44 | 7.44 | |
| Total | 151 | 42 | 19 | 212 |

Chi-Sq = 0.259 +  7.072 + 6.340 +
          0.403 + 10.991 + 9.853 = 34.918
DF = 2,       P-Value = 0.000

a. State the null and alternative hypotheses being tested in this situation.

b. State a conclusion for the test, using $\alpha = .05$.

c. Using the results shown in the Minitab output, identify the cells with large "contributions to chi-square." Explain in words that would be understood by someone with no training in statistics why the contributions are so large for those cells.

15.49 In the 1996 General Social Survey, participants were asked, "Should divorce in this country be easier or more difficult to obtain than it is now?" The results are shown in the following table and Minitab output.

**Data for Exercise 15.49**

| Sex | Easier | More Difficult | Stay Same | Total |
|---|---|---|---|---|
| Male | 247 | 413 | 151 | 811 |
| Female | 280 | 547 | 206 | 1033 |
| Total | 527 | 960 | 357 | 1844 |

*Data source: 1996 SDA archive at http://csa.berkeley.edu:7502/.*

| | Easier | More Diff. | Same | Total |
|---|---|---|---|---|
| M | 247 | 413 | 151 | 811 |
| | 231.78 | 422.21 | 157.01 | |
| F | 280 | 547 | 206 | 1033 |
| | 295.22 | 537.79 | 199.99 | |
| Total | 527 | 960 | 357 | 1844 |

Chi-Sq = 1.000 + 0.201 + 0.230 +
      0.785 + 0.158 + 0.181 = 2.554
DF = 2,    P-Value = 0.279

a. Carry out the five steps for a chi-square test for this situation.

b. Draw a picture of the appropriate chi-square distribution, and shade the region for the $p$-value for this test.

15.50 Refer to Exercise 15.49. Suppose the same question were to be asked in many independent surveys. Assuming that the null hypothesis is true, into what range should the test statistic fall about 95% of the time, where 0 is at the lower end of the range?

15.51 Explain why a chi-square test statistic cannot be negative.

15.52 Students ($n = 183$) were asked to identify their own eye color as well as the eye color to which they are most attracted. The results are shown here:

**Eye Color and Attraction**

| Own Eyes | Eyes Attracted To | | | | |
|---|---|---|---|---|---|
| | Brown | Blue | Hazel | Green | Total |
| Brown | 30 | 22 | 6 | 13 | 71 |
| Blue | 15 | 37 | 3 | 11 | 66 |
| Hazel | 4 | 12 | 7 | 7 | 30 |
| Green | 4 | 8 | 1 | 3 | 16 |
| Total | 53 | 79 | 17 | 34 | 183 |

a. The Minitab results of a chi-square analysis follow. On the basis of these results, explain why a chi-square analysis is not appropriate.

| | Brown | Blue | Hazel | Green | Total |
|---|---|---|---|---|---|
| Brown | 30 | 22 | 6 | 13 | 71 |
| | 20.56 | 30.65 | 6.60 | 13.19 | |
| Blue | 15 | 37 | 3 | 11 | 66 |
| | 19.11 | 28.49 | 6.13 | 12.26 | |
| Hazel | 4 | 12 | 7 | 7 | 30 |
| | 8.69 | 12.95 | 2.79 | 5.57 | |
| Green | 4 | 8 | 1 | 3 | 16 |
| | 4.63 | 6.91 | 1.49 | 2.97 | |
| Total | 53 | 79 | 17 | 34 | 183 |

Chi-Sq = 4.331 + 2.441 + 0.054 + 0.003 +
      0.886 + 2.541 + 1.599 + 0.130 +
      2.530 + 0.070 + 6.369 + 0.365 +
      0.087 + 0.173 + 0.159 + 0.000 = 21.738
DF = 9,    P-Value = 0.010
4 cells with expected counts less than 5.0

b. Combine the categories for "hazel" and "green," and rerun the analysis. Carry out all five steps of the hypothesis test.

c. Refer to the results from part (b). Identify the cells that have the largest "contributions to chi-square," and explain why the results in those cells support the alternative hypothesis.

15.53 Refer to Exercise 15.52. Using the original data (not combining green and hazel), construct a contingency table for the two variables:

*Explanatory variable:* Eye color.
*Response variable:* Finds own eye color most attractive, yes or no.

a. Conduct a chi-square test to determine if these two variables are related. Specify all five steps of the test.

b. Determine which eye color(s) contributes the most to the chi-square statistic. Explain whether that eye color(s) is more attractive to people who have it as their own color than would be expected or less so.

15.54 Gillespie (1999) and Chambers (2000) reported on two Gallup polls, taken in August 1999 and August 2000 using independent samples, which asked parents the question "How satisfied are you with the quality of education your oldest child is receiving? Would you say completely satisfied, somewhat satisfied, somewhat dissatisfied or completely dissatisfied?" The results of the two polls are shown in the following table:

**Results of Gallup Polls Taken in 1999 and 2000**

| | August 24–26, 1999 | August 24–27, 2000 |
|---|---|---|
| Completely Satisfied | 125 | 87 |
| Somewhat Satisfied | 155 | 133 |
| Somewhat Dissatisfied | 41 | 34 |
| Completely Dissatisfied | 7 | 17 |
| Just Starting School (Volunteered Answer) | 7 | 11 |
| No Opinion | 3 | 0 |
| Total | 338 | 282 |

a. Refer to the technical note on page 638 about the difference between tests for homogeneity and tests for independence. Which is more appropriate for this situation, for comparing the results of the survey for the two years? Specify the appropriate null and alternative hypotheses.

b. Ignore the last two categories (Just Starting School; No Opinion). Carry out the five steps for a chi-square test to determine whether opinions differed in 1999 and 2000. Be sure to identify the appropriate populations represented by these data.

## Dataset Exercises

**Statistics⊖Now**™ Datasets **are required** to solve these exercises and can be found at **http://1pass.thomson.com** or on your CD.

15.55 For this exercise, use the **GSS-02** dataset. The variable *owngun* indicates whether the respondent owns a gun or not, and the variable *polparty* contains the respondent's political party preference. Is there a significant relationship between *owngun* and *polparty?*

15.56 For this exercise, use the **UCDavis2** dataset. Respondents were categorized as male or female (*Sex*) and were asked whether they typically sit in the front, middle, or back of the classroom (*Seat*).

a. Carry out the five steps to determine whether there is a significant relationship between these two variables.

b. The *p*-value for your test in part (a) should be about .02. Explain in words which cells contribute the most to the chi-square statistic and how those cells support the alternative hypothesis.

15.57 For this exercise, use the **UCDavis2** dataset. Two variables that were measured were whether the respondent was left- or right-handed (*Hand*) and whether the respondent finds it easier to make friends with people of the same or opposite sex (*Friends*). Carry out the five steps to determine whether there is a significant relationship between those variables.

15.58 For this exercise, use the **UCDavis2** dataset. Identify two variables for which it would be of interest to you to test whether or not there is a relationship. Carry out the five steps of the chi-square test. If one or both of the variables are quantitative, create reasonable categories. For instance, you could classify students as nondrinkers, moderate drinkers, or heavy drinkers using the variable *Alcohol.* Make sure you explain what variables you used and any recoding you did.

**Statistics⊖Now**™ Preparing for an exam? Assess your progress by taking the post-test at **http://1pass.thomson.com.**

**⊽Mentor**™
Do you need a live tutor for homework problems? Access vMentor at **http://1pass.thomson.com** for one-on-one tutoring from a statistics expert.

# 16

*Is GPA related to where a student likes to sit?*

See Example 16.1 *(p. 670)*

Nick Daly/Photonica/Getty Images

# Analysis of Variance

> *Analysis of variance* is a versatile tool for analyzing how the mean value of a quantitative response variable is affected by one or more categorical explanatory factors. For instance, it can be used to compare the mean weight loss for three different weight-loss programs or the mean testosterone levels of men in seven different occupations.

Suppose that a researcher wants to compare the mean weight loss for three different weight-loss programs or that another researcher wants to compare the mean testosterone levels of men in seven different occupations. What statistical methods can be used to make the desired comparisons? In the data analysis, the researchers might, at some point, use confidence intervals and significance tests to compare two means at a time (e.g., compare mean testosterone levels of occupations 1 and 2, occupations 1 and 3, and so on). Usually, however, an important first step in the analysis of more than two means is to do a significance test to determine if there are any differences at all among the population means being compared. The significance test for doing this is part of a procedure called the **analysis of variance,** which is also sometimes referred to as **ANOVA.**

In this chapter, we focus on the comparison of the means of more than two populations. When different values or levels of a single categorical explanatory variable (weight-loss programs, for instance) define the populations being compared, the ANOVA procedure is called **one-way analysis of variance.** In general, *analysis of variance* is a versatile tool for analyzing how the mean value of a quantitative response variable is related to one or more categorical explanatory factors. While most of the chapter is about one-way analysis of variance, the final section gives an overview of the concepts of *two-way analysis of variance*, which is used to examine the effects of two categorical explanatory variables on the mean value of a quantitative response variable. ▮

## 16.1  Comparing Means with an Anova *F*-Test

When we compare the means of populations represented by independent samples of a quantitative response variable, a null hypothesis of interest is that all means have the same value. An alternative hypothesis is that the means are not all equal. Notice that this alternative hypothesis does not require that all means must differ from each other. The alternative would be true, for example, if only one of the means were different from the others. If $k$ = the number of populations, the null and alternative hypotheses for comparing population means can be written as

$H_0: \mu_1 = \mu_2 = \ldots = \mu_k$
$H_a$: The population means are not all equal.

An **F-statistic** that arises from a *one-way analysis of variance* of the sample data is used to test the hypotheses about the population means, and the significance test is called an **F-test.** The $F$-statistic is sensitive to differences among a set of sample means. The greater the variation among the sample means, the larger is the value of the test statistic. The smaller the variation among the observed means, the smaller the value of the test statistic. In this section, we are concerned with the general ideas of the $F$-test. The specific details and formulas are given in the next section.

Conceptually, the $F$-statistic can be viewed as follows:

$$F = \frac{\text{Variation among sample means}}{\text{Natural variation within groups}}$$

The variation among sample means is 0 if all $k$ of the sample means are exactly equal and gets larger the more spread out they are. If that variation is large enough, it is evidence that at least one of the $k$ population means is different from the other means. In that event, the null hypothesis should be rejected.

The denominator of the $F$-statistic, the natural variation within groups, provides a yard-stick for determining whether the numerator is large enough to reject the null hypothesis. Much like the standard error in a $z$-statistic, it standardizes the numerator so that the $p$-value can be found from common tables. In fact, the denominator of the $F$-statistic is simply a pooled estimate of the variance within each group, a fact that will be discussed in more detail in the next section.

To find the $p$-value, which is the probability that the computed $F$-statistic would be as large as it is (or larger) if the null hypothesis is true, a probability distribution called the **F-distribution** is used. As with all other significance tests, the null hypothesis is rejected if the $p$-value is as small or smaller than the desired level of significance (usually $\alpha = .05$). When the null hypothesis is rejected, the conclusion is that the population means do not all have the same value.

**Example 16.1**

Statistics⊘Now™

Watch a video example at **http://1pass.thomson.com** or on your CD.

For software help, download your Minitab, Excel, TI-83, SPSS, R, and JMP manuals from **http://1pass.thomson.com**, or find them on your CD.

**Classroom Seat Location and Grade Point Average** Is it true that the best students sit in the front of a classroom, or is that a false stereotype? In surveys done in two statistics classes at the University of California at Davis, students reported their grade point averages and also answered the question, "Where do you typically sit in a classroom (front, middle, back)?" In all, 384 students gave valid responses to both questions, and among these students, 88 said that they typically sit in the front, 218 said they typically sit in the middle, and 78 said they typically sit in the back.

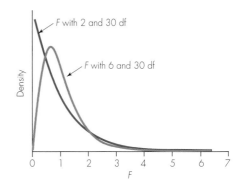

**Figure 16.3** ▌ An $F$-distribution with 2 and 30 df and an $F$-distribution with 6 and 30 df

# Determining the $p$-Value

The $p$-value for the $F$-test is the area to the right of the value of the $F$-statistic, under an $F$-distribution with $k - 1$ and $N - k$ degrees of freedom. All statistical software programs incorporate analysis of variance procedures, and the $p$-value will be reported as part of the output. In situations where an $F$-statistic has been computed without statistical software, either a statistical calculator or the Excel command FDIST (*value, numerator df, denominator df*) can be used to find the $p$-value. It is also possible to find a range for the $p$-value using tabled values for the appropriate $F$-distribution.

Table A.4 in the Appendix gives **critical values** for two different levels of significance, $\alpha = .05$ and $\alpha = .01$. If the observed $F$-statistic is greater than the critical value for a particular level of significance, the result is statistically significant at that level. The region of $F$-statistic values greater than the critical value is called the **rejection region** for the test.

## Using Table A.4 to Judge Statistical Significance

First, determine values for the numerator degrees of freedom and the denominator degrees of freedom. Possible values for the numerator degrees of freedom are given as column headings across the top of Table A.4. Possible values for the denominator degrees of freedom are given as row labels in the leftmost column.

- If the value of the $F$-statistic is greater than the critical value in the $\alpha = .05$ portion of the table, the result is statistically significant at the .05 significance level. In this case, the $p$-value is less than .05, which may be written $p < .05$.

- If the value of the $F$-statistic is greater than the critical value in the $\alpha = .01$ portion of the table, the result is statistically significant at the .01 significance level. In this case, the $p$-value is less than .01, which may be written $p < .01$.

- If the value of the $F$-statistic is between the critical values given for $\alpha = .05$ and $\alpha = .01$, the $p$-value is between .05 and .01 ($.01 < p < .05$). In this case, the result is statistically significant for $\alpha = .05$ but not for $\alpha = .01$.

As an example, suppose the $F$-statistic for comparing four means is $F = 4.26$ and numerator df $= 3$ and denominator df $= 20$. In Table A.4, for $\alpha = .05$, the critical value is given as 3.10 (in the column labeled "3" and row labeled "20"). Because the observed $F = 4.26$ is greater than the critical value $= 3.10$, the result is statistically significant at the $\alpha = .05$ level. The critical value for the $\alpha = .01$ is given as 4.94 in Table A.4. Because 4.26 is not greater than this critical value, the result is not significant at the 0.01 level. Thus, the $p$-value is between .01 and .05.

**Example 16.5**

**The $p$-Value for the Testosterone and Occupational Choice Example**  The authors of the study described in Example 16.4 (p. 674) reported that for comparing the mean testosterone levels of seven occupations the value of the $F$-statistic was $F = 2.5$. There were $k = 7$ occupational groups and $N = 66$ total observations. Thus, for the $F$-distribution that is used to find the $p$-value, numerator df $= k - 1 = 7 - 1 = 6$ and denominator df $= N - k = 66 - 7 = 59$.

Figure 16.4 illustrates the $p$-value, which is the area to the right of 2.5 under an $F$-distribution with 6 and 59 df. In Excel, FDIST(*value, numerator df, denominator df*) can be used to find the probability of an $F$-statistic as large as or larger than the specified value. In this case, FDIST (2.5,6,59) provides the information that the $p$-value is .032. Using Table A.4, we learn that the approximate critical value is 2.25 for $\alpha = .05$. This is the value given for numerator df $= 6$ and denominator df $= 60$. Because 2.5 is larger than this critical value, the result is statistically significant for the .05 significance level, and the $p$-value is smaller than .05.

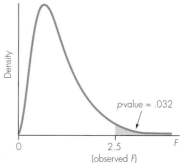

**Figure 16.4** ▮ The $p$-value for Example 16.2 is the area to the right of $F = 2.5$ under an $F$-distribution with 6 and 59 degrees of freedom

## Multiple Comparisons

The term **multiple comparisons** is used when two or more comparisons are made to examine the specific pattern of differences among means. The most commonly analyzed set of multiple comparisons is the set of all pairwise comparisons among population means. In Example 16.1 (p. 670), about GPA and classroom seat location, the possible pairwise comparisons are Front versus Middle, Middle versus Back, and Front versus Back. In Example 16.2 (p. 672), about occupation and testosterone level, the mean testosterone levels for 21 different pairs of occupations can be compared (actor versus minister, football player versus professor, and so on).

Either of two equivalent approaches can be used to make inferences about pairs of means. For each pair, a significance test could be done to determine whether the two means significantly differ. Or a confidence interval for the difference in each pair of means could be computed and then interpreted in terms of statistical significance. If a confidence interval for a difference does not include the value 0, there is a statistically significant difference.

Remember that when the $\alpha = .05$ significance level is used, about 1 in 20 independent statistical tests will achieve statistical significance even if all null hypotheses are true. In other words, when many statistical tests are done there is an increased risk of making at least one type 1 error (erroneously rejecting a null hypothesis). Consequently, several procedures have been developed to control the overall **family type 1 error rate** or the overall **family confidence level** when inferences for a set (family) of multiple comparisons are done.

---

*definition*

- A **family error rate** for a set of significance tests is the probability of making one or more type 1 errors (erroneously rejecting a null hypothesis) when more than one significance test is done.

- A **family confidence level** for a procedure used to create a set of confidence intervals is the proportion of times that all intervals in the set capture their true parameter values.

---

**Tukey's procedure** is one of several available procedures that control the family type 1 error rate (and family confidence level) for multiple comparisons between pairs of population means. If the family error rate is not a concern, **Fisher's procedure** may be used. In Fisher's procedure as it is used for confidence intervals, a specified confidence level (say, 95%) is specified for each separate comparison of two means.

**Example 16.6**

Statistics⬡Now™

Watch a video example at **http://1pass.thomson.com** or on your CD.

**Pairwise Comparisons of GPAs Based on Seat Locations** Example 16.1 (p. 670) was about GPA and preferred classroom seat location for college students. Figure 16.5 shows Minitab output for pairwise comparisons of the mean GPA by seat location. The family confidence level used for the Tukey procedure was .95, while for Fisher's procedure a .95 confidence level was used

**Figure 16.5** ▮ Confidence intervals for pairwise differences in the GPA and seat location example

Tukey 95% Simultaneous Confidence Intervals

Seat = Back subtracted from:

| Seat | Lower | Center | Upper |
|------|-------|--------|-------|
| Middle | −0.1028 | 0.0659 | 0.2347 |
| Front | 0.0846 | 0.2835 | 0.4824 |

Seat = Middle subtracted from:

| Seat | Lower | Center | Upper |
|------|-------|--------|-------|
| Front | 0.0561 | 0.2176 | 0.3791 |

Fisher 95% Individual Confidence Intervals

Seat = Back subtracted from:

| Seat | Lower | Center | Upper |
|------|-------|--------|-------|
| Middle | −0.0759 | 0.0659 | 0.2077 |
| Front | 0.1164 | 0.2835 | 0.4506 |

Seat = Middle subtracted from:

| Seat | Lower | Center | Upper |
|------|-------|--------|-------|
| Front | 0.0819 | 0.2176 | 0.3533 |

for each individual interval. For both methods, the output gives a confidence interval for the difference between population means for each of the three possible differences between two locations (back versus middle, back versus front, and middle versus back). The numbers given under "Lower" and "Upper" are the lower and upper ends of a confidence interval for the difference between population means for the two locations. (The value under "Center" is the difference between sample means.) For example, in the results for the Tukey procedure, the "Lower" and "Upper" values for "Back subtracted from Front" are 0.0846 and 0.4824, respectively. Thus, a confidence interval for the difference in means ($\mu_{\text{Front}} - \mu_{\text{Back}}$) is 0.0846 to 0.4824. In the results given for the Fisher procedure, the confidence interval for ($\mu_{\text{Front}} - \mu_{\text{Back}}$) is 0.1164 to 0.4506.

In general, two means are significantly different if the confidence interval for the difference does not cover 0. Notice that the two procedures lead to the same conclusions about statistical significance in this example. In both procedures, the only confidence interval that covers 0 is the one for ($\mu_{\text{Middle}} - \mu_{\text{Back}}$). We conclude that for students who sit in the front, the population mean GPA is different from the population mean for students who sit in the middle and the back of the room. We are not able to say that there is a difference between the population means of those who sit in the middle and those who sit in the back.

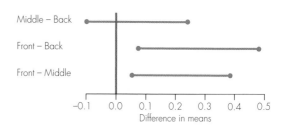

**Figure 16.6** ▌ Confidence intervals for differences between pairs of means in the GPA and seat location example (Tukey procedure)

Figure 16.6 is a graph of the confidence intervals found using the Tukey method. Each interval estimates the difference between the population mean GPAs for two different seat locations. A vertical line is drawn at 0, the value that would indicate no difference between two means. Notice that the confidence interval for ($\mu_{\text{Middle}} - \mu_{\text{Back}}$) covers the line drawn at 0, whereas confidence intervals for ($\mu_{\text{Front}} - \mu_{\text{Back}}$) and ($\mu_{\text{Front}} - \mu_{\text{Middle}}$) do not. ▪

16.1 Exercises are on page 693.

*in summary*

## Concepts for Comparing More Than Two Population Means

- *One-way analysis of variance* is used to compare more than two population means.
- The *null and alternative hypotheses* used for comparing $k$ population means can be written as follows:

    $H_0$: $\mu_1 = \mu_2 = \ldots = \mu_k$
    $H_a$: The means are not all equal.

- The *test statistic* is an *F*-statistic which can be expressed conceptually as follows:

$$F = \frac{\text{Variation among sample means}}{\text{Natural variation within groups}}$$

- An *F*-distribution is used to determine the *p*-value that is used as the basis for reaching a conclusion about the hypotheses.

---

| MINITAB *tip* | **One-Way Analysis of Variance** |
|---|---|

- Use **Stat>ANOVA>One-way** if values of the response variable are in one column (*Response*) and values or names for groups are in a second column (*Factor*). In the dialog box, the ***Comparisons*** button is used for pairwise comparisons.

- Use **Stat>ANOVA>One-way (unstacked)** if the data for different groups are stored in separate columns. Specify the separate columns for the groups in the dialog box. There is not a pairwise comparisons capability in this procedure in Versions 13 or lower, but there is in Version 14.

---

# 16.2  Details of One-Way Analysis of Variance

In this section, we provide some computational details for the *F*-statistic and give the necessary conditions for using the statistic. In practice, statistical software is nearly always used to do a one-way analysis of variance, so it is not necessary to learn formulas to do the computations "by hand." The formulas, however, provide useful insights into the concepts of analysis of variance and how the *F*-test is constructed.

## The Analysis of Variance Table

A fundamental concept in one-way analysis of variance is that the variation among the data values in the overall sample can be separated into (1) differences between group means, and (2) natural variation among observations within a group. The calculations and theory for analysis of variance stem from the fact that for a particular way of measuring variation, the sum of the variation between group means and the variation among observations within groups equals the total variation. Expressed as an equation, the relationship is

Total variation = Variation between groups + Variation within groups

An **analysis of variance table,** like the one shown in Figure 16.1 for Example 16.1 (p. 670), is used to display information about the sources of variation in the response variable. The *F*-statistic that is used to compare population means measures the relative size of the variation between group means and the natural variation within groups.

Example 16.7

**Comparison of Weight-Loss Programs**  Suppose that a researcher does a randomized experiment to compare the mean weight loss for three different programs for losing weight, and the observed weight losses after three months are as follows:

| Program 1 | Program 2 | Program 3 |
|-----------|-----------|-----------|
| 7 | 9 | 15 |
| 9 | 11 | 12 |
| 5 | 7 | 18 |
| 7 | | |

A look at the dotplot of the data shown in Figure 16.7 shows that over the total dataset, the weight losses ranged from 5 to 18 pounds. The sample means for the three programs are $\bar{x}_1 = 7, \bar{x}_2 = 9, \bar{x}_3 = 15$. While there is some variation among individuals within each program, generally it appears that differences *between* weight-loss programs account for much of the variation in the total dataset. The weight losses for program 3 in particular are noticeably greater than the weight losses for the other two programs.

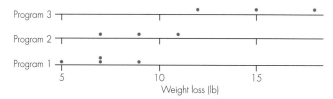

**Figure 16.7** ▌ Dotplot for Example 16.7

## Measuring Variation Between Groups

The variation between group means is measured with a weighted sum of squared differences between the sample means (the $\bar{x}_i$) and $\bar{x}$, the overall mean of all data. Each squared difference is multiplied by the appropriate group sample size, $n_i$, in this sum. This quantity is called **sum of squares for groups** or **SS Groups.** A formula is

$$\text{SS Groups} = n_1(\bar{x}_1 - \bar{x})^2 + n_2(\bar{x}_2 - \bar{x})^2 + \cdots + n_k(\bar{x}_k - \bar{x})^2 = \sum_{\text{groups}} n_i(\bar{x}_i - \bar{x})^2$$

The numerator of the $F$-statistic for comparing means is called the **mean square for groups** or **MS Groups,** and it is calculated as follows:

$$\text{MS Groups} = \frac{\text{SS Groups}}{k - 1}$$

## Measuring Variation Within Groups

To measure the variation among individuals within groups, find the sum of squared deviations between data values and the sample mean in each group, and then add these quantities. This is called the **sum of squared errors** or **SSE.**

A formula in terms of sample standard deviations is

$$\text{SS Error} = (n_1 - 1)s_1^2 + (n_2 - 1)s_2^2 + \cdots + (n_k - 1)s_k^2 = \sum_{\text{groups}} (n_i - 1)s_i^2$$

The denominator of the $F$-statistic is called the **mean square error** or **MSE,** and it is calculated as follows:

$$\text{MSE} = \frac{\text{SSE}}{N - k} = \frac{(n_1 - 1)s_1^2 + (n_2 - 1)s_2^2 + \cdots + (n_k - 1)s_k^2}{n_1 + n_2 + \cdots + n_k - k}$$

Notice that MSE is just a weighted average of the sample variances for the $k$ groups. In fact, if all $n_i$ are equal, MSE is simply the average of the $k$ sample variances. The square root of MSE, denoted by $s_p$ and called the **pooled standard deviation,** estimates the population standard deviation of the response variable. Recall that the populations being compared are assumed to have equal standard deviations.

*definition*  |  The **pooled standard deviation** $s_p = \sqrt{\text{MSE}}$ estimates the population standard deviation, $\sigma$. This assumes that each population has the same value of $\sigma$.

## Measuring Total Variation

The total variation in the data from all samples combined is measured by computing the sum of squared deviations between data values and the mean of all the data. This quantity is referred to as the **total sum of squares** or **SS Total.** The total sum of squares may also be referred to as SSTO. A formula for the sum of squared differences from the overall mean is

$$\text{SS Total} = \sum_{\text{values}} (x_{ij} - \bar{x})^2$$

where $x_{ij}$ represents the $j$th observation within the $i$th group, and $\bar{x}$ is the mean of all observed data values. Notice that SS Total would be the numerator of the sample variance (with denominator $N - 1$) if all the data were combined and treated as a single sample.

The relationship between SS Total, SS Groups, and SS Error is

$$\text{SS Total} = \text{SS Groups} + \text{SS Error}$$

This additive relationship is useful computationally. For example, if the values of SS Total and SS Groups are known, then SS Error can be calculated as

$$\text{SS Error} = \text{SS Total} - \text{SS Groups}$$

**Example 16.8**  |  **Analysis of Variation Among Weight Losses**  For Example 16.7, the $k = 3$ sample means are $\bar{x}_1 = 7, \bar{x}_2 = 9$, and $\bar{x}_3 = 15$, and the mean of all $N = 10$ observations is $\bar{x} = 10$.

$$\begin{aligned}
\text{SS Groups} &= n_1(\bar{x}_1 - \bar{x})^2 + n_2(\bar{x}_2 - \bar{x})^2 + n_3(\bar{x}_3 - \bar{x})^2 \\
&= 4(7 - 10)^2 + 3(9 - 10)^2 + 3(15 - 10)^2 = 114
\end{aligned}$$

$$\text{MS Groups} = \frac{\text{SS Groups}}{k-1} = \frac{114}{3-1} = 57$$

$$\begin{aligned}
\text{SS Total} = {}&(7-10)^2 + (9-10)^2 + (5-10)^2 + (7-10)^2 + (9-10)^2 \\
&+ (11-10)^2 + (7-10)^2 + (15-10)^2 + (12-10)^2 \\
&+ (18-10)^2 = 148
\end{aligned}$$

$$\text{SS Error} = \text{SS Total} - \text{SS Groups} = 148 - 114 = 34$$

$$\text{MSE} = \frac{\text{SS Error}}{N-k} = \frac{34}{10-3} = 4.857$$

$$F = \frac{\text{MS Groups}}{\text{MSE}} = \frac{57}{4.857} = 11.74 \qquad \text{with 2 and 7 df}$$

Figure 16.8 shows Minitab output for this example. The analysis of variance table format used is standard for most software. The line labeled "Programs" gives information about SS Groups and MS Groups and includes the $F$-statistic and the $p$-value. The label for this line is the name the user gives the explanatory variable when the dataset is created. In some instances, "Factor" may be used as a generic label for the explanatory variable. The line labeled "Error" gives information about SS Error and MSE. At the bottom of the output, the pooled standard deviation is given as 2.204. This can be computed as $\sqrt{\text{MSE}} = \sqrt{4.86}$.

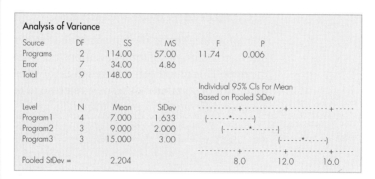

**Figure 16.8** ▌ Minitab output for Example 16.8

*in summary*    ## General Format of a One-Way ANOVA Table

The general format of the one-way analysis of variance table is

| Source | df | SS | MS | F |
|--------|-----|-----|-----|-----|
| Between groups (due to factor) | $k-1$ | $\text{SS Groups} = \sum_{\text{groups}} n_i(\bar{x}_i - \bar{x})^2$ | $\dfrac{\text{SS Groups}}{k-1}$ | $F = \dfrac{\text{MS Groups}}{\text{MSE}}$ |
| Error (within groups) | $N-k$ | $\text{SSE} = \sum_{\text{groups}} (n_i - 1)s_i^2$ | $\dfrac{\text{SSE}}{N-k}$ | |
| Total | $N-1$ | $\text{SSTO} = \sum_{\text{values}} (x_{ij} - \bar{x})^2$ | | |

**Example 16.9**

**Top Speeds of Supercars** Kitchens (1998, p. 783) gives data gathered by *Car and Driver* magazine on the top speeds of five supercars from five different countries: Acura NSX-T from Japan, Ferrari F355 from Italy, Lotus Esprit S4S from Great Britain, Porsche 911 Turbo from Germany, and Dodge Viper RT/10 from the United States. The data represent the top speeds for six runs on each car, using as much distance as necessary without exceeding the engine's redline. To cancel grade or wind effects, there were three runs in each direction on the test facility. Figure 16.9 includes a dotplot of the data, as well as output for a one-way analysis of variance that compares the five cars. Some important features of the results shown in the figure are as follows:

- The *p*-value is .000, so we can reject the null hypothesis that the population mean speeds are the same for all five cars.
- $F = 25.15$. The calculation was $F = $ MS Cars/MS Error $= 364.1/14.5$. An $F$-distribution with 4 and 25 df was used to determine the *p*-value.
- The necessary conditions for doing an ANOVA $F$-test are present. The data are not skewed and there are no extreme outliers. The largest sample standard deviation (5.02 for Viper) is not more than twice as large as the smallest standard deviation (2.92 for Acura).
- MS Error $= 14.5$ is an estimate of the variance of the top speed for the hypothetical distribution of all possible runs with one car. The estimated standard deviation for each car is $\sqrt{\text{MS Error}} = \sqrt{14.5} = 3.81$. This value is given as the pooled standard deviation at the bottom of the output.
- On the basis of the sample means and corresponding confidence intervals for population means, the Porsche and Ferrari seem to be faster on average than the other three cars.

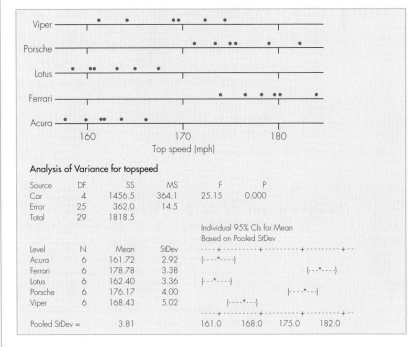

**Figure 16.9** ❚ Dotplot and analysis of variance output for Example 16.9

# Computation of 95% Confidence Intervals for the Population Means

It is informative to examine and compare confidence intervals for the population means. In Figure 16.9 (p. 683) we see that the Minitab output for one-way analysis of variance includes a graph showing a 95% confidence interval for each population mean. These intervals are computed by using the same general format we learned in Chapter 11 for a confidence interval for a mean:

$$\text{Sample mean} \pm \text{Multiplier} \times \text{Standard error}$$

The sample mean for the $i$th group is $\bar{x}_i$. For the $i$th sample mean, the standard error is s.e.$(\bar{x}_i) = s_p/\sqrt{n_i}$. Notice that the pooled standard deviation is used regardless of group, but the sample size is specific to the group. The "multiplier" is determined by using a $t$-distribution as described in Section 11.1, but in this situation, df $= N - k$.

---

**formula**  |  **Confidence Interval for a Population Mean**

In one-way analysis of variance, a **confidence interval for a population mean** $\mu_i$ is

$$\bar{x}_i \pm t^* \frac{s_p}{\sqrt{n_i}}.$$

where $s_p = \sqrt{\text{MSE}}$ and $t^*$ is such that the confidence level is the probability between $-t^*$ and $t^*$ in a $t$-distribution with $N - k$ degrees of freedom.

---

**Example 16.10** | **95% Confidence Intervals for Mean Car Speeds**  The pooled deviation for Example 16.9 (p. 683) is $s_p = 3.81$ mph, a value displayed at the bottom of Figure 16.9. For each car, $n_i = 6$, so the standard error of the mean for any car is s.e.$(\bar{x}_i) = 3.81/\sqrt{6}$. The df for the $t^*$ multiplier are $N - k = 30 - 5 = 25$. From the table of $t^*$ multipliers (Table A.2), we learn that for a .95 confidence level and df $= 25$, the multiplier is $t^* = 2.06$. A 95% confidence interval for any mean is $\bar{x}_i \pm 2.06(3.81/\sqrt{6})$, or $\bar{x}_i \pm 3.02$. For example, the 95% confidence interval for the Acura is $161.72 \pm 3.02$ mph, for the Ferrari it is $178.78 \pm 3.02$ mph,

16.2 Exercises are on page 696.  |  and so on. ■

---

***thought question 16.2***  In Example 16.9, each 95% confidence interval had the same width. Why did this happen? When would the 95% confidence intervals have different widths?*

---

*****HINT:** In the formula given just before Example 16.10, which of the quantities affecting interval width may or may not differ from group to group?

*in summary*    Steps for the *F*-Test for Comparing Several Means

**Step 1:** *Determine the* null *and* alternative *hypotheses.*

Null hypothesis:       $H_0: \mu_1 = \mu_2 = \ldots = \mu_k$
Alternative hypothesis:   $H_a$: The $\mu$'s are not all equal.

**Step 2:** *Verify necessary data conditions, and if met, summarize the data into an appropriate* test statistic. The *F*-statistic can be used if the data are not extremely skewed, there are no extreme outliers, and the group standard deviations are not markedly different. A criterion for standard deviations is that the largest of the sample standard deviations should not be more than twice as large as the smallest of the sample standard deviations. The test statistic is

$$F = \frac{\text{MS Groups}}{\text{MS Error}} = \frac{\dfrac{\sum n_i(\bar{x}_i - \bar{x})^2}{k-1}}{\dfrac{\sum(n_i-1)s_i^2}{N-k}}$$

**Step 3:** *Assuming that the null hypothesis is true, find the p-value.* Using the *F*-distribution with numerator df $= k - 1$ and denominator df $= N - k$, the *p*-value is the area in the tail to the right of the test statistic *F*.

**Step 4:** *Decide whether or not the result is* statistically significant *based on the p-value.* Choose a significance level ("alpha"); standard is $\alpha = .05$. The result is statistically significant if the *p*-value $\le \alpha$.

**Step 5:** *Report the conclusion in the context of the situation.* In analysis of variance, when the null hypothesis is rejected these steps are usually followed by a multiple comparison procedure to determine which group means are significantly different from each other.

# 16.3 Other Methods for Comparing Populations

You probably will not be surprised to learn that the necessary conditions for using an analysis of variance *F*-test do not hold for all datasets. In this section, we discuss methods that can be used when one or both of the assumptions about equal population standard deviations and normal distributions are violated. It is important to remember that no inference method is appropriate if the observed data do not represent the population for the question of interest.

**Example 16.11**   **Drinks per Week and Seat Location** In the two surveys described for Example 16.1 (p. 670), students were asked, "How many alcoholic beverages do you consume each week?" Let's compare responses for the same three groups

Statistics⊝Now™

Watch a video example at **http://
1pass.thomson.com** or on your CD.

compared in Example 16.1: students who typically sit in the front, middle, or back of a classroom. Table 16.1 contains summary statistics for the three seat locations and Figure 16.10 displays a boxplot of the data. The sample sizes are a bit different here than in Example 16.1 because some students did not give a response to every question.

**Table 16.1** Summary Statistic by Seat Location for Number of Alcoholic Beverages per Week

| Location | $n$ | Mean | Median | s.d. |
|----------|-----|------|--------|------|
| Front | 87 | 1.6 | 0 | 3.4 |
| Middle | 207 | 3.9 | 1 | 6.5 |
| Back | 79 | 8.5 | 5 | 10.5 |

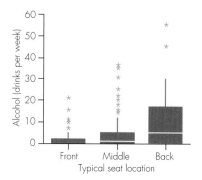

**Figure 16.10** ▮ Boxplot by seat location for number of alcoholic beverages per week

The information in Figure 16.10 and Table 16.1 reveals that the students who typically sit in the back report drinking more alcohol than the other students do. The information in the figure and table also shows that the group standard deviations differ, and the data appear to be skewed. The boxplot shows much more variation in the responses of the back of the classroom group than in the other two groups. In Table 16.1, we see that the standard deviation in the "back" group (10.5) is more than twice as large as the standard deviation in the "front" group (3.4). Also, the mean is greater than the median in each group, which is evidence of skewness in the data. The necessary conditions for doing an analysis of variance are violated in this dataset. We will use other methods to analyze these data in Examples 16.12 and 16.13. ▮

## Hypotheses About Medians

When the observed data are skewed or when extreme outliers are present, it usually is better to analyze the median rather than the mean. This is also what should usually be done if the response variable is an ordinal variable. When several population medians are compared, null and alternative hypotheses of interest are

$H_0$: Population medians are equal.
$H_a$: Population medians are not all equal.

The notation used for a population median varies from author to author. The most commonly used symbol is $\eta$ (pronounced "eta"). With this notation, the null hypothesis about the medians of $k$ populations is $H_0$: $\eta_1 = \eta_2 = \ldots = \eta_k$. The letter M is generally used to denote a sample median. In considering several samples, the median of the $i$th sample would be written $M_i$.

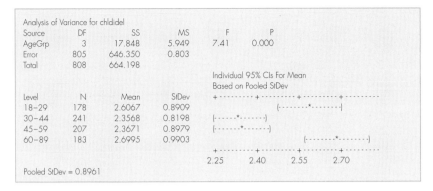

**Output for Exercise 16.7**

16.7 ◆ In the 2002 General Social Survey, a randomly selected sample of U.S. adults was asked what they think is the ideal number of children for a couple to have. The output for this exercise is for a one-way analysis of variance that compares the mean response in four age groups: 18–29, 30–44, 45–59, and 60+. (*Data source:* **GSS-02** dataset on the CD for this book.)
  a. Write null and alternative hypotheses for this problem. State the hypotheses in words and also state them using mathematical notation. To what populations do these hypotheses apply?
  b. What values are given in the output for the *F*-statistic and the *p*-value? What conclusion can be reached about the hypotheses?
  c. Based on the 95% confidence intervals given for the population means, describe the differences among the age groups.
  d. Is the assumption of equal population standard deviations reasonable for these data? Briefly explain.

16.8 Refer to Exercise 16.7 about the ideal number of children for a couple to have. The following output is for a Tukey procedure done to examine all pairwise comparisons of the four age groups. A 95% family confidence level was used.

Tukey 95% Simultaneous Confidence Intervals
All Pairwise Comparisons among Levels of agegroup

agegroup = 18–29 subtracted from:

| agegroup | Lower | Center | Upper |
|---|---|---|---|
| 30–44 | −0.4772 | −0.2499 | −0.0226 |
| 45–59 | −0.4747 | −0.2396 | −0.0045 |
| 60+ | −0.1494 | 0.0927 | 0.3348 |

agegroup = 30–44 subtracted from:

| agegroup | Lower | Center | Upper |
|---|---|---|---|
| 40–59 | −0.2077 | 0.0103 | 0.2283 |
| 60+ | 0.1171 | 0.3426 | 0.5681 |

agegroup = 45–59 subtracted from:

| agegroup | Lower | Center | Upper |
|---|---|---|---|
| 60+ | 0.0989 | 0.3323 | 0.5657 |

a. Which pairs of age groups exhibit statistically significant differences? In one or two sentences, summarize the differences in mean response for the four age groups.
b. Briefly explain what it means to say that a 95% family confidence level was used for the six confidence intervals shown in the output for this problem.

16.9 To evaluate a standard test of the flammability of fabric for children's sleepwear, the American Society for Testing Materials had five laboratories each test 11 pieces of the same type of fabric. The response variable is the length of the char mark made when the fabric sample is held over a flame for a specified time period. Ideally, all labs using the standardized test to test the same fabric should observe about the same value for this response variable. Ryan and Joiner (2001, p. 269) give the following data for the observed char lengths. (The measurement unit was not specified by Ryan and Joiner.)

*Lab 1:* 2.9  3.1  3.1  3.7  3.1  4.2  3.7  3.9  3.1  3.0  2.9

*Lab 2:* 2.7  3.4  3.6  3.2  4.0  4.1  3.8  3.8  4.3  3.4  3.3

*Lab 3:* 3.3  3.3  3.5  3.5  2.8  2.8  3.2  2.8  3.8  3.5  3.8

*Lab 4:* 3.3  3.2  3.4  2.7  2.7  3.3  2.9  3.2  2.9  2.6  2.8

*Lab 5:* 4.1  4.1  3.7  4.2  3.1  3.5  2.8  3.5  3.7  3.5  3.9

a. Graph the data in a way that is useful for comparing the laboratories. Describe any differences among laboratories that can be identified in your graph.
b. Use the graph drawn for part (a) to assess whether the necessary conditions are present for using one-way analysis of variance to compare the five laboratories.

16.10 Refer to Exercise 16.9 about testing the flammability of children's sleepwear. Output for a one-way analysis of variance of the data is shown on the next page.
a. Write null and alternative hypotheses for comparing the mean char lengths for the five labs. State the hypotheses in words, and also state them using mathematical notation. To what populations do these hypotheses apply?

```
Analysis of Variance for CharLng
Source      DF        SS        MS        F        P
Lab         4         2.987     0.747     4.53     0.003
Error       50        8.233     0.165
Total       54        11.219

                                          Individual 95% CIs For Mean
                                          Based on Pooled StDev
Level    N       Mean      StDev      -- + --------- + --------- + --------- + ----
1        11      3.3364    0.4523        (-------*------)
2        11      3.6000    0.4604            (-------*------)
3        11      3.3000    0.3715        (-------*------)
4        11      3.0000    0.2864     (-------*------)
5        11      3.6455    0.4321          (-------*------)
                                       -- + --------- + --------- + --------- + ----
Pooled StDev = 0.4058                   2.80      3.15      3.50      3.85
```

**Output for Exercise 16.10**

b. Are there statistically significant differences among the labs? Justify your answer using the F-statistic and p-value found in the output.

c. Draw a sketch illustrating how the p-value was determined for this problem. See Figure 16.4 (p. 676) for guidance.

d. Based on the sample means and the 95% confidence intervals given for the population means, describe the differences among the five laboratories. Be specific about which laboratories appear to have different mean values from the others.

16.11 Koopmans (1987, p. 93) gave data on the testosterone levels (measured as milligrams per 100 milliliters of blood samples) of 46 women classified into three occupational groups: (1) not employed, (2) employed in job not requiring an advanced degree, and (3) employed in job requiring advanced degree. The data are displayed in the dotplot below.

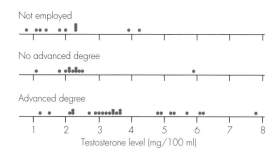

a. Write null and alternative hypotheses for comparing the mean testosterone levels in the three groups. Use proper notation.

b. What does the dotplot indicate about the necessary conditions for doing a one-way analysis of variance of the data? Do any conditions appear to be violated? If so, which condition(s)?

## Section 16.2

16.12 ● Give a value for each of the missing elements in the following analysis of variance table:

| Source | df | SS | MS | F |
|---|---|---|---|---|
| Between Groups | 5 | 40 | — | — |
| Error | 10 | 60 | — | |
| Total | — | | — | |

16.13 ● Give a value for each of the missing elements in the following analysis of variance table:

| Source | df | SS | MS | F |
|---|---|---|---|---|
| Between Groups | 2 | 10 | — | — |
| Error | — | — | — | |
| Total | 30 | 300 | | |

16.14 ● Refer to Exercises 16.4 and 16.5 about female student height and preferred classroom seating location. Sample sizes, means, and standard deviations by seating location are as follows:

| Location | N | Mean | Std.Dev. |
|---|---|---|---|
| Front | 38 | 63.86 | 2.09 |
| Middle | 93 | 64.80 | 2.80 |
| Back | 22 | 66.55 | 2.76 |
| All | 153 | 64.81 | |

a. Show calculations verifying that SS Groups is approximately 101.

b. Use the formula SS Error = $\sum_{\text{groups}} (n_i - 1)s_i^2$ to verify that SS Error is approximately 1043.

**c.** The value of df for error is 150. Find the value of MSE, and find the value of the pooled standard deviation $s_p$.

**16.15** ● Suppose that SS Groups = 0 in an analysis of variance. What would this indicate about the sample means?

**16.16** Thirty male college students were randomly divided into three groups of 10, and the groups received different doses of caffeine (0, 100, and 200 mg). Two hours after consuming the caffeine, each participant tapped a finger as rapidly as possible, and the number of taps per minute was recorded. The data are displayed in a comparative boxplot in the following figure. (*Data source:* Hand et. al., 1994, dataset 50.)

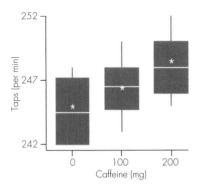

**a.** Write null and alternative hypotheses for comparing the mean taps per minute in the three caffeine amount groups. Use proper notation.

**b.** Using the figure, evaluate the necessary conditions for doing a one-way analysis of variance to compare the mean taps per minute for the three caffeine amount groups.

**16.17** Refer to Exercise 16.16 about finger tapping and caffeine amount. Here is a partial analysis of variance table for the problem with some elements intentionally omitted:

| Source | df | SS | MS | F | p |
|---|---|---|---|---|---|
| Caffeine | — | 61.40 | — | — | 0.006 |
| Error | — | — | — | | |
| Total | 29 | 195.50 | | | |

**a.** Write out the completed analysis of variance table. Indicate how you determined a value for each element that is not shown in the table.

**b.** Calculate the pooled standard deviation, $s_p$. Describe what parameter is estimated by this statistic.

**c.** Based on the p-value given in the ANOVA table, what conclusion can be reached about the mean taps per minute in the population for the three caffeine amounts?

**16.18** Refer to Exercises 16.9 and 16.10 about testing the flammability of children's sleepwear. Assuming that the equal standard deviations assumption is valid, calculate 95% confidence intervals for each of $\mu_1$, $\mu_2$, $\mu_3$, $\mu_4$, and $\mu_5$. The output for Exercise 16.10 shows the sample means and the pooled standard deviation.

**16.19** Suppose that four soil treatments that might be useful for improving the yield of alfalfa grown in fields are compared in an experiment. Each treatment is used in six fields, and the response variable is the crop yield per acre. Suppose further that SS Groups = 150 and SS Error = 200.

**a.** Write null and alternative hypotheses for comparing the mean crop yields in the four groups. Use proper notation.

**b.** Calculate the F-statistic for comparing the mean yield of the four treatments.

**c.** What are the degrees of freedom for the F-statistic?

**d.** The correct p-value is about .01. Draw a sketch illustrating how the p-value is found.

**e.** Based on the p-value given in part (d), what conclusion can be reached about the mean yields for the four soil treatments?

**16.20** Refer to Exercise 16.19. Use the information given or determined in that exercise to write out a completed analysis of variance table.

**16.21** Suppose that three drugs used to reduce cholesterol are compared in a randomized experiment in which three people use each drug for a month. The data for the reductions in cholesterol level for the N = 9 participants are as follows:

| Drug 1 | Drug 2 | Drug 3 |
|---|---|---|
| 6 | 10 | 9 |
| 4 | 14 | 12 |
| 2 | 9 | 6 |

**a.** Calculate the overall mean, $\bar{x}$, and the group means, $\bar{x}_1$, $\bar{x}_2$, and $\bar{x}_3$.

**b.** Calculate SS Groups, the sum of squares for groups.

**c.** Calculate SS Total, the total sum of squares.

**d.** Determine SS Error, the sum of squared errors.

**e.** Calculate the F-statistic for comparing the mean cholesterol reductions for the three drugs. What are the degrees of freedom for this statistic?

## Section 16.3

**16.22** ● Refer to Exercise 16.3 about hours of watching television in a typical day and the highest educational degree attained for respondents in the 1993 General Social Survey. Minitab output for a Mood's Median Test of the same data follows.

```
Mood median test for tvhours

Chi-Square = 138.16      DF = 4      P = 0.000

degree        N<=      N>      Median
Bachelor      182      70      2.00
Graduate       94      24      2.00
HighScho      409     425      3.00
JunCol         62      34      2.00
NotHS          90     201      4.00

Overall median = 2.00
```

a. In the context of this situation, write the null and alternative hypotheses that are tested by the Mood's Median Test.
b. What is the *p*-value for testing the hypotheses written in part (a)? What conclusion can be reached about the hypotheses?
c. Which group had the highest median hours of television watched in a typical day? What was the median value for that group?
d. The overall median for the sample is 2 hours of watching television in a typical day. What percentage of respondents in the high school degree group said that they watch television more than 2 hours in a typical day? What percentage of those who did not get a high school degree watch more than 2 hours of television in a typical day?

**16.23** ● A sample of $n = 729$ college students is asked to rate how much they like various types of music on a scale of 1 to 6 on which 1 = don't like at all and 6 = like a lot. The students also were asked whether their hometown was a big city, rural, a small town, or suburban. Results for a Kruskal–Wallis test comparing ratings of reggae music for the four types of hometown are given in the output for this exercise.

```
Kruskal–Wallis Test on Reggae

Hometown     N      Median     Ave Rank      Z
Big city     89     5.000      476.2       5.32
Rural        96     3.000      335.2      -1.49
Small town  176     3.000      360.3      -0.34
Suburban    368     3.000      348.2      -2.18
Overall     729                365.0

H = 29.17     DF = 3     P = 0.000
```

**Output for Exercise 16.23**

a. Write null and alternative hypotheses for comparing the four hometowns with regard to the rating of reggae music.
b. What *p*-value is given in the output? Explain what conclusion can be made based on this *p*-value.
c. For which type of hometown was the median rating of reggae the highest? What was the median rating in that type?

**16.24** ● In the survey that was described in Exercise 16.23, students were asked how many times they pray per week and what they felt was the importance of religion in their life (very important, fairly important, or not very). Descriptive statistics comparing number of times praying per week for the three religious importance groups are as follows:

| Importance | N | Mean | Median | StDev |
|---|---|---|---|---|
| Very | 169 | 9.57 | 7 | 8.64 |
| Fairly | 312 | 4.76 | 3 | 2.15 |
| Not very | 252 | 0.77 | 0 | 1.83 |

Why would it be preferable to compare the prayer frequency for the three religious importance groups using a nonparametric technique rather than analysis of variance? (*Hint:* Which of the necessary conditions for using an *F*-statistic to compare means are violated by these data?)

**16.25** A sample of $n = 235$ college students was asked, "Consider how important a person's personality and looks (attractiveness) are to you. Rate this importance on a scale of 1 (personality is most important) to 25 (looks are most important)."

a. Explain whether the response is a quantitative variable, a categorical variable, or an ordinal variable.
b. Results for a Mood's Median Test that compares the responses of men and women are shown below in the Minitab output for this exercise. What are the null and alternative hypotheses of the Mood's Median Test in this instance? Make your answer specific to this situation.
c. What is the overall median for all responses? What percentage of the women gave an answer that was less than or equal to the overall median? What percentage of the men gave an answer that was less than or equal to the overall median?

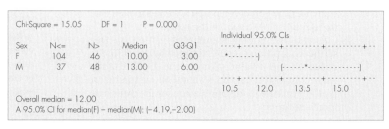

**Output for Exercise 16.25**

d. What conclusion can be reached about the difference between men and women with regard to the relative importance of looks versus personality? Explain how you made this decision, and be specific about the populations to which you think the conclusion applies.

**16.26** Refer to Exercise 16.25. The *p*-value is given as P = 0.000, and the value of the chi-square statistic is given as Chi-Square = 15.05. Explain the connection between the *p*-value and the chi-square statistic.

**16.27** Refer to Exercise 16.11 about women's testosterone levels. Results are shown in the accompanying output for a Kruskal–Wallis test done to compare the testosterone levels of the three groups of women.

| group | N | Median | Ave Rank | Z |
|---|---|---|---|---|
| Adv Degr | 24 | 3.400 | 30.1 | 3.46 |
| No Adv D | 11 | 2.200 | 17.6 | −1.67 |
| Not Empl | 11 | 2.000 | 15.1 | −2.38 |
| Overall | 46 | | 23.5 | |

H = 12.19   DF = 2   P = 0.002
H = 12.21   DF = 2   P = 0.002 (adjusted for ties)

**a.** Write null and alternative hypotheses for the Kruskal–Wallis test. Make your answer specific to this situation.

**b.** Based on the output, what conclusion can be reached about testosterone levels in the three groups? Justify your answer using information given in the output.

c. What are the sample medians for the three groups? Based on these medians, and the appearance of the dotplot in the figure for Exercise 16.11, describe the difference among the three groups.

## Section 16.4

**16.28** For each scenario, explain whether an interaction is described or not. If there is an interaction, what are the two variables that interact?

**a.** For women, there was no difference between the mean hours spent studying per week for members and nonmembers of Greek organizations. For men, there was a five-hour difference between mean hours spent studying per week for members of Greek organizations versus nonmembers.

b. The difference between the mean survival times for the two treatments was about six months in each patient age group.

**16.29** Explain whether the following statement is an example of a main effect or an interaction: The mean number of classes missed per week was significantly lower for students who think religion is very important than it was for students who think religion is either fairly or not very important.

**16.30** Exercise 16.23 described a survey in which college students were asked to rate various types of music on a 1 to 6 scale on which 1 = don't like and 6 = like a lot. The figure for this exercise is an interaction plot showing how the mean rating of rock music relates to hometown and gender.

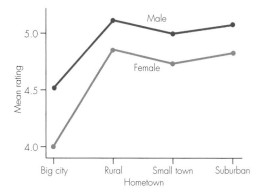

**a.** Which gender liked rock music better?

**b.** What is the evidence of a possible main effect of hometown?

**c.** Discuss whether you think there is evidence of a possible interaction between gender and hometown.

**16.31** The figure for this exercise shows how mean hours slept the previous night is related to gender and preferred classroom seat location for a sample of *n* = 173 college students (*Data source:* **UCDavis** dataset on the CD for this book.)

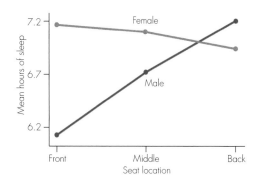

a. For males, describe how mean hours slept the previous night is related to preferred classroom seat location.

b. For females, describe how mean hours slept the previous night is related to preferred classroom seat location.

c. Explain why the plot is evidence of a possible two-way interaction between gender and seat location.

**16.32** Give an example not given in this chapter in which two-way analysis of variance would be used to make a comparison of means. Be specific about what the response variable of interest is, what the explanatory variables are, and what specific levels or categories of the explanatory variables are used.

**16.33** In Example 16.1 about mean GPA and typical classroom seat location, the finding was that there were differences among the mean grade point averages for the three locations. Suppose that the sex of the students is also considered along with the seat location.

**a.** What would be indicated about the difference in the mean GPA of men and women if there were an interaction between student gender and seat location?

**b.** What would be indicated about the difference in the mean GPA of men and women if there were not an interaction between student gender and seat location?

**16.34** Students in a statistics class gave information on their height and weight and also reported their perception of their weight (about right, underweight, overweight). The mean body mass index (BMI) is shown in the following table, along with sample size, for each combination of student gender and perception of weight category. (BMI is calculated as Weight in kilograms/(Height in meters)$^2$.)

**Mean Body Mass Index by Gender and Perception of Weight**

| | Perception of Weight | | |
| --- | --- | --- | --- |
| | Underweight | About Right | Overweight |
| **Females** | 19.3 ($n = 7$) | 21.1 ($n = 107$) | 24.4 ($n = 31$) |
| **Males** | 20.1 ($n = 14$) | 23.9 ($n = 56$) | 27.8 ($n = 13$) |

a. Draw an interaction plot of the means. Figures 16.13 (p. 690) and 16.14 (p. 691) may provide guidance for how to do this.

b. Based on the given data, do you think gender and perception of weight are interacting variables? Briefly explain why or why not.

**16.35** Ryan and Joiner (2001) give data collected in an experiment done to examine the effects of storage temperature on the rot that occurs in stored potatoes. Potatoes were injected with bacteria known to cause potato rot, and three different bacteria amounts (low, medium, and high) were used. For each bacteria amount, half of the injected potatoes were stored at 10°C, while the others were stored at 16°C. The response variable was the diameter (mm) of the rot in a potato after being stored. The mean rot diameters for all combinations of bacteria amount and temperature are shown in the table at the top of the next column. For each combination, the sample size is 9.

**Mean Diameter of Potato Rot by Storage Temperature and Bacteria Amount**

| | Temperature | |
| --- | --- | --- |
| Bacteria | 10°C | 16°C |
| **Low** | 3.6 | 7.0 |
| **Medium** | 4.8 | 13.6 |
| **High** | 8.0 | 19.6 |

a. Draw an interaction plot of the means. Figures 16.13 (p. 690) and 16.14 (p. 691) may provide guidance for how to do this.

b. What is the evidence that there may be an interaction between bacteria amount and storage temperature? Briefly describe the pattern of this possible interaction.

## Chapter Exercises

**16.36** Give an example not already given anywhere in this chapter in which one-way analysis of variance could be used to make a comparison of means. Be specific about what the response variable of interest is and what groups are compared.

**16.37** Refer to Figure 16.1 (p. 671) for Example 16.1 about GPA by classroom seat location. Explain why the 95% confidence intervals for the mean GPA in the three seat locations have unequal widths.

**16.38** Students in grades 3 to 8 in 11 public schools in Ohio were asked how often they engage in violent behaviors and how often they watch television (Singer et al., 1998). A violent behaviors score was calculated by adding frequency scores (0 = never, 3 = almost every day) for five different violent acts: threatening, hitting before being hit, hitting after being hit, beating up, and attacking with knife. For five groups determined by self-reported hours of television, the sample size, sample mean, and sample standard deviation for the violent behaviors scores are as follows:

| Daily Hours of Television | $n_i$ | $\bar{x}_i$ | $s_i$ |
| --- | --- | --- | --- |
| <1 | 227 | 2.94 | 2.73 |
| 1–2 | 526 | 2.59 | 2.44 |
| 3–4 | 666 | 2.87 | 2.36 |
| 5–6 | 310 | 3.10 | 2.45 |
| >6 | 488 | 4.03 | 2.81 |

**a.** Write null and alternative hypotheses for comparing the mean violent behaviors score for the five television-watching groups. State the hypotheses in words, and also state them using mathematical notation.

**b.** What do you think are the populations to which the results of an analysis of variance $F$-test would apply?

**c.** The researchers did a one-way analysis of variance and reported the results as $F = 22.95$, $p < .001$. What does this result indicate about the null and alternative hypotheses for this problem?

**d.** Notice that the highest mean violent behaviors score is in the group that watches the most television. Explain why this cannot necessarily be interpreted to mean that watching a lot of television causes violent behavior in children. What is another explanation for the observed result?

**16.39** Refer to Exercise 16.38.

**a.** On the basis of the information given, which of the necessary conditions for doing a one-way analysis of variance appear to be satisfied? What is the evidence of this?

**b.** For children who watch television less than 1 hour per day, use the Empirical Rule (see Section 2.7) to estimate the interval in which the middle 95% of scores would fall if the data were bell-shaped. Considering that 0 is the lowest possible violent behaviors score, explain why the result of your calculation is evidence that the data either are skewed or contain outliers.

**c.** Is the difficulty noted in part (b) also present in the other television-watching groups? Show calculations that support your answer.

**16.40** Refer to Exercises 16.38 and 16.39. The investigators reported results separately for boys and girls in the five television-watching groups. In this two-way classification, the approximate observed mean violent behaviors scores were as follows:

| | \<1 | 1–2 | 3–4 | 5–6 | 6+ |
|---|---|---|---|---|---|
| **Hours of Daily Television Watching** | | | | | |
| **Boys** | 3.7 | 3.0 | 3.4 | 3.4 | 4.1 |
| **Girls** | 2.3 | 2.3 | 2.4 | 2.8 | 3.9 |

Explain whether there is an interaction between sex and amount of television watching or not. If there is an interaction, describe the pattern of the interaction.

**16.41** Refer to Example 16.9 (p. 683) about the top speeds of supercars. The accompanying output (top of next column) shows Fisher's procedure for pairwise comparisons, with a 95% confidence level for each individual confidence interval. Interpret the output. Specifically, use the output to describe the statistically significant differences among the mean top speeds of the cars. Look back at Example 16.9 to see which were the fastest and slowest cars.

Fisher 95% Individual Confidence Intervals
All Pairwise Comparisons

Acura subtracted from:

| | Lower | Center | Upper |
|---|---|---|---|
| Ferrari | 12.542 | 17.067 | 21.592 |
| Lotus | −3.842 | 0.683 | 5.208 |
| Porsche | 9.925 | 14.450 | 18.975 |
| Viper | 2.192 | 6.717 | 11.242 |

Ferrari subtracted from:

| | Lower | Center | Upper |
|---|---|---|---|
| Lotus | −20.908 | −16.383 | −11.858 |
| Porsche | −7.142 | −2.617 | 1.908 |
| Viper | −14.875 | −10.350 | −5.825 |

Lotus subtracted from:

| | Lower | Center | Upper |
|---|---|---|---|
| Porsche | 9.242 | 13.767 | 18.292 |
| Viper | 1.508 | 6.033 | 10.558 |

Porsche subtracted from:

| | Lower | Center | Upper |
|---|---|---|---|
| Viper | −12.258 | −7.733 | −3.208 |

**Output for Exercise 16.41**

**16.42** At a private university in southern California, 334 freshman (197 males and 137 females) reported their frequency of drinking alcoholic beverages and also completed a test of personality and psychological characteristics (Ichimaya and Kruse, 1998). One response variable was the score on a 12-item test of conscientiousness for which a high score indicates dependability. Summary statistics for three groups defined by frequency of binge drinking (five or more drinks in one session) are as follows:

| Drinking Group | $n_i$ | $\bar{x}_i$ | $s_i$ |
|---|---|---|---|
| **Nonbinge** | 162 | 32.59 | 7.15 |
| **Occasional Binge** | 67 | 29.93 | 6.43 |
| **Frequent Binge** | 105 | 30.10 | 6.45 |

**a.** Write null and alternative hypotheses for comparing the mean conscientiousness score for the three groups. Write the hypotheses in words, and also write the null hypothesis using proper statistical notation.

**b.** What do you think are the populations to which the results of an analysis of variance $F$-test would apply?

**c.** The researchers did a one-way analysis of variance and reported the results as $F = 5.95$, $p < .0005$. What does this result indicate about the null and alternative hypotheses for this problem? (*Note:* The *p*-value reported by the researchers is incorrect. The correct *p*-value is $p = .003$.)

d. Using the reported sample means, describe the differences among the mean conscientiousness scores for the three groups. Explain whether you think any observed differences have practical significance or not.

e. Is it reasonable to assume that population standard deviations are equal? Why or why not?

**16.43** Refer to Exercise 16.42. Use formulas in Section 16.2 to compute SS Error, MSE, and the pooled standard deviation, $s_p$.

**16.44** Refer to Exercise 16.16 about finger tapping and amount of caffeine consumed. Notice that caffeine amount is a quantity, so it could be analyzed as a quantitative variable. What statistical technique other than one-way analysis of variance could be used to analyze the relationship between taps per minute and caffeine amount? What would be an advantage of using that technique to analyze this relationship? (*Tip:* It might help to look again at the boxplot for Exercise 16.16.)

## Dataset Exercises

**Statistics☐Now**™  Datasets **are required** to solve these exercises and can be found at **http://1pass.thomson.com** or on your CD.

16.45 Use the dataset **wineratings.** The variable *Quality* gives ratings of overall quality for 38 samples of Pinot Noir wine made in three different wine-producing regions. The variable **Region** gives the region (1, 2, or 3).

a. Create a comparative boxplot to compare overall quality ratings in the three different regions. Describe any regional differences.

b. Determine and compare sample means of overall quality ratings for the three regions. Find sample standard deviations for the three regions.

c. Use software to do a one-way analysis of variance to compare overall quality ratings for the three regions. State null and alternative hypotheses, give the *p*-value, and clearly state a conclusion about the regions.

16.46 Use the dataset **wineratings.** The variable *Aroma* gives ratings of aroma for 38 samples of Pinot Noir wine made in three different wine-producing regions. The variable *Region* gives the region for a sample.

a. Determine and compare sample mean aroma ratings for the three regions.

b. Create a comparative boxplot to compare aroma ratings in the three different regions. Describe any regional differences.

c. Use software to do a one-way analysis of variance to compare aroma ratings for the three regions. State null and alternative hypotheses, give the *p*-value, and clearly state a conclusion about the regions.

16.47 Use the dataset **Student0405,** which contains data that were gathered in surveys of statistics classes at a large university in the United States during 2004 and 2005. The variable *MissClass* gives student estimates of how many classes they typically miss in a week. The variable *Seat* gives responses to the question "Where do you prefer to sit in a classroom (front, middle, back)?"

a. Determine and compare sample means of missed classes per week for the three seat locations.

b. Use a comparative boxplot to compare missed classes for the three seat locations. Describe any possible differences.

c. Do a one-way analysis of variance *F*-test to compare mean missed classes for the seat locations. State null and alternative hypothesis, give the *p*-value, and clearly state a conclusion about the locations.

d. Explain whether the data satisfy the necessary conditions for using an *F*-test to compare population means.

16.48 Use the dataset **Student0405,** which contains data that were gathered in surveys of statistics classes at a large university in the United States during 2004 and 2005. The variable *StudyHrs* gives student estimates of how many hours they typically study in a week. The variable *ReligImp* gives responses to the question "How important is religion in your own life (very, fairly, not very)?"

a. Graphically compare study hours for the three religious importance groups. Describe the results.

b. Use software to do a one-way analysis of variance to compare mean weekly hours of study for the three religious importance groups. State null and alternative hypotheses, give the *p*-value for the *F*-test, and clearly state a conclusion about whether population means differ or not.

c. Explain whether the data satisfy the necessary conditions for using an *F*-test to compare population means.

16.49 Refer to Exercise 16.48. Compare study hours for the three religious importance groups using either a Mood's Median test or a Kruskal–Wallis test. Give the output as part of your answer, and state a clear conclusion in the context of this situation.

16.50 Use the **GSS-02** dataset. The variable *degree* gives the highest educational degree achieved (five categories) for respondents in the 2002 General Social Survey, and *tvhours* gives responses for hours of watching television in a typical day.

a. Find mean and median weekly television watching amounts for each educational degree group. Describe any differences among the groups.

b. Use either a Kruskal–Wallis test or a Mood's Median test to compare the amount of television watching in the five educational degree groups. Give null and alternative hypotheses, give a *p*-value for the test you chose to do, and state a conclusion in the context of this situation.

The effect of the anti-tobacco efforts has been fewer smokers and fewer deadly cases of cancer related to smoking, health officials said. August and Pechacek both said they expect the trend to continue.

August said that means there will be up to 4,000 fewer lung cancer cases in California this year and about 2,000 fewer deaths.

In its report, the CDC compared cancer registries in California, Connecticut, Hawaii, Iowa, New Mexico and Utah, as well as Seattle, Atlanta and Detroit.

In 1988, the lung cancer rate in California was 72 cases per 100,000 people, slightly higher than that of the other regions studied. By 1997, California's rate had dropped to about 60 per 100,000.

The CDC averages the statistics for the first two years and the last two years studied to arrive at an accurate representation, health officials said. The numbers the CDC used were 71.9 for 1988 and 70.3 for 1989, averaging to 71.1; and 62.2 for 1996 and 60.1 for 1997, averaging to 61.15. That computes to a 14 percent drop in lung cancer cases.

While lung cancer rates for women in the other regions rose 13 percent, the rate for California women dropped 4.8 percent. Among California men, lung cancer rates dropped 23 percent, compared with a 13 percent drop among men elsewhere.

Dr. David Burns, a volunteer with the American Lung Association in California, said: "This is an accomplishment of Proposition 99 money being invested wisely by the state to help people change their smoking behavior."

*Source:* http://dailynews.yahoo.com/htx/ap/20001130/hl/cancer_study_2.html.

Suppose you were a lawmaker in another state and read these results. Would you attempt to enact legislation similar to that in California? The evidence is certainly tempting, but there are some statistical issues that need to be considered. First, these results are based on observational data. California was not randomly chosen to apply tough antismoking laws. People in California may be more health conscious than people in other states. California has other strict air pollution legislation that was enacted in the same time period, such as tough maintenance and inspection laws for vehicle emissions. California industry has changed over the years, favoring high-tech industries that are likely to employ younger, healthier workers.

If you were a lawmaker, you would need to weigh consequences of possible decisions. For instance, as mentioned, California has had an active antismoking education program. The consequences of launching such a program would presumably not be harmful, so a decision to do so would have positive or neutral consequences. On the other hand, a decision to ban smoking in bars and restaurants would involve more controversial consequences. ■

Chapter Exercises are on page 721.

# 17.5 Understanding Our World

Sometimes statistical studies are done to help us understand ourselves and our world, without involving any decisions. There are thousands of academic journals containing the results of statistical studies, most of which will never be used to change personal or societal decisions or behavior. Scan any major news source for a few weeks and you will find reports of many interesting studies that are done to help us understand the world. Many of these studies are exploratory in nature and have results that are controversial. That's part of why they make interesting news.

**Example 17.6**

**Is It Wining or Dining That Helps French Hearts?** French citizens have lower rates of heart disease than people in neighboring countries, and some recent speculation has focused on red wine as a partial cause. The following article discusses some other possible explanations. Notice that no one is likely to change laws or behavior as a result of this research, it is simply interesting information.

Friday May 28 1:24 PM ET (1999)

*Wine May Not Explain France's Lower Heart Disease Rate*

NEW YORK, May 28 (Reuters Health)—Previous research has suggested that red wine consumption may explain why the French have a much lower rate of heart disease than other nations. But in a report published this week in the British Medical Journal, two UK experts dispute this theory, suggesting instead that the reduced risk lies in the fact that the French diet traditionally contains less animal fat.

And because the levels of animal fat and cholesterol in the French diet have risen in the last 15 years, their explanation suggests that it is only a matter of time before the heart disease rate in France catches up with the rate in Britain, where heart disease mortality is about four times higher.

Dr. Malcolm Law and Professor Nicholas Wald from the Wolfson Institute of Preventive Medicine in London, UK, suggest that the tendency of French doctors to attribute heart disease deaths to other causes "could account for about 20% of the difference," according to the report.

Law and Wald do give red wine some credit as a preventive factor, but say that this effect is small. The high consumption of red wine in France, they write, "explains (less than 5%) of the difference."

On the other hand, they provide considerable statistical evidence to support their "time lag" theory. "Mortality from . . . heart disease," the authors write, "was strongly associated with past animal fat consumption . . . and past (serum cholesterol) values, but not with recent values."

"Animal fat consumption and serum cholesterol concentration have been similar in France and Britain for a relatively short time—about 15 years," write Law and Wald. "Serum cholesterol concentration in 1970 was . . . lower in France than in Britain . . . and this explains most of its lower mortality from heart disease," they suggest.

Several related editorials question Law and Wald's hypothesis. Meir Stampfer and Eric Rimm from Harvard University, Boston, Massachusetts, write, "Obviously, other factors must play a role. . . . We think it more likely that the difference in coronary mortality rests on behavioral (especially dietary) differences that have not received adequate attention."

In another commentary, D.J.P. Barker from the University of Southampton in Southampton, UK, writes that "recent trends in coronary heart disease are only weakly related to trends in serum cholesterol" and argues that "coronary heart disease originates in utero, through adaptations that the fetus makes to undernutrition."

Finally, Johan Mackenbach and Anton Kunst from Erasmus University in Rotterdam, the Netherlands, contend that "heterogeneity of populations should be taken into account."

But Law and Wald dismiss these arguments. "We believe," they write in response, "that the time lag explanation is the major reason and that the alternative explanations offered in the commentaries are quantitatively unimportant."

*Source: British Medical Journal* (1999), 318:1471–1480. ■

**Example 17.7**

**Give Her the Car Keys** Memory is a fascinating ability, and most of us wish we had more of that ability. Studies about memory may eventually help us find ways to improve it. In the meantime, the following study is simply interesting. If it leads to any decision at all, it would be related to who thinks they should get the car keys and who actually should. We'll let you figure it out.

Thursday March 26, 1998 1:30 PM EST    Yahoo Health News

*Women Remember Item Location Better*

NEW YORK (Reuters)—Women are better at remembering where objects are than men, but show less confidence in their memory, according to a study.

"When it comes to memory, women have more skill than confidence, and men have more confidence than skill," concludes Dr. Robin West, a University of Florida (UF) psychology professor and researcher.

West, along with UF psychology graduate student Duana Welch, compared the spatial memories of over 300 healthy men and women of various ages in a study funded by the National Institutes of Mental Health.

They used specially-designed computer tests that had each participant place 20 common household "objects" into one of 12 "rooms" (e.g., kitchen, bedroom, bathroom) depicted on the screen. After a 40-minute interval, each participant was asked to remember the location of each object.

According to Welch, the test focused on "something that has very important day-to-day meaning for people; that is, the ability to find where you put something."

The results? West and Welch say they found "younger adults performing better than older adults, and women performing better than men." In fact, women scored an average of 14.4 points on the test, compared with the men's average score of 13.5.

Those scores didn't meet male expectations of their own memory skills, the researchers say. In a statement issued by the university, West and Welch say that when questioned, "men tended to overestimate their ability to remember object locations," while women tended to underestimate their powers of recall. The study authors note these findings "are in keeping with other literature suggesting that women are not as confident about their cognitive (intellectual) abilities as they should be."

Those types of gender-based insecurities may be ending, however. In a previous study involving married couples, West discovered that gender-specific differences in memory-confidence disappeared among younger (ages 35 to 45) couples. West credits those results on "historical changes in the beliefs and attitudes of our culture. The older women grew up in a time when men were thought to have more mental

Chapter Exercises are on page 721.

ability than women. Such beliefs—learned in childhood and reinforced over time—are hard to change." ■

# 17.6 Getting to Know You

As social beings, we are curious about how others think and behave. What do they do with their time? Are we in the majority with our opinions on controversial issues? At what age do people typically get married? What proportion of the population is left-handed? What proportion is gay or lesbian? Are people basically honest? Many of these questions can be answered by surveying random or representative samples.

Most national governments have agencies that collect samples to answer some of these questions on a routine basis. The U.S. government collects diverse and numerous statistics on the behavior and opinions of its people. The website http://www.fedstats.gov, which is headlined "The gateway to statistics from over 100 U.S. Federal agencies," provides links to all kinds of information about the U.S. population.

**Example 17.8**

**Lifestyle Statistics from the Census Bureau** The U.S. Census Bureau collects voluminous data on many aspects of American life. The ongoing "Current Population Survey" polls a random sample of U.S. households on a wide variety of topics. From this survey, trends in lifestyle decisions can be tracked. Here are a few examples of trends reported in 1995 ("Marital Status and Living Arrangements," Arlene F. Saluter, http://www.census.gov/population/www/pop-profile/msla.html):

- "The estimated median age at first marriage is higher than ever before. In 1994, the median age at first marriage was 26.7 years for men and 24.5 years for women, approximately $3\frac{1}{2}$ years higher than the median age in 1970 (23.2 years for men and 20.8 years for women).

- An unmarried-couple household is composed of two unrelated adults of the opposite sex (one of whom is the householder) who share a housing unit with or without the presence of children under 15 years old. There were 7 unmarried couples for every 100 married couples in 1994, compared with only 1 for every 100 in 1970. About one-third had children under 15 years old present in the home.

- Children living with one parent (18.6 million) represented 27 percent of all children under 18 years old in 1994, up from 12 percent in 1970. The majority lived with their mother, but an increasing proportion lived with their father. In 1994, 12 percent of the children in a one-parent situation lived with their father, up from 9 percent in 1970."

Notice that these statistics are presented as if they were known for the population, but in fact, they are based on random samples of various sizes. Consequently, they have a margin of error that could be computed from knowing the sample size. ■

**Example 17.9**

**In Whom Do We Trust?**   If you want to know about the opinions of Americans on almost any topic, visit the website for the Gallup Organization (http://www.gallup.com) and type in the key word of your choice. (Unfortunately, Gallup now charges for this service.) For instance, a news release on November 27, 2000, was headlined "Nurses Remain at Top of Honesty and Ethics Poll, Car Salesmen Still Seen as Least Honest and Ethical." The accompanying story (Darren K. Carlson, http://www.gallup.com/poll/releases/pr001127.asp) reported on a poll that has been taken yearly since 1977. Respondents were asked, "Please tell me how you would rate the honesty and ethical standards of people in these different fields—very high, high, average, low, or very low" and were then given a list of several dozen occupations, in random order. Some of the results highlighted in the news release were

- "For the second year in a row, Gallup's survey on honesty and ethics in professions finds that the American public rates nursing as the field with the highest standards of honesty and ethics. Almost eight in 10 Americans —79%—say nurses have 'very high' or 'high' ethical standards. Pharmacists finish second with 67%. Pharmacists had consistently finished first in the survey, until nurses were added to the list in 1999.

- Only one profession is rated as having 'low' or 'very low' standards by a majority of Americans: car sales.

- Americans remain skeptical of the honesty and ethics of their elected officials. As with every previous year, less than half the public said any elected official on Gallup's list had 'very high' or 'high' honesty and ethics.

- The long-term trends of Gallup's honesty and ethics survey reveal a growing skepticism among the American public regarding the ethics of news professionals and lawyers. The three news professions surveyed all remain significantly lower than when they were first placed on the list. Twenty-one percent of the public says journalists are honest and ethical, a percentage that has declined steadily from its debut of 33% in 1977.

- Lawyers are often the punch line to jokes regarding professional ethics and honesty. They debuted on Gallup's survey list in 1977 with 26%. Lawyers have not been rated as having high ethical standards by more than 20% of the public since 1991, and the latest poll shows a rating of 17%."

College teachers were rated highly by 59%, and grade school and high school teachers were rated highly by 62%. And, of course, to interpret the results correctly, you need to know that "the results reported here are based on telephone interviews with a randomly selected national sample of 1,028 adults, 18 years and older, conducted November 13–15, 2000. For results based on the whole sample, one can say with 95 percent confidence that the maximum error attributable to sampling and other random effects is plus or minus 3 percentage points." ∎

Chapter Exercises are on page 721.

# 17.7  Words to the Wise

We hope that by now you realize that you are likely to encounter statistical studies in your personal and professional life and that they can provide useful information. We also hope you realize that results from such studies can be misleading if you don't interpret them wisely.

Throughout this book we have presented warnings and examples of how statistical studies are often misinterpreted. The ten guiding principles presented here are a synthesis of those warnings. Keeping these principles in mind when you read statistical studies will help you gain wisdom about the world around you while maintaining a healthy dose of skepticism about legitimate conclusions.

*in summary*       ## Ten Guiding Principles

1. A representative sample can be used to make inferences about a larger population, but descriptive statistics are the only useful results for an unrepresentative sample.
2. Cause and effect can be inferred from randomized experiments but generally not from observational studies, in which confounding variables are likely to cloud the interpretation.
3. A conservative estimate of *sampling* error in a survey is the margin of error, $1/\sqrt{n}$. This provides a bound on the difference between the true proportion and the sample proportion that holds for at least 95% of properly conducted surveys.
4. The margin of error does *not* include nonsampling error, such as errors due to biased wording, nonresponse, and so on.
5. When the individuals measured constitute the whole population, there is no need for statistical inference because the truth is known.
6. A significance test based on a very large sample is likely to produce a statistically significant result even if the true value is close to the null value. In such cases, it is wise to examine the magnitude of the parameter with a confidence interval to determine whether the result has practical importance.
7. A significance test based on a small sample may not produce a statistically significant result even if the true value differs substantially from the null value. It is important that the null hypothesis not be "accepted" on this basis.
8. When deciding how readily to reject the null hypothesis (what significance level to use), it is important to consider the consequences of type 1 and type 2 errors. If a type 1 error has serious consequences, the level of significance should be small. If a type 2 error is more serious, a higher level of significance should be used.
9. A study that examines many hypotheses could find one or more statistically significant results just by chance, so you should try to find out how many tests were conducted when you read about a significant result. For instance, with a level of significance of .05, about 1 test in 20 should result in statistical significance when all of the null hypotheses are correct and the tests are independent. It is common in large studies to find that one test attracts media attention, so it is important to know whether

that test was the only one out of many conducted that achieved statistical significance.

10. You will sometimes read that researchers were surprised to find "no effect" and that a study "failed to replicate" an earlier finding of statistical significance. In that case, consider two possible explanations. One is that the sample size was too small and the test had low power. The other possibility is that the result in the first study was based on a type 1 error. This is particularly likely if the effect was moderate and was part of a larger study that covered multiple hypotheses.

*If you still think statistics is a boring and useless subject, check to make sure you are still breathing, your heart is beating, and your mind has not turned off!*

# Exercises

● Denotes basic skills exercises

◆ Denotes dataset is available in StatisticsNow at **http://1pass.thomson.com** or on your CD but is **not required** to solve the exercise.

**Bold-numbered exercises** have answers in the back of the text and fully worked solutions in the Student Solutions Manual.

**Statistics⬢Now™** Go to the StatisticsNow website at **http://1pass.thomson.com** to:
• Assess your understanding of this chapter
• Check your readiness for an exam by taking the Pre-Test quiz and exploring the resources in the Personalized Learning Plan

## Chapter Exercises

*Exercises 1 to 15 are based on the following three reports taken from news organization websites. In each case, the headline and a few sentences from the news story are provided. The original source for Studies #2 and #3 are on the CD for this book*

**Study #1:** "*Calcium Lowers Blood Pressure in Black Teens:* In the new study, researchers followed 116 African-American teens who were enrolled in grades 10 through 12 at three high schools in Los Angeles. Over the course of eight weeks, the study participants were given either a placebo or a calcium supplement. They were also asked to fill out food questionnaires. Researchers measured the teens' blood pressure at the outset of the study and two, four and eight weeks later." (Linda Carroll, Medical Tribune News Service, Sept. 7, 1998; study published in the *American Journal of Clinical Nutrition* (1998), Vol. 68, 648–655.)

**Study #2:** "*Daily Two-Mile Walk Halves Death Risk:* Between 1980 and 1982, multicenter researchers in the Honolulu Heart Program studied 707 nonsmoking, retired men, aged 61 to 81 years, and collected mortality data on these men over the following 12 years. During the study, 208 of the men died. The study results show that while 43.1% of men who walked less than one mile per day died, only half this figure—21.5%—of the men who walked more than two miles per day died." (Reuters News, Jan. 8, 1998; study published in *The New England Journal of Medicine* (1998), Vol. 338, 94–99.)

**Study #3:** "*Tea Doubles Chance Of Conception:* Tea for two might make three, according to a new study. The investigators asked 187 women, each of whom said they were trying to conceive, to record information concerning their daily dietary intake over a one-year period. According to the researchers, an analysis of the data suggests drinking one-half cup or more of tea daily approximately doubled the odds of conception per cycle, compared with non-tea drinkers. The mechanism behind tea's apparent influence on fertility remains unclear. But the investigators point out that, on average, tea drinking is associated with a 'preventive or healthier' lifestyle, which might encourage conception." (Reuters News, Feb. 27, 1998; study published in the *American Journal of Public Health* (1998), Vol. 88, No. 2: 270–274.)

**17.1** Each of the three headlines implies a causal relationship between an explanatory and response variable. For each of the three studies, describe the explanatory and response variables.

17.2 Do you think Study #1 was a randomized experiment or an observational study? Explain. Based on your answer, explain whether or not the headline is justified.

17.3 Do you think Study #2 was a randomized experiment or an observational study? Explain. Based on your answer, explain whether or not the headline is justified.

**17.4** Do you think Study #3 was a randomized experiment or an observational study? Explain. Based on your answer, explain whether or not the headline is justified.

17.5 Give an example of a possible confounding variable for Study #2. Explain why you think it fits the definition of a confounding variable. (See p. 120.)

**17.6** Give an example of a possible confounding variable for Study #3. Explain why you think it fits the definition of a confounding variable. (See p. 120.)

17.7 Can the results of Study #1 be applied to a larger population? If so, to what population? If not, why not?

17.8 Can the results of Study #2 be applied to a larger population? If so, to what population? If not, why not?

**17.9** Can the results of Study #3 be applied to a larger population? If so, to what population? If not, why not?

17.10 In Study #1, participants were also asked to fill out a food questionnaire. Another quote from the news article was: "Overall, diastolic blood pressure—the second of the two numbers given when blood pressure is measured—dropped about two points in the group given calcium supplements. But the drop was almost five points in teens with a particularly low intake of calcium-containing foods."
   a. Why do you think the researchers wanted the information provided by the food questionnaire?
   b. Is "intake of calcium-containing foods" a confounding variable or an interacting variable in this study? Explain.

17.11 Here are some additional quotes from the article for Study #2. Comment on each quote using issues covered in this chapter.
   a. "The investigators also report that 'when the distance walked is increased by one mile per day . . . the risk of death can be reduced by 19%.'"
   b. "Cancer was the most common cause of death in the men studied. The research team found that 13.4% of the men who walked less than one mile per day died of cancer, but only 5.3% of those who walked more than two miles per day died of cancer. Even when age was included in the analysis, this result held."
   c. "Walking also appeared to protect against dying from heart disease and stroke, but the findings were not statistically significant, meaning that they could have occurred by chance."

17.12 The article for Study #3 also noted: "The study results also show that the women who consumed the most tea 'had significantly higher energy and fat intakes' than women who drank less tea."
   a. Do you think the word *significantly* in this quote refers to statistical significance? If not, why not? If so, describe in words the null and alternative hypotheses you think the researchers tested to reach that conclusion.
   b. Are energy and fat intake confounding variables or interacting variables in this study? Explain.

17.13 The article for Study #3 also said: "The California researchers say they found no significant association between coffee intake and fertility, but 'in later cycles . . . there was a suggestion that both total caffeine and

coffee were associated with a nonsignificant reduction in fertility,' a finding in keeping with previous research which pointed to the possibility that coffee or caffeine might lower conception rates."
   a. The phrase "*no* significant association between coffee intake and fertility" seems to contradict the "suggestion that both total caffeine and coffee were associated with a nonsignificant reduction in fertility." Explain this apparent contradiction.
   b. Refer to item 10 in Section 17.7, and discuss it in the context of this quote.

17.14 The two main variables measured in Study #1 were the treatment given to the participant and the participant's blood pressure.
   a. For each of the two variables, determine whether it was recorded as a categorical variable or a quantitative variable. If it was a categorical variable, specify the categories.
   b. Researchers measured the participants' blood pressure at the beginning of the study and two, four, and eight weeks later. If they wanted to determine whether blood pressure had dropped significantly from the beginning of the study to eight weeks later for the participants in the calcium group, would they use a paired $t$-test or a test for the difference in means for independent samples? Explain.
   c. Refer to part (b). Write the null and alternative hypotheses that the researchers would use.
   d. If the researchers wanted to examine the change in blood pressure after eight weeks of taking calcium compared to eight weeks of taking a placebo, what hypothesis-testing method would they use?
   e. Refer to part (d). Write the null and alternative hypotheses that the researchers would use.

**17.15** Refer to the headline in Study #2 and to the statement "While 43.1% of the men who walked less than one mile per day died, only half this figure—21.5%—of the men who walked more than two miles per day died."
   a. What two categorical variables measured on each participant are used in making that statement? What are the categories for each one?
   b. If the researchers wanted to determine whether the result was statistically significant, what inference procedure would be appropriate? Be specific.

17.16 Find an example of an observational study in the news and answer these questions about it:
   a. Explain why it fits the definition of an observational study rather than a randomized experiment.
   b. Describe an explanatory and response variable considered in the study.
   c. Using the variables in part (b), give an example of a possible confounding variable and explain why it fits the definition of a confounding variable.
   d. Write two headlines to accompany the study, one that is correct and one that implies a conclusion that is not justified. Explain which is which and why. (You may use the actual headline accompanying the story as one of these.)

**Table A.1** Standard Normal Probabilities (for $z > 0$)

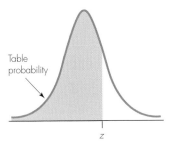

Table probability

| z | .00 | .01 | .02 | .03 | .04 | .05 | .06 | .07 | .08 | .09 |
|---|---|---|---|---|---|---|---|---|---|---|
| 0.0 | .5000 | .5040 | .5080 | .5120 | .5160 | .5199 | .5239 | .5279 | .5319 | .5359 |
| 0.1 | .5398 | .5438 | .5478 | .5517 | .5557 | .5596 | .5636 | .5675 | .5714 | .5753 |
| 0.2 | .5793 | .5832 | .5871 | .5910 | .5948 | .5987 | .6026 | .6064 | .6103 | .6141 |
| 0.3 | .6179 | .6217 | .6255 | .6293 | .6331 | .6368 | .6406 | .6443 | .6480 | .6517 |
| 0.4 | .6554 | .6591 | .6628 | .6664 | .6700 | .6736 | .6772 | .6808 | .6844 | .6879 |
| 0.5 | .6915 | .6950 | .6985 | .7019 | .7054 | .7088 | .7123 | .7157 | .7190 | .7224 |
| 0.6 | .7257 | .7291 | .7324 | .7357 | .7389 | .7422 | .7454 | .7486 | .7517 | .7549 |
| 0.7 | .7580 | .7611 | .7642 | .7673 | .7704 | .7734 | .7764 | .7794 | .7823 | .7852 |
| 0.8 | .7881 | .7910 | .7939 | .7967 | .7995 | .8023 | .8051 | .8078 | .8106 | .8133 |
| 0.9 | .8159 | .8186 | .8212 | .8238 | .8264 | .8289 | .8315 | .8340 | .8365 | .8389 |
| 1.0 | .8413 | .8438 | .8461 | .8485 | .8508 | .8531 | .8554 | .8577 | .8599 | .8621 |
| 1.1 | .8643 | .8665 | .8686 | .8708 | .8729 | .8749 | .8770 | .8790 | .8810 | .8830 |
| 1.2 | .8849 | .8869 | .8888 | .8907 | .8925 | .8944 | .8962 | .8980 | .8997 | .9015 |
| 1.3 | .9032 | .9049 | .9066 | .9082 | .9099 | .9115 | .9131 | .9147 | .9162 | .9177 |
| 1.4 | .9192 | .9207 | .9222 | .9236 | .9251 | .9265 | .9279 | .9292 | .9306 | .9319 |
| 1.5 | .9332 | .9345 | .9357 | .9370 | .9382 | .9394 | .9406 | .9418 | .9429 | .9441 |
| 1.6 | .9452 | .9463 | .9474 | .9484 | .9495 | .9505 | .9515 | .9525 | .9535 | .9545 |
| 1.7 | .9554 | .9564 | .9573 | .9582 | .9591 | .9599 | .9608 | .9616 | .9625 | .9633 |
| 1.8 | .9641 | .9649 | .9656 | .9664 | .9671 | .9678 | .9686 | .9693 | .9699 | .9706 |
| 1.9 | .9713 | .9719 | .9726 | .9732 | .9738 | .9744 | .9750 | .9756 | .9761 | .9767 |
| 2.0 | .9772 | .9778 | .9783 | .9788 | .9793 | .9798 | .9803 | .9808 | .9812 | .9817 |
| 2.1 | .9821 | .9826 | .9830 | .9834 | .9838 | .9842 | .9846 | .9850 | .9854 | .9857 |
| 2.2 | .9861 | .9864 | .9868 | .9871 | .9875 | .9878 | .9881 | .9884 | .9887 | .9890 |
| 2.3 | .9893 | .9896 | .9898 | .9901 | .9904 | .9906 | .9909 | .9911 | .9913 | .9916 |
| 2.4 | .9918 | .9920 | .9922 | .9925 | .9927 | .9929 | .9931 | .9932 | .9934 | .9936 |
| 2.5 | .9938 | .9940 | .9941 | .9943 | .9945 | .9946 | .9948 | .9949 | .9951 | .9952 |
| 2.6 | .9953 | .9955 | .9956 | .9957 | .9959 | .9960 | .9961 | .9962 | .9963 | .9964 |
| 2.7 | .9965 | .9966 | .9967 | .9968 | .9969 | .9970 | .9971 | .9972 | .9973 | .9974 |
| 2.8 | .9974 | .9975 | .9976 | .9977 | .9977 | .9978 | .9979 | .9979 | .9980 | .9981 |
| 2.9 | .9981 | .9982 | .9982 | .9983 | .9984 | .9984 | .9985 | .9985 | .9986 | .9986 |
| 3.0 | .9987 | .9987 | .9987 | .9988 | .9988 | .9989 | .9989 | .9989 | .9990 | .9990 |
| 3.1 | .9990 | .9991 | .9991 | .9991 | .9992 | .9992 | .9992 | .9992 | .9993 | .9993 |
| 3.2 | .9993 | .9993 | .9994 | .9994 | .9994 | .9994 | .9994 | .9995 | .9995 | .9995 |
| 3.3 | .9995 | .9995 | .9995 | .9996 | .9996 | .9996 | .9996 | .9996 | .9996 | .9997 |
| 3.4 | .9997 | .9997 | .9997 | .9997 | .9997 | .9997 | .9997 | .9997 | .9997 | .9998 |

In the Extreme (for $z > 0$)

| z | 3.09 | 3.72 | 4.26 | 4.75 | 5.20 | 5.61 | 6.00 |
|---|---|---|---|---|---|---|---|
| Probability | .999 | .9999 | .99999 | .999999 | .9999999 | .99999999 | .999999999 |

S-PLUS was used to determine information for the "In the Extreme" portion of the table.

**Table A.2**  *t\** Multipliers for Confidence Intervals and Rejection Region Critical Values

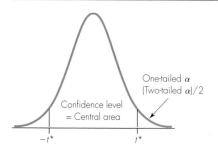

One-tailed $\alpha$
(Two-tailed $\alpha$)/2

Confidence level
= Central area

$-t^*$     $t^*$

| df | .80 | .90 | .95 | .98 | .99 | .998 | .999 |
|---|---|---|---|---|---|---|---|
| | | | | Confidence Level (*k%*) | | | |
| 1 | 3.08 | 6.31 | 12.71 | 31.82 | 63.66 | 318.31 | 636.62 |
| 2 | 1.89 | 2.92 | 4.30 | 6.96 | 9.92 | 22.33 | 31.60 |
| 3 | 1.64 | 2.35 | 3.18 | 4.54 | 5.84 | 10.21 | 12.92 |
| 4 | 1.53 | 2.13 | 2.78 | 3.75 | 4.60 | 7.17 | 8.61 |
| 5 | 1.48 | 2.02 | 2.57 | 3.36 | 4.03 | 5.89 | 6.87 |
| 6 | 1.44 | 1.94 | 2.45 | 3.14 | 3.71 | 5.21 | 5.96 |
| 7 | 1.41 | 1.89 | 2.36 | 3.00 | 3.50 | 4.79 | 5.41 |
| 8 | 1.40 | 1.86 | 2.31 | 2.90 | 3.36 | 4.50 | 5.04 |
| 9 | 1.38 | 1.83 | 2.26 | 2.82 | 3.25 | 4.30 | 4.78 |
| 10 | 1.37 | 1.81 | 2.23 | 2.76 | 3.17 | 4.14 | 4.59 |
| 11 | 1.36 | 1.80 | 2.20 | 2.72 | 3.11 | 4.02 | 4.44 |
| 12 | 1.36 | 1.78 | 2.18 | 2.68 | 3.05 | 3.93 | 4.32 |
| 13 | 1.35 | 1.77 | 2.16 | 2.65 | 3.01 | 3.85 | 4.22 |
| 14 | 1.35 | 1.76 | 2.14 | 2.62 | 2.98 | 3.79 | 4.14 |
| 15 | 1.34 | 1.75 | 2.13 | 2.60 | 2.95 | 3.73 | 4.07 |
| 16 | 1.34 | 1.75 | 2.12 | 2.58 | 2.92 | 3.69 | 4.01 |
| 17 | 1.33 | 1.74 | 2.11 | 2.57 | 2.90 | 3.65 | 3.97 |
| 18 | 1.33 | 1.73 | 2.10 | 2.55 | 2.88 | 3.61 | 3.92 |
| 19 | 1.33 | 1.73 | 2.09 | 2.54 | 2.86 | 3.58 | 3.88 |
| 20 | 1.33 | 1.72 | 2.09 | 2.53 | 2.85 | 3.55 | 3.85 |
| 21 | 1.32 | 1.72 | 2.08 | 2.52 | 2.83 | 3.53 | 3.82 |
| 22 | 1.32 | 1.72 | 2.07 | 2.51 | 2.82 | 3.50 | 3.79 |
| 23 | 1.32 | 1.71 | 2.07 | 2.50 | 2.81 | 3.48 | 3.77 |
| 24 | 1.32 | 1.71 | 2.06 | 2.49 | 2.80 | 3.47 | 3.75 |
| 25 | 1.32 | 1.71 | 2.06 | 2.49 | 2.79 | 3.45 | 3.73 |
| 26 | 1.31 | 1.71 | 2.06 | 2.48 | 2.78 | 3.43 | 3.71 |
| 27 | 1.31 | 1.70 | 2.05 | 2.47 | 2.77 | 3.42 | 3.69 |
| 28 | 1.31 | 1.70 | 2.05 | 2.47 | 2.76 | 3.41 | 3.67 |
| 29 | 1.31 | 1.70 | 2.05 | 2.46 | 2.76 | 3.40 | 3.66 |
| 30 | 1.31 | 1.70 | 2.04 | 2.46 | 2.75 | 3.39 | 3.65 |
| 40 | 1.30 | 1.68 | 2.02 | 2.42 | 2.70 | 3.31 | 3.55 |
| 50 | 1.30 | 1.68 | 2.01 | 2.40 | 2.68 | 3.26 | 3.50 |
| 60 | 1.30 | 1.67 | 2.00 | 2.39 | 2.66 | 3.23 | 3.46 |
| 70 | 1.29 | 1.67 | 1.99 | 2.38 | 2.65 | 3.21 | 3.44 |
| 80 | 1.29 | 1.66 | 1.99 | 2.37 | 2.64 | 3.20 | 3.42 |
| 90 | 1.29 | 1.66 | 1.99 | 2.37 | 2.63 | 3.18 | 3.40 |
| 100 | 1.29 | 1.66 | 1.98 | 2.36 | 2.63 | 3.17 | 3.39 |
| 1000 | 1.282 | 1.646 | 1.962 | 2.330 | 2.581 | 3.098 | 3.300 |
| Infinite | 1.281 | 1.645 | 1.960 | 2.326 | 2.576 | 3.090 | 3.291 |
| Two-tailed $\alpha$ | .20 | .10 | .05 | .02 | .01 | .002 | .001 |
| One-tailed $\alpha$ | .10 | .05 | .025 | .01 | .005 | .001 | .0005 |

Note that the *t*-distribution with infinite df is the standard normal distribution.

**Table A.3** One-Sided p-Values for Significance Tests Based on a t-Statistic

- A p-value in the table is the area to the right of t.
- Double the value if the alternative hypothesis is two-sided (not equal).

| | | | | Absolute Value of t-Statistic | | | | |
|---|---|---|---|---|---|---|---|---|
| df | 1.28 | 1.50 | 1.65 | 1.80 | 2.00 | 2.33 | 2.58 | 3.00 |
| 1 | .211 | .187 | .173 | .161 | .148 | .129 | .118 | .102 |
| 2 | .164 | .136 | .120 | .107 | .092 | .073 | .062 | .048 |
| 3 | .145 | .115 | .099 | .085 | .070 | .051 | .041 | .029 |
| 4 | .135 | .104 | .087 | .073 | .058 | .040 | .031 | .020 |
| 5 | .128 | .097 | .080 | .066 | .051 | .034 | .025 | .015 |
| 6 | .124 | .092 | .075 | .061 | .046 | .029 | .021 | .012 |
| 7 | .121 | .089 | .071 | .057 | .043 | .026 | .018 | .010 |
| 8 | .118 | .086 | .069 | .055 | .040 | .024 | .016 | .009 |
| 9 | .116 | .084 | .067 | .053 | .038 | .022 | .015 | .007 |
| 10 | .115 | .082 | .065 | .051 | .037 | .021 | .014 | .007 |
| 11 | .113 | .081 | .064 | .050 | .035 | .020 | .013 | .006 |
| 12 | .112 | .080 | .062 | .049 | .034 | .019 | .012 | .006 |
| 13 | .111 | .079 | .061 | .048 | .033 | .018 | .011 | .005 |
| 14 | .111 | .078 | .061 | .047 | .033 | .018 | .011 | .005 |
| 15 | .110 | .077 | .060 | .046 | .032 | .017 | .010 | .004 |
| 16 | .109 | .077 | .059 | .045 | .031 | .017 | .010 | .004 |
| 17 | .109 | .076 | .059 | .045 | .031 | .016 | .010 | .004 |
| 18 | .108 | .075 | .058 | .044 | .030 | .016 | .009 | .004 |
| 19 | .108 | .075 | .058 | .044 | .030 | .015 | .009 | .004 |
| 20 | .108 | .075 | .057 | .043 | .030 | .015 | .009 | .004 |
| 21 | .107 | .074 | .057 | .043 | .029 | .015 | .009 | .003 |
| 22 | .107 | .074 | .057 | .043 | .029 | .015 | .009 | .003 |
| 23 | .107 | .074 | .056 | .042 | .029 | .014 | .008 | .003 |
| 24 | .106 | .073 | .056 | .042 | .028 | .014 | .008 | .003 |
| 25 | .106 | .073 | .056 | .042 | .028 | .014 | .008 | .003 |
| 26 | .106 | .073 | .055 | .042 | .028 | .014 | .008 | .003 |
| 27 | .106 | .073 | .055 | .042 | .028 | .014 | .008 | .003 |
| 28 | .106 | .072 | .055 | .041 | .028 | .014 | .008 | .003 |
| 29 | .105 | .072 | .055 | .041 | .027 | .013 | .008 | .003 |
| 30 | .105 | .072 | .055 | .041 | .027 | .013 | .008 | .003 |
| 40 | .104 | .071 | .053 | .040 | .026 | .012 | .007 | .002 |
| 50 | .103 | .070 | .053 | .039 | .025 | .012 | .006 | .002 |
| 60 | .103 | .069 | .052 | .038 | .025 | .012 | .006 | .002 |
| 70 | .102 | .069 | .052 | .038 | .025 | .011 | .006 | .002 |
| 80 | .102 | .069 | .051 | .038 | .024 | .011 | .006 | .002 |
| 90 | .102 | .069 | .051 | .038 | .024 | .011 | .006 | .002 |
| 100 | .102 | .068 | .051 | .037 | .024 | .011 | .006 | .002 |
| 1000 | .100 | .067 | .050 | .036 | .023 | .010 | .005 | .001 |
| Infinite | .1003 | .0668 | .0495 | .0359 | .0228 | .0099 | .0049 | .0013 |

Note that the t-distribution with infinite df is the standard normal distribution.

**Table A.4**  Critical Values for F-Test ($\alpha = .05$)

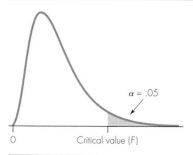

$\alpha = .05$

0    Critical value (F)

| | | | | | Numerator df | | | | | |
|---|---|---|---|---|---|---|---|---|---|---|
| Denom df | 1 | 2 | 3 | 4 | 5 | 6 | 7 | 8 | 9 | 10 |
| 1 | 161.45 | 199.50 | 215.71 | 224.58 | 230.16 | 233.99 | 236.77 | 238.88 | 240.54 | 241.88 |
| 2 | 18.51 | 19.00 | 19.16 | 19.25 | 19.30 | 19.33 | 19.35 | 19.37 | 19.38 | 19.40 |
| 3 | 10.13 | 9.55 | 9.28 | 9.12 | 9.01 | 8.94 | 8.89 | 8.85 | 8.81 | 8.79 |
| 4 | 7.71 | 6.94 | 6.59 | 6.39 | 6.26 | 6.16 | 6.09 | 6.04 | 6.00 | 5.96 |
| 5 | 6.61 | 5.79 | 5.41 | 5.19 | 5.05 | 4.95 | 4.88 | 4.82 | 4.77 | 4.74 |
| 6 | 5.99 | 5.14 | 4.76 | 4.53 | 4.39 | 4.28 | 4.21 | 4.15 | 4.10 | 4.06 |
| 7 | 5.59 | 4.74 | 4.35 | 4.12 | 3.97 | 3.87 | 3.79 | 3.73 | 3.68 | 3.64 |
| 8 | 5.32 | 4.46 | 4.07 | 3.84 | 3.69 | 3.58 | 3.50 | 3.44 | 3.39 | 3.35 |
| 9 | 5.12 | 4.26 | 3.86 | 3.63 | 3.48 | 3.37 | 3.29 | 3.23 | 3.18 | 3.14 |
| 10 | 4.96 | 4.10 | 3.71 | 3.48 | 3.33 | 3.22 | 3.14 | 3.07 | 3.02 | 2.98 |
| 11 | 4.84 | 3.98 | 3.59 | 3.36 | 3.20 | 3.09 | 3.01 | 2.95 | 2.90 | 2.85 |
| 12 | 4.75 | 3.89 | 3.49 | 3.26 | 3.11 | 3.00 | 2.91 | 2.85 | 2.80 | 2.75 |
| 13 | 4.67 | 3.81 | 3.41 | 3.18 | 3.03 | 2.92 | 2.83 | 2.77 | 2.71 | 2.67 |
| 14 | 4.60 | 3.74 | 3.34 | 3.11 | 2.96 | 2.85 | 2.76 | 2.70 | 2.65 | 2.60 |
| 15 | 4.54 | 3.68 | 3.29 | 3.06 | 2.90 | 2.79 | 2.71 | 2.64 | 2.59 | 2.54 |
| 16 | 4.49 | 3.63 | 3.24 | 3.01 | 2.85 | 2.74 | 2.66 | 2.59 | 2.54 | 2.49 |
| 17 | 4.45 | 3.59 | 3.20 | 2.96 | 2.81 | 2.70 | 2.61 | 2.55 | 2.49 | 2.45 |
| 18 | 4.41 | 3.55 | 3.16 | 2.93 | 2.77 | 2.66 | 2.58 | 2.51 | 2.46 | 2.41 |
| 19 | 4.38 | 3.52 | 3.13 | 2.90 | 2.74 | 2.63 | 2.54 | 2.48 | 2.42 | 2.38 |
| 20 | 4.35 | 3.49 | 3.10 | 2.87 | 2.71 | 2.60 | 2.51 | 2.45 | 2.39 | 2.35 |
| 21 | 4.32 | 3.47 | 3.07 | 2.84 | 2.68 | 2.57 | 2.49 | 2.42 | 2.37 | 2.32 |
| 22 | 4.30 | 3.44 | 3.05 | 2.82 | 2.66 | 2.55 | 2.46 | 2.40 | 2.34 | 2.30 |
| 23 | 4.28 | 3.42 | 3.03 | 2.80 | 2.64 | 2.53 | 2.44 | 2.37 | 2.32 | 2.27 |
| 24 | 4.26 | 3.40 | 3.01 | 2.78 | 2.62 | 2.51 | 2.42 | 2.36 | 2.30 | 2.25 |
| 25 | 4.24 | 3.39 | 2.99 | 2.76 | 2.60 | 2.49 | 2.40 | 2.34 | 2.28 | 2.24 |
| 26 | 4.23 | 3.37 | 2.98 | 2.74 | 2.59 | 2.47 | 2.39 | 2.32 | 2.27 | 2.22 |
| 27 | 4.21 | 3.35 | 2.96 | 2.73 | 2.57 | 2.46 | 2.37 | 2.31 | 2.25 | 2.20 |
| 28 | 4.20 | 3.34 | 2.95 | 2.71 | 2.56 | 2.45 | 2.36 | 2.29 | 2.24 | 2.19 |
| 29 | 4.18 | 3.33 | 2.93 | 2.70 | 2.55 | 2.43 | 2.35 | 2.28 | 2.22 | 2.18 |
| 30 | 4.17 | 3.32 | 2.92 | 2.69 | 2.53 | 2.42 | 2.33 | 2.27 | 2.21 | 2.16 |
| 40 | 4.08 | 3.23 | 2.84 | 2.61 | 2.45 | 2.34 | 2.25 | 2.18 | 2.12 | 2.08 |
| 50 | 4.03 | 3.18 | 2.79 | 2.56 | 2.40 | 2.29 | 2.20 | 2.13 | 2.07 | 2.03 |
| 60 | 4.00 | 3.15 | 2.76 | 2.53 | 2.37 | 2.25 | 2.17 | 2.10 | 2.04 | 1.99 |
| 70 | 3.98 | 3.13 | 2.74 | 2.50 | 2.35 | 2.23 | 2.14 | 2.07 | 2.02 | 1.97 |
| 80 | 3.96 | 3.11 | 2.72 | 2.49 | 2.33 | 2.21 | 2.13 | 2.06 | 2.00 | 1.95 |
| 90 | 3.95 | 3.10 | 2.71 | 2.47 | 2.32 | 2.20 | 2.11 | 2.04 | 1.99 | 1.94 |
| 100 | 3.94 | 3.09 | 2.70 | 2.46 | 2.31 | 2.19 | 2.10 | 2.03 | 1.97 | 1.93 |
| 200 | 3.89 | 3.04 | 2.65 | 2.42 | 2.26 | 2.14 | 2.06 | 1.98 | 1.93 | 1.88 |
| 500 | 3.86 | 3.01 | 2.62 | 2.39 | 2.23 | 2.12 | 2.03 | 1.96 | 1.90 | 1.85 |
| 1000 | 3.85 | 3.00 | 2.61 | 2.38 | 2.22 | 2.11 | 2.02 | 1.95 | 1.89 | 1.84 |

**Table A.4**  Critical Values for *F*-Test ($\alpha = .01$)

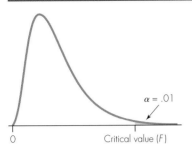

$\alpha = .01$

0        Critical value (*F*)

| | Numerator df | | | | | | | | | |
|---|---|---|---|---|---|---|---|---|---|---|
| *Denom df* | *1* | *2* | *3* | *4* | *5* | *6* | *7* | *8* | *9* | *10* |
| 1 | 4052 | 4999 | 5404 | 5624 | 5764 | 5859 | 5928 | 5981 | 6022 | 6056 |
| 2 | 98.50 | 99.00 | 99.16 | 99.25 | 99.30 | 99.33 | 99.36 | 99.38 | 99.39 | 99.40 |
| 3 | 34.12 | 30.82 | 29.46 | 28.71 | 28.24 | 27.91 | 27.67 | 27.49 | 27.34 | 27.23 |
| 4 | 21.20 | 18.00 | 16.69 | 15.98 | 15.52 | 15.21 | 14.98 | 14.80 | 14.66 | 14.55 |
| 5 | 16.26 | 13.27 | 12.06 | 11.39 | 10.97 | 10.67 | 10.46 | 10.29 | 10.16 | 10.05 |
| 6 | 13.75 | 10.92 | 9.78 | 9.15 | 8.75 | 8.47 | 8.26 | 8.10 | 7.98 | 7.87 |
| 7 | 12.25 | 9.55 | 8.45 | 7.85 | 7.46 | 7.19 | 6.99 | 6.84 | 6.72 | 6.62 |
| 8 | 11.26 | 8.65 | 7.59 | 7.01 | 6.63 | 6.37 | 6.18 | 6.03 | 5.91 | 5.81 |
| 9 | 10.56 | 8.02 | 6.99 | 6.42 | 6.06 | 5.80 | 5.61 | 5.47 | 5.35 | 5.26 |
| 10 | 10.04 | 7.56 | 6.55 | 5.99 | 5.64 | 5.39 | 5.20 | 5.06 | 4.94 | 4.85 |
| 11 | 9.65 | 7.21 | 6.22 | 5.67 | 5.32 | 5.07 | 4.89 | 4.74 | 4.63 | 4.54 |
| 12 | 9.33 | 6.93 | 5.95 | 5.41 | 5.06 | 4.82 | 4.64 | 4.50 | 4.39 | 4.30 |
| 13 | 9.07 | 6.70 | 5.74 | 5.21 | 4.86 | 4.62 | 4.44 | 4.30 | 4.19 | 4.10 |
| 14 | 8.86 | 6.51 | 5.56 | 5.04 | 4.69 | 4.46 | 4.28 | 4.14 | 4.03 | 3.94 |
| 15 | 8.68 | 6.36 | 5.42 | 4.89 | 4.56 | 4.32 | 4.14 | 4.00 | 3.89 | 3.80 |
| 16 | 8.53 | 6.23 | 5.29 | 4.77 | 4.44 | 4.20 | 4.03 | 3.89 | 3.78 | 3.69 |
| 17 | 8.40 | 6.11 | 5.19 | 4.67 | 4.34 | 4.10 | 3.93 | 3.79 | 3.68 | 3.59 |
| 18 | 8.29 | 6.01 | 5.09 | 4.58 | 4.25 | 4.01 | 3.84 | 3.71 | 3.60 | 3.51 |
| 19 | 8.18 | 5.93 | 5.01 | 4.50 | 4.17 | 3.94 | 3.77 | 3.63 | 3.52 | 3.43 |
| 20 | 8.10 | 5.85 | 4.94 | 4.43 | 4.10 | 3.87 | 3.70 | 3.56 | 3.46 | 3.37 |
| 21 | 8.02 | 5.78 | 4.87 | 4.37 | 4.04 | 3.81 | 3.64 | 3.51 | 3.40 | 3.31 |
| 22 | 7.95 | 5.72 | 4.82 | 4.31 | 3.99 | 3.76 | 3.59 | 3.45 | 3.35 | 3.26 |
| 23 | 7.88 | 5.66 | 4.76 | 4.26 | 3.94 | 3.71 | 3.54 | 3.41 | 3.30 | 3.21 |
| 24 | 7.82 | 5.61 | 4.72 | 4.22 | 3.90 | 3.67 | 3.50 | 3.36 | 3.26 | 3.17 |
| 25 | 7.77 | 5.57 | 4.68 | 4.18 | 3.85 | 3.63 | 3.46 | 3.32 | 3.22 | 3.13 |
| 26 | 7.72 | 5.53 | 4.64 | 4.14 | 3.82 | 3.59 | 3.42 | 3.29 | 3.18 | 3.09 |
| 27 | 7.68 | 5.49 | 4.60 | 4.11 | 3.78 | 3.56 | 3.39 | 3.26 | 3.15 | 3.06 |
| 28 | 7.64 | 5.45 | 4.57 | 4.07 | 3.75 | 3.53 | 3.36 | 3.23 | 3.12 | 3.03 |
| 29 | 7.60 | 5.42 | 4.54 | 4.04 | 3.73 | 3.50 | 3.33 | 3.20 | 3.09 | 3.00 |
| 30 | 7.56 | 5.39 | 4.51 | 4.02 | 3.70 | 3.47 | 3.30 | 3.17 | 3.07 | 2.98 |
| 40 | 7.31 | 5.18 | 4.31 | 3.83 | 3.51 | 3.29 | 3.12 | 2.99 | 2.89 | 2.80 |
| 50 | 7.17 | 5.06 | 4.20 | 3.72 | 3.41 | 3.19 | 3.02 | 2.89 | 2.78 | 2.70 |
| 60 | 7.08 | 4.98 | 4.13 | 3.65 | 3.34 | 3.12 | 2.95 | 2.82 | 2.72 | 2.63 |
| 70 | 7.01 | 4.92 | 4.07 | 3.60 | 3.29 | 3.07 | 2.91 | 2.78 | 2.67 | 2.59 |
| 80 | 6.96 | 4.88 | 4.04 | 3.56 | 3.26 | 3.04 | 2.87 | 2.74 | 2.64 | 2.55 |
| 90 | 6.93 | 4.85 | 4.01 | 3.53 | 3.23 | 3.01 | 2.84 | 2.72 | 2.61 | 2.52 |
| 100 | 6.90 | 4.82 | 3.98 | 3.51 | 3.21 | 2.99 | 2.82 | 2.69 | 2.59 | 2.50 |
| 200 | 6.76 | 4.71 | 3.88 | 3.41 | 3.11 | 2.89 | 2.73 | 2.60 | 2.50 | 2.41 |
| 500 | 6.69 | 4.65 | 3.82 | 3.36 | 3.05 | 2.84 | 2.68 | 2.55 | 2.44 | 2.36 |
| 1000 | 6.66 | 4.63 | 3.80 | 3.34 | 3.04 | 2.82 | 2.66 | 2.53 | 2.43 | 2.34 |

**Table A.5** Chi-Square Distribution

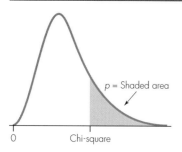

$p$ = Shaded area

0    Chi-square

| | | | | $p$ = Area to Right of Chi-Square Value | | | | | |
|---|---|---|---|---|---|---|---|---|---|
| df | .50 | .25 | .10 | .075 | .05 | .025 | .01 | .005 | .001 |
| 1 | 0.45 | 1.32 | 2.71 | 3.17 | 3.84 | 5.02 | 6.63 | 7.88 | 10.83 |
| 2 | 1.39 | 2.77 | 4.61 | 5.18 | 5.99 | 7.38 | 9.21 | 10.60 | 13.82 |
| 3 | 2.37 | 4.11 | 6.25 | 6.90 | 7.81 | 9.35 | 11.34 | 12.84 | 16.27 |
| 4 | 3.36 | 5.39 | 7.78 | 8.50 | 9.49 | 11.14 | 13.28 | 14.86 | 18.47 |
| 5 | 4.35 | 6.63 | 9.24 | 10.01 | 11.07 | 12.83 | 15.09 | 16.75 | 20.51 |
| 6 | 5.35 | 7.84 | 10.64 | 11.47 | 12.59 | 14.45 | 16.81 | 18.55 | 22.46 |
| 7 | 6.35 | 9.04 | 12.02 | 12.88 | 14.07 | 16.01 | 18.48 | 20.28 | 24.32 |
| 8 | 7.34 | 10.22 | 13.36 | 14.27 | 15.51 | 17.53 | 20.09 | 21.95 | 26.12 |
| 9 | 8.34 | 11.39 | 14.68 | 15.63 | 16.92 | 19.02 | 21.67 | 23.59 | 27.88 |
| 10 | 9.34 | 12.55 | 15.99 | 16.97 | 18.31 | 20.48 | 23.21 | 25.19 | 29.59 |
| 11 | 10.34 | 13.70 | 17.28 | 18.29 | 19.68 | 21.92 | 24.73 | 26.76 | 31.26 |
| 12 | 11.34 | 14.85 | 18.55 | 19.60 | 21.03 | 23.34 | 26.22 | 28.30 | 32.91 |
| 13 | 12.34 | 15.98 | 19.81 | 20.90 | 22.36 | 24.74 | 27.69 | 29.82 | 34.53 |
| 14 | 13.34 | 17.12 | 21.06 | 22.18 | 23.68 | 26.12 | 29.14 | 31.32 | 36.12 |
| 15 | 14.34 | 18.25 | 22.31 | 23.45 | 25.00 | 27.49 | 30.58 | 32.80 | 37.70 |
| 16 | 15.34 | 19.37 | 23.54 | 24.72 | 26.30 | 28.85 | 32.00 | 34.27 | 39.25 |
| 17 | 16.34 | 20.49 | 24.77 | 25.97 | 27.59 | 30.19 | 33.41 | 35.72 | 40.79 |
| 18 | 17.34 | 21.60 | 25.99 | 27.22 | 28.87 | 31.53 | 34.81 | 37.16 | 42.31 |
| 19 | 18.34 | 22.72 | 27.20 | 28.46 | 30.14 | 32.85 | 36.19 | 38.58 | 43.82 |
| 20 | 19.34 | 23.83 | 28.41 | 29.69 | 31.41 | 34.17 | 37.57 | 40.00 | 45.31 |
| 21 | 20.34 | 24.93 | 29.62 | 30.92 | 32.67 | 35.48 | 38.93 | 41.40 | 46.80 |
| 22 | 21.34 | 26.04 | 30.81 | 32.14 | 33.92 | 36.78 | 40.29 | 42.80 | 48.27 |
| 23 | 22.34 | 27.14 | 32.01 | 33.36 | 35.17 | 38.08 | 41.64 | 44.18 | 49.73 |
| 24 | 23.34 | 28.24 | 33.20 | 34.57 | 36.42 | 39.36 | 42.98 | 45.56 | 51.18 |
| 25 | 24.34 | 29.34 | 34.38 | 35.78 | 37.65 | 40.65 | 44.31 | 46.93 | 52.62 |
| 26 | 25.34 | 30.43 | 35.56 | 36.98 | 38.89 | 41.92 | 45.64 | 48.29 | 54.05 |
| 27 | 26.34 | 31.53 | 36.74 | 38.18 | 40.11 | 43.19 | 46.96 | 49.65 | 55.48 |
| 28 | 27.34 | 32.62 | 37.92 | 39.38 | 41.34 | 44.46 | 48.28 | 50.99 | 56.89 |
| 29 | 28.34 | 33.71 | 39.09 | 40.57 | 42.56 | 45.72 | 49.59 | 52.34 | 58.30 |
| 30 | 29.34 | 34.80 | 40.26 | 41.76 | 43.77 | 46.98 | 50.89 | 53.67 | 59.70 |

# References

Abelson, R. P., E. F. Loftus, and A. G. Greenwald (1992). "Attempts to improve the accuracy of self-reports of voting," in *Questions about Questions*, J. M. Tanur (ed.), New York: Russell Sage Foundation, pp. 138–153.

American Statistical Association (1999). Ethical guidelines for statistical practice; available at http://www.amstat.org/profession/.

Anastasi, Anne (1988). *Psychological Testing*, 6th ed., New York: Macmillan.

Anderson, M. J., and S.E. Fienberg (2001). *Who Counts? The Politics of Census-Taking in Contemporary America*, New York: Russell Sage Foundation.

Araf, J. (1994). "Leave Earth or perish: Sagan," *China Post*, December 14, 1994, p. 4.

Ashenfelter, Orley (1994). "Report on expected absentee ballot," revised March 29, 1994. Unpublished report submitted to Judge Clarence Newcomer, U.S. District Court, Eastern District of Pennsylvania, March 30, 1994.

Baird, D. D., and A. J. Wilcox (1985). "Cigarette smoking associated with delayed conception," *Journal of the American Medical Association*, Vol. 253, pp. 2979–2983.

Bem, D., and C. Honorton (1994). "Does psi exist? Replicable evidence for an anomalous process of information transfer," *Psychological Bulletin*, Vol. 115, No. 1, pp. 4–18.

Bickel, P. J., E. A. Hammel, and J. W. O'Connell (1975). "Sex bias in graduate admissions: Data from Berkeley," *Science*, Vol. 187, pp. 298–304.

Borzekowski, D. L. G, T. N. Robinson, and J. D. Killen (2000). "Does the camera add 10 pounds?: Media use, perceived importance of appearance, and weight concerns among teenage girls," *Journal of Adolescent Health*, Vol. 26, pp. 36–41.

Bryson, M. C. (1976). "The *Literary Digest* poll: Making of a statistical myth," *American Statistician*, Vol. 30, pp. 184–185.

Capron, C., and M. Duyme (1991). "Children's IQs and SES of biological and adoptive parents in a balanced cross-fostering study," *European Bulletin of Cognitive Psychology*, Vol. 11, No. 3, pp. 323–348.

Carpenter, J., J. Hagemaster, and B. Joiner (1998). Letter to the editor, *Journal of the American Medical Association*, Vol. 280, No. 22, p. 1905.

Carter, C. L., D. Y Jones, A. Schatzkin, and L. A. Brinton (1989). "A prospective study of reproductive, familial, and socioeconomic risk factors for breast cancer using NHANES I data," *Public Health Reports*, Vol. 104, January–February 1989, pp. 45–49.

Casella, George, and Roger L. Berger (2002). *Statistical Inference*, 2nd ed., Belmont, CA: Brooks/Cole.

Chambers, Chris (2000). "Americans are overwhelmingly happy and optimistic about the future of the United States," Gallup Poll Press Release, October 13, 2000, www.gallup.com/poll/releases/pr001013.asp.

——— (2000). "Americans dissatisfied with U.S. education in general, but parents satisfied with their kids' schools," Gallup Poll Release, Sept. 5, 2000 (www.gallup.com).

Citro, C. F. and J. L. Norwood (eds.) (1997). *The Bureau of Transportation Statistics: Priorities for the Future*, Washington, D.C.: National Academy Press.

Clark, H. H., and M. F. Schober (1992). "Asking questions and influencing answers," in *Questions about Questions*, J. M. Tanur (ed.), New York: Russell Sage Foundation, pp. 15–48.

Cohen, J. (1988). *Statistical Power Analysis for the Behavioral Sciences*, 2nd ed., Hillsdale, NJ: Lawrence Erlbaum Associates.

Cohn, Victor, and Lewis Cope (2001). *News and Numbers: A Guide to Reporting Statistical Claims and Controversies in Health and Other Fields*, 2nd ed., Ames, IA: Iowa State University Press.

Cole, C. R., J. M. Foody, E. H. Blackstone, and M. S. Lauer (2000). "Heart rate recovery after submaximal exercise as a predictor of mortality in a cardiovascularly healthy cohort," *Annals of Internal Medicine*, Vol. 132, pp. 552–555.

Coren, S., and D. Halpern (1991). "Left-handedness: A marker for decreased survival fitness," *Psychological Bulletin*, Vol. 109, No. 1, pp. 90–106.

Crossen, Cynthia (1994). *Tainted Truth*, New York: Simon and Schuster.

Dabbs, J. M., D. de LaRue, and P. Williams (1990). "Testosterone and occupational choice: Actors, ministers, and other men," *Journal of Personality and Social Psychology*, Vol. 59, No. 6, pp. 1261–1265.

Davidson, R. J., J. Kabat-Zinn, J. Schumacher, M. Rosenkranz, D. Muller, S. F. Santorelli, F. Urbanowski, A. Harrington, K. Bonus, and J. F. Sheridan (2003). "Alterations in brain and immune function produced by mindfulness meditation," *Psychosomatic Medicine*, Vol. 65, pp. 564–570.

Davis, Robert (1998). "Prayer can lower blood pressure," *USA Today*, August 11, 1998, p. 1D.

Devore, J., and R. Peck (1993). *Statistics: The Exploration and Analysis of Data*, 2nd ed., Belmont, CA: Duxbury Press.

Diaconis, P., and F. Mosteller (1989). "Methods for studying coincidences," *Journal of the American Statistical Association*, Vol. 84, pp. 853–861.

Eddy, D. M. (1982). "Probabilistic reasoning in clinical medicine: Problems and opportunities," Chapter 18 in D. Kahneman, P. Slovic, and A. Tversky (eds.), *Judgment under Uncertainty: Heuristics and Biases*, Cambridge, U.K.: Cambridge University Press.

Edwards, Gary, and Josephine Mazzuca (2000). "About four-in-ten Canadians accepting of same sex marriages, adoption," Gallup Poll Release, March 7, 2000 (www.gallup.com).

Elias, Marilyn (2001). "Web use not always a downer. Study disputes link to depression," *USA Today*, July 23, 2001, p. 1D.

Ellsworth, P. C., J. M. Carlsmith, and A. Henson (1972). "The stare as a stimulus to flight in human subjects: A series of field experiments," *Journal of Personality and Social Psychology*, Vol. 21, pp. 302–311.

Faigenbaum, A. D., W. L. Westcott, R. LaRosa Loud, and C. Long (1999). "The effects of different resistance training protocols on muscular strength and endurance development in children," *Pediatrics Electronic Article*, Vol. 104, No. 1 (July), p. e5.

Fichtenberg, C. M., and S. A. Glantz (2000). "Association of the California Tobacco Control program with declines in cigarette consumption and mortality from heart disease," *New England Journal of Medicine*, Vol. 343, No. 24, 1772–1777.

Fletcher, S. W., B. Black, R. Harris, B. K. Rimer, and S. Shapiro (1993). "Report of the international workshop on screening for breast cancer," *Journal of the National Cancer Institute*, Vol. 85, No. 20, Oct. 20, 1993, pp. 1644–1656.

Freedman, D., R. Pisani, R. Purves, and A. Adhikari (1991). *Statistics*, 2nd ed., New York: W. W. Norton.

Freinkel, A. (1998). Letter to the editor, *Journal of the American Medical Association*, Vol. 280, No. 22, p. 1905.

French, J. R. P (1953). "Experiments in field settings," in L. Festinger and D. Katz (eds.), *Research Method in the Behavioral Sciences*, New York: Holt, pp. 98–135.

Gastwirth, J. L. (1988). *Statistical Reasoning in Law and Public Policy*, New York: Academic Press.

Gastwirth, J. L. (1988). *Statistical Reasoning in Law and Public Policy, Vol. 2: Tort Law, Evidence and Health*, Boston: Academic Press.

Gehring, W. J., and A. R. Willoughby (2000). "The medial frontal cortex and the rapid processing of monetary gains and losses," *Science*, Vol. 295, pp. 2279–2284.

Gillespie, Mark (1999). "Local schools get passing grades: Americans support higher pay for teachers," Gallup Poll Release, Sept. 8, 1999.

Griffiths, R. R., and E. M. Vernotica (2000). "Is caffeine a flavoring agent in cola soft drinks?" *Arch. Fam. Med.*, Vol. 9, pp. 727–734.

Gustafson, R., and H. Kallmen (1990). "Alcohol and Color Word Test," *Perceptual and Motor Skills*, Vol. 71, pp. 99–105.

Hand, D. J., F. Daly, A. D. Lunn, K, J. McConway, and E. Ostrowski (1994). *A Handbook of Small Data Sets*, London: Chapman and Hall.

Harmon, Amy (1998). "Sad, lonely world discovered in cyberspace," *New York Times*, August 30, 1998, p. A3.

Hedges, L. V., and I. Olkin (1985). *Statistical Methods for Meta-Analysis*, New York: Academic Press.

Hite, Shere (1987). *Women and Love*, New York: Knopf.

Holbrook, M. B., and R. M. Schindler (1989). "Some exploratory findings on the development of musical tastes," *Journal of Consumer Research*, Vol. 16, pp. 119–124.

Horne, J. A., L. A. Reyner, and P. R. Barrett (2003). "Driving impairment due to sleepiness is exacerbated by low alcohol intake," *Journal of Occupational and Environmental Medicine*, Vol. 60, pp. 689–692.

Howell, David C. (1997). *Statistical Methods for Psychology*, 4th ed., Belmont, CA: Duxbury.

Hurt, R., L. Dale, P. Fredrickson, C. Caldwell, G. Lee, K. Offord, G. Lauger, Z. Maruisic, L. Neese, and T. Lundberg (1994). "Nicotine patch therapy for smoking cessation combined with physician advice and nurse follow-up," *Journal of the American Medical Association*, Vol. 271, No. 8, (Feb. 23), pp. 595–600.

Ichimaya, M. A., and M. I. Kruse (1998). "The social contexts of binge drinking among university freshmen," *Journal of Alcohol and Drug Education*, Vol. 44, pp. 18–33.

Iman, R. L. (1994). *A Data-Based Approach to Statistics*, Belmont, CA: Wadsworth.

Jorenby, D. E., S. J. Leischow, M. A. Nides, S. I. Rennard, J. A. Johnston, A. R. Hughes, S. S. Smith, M. L. Muramoto, D. M. Daughton, K. Doan, M. C. Fiore, and T. B. Baker (1999). "A controlled trial of sustained-release bupropion, a nicotine patch, or both for smoking cessation," *New England Journal of Medicine*, Vol. 340, No. 9, pp. 685–691.

Kalichman, S. C. (1986). "Horizontality as a function of sex and academic major," *Perceptual and Motor Skills*, Vol. 63, pp. 903–908.

Kawachi, I., D. Sparrow, A. Spiro III, P. Vokonas, and S. T. Weiss (1994). "A prospective study of anger and coronary heart disease: The normative aging study," *Circulation*, Vol. 94, pp. 2090–2095.

Kiener, Robert (accessed June 3, 2005). "How honest are we?" *Reader's Digest Canada* (http://www.readersdigest.ca/mag/1997/03/think_01.html).

Kitchens, L. J. (1998). *Exploring Statistics: A Modern Introduction to Data Analysis and Inference*, 2nd ed., Boston: Duxbury Press.

Kohlmeier, L., G. Arminger, S. Bartolomeycik, B. Bellach, J. Rehm, and M. Thamm (1992). "Pet birds as an independent risk factor for lung cancer: Case-control study," *British Medical Journal*, Vol. 305, pp. 986–989.

2.86 a. Somewhat unusual. Only about 2.5% of adult males will have a smaller measurement because 52 cm is two standard deviations below the mean.

2.90 a. Population. b. Population, 14.77.

2.94 a. Would hold without the two outliers; should still be close. b. Yes, range is 10.75 cm, close to 6 standard deviations. Expect between 4 and 6 standard deviations.

2.98 a. $z = -0.5$; .3085. b. $z = 2.5$; .9938.

2.100 50, 50, 50, 50, 50, 50, 50; no.

2.103 a. Mean = 51.47 years; standard deviation = 8.92 years (population standard deviation = 8.85). b. Range is 42 years, 4.7 standard deviations, so it holds. c. $z$ for youngest CEO is $-2.18$, $z$ for oldest CEO is 2.53; about as expected from the Empirical Rule.

2.110 a. This is a personal preference; some may prefer a large family. b. An outlier in the high direction.

2.111 a. Categorical. b. Quantitative.

2.113 a. Yes; night light use, for example. b. No.

2.117 a. Set 2—it covers a much wider range of heights.

2.119 a. Amount of beer consumed per unit of time (week, etc.) and systolic blood pressure. b. Calories of protein consumed per day on average and whether or not they had colon cancer.

2.122 Do the students who studied the most last week tend to be the students with the highest grade point averages?

## Chapter 3

3.3 a. Yes. The heights of women in the class probably are similar to the heights of all women at the college. b. No. The sample of daycare parents is likely to be more supportive than the general population.

3.4 a. All registered voters in the community.

3.9 c. Nonresponse bias.

3.11 For example, testing manufactured parts and the test damages the products.

3.13 a. Selection bias. b. Nonresponse bias.

3.15 a. Does not hold. Professional basketball players do not represent any larger population. b. Probably holds. Students taking statistics most likely have representative pulse rates.

3.17 No. All possible sets of four songs do not have the same probability to be a sample. For example, a sample consisting of the first four songs on the first CD is impossible.

3.19 .014 or 1.4%.

3.24 b. 40% ± 3.15% or 36.85% to 43.15%.

3.28 a. .49 ± .03 = .46 to .52. b. .47 ± .03 = .44 to .50.

3.37 a. Use 01 to 49 as is and subtract 50 from 51 to 99 to get 15, 23, 20, 21, 29, 5. b. It doesn't matter.

3.40 Simple random sample.

3.43 a. Stratified sample: Use the three types of schools as strata. Create a list of all students for each of the three strata; draw a simple random sample from each of the three lists. b. Cluster sample: Use individual schools or individual classes as clusters. Take a random sample of clusters; measure all students in those clusters. c. Simple random sample: Obtain a list of all students in the classes at all schools; take a simple random sample from that combined list.

3.46 Cluster sample because a sample of exchanges is found and then only numbers within those exchanges are sampled.

3.48 a. All taxpayers. b. Parents of all the school children.

3.52 a. Dentists who subscribe to one of two dental magazines. Yes, because not all dentists subscribe to those two magazines. b. Nonresponse. c. Send a reminder or call those who didn't respond.

3.61 Anonymous testing.

3.64 Any two questions in which one changes the way respondents would think about the other. An example: "Are you aware that over 30% of homeless people in this city are mothers with children?" and "Do you think more public money should be used to help homeless people?"

3.66 Desire to please; confidentiality and anonymity.

3.69 a. Open-form question. b. Yes, because of all the publicity he had just received. c. Probably lower.

3.76 Any list that contains the two-digit strings in any order: 00, 07, 15, 19, 24, 33, 44, 51, 65, 99. For instance, 24190 03351 99076 54415.

3.79 a. Put an ad in the local paper asking people to fill out the survey. b. Ask "Don't you agree that there is too much trash in our streets and that more public trash containers are needed?"

3.85 a. All students at the university. b. All students enrolled in statistics classes. c. The 500 students to whom the survey was mailed.

3.87 The sample sizes or margin of error. The 20% versus 25% may be within the margin of error for the surveys.

3.89 The quickie poll would probably be most representative.

3.91 Nonresponse.

3.92 b. Send the survey to a legitimate random sample, but make the questions so outrageous that only those who support the position would respond; others would not take it seriously.

3.96 No, only those who are likely to vote should be used.

3.99 Larger; the sample size for Republicans only would be smaller.

3.101 a. .078; .66 ± .078 or 58.2% to 73.8%. b. .066; .38 ± .066 or 31.4% to 44.6%. c. Yes, intervals do not overlap.

## Chapter 4

4.3 c. Observational study.

4.4 a. Response variable is daughter's height.

4.7 c. Randomized experiment. d. Observational study.

4.9 a. Probably not, because long-term meditation is a matter of choice, not easily randomly assigned. b. Yes, volunteers could be randomly assigned to attend the program or not.

4.10 a. The explanatory variable is long-term practice of meditation; the response is blood pressure.

4.11 a. Amount of salt and sodium in the diet or other diet-related factors.

4.20 a. Yes b. No

4.25 Completely randomized design.

4.29 a. Single-blind; customers knew what plan they had, but the technician did not.

4.30 a. Matched pairs.

**4.31**  a.  The explanatory variable is the rate plan; the response is usage during peak hours.

**4.32**  a.  A control group was used (matched with the treatment group); a placebo was not.

**4.38**  a.  Yes.  b.  No.  c.  No.

**4.42**  a.  Observational study.  b.  The explanatory variable was whether or not the couple owned a pet, and the response variables were marriage satisfaction and stress levels.  c.  An example is the amount of business travel they do. Couples who travel frequently may be less likely to own pets and also have more stress.

**4.46**  a.  Interacting variable.

**4.49**  The explanatory variable is taking the new medication or not; the response variable is weight gain or loss; the interacting variable is gender.

**4.53**  a.  Yes.  b.  Yes.  c.  No.

**4.57**  c.  Confounding variables and the implication of causation (e.g., current diet). Relying on memory; it would be hard for people to remember how high in fat their childhood diet was.

**4.60**  Observational study.

**4.63**  Postmenopausal women who did not use hormone therapy at all.

**4.67**  Extending results inappropriately. If women today take lower levels of hormones than the women in the study, these results might not apply to women currently on hormone therapy.

**4.72**  This is an example of an interacting variable, since the effect on self-esteem of thinking about their bad hair is different for men than for women.

**4.78**  a.  Individual unit is a tomato plant. Two variables were the number of tomatoes produced and whether the tomato plant was raised in full sunlight or partial shade.

**4.80**  a.  Yes.  b.  Yes.  c.  No.

**4.87**  a.  A random sample was used, so results can be extended to all customers.

## Chapter 5

**5.1**  a.  Negative association.  b.  No association.

**5.2**  a.  Negative association.  b.  Roughly linear.  c.  Highest average math SAT is about 600; fewer than 5% took the test.  d.  Lowest average math SAT is about 475; about 60% took the test.

**5.3**  a.  Yes, both variables are quantitative.

**5.7**  a.  The speed of the car is the explanatory variable, and the stopping distance is the response variable.  b.  Positive association, somewhat linear but possibly curvilinear.

**5.9**  a.  Average weight = $-250 + 6(70) = 170$ lb.  b.  On average, weight increases 6 pounds per each 1-inch increase in height.

**5.11**  b.  Predicted average = $575 - 1.11(8) = 566.12$.  c.  Residual = $573 - 566.12 = 6.88$.

**5.12**  Deterministic relationship.

**5.14**  a.  Height increases an average of 0.7 inch for each centimeter increase in handspan.  b.  65.1 in.  c.  Residual = $66.5 - 65.1 = 1.4$ in.

**5.19**  a.  For line 1, SSE = 10; for line 2, SSE = 4.  b.  Line 2 is better because SSE is smaller.

**5.22**  The two variables do not have a linear association.

**5.23**  a.  Chest girth increases as length increases.  b.  The value of $r$ is high, so the relationship apparently is strong.  c.  $r = +0.82$.

**5.26**  It would still be 0.95. Changing the units of measurement does not change the correlation.

**5.29**  a.  Graph 2 is the strongest; Graph 3 is the weakest.

**5.31**  b.  Positive, because larger cities have more of both.  c.  Negative, because strength decreases with age.

**5.40**  a.  Estimated stopping distance when speed is 80 mph is $-44.2 + 5.7(80) = 411.8$ ft. No, this is probably not an accurate estimate. The relationship is probably not linear.  b.  From the plot, the relationship looks curvilinear. Estimated stopping distance is about 500 ft.

**5.42**  Men and women should not be combined. Women probably have higher scores but lower heights.

**5.48**  Both variables will have relatively high values in the winter and relatively low values in the summer.

**5.49**  Gender differences create the observed negative correlation. Females tend to be shorter and to have more ear pierces.

**5.68**  a.  For a man, $53 + 0.7(140) = 151$ pounds. For a woman, $44 + 0.6(140) = 128$ pounds. The man wants to weigh 11 pounds more; the woman wants to weigh 12 pounds less.  b.  No, there are not actual weights close to 0.  c.  Yes. For each increase of 1 pound in actual weight, the slope provides the estimated average increase in ideal weight, which is 0.6 pound for women and 0.7 pound for men.

**5.70**  b.  $-0.312$  c.  The correlation for hardcover books only is 0.348. The correlation for softcover books only is 0.637.

**5.72**  a.  Perhouse = $44.4 - (0.02095)$(Year). The estimated number of persons per household in 2010 is $44.4 - (0.02095)(2010) = 2.291$ persons.  b.  Slope = $-0.02095$, indicating that the average number of people per household decreases by about 0.02 per year, or about 0.2 every 10 years.

**5.80**  b.  Diff = $-52.5 + 0.312$ (Actual). For men who weigh 150 pounds, the estimated average difference is $-52.5 + 0.312(150) = -5.7$ pounds. This is actual $-$ ideal, so they want to weigh more.  d.  $r^2 = .353$ or 35.3%.

## Chapter 6

**6.1**  a.  $1322/2057 = 64.3\%$; row percent.  b.  $83/548 = 15.1\%$; column percent.  c.  $285/2858 = 10.1\%$.

**6.3**  b.  $170/300 = 56.7\%$. Among females, 56.7% are freshmen.

**6.5**  a.  Yes, both variables are categorical.  c.  No, both variables are quantitative.

**6.6**  No. Totals are given for categories of each variable, but counts for combinations are not provided.

**6.9**  a.

| Anger? | Heart Disease | No Heart Disease | Total |
|---|---|---|---|
| No Anger | 8 | 191 | 198 |
| Most Anger | 59 | 500 | 559 |
| Total | 67 | 691 | 758 |

b.  $8/199 = 4.0\%$.  c.  $59/559 = 10.6\%$.

6.12 a. 1.0
6.13 a. 10/100 = .1.
6.14 a. .1/.05 = 2.
6.15 a. 2.11
6.19 1.4
6.22 a. Relative risk is .106/.04 = 2.65. b. Percent increase in risk is 100% × (10.6 − 4.0)/4.0 = 165%.
6.25 a. Aged 18 to 29, 13.9%; aged 30 or over, 10.6%.
b. Relative risk is .139/.106 = 1.3. Adults under 30 are 1.3 times as likely to report seeing a ghost as are adults 30 years old or older.
6.31 Probably not. The relationship between blood pressure and religious activity would need to be reversed when separate categories of a third variable are considered.
6.33 Occupation; men may be more likely to have jobs that require long-distance driving.
6.37 a. Response variable is outcome (successful or not).
c. Compare treatments separately within the two initial severity groups.
6.38 a. Null: Gender and opinion on death penalty are not related. Alternative: Gender and opinion on death penalty are related. b. Conclude that gender and opinion are related because p-value < .05.
6.39 For males, oppose = 197.31, favor = 433.69. For females, oppose = 211.69, favor = 465.31.
6.43 b. Do not reject the null hypothesis because the p-value is greater than .05. f. Reject the null hypothesis because the chi-square statistic is greater than 3.84.
6.46 a. Yes. b. Yes. c. No.
6.49 a. Null hypothesis is that there is no relationship between gender and opinions about capital punishment. Alternative hypothesis is that there is a relationship.
b. The relationship is not statistically significant because 0.19 > 0.05.
6.50 a. Chi-square statistic = 1.714.
6.55 a. Yes, there appears to be a relationship based on the differences among the following percents:

| Pierces | % with Tattoo |
|---|---|
| 2 or less | 40/538 = 7.4% |
| 3 or 4 | 58/432 = 13.4% |
| 5 or 6 | 77/279 = 27.6% |
| 7 or more | 53/126 = 42.1% |

c. (77 + 53)/228 = 57% of women with a tattoo have five or more ear pierces. (202 + 73)/1147 = 24% of women with no tattoo have five or more ear pierces.
d. 228/1375 = 16.6%. e. 498/1375 = 36.2%.

6.57 a.

|  | Cancer | Control | Total |
|---|---|---|---|
| Bird | 98 | 101 | 199 |
| No Bird | 141 | 328 | 469 |
| Total | 239 | 429 | 668 |

b. Bird owners: 98/199 = .4925, or 49.25%. Nonowners: 141/469 = .301, or 30.1%. c. No. This was a case control study in which the researchers purposely sampled more lung cancer patients than would naturally occur in a group of 668 people. The sample percentages falling into the cancer category are much higher than you would see in a random sample. d. The relative risk for bird owners is .4925/.301 = 1.64. e. The "baseline" risk of lung cancer for people like yourself. f. No. This is an observational study, and there may be confounding factors that explain the results. g. Compare the smoking habits of the bird owners and nonowners and see whether or not the bird owners generally smoke more.
6.61 Chi-square = 4.817, p-value = .028; relationship is statistically significant.
6.65 If we use the approximate odds given in Exercise 6.64, the odds ratio is 24/8.5 = 2.82. The precise odds ratio is found by taking (59 × 191)/(8 × 500) = 2.817. The odds of remaining free of heart disease versus getting heart disease for men with no anger are about 2.8 times the odds of those events for men with the most anger.
6.68 a. For men, the odds for Program A are 400 to 250, or about 1.6 to 1. For women, the odds are 50 to 25, or 2 to 1. The odds ratio for women compared to men is therefore 2/1.6, or 1.25. The odds of being admitted versus being denied admittance for women are about 1.25 times what they are for men, so women have better odds. b. For men, the odds for the combined programs are 450 to 550, or about .8 to 1. For women, the odds are 175 to 325, or about .54 to 1. The odds ratio (women to men) is therefore about .54/.8 = .675. The odds of being admitted versus being denied admittance for women are about 2/3 of what they are for men, so women have worse odds. c. The last sentence in the solutions for parts (a) and (b) can be used for this purpose.

## Chapter 7

7.1 Random circumstance: Domestic flight is on time or not. The probability is .761 that flight is on time.
7.5 1/125
7.6 c. Yes. d. No.
7.9 a. Relative frequency; probably determined by observing the number injured by lightning for many years, divided by the average population in those years.
b. Personal; based on her assessment of the health of the plants and growing conditions.
7.10 Random circumstances: Traffic light is green or not. Determine probability by observing the proportion of time it's green over many months of driving to work.
7.12 Play over and over again, and record how many games he or she won out of the total number of games played.
7.13 Example: the probability that the United States President will be assassinated this year.
7.16 17/172.
7.18 a. Yes. b. No.
7.21 a. Yes. b. No.
7.24 a. Yes, red die cannot be both 3 and 6. b. No, red die can be 3 in the same toss that green die is 6.
7.25 a. No, knowing A occurred changes P(B) to 0.
b. Yes, knowing the red die is a 3 does not change P(B).
7.29 No, knowing the woman's age changes the probability that she is fertile.

7.32  a.  Getting at least one tail.    b.  7/8
7.35  a.  No.    b.  .8
7.38  a.  11/12    b.  11/12    c.  121/144    d.  23/144
7.41  a.  Long-run relative frequency of these flights arriving on time to get to the ship.    b.  No, in Miami, bad weather would increase the chances of being late for both.    c.  .72    d.  .26
7.43  a.  .0625    b.  .0577    c.  .5625    d.  .5577
7.44  b.  2/3    c.  1/2
7.47  .085
7.50  (1/13)(48/51)
7.53  $P$(calculus student) = .4; $P$(Junior|calculus student) = .15/.4 = .375.
7.56  45/10000
7.58  a.  8/50    b.  7/31    c.  1/50
7.60  Use only two digits, such as 8 and 9, to represent correct guesses.
7.63  a.  .12    b.  .4615
7.66  They are equally likely.
7.70  a.  1/365    b.  .0136    c.  No, that specific event has low probability, but it is not unlikely that someone in the class would have a birthday matching a teammate's family member's birthday.
7.72  No, birth outcomes are independent.
7.76  Not surprising; $1 - P$(no match) = .1829.
7.79  a.  .27    b.  .86    c.  .0225    d.  .0064    e.  .0145
7.82  a.  40%    b.  10%    c.  31%
7.86  a.  .5 (problem stated that no responses equal median). b.  .0625    c.  No, they would not be independent.
7.89  a.  .1225    b.  No, their statuses are not independent. c.  Relative frequency.    d.  Statement must be about a separate randomly selected foreign-born person in each of the two years.
7.92  485/1669
7.95  238/612
7.97  .4125

## Chapter 8

8.2  a.  Discrete.    b.  Continuous.    c.  Discrete. d.  Discrete.
8.5  Continuous example: amount of rainfall during the event; $S$ includes the range from 0 to the maximum possible rain in one day. Discrete example: number of emergency vehicles with sirens driving by during event; $S = \{0,1,2, \ldots , k\}$, where $k$ is the logical maximum for the situation.
8.6  a.  3/6
8.9

| $k$ | 0 | 1 | 2 | 3 |
|---|---|---|---|---|
| $P(X = k)$ | 1/8 | 3/8 | 3/8 | 1/8 |

8.10  a.  .80
  d.

| $k$ | 0 | 1 | 2 | 3 | 4 |
|---|---|---|---|---|---|
| $P(X \leq k)$ | .14 | .51 | .80 | .95 | 1 |

8.14  a.  Sample space events = G, BG, BBG, BBBG, BBBB.
8.18  a.  $30.    b.  $0 and $100 with probabilities .9 and .1.
8.20  −$3.95.
8.23  $E(X)$ = 1.5.

8.26  Standard deviation = $29.67.
8.29  $E(X)$ = 1.69; not a possible value, so meaningful only as a long-run average.
8.32  a.  Yes, 200, .5.    b.  100
8.34  a.  $n$ = 30, $p$ = 1/6.
8.35  a.  5
8.38  a.  The probability of success does not remain the same from one trial to the next.    b.  There are not a pre-specified number of trials.    c.  Outcomes are not independent from one trial to the next; probability of success changes.
8.41  a.  .2051    b.  .3504
8.42  b.  The desired probability is $P(X \geq 110)$, $n$ is 500, and $p$ is .20.
8.46  a.  $X$ is a uniform random variable. b.  $f(x)$ = 1/100 for any $x$ between 0 and 100; $f(x)$ = 0 for any $x$ not between 0 and 100. c.  $P(X \leq 15$ seconds) = $(15 - 0)/100$ = .15.
8.49  b.  $z$ = −1.0.
8.50  b.  .3632    c.  .6368
8.53  a.  .0808
8.55  a.  $P(Z \leq -0.5)$ = .3085.
8.57  a.  −1.96
8.59  a.  .3665
8.60  $P(Z \leq -.675)$ is about .25, so about 25% of the women are shorter than $-.675 \times 2.7 + 65$ = 63.2 inches.
8.63  a.  .0918 (for $z$ = −1.33)    b.  .0228    c.  .3830    d.  .0023
8.66  a.  .0571    b.  .0749
8.70  a.  400 seconds.    b.  20 seconds.    c.  .0228
8.73  b.  $X + Y$ is normal with mean = 150, variance = 325, standard deviation = 18.03.
8.76  $P(Z < -1.21)$ = .1131.
8.78  .0918 (for $z$ = −1.33).
8.80  Probably not. It's a mixture of two bell-shaped distributions with different means.
8.83  a.  $X$ is −$2, or $70; $P(X = -\$2)$ = 37/38, $P(X = \$70)$ = 1/38.    b.  $E(X)$ = −$4/38 = −$0.1053, players lose about 10 cents per $2 bet over the long run.
8.86  a.  Probability of success is not the same for all trials. b.  Outcomes are not independent from one trial to the next.
8.89  a.  It makes sense to assume independence unless the children influence each others' choices or some other common factor leads them both to pick a long or short story.    b.  $P(Z > 1.77)$ = .0384.

## Chapter 9

9.1  a.  Statistic.    b.  Parameter.    c.  Statistic. d.  Parameter.
9.4  Population parameter.
9.7  a.  Parameters; they represent the population values. b.  Yes, they know the means exactly, and the mean for males is $1500 higher.
9.9  a.  $\hat{p}$    b.  $p$    c.  $\hat{p}$
9.11  a.  What proportion of parents in the school district support the new program?    b.  Parameter is $p$ = proportion of parents in school district who support the new program.    c.  Sample estimate is $\hat{p} = 104/300$ = .347.

9.14   a.   How much difference is there between the mean number of CDs owned in the populations of male and female students?   b.   Parameter $= \mu_1 - \mu_2$, where $\mu_1 =$ population mean for males and $\mu_2 =$ population mean for females.   c.   $\bar{x}_1 - \bar{x}_1 = 110 - 90 = 20$.

9.16   One research question of interest would be "What is the mean difference between hours spent studying and hours spent socializing?" The parameter is $\mu_d =$ mean difference for all students at the university. The sample estimate is $\bar{d} =$ mean difference for the 100 students in the sample.

9.17   Paired data. Two variables are measured on each individual.

9.22   a.   Parameter is $p =$ proportion of all teens who go out on dates that would say that they have been out with someone of another race or ethnic group. Sample statistic is $\hat{p} = .57$.

9.24   a.   $p$   b.   $\hat{p}$

9.27   a.   The same, it would be $\mu$.   b.   The same, approximately normal.   c.   The same.

9.29   a.   Not likely.   c.   Likely.

9.31   a.   Mean $= .5$; s.d. $=$ s.d. $= \sqrt{\dfrac{(.5)(.5)}{400}} = .025$.
     b.   Mean $= .5$; s.d. $= .0125$.

9.33   a.   $.55$   b.   $\sqrt{\dfrac{(.55)(.45)}{100}} = .0497$, or about $.05$.
     c.   $.55 \pm 3(.05)$, or between about $.40$ and $.70$.

9.35   a.   $\hat{p} = 300/500 = .60$.   b.   s.e.$(\hat{p}) = .022$.

9.39   $.05 \pm 2\sqrt{\dfrac{(.05)(.95)}{400}}$, or $.03$ to $.07$.

9.42   a.   $.20$   b.   $.18$   c.   $.0384$   d.   $.20$   e.   $.04$

9.46   $p_1 - p_2 = .30 - .36 = -0.06$.

9.49   c.   No. The sample size condition is not met. With a total of 10 from each method, there could not be at least 10 that succeed and 10 that fail.

9.52   a.   Yes, it is $p_1 - p_2 = 0$.   b.   Yes; approximately normal.   c.   Yes.   d.   No.

9.55   a.   Yes, this falls into Situation 1 described in Section 9.6.   b.   No. For a skewed population, the sample size is too small.   c.   Yes, this falls into Situation 2.

9.56   a.   6   b.   3   c.   s.d.$(\bar{x})$ decreases when sample size is increased.

9.58   a.   $\bar{x}$ and $s$.   b.   s.e.$(\bar{x}) = 47.7/\sqrt{28} = 9.014$.

9.61   b.   Approximately normal with a mean of 210 pounds and a standard deviation of 3.95.   d.   $P(z > 2.53) = .0057$.

9.62   Standard deviation of the sampling distribution is $\sigma/\sqrt{n}$. Standard error of the mean replaces population $\sigma$ with the sample version $s$ and is more commonly used because $\sigma$ is rarely known.

9.65   a.   Two independent samples.

9.68   d.   Approximately normal, mean $= 0$,
     s.d.$(\bar{d}) = \dfrac{5}{\sqrt{12}} = 1.44$.

9.71   a.   5   b.   1.302 inches.

9.74   a.   0   d.   No.   e.   No.

9.76   Mean $\mu$ and standard deviation $\sigma$.

9.78   b.   1.84

9.80   b.   2.58

9.83   b.   7.01

9.85   a.   $z = 2$.   b.   $z = 2.828$.

9.88   a.   $-1.0$ and $+1.6$.

9.89   a.   $z = 1$.   b.   $z = -1$.

9.92   a.   $t = -1$; df $= 16 - 1 = 15$.

9.94   a.   $t = 0.417$.   b.   $z = 0.5$.   c.   $t = -3$.

9.98   a.   $t$-distribution.   b.   df $= 5$.   c.   Normal distribution.

9.99   a.   $t = 2.5$.

9.105   b.   $1/10$   d.   $R = 1$ is most likely, and $R = 4$ is least likely.

9.107   a.   0, 1, 2, 3, 4.

9.108   a.   $1/2$   b.   Integers from 0 to 6.   c.   $(1/2)^6 = 1/64$.   d.   $B$ is binomial with $n = 6$ and $p = 1/2$; probabilities for $B = 0, 1, 2, 3, 4, 5, 6$ are $1/64$, $6/64$, $15/64$, $20/64$, $15/64$, $6/64$, and $1/64$, respectively.

9.110   a.   The population of possible net gains is highly skewed (not bell-shaped), and the sample size is small.   b.   $-\$0.35$; $\$9.38$

9.112   Because of the symmetry of the situation, it should be a mirror image of the sampling distribution of $H$, the highest number, given in Figure 9.12, so it should be highly skewed to the right.

9.115   a.   Binomial; $n = 10$, $p = .5$.   b.   5   c.   1.58

9.122   Actual population exists (students at your college); fixed proportion are left-handed; $n = 200$ is large enough for number of left-handers to be sufficient.

9.125   b.   Yes; the sample proportion is almost surely in the range $.447$ to $.553$, so $.70$ is unlikely.

9.129   Approximately normal with a mean of 105 and a standard deviation of 1.5.

9.132   a.   Approximately normal, mean $= .20$, standard deviation $= \sqrt{.2 \times .8/2500} = .008$.   b.   Not likely. We expect virtually all samples to have between 17.6% and 22.4% of viewers.

9.135   a.   No, sample size is too small.   b.   Yes.   c.   No, days in January and February only do not constitute a random sample.   d.   Yes.

9.139   225

## Chapter 10

10.1   c.   Sample proportion, $.55$.

10.4   a.   Choice ii, it could be 171 again.

10.7   a.   $36\% \pm 3.5\%$, or $32.5\%$ to $39.5\%$. With 95% confidence, we can say that the percentage of American adults who feel that they don't get enough sleep is between 32.5% and 39.5%.   b.   If college students generally differ from others on this question, then this sample will not correctly estimate the percent for students.

10.9   About $0.95 \times 200 = 190$ intervals will capture the population proportion, so about 10 intervals will not.

10.11   A confidence interval should not be computed because this is not a randomly selected sample, nor is it likely to be a representative sample. It is a self-selected sample, and this type of sample often is biased toward a particular trait or opinion.

10.12   5%

10.14   a.   $p =$ the proportion of the population that suffers from allergies.

10.16   a.   $.016$   b.   $.558$ to $.622$.

10.19   a.   $\hat{p} = \dfrac{166}{200} = .83$.   b.   $.83 \pm 1.96\sqrt{\dfrac{(.83)(1 - .83)}{200}}$, or

.778 to .882. With 95% confidence, we can say that in the population of patients with this disease, the proportion who would be cured with this treatment is in this interval.

10.20 $.83 \pm 2.33\sqrt{\dfrac{(.83)(.17)}{200}}$, or $.83 \pm .062$. The 98% confidence interval is wider than the 95% interval.

10.23 a. .90   b. .95   c. .98   d. .99

10.24 a. $\hat{p} = \dfrac{220}{400} = .55$.   b. s.e.$(\hat{p}) = \sqrt{\dfrac{.55(1 - .55)}{400}} = .025$.
c. $.55 \pm (1.96)(.025)$, or .501 to .599.
d. $.55 \pm (2.33)(.025)$, or .492 to .608.

10.28 $.19 \pm 1.645\sqrt{\dfrac{.19(1 - .19)}{300}}$.

10.29 $z^* = 1.28$.

10.32 a. 99%

10.34 The probability is .95 that the difference between sample and population proportions (or percentages) is less than the margin of error, so the probability that the difference is greater than the margin of error (3% here) is $1 - .95 = .05$.

10.35 a. $2\sqrt{\dfrac{.56(1 - .56)}{100}} = .0993$.   b. .0496

10.37 a. $\dfrac{1}{\sqrt{501}} = .045$, or 4.5%. They apparently rounded up to 5%.

10.38 a. This can be calculated as $\sqrt{\dfrac{(.03)(.97)}{883}} = .00574$.
b. $2\sqrt{\dfrac{(.03)(.97)}{883}} = .011$.

10.41 a. Yes, there are two independent samples with sufficiently large sample sizes.   b. No, the sample sizes are too small.   c. No, the situation is about the difference in two means, not two proportions.

10.45 b. $(.466 - .380) \pm 2\sqrt{\dfrac{.466(1 - .466)}{4594} + \dfrac{.380(1 - .380)}{7076}}$.

10.47 This is a difference between the proportions in two different categories of a variable in the same sample. The method in Section 10.4 is for a difference between proportions with the same trait in two independent samples.

10.50 a. 55% ± 3%, or 52% to 58%.   b. The confidence interval is entirely above 50%, so it is reasonable to say that more than 50% of the population thinks their weight is about right.

10.51 The 28% claim is not reasonable because it is not within the confidence interval.

10.54 a. 95% confidence intervals for population percentages in the two years do not overlap. It is reasonable to conclude that the population percent is higher in 2002.

10.56 b. While the sample is not a random sample, for the question of interest it may provide a suitable estimate of the proportion that is left-handed in the population. It doesn't seem that this sample of 400 students should be biased toward having a different proportion that is left-handed than in the population.

10.59 a. Observational study. The researchers did not assign individuals to these groups.   b. There may be other differences between the two groups that could account for

the difference. For instance, police officers may have worse diets or may smoke more.

10.63 c. 71% ± 5% or 66% to 76%. We can be 95% confident that somewhere between 66% and 76% of all adult Americans would answer that they are for the death penalty when asked Question 1.   d. 56% ± 5% which is 51% to 61%. We are 95% confident that between 51% and 61% of all adult Americans would respond "death penalty" when asked Question 2.

10.66 a. −0.07 to 0.04   b. The confidence interval contains 0, so we cannot conclude that there is a difference in the population proportions.

10.69 a. $(.63 - .29) \pm 2\sqrt{\dfrac{(.63)(1 - .63)}{300} + \dfrac{(.29)(1 - .29)}{300}}$.

10.72 c. A 99% confidence interval is about .090 to .302, computed as $.1965 \pm (2.576)(.0412)$.

## Chapter 11

11.2 Are you asking a question about children in this school only, or are you using them to represent a larger population?

11.4 b. $\bar{x}_1 - \bar{x}_2 = 22.5 - 16.3 = 6.2$ days.

11.7 a. Two independent samples.   b. Paired data.   c. Two independent samples.

11.10 a. The two populations are women who got married in 1970 and women who got married in 2000. The response variable is the woman's age when she got married. The question of interest is whether the average age at marriage for women changed between 1970 and 2000.

11.12 s.e.$(\bar{x}) = \dfrac{2}{\sqrt{64}} = 0.25$ cm. Roughly, for samples of this size, the average difference between the sample and population means is 0.25 cm.

11.14 Calculate the difference for each person. The standard error of the sample mean difference is calculated using the formula $s/\sqrt{n}$, where $s$ = standard deviation of the sample of differences and $n = 50$.

11.16 Take a large enough sample or samples.

11.19 a. 2.13

11.21 a. $76 \pm 2.31\dfrac{6}{\sqrt{9}}$.   e. $100 \pm 2.13\dfrac{8}{\sqrt{16}}$.

11.23 a. $t^* = 2.52$ (df = 21).   b. $t^* = 2.13$ (df = 4).

11.26 The confidence interval is about 7.04 to 9.16 days, computed as $8.1 \pm 2.82 \times \dfrac{1.8}{\sqrt{23}}$.

11.28 The calculation is $\bar{x} \pm 2 \times s$. Rounding to the nearest dollar, about 95% had textbook expenses in the interval $285 ± (2)($96), which is $93 to $487.

11.31 a. s.e.$(\bar{x}) = \dfrac{s}{\sqrt{n}} = \dfrac{2.7}{\sqrt{81}} = 0.3$ inch. Approximate 95% confidence interval is $64.2 \pm (2 \times 0.3)$, or 63.6 to 64.8 inches.

11.33 The interval is 27 to 28 cm, computed as $27.5 \pm 2 \times 0.25$. With approximate 95% confidence, we can say that in the population of men represented by this sample the mean foot length is between 27 and 28 cm.

11.36 a. $5.36 \pm (2.06)(3.05)$ or −0.92 to 11.64.   b. No, the interval covers 0.   c. Two different variables are measured for each individual.

11.38 No. Any value in the interval is a plausible value for the difference.

11.41 a. The population means differ.   b. We cannot say that the population means differ.

11.42 a. $(11.6 - 10.7) \pm 2.07\sqrt{\dfrac{3.39^2}{16} + \dfrac{2.59^2}{10}}$.

11.45 a. 2.2 cm.   b. $\sqrt{\dfrac{2.4^2}{36} + \dfrac{1.8^2}{36}} = 0.5$.

c. $2.2 \pm (2 \times 0.5)$, or 1.2 to 3.2 cm.

11.48 The variances (or standard deviations) in the two populations are equal.

11.50 $s_p = 1.7$. The 95% confidence interval is

$(8.1 - 4.5) \pm 2.01\sqrt{\dfrac{1.7^2}{23} + \dfrac{1.7^2}{25}}$.

11.52 c. The interval is approximately $72.8 \pm 2\dfrac{72.2}{\sqrt{204}}$, or

$72.8 \pm 10.1$.

11.54 Both intervals are entirely above 1, so it is reasonable to conclude that risk of death is greater in the abnormal heart rate recovery group than in the normal heart rate group.

11.63 a. The interval will be wider for 95% confidence.
b. The interval will become more narrow because the standard error will decrease.   c. Assuming that the standard deviation stays the same, the interval width stays the same.

11.66 a. On each team, the ninth weight is the weight of the coxswain, who gives instructions about the rowing cadence but does not row. A coxswain's weight is much less than the weights of the rowers, so there are two substantial outliers.

11.69 b. Paired data.

11.71 d. The formula for a confidence interval for the difference between two independent means was used.

The calculation is $(2.37 - 1.95) \pm 2\sqrt{\dfrac{1.87^2}{59} + \dfrac{1.51^2}{116}}$.

11.73 a. A "success" is an interval that captures the population mean.   b. .90   e. $.90^{100}$

## Chapter 12

12.1 a. All babies.   b. $p$ = proportion of babies in the population born during the 24-hour period surrounding a

full moon.   c. $H_0: p = \dfrac{1}{29.53}$.

12.4 a. 21-year-old men and women in the United States.
b. Researchers are interested in the hypothesis that equal proportions of 21-year-old men and women in the United States have high school diplomas.

12.7 An example is the hypothesis that equal proportions of men and women are unemployed. The populations are men and women, and the proportion of interest is the proportion who are unemployed.

12.10 a. No.   b. Yes.   c. No.

12.12 a. There are five choices, so $p = 1/5$, or .2.
b. $H_0: p = .2$, $H_a: p < .2$.

12.14 Before they look at the data.

12.16 a. The population is all school-aged children in the state; $p$ is the proportion of them living with one or more grandparents.

12.18 b. 32

12.20 a. Do not reject the null hypothesis.   b. Reject the null hypothesis.

12.22 a. There is not sufficient evidence to conclude that smokers are more likely to get the disease.

12.26 a. Cannot reject the null hypothesis.   b. Reject the null hypothesis (or accept the alternative hypothesis).
c. Reject the null hypothesis (or accept the alternative hypothesis).

12.29 a. False.   b. True.   c. False.   d. False.

12.31 a. Correct decision.   b. Type 2 error.

12.34 a. Example: a diagnosis that leads to major surgery.
b. Example: an infection that could be cured by antibiotics but gets very serious if untreated.

12.36 c. Type 1 is probably more serious: voting for the bill that implements substantial change when actually a majority do not support it.   d. Type 1.

12.41 a. .0358   b. .0228   c. .1379   d. .00001

12.43 a. Yes.   b. No.   c. No.   d. No.

12.45 b. $-2.68$

12.46 b. .0037

12.47 Null standard error uses the null value $p_0$, and standard error uses the sample proportion.

12.49 a. No, don't know if it was a random sample.
b. Step 1: $H_0: p = .10$, $H_a: p > .10$.
Step 2: Sample size condition is met; see part (a) for other. Sample proportion is .12,

$$z = \dfrac{.12 - .1}{\sqrt{\dfrac{.1(.9)}{150}}} = 0.82$$

Step 3: Exact $p$-value = .242; $p$-value for $z$ is .206 (or .207).
Step 4: Cannot reject the null hypothesis; result is not statistically significant.
Step 5: Cannot conclude that artists are more likely to be left-handed than the general population.

12.51 a. No, $.07 > .05$.   b. Yes.

12.55 a. Probability of 17 to 50 successes for a binomial with $n = 50$ and $p = .25$.

12.58 a. .0401   b. .9599

12.59 a. Reject the null hypothesis or accept the alternative hypothesis.   b. Reject the null hypothesis or accept the alternative hypothesis.

12.60 c. $z < -1.645$; do not reject the null hypothesis.
d. $z > 1.645$; do not reject the null hypothesis.

12.63 $z = 3.26$, $p = .001$; reject the null hypothesis and conclude that a higher proportion of women would claim that they would return the money.

12.65 $z = 6.26$, $p$-value $< .000000001$. We can reject the null hypothesis. We can conclude that in the population(s) represented by the sample(s), the proportion of women who would say there should be "more strict" laws covering the sale of firearms is higher than the proportion of men who would say this.

12.67 $z = 2.40$, $p$-value $\approx .008$. Reject the null hypothesis. The conclusion is that in the population(s) represented by the sample(s), a higher proportion of women than men would claim that they would date someone with a great personality even if they did not find the person physically attractive.

12.70  a. .0725   d. .0145

12.72  Very large.

12.74  a. 1000   b. .45   c. .45

12.78  Probably the finding was that there was no *significant* difference, in the statistical meaning, not the ordinary-language meaning. It could be because there really is no difference or because the sample was too small to detect it and a type 2 error was made.

12.80  b. No, this is not a contradiction. There is not much (if any) practical importance to the observed difference in incidence of drowsiness (6% versus 8%), but the large sample sizes led to a *statistically* significant difference.

12.84  Step 1: $H_0: p = .10$, $H_a: p \neq .10$.
       Step 2: Sample conditions are met. Sample proportion is 20/240 = .0833,
       $$z = \frac{.0833 - .1}{\sqrt{\frac{.1(.9)}{240}}} = -0.86$$
       Step 3: Exact $p$-value = .395; $p$-value for $z$ is .3898.
       Step 4: Cannot reject the null hypothesis; result is not statistically significant.
       Step 5: Cannot conclude that proportion of UCD students who are left-handed differs from the national proportion.

12.87  a. Population parameter = proportion of all people who suffer from this chronic pain who would experience temporary relief if taking the new medication. Hypotheses are $H_0: p = .7$, $H_a: p > .7$.

12.89  a. .1470   b. .0375

12.92  $z > 0$.

12.94  b. Population; only this class is of interest.

12.97  Assume that students represent a random sample of a population of interest on this question and $p$ = proportion of population who would choose first drink. $H_0: p = .5$, $H_a: p > .5$. Both 86 and 73 are large enough to meet the conditions. $z = 1.03$, $p$-value = .1515, not statistically significant. Cannot conclude that the first drink presented is more likely to be selected.

12.99  Larger samples should be used. Power for samples of 500 is only .2768.

12.100  b. $z = 3.19$; samples are large enough and independent, $p$-value = .001, conclude the population proportions were significantly different.

12.102  b. Type 1, people would take aspirin although it doesn't help; type 2, people would not take aspirin and heart attacks that could be prevented would not be; type 2 is more serious.

12.104  $z = 0.98$, $p$-value = .1635 (exact $p$-value = .189); do not reject $H_0$; cannot guess at significantly better than chance level.

12.107  a. $z = 0.62$, $p$-value = .266; do not reject $H_0$ for $\alpha = .05$.   b. $z = 4.59$; yes.

12.110  $z = 2.51$, $p$-value = .012; conclude that proportions were significantly different.

## Chapter 13

13.2  a. Difference between the means of two populations.

13.3  a. $\mu_1 - \mu_2$, where $\mu_1$ is the mean height of the population of 25-year-old women and $\mu_2$ is the mean height of the population of 45-year-old women.

13.4  a. $H_0: \mu_1 - \mu_2 = 0$.

13.9  Yes. The hypotheses can (and should) be specified.

13.11  a. 1.11

13.13  a. Rejection region is $t \geq 1.72$; reject $H_0$.   c. Rejection region is $|t| \geq 2.09$; reject $H_0$.   g. Rejection region is $t \leq -1.83$; do not reject $H_0$.

13.16  a. $H_0: \mu = 72$ beats per minute, $H_a: \mu \neq 72$. Since $n = 57$ is large, the test can proceed. Test statistic is $t = -1.20$, df = 56, exact $p$-value = 2(.1176) = .2352. Cannot reject the null hypothesis; mean male pulse rate is not significantly different from 72.

13.18  a. No. Even one standard deviation below the mean is negative.   b. Yes, $n$ is large.

13.21  $\bar{x} = 63.86$, $s = 2.086$, $t = -3.38$, df = 37, $p$-value = .001.

13.24  b. $t = -1.89$, df = 49, $.05 < p < .078$. Exact $p = .064$.

13.25  $t = 3.4$; reject the null hypothesis.

13.29  c. $t = \frac{8 - 10}{4/\sqrt{20}} = -2.236$.   d. df = 19, $.015 < p < .03$. Exact $p = .0188$.

13.32  b. $H_a: \mu_1 - \mu_2 < 0$.   c. $H_a: \mu_1 - \mu_2 < 0$.

13.34  a. 1.414

13.36  Plots show necessary conditions are satisfied, $t = 0.43$; don't reject the null hypothesis.

13.39  a. Unpooled $t = 3.97$, pooled $t = 3.96$; reject the null hypothesis.   b. Can use pooled because sample sizes and standard deviations are close.

13.42  a. From Minitab: $t = 1.89$, df = 13, $p$-value = .041, reject the null hypothesis.   b. From Minitab: $t = 1.96$, df = 18, $p$-value = .033, reject the null hypothesis.

13.44  a. No.   c. No.   e. Yes.

13.47  a. Reject $H_0$ for $\alpha = .05$.   b. Do not reject $H_0$ for $\alpha = .025$.

13.49  a. Can't tell. The entire interval could be greater than 10, or it could include 10.

13.50  a. $\mu_1 - \mu_2$, where $\mu_1$ and $\mu_2$ are the mean running times for the 50-yard dash for the populations of first-grade boys and girls, respectively.

13.53  a. Estimate the difference in proportions of college men and women who would answer yes, $p_1 - p_2$, and test whether $p_1 - p_2 = 0$. May also want to estimate $p_1$ and $p_2$ separately.

13.55  a. 68% of the differences fall in the range $-2$ to $+6$.   c. .695

13.57  a. .632

13.60  a. 0.23   d. No.

13.62  Items 2, 4, and 5.

13.65  a. All except item 5 rely on knowing the $p$-value.   b. Items 2, 3, and 4.

13.66  a. Parameter = proportion of adult Americans between the ages of 18 and 30 who think the use of marijuana should be legalized; null value = .5.   c. Parameter = difference in mean ages of death for left- and right-handed people; null value = 0.

13.70  a. $H_0: \mu_d = 0$, $H_a: \mu_d > 0$, $t = 7.12$, df = 169, $p$-value $\approx 0$.

13.73  Reject the null hypothesis using $\alpha = .025$.

13.77  a. Type 1, patients are retested perhaps without cause; type 2, patients need to be retested but are not, so the cholesterol readings may be too high.   b. Type 2 seems more serious because a medical decision is not made by using the most accurate data.

## Chapter 14

14.1   a. $\hat{y} = 119 - 1.64(40) = 53.4$.
    b. $y - \hat{y} = 50 - 53.4 = -3.4$.

14.2   a. $\beta_0$ = intercept; $\beta_1$ = slope.

14.4   a. Slope = 0.9. Mean blood pressure increases 0.9 per one-year increase in age.

14.6   b. $b_1$

14.9   a. $r^2 = \dfrac{\text{SSTO} - \text{SSE}}{\text{SSTO}} = \dfrac{500 - 300}{500} = .40$.

14.10  a. $\hat{y}_i = 513.79$, or about 514 ft.   b. $e_i = 11.21$, or about 11 ft.   c. Approximately $514 \pm 100$ ft.   d. 650 ft is more than two standard deviations from the mean distance for drivers 21 years old, so it would be unusual.

14.11  a. $s = 2.837$ is roughly the average deviation of $y$-values from the sample regression line.   b. Geographic latitude explains 91.7% of the variation in mean April temperatures.

14.13  b. SSE = 18.   c. $s = \sqrt{\dfrac{18}{6 - 2}} = 2.12$.

14.14  a. The $p$-value is not less than .05, so the result is not statistically significant.   b. .552

14.17  b. 0.165

14.18  d. The result may not have practical significance. The slope indicates only a small increase in television watching per one-year increase in age, and $R^2$ is small, so the relationship is weak.

14.20  a. $H_0: \rho = 0$ versus $H_a: \rho \neq 0$. It may be that in reality, the researchers' alternative hypothesis was $\rho > 0$, but computer software routinely provides a $p$-value for the two-sided alternative. The researchers have decided to reject the null hypotheses.

14.21  a. Prediction interval for a value of $y$.   b. Confidence interval for the mean.

14.22  a. Grade point average of an individual who misses two classes per week.   b. Mean grade point average of all who miss two classes per week.

14.25  a. $\hat{y}_i = 3.59 + (.967)(40)$.   d. The width of the prediction interval reflects the variation among the individual ages of husbands married to women 40 years old. The confidence interval is narrow because, with $n = 170$, the mean age can be estimated with good precision.

14.26  $42.258 \pm (2 \times 0.313)$, which gives approximately the interval provided by Minitab.

14.28  a. Conditions 2 and 4.   b. Conditions 1, 2, and 3.   c. Condition 5.   d. Conditions 1, 2, and 3.

14.30  There is an outlier; aside from the outlier, the distribution might be normal.

14.31  Condition 4 seems to hold. It's not unreasonable to assume that the residuals have a normal distribution.

14.33  a. There appears to be an outlier, so Condition 2 is violated.

14.34  c. The plot will have a U shape.

14.35  a. $b_1 = -0.269$.   b. $-0.26917 \pm (2 \times 0.06616)$, or about $-0.401$ to $-0.137$.   c. $H_0: \beta_1 = 0$ versus $H_a: \beta_1 \neq 0$.   d. $t = -4.07$, df = 114.

e. The relationship is statistically significant ($p$-value = 0.000).

14.37  $b_0 = 7.56$. This is the mean hours of sleep for students who studied 0 hours.

14.38  Conditions 1, 2, and 3 hold. The form of the equation is probably correct, there are no outliers, and the standard deviation is about the same for all values of $x$.

14.39  (3.746, 9.750). Of students in the population who studied 3 hours, about 95% slept between 3.746 and 9.75 hours. Or the probability is .95 that the hours of sleep will be in the given interval for a student randomly selected from the population of students who studied 3 hours.

14.41  b. $\hat{y} = 25.38 + 0.7069x$.   c. $\hat{y} = 78.4$.   d. 72.97 to 83.83.

14.43  a. $R^2 = 44.7\%$. The father's height explains about 44.7% of the variation in son's height.   b. $r = \sqrt{0.447} = 0.67$.

14.49  a. The two observations with the greatest chest girths could be outliers, or there might be a gently curving pattern.   b. $\hat{y} = -206.9 + 10.76x$.   c. $R^2 = 95.1\%$.   d. $\hat{y}_i = 223.5$.   e. 95% prediction interval is (180.47, 266.49).   f. 95% confidence interval is (211.41, 235.55).

## Chapter 15

15.1   a. Appropriate. Both variables are categorical.
    b. Not appropriate. Both variables are quantitative.

15.4   a. $p$-value = .05.   b. $.10 < p$-value $< .25$ (.153).

15.5   b. 16.81 (df = 6).

15.7   a. No, $\chi^2 < 6.63$.   b. No, $\chi^2 < 6.63$.   c. Yes, $\chi^2 > 13.28$.   d. No, $\chi^2 < 30.58$.

15.9   a. $H_0$: Gender and type of class taken are not related for the population of students. $H_a$: Gender and type of class taken are related for the population of students.   b. df = 1.   c. 1.258   d. $p$-value = .262.   e. Do not reject the null hypothesis for $\alpha = .05$. The relationship between gender and type of class taken is not statistically significant.

15.11  c. $H_0$: Opinion about death penalty and opinion about marijuana legalization are not related; $H_a$: Opinion about death penalty and opinion about marijuana legalization are related.   d. $\chi^2 = .287$, df = 1, $p$-value = .592.

15.12  For expected counts, the proportion with an ear infection is .22 within each treatment.

15.15  f. We're not telling. Ask an engineer.

15.17  $\chi^2 = 807(191 \times 171 - 341 \times 104)^2 / (532 \times 275 \times 295 \times 512) = 0.287$.

15.18  a. The researcher probably thought that short students would be more likely to be bullied. That is a one-sided hypothesis, so a $z$-test is appropriate.   c. $z = 3.02$, $p$-value = .001.

15.21  No. The response variable has three categories.

15.24  a. Given that 2 out of 10 participants have reduced pain, what is the probability that both of them would be in the magnet-treated group?

15.26  a. 100 for each of the three categories.

15.28  a. No.  b. Yes.  c. Yes.  d. No.  e. No.  f. No.

15.30  a.  $H_0$: $p_1 = .3$, $p_2 = .6$, $p_3 = .1$ where $p_1$, $p_2$, and $p_3$ are the population proportions who drove, biked, and used another way that day, respectively.   c.  6.67

15.32  $\chi^2 = 1.766$, df = 2, $.25 <$ p-value $< .50$ (.413).

15.34  a.  HH, TH, HT, TT, $X = 0$, $p = 1/4$; $X = 1$, $p = 1/2$; $X = 2$, $p = 1/4$.   b.  $H_0$: $p_0 = .25$, $p_1 = .50$, $p_2 = .25$.   c.  $\chi^2 = 7.2$, df = 2, $.025 <$ p-value $< .05$ (.027).

15.36  $\chi^2 = 15.8$, df = 5, $.005 <$ p-value $< .01$. Reject $H_0$: proportions are as stated at M&M website, conclude at least two proportions are not as stated.

15.39  b.  $H_0$: Number of ear pierces and likelihood of having a tattoo are not related, $H_a$: Number of ear pierces and likelihood of having a tattoo are related.   c.  $\chi = 119.279$, df = 3, p-value $< .001$. Decide $H_a$.

15.42  a.  $n = 10$, $p = .5$.

c.

| Number Correct | Null Probabilities |
|---|---|
| $\leq 2$ | .0547 |
| 3 | .1172 |
| 4 | .2051 |
| 5 | .2461 |
| 6 | .2051 |
| 7 | .1172 |
| $\geq 8$ | .0547 |

15.44  a.  The conditions are met.   b.  $\chi^2 = 157$, df = 12.

15.47  a.  $\chi^2 = 0.267$, df = 1, not significant.   b.  $\chi^2 = 13.025$, df = 1, significant.   c.  $\chi^2 = 6.748$, df = 1, significant.   d.  $\chi^2 = 2.356$, df = 1, not significant.

15.50  Between 0 and 5.99.

15.52  b.  $\chi^2 = 15.5$, df = 4, $.001 <$ p-value $< .005$. There is a statistically significant relationship.   c.  The largest contributions occur in cells where student eye color and eye color student finds attractive are both blue or both brown.

15.54  a.  Homogeneity.

## Chapter 16

16.1  a.  Appropriate.   d.  Not appropriate. Not a comparison of independent groups.

16.2  a.  Mean body weight is equal for the populations in the three age groups; $H_0$: $\mu_1 = \mu_2 = \mu_3$.   b.  Mean body weight is not equal for the populations in the three age groups.

16.5.  a.  1.018 to 4.362; interval does not cover 0 so it's reasonable to say means differ.

16.6  a.  Reject the null; F-statistic is greater than 3.89, the critical value.

16.8  a.  18–29 and 60+ age groups differ from the other two age groups.   b.  95% confident that all six intervals capture the corresponding population parameters.

16.9  b.  Variations is about the same for each lab and there are no outliers.

16.10  a.  $H_0$: $\mu_1 = \mu_2 = \mu_3 = \mu_4 = \mu_5$ versus $H_a$: Not all $\mu_i$ the same.   b.  Yes. $F = 4.53$, p-value = .003.   c.  p-value is area to the right of 4.53 under a skewed curve.   d.  The mean for lab 4 is lower than the means for labs 2 and 5. Other comparisons are more difficult to judge.

16.12  Completed table is

| Source | df | SS | MS | F |
|---|---|---|---|---|
| Between groups | 5 | 40 | 8 | 1.333 |
| Error | 10 | 60 | 8 | |
| Total | 15 | 100 | | |

16.14  c.  MSE = 1043/150 = 6.953; $s_p = \sqrt{6.953} = 2.637$.

16.17  a.

| Source | df | SS | MS | F | p |
|---|---|---|---|---|---|
| Caffeine | 2 | 61.40 | 30.70 | 6.18 | 0.006 |
| Error | 27 | 134.10 | 4.97 | | |
| Total | 29 | 195.50 | | | |

b.  $s_p = \sqrt{4.97} = 2.23$.

16.18  Each interval has the form $\bar{x}_i \pm 2.01 \dfrac{0.4058}{\sqrt{11}}$, or $\bar{x}_i \pm 0.246$.

16.19  b.  $F = 50/10 = 5$.   c.  df = 3 and 20.

16.21  a.  $\bar{x} = 8$, $\bar{x}_1 = 4$, $\bar{x}_2 = 11$, $\bar{x}_3 = 9$.   b.  SS Groups = 78.   c.  SS Total = 118.   d.  SSE = $118 - 78 = 40$.

16.23  a.  Null is that median ratings would be equal for populations of students from the four types of hometowns; alternative is that median ratings would not be equal for the four types of hometowns.   b.  p-value = .000; conclude that median ratings would not be equal.   c.  Big city, median = 5.

16.25  a.  Ordinal.

16.27  a.  $H_0$: Medians are the same for the three populations of occupational groups; $H_a$: Population medians are not all the same.   b.  Decide in favor of $H_a$ (p-value = .002).

16.28  a.  Interaction; amount of difference between members and nonmembers of Greek organizations depends on gender.

16.30  a.  Males.   b.  Students from big cities rate rock music lower than students from other types of hometowns.   c.  Pattern of relationship between rating and hometown is about the same for males and females, so the interaction is weak or nonexistent.

16.33  a.  The amount of difference between mean GPA of men and women would depend on seat location.   b.  The amount of difference between men's and women's mean GPAs would be the same in each seat location.

16.35  b.  As the bacteria amount increases, the difference in mean rot for 10°C and 16°C increases.

16.38  a.  $H_0$: $\mu_1 = \mu_2 = \mu_3 = \mu_4 = \mu_5$ versus $H_a$: Not all $\mu_i$ the same.   b.  Population might be all public school children of this age in the United States.   c.  Decide in favor of $H_a$. Not all the means are the same.   d.  Perhaps children who engage in more violent behaviors also like to watch more television.

16.40  There may be interaction. The difference between boys and girls is smaller in the 6+ hours of television group than it is in other television watching groups.

16.43  SSE = 15,286, $s_p = \sqrt{46.18} = 6.796$.

16.44  Regression methods could be used. Both variables are quantitative.

## Chapter 17

**17.1** Study #1; the explanatory variable is taking calcium or placebo, and the response variable is blood pressure.

**17.4** Observational study; the headline is not justified.

**17.6** Drinking coffee or not. People who don't drink coffee may be more likely to drink tea and may also be more likely to conceive for other reasons.

**17.9** Probably. We are not told who was in the sample, but there is no indication that it was an unrepresentative group with regard to diet and conception.

**17.12** a. Statistical significance; the null hypothesis is that energy and fat intake are not related to tea consumption. b. Confounding variables.

**17.13** a. There was an association, but it was not strong enough to be *statistically* significant for the sample size used.

**17.15** a. Amount walked per day (less than 1 mile, between 1 and 2 miles, more than 2 miles) and whether the man died during the study (yes, no).

**17.18** a. No, the entire population was already measured. b. Yes, assuming that the mechanism causing geyser eruptions stays the same.

## Supplemental Topic 1

**S1.2** Yes, $X$ is a hypergeometric random variable with $N = 250$, $n = 5$, $a = 100$.

**S1.5** $N = 100$, $n = 12$, $a = 40$.

**S1.6** $X$ is an integer between 0 and the smaller of $(a, n)$, so in this case, it is an integer between 0 and 12.

**S1.9** $X$ = number of winning numbers between 1 and 31; $X$ is a hypergeometric random variable with $N = 51$, $n = 6$, and $a = 31$. $P(X = 6) = .0409$.

**S1.13** c. $X$ is a binomial random variable; $\mu = np = 40$ and $\sigma = 5.66$. d. $X$ is a hypergeometric random variable; $\mu = np = 20$ and $\sigma = 2.836$.

**S1.16** .0067

**S1.19** $P(X \geq 20) = .0035$.

**S1.22** a. $\mu = 30 \times 2 = 60$.

**S1.23** a. Mean = variance = 4, standard deviation = 2.

**S1.24** a. $e^{-4} = .0183$.

**S1.27** a. $n = 200$, $p = .01$. b. $X$ is approximately a Poisson random variable with $\mu = np = 2$, so the Poisson formula can be used to find probabilities related to $X$. c. $P(X = 0) = (.99)^{200} = .1340$. d. Approximate $P(X = 0) = e^{-2} = .1353$. The answers are very similar. There is about a 13.5% chance of no defective candies in a bag.

**S1.31** Mean = $np_i = 30(1/6) = 5$, variance = $np_i(1 - p_i) = 30(1/6)(5/6) = 4.167$.

**S1.34** Yes, this is a multinomial experiment. There are 1000 "trials" with three possible outcomes on each one, and the outcomes are independent from one person to the next.

## Supplemental Topic 2

**S2.1** a. Yes. $H_0$: $\eta$ = median pulse = 72, $H_a$: $\eta$ = median pulse > 72.

**S2.4** b. Sample median = $355. c. $S^+ = 4$, $S^- = 12$, $n_a = 16$. d. Find $P(X \leq 4)$ in a binomial distribution with $n = 16$ and $p = .5$. e. Reject $H_0$, conclude that

the median is less than $400. Exact $p$-value = .0384. With normal approximation but no continuity correction, $z = -2$, $p$-value = .0228. With continuity correction, $z = -1.75$, $p$-value = .0401.

**S2.7** a. $H_0$: Distribution of pulse rates is the same for populations of those who do and do not exercise. $H_a$: Pulse rates tend to be higher than for those who do not exercise (or lower for those who do).

b. $W = 103.5$. c. $\mu_W = 8\dfrac{1 + 20}{2} = 84$,

$\sigma_W = \sqrt{\dfrac{8(12)(8 + 12 + 1)}{12}} = 12.96$.

d. $z = \dfrac{103.5 - 84}{12.96} = 1.50$,

$p$-value = $P(Z \geq 1.50) = .0668$. With continuity correction, $z = 1.466$, $p$-value = .0713. e. Cannot reject $H_0$: cannot say those who do not exercise tend to have higher pulse rates.

**S2.9** There is an obvious outlier in the diet plan group (value = 35).

**S2.14** a. $T^+ = 7.5 + 15 + 3 + 1 = 26.5$.

b. $z = \dfrac{26.5 - 68}{19.339} = -2.12$.

c. $p$-value = $P(Z < -2.12) = .017$. Reject $H_0$, conclude that in the population, the median amount spent is less than $400.

**S2.18** c. Median = 0 doesn't mean that the two measurements are equal for all people. Median is $\approx 0$ if forearm is larger for half of all people and foot is longer for the other half. Also, failing to reject $H_0$ might be a type 2 error.

**S2.19** a. $\bar{R}_1 = 10$, $\bar{R}_2 = 12.8333$, $\bar{R}_3 = 5.6667$.

b. $H = \dfrac{12}{18(18 + 1)}[6(10 - 9.5)^2 + 6(12.8333 - 9.5)^2 + 6(5.6667 - 9.5)^2] \approx 5.49$.

**S2.21** a. $H_0$: Distribution of weekly times of exercising is the same for each reason. $H_a$: Weekly times of exercising tends to be larger for at least one reason than for other reasons.

## Supplemental Topic 3

**S3.4** Necessary conditions look to be supported. The plot shows a roughly horizontal band of points with no obvious pattern.

**S3.6** a. $\hat{y}_i = 20.8 - 0.415(49) + 1.24(56) = 69.905$. b. $e_i = y_i - \hat{y}_i = 64 - 69.905 = -5.905$.

**S3.8** Intercept estimates the average height of male students whose mothers and fathers have height = 0. This is impossible.

**S3.9** 0.539 inch (or 0.53897). This is the value of $b_2$.

**S3.10** a. .031. Reject H, conclude that the student height and mother's height are related.

b. $t = \dfrac{0.19294}{0.08777} = 2.20$.

**S3.15** a. $H_0$: $\beta_1 = \beta_2 = 0$; $H_a$: At least one of $\beta_1$ and $\beta_2$ is $\neq 0$. b. Reject $H_0$ ($F = 15.19$, $p$-value = 0.000). Conclude that left arm length is related to at least one of left foot length and/or right foot length.

**S3.17** a. $s = 1.796$ cm. Roughly, the average absolute size of the deviations between actual and predicted arm lengths is about 1.796 cm.   b. $R^2 = 36.9\%$. The foot length variables explain 36.9% of the variation among arm lengths.

S3.18   a.   $b_0 = 119.58$, $b_1 = 0.9636$, $b_2 = -1.0903$, $b_3 = 2.6295$, $b_4 = -1.2080$, $b_5 = -0.6413$.

S3.21   a.   $y_i = \beta_0 + \beta_1 x_{i1} + \beta_2 x_{i2} + \epsilon_i$, where we assume that $\epsilon_i$ are normally distributed with mean 0 and standard deviation $\sigma$.

## Supplemental Topic 4

S4.1   a.   Response variable is weight loss after two months. Factor A is whether the person smokes or not, $a = 2$. Factor B is the exercise program, $b = 3$.

S4.5   a.   $\mu_{12} = 8.0$ hours.   b.   $\mu_{..} = 28.4/4 = 7.1$ hours.
c.   $\mu_{1.} = 7.0$ hours, $\mu_{2.} = 7.2$ hours.   d.   $\mu_{.1} = 6.5$ hours; $\mu_{.2} = 7.7$ hours.

S4.8   $\mu_{11} = \mu_{..} + \alpha_1 + \beta_1 + \alpha\beta_{11} = 7.1 - 0.1 - 0.6 - 0.4 = 6.0$ hours.

S4.10   a.   Yes. The mean DDT level appears to be much higher in the Arctic region than in the United States or Canada.

S4.11   a.   $H_0$: The population mean DDT levels in falcons are the same in the Arctic, the United States, and Canada. $H_a$: For at least one of the three sites, the population mean DDT level is different.

S4.12   a.   $p$-value $= .000$; reject the null hypothesis; conclude that mean DDT levels differ across sites.

S4.14   For Factor A, F = 8892.70/3.44 = 2585.1.

S4.17   c.   $H_0$: The difference in mean commute times for the three routes is the same across days. $H_a$: The difference in mean commute times for the three routes depends on the day.

S4.18   c.   $F = 8.40$, df $= (4, 81)$. From Table A.4, $p$-value $< .01$ because $8.40 > 3.56$. Reject $H_0$. Conclude that there is an interaction, so the difference in mean commute times for the routes depends on the day of the week category.

S4.22   a.   1, based on subtracting the given df values from the total value of 53.   b.   225, found as SS/df $= 450/2$.
c.   840 because df $= 1$.

S4.24   a.   From knowing df $= 2$, $a = 3$.

S4.25   c.   Degrees of freedom for groups $=$ Total df $-$ Error df $= 53 - 48 = 5$. Also, there are six groups, so df $= 6 - 1 = 5$.
d.   Degrees of freedom for error is the same, 48.

## Supplemental Topic 5

S5.4   a.   This would be a good idea.

S5.7   No, the intervals overlap, and there is no clear choice. One method produces a slightly higher mean but also has more variability.

S5.10   This condition was not met. The 9-year-old girl was in immediate proximity to the practitioners, and she knew which hand she was hovering over with her hand.

S5.13   The main problem is that participants were deceived into thinking that they were hurting others. But there would have been no way to conduct the experiment if they had been told the truth.

S5.17   The problem is that they chose the confidence level to present the outcome they desired. They should have reported the standard 95% confidence interval.

S5.19   The complaint raised by the letter is that the study was not measuring what it purported to measure.

# Credits

This page constitutes an extension of the copyright page. We have made every effort to trace the ownership of all copyrighted material and to secure permission from copyright holders. In the event of any question arising as to the use of any material, we will be pleased to make the necessary corrections in future printings. Thanks are due to the following authors, publishers, and agents for permission to use the material indicated.

## Photos

**xxvi** © DiMaggio/Kalish/Corbis; **12** © Tony Dizinno/Allsport Concepts/Getty Images; **70** © Roy Morsch/zefa/Corbis; **116** © Royalty-free/Corbis; **150** © Royalty-free/Corbis; **192** © Simon Wilkinson/The Image Bank/Getty Images; **228** © Mario Tama/Getty Images; **278** © Ron Sherman/Stone/Getty Images; **330** © Joos Mind/Stone/Getty Images; **400** © Peter Scholey/Photographer's Choice/Getty Images; **442** © Geoffrey Clifford/The Image Bank/Getty Images; **494** © Laurence Mouton/PhotoAlto/Getty Images; **550** © Photodisc/Getty Images; **598** © Royalty-free/Corbis; **634** © John Giustina/Iconica/Getty Images; **668** © Nick Daly/Photonica/Getty Images; **704** © Photodisc/Getty Images; **S1-1** © Rudy Gutierrez/America 24-7/Getty Images; **S2-1** © Ian Shaw/Stone/Getty Images; **S3-1** © Royalty-free/Corbis; **S4-1** © Wide Group/Iconica/Getty Images; **S5-1** © Zia Soleil/Iconica/Getty Images

## Text

*Chapter 1* **6** Case Study 1.7 (excerpt): "Sad, Lonely World Discovered in Cyberspace," by A. Harmon, *New York Times*, August 30, 1998, p. A3. Reprinted with permission of the New York Times Company.

*Chapter 2* **38** Example 2.11 (excerpt): From Amy Goldwitz, *The Davis Enterprise*, March 4, 1998, p. A1. Reprinted by permission of *The Davis Enterprise*.

*Chapter 3* **100** Example 3.18 (excerpt): Reprinted by permission of *The Davis Enterprise*.

*Chapter 4* **125** (excerpt): *Sacramento Bee*, July 7, 1999, p. G5. Reprinted by permission of the *Sacramento Bee*. **137** Example 4.4 (excerpt): From "Study: Age doesn't sap memory," *Sacramento Bee*, Kathryn Doré Perkins, July 7, 1999, pp. A1 and A10. Reprinted by permission of the *Sacramento Bee*. **122–123** Case Study 4.1 (excerpt): From Sternberg, Steve, "Lead Exposure Linked to Bad Teeth in Children," *USA Today*, 23 June 1999, Internet. Reprinted with permission.

*Chapter 10* **421** Example 10.7 (excerpt): Copyright © 1997 The Gallup Organization. **433** Exercise 10.7: Copyright © 2000 The Gallup Organization.

*Chapter 12* **537** Case Study 5.1 (excerpt): From "Probability experts may decide Pennsylvania vote," by P. Passell, *New York Times*, April 11, 1994, A15. Reprinted with permission of the New York Times Company. **537** Case Study 5.1 (excerpt): From "Sad, Lonely World Discovered in Cyberspace," by A. Harmon, *New York Times*, August 30, 1998, p. A3. Reprinted by permission of the New York Times Company. **544** Exercise 12.65: Copyright © 2000 The Gallup Organization. **547** Exercise 12.100: Copyright © 2000 The Gallup Organization. **548** Exercise 12.106: Copyright © 2000 The Gallup Organization.

*Chapter 15* **660** Exercise 15.13: Copyright © 1999 The Gallup Organization. **666** Exercise 15.54: Copyright © 1999, 2000 The Gallup Organization.

# Index

# t* Multipliers for Confidence Intervals and Rejection Region Critical Values

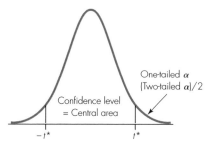

One-tailed $\alpha$
(Two-tailed $\alpha$)/2

Confidence level
= Central area

$-t^*$     $t^*$

| df | .80 | .90 | .95 | .98 | .99 | .998 | .999 |
|---|---|---|---|---|---|---|---|
| | | | | Confidence Level | | | |
| 1 | 3.08 | 6.31 | 12.71 | 31.82 | 63.66 | 318.31 | 636.62 |
| 2 | 1.89 | 2.92 | 4.30 | 6.96 | 9.92 | 22.33 | 31.60 |
| 3 | 1.64 | 2.35 | 3.18 | 4.54 | 5.84 | 10.21 | 12.92 |
| 4 | 1.53 | 2.13 | 2.78 | 3.75 | 4.60 | 7.17 | 8.61 |
| 5 | 1.48 | 2.02 | 2.57 | 3.36 | 4.03 | 5.89 | 6.87 |
| 6 | 1.44 | 1.94 | 2.45 | 3.14 | 3.71 | 5.21 | 5.96 |
| 7 | 1.41 | 1.89 | 2.36 | 3.00 | 3.50 | 4.79 | 5.41 |
| 8 | 1.40 | 1.86 | 2.31 | 2.90 | 3.36 | 4.50 | 5.04 |
| 9 | 1.38 | 1.83 | 2.26 | 2.82 | 3.25 | 4.30 | 4.78 |
| 10 | 1.37 | 1.81 | 2.23 | 2.76 | 3.17 | 4.14 | 4.59 |
| 11 | 1.36 | 1.80 | 2.20 | 2.72 | 3.11 | 4.02 | 4.44 |
| 12 | 1.36 | 1.78 | 2.18 | 2.68 | 3.05 | 3.93 | 4.32 |
| 13 | 1.35 | 1.77 | 2.16 | 2.65 | 3.01 | 3.85 | 4.22 |
| 14 | 1.35 | 1.76 | 2.14 | 2.62 | 2.98 | 3.79 | 4.14 |
| 15 | 1.34 | 1.75 | 2.13 | 2.60 | 2.95 | 3.73 | 4.07 |
| 16 | 1.34 | 1.75 | 2.12 | 2.58 | 2.92 | 3.69 | 4.01 |
| 17 | 1.33 | 1.74 | 2.11 | 2.57 | 2.90 | 3.65 | 3.97 |
| 18 | 1.33 | 1.73 | 2.10 | 2.55 | 2.88 | 3.61 | 3.92 |
| 19 | 1.33 | 1.73 | 2.09 | 2.54 | 2.86 | 3.58 | 3.88 |
| 20 | 1.33 | 1.72 | 2.09 | 2.53 | 2.85 | 3.55 | 3.85 |
| 21 | 1.32 | 1.72 | 2.08 | 2.52 | 2.83 | 3.53 | 3.82 |
| 22 | 1.32 | 1.72 | 2.07 | 2.51 | 2.82 | 3.50 | 3.79 |
| 23 | 1.32 | 1.71 | 2.07 | 2.50 | 2.81 | 3.48 | 3.77 |
| 24 | 1.32 | 1.71 | 2.06 | 2.49 | 2.80 | 3.47 | 3.75 |
| 25 | 1.32 | 1.71 | 2.06 | 2.49 | 2.79 | 3.45 | 3.73 |
| 26 | 1.31 | 1.71 | 2.06 | 2.48 | 2.78 | 3.43 | 3.71 |
| 27 | 1.31 | 1.70 | 2.05 | 2.47 | 2.77 | 3.42 | 3.69 |
| 28 | 1.31 | 1.70 | 2.05 | 2.47 | 2.76 | 3.41 | 3.67 |
| 29 | 1.31 | 1.70 | 2.05 | 2.46 | 2.76 | 3.40 | 3.66 |
| 30 | 1.31 | 1.70 | 2.04 | 2.46 | 2.75 | 3.39 | 3.65 |
| 40 | 1.30 | 1.68 | 2.02 | 2.42 | 2.70 | 3.31 | 3.55 |
| 50 | 1.30 | 1.68 | 2.01 | 2.40 | 2.68 | 3.26 | 3.50 |
| 60 | 1.30 | 1.67 | 2.00 | 2.39 | 2.66 | 3.23 | 3.46 |
| 70 | 1.29 | 1.67 | 1.99 | 2.38 | 2.65 | 3.21 | 3.44 |
| 80 | 1.29 | 1.66 | 1.99 | 2.37 | 2.64 | 3.20 | 3.42 |
| 90 | 1.29 | 1.66 | 1.99 | 2.37 | 2.63 | 3.18 | 3.40 |
| 100 | 1.29 | 1.66 | 1.98 | 2.36 | 2.63 | 3.17 | 3.39 |
| 1000 | 1.282 | 1.646 | 1.962 | 2.330 | 2.581 | 3.098 | 3.300 |
| Infinite | 1.281 | 1.645 | 1.960 | 2.326 | 2.576 | 3.090 | 3.291 |
| Two-tailed $\alpha$ | .20 | .10 | .05 | .02 | .01 | .002 | .001 |
| One-tailed $\alpha$ | .10 | .05 | .025 | .01 | .005 | .001 | .0005 |

Note that the t-distribution with infinite df is the standard normal distribution.

## Standard Normal Probabilities (for z < 0)

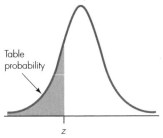

Table probability

| z | .00 | .01 | .02 | .03 | .04 | .05 | .06 | .07 | .08 | .09 |
|------|------|------|------|------|------|------|------|------|------|------|
| −3.4 | .0003 | .0003 | .0003 | .0003 | .0003 | .0003 | .0003 | .0003 | .0003 | .0002 |
| −3.3 | .0005 | .0005 | .0005 | .0004 | .0004 | .0004 | .0004 | .0004 | .0004 | .0003 |
| −3.2 | .0007 | .0007 | .0006 | .0006 | .0006 | .0006 | .0006 | .0005 | .0005 | .0005 |
| −3.1 | .0010 | .0009 | .0009 | .0009 | .0008 | .0008 | .0008 | .0008 | .0007 | .0007 |
| −3.0 | .0013 | .0013 | .0013 | .0012 | .0012 | .0011 | .0011 | .0011 | .0010 | .0010 |
| −2.9 | .0019 | .0018 | .0018 | .0017 | .0016 | .0016 | .0015 | .0015 | .0014 | .0014 |
| −2.8 | .0026 | .0025 | .0024 | .0023 | .0023 | .0022 | .0021 | .0021 | .0020 | .0019 |
| −2.7 | .0035 | .0034 | .0033 | .0032 | .0031 | .0030 | .0029 | .0028 | .0027 | .0026 |
| −2.6 | .0047 | .0045 | .0044 | .0043 | .0041 | .0040 | .0039 | .0038 | .0037 | .0036 |
| −2.5 | .0062 | .0060 | .0059 | .0057 | .0055 | .0054 | .0052 | .0051 | .0049 | .0048 |
| −2.4 | .0082 | .0080 | .0078 | .0075 | .0073 | .0071 | .0069 | .0068 | .0066 | .0064 |
| −2.3 | .0107 | .0104 | .0102 | .0099 | .0096 | .0094 | .0091 | .0089 | .0087 | .0084 |
| −2.2 | .0139 | .0136 | .0132 | .0129 | .0125 | .0122 | .0119 | .0116 | .0113 | .0110 |
| −2.1 | .0179 | .0174 | .0170 | .0166 | .0162 | .0158 | .0154 | .0150 | .0146 | .0143 |
| −2.0 | .0228 | .0222 | .0217 | .0212 | .0207 | .0202 | .0197 | .0192 | .0188 | .0183 |
| −1.9 | .0287 | .0281 | .0274 | .0268 | .0262 | .0256 | .0250 | .0244 | .0239 | .0233 |
| −1.8 | .0359 | .0351 | .0344 | .0336 | .0329 | .0322 | .0314 | .0307 | .0301 | .0294 |
| −1.7 | .0446 | .0436 | .0427 | .0418 | .0409 | .0401 | .0392 | .0384 | .0375 | .0367 |
| −1.6 | .0548 | .0537 | .0526 | .0516 | .0505 | .0495 | .0485 | .0475 | .0465 | .0455 |
| −1.5 | .0668 | .0655 | .0643 | .0630 | .0618 | .0606 | .0594 | .0582 | .0571 | .0559 |
| −1.4 | .0808 | .0793 | .0778 | .0764 | .0749 | .0735 | .0721 | .0708 | .0694 | .0681 |
| −1.3 | .0968 | .0951 | .0934 | .0918 | .0901 | .0885 | .0869 | .0853 | .0838 | .0823 |
| −1.2 | .1151 | .1131 | .1112 | .1093 | .1075 | .1056 | .1038 | .1020 | .1003 | .0985 |
| −1.1 | .1357 | .1335 | .1314 | .1292 | .1271 | .1251 | .1230 | .1210 | .1190 | .1170 |
| −1.0 | .1587 | .1562 | .1539 | .1515 | .1492 | .1469 | .1446 | .1423 | .1401 | .1379 |
| −0.9 | .1841 | .1814 | .1788 | .1762 | .1736 | .1711 | .1685 | .1660 | .1635 | .1611 |
| −0.8 | .2119 | .2090 | .2061 | .2033 | .2005 | .1977 | .1949 | .1922 | .1894 | .1867 |
| −0.7 | .2420 | .2389 | .2358 | .2327 | .2296 | .2266 | .2236 | .2206 | .2177 | .2148 |
| −0.6 | .2743 | .2709 | .2676 | .2643 | .2611 | .2578 | .2546 | .2514 | .2483 | .2451 |
| −0.5 | .3085 | .3050 | .3015 | .2981 | .2946 | .2912 | .2877 | .2843 | .2810 | .2776 |
| −0.4 | .3446 | .3409 | .3372 | .3336 | .3300 | .3264 | .3228 | .3192 | .3156 | .3121 |
| −0.3 | .3821 | .3783 | .3745 | .3707 | .3669 | .3632 | .3594 | .3557 | .3520 | .3483 |
| −0.2 | .4207 | .4168 | .4129 | .4090 | .4052 | .4013 | .3974 | .3936 | .3897 | .3859 |
| −0.1 | .4602 | .4562 | .4522 | .4483 | .4443 | .4404 | .4364 | .4325 | .4286 | .4247 |
| −0.0 | .5000 | .4960 | .4920 | .4880 | .4840 | .4801 | .4761 | .4721 | .4681 | .4641 |

### In the Extreme (for z < 0)

| z | −3.09 | −3.72 | −4.26 | −4.75 | −5.20 | −5.61 | −6.00 |
|-------------|-------|-------|--------|---------|----------|-----------|------------|
| Probability | .001 | .0001 | .00001 | .000001 | .0000001 | .00000001 | .000000001 |

S-PLUS was used to determine information for the "In the Extreme" portion of the table.